LYSOZYME

Proceedings of the Lysozyme Conference held on the occasion of the 50th anniversary of the discovery of lysozyme by Sir Alexander Fleming. Sponsored by the Institute of Cancer Research, College of Physicians and Surgeons, Columbia University, New York, New York. Held at Arden House, Harriman, New York, October 29–31, 1972.

LYSOZYME

EDITED BY

ELLIOTT F. OSSERMAN

Institute of Cancer Research and Department of Medicine
College of Physicians and Surgeons
Columbia University
New York, New York

ROBERT E. CANFIELD

Department of Medicine
College of Physicians and Surgeons
Columbia University
New York, New York

SHERMAN BEYCHOK

Departments of Biological Sciences and Chemistry
Columbia University
New York, New York

ACADEMIC PRESS New York and London 1974

A Subsidiary of Harcourt Brace Jovanovich, Publishers

ACADEMIC PRESS, INC.
111 Fifth Avenue, New York, New York 10003

United Kingdom Edition published by
ACADEMIC PRESS, INC. (LONDON) LTD.
24/28 Oval Road, London NW1

Library of Congress Cataloging in Publication Data

Lysozyme Conference, Arden House, 1972.
    Lysozyme.

    Sponsored by the Institute of Cancer Research,
Columbia University.
    Bibliography: p.
    1.   Lysozyme—Congresses.   I.   Osserman, Elliott F.,
ed.   II.   Canfield, Robert E., ed.   III.   Beychok,
Sherman, ed.   IV.   Columbia University. Institute of
Cancer Research.   V.   Title   [DNLM: 1.   Muramidase—
Congresses.   QU135 L995L 1972]
QP609.L9L95 1972   574.1'9256   73-9442
ISBN 0—12—528950—2

Contents

v

Section 3 BIOLOGICAL AND CLINICAL STUDIES

Section 4 LYSOZYME BIBLIOGRAPHY AND INDEX

The Fleming–Lysozyme Medal

This medal was created to commemorate the fiftieth anniversary of the discovery of lysozyme by Sir Alexander Fleming and to mark the occasion of the Lysozyme Conference held at Arden House, Harriman, New York, October 29–31, 1972, under the auspices of the Institute of Cancer Research of the College of Physicians and Surgeons of Columbia University. This medal was presented at the meeting to Lady Amalia Fleming, Honorary Chairman of the conference.

The portrait on the obverse side of the medal shows Sir Alexander at approximately the time (1922) of the lysozyme discovery. Under the microscope to the right of the portrait are two preparations of *M. lysodeikticus*. The upper field shows the untreated bacteria, and the lower shows the organisms after exposure to lysozyme, markedly swollen and with distorted cell walls. This representation of the morphologic effect of lysozyme was taken from an illustration in Fleming's original report of the lysozyme discovery (*Proc. Roy. Soc. London* **B93**: 306, 1922).

To the left of the portrait is the well-known dictum of Pasteur, "Chance favors the prepared mind." This was frequently cited by Sir Alexander, and assuredly applied to his own life and contributions.

On the reverse side of the medal, lower center, is another illustration adapted from Fleming's original article showing a culture plate with a heavy growth of *M. lysodeikticus*. In the middle of this bacterial growth, a small well has been cut in the agar and filled with tears. After a period of several hours a zone of clearing due to bacterial lysis was evident around the tear sample. The labeling of *"M. lysodeikticus"* and "lysozyme (tears)" is in Fleming's own hand.

A part of the three-demensional molecular model of hen egg-white lysozyme, representing the present state of knowledge of the enzyme, surrounds Fleming's initial culture plate. The portion shown is the active site of the molecule as defined by the X-ray crystallographic studies of David Phillips and his associates. The four circles represent the oxygens of the aspartic 52 and glutamic 35 residues which specifically accomplish the

xi

hydrolytic cleavage of the lysozyme substrate. This molecular model is representative of the wealth of information concerning Fleming's lysozyme which has been assembled through the efforts of a great many investigators in the fifty years since Fleming's discovery.

The medal was created by the distinguished sculptor Abram Belskie, renowned for his many works relating to landmarks of medicine and science and their discoverers. Born in London, Mr. Belskie spent several years in Scotland apprenticed to William Petrie. Thus, he felt a particular kindred to Scottish-born Sir Alexander Fleming. After coming to the United States in 1939, Mr. Belskie worked extensively with the brilliant sculptor, John Gregory.

Mr. Belskie is the recipient of many prizes and honors, including the Sir John Edward Burnett prize, the Lindsay Morris prize, and the Golden Anniversary prize of the Allied Artists of America. He is a fellow of the National Academy of Arts and the National Sculpture Society.

Mr. Belskie resides and has his studio in Closter, New Jersey.

ELLIOTT F. OSSERMAN

Personal Recollections of Lysozyme and Fleming

It is well known that Fleming is famous for his discovery of penicillin and indeed for its value in the treatment of diseases and for starting the era of the antibiotics. Fleming's first paper on penicillin is already extremely important and "a milestone in the history of medical progress," as the Editor of the *British Medical Journal* has called it (1). However, seen from a purely scientific point of view, the discovery of the substance to which Fleming gave the Greek name "lysozyme" is quite possibly greater. At the time of his lysozyme studies, Fleming himself, the introvert, was unable to hide his excitement. His rich imagination and the dreams which he kept so cautiously to himself overpowered him, and he expounded a stream of wonderful hypotheses. The six papers (2–7) which he wrote about lysozyme show perhaps more than any of his other work the brilliance of his mind.

Fleming had two great reasons to cherish lysozyme. It was the first antiseptic he had studied during long years of hard work and search which fulfilled what he required from a bactericidal substance: that it be *selectively more lethal to bacteria than to the host cells*. And what was more and dearer to Fleming's heart, lysozyme was a constituent of human cells and part of the whole body's natural resistance mechanism. As a pupil of Almroth Wright, Fleming had high respect for natural defense mechanisms, and he kept this respect to the end in spite of the miraculous achievements of his penicillin and all the other antibiotics.

It would be right to say that lysozyme made the discovery of penicillin easier, although my own feeling is that it might have been better if penicillin had come first. Ironically, the reading before a learned audience of colleagues of both these extremely important papers was received with the same icy silence and indifference. Both substances have very much in common. They were discovered in almost an identical way. To study penicillin, Fleming used the methods he had devised to study lysozyme. And what is particularly important, and perhaps more so to me, both discoveries show the two great qualities of Fleming's mind. The first characteristic was that he would immediately understand the implications of a chance phenomenon and go straight to the right conclusion. Other scientists before Fleming had noticed the bactericidal power of egg white (8), and much work had been done on that of leucocytes. It is probable that these scientists were investigating the actions of lysozyme but, as Fleming said in his Presidential Address to the Royal Society in 1932, "All these authors considered that the antibacterial phenomena they observed were peculiar to the substance with which they were working—leucocytes or egg white—and none of them apparently had any inkling that the lytic element was widely distributed throughout the animal and vegetable Kingdom" (9).

A drop from the nose of Fleming who had a cold fell onto an agar plate where large yellow colonies of a contaminant had grown, and lysozyme was discovered. He made this important discovery because when he saw that the colonies of the contaminants were fading, his mind went straight to the right cause of the phenomenon he was observing: that the drop from his nose contained a lytic substance. And, also immediately, he thought that this substance might be present in many secretions and tissues of the body. And, he found that this was so—the substance was in tears, saliva, leucocytes, skin, fingernails, mothers' milk—thus, very widely distributed in animals and also in plants.

The circumstances of the discovery of penicillin were very similar. Other scientists, and I believe his friend the Belgian Professor Gzatia was among them, had seen colonies of bacteria fading around a chance mould contaminant, but they did not realize the possible significance of what they were seeing—and they did not discover penicillin.

This great ability which Fleming possessed to understand immediately the meaning and cause of things often impressed me. If I had a problem which to me seemed difficult and confused and I felt like I was helplessly searching my way in a thick wood, he used to find the solution immediately, giving me the impression that he was looking at this inscrutable forest from above and could easily and clearly see the tortuous little path out, and how and where it led.

Fleming's second important characteristic was that his mind was not cluttered or closed to new findings and developments by the beliefs his previous experiments had led him to. And this is proved by the way he threw overboard all these beliefs when he discovered lysozyme and after it penicillin. Hundreds of his neat, conclusive experiments had shown that all antibacterial substances in use at the time, and in the way they were used, did more harm than good. All were killing the bacteria only in concentrations which destroyed the leucocytes. By contrast, lysozyme was harmless to leucocytes and killed some bacteria. Its bactericidal action on the pathogenic bacteria was very weak, which also was natural because otherwise the bacteria would not have been able to establish themselves in the body and be pathogenic; they would have been killed by the lysozyme. Fleming thought that the definition of a pathogenic germ was just that— that it was resistant to lysozyme, at least to its usual concentration in the body. Having found that in egg white the concentration of the lytic element was much greater, a hundred times higher than that in the tears—which he had found to be the richest in lysozyme in the human body—he tested the toxicity of this higher concentration with his usual comparative tests on leucocytes and bacteria. With delight he found that in this higher concentration, lysozyme was still absolutely harmless to the leucocytes while it had a marked effect on some pathogenic bacteria. Fleming then injected egg-white solution intravenously in a rabbit and found that it did not upset the rabbit, while it markedly enhanced the bactericidal power of its blood. He tried to obtain the lytic element in pure form to inject it into the blood stream. But neither he nor anyone else in Almroth Wright's team was a chemist. Almroth Wright despised chemists and would not have one in his laboratory! And so, having nothing better to use in the blood stream, Fleming tried diluted egg white and wrote his conclusion: "It is possible that in cases of generalized infection with a microbe susceptible to the bacteriolytic action of egg white . . . the intravenous injection of a solution of egg white might be beneficial . . ." (5). Yet, earlier, in his 1919 Hunterian Lecture, he had dismissed the possibility of benefit from any of the antibacterial substances then known. "It seems a pity," he had said, "that the surgeon should wish to share his glory with a chemical antiseptic of more than doubtful utility . . ." (10).

Exactly the same thing happened with the biological testing of penicillin. Fleming tested the broth culture of the mould for toxicity in his usual way and found that the antibacterial substance it contained, although very powerful on pathogenic bacteria, did not interfere with the action of leucocytes. He then gave intravenous injections of the broth culture filtrate to rabbits and found that it did not upset the animals more than pure broth (11). Fleming once again dismissed his previous convictions regarding the "doubtful utility" of "chemical antiseptics." Actually, earlier in the *same year* that he discovered penicillin, while studying a mercury compound which seemed to have some possibilities, he had written, "There is little chance of finding any general antiseptic capable of killing bacteria in the blood stream, though there is some hope that chemicals may be produced with special affinities for special bacteria which may be able to destroy these in the blood, although they may be quite without action on other, and it may be, closely allied bacteria" (12).

With penicillin he had this chemical, he had this antibacterial substance; a mould was producing it. Again he tried to extract the active element pure from the culture but, as with lysozyme, he failed again. And so did all the chemists he asked to try to do it. As he obviously could not inject a broth culture into the blood stream of humans, he tried giving to patients who were dying from septicemia (because only then would the physicians allow him to do so), a *milk* culture of the mould to eat. It looked like Stilton cheese, although certainly not tasting as good. This therapeutic effort sometimes brought slight, short-lasting, unreliable beneficial results. To be used, penicillin and lysozyme would have to wait until they were purified.

With the discovery of lysozyme, Fleming's mind gave birth to wonderful hypotheses, and most of them he proved. Lysozyme, he thought, was the natural defence of the organism, the defence Nature had provided to all living organisms: human, animal, and vegetable. Perhaps at some primeval time no germ could establish itself and cause a disease: they were all sensitive to lysozyme. He thought that pathogenic germs, as we know them, were in fact pathogenic because they had become resistant to the action of lysozyme; that is, they are probably the descendants of resistant mutants of sensitive bacteria. Fleming himself produced such resistant mutants *in vitro,* and using them he proved that the intracellular digestion of bacteria was related to the action of lysozyme in these cells (7). He thought that the different pathogenicity of bacteria to different animals might be due to strain differences in the quantity and quality of lysozyme. He thought that lysozyme *should be* in greater concentration in these parts of the body more exposed to infections or lacking other forms of natural protection. And he proved that this was so.

Years earlier, during World War I, Fleming had struggled to find an efficient antiseptic which would help prevent and cure the wound infections. At that time he had dreamt that Nature (this Nature he believed in so much) *must* have provided every living thing with an effective defence mechanism which would protect it *in all its parts.* With the discovery of lysozyme, Fleming believed that he had found this primeval general natural defence mechanism. He had also found something much greater: he had found *Hope.* Lysozyme proved that a substance did exist which had selective bactericidal action while being harmless to human cells. So this, which up to that time had been considered impossible, *was* possible. Fleming thought that if nothing else could be found, lysozyme, purified and many times more potent, might be an answer—might be a help.

Fleming believed in lysozyme. He believed that it was bound to have a great future; he had discovered it and had done marvelous work on it. Other scientists would follow the path he had opened. Others would purify it, advance it. With absolute confidence he used to say, "We shall hear more about lysozyme."

Your work has made Fleming's prophecy come true.

LADY AMALIA FLEMING

Cheyne Walk, London

BIBLIOGRAPHY

1. Fleming, A. The discovery of penicillin. *Brit. Med. J.* **1:** 711, 1955.
2. Fleming, A. On a remarkable bacteriolytic element found in tissues and secretions. *Proc. Roy. Soc. London* **B93:** 306, 1922.
3. Fleming, A., and Allison, V. D. Further observations on a bacteriolytic element found in tissues and secretions. *Proc. Roy. Soc. London* **B94:** 142, 1922.
4. Fleming, A., and Allison, V. D. Observations on a bacteriolytic substance—lysozyme—found in secretions and tissues. *Brit. J. Exp. Path.* **3:** 252, 1922.
5. Fleming, A., and Allison, V. D. On the antibacterial power of egg white. *Lancet* **1:** 1303, 1924.
6. Fleming, A., and Allison, V. D. On the specificity of the protein of human tears. *Brit. J. Exp. Path.* **6:** 87, 1925.
7. Fleming, A., and Allison, V. D. On the development of strains of bacteria resistant to lysozyme action and the relation of lysozyme action to intracellular digestion. *Brit. J. Exp. Path.* **8:** 214, 1927.
8. Rettger, L. F., and Sperry, J. A. The antiseptic and bactericidal properties of egg white. *J. Med. Res.* **26:** 55, 1912.
9. Fleming, A. Lysozyme. *Proc. Roy. Soc.* **B26:** 1, 1932.
10. Fleming, A. The action of chemical and physiological antiseptics in a septic wound (1919 Hunterian Lecture). *Brit. J. Surg.* **7:** 99, 1919.
11. Fleming, A. On the antibacterial action of cultures of penicillium with special reference to their use in the isolation of *B. influenza. Brit. J. Exp. Path.* **10:** 226, 1929.
12. Fleming, A. The bactericidal power of human blood and some methods of altering it. *Proc. Roy. Soc. Med.* **21:** 839, 1928.

List of Contributors and Participants

Numbers in parentheses indicate the pages on which the authors' contributions begin.

MATTEO ADINOLFI[1] (463), Institute of Cancer Research and Department of Medicine, College of Physicians and Surgeons, Columbia University, New York, New York

VAGN ANDERSEN (307), Division of Hematology, Department of Medicine, A. Rigshospitalet, University Hospital of Copenhagen, Copenhagen, Denmark

HARRIETT ANSARI, Institute of Cancer Research, College of Physicians and Surgeons, Columbia University, New York, New York

NORMAN ARNHEIM (81, 153), Department of Biochemistry, State University of New York at Stony Brook, Stony Brook, New York

RUTH ARNON (105), Department of Chemical Immunology, The Weizmann Institute of Science, Rehovot, Israel

H. ASAMER (373), Department of Medicine, University of Innsbruck, Innsbruck, Austria

S. K. BANERJEE (251), University of Arizona, Tucson, Arizona

S. H. BANYARD[2] (71), Laboratory of Molecular Biophysics, Oxford University, Oxford, England

I. BERNIER (31), Laboratory of Biochemistry, University of Paris, Paris, France

J. BERTHOU (31), Laboratory of Biochemistry, University of Paris, Paris, France

SHERMAN BEYCHOK (165, 281), Departments of Biological Sciences and Chemistry, Columbia University, New York, New York

STEVEN BIRKEN, Department of Medicine, College of Physicians and Surgeons, Columbia University, New York, New York

C. C. F. BLAKE (71), Laboratory of Molecular Biophysics, Oxford University, Oxford, England

BENJAMIN BONAVIDA (143), Department of Microbiology and Immunology, University of California, Los Angeles, California

H. BRAUNSTEINER (373), Department of Medicine, University of Innsbruck, Innsbruck, Austria

KEITH BREW (55), Department of Biochemistry, University of Leeds, Leeds, England

C. F. BREWER (239), Albert Einstein College of Medicine, Bronx, New York

I. D. CAMPBELL (219), Department of Biochemistry, Oxford University, Oxford, England

ROBERT E. CANFIELD (3, 63), Department of Medicine, College of Physicians and Surgeons, Columbia University, New York, New York

O. CASTRO (335), Department of Medicine, Yale University School of Medicine, New Haven, Connecticut

D. CHARLEMAGNE (31), Laboratory of Biochemistry, University of Paris, Paris, France

[1] Present address: Pediatric Research Unit, Guy's Hospital Medical School, London, England.

[2] Present address: Edgenössische Technische Hochschule, Laboratorium für Organische Chemie, Zurich, Switzerland.

xix

LORRAINE M. CLAUSS[3] (269), Department of Biochemistry, University of Minnesota Medical School, Minneapolis, Minnesota

JANNA C. COLLINS (63), Department of Pediatrics, Babies Hospital, Columbia-Presbyterian Medical Center, New York, New York

ARTHUR M. DANNENBERG, JR., Department of Radiological Science, School of Hygiene and Public Health, Johns Hopkins University, Baltimore, Maryland

C. M. DOBSON (219), Department of Inorganic Chemistry, Oxford University, Oxford, England

DAVID DOLPHIN (229), Department of Chemistry, Harvard University, Cambridge, Massachusetts

J. A. DONADIO (335), Department of Medicine, Yale University School of Medicine, New Haven, Connecticut

JOANNA ECONOMIDOU, Blood Research Laboratory, Hellenic Red Cross, Athens, Greece

REUBEN EISENSTEIN (399), Department of Pathology, Rush Medical College, Chicago, Illinois

YUVAL ESHDAT (195), Department of Biophysics, The Weizmann Institute of Science, Rehovot, Israel

MEHDI FARHANGI (379), Institute of Cancer Research, and Department of Medicine, College of Physicians and Surgeons, Columbia University, New York, New York

A. FAURE (31), Laboratory of Biochemistry, University of Paris, Paris, France

STUART C. FINCH (335, 359, 391), Department of Medicine, Yale University School of Medicine, New Haven, Connecticut

ROBERT E. FISCHEL (471), Institute of Cancer Research and Department of Medicine, College of Physicians and Surgeons, Columbia University, New York, New York

ISABELLA LAM FUNG, Institute of Cancer Research, College of Physicians and Surgeons, Columbia University, New York, New York

JACOB FURTH, Institute of Cancer Research, College of Physicians and Surgeons, Columbia University, New York, New York

J. B. L. GEE (391), Department of Internal Medicine, Yale University Lung Research Center, Yale University School of Medicine, New Haven, Connecticut

JEAN-MARIE GHUYSEN (185), Service de Microbiologie, Faculté de Médecine, Institut de Botanique, Université de Liège, Sart-Tilman, Liège, Belgique

HARRIET S. GILBERT (355), Department of Medicine, Mount Sinai School of Medicine of the City University of New York, New York, New York

ALAN GLYNN, Department of Bacteriology, St. Mary's Hospital Medical School, University of London, London, England

JOEL S. GREENBERGER (385), Peter Bent Brigham Hospital, Harvard Medical School, Boston, Massachusetts

ROBERT A. GREENWALD (411), Department of Medicine, Long Island Jewish-Hillside Medical Center, New Hyde Park, New York

A. P. GROLLMAN (239), Albert Einstein College of Medicine, Bronx, New York

RICHARD J. GUALTIERI (281), Departments of Biological Sciences and Chemistry, Columbia University, New York, New York

JAMES HALPER (471), Institute of Cancer Research and Department of Medicine, College of Physicians and Surgeons, Columbia University, New York, New York

NIELS EBBE HANSEN (307), Division of Hematology, Department of Medicine A, Rigshospitalet, University Hospital of Copenhagen, Copenhagen, Denmark

JOHN H. HASH (95), Department of Microbiology, Vanderbilt University School of Medicine, Nashville, Tennessee

J. HERMANN (31), Laboratory of Biochemistry, University of Paris, Paris, France

ROBERT L. HILL (55), Department of Biochemistry, Duke University Medical Center, Durham, North Carolina

NEWTON E. HYSLOP, JR., (449), Department of Medicine, Harvard Medical School, and Medical Service (Infectious Disease Unit), Massachusetts General Hospital, Boston, Massachusetts

[3] Present address: American Institute of Baking, Chicago, Illinois.

TAKASHI ISOBE, Institute of Cancer Research, College of Physicians and Surgeons, Columbia University, New York, New York

ERIC R. JOHNSON (269), Department of Biochemistry, University of Minnesota Medical School, Minneapolis, Minnesota

J. JOLLÈS (31), Laboratory of Biochemistry, University of Paris, Paris, France

P. JOLLÈS (31), Laboratory of Biochemistry, University of Paris, Paris, France

ALAN S. JOSEPHSON (411), Department of Medicine, Downstate Medical Center, State University of New York, Brooklyn, New York

PETER C. KAHN, Department of Biological Sciences, Columbia University, New York, New York

SANDRA KAMMERMAN, Department of Medicine, New York University School of Medicine, New York, New York

HANS KARLE (307), Division of Hematology, Department of Medicine A, Rigshospitalet, University Hospital of Copenhagen, Copenhagen, Denmark

A. KATZ (419), Immunoprotein Research Laboratory of the University of Toronto, Rheumatic Disease Unit, The Wellesley Hospital, and the Department of Pathology, St. Michael's Hospital, Toronto, Canada

FRIEDRICK KATZ, Psychiatric Division, Bellevue Hospital, New York, New York

KATHRYN C. KERN (449), Infectious Disease Unit, Massachusetts General Hospital, Boston, Massachusetts

MATTI KLOCKARS[4] (471), Institute of Cancer Research, College of Physicians and Surgeons, Columbia University, New York, New York

FREDERICK W. KRAUS, Department of Microbiology, University of Alabama, Birmingham, Alabama

I. KREGAR (251), Department of Biochemistry, J. Stefan Institute, University of Ljubljana, Ljubljana, Yugoslavia

KLAUS E. KUETTNER (399), Departments of Orthopedic Surgery and Biochemistry, Rush-Presbyterian-St. Luke's Medical Center, Chicago, Illinois

S. LAPANJE (251), Department of Chemistry, University of Ljubljana, Ljubljana, Yugoslavia

GUSTAVE LIENHARD, Department of Biochemistry, Dartmouth Medical School, Hanover, New Hampshire

M. E. LIPPMAN (335), Department of Internal Medicine, Yale University School of Medicine, New Haven, Connecticut

JOHN N. LOEB (463), Department of Medicine, College of Physicians and Surgeons, Columbia University, New York, New York

THOMAS MAACK (321), Department of Physiology, Cornell University Medical College, New York, New York

JEFFREY McKELVEY, Department of Anatomy, University of Connecticut Health Center, Storrs, Connecticut

JORGEN MALMQUIST (347), Department of Medicine, University of Lund, Malmö General Hospital, Malmö, Sweden

D. MARCUS (239), Albert Einstein College of Medicine, Bronx, New York

PAUL A. MARKS, Department of Medicine, College of Physicians and Surgeons, Columbia University, New York, New York

ELCHANAN MARON[5] (121), Department of Chemical Immunology, The Weizmann Institute of Science, Rehovot, Israel

JIRI MESTECKY, Department of Microbiology, University of Alabama, Birmingham, Alabama

KARL MEYER, Department of Chemistry, Yeshiva University, New York, New York

ALEXANDER MILLER (143), Department of Bacteriology, University of California, Los Angeles, California

WILLIAM C. MOLONEY (385), Peter Bent Brigham Hospital, Harvard Medical School, Boston, Massachusetts

BETTY ROSE MOORE (493), Institute of Cancer Research, College of Physicians and Surgeons, Columbia University, New York, New York

---

[4] Present address: Fourth Department of Medicine, University of Helsinki, Helsinki, Finland.

[5] Deceased.

FRANK J. MORGAN (81), Department of Medicine, College of Physicians and Surgeons, Columbia University, New York, New York

RODERICK S. MULVEY (281), Departments of Biological Sciences and Chemistry, Columbia University, New York, New York

LOUIS H. MUSCHEL, Research Department, American Cancer Society, National Office, New York, New York

M. A. OGRYZLO (419), Department of Medicine, University of Toronto, Toronto, Ontario, Canada

ELLIOTT F. OSSERMAN (303, 379, 463, 471, 493), Institute of Cancer Research and Department of Medicine, College of Physicians and Surgeons, Columbia University, New York, New York

R. S. PASCUAL (391), Department of Internal Medicine, Yale University Lung Research Center, Yale University School of Medicine, New Haven, Connecticut

STEVEN L. PATT (229), Department of Chemistry, Harvard University, Cambridge, Massachusett

PASQUALE E. PERILLIE[6] (335, 359, 391), Department of Medicine, Yale University School of Medicine, New Haven, Connecticut

J. P. PÉRIN (31), Laboratory of Biochemistry, University of Paris, Paris, France

D. C. PHILLIPS (9), Laboratory of Molecular Biophysics, Department of Zoology, Oxford University, Oxford, England

MIROSLAV D. POULIK, Department of Immunochemistry, Wayne State School of Medicine, Detroit, Michigan

E. M. PRAGER (127), Department of Biochemistry, University of California, Berkeley, California

W. PRUZANSKI (419), Immunoprotein Research Laboratory of the University of Toronto, Rheumatic Disease Unit, The Wellesley Hospital, and the Department of Pathology, St. Michael's Hospital, Toronto, Ontario, Canada

ROY J. RIBLET (89), The Salk Institute for Biological Studies, San Diego, California

DAVID S. ROSENTHAL (385), Peter Bent Brigham Hospital, Harvard Medical School, Boston, Massachusetts

J. A. RUPELY (251), Department of Chemistry, University of Arizona, Tucson, Arizona

J. SAINT-BLANCARD (31), Laboratory of Biochemistry, University of Paris, Paris, France

MILTON R. J. SALTON, Department of Microbiology, New York University School of Medicine, New York, New York

F. SCHMALZL (373), Department of Medicine, University of Innsbruck, Innsbruck, Austria

GEBHARD F. B. SCHUMACHER (427), Section of Reproductive Biology, Department of Obstetrics and Gynecology, The University of Chicago, Chicago, Illinois

ROBERT J. SCIBIENSKI[7] (143), Department of Bacteriology, University of California, Los Angeles, California

DUANE SEARS, Department of Biological Sciences, Columbia University, New York, New York

BEATRICE SEEGAL, Department of Microbiology, Columbia University, New York, New York

ELI SERCARZ (143), Department of Bacteriology, University of California, Los Angeles, California

NATHAN SHARON (195), Department of Biophysics, The Weizmann Institute of Science, Rehovot, Israel

A. F. SHRAKE (251), Department of Chemistry, University of Arizona, Tucson, Arizona

DANIEL SIGULEM (321), Department of Physiology, Cornell University Medical College, New York, New York

ARTHUR T. SKARIN, Department of Medicine, Harvard Medical School, Boston, Massachusetts

JOAN H. SOBEL (63), Department of Medicine, College of Physicians and Surgeons, Columbia University, New York, New York

---

[6] Present address: Department of Medicine, Bridgeport Hospital, Bridgeport, Connecticut.

[7] Present address: Department of Medical Microbiology, School of Medicine, University of California, Davis, California.

ALAN SOLOMON, University of Tennessee Memorial Research Center and Hospital, Knoxville, Tennessee

ALKIS J. SOPHIANOPOULOS, Department of Biochemistry, Emory University, Atlanta, Georgia

NINO SORGENTE (399), Department of Orthopedic Surgery, Rush Medical College, Chicago, Illinois

HOWARD M. STEINMAN (55), Department of Biochemistry, Duke University Medical Center, Durham, North Carolina

H. STERNLICHT (239), Bell Laboratories, Murray Hill, New Jersey

JOAN A. STRATTON (143), Department of Radiology, Harbor General Campus, University of California, Torrance, California

JACK L. STROMINGER (169), Department of Biochemistry and Molecular Biology, Harvard University; The Biological Laboratories, Cambridge, Massachusetts

I. D. A. SWAN (71), The Astbury Department of Biophysics, University of Leeds, Leeds, England

BRIAN D. SYKES (229), Department of Chemistry, Harvard University, Cambridge, Massachusetts

DONALD J. TIPPER (169), Department of Microbiology, University of Massachusetts Medical School, Worcester, Massachusetts

V. TURK (251), Department of Biochemistry, J. Stefan Institute, University of Ljubljana, Ljubljana, Yugoslavia

JAN WALDENSTRÖM, Department of Medicine. University of Lund, Malmö General Hospital, Malmö, Sweden

W. ALLAN WALKER (449), Gastrointestinal Unit, Pediatric Service, Massachusetts General Hospital, Boston, Massachusetts

BABETTE WEKSLER, Department of Medicine, Cornell University Medical College, New York, New York

DONALD B. WETLAUFER (269), Department of Biochemistry, University of Minnesota Medical School, Minneapolis, Minnesota

R. J. P. WILLIAMS (219), Department of Inorganic Chemistry, Oxford University, Oxford, England

A. C. WILSON (127), Department of Biochemistry, University of California, Berkeley, California

A. V. XAVIER[8] (219), Department of Inorganic Chemistry, Oxford University, Oxford, England

[8] Present address: Centro Estudos Quimica Nuclear, Instituto Superior Tecnico, Lisboa, Portugal.

Preface

1972 marked the fiftieth anniversary of Alexander Fleming's first paper describing the discovery of lysozyme.* The Conference on which this work is based was organized to commemorate the lysozyme discovery and to provide a forum for surveying at least a part of the vast amount of information concerning lysozyme which has been gathered in the intervening fifty years by many hundreds of laboratory and clinical investigators. The organizers of the Conference were particularly pleased that Lady Amalia Fleming agreed to serve as Honorary Chairman. Some personal recollections of her late husband and his convictions regarding the importance of lysozyme are presented on p. xiii.

The diversity of interests represented by the participants in the Conference—biophysics, chemistry, microbiology, physiology, and clinical medicine—testifies to the enormous scope and wide ramifications of lysozyme research.

As this work clearly documents, we now know a great deal about the chemistry and structure of several lysozymes, interactions with substrates and inhibitors, distribution in tissues, and changes associated with various disease states. Despite this wealth of information, however, we almost certainly do *not know* the full range of *functions* of the diverse "lysozyme systems" as they occur throughout the plant and animal kingdoms. Thus, it would seem that Nature has employed this basic enzyme system for different purposes in different settings. Starting with Fleming's own studies, much of the work on lysozyme has been focused on its antibacterial activities, and these functions are now well documented and apparently well established. But the virtual ubiquity of lysozymes and the paradoxically high concentrations in certain sites such as cartilage, genital secretions, and the eggs of fish, amphibia, and birds already suggested to Fleming and subsequently to many other investigators that lysozyme might serve other functions besides its antibacterial role. This critical point is considered by

*Fleming, A. On a remarkable bacteriolytic element found in tissues and secretions. *Proc. Roy. Soc. London* **B93**: 306, 1922.

several of the contributors to this volume, and some preliminary evidence is presented indicating that lysozyme may, in fact, significantly alter certain mammalian cell constituents. If this proves to be correct, an entirely new and potentially very important area of lysozyme research will have been opened, thus validating Fleming's prediction, as recalled by Lady Fleming, that "we shall hear more about lysozyme."

Obviously, in our search for further insights into the possible biological functions of lysozyme, additional structural and biochemical data will be extremely useful—if we know what a particular molecule (enzyme) *can* do, it should be easier to determine what it *is* doing in a particular setting. In great measure, this was one of the primary goals of the Conference.

The limitations of time at the Conference and space in this volume, unfortunately, necessitated the exclusion of many important topics including the bacteriophage and plant lysozymes. It is hoped that these subjects can be considered at subsequent lysozyme conferences.

We wish to express our deep appreciation to the Damon Runyon Memorial Fund, Hoffman-La Roche, Inc., and Kallestad Laboratories, Inc., for their generous support of the Conference, and our indebtedness to Mrs. Vivian Meyer and Mr. James Quirk for their invaluable efforts in organizing the meeting.

ELLIOTT F. OSSERMAN

Note on Nomenclature

For the past several years, there has been considerable confusion and disagreement regarding the terms "lysozyme" and "muramidase." It seemed appropriate, therefore, that this question be considered by the participants in the Lysozyme Conference since many of the more active investigative groups in this field were represented.

The following recommendation was drafted at the Conference by an *ad hoc* committee (A. A. Glynn, Ida Bernier, S. C. Finch, M. R. J. Salton):

> The present position is that while the Commission on Enzyme Nomenclature in 1961 decided that E.C. 3.2.1.17 should be called muramidase, it reversed this decision in 1964. The recommendation now reads:
>
> E.C. 3.2.1.17. Systematic name: mucopeptide *N*-acetylmuramylhydrolase; recommended trivial name: mucopeptide glucohydrolase, lysozyme; other name not recommended: muramidase.
>
> This recommendation should be followed since otherwise confusion in indexing will result.
>
> The term lysozyme should include only enzymes which are β-1,4-glycan-hydrolases. They may be of animal, plant, or microbial origin. In a full description, the species of origin should be given and any allelic or other variations indicated.
>
> The primary identification of a lysozyme is usually made by determining its lytic activity against *Micrococcus lysodeikticus* (NCTC (2665). Chitinases have no action on this microorganism and are, therefore, excluded.

This statement was unanimously approved by the Conference participants.

BIBLIOGRAPHY

"Comprehensive Biochemistry," 1st edition (M. Florkin and E. H. Stotz, eds.), Vol. 13, p. 106. Elsevier, Amsterdam, 1964.
"Comprehensive Biochemistry," 2nd edition (M. Florkin and E. H. Stotz, eds.), Vol. 13, p. 138. Elsevier, Amsterdam, 1965.

SECTION

1

STRUCTURE OF LYSOZYMES

Introduction

R. E. CANFIELD

The contributions to this work represent a balance between a review of existing knowledge about lysozyme and new contributions that report recently completed research. In organizing the contents it seemed wisest to begin by focusing on the chemical nature of lysozymes. Therefore, the aim of this first section is to review the status of our knowledge of the structure of various lysozymes to provide a basis for later discussions of the enzyme's mechanism of action and its biological role.

The papers dealing with structural aspects of the lysozyme molecule have been divided into those concerned with primary structure, those concerned with three-dimensional structure as determined by X-ray crystallography, and finally a group of immunochemical studies which detail the structural aspects of this protein from the perspective of an antibody.

Primary Structure

Avian Lysozymes

Nearly 30 years elapsed following Fleming's discovery of lysozyme before serious studies of the enzyme's structure began. As techniques of protein chemistry evolved, the ready availability of purified hen egg-white lysozyme made this protein ideal for the application of new techniques including those whereby amino acid sequence information could be obtained. Some of the earliest structural studies were conducted in the laboratory of Professor Claude Fromageot in Paris (1, 2) and, following his untimely death, this work was carried to completion by Pierre Jollès and his colleagues. An independent determination of the primary structure of hen egg-white lysozyme was also begun in the laboratory of Dr. Christian B. Anfinsen, and both of these studies were concluded 10 years ago (3, 4). Thus one may view the elucidation of the primary structure of hen egg-white lysozyme as the beginning of a decade of progress in the knowledge of the chemistry of these molecules. The group with Professor Jollès in Paris has been the most active in determining the primary structure of lysozymes and, therefore, they were asked to review the current status of this subject.

The completion of a high resolution X-ray crystallographic analysis of hen egg-white lysozyme in 1965, with its implications for the enzyme's mechanism of action, stimulated a number of groups to study the primary structure of other lysozymes. It was hoped that evolutionary variations in structure around the active site would permit one to note those features common to all lysozymes which might be essential for enzyme activity and also to note those dissimilar features that might explain differences in the observed characteristics of the action of different lysozymes. In general, the other avian lysozymes have proved to be quite similar to chicken lysozyme as described in the paper by Jollès and also in a later paper in this volume by Wilson. One exception appears to be the lysozyme in the egg white from White Embden geese. When compared to chicken lysozyme, it is found to have a very different amino acid composition, a dissimilar pattern of tryptic peptides, and a much higher specific activity than hen egg-white lysozyme (5). These results, coupled with the suggestion of marked immunologic differences (6) and different substrate-inhibition characteristics (7), led us to pursue amino acid sequence studies which have indicated that White Embden goose egg-white lysozyme probably bears no homologous relationship to the hen egg-white group of lysozymes (8). Arnheim has shown that both of these genetically different types of lysozymes are present in the egg white of the black swan (6), and the most recent chemical results are presented in a paper in this volume by Morgan and Arnheim. In a separate paper Arnheim reviews some additional findings. All of the evidence suggests that the goose-type lysozyme possesses a distinctly different architectural solution from that of the hen-type lysozyme to the problem of catalyzing and hydrolysis of β-$(1 \rightarrow 4)$-glycosidic linkages between N-acetylglucosamine and N-acetylmuramic acid in bacterial cell walls.

As discussed in the paper by Hash, a protein with lysozyme activity is synthesized by the fungus *Chalaropsis*. Sufficient details of the amino acid sequence of the protein are reported to indicate that this enzyme must be representative of a third class of lysozymes that have appeared in evolution with no homologous relationship to either the hen-type or the goose-type lysozymes. The plant lysozymes (9) and the bacteriophage lysozymes (10) appear to provide two more architectural solutions for catalyzing the hydrolysis of these β-(1 → 4)-glycosidic bonds, but these subjects are not reviewed in this work.

α-Lactalbumins

Shortly after the publication of the primary structure of hen egg-white lysozyme, Brew, Vanaman, and Hill recognized the presence of an homologous relationship between the structure of this lysozyme and that of bovine α-lactalbumin (11). In recent years these investigators and their co-workers have carried out extensive studies of the primary structures of several α-lactalbumins and the paper by Hill *et al.* reviews these data together with some discussion of their significance.

Mammalian Lysozymes

Fleming described the presence of lysozyme activity in a number of mammalian tissues, but the purification of large quantities of enzyme was less readily achieved than that of hen egg-white lysozyme. Then in 1966 Osserman made the important observation that patients with myelomonocytic leukemia often excrete large quantities of this enzyme in their urine, with some patients excreting in excess of 1 gm per day (12). This provided an excellent source of highly purified enzyme and we have included a brief notation concerning our studies with human leukemia lysozyme in a subsequent paper. The structural data for human milk lysozyme obtained by Jollès indicate that it is the same as the leukemic enzyme and that there is a high degree of homology between the human and the hen egg-white group of lysozymes. The data presented by Riblit in this volume indicate that lysozyme from the mouse is also homologous with the hen egg-white group of lysozymes.

X-Ray Crystallography

In the 50 years since Fleming's discovery of lysozyme, undoubtedly the single most significant contribution to our knowledge concerning this enzyme was the X-ray crystallographic analysis performed by Phillips and his colleagues when they were at The Royal Institution in London. These crystallographic studies of lysozyme and the observations of interactions with substrates and inhibitors are reviewed in the paper by Phillips. In addition, recent results of a high resolution X-ray crystallographic study of human leukemia lysozyme by Banyard, Blake, and Swan are included. These details of the three-dimensional structures of hen egg-white lysozyme and human lysozyme provide the structural basis for the discussion of enzymatic and nonenzymatic behavior in the subsequent sections.

Immunochemical Studies

Immunochemists have taken advantage of the growing knowledge of the primary and tertiary structures of lysozymes from many different species to employ these molecules, or peptides derived from them, as model antigens. Arnon and her co-workers at the Weizmann Institute have made many original contributions to this field, particularly with studies of the so-called loop peptide of hen egg-white lysozyme, i.e., residues 60–83, and these are reviewed in her paper. A further application of immunologic studies of the loop region of avian and human lysozymes is described in the paper by Maron.

The correlation between immunologic cross-reactivity and alterations in the primary structure of different lysozymes is a matter of considerable interest. Wilson reviews his earlier work in this area with bird lysozymes and has added in his paper some recent data concerning the cross-reactivity of primate lysozymes. Arnheim has detailed the application of immunologic techniques in the study of multiple genes for lysozyme with the goose-type and chicken-type lysozymes that were noted earlier. Finally, Sercarz *et al.* report an interesting application of the use of lysozyme as a model antigen in studies of the regulation of the immune response, and their paper contains some additional data describing the cross-reactivity of gallinaceous lysozymes with human lysozyme.

References

1. Acher, R., Jutisz, M., Fromageot, C., *Biochim Biophys. Acta* **8,** 442 (1952).
2. Acher, R., Laurila, U. R., Thaureaux, J., and Fromageot, C., *Biochim. Biophys. Acta* **14,** 151 (1954).
3. Jollès, J., Jauregui-Adell, J., Bernier, I., and Jollès, P., *Biochim. Biophys. Acta* **78,** 668 (1963).
4. Canfield, R. E., *J. Biol. Chem.* **238,** 2698 (1963).

5. Canfield, R. E., and McMurry, S., *Biochem. Biophys. Res. Commun.* **26,** 38 (1967).
6. Arnheim, N., and Steller, R., *Arch. Biochem. Biophys.* **141,** 656 (1970).
7. Saint-Blancard, J., Henique, Y., Ducasse, D., Dianoux, A. C., and Jollès, P., *Bull. Soc. Chim. Biol. (Paris)* **50,** 1783 (1968).
8. Canfield, R. E., Kammerman, S., Sobel, J. H., and Morgan, F. J., *Nature (Lond.)* **232,** 16 (1971).
9. Howard, J. B., and Glazer, A. N., *J. Biol. Chem.* **244,** 1399 (1969).
10. Tsugita, A., *in* "The Enzymes" (P. D. Boyer, ed.), 3rd ed., Vol. 5, pp. 344–411. Academic Press, New York, 1971.
11. Brew, K., Vanaman, T. C., and Hill, R. L., *J. Biol. Chem.* **242,** 3747 (1967).
12. Osserman, E. F., and Lawlor, D. P., *J. Exp. Med.* **124,** 921 (1966).

Crystallographic Studies of Lysozyme and Its Interactions with Inhibitors and Substrates

D. C. PHILLIPS

The crystallography of lysozyme began in the 1930's at Oxford; E. P. Abraham, a member of the famous penicillin team, and Robert Robinson succeeded in crystallizing the hen egg-white enzyme (1). They tried to improve these crystals in order to start Oxford crystallographers working on them, but had difficulty in reproducing the crystallization for reasons that are now obscure, though they may have been connected with the use of Chinese eggs as the source of the enzyme. It may have been that the lysozyme was not in very good condition by the time it reached Oxford.

It was not long, however, before Alderton and Fevold (2) succeeded in crystallizing the enzyme in a rather straightforward way, and it was the Abraham and Robinson crystal form, as produced by the Alderton and Fevold method, that we began to study at The Royal Institution in about 1960.

This was after the successful work on myoglobin (32) and it was some time after other crystallographers had begun studying lysozyme. The problem was introduced to our group at The Royal Institution by a visiting Argentinian, Roberto Poljak, who is now working on immunoglobulins at Johns Hopkins, and for some reason things went extremely well for us, so that by 1965, thanks to the efforts of Poljak, who left us in 1962, and

then of C. C. F. Blake, A. C. T. North, V. R. Sarma, D. F. Koenig, and
G. A. Mair, we succeeded in producing an electron density map of the enzyme, part of which is shown in Fig. 1 (10).

  Figure 1 shows the type of information that crystallographers produce
about protein molecules. In 1965, of course, it was the first electron density
map of an enzyme, and, happily, it was clear enough for us to be able to
see the main outline of the structure without very much difficulty. But I
must add immediately that interpreting the image was made much more
straightforward—in fact, perhaps I should say "was made possible"—by
the fact that at about the same time Pierre Jollès in Paris and R. E. Canfield
at NIH, and more recently in New York, with their colleagues (16, 18,

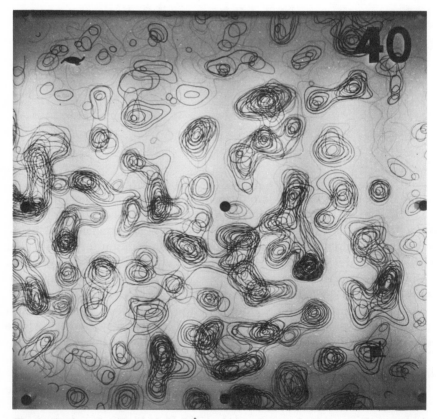

Fig. 1.   Sections Z = 40–44 of the 2-Å resolution electron density map. Contours at intervals of 0.25 eA⁻³. The heavy peak at lower right corresponds to part of a disulfide bridge
(residues 30–115) which continues into sections above those shown. The feature to the left
of this peak represents the side chain of Phe 34.

28, 30), had determined the amino acid sequence of the enzyme, so that we were able by looking at this map, which shows the electron density at less than full atomic resolution, to identify the features in it in relationship to the sequence.

The part of the map shown in Fig. 1 is, in fact, the part that we first plotted and were able immediately to relate to the sequence. The procedure was very simple. We looked for the highest electron density in the map, which happens to be bottom right of center in these sections. Knowing something through the work of Pauling and Corey and many others about possible protein conformations, we were then able to postulate immediately that this feature corresponds to a disulfide bridge, where two sulfur atoms are located, and that it is related above to a run of polypeptide chain which is clearly in the α-helical conformation.

This may not be too clear in the figure, but in the three-dimensional map it really is rather obvious. Furthermore, one can count amino acid residues from the disulfide bridge, around one turn of the helix, to a feature to its left, which is a flat, platelike region of electron density clearly linked back to the main polypeptide chain and corresponding to the side chain of a phenylalanine residue. The map shows that these two residues, the half cystine and the phenylalanine, are four residues apart in the amino acid

	1	2	3	4	5	6	7	8	9	10	11	12	13	14	15	16	17	18	19
HLL	LYS	VAL	PHE	GLU	ARG	CYS	GLU	LEU	ALA	ARG	THR	LEU	LYS	ARG	LEU	GLY	MET	ASP	GLY
HEL	LYS	VAL	PHE	GLY	ARG	CYS	GLU	LEU	ALA	ALA	ALA	MET	LYS	ARG	HIS	GLY	LEU	ASP	ASN
BAL	GLU	GLN	LEU	THR	LYS	CYS	GLU	VAL	PHE	ARG	GLU	LEU	LYS	ASP	...	...	LEU	LYS	GLY

	20	21	22	23	24	25	26	27	28	29	30	31	32	33	34	35		36	37
HLL	TYR	ARG	GLY	ILE	SER	LEU	ALA	ASN	TRP	MET	CYS	LEU	ALA	LYS	TRP	GLU	...	SER	GLY
HEL	TYR	ARG	GLY	TYR	SER	LEU	GLY	ASN	TRP	VAL	CYS	ALA	ALA	LYS	PHE	GLU	...	SER	ASN
BAL	TYR	GLY	GLY	VAL	SER	LEU	PRO	GLU	TRP	VAL	CYS	THR	THR	...	PHE	HIS	THR	SER	GLY

	38	39	40	41	42	43	44	45	46	47	48	49	50	51	52	53	54	55	56
HLL	TYR	ASN	THR	ARG	ALA	THR	ASN	TYR	ASN	ALA	GLY	ASP	ARG	SER	THR	ASP	TYR	GLY	ILE
HEL	PHE	ASN	THR	GLN	ALA	THR	ASN	ARG	ASN	THR	...	ASP	GLY	SER	THR	ASP	TYR	GLY	ILE
BAL	TYR	ASP	THR	GLU	ALA	ILE	VAL	GLN	ASN	...	...	ASN	GLN	SER	THR	ASP	TYR	GLY	LEU

	57	58	59	60	61	62	63	64	65	66	67	68	69	70	71	72	73	74	75
HLL	PHE	GLN	ILE	ASN	SER	ARG	TYR	TRP	CYS	ASN	ASP	GLY	LYS	THR	PRO	GLY	ALA	VAL	ASN
HEL	LEU	GLN	ILE	ASN	SER	ARG	TRP	TRP	CYS	ASN	ASP	GLY	ARG	THR	PRO	GLY	SER	ARG	ASN
BAL	PHE	GLN	ILE	ASN	ASN	LYS	ILE	TRP	CYS	LYS	ASP	ASP	GLN	ASN	PRO	HIS	SER	SER	ASN

	76	77	78	79	80	81	82	83	84	85	86	87	88	89	90	91	92	93	94
HLL	ALA	CYS	HIS	LEU	SER	CYS	SER	ALA	LEU	LEU	GLN	ASP	ASN	ILE	ALA	ASP	ALA	VAL	ALA
HEL	LEU	CYS	ASN	ILE	PRO	CYS	SER	ALA	LEU	LEU	SER	SER	ASP	ILE	THR	ALA	SER	VAL	ASN
BAL	ILE	CYS	ASN	ILE	SER	CYS	ASP	LYS	PHE	LEU	ASN	ASN	ASP	LEU	THR	ASN	ASN	ILE	MET

	95	96	97	98	99	100	101	102	103	104	105	106	107	108	109	110	111	112	113
HLL	CYS	ALA	LYS	ARG	VAL	VAL	ARG	ASP	PRO	GLN	GLY	ILE	ARG	ALA	TRP	VAL	ALA	TRP	ARG
HEL	CYS	ALA	LYS	LYS	ILE	VAL	SER	ASP	GLY	ASN	GLY	MET	ASN	ALA	TRP	VAL	ALA	TRP	ARG
BAL	CYS	VAL	LYS	LYS	ILE	LEU	...	ASP	LYS	VAL	GLY	ILE	ASN	TYR	TRP	LEU	ALA	HIS	LYS

	114	115	116	117	118	119	120	121	122	123	124	125	126	127	128		129	130
HLL	ASN	ARG	CYS	GLN	ASN	ARG	ASP	VAL	ARG	GLN	TYR	VAL	GLN	GLY	CYS	...	GLY	VAL
HEL	ASN	ARG	CYS	LYS	GLY	THR	ASP	VAL	GLN	ALA	TRP	ILE	ARG	GLY	CYS	...	ARG	LEU
BAL	ALA	LEU	CYS	SER	GLU	LYS	LEU	ASP	GLN	...	TRP	LEU	...	...	CYS	GLU	LYS	LEU

Fig. 2. Amino acid sequences of human leukemic lysozyme (HLL), hen egg-white lysozyme (HEL), and bovine α-lactalbumin (BAL). Figure kindly provided by Dr. R. E. Canfield.

sequence. We therefore looked straightaway at the sequence which is shown in Fig. 2 and found only one place in which there was this relationship between a disulfide bridge and a phenylalanine, at residue Cys 30 and residue Phe 34.

In this way we were able to locate where we were in the polypeptide chain, and then within a few days to follow it around the molecule, matching the electron density to the amino acid sequence and building a model of the enzyme. Very soon Sir Lawrence Bragg was able to draw a freehand sketch of the layout of the polypeptide chain in the molecule. His sketch is shown in Fig. 3. This drawing shows that we located some significant lengths of α-helix and otherwise a rather random-looking arrangement of the polypeptide chain.

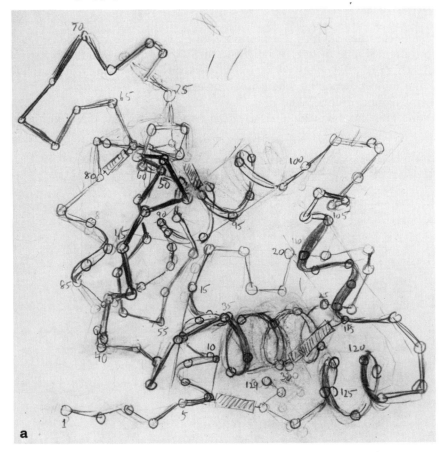

Fig. 3. Drawing by W. L. Bragg of the course of the main polypeptide chain in hen eggwhite lysozyme: (a) original pencil sketch; (b) copy prepared for publication by S. J. Cole.

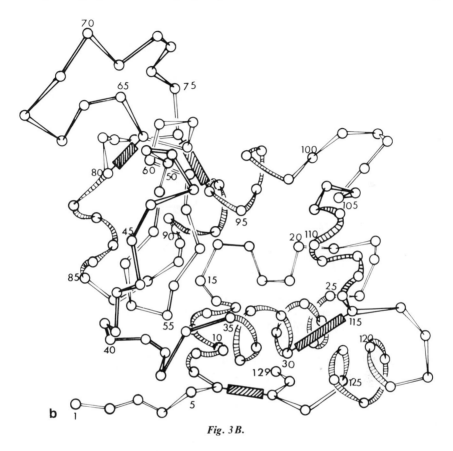

Fig. 3 B.

Subsequent work has enabled us to improve that description. We now see the elements of secondary structure in the enzyme, as shown in Fig. 4. This diagram illustrates the hydrogen bonding in the helices and also the fact that there is a somewhat distorted region of mainly antiparallel β-pleated sheet in the molecule. These are the two elements of protein conformation that had been predicted by Pauling and Corey in 1951 (44, 45).

We are still in the course of refining the structure, in the hope eventually of being able to define the atomic positions in the enzyme molecule to within less than 0.1 Å. This work is being done in collaboration with R. Diamond and M. Levitt in the MRC Molecular Biology Laboratory at Cambridge. Starting with the atomic coordinates of our model, taken by hand from the electron density, Diamond has performed a series of real space refinements, systematic attempts by means of an objective computer program, to derive a model which fits the observed electron density as well

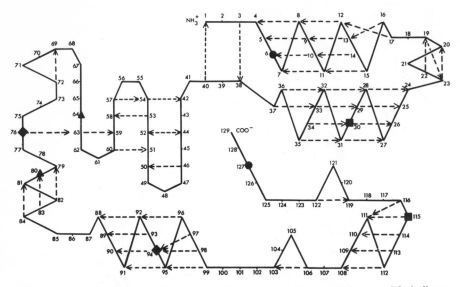

Fig. 4. Secondary structure in hen egg-white lysozyme. Arrows, NH ----> CO, indicate hydrogen bonds between main chain groups; symbols, circles, squares, triangles, and diamonds, show the pairing of cysteine residues in disulfide bridges.

as possible while remaining consistent with all that we know about the stereochemistry of protein molecules (20, 21).

At the same time, Levitt (36) has been refining the crystallographic model to give a minimum energy conformation, using our best knowledge of the energetics of interactions between different groups in a protein molecule. These two approaches lead to encouragingly similar descriptions of the molecule in which the atomic positions can be specified much more closely than before.

Further refinement of this structure will involve recalculation of the electron density map using information derived from our current model of the structure. Such work is already in progress together with extension of the study to a higher level of resolution by the collection of additional diffraction data at higher angles. Unfortunately a limit is set to this approach by the fact that these tetragonal crystals of hen egg-white lysozyme do not diffract well at angles corresponding to interplanar spacings less than 2 Å. Consequently a more promising approach may be to concentrate attention for high resolution studies on the triclinic crystals first described by Stein-rauf (54) which do diffract well at high angles and make possible an analysis at near 1 Å resolution. The structure of this crystal form has been determined at low resolution by comparison with the known tetragonal crystal structure (31) and L. H. Jensen in Seattle is now engaged in extending the analysis to the highest attainable level of resolution.

That is all I wish to say at this time about current crystallographic studies of the native enzyme structure, but it will be clear that attempts are being made to define the structure and atomic positions more closely. This information can then be used to further our understanding of the activity of this enzyme and also, of course, of how it is folded and how the folded structure is stabilized.

Of course, Fleming (23) himself discovered the gram-positive species of bacteria, *Micrococcus lysodeikticus*, that is particularly susceptible to the action of the enzyme and which is still used universally in assaying its activity. It was not until much later that Salton (51) demonstrated that the substrate was located entirely within the bacterial cell wall and it was another 10 years before its chemical constitution was fully established. Important early experiments by Karl Meyer, Ernst Chain, and their colleagues (22, 39, 40) showed that lysozyme releases *N*-acetylamino sugars from *M. lysodeikticus*, but the first indication of the type of linkage attacked by lysozyme came when Berger and Weiser (8) showed that lysozyme also degrades chitin, the linear polymer of *N*-acetylglucosamine (GlcNAc). They concluded that the enzyme has β-$(1 \to 4)$-glucosaminidase activity. Subsequently, Salton and Ghuysen (52) isolated a tetrasaccharide from lysozyme digests of *M. lysodeikticus* cell walls and showed that it contained equimolar amounts of *N*-acetylglucosamine (GlcNAc) and *N*-acetylmuramic acid (MurNAc). On incubation with lysozyme the tetrasaccharide yielded a disaccharide of GlcNAc and MurNAc as the only product. It was deduced that the tetrasaccharide was a dimer of this disaccharide in which the two units are joined by the lysozyme-sensitive β-$(1 \to 4)$ linkage. The structure originally assigned to the disaccharide was *N*-acetylglucosaminyl-β-$(1 \to 6)$-*N*-acetylmuramic acid, but, shortly before our crystallographic analysis of the lysozyme structure came to fruition, Jeanloz *et al.* (25) synthesized this β-$(1 \to 6)$-disaccharide and found its properties to be different from those of the cell wall disaccharide. They suggested that the linkage in the natural dimer is β-$(1 \to 4)$ and hence that the cell wall tetrasaccharide has the structure GlcNAc-β-$(1 \to 4)$-MurNAc-β-$(1 \to 4)$-GlcNAc-β-$(1 \to 4)$-MurNAc, which is shown in Fig. 5 together with the lysozyme-sensitive linkage.

From these results it appeared that long-chain mucopolysaccharides comprising alternative residues of GlcNAc and MurNAc joined by β-$(1 \to 4)$ linkages are important constituents of the bacterial cell walls. It is clear that studies involving lysozyme have played a leading part in the analysis of cell wall structures. This topic is discussed further in this book by Strominger and by Ghuysen who have played important parts in the work. It may be appropriate to remark at this point, however, that in concentrating on this aspect of the activity of lysozyme, I certainly do not wish to give the impression that lysozyme has only this particular glucosaminidase activity or

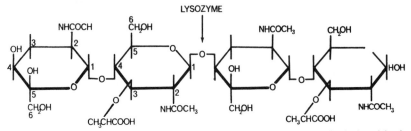

Fig. 5. Cell wall tetrasaccharide with the β-(1→4)-glycosidic linkage hydrolyzed by lysozyme, shown by an arrow. The formula is drawn unconventionally to resemble more closely the actual atomic arrangement in which acetamido groups from adjacent sugar residues project in opposite directions.

that we know precisely what its major biological role is. This work, almost for the first time, brings out clearly the growing diversity of lysozyme studies and, hopefully it will help to clarify what part the enzyme plays in other situations.

The structure of the cell wall mucopolysaccharide is very similar to the structure of chitin, differing from it only through the inclusion of MurNAc residues with their characteristic lactyl side groups, so that the effects of lysozyme upon chitin and upon the cell wall polymer seem likely to be very closely related. One problem, solved by crystallographic studies, was to understand the specificity of lysozyme for the linkages MurNAc-β-(1→4)-GlcNAc, but otherwise studies of the interactions of lysozyme with oligomers derived from chitin seemed likely to be illuminating and such studies have played a major part in our crystallographic investigations. These experiments were prompted in particular by the observations of Wenzel *et al.* (57) that lysozyme promotes the cleavage of the tri-*N*-acetyl-chitotriose (tri-GlcNAc) with the release of mono- and disaccharides and, most importantly, that the products inhibit the activity of the enyzme competitively.

These results (together with the observation that the trisaccharide is a very poor substrate and acts itself as a competitive inhibitor of the enzyme activity) suggested some rather simple and direct crystallographic studies of the ways in which the sugar inhibitor molecules bind to the enzyme (26). Our hope was, of course, that the inhibitors would prove to bind individually in ways related to their binding as components of the substrate and, as I shall indicate, this hope appears to be realized. Happily our lysozyme crystals, like all known protein crystals, include a high proportion of liquid so that relatively small sugar molecules, such as *N*-acetylglucosamine and its short oligomers, can diffuse quite readily into them. Since the arrangement of the protein molecules is not generally upset by incorporation of the sugars into the crystal structure (i.e., since the inhibitor-containing crystals are isomorphous with the native crystals), it is a simple matter to determine the change in structure brought about by the sugar. When the diffraction pattern of the derivative crystal is measured, the difference in electron density between derivative and native structure can then be calculated from the observed changes in the intensities of the X-ray reflections and their known phases, since the latter have been determined already in the original analysis of the native structure (e.g., 46).

The kind of result that we produced in this way (9) is shown in Fig. 6. This map shows a part of the difference in electron density between lysozyme crystals and similar crystals into which *N*-acetylglucosamine had been diffused. It happens that both of the anomeric forms of *N*-acetylglucosamine, the α-form in which the anomeric hydroxyl is axial, and the β-form in which it is equatorial, bind to lysozyme in these crystals in related but different ways. The part of the electron density map shown in Fig. 6 shows the α-form of the molecule, characterized by a lobe of electron density protruding to the right and representing the α-hydroxyl group. Electron density corresponding to β-GlcNAc appears elsewhere in the map.

Although the principal features in the difference electron density map correspond to the bound inhibitor molecules, there are some other features, both positive and negative, which show where water molecules, for example, have been displaced or where various groups on the enzyme have moved when the inhibitor was bound. By careful analysis of such maps, therefore, we can locate the inhibitors precisely in relation to the enzyme molecule and define any conformational changes that have taken place in the enzyme itself as a result of the inhibitor binding.

Thanks to experiments of this kind, we have been able to study the interactions between lysozyme and a number of inhibitors related to the substrate. The key study (9) was of the interaction between lysozyme and tri-*N*-acetylchitotriose, illustrated in Fig. 7. Here we see that the trisaccharide fits very neatly and specifically into a groove in the enzyme surface which actually extends about twice as far as is needed to accommodate the trisaccharide. This was the principal observation that we had in early 1966 and that led us to a plausible model of the lysozyme enzyme-substrate complex.

Although the trisaccharide is hydrolyzed by lysozyme, under certain circumstances (57), we knew that this particular complex is quite stable. We wondered, therefore, whether this nonproductive complex might form part of a more extensive and productive enzyme-substrate complex. There were two clues. The first, as has been mentioned, was that the trisaccharide fills up only half of the groove in the enzyme surface; there is room for more saccharide units below the trisaccharide (Fig. 7). The second clue was provided by John Rupley's (50) observations of the effect of lysozyme on chitin oligomers (Fig. 8). Figure 8 shows the products obtained by hydrolysis of the dimer, trimer, tetramer, pentamer, and hexamer derived from chitin. The results are rather complicated for the trimer and tetramer, but they become a little more clear for the pentamer, and clearer again for the hexamer, which is cleaved only between the fourth and fifth sugar residues. They suggest that the binding site for the enzyme might be able to accommodate six sugar residues, and that the site of catalysis then is between the fourth and fifth.

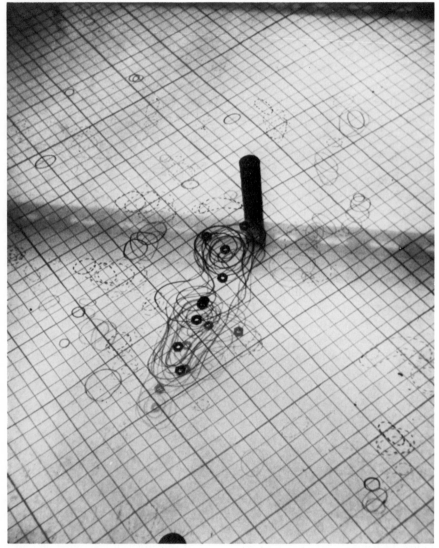

Fig. 6. Map showing the difference in electron density between a lysozyme molecule with bound α-*N*-acetylglucosamine and one without. The axial α-hydroxyl group projects to the right of the density representing the sugar ring and the acetamido group points toward the top of the map near the central support. Nuts indicate the atomic positions.

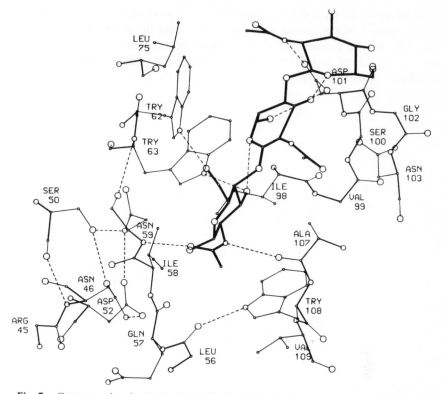

Fig. 7.   Computer drawing by A. C. T. North of the binding to lysozyme of tri-*N*-acetyl-chitotriose. Sugar molecule represented by heavy lines, hydrogen bonds by broken lines, and oxygen atoms by open circles.

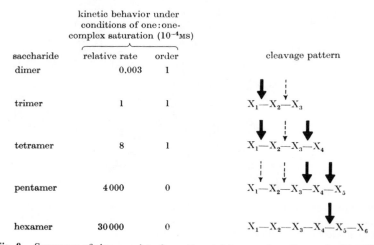

saccharide	relative rate	order	cleavage pattern
dimer	0.003	1	
trimer	1	1	$X_1 \!-\! X_2 \!-\! X_3$
tetramer	8	1	$X_1 \!-\! X_2 \!-\! X_3 \!-\! X_4$
pentamer	4 000	0	$X_1 \!-\! X_2 \!-\! X_3 \!-\! X_4 \!-\! X_5$
hexamer	30 000	0	$X_1 \!-\! X_2 \!-\! X_3 \!-\! X_4 \!-\! X_5 \!-\! X_6$

kinetic behavior under conditions of one:one-complex saturation (10^{-4}MS)

Fig. 8.   Summary of cleavage data from *N*-acetylglucosamine oligosaccharides (50).

Stimulated by these findings, and by our knowledge of the structure, we went on with model building to see if we could fit in another three sugar residues. Figure 9 shows the result. The three sugar residues whose binding we have observed in the trisaccharide complex are marked here A, B, and C and three more that we were able to fit into the vacant part of the cleft in

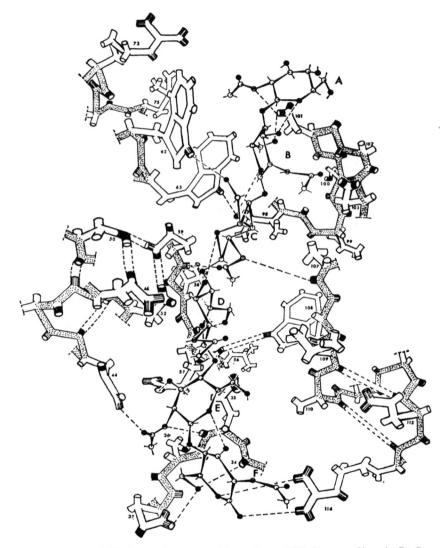

Fig. 9. Proposed binding to lysozyme of hexa-*N*-acetylchitohexaose. Sites A, B, C are as observed in the binding of tri-*N*-acetylchitotriose (Fig. 7). Sites D, E, F were inferred originally from model building (drawing by Mrs. W. J. Browne).

the enzyme, guided by reasonable stereochemical considerations, are labeled D, E, and F. Sites B, D, and F in this model have room for MurNAc residues, with their bulky lactyl side chains, whereas sites C and E do not and consideration of the enzyme's specificity against cell wall substrates leads immediately to location of the catalytic site between residues D and E (47, 48), a result that is happily consistent with Rupley's findings (Fig. 8). The model also explains all other known features of lysozyme specificity, such as the protection afforded by *O*-acetylation at position 6 of the MurNAc residues (14) which cannot then bind in site D.

There was, however, one problem in this model building in that fitting the fourth sugar residue, called here D, turned out to be difficult because there were impossibly close interactions between the enzyme and the C-6–O-6 group on this sugar residue as judged by measurements from the model. This meant either that it could not fit in this particular position, or that the conformation of either the sugar residue D, or the enzyme, or both had to change in order to allow it to fit. We chose the simplest course of changing the conformation of the sugar in site D from the usual chair conformation to a conformation in which the 6-hydroxymethyl group is rather more axial than equatorial. At this point I must mention that Levitt (35) has suggested recently on the basis of preliminary energy calculations from the refined atomic coordinates, that comparatively small perturbations of the enzyme structure permit binding of an undistorted GlcNAc residue in site D. This is now being investigated further, with special reference to the new experimental data on xylose and lactone binding discussed below.

At the beginning of 1966 we wondered whether the structure of our model told us anything about catalysis. Happily it was possible, fairly straightforwardly, to suggest a plausible mechanism of action for the enzyme, which involves the participation of the two carboxylic acid residues, glutamic acid 35 and aspartic acid 52 (Fig. 9) disposed on either side of the critical glycosidic linkage between sugar residues D and E. The factors that we suggested might be involved, and we were helped at this stage by consultations with Charles Vernon (56) and by knowledge of which bond is broken (50), were proton transfer from glutamic 35 to the glycosidic oxygen, leading to the breaking of the (C-1)–oxygen bond and the formation of a carbonium ion at C-1 of residue D which is promoted and stabilized by the negatively charged aspartate 52 and promoted also by the conformation of sugar residue D that is required for binding at that site.

These proposals for the enzyme mechanism put forward in 1966 have stimulated since then a considerable amount of work and, naturally among the various people working to see what additional evidence might be produced, we have continued ourselves to see whether we could get closer crystallographically to studying directly the enzyme substrate complex. It is not possible to review here all of the studies that have been made by many people but we may note that there is now a wealth of evidence (e.g., 19, 27) for the crystallographic model and, in particular, that energy has to be expended in order to bind a sugar residue in site D, no matter how that energy may be utilized. What follows is an incomplete summary of the ways in which we have attempted to arrive at additional crystallographic evidence.

Figure 10 shows in a more schematic way the factors which we suggested are involved in the catalysis, and, in particular, it emphasizes that the difficulty in binding sugar residue D arises because there seems not to be enough room to accommodate the 6-hydroxymethyl group, C-6–O-6 here. This suggested to various people that removal or modification of this group might affect the susceptibility of the substrate to attack by lysozyme. Osawa (42) found that lysozyme does not release nitrophenol from GlcNAc oligosaccharides containing nitrophenol-6-deoxy-GlcNAc at the glycoside terminus. Raftery and Rand-Meir (49) have shown a similar effect with oligosaccharides containing *p*-nitrophenylxylose at the glycoside terminus.

Prompted by this thought John Moult (7, 14) studied the binding to lysozyme of *N*-acetylglucosaminyl-*β*-(1 → 4)-xylose. He succeeded in binding this disaccharide in the crystals, and the result is shown in Fig. 11. Here again we see the typical binding of an *N*-acetylglucosaminyl residue in site C, guided there by the acetamido side chain which binds very specifically in this site (see Fig. 9) making good hydrogen bonds. But here, for the first time, in these crystals we actually saw a sugar binding in our proposed site D. It is bound, as far as we can tell in the rather fuzzy difference electron density map, in a somewhat different orientation than we suggested for a sugar

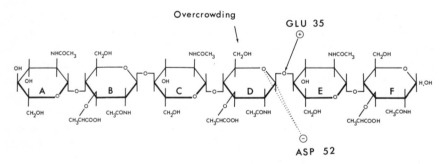

Fig. 10. Schematic diagram illustrating the stereospecificity of lysozyme and its proposed mechanism of action. Six sugar residues participate in formation of the enzyme-substrate complex with the groups shown along the upper edge of the molecule directed into the active site cleft of the enzyme. The sugar residue in site D is distorted from the chair conformation. This distortion and the action of amino acid residues Glu 35 and Asp 52 promote the cleavage of the C-1–O bond in the glycosidic linkage between sugar residues D and E.

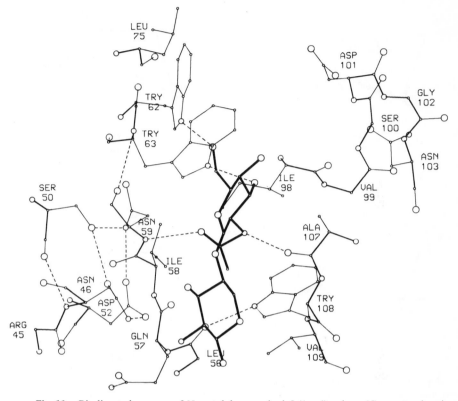

Fig. 11. Binding to lysozyme of *N*-acetylglucosaminyl-*β*-(1 → 4)-xylose. (Computer drawing by A. C. T. North.)

residue in the substrate, being rotated a little so that the oxygen in the 2 position approaches aspartic acid 52 more closely, and the hydrogen replacing C-6 is rather more to the right than we expected. Nevertheless this was encouraging support for our substrate model and we sought ways to investigate it further.

A different line of thought was to consider the likely course of the catalyzed reaction and to wonder what kind of molecule might be devised that would look like the substrate as it bound to the enzyme. Here we recalled that, in proposing the mechanism in discussion with Charles Vernon (56), we had taken into account the proposal of Lemieux and Huber (34) that, in the course of a reaction involving an intermediate with a carbonium-ion at C-1, a hexose ring is likely to take up a conformation in which the C-2, C-1, O-5, and C-5 atoms lie in a plane. Lemieux and Huber suggested that in their particular chemical studies the half-chair conformation illustrated in Fig. 12 is likely to be adopted but we may note at this point that co-

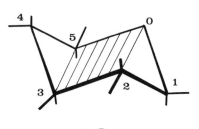

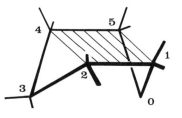

$${}^{4}C_{1}$$

$$B_{3,0}$$

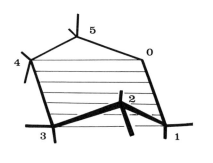

$${}^{2,5}B$$

$${}^{4}H_{3}$$

Fig. 12. Conformations of β-D-glucose. Reference planes are shown hatched (55) and hydroxyl groups by the longer bonds at each ring carbon.

planarity of atoms around the partial double bond C-1–O-5 is achieved also in the boat conformation variously described as B2 (49a) or ${}^{2,5}B$ (55) in which the atoms C-2 and C-5 lie out of the plane of the remaining gluco-pyranose ring atoms.

These considerations led Leaback (33) to suggest that an aldono-1:5 lactone would be likely to act as a competitive inhibitor of enzymic reactions involving C-1–carbonium ion intermediates, a proposition that is consistent with the body of evidence produced by Levvy and others for the inhibition of glycosidases by aldonolactones (e.g., 37). They also prompted us to investigate the interactions of lactones with lysozyme. The reason is illustrated in Fig. 13 which shows the bond rearrangement that is expected when a carbonium ion at C-1 shares its positive charge with the ring oxygen and the analogous rearrangement when the carbonyl oxygen of the corresponding lactone takes up a negative charge and its ring oxygen acquires a complementary positive charge.

In Oxford, we have made numerous attempts to bind small lactones to the enzyme (6), including the monosaccharide 2-acetamido-2-deoxygluconolactone derived from GlcNAc, but these experiments have not led to helpful results (in the present connection) because the acetamido group, when present, always leads to binding in site C (Fig. 9) and gluconolactone itself binds anomalously in the 1:4-form (6, 7). Progress was made only on Secemski and Lienhard's (53) synthesis of the tetrasaccharide lactone by oxidation of tetra-*N*-acetylchitotetraose. This compound (see also, 38) has an association constant for binding to lysozyme some 40 times larger than that for tetra-*N*-acetylchitotetraose itself at 25°C and pH 5.0. We were therefore delighted to accept Lienhard's generous gift of material for a crystallographic study which is being conducted in Oxford by L. N. Johnson, P. A. Machin, and R. Tjian. I am able to include here only a brief preliminary account of the results that are being obtained.

We do indeed see the tetrasaccharide mentioned above bound in sites A, B, C, and D with the residues in sites A, B, and C very much the same as those observed in the trisaccharide. What then are the conformation and orientation of the lactone in site D?

In discussion of these aspects of our findings it must be remembered that the electron density is somewhat ill defined, very similar in resolution to the electron density shown in Fig. 6 that corresponds to the binding of α-*N*-acetylglucosamine. It shows quite clearly, however, a structured slab of electron density corresponding to the sugar ring, with a well-defined lobe projecting from it that corresponds, most probably, to an electron-dense axial group. When we come to build the lactone into this electron density it becomes clear immediately that the axial group must be the 6-hydroxymethyl side chain. Density is also observed that fits equatorial acetamido- and hydroxyl-groups at the 2- and 3-positions, respectively, and the glycosidic linkage to the sugar in site C.

We can note immediately that an axial C-6 is not compatible with the standard half-chair conformation in which this group is quasi-equatorial (24), but related conformations are available for consideration. Before turning to them, however, we can observe that O-4 must be equatorial in

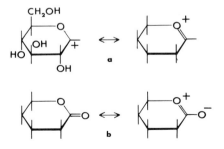

Fig. 13. Resonance hybrid structure of (a) *N*-acetylglucosamine incorporating a carbonium ion; and (b) *N*-acetylglucosaminono-(1, 5)-lactone.

any acceptable conformation, since the glycosidic linkage between the sugar residues in sites C and D appear from the electron density to be linked in the same way as the residues in sites B and C and those in sites A and B (i.e., with the glycosidic oxygen equatorial to each of the linked sugars). Furthermore the lactone ring is expected to include a planar trigonal arrangement of bonds at C-1.

The most obvious way of modifying the conformation of an isolated D-glucopyranose ring to make C-6 axial is to move C-5 across the rough plane of O-5, C-1, C-3, and C-4 so as to give the boat conformation $^{2,5}B$ (55). Encouragingly, as we noted earlier, the atoms C-2, C-1, O-5, and C-5 lie in a plane in this conformation but O-4 is axial and cannot participate in the required glycosidic linkage between the sugars in sites C and D. This dilemma directs our attention to the boat $B_{3,0}$, in which the atoms C-3 and O-5 are displaced in the same direction from the approximate plane of the atoms C-1, C-2, C-4, and C-5, and we note that inclusion of a planar-trigonal C-1 in this conformation, as required by the lactone, tends to reduce the extent to which O-5 lies out of the plane of C-1, C-2, C-4, and C-5. The $B_{3,0}$ boat, adapted in this rough way to accommodate the lactone, approaches a conformation in which C-1, C-2, C-4, C-5, and O-5 lie in a plane, i.e., a sofa conformation (3).

At the present stage of the analysis this conformation appears to fit the observed electron density in site D reasonably well but I must emphasize that this is a preliminary conclusion and that further refinement may well show that some related conformation, for example, a skew boat on the pathway between $^{2,5}B$ and $B_{3,0}$, gives even better agreement. Such possibilities are being investigated.

When we consider the constraints upon our model building in 1966 it is perhaps not very suprising to find that we arrived at a very similar conformation for a sugar residue bound in site D as part of the proposed enzyme substrate complex (see Fig. 9) (47). The constraints again included the apparent necessity for C-6 to be axial (to avoid close contacts with the enzyme molecule) and O-4 to be equatorial (again to make a satisfactory glycosidic linkage with the sugar residue in site C). Instead of having to accommodate a planar lactone group we had to make O-1 equatorial also in order to achieve a satisfactory glycosidic linkage with the proposed sugar residue in site E. The combined effect of these constraints on the sugar residue in site D of the proposed substrate model was to make atoms C-5, O-5, C-1, C-2, and, somewhat less closely, C-4 approach coplanarity. This conformation was described, rather loosely, as being related to the half-chair, and it is very similar to the conformation that now appears best to fit the observed electron density for Lienhard's lactone.

When we turn to consider the orientation of the lactone ring with respect to the enzyme we find that the results do not agree so closely with our proposals from model building. In our model (Fig. 9), O-6 is tucked underneath C-6 and participates in a hydrogen bond to a main chain carbonyl group at residue 57 in the back of the cleft while the acetamido group points straight out and does not interact strongly with any group on the enzyme. The lactone, however, appears to bind with O-6 further over to the right, where it may interact either with glutamic acid 35 or with a main chain carbonyl group at residue 108. The sugar ring seems to be twisted around slightly from the orientation proposed in Fig. 9, so that the acetamido group approaches more closely asparagine 46 which may be in hydrogen bond contact with the acetamido NH. This is clearly similar to the way in which a xylose residue appears to bind in this site (Fig. 11).

In considering the relationship of these results to the activity of the enzyme we must take these two new interactions into account. We must also remember the carbonyl oxygen of residue 57 which no longer appears to interact with O-6 but is little more than 3 Å from the ring oxygen O-5. O-5 and C-1 are still within 3 Å of aspartic acid 52, however, as we suggested they would be, though there are indications from the map that aspartic acid 52 has moved slightly away from the inhibitor position in relation to the rest of the enzyme structure.

In a preliminary way, then, the results of studying Lienhard's lactone are encouragingly similar to the proposals that we made about the enzyme-substrate interaction, but there are one or two intriguing differences in detail in the way that it interacts, and these we are still trying to define as closely as possible before we embark upon consideration of their implications for the mechanism.

It is unfortunate that we have not yet been able to observe the binding of sugars at our proposed sites E and F and the reason for this is that, in these particular crystals of the hen egg-white enzyme, enzyme molecules come rather close together and interact in the region of these sites, so that there is no room for part of the substrate to fit there without breaking up the crystals themselves. One of our reasons, therefore, for studying other crystal forms of the hen egg enzyme and other species of lysozyme is that we would like to find a crystal form in which sites E and F are accessible. This may be the case with the crystals of the human enzyme which Colin Blake and his colleagues have been studying at Oxford (Banyard, Blake, and Swan, Chapter 6, this volume), studies which were stimulated by Elliott Osserman (43), who kindly provided us with the material.

We have also been looking recently at crystals of the baboon enzyme (5) using material provided by Buss (15). Again in these crystals it looks as though sites E and F may be available. These crystals are closely related to the crystals of the human enzyme, though not closely enough for the two to be regarded as being isomorphous.

These studies of lysozyme from different species are going on partly, therefore, with a view to extending our understanding of the enzyme activity. But we are also interested, of course, in the relationships between these different structures. We are interested in knowing to what extent changes in the amino acid sequences of these molecules affect their three-dimensional structures and in learning, as are so many others (see, for example, the chapters by Morgan and Arnheim, Maron, and Wilson in this volume), how the enzyme has evolved.

These interests have naturally led us also to think about other problems, in particular to wonder how widespread, without our yet knowing it, is the lysozyme structure in nature. Here, of course, I allude to the fascinating work on α-lactalbumin (11; see also Hill, Steinman, and Brew, this volume) which certainly is chemically closely related to lysozyme (see Fig. 2) (12) and may well be closely similar to lysozyme in its three-dimensional structure also (13).

Unfortunately the structure of α-lactalbumin has not yet been determined though we are still working on it intensively and there is now good hope that our studies of baboon α-lactalbumin will be successful. Our studies of this protein so far illustrate very well the facts that even simple protein structures cannot be determined unless satisfactory crystals are available (4) and that a survey of species is often the most direct way of circumventing the problems of crystallization. Such surveys of different species, of course, lead us directly into the study of evolution at the molecular level. It seems appropriate for me to end with a brief remark about that particular problem since it is one that has attracted many diverse scientists to the study of Fleming's lysozyme.

Hugh MacKenzie in Canberra has recently been investigating the milk proteins of animals unique to the Australian continent and, in the course of this work, he has extracted lysozyme and α-lactalbumin from the echidna, or spiny anteater. His preliminary finding is that one of the subspecies of echidna produces a lysozyme which also has α-lactalbumin activity. The technical difficulties of the experiments involved in this study must not be underestimated; but, whether these early results are confirmed or not, it is evident that the evolution of lysozyme and α-lactalbumin activity poses a biological problem that Alexander Fleming would have relished had he, happily, still been with us.

Acknowledgments

I am much indebted to my colleagues for their vital contributions to these studies and for permission to discuss unpublished results; to Drs. G. Lienhard, J. A. Rupley, and N. Sharon for generous gifts of saccharides; and to the Medical Research Council for supporting this work from its inception.

References

1. Abraham, E. P., and Robinson, R., *Nature (Lond.)* **140,** 24 (1937).
2. Alderton, G., and Fevold, H. L., *J. Biol. Chem.* **164,** 1 (1946).
3. Anet, E. F. L. J., *Carbohyd. Res.* **1,** 348 (1966).
4. Aschaffenburg, R., Fenna, R. E., and Phillips, D. C., *J. Mol. Biol.* **67,** 529 (1972).
5. Aschaffenburg, R., Fenna, R. E., and Phillips, D. C., in preparation (1973).
6. Beddell, C. R., D.Phil. thesis, Oxford University, 1970.
7. Beddell, C. R., Moult, J., and Phillips, D. C., *in* "Molecular Properties of Drug Receptors" (R. Porter and M. O'Connor, eds.) p. 85. Churchill, London, 1970.
8. Berger, L. R., and Weiser, R. S., *Biochim. Biophys. Acta* **26,** 517 (1959).
9. Blake, C. C. F., Johnson, L. N., Mair, G. A., North, A. C. T., Phillips, D. C., and Sarma, V. R., *Proc. Roy. Soc., Lond.* **B167,** 378 (1967).
10. Blake, C. C. F., Koenig, D. F., Mair, G. A., North, A. C. T., Phillips, D. C., and Sarma, V. R., *Nature (Lond.)* **206,** 757 (1965).
11. Brew, K., and Campbell, P. N., *Biochem. J.* **102,** 258 (1967).
12. Brew, K., Vanaman, T. C., and Hill, R. L., *J. Biol. Chem.* **242,** 3747 (1967).
13. Browne, W. J., North, A. C. T., Phillips, D. C., Brew, K., Thomas, C., and Hill, R. L., *J. Mol. Biol.* **42,** 65 (1969).
14. Brumfitt, W., *Br. J. Exp. Pathol.* **40,** 441 (1959).
15. Buss, D. H., *Biochim. Biophys. Acta* **236,** 587 (1971).
16. Canfield, R. E., *J. Biol. Chem.* **238,** 2098 (1963).
17. Canfield, R. E., Kammerman, S., Sobel, J. N., and Morgan, F. J., *Nature (Lond.)* **232,** 16 (1971).
18. Canfield, R. E., and Liu, A. K., *J. Biol. Chem.* **240,** 1997 (1965).
19. Chipman, D., and Sharon, N., *Science* **165,** 454 (1969).
20. Diamond, R., *Acta Crystallogr. [A] (Kbh.)* **27,** 436 (1971).
21. Diamond, R., *J. Mol. Biol.* (1973) (in press).
22. Epstein, L. A., and Chain, E. *Br. J. Exp. Pathol.* **21,** 339 (1940).
23. Fleming, A., *Proc. Roy. Soc., Lond.* **B93,** 306 (1922).
24. Hackert, M. L., and Jacobson, R. A., *Acta Crystallogr.* **B27,** 203 (1971).
25. Jeanloz, R. W., Sharon, N., and Flowers, H. M., *Biochem. Biophys. Res. Commun.* **13,** 20 (1963).
26. Johnson, L. N., and Phillips, D. C., *Nature (Lond.)* **206,** 761 (1965).
27. Johnson, L. N., Phillips, D. C., and Rupley, J. A., *Brookhaven Symp. Biol.* **21,** 120 (1968).
28. Jollès, J., Jauregui-Adell, J., Bernier, I., and Jollès, P., *Biochim. Biophys. Acta* **78,** 668 (1963).

29. Jollès, P., *Chimia* **25**, 1 (1971).

30. Jollès, P., Jauregui-Adell, J., and Jollès, J., *C. R. Acad. Sci. (Paris)* **258**, 3926 (1964).

31. Joynson, M., North, A. C. T., Sarma, V. R., Dickerson, R. E., and Steinrauf, L. K., *J. Mol. Biol.* **50**, 137 (1970).

32. Kendrew, J. C., Dickerson, R. E., Strandberg, B. E., Hart, R. G., Davies, D. R., Phillips, D. C., and Shore, V. C. *Nature (Lond.)* **185**, 442 (1960).

33. Leaback, D. H., *Biochem. Biophys. Res. Commun.* **32**, 1025 (1968).

34. Lemieux, R. H., and Huber, G., *Can. J. Chem.* **33**, 128 (1955).

35. Levitt, M., Ph.D. Thesis, Cambridge University, 1972.

36. Levitt, M., *J. Mol. Biol.* (1973) (in press).

37. Levvy, G. A., and Snaith, S. M., *Adv. Enzymol.* **36**, 151 (1972).

38. Lienhard, G. E., Secemski, I. I., Koehler, K. A., and Lindquist, R. N., *Symp. Quant. Biol.* **36** (1972).

39. Meyer, K., Hahnel, E., and Steinberg, A., *J. Biol. Chem.* **163**, 733 (1946).

40. Meyer, K., Thompson, R., Palmer, J. W., and Khorazo, D., *J. Biol. Chem.* **113**, 303 (1936).

41. Moult, J., D.Phil. thesis, Oxford University, 1970.

42. Osawa, T., *Carbohyd. Res.* **7**, 217 (1968).

43. Osserman, E. F., and Lawlor, D. P., *J. Exp. Med.* **124**, 921 (1966).

44. Pauling, L., and Corey, R. B., *Proc. Natl. Acad. Sci. U.S.* **37**, 729 (1951).

45. Pauling, L., Corey, R. B., and Branson, H. R., *Proc. Natl. Acad. Sci. U.S.* **37**, 205(1951).

46. Phillips, D. C., *Adv. Res. Diffraction Methods* **2**, 75 (1966).

47. Phillips, D. C., *Sci. Am.* **215**, 78 (1966).

48. Phillips, D. C., *Proc. Natl. Acad. Sci. U.S.* **57**, 484 (1967).

49. Raftery, M. A., and Rand-Meir, T., *Biochemistry* **7**, 3281 (1968).

49a. Reeves, R. E., *Adv. Carbohyd. Res.* **6**, 107 (1951).

50. Rupley, J. A., *Proc. Roy. Soc., Lond.* **B167**, 416 (1967).

51. Salton, M. J. R., *Nature (London)* **170**, 746 (1952).

52. Salton, M. J. R., and Ghuysen, J. M., *Biochim. Biophys. Acta* **36**, 552 (1959).

53. Secemski, I. I., and Lienhard, G. E., *J. Am. Chem. Soc.* **93**, 3549 (1971).

54. Steinrauf, L. K., *Acta Crystallogr. (Kbh.)* **12**, 77 (1959).

55. Stoddart, J. F., "Stereochemistry of Carbohydrates." Wiley (Interscience), New York, 1971.

56. Vernon, C. A., *Proc. Roy. Soc., Lond.* **B167**, 389 (1967).

57. Wenzel, M., Lenk, H' P., and Schutte, E., *Hoppe Seylers Z. Physiol. Chem.* **327**, 13 (1962).

From Lysozymes to Chitinases:
Structural, Kinetic, and Crystallographic Studies

P. JOLLÈS, I. BERNIER, J. BERTHOU, D. CHARLEMAGNE,
A. FAURE, J. HERMANN, J. JOLLÈS, J.-P. PÉRIN, AND
J. SAINT-BLANCARD

Introduction

Lysozyme (EC 3.2.1.17) was discovered by Fleming in 1922. In his first paper devoted to this "remarkable bacteriolytic element," Fleming noted (1): "It was found that nasal mucus contained a large amount of lysozyme, and it was later found that tears and sputum were very potent in their lytic action. It was also found that this property was possessed by a very large number of the tissues and organs of the body." Do all these human tissues and organs contain the same or different forms of lysozyme? The problem of possible multiple forms of an enzyme in the same individual was thus stated. Concerning the distribution of lysozymes in nature, Fleming noted (1): "Only a limited amount of work has been done in this direction, but it is sufficient to show that lysozyme is very widespread in nature." This observation was the starting point for studies on the evolution of lysozymes. This latter expression was for the first time employed in 1932 by Fleming (2): "It has been shown, however, that the lysozyme of different tissues and secretions has quite varied antibacterial powers toward different microbes, and it seems that there are some differences in the antibacterial ferments (which we may call lysozymes) of different tissues whereby the bacterial affinities may be very different."

The story of hen egg-white lysozyme is remarkable from many points of view. The establishment of its primary structure (3, 4) preceded the study of its mode of action and even now its biological role is not completely known. Extensive studies were devoted to the relationship between structure and activity and to its action on different substrates. Furthermore, as a result of intensive X-ray crystallographic investigations, hen lysozyme became the first enzyme for which the complete three-dimensional structure was determined; for the first time a detailed catalytic mechanism was proposed (5).

In the last few years considerable interest has been directed, in our own laboratory, toward the comparative structural, evolutionary, kinetic, immunologic, and crystallographic behavior of lysozymes of diverse origins. Some of our results are summarized in this communication and, in addition, we are reporting recent data on the crystallization of hen egg-white lysozyme at high temperatures (up to 60°C). A phase transition has been observed; the classic tetragonal form was not stable above 25°–30°C and was transformed into an orthorhombic form.

Structural Studies

Primary Structures and Disulfide Bonds

Unlike the situation found for cytochromes c and myoglobins, where primary structure data of these proteins have been determined for a large number of species, only about 10 lysozymes of diverse origins have been detailed with respect to amino acid sequence. In our laboratory, following the elucidation of the primary structure of hen lysozyme, the complete sequences of two duck egg-white (6, 7), guinea hen egg-white (8), human milk (9, 10), and baboon milk lysozymes (11) were established (Table I). The reduced alkylated enzymes were subjected to digestion with trypsin. The structures of all the tryptic peptides were established. Alignment of the tryptic peptides into a single chain was determined by locating overlapping regions that contained lysine or arginine and these were obtained mainly by digestion of the lysozymes with chymotrypsin. The position of the disulfide bonds was studied in detail for the hen and duck enzymes. In one case (baboon lysozyme) partial sequence information was obtained using an automated sequencer.

Three-Dimensional Structure

Following the studies of Blake *et al.* (12, 13) on hen and human leukemic lysozymes, Berthou, Laurent, and Jollès (14) decided to establish the tertiary structure of the two histidine-free duck lysozymes. Single crystals of duck egg-white lysozyme II suitable for high-resolution X-ray diffraction studies were obtained at 4°, 20°, and 37°C (3 × 1 × 1 mm^3 in a few days). The crystals were monoclinic; the space group was P2$_1$ and unit cell dimensions were $a = 28.5$; $b = 66.2$; $c = 32$; $\beta = 113°$. An Okl 30°C precession photograph showed that the diffraction maxima remained fairly intense out to a spacing of at least 1.4 Å (Fig. 1). Suitable heavy atom derivatives have been prepared. Since the ions diffuse slowly, we performed the diffusion at 37°C in light of observations that the crystals grew in the same system up to this temperature. Two mercury derivatives gave very similar difference-Patterson maps. For one of them (Hg acetate), the 010 projection has been fully interpreted. The map showed many peaks and one of them seemed particularly important. For this main peak, the x and z coordinates could be deduced and used for the calculation of a difference-Fourier map. On this map three secondary sites could be recognized; their coordinates were measured and occupancies roughly estimated. The four sites so determined allowed a complete interpretation of the difference-Patterson map.

Multiple Forms of Lysozyme

Only a single type of lysozyme was found in hen, guinea hen, and goose egg white as well as in *normal* human tissues and secretions. In the latter, lysozymes were purified from tears, leukocytes, plasma, milk, and placenta, and the different enzymes were found to have the same chromatographic and electrophoretic behaviors as well as the same amino acid compositions (15).

Multiple lysozymes have been found in the duck egg white (16), and in the tissues or secretions of patients suffering from leukemia (17).

Duck Lysozymes

Lysozyme was purified from pooled samples of Peking, Kaki, and "wild-type" duck egg whites by gel filtration and ion-exchange chromatography (7, 16). In each case, at least three peaks called I, II, and III containing material active against *Micrococcus lysodeikticus* were characterized by ion-exchange chromatography on Amberlite CG-50. From peaks II and III, histidine-free duck lysozymes were isolated and their structures determined (6, 7). Six replacements were noted between these two enzymes. Their electrophoretic (16) and immunologic (18) behaviors as well as their behavior in the presence of inhibitors (19) were slightly different. Analysis by starch-gel electrophoresis of the partially purified lysozymes present in 44 individual duck eggs strongly suggests that the multiple forms of duck

TABLE I

Primary Structure of Human Milk Lysozyme (Hu) (9,10) and Comparison with Hen (H) (3,3a), Duck II (DII) (6), Duck III (DIII) (7), and Guinea Hen (G) (8) Egg-White Lysozymes and Bovine α-Lactalbumin (L) (26)[a]

	1	2	3	4	5	6	7	8	9	10	11	12	13	14	15	16
Hu 1	Lys	Val	Phe	Glu	Arg	Cys	Glu	Leu	Ala	Arg	Thr	Leu	Lys	Arg	Leu	Gly
H 1				Gly						Ala	Ala	Met			His	
DII 1			Tyr	Ser						Ala	Ala	Met				
DIII 1			Tyr	Glu						Ala	Ala	Met				
G 1				Gly						Ala	Ala	Met			His	
L 1	Glu	Gln	Leu	Thr	Lys			Val	Phe		Glu		—[b]	—		Asp

	17	18	19	20	21	22	23	24	25	26	27	28	29	30	31	32
Hu 17	Met	Asp	Gly	Tyr	Arg	Gly	Ile	Ser	Leu	Ala	Asn	Trp	Met	Cys	Leu	Ala
H 17	Leu		Asn				Tyr			Gly			Val		Ala	
DII 17	Leu		Asn				Tyr			Gly			Val		Ala	
DIII 17	Leu		Asn				Tyr			Gly			Val		Ala	
G 17	Leu		Asn				Tyr			Gly			Val		Ala	
L 15	Leu	Lys		Gly			Val			Pro	Glu		Val		Thr	Thr

	33	34	35	36	37	38	39	40	41	42	43	44	45	46	47	48
Hu 33	Lys	Trp	Glu	Ser	Gly	Tyr	Asn	Thr	Arg	Ala	Thr	Asn	Tyr	Asn	Ala	Gly
H 33		Phe			Asn	Phe			Gln				Arg		Thr	—
DII 33	Asn	Tyr		Ser		Phe			Gln				Arg		Thr	—
DIII 33	Asn	Tyr		Ser		Phe			Gln				Arg		Thr	—
G 33		Phe			Asn	Phe		Ser	Gln				Arg		Thr	—
L 31	—	Phe	His				Asp		Glu	Ile	Val	Glu			—	—

Note (L 31): a branch "\ /" leads to **Thr** shown below the His/Asp region.

	49	50	51	52	53	54	55	56	57	58	59	60	61	62	63	64
Hu 49	Asp	Arg	Ser	Thr	Asp	Tyr	Gly	Ile	Phe	Gln	Ile	Asn	Ser	Arg	Tyr	Trp
H 48	Gly								Leu					Trp		
DII 48	Gly								Leu	Glu				Trp		
DIII 48	Gly								Leu	Glu				Trp		
G 48	Gly							Val	Leu					Trp		
L 46	Asn	Gln							Leu					Asn	Lys	Ile

	65	66	67	68	69	70	71	72	73	74	75	76	77	78	79	80
Hu 65	Cys	Asn	Asp	Gly	Lys	Thr	Pro	Gly	Ala	Val	Asn	Ala	Cys	His	Leu	Ser
H 64					Arg				Ser	Arg	Leu		Asn	Ile	Pro	
DII 64		Asp	Asn						Ser	Lys			Gly	Ile	Pro	
DIII 64		Asp	Asn					Arg		Lys			Gly	Ile	Pro	
G 64					Arg				Ser	Arg	Leu		Asn	Ile	Pro	
L 62	Lys	Asn	Asp	Gln	Asp			His	Ser	Ser		Ile		Asn	Ile	

	81	82	83	84	85	86	87	88	89	90	91	92	93	94	95	96
Hu 81	Cys	Ser	Ala	Leu	Leu	Gln	Asp	Asn	Ile	Ala	Asp	Ala	Val	Ala	Cys	Ala
H 80					Ser	Ser	Asp		Thr	Ala	Ser		Asn			
DII 80			Val		Arg	Ser	Asp		Thr	Glu			Arg			
DIII 80			Val		Arg	Ser	Asp		Thr	Glu			Lys			
G 80				Gln	Ser	Ser	Asp		Thr	Ala	Thr	Ala	Asn			
L 78	Asp	Lys	Phe		Asn	Asn	Asp	Leu	Thr	Asn	Asn	Ile	Met		Val	

TABLE I (*contd*)

Label	No.																	
Hu	97	Lys	Arg	Val	Val	Arg	Asp	Pro	Gln	Gly	Ile	Arg	Ala	Trp	Val	Ala	Trp	
H	96		Lys	Ile			Ser		Gly	Asp		Met	Asn					
DII	96			Ile			Ser		Gly	Asp		Met	Asn					
DIII	96			Ile			Ser		Gly	Asp		Met	Asn					
G	96		Lys	Ile			Ser		Gly	Asp		Met	Asn					
L	94		Lys	Ile	Leu	—	Lys	Val					Asn	Tyr		Leu	His	

Label	No.																			
Hu	113	Arg	Asn	Arg	Cys	Gln	Asn	Arg	Asp	Val	Arg	Gln	Tyr	Val	Gln	Gly	Cys	Gly	Val	
H	112				Lys	Gly	Thr				Gln	Ala	Trp	Ile	Arg			Arg	Leu	
DII	112				Arg	Gly	Thr				Ser	Lys	Trp	Ile	Arg			Arg	Leu	
DIII	112				Lys	Gly	Thr				Ser	Arg	Trp	Ile	Arg			Arg	Leu	
G	112		Lys	His	Lys	Gly	Thr					Val	Trp	Ile	Lys			Arg	Leu	
L	109	Lys	Ala	Leu		Ser	Glu	Lys	Leu	Asp	Gln	—	Trp	Leu	—	—		Lys	Leu	

\ /
Glu

[a] Only replacements are indicated.
[b] — indicates deletion.

lysozyme are due to multiple alleles at a single locus (20). According to Arnheim and Steller (21), two nonallelic genes for egg-white lysozyme appear to exist in the black swan, *Cygnus atratus*. The existence of residue 37 in duck lysozyme III under two forms (Ser 70%; Gly 30%) should also be mentioned in this regard (7).

Human Leukemic Lysozymes

Ion-exchange chromatography on Amberlite CG-50 permitted the characterization of only one biologically active peak from the leukocytes of normal persons and two distinct active peaks from the leukocytes of patients with chronic myelogenous leukemia (CML). A similar observation was made with the lysozyme isolated by this technique from the urine of patients with monocytic and monomyelocytic leukemias (15, 17, 22). The first active peak from CML patients was eluted in the same volume as the normal enzyme and the second one had a more basic behavior on the column. Studies of the molecular weight, amino acid composition, UV spectrum, electrophoretic behavior, the NH_2-terminus, and action against chitopentaose did not distinguish the two enzyme peaks. However, their behavior on rechromatography on Amberlite CG-50 suggested a difference between their nitrogen amide content, as discussed in detail by Périn and Jollès (22).

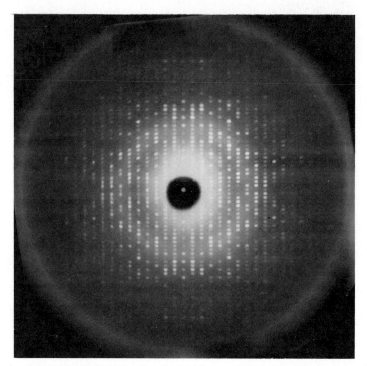

Fig. 1. A 30°C precession photograph of Okl zone of duck egg-white lysozyme II.

Robinson *et al.* (23) have suggested that a biological function of glutaminyl and asparaginyl residues in proteins may be to effect a timed alteration of chemical structure through deamidation. Since the CML leukocytes include a large quantity of young cells of the myelogenous series which are not fully differentiated, the lysozyme they contain may be at a less advanced lifetime than the enzyme of normal mature cells; thus this latter might be less amidated.

The structure of urine leukemia lysozyme, purified on BioRex 70 and CM-32, was established by Canfield *et al.* (24).

Conclusion

It is worth mentioning that (during the structural studies of hen lysozyme) Jollès observed that residue 118 in hen egg-white lysozyme, Thr was in part ($<5\%$) present as Ser.

Phylogenetic Studies

Close Structural Homology between Human and Several Bird Egg-White Lysozymes

As shown in Table I, human milk lysozyme is apparently identical to the other human lysozymes that have been studied and can be ranged among the group of lysozymes, including the enzymes from many bird egg whites, with the one exception noted below. All these enzymes have considerable homology. In order to optimize homologous relationships, an insertion was suggested in the sequence of human lysozyme to maximize alignment with bird lysozymes. The most probable position for this insertion appears to be residue 48 based upon homology and following X-ray crystallographic data presented by Blake and Swan (13). All the lysozymes mentioned contain 129 (bird) or 130 (human) amino acid residues (MW, around 15,000) and between 70 and 80 of these residues are in identical positions. The half-cystines as well as the two acidic residues essential for catalytic activity (Glu 35, Asp 52 for bird and 53 for human lysozymes) appear in identical positions. The same is true for tryptophan residues (all Trp residues for bird, 4 of 5 for human lysozymes) and for residues involved in hydrogen bonds playing a role in holding the folded chain together (Ser 24, Trp 28, Ser 36, Asn 39). The most constant regions among the various lysozymes are situated between (a) residues 49–72 for human and 48–71 for bird lysozymes (in part β-pleated sheet region) where the only acceptable substitutions are Ile for Val, Phe for Leu, Tyr for Trp, Lys for Arg; (b) residues 108–113 for human and 107–112 for bird lysozymes (sequence Ala-Trp-Val-Ala-Trp-Arg). In general, the NH_2-terminal ends of the enzymes show more homology than the COOH-terminal parts. Residue 123 for human and 122 for bird lysozymes are particularly exposed to changes in all the enzymes studied.

Despite the common elements in the primary structures of all these lysozymes (Table II) (3, 6–8, 10, 24a, 24b), immunologic studies as well as studies of their action on various substrates have indicated important differences (25).

Human and Bird Lysozymes and α-Lactalbumin

It is worth pointing out that bovine α-lactalbumin is a protein closely related to all the lysozymes, as indicated in Table I. The suggestion made by Brew *et al.* (26) that these molecules evolved from a common ancestral gene seems thus to be confirmed.

TABLE II

Number of Amino Acid Replacements between Hen (H), Guinea Hen (G), Duck II (DII), Duck III (DIII), Turkey (T), Quail (Q), and Human Milk (Hu) Lysozymes[a]

Lysozyme	Ref.	H	G	DII	DIII	T	Q	Hu[b]
H	3	—	10	19	20	7	6	50
G	8	10	—	26	27	16	15	55
DII	6	19	26	—	6	19	22	48
DIII	7	20	27	6	—	20	23	47
T	24a	7	16	19	20	—	10	51
Q	24b	6	15	22	23	10	—	52
Hu	10	50[b]	55[b]	48[b]	47[b]	51[b]	52[b]	—

[a] The changes due to a difference in the nitrogen amide content have not been taken into account in this table.
[b] With 1 insertion.

Goose Egg-White Lysozyme

Goose lysozyme is quite different (Table III) from other bird egg-white lysozymes studied in our laboratory (19, 27) or in others (24), including the duck enzymes. This is rather surprising since the duck and goose are in the same bird family (Anatidae). Goose lysozyme is heat labile, probably because of its low cystine and tryptophane contents; it is not inhibited by the same molecules that usually inhibit other lysozymes such as N-acetylglucosamine (GlcNAc), and it does not digest low molecular weight substrates (28). Preliminary structural studies of Canfield *et al.* (24) confirm our earlier suggestion that goose lysozyme is representative of a class of lysozymes different from the known avian enzymes.

Lysozymes from Invertebrates

We have conducted detailed studies of lysozymes isolated from a large number of invertebrates by our general purification procedure. Traces of the enzyme were found in nearly all the animals, and larger amounts were found only in insects, echinoderms, and annelids (Fig. 2). We investigated the lysozymes from *Periplaneta americana* (29), *Asterias rubens* (30), and *Nephthys hombergii* (31), and all these enzymes could be arranged in the class that includes hen egg-white and human milk lysozymes.

Plant Lysozymes

Fleming detected the presence of lysozyme in many plants, notably in their floral tissues (1). A number of roots and tubers were also examined (2).

TABLE III

Some Differences between Hen and Goose Egg-White Lysozymes[a,b]

	Hen	Goose
Ion-exchange chromatography on Amberlite CG-50		
pH	7.18	7.40
Molarity	0.2	0.2
Amino acids/mole	129	128 ± 6
Trp/mole	6	3
(Cys–)/mole	8	4
C-terminal amino acid	Leu	Tyr
N-terminal amino acid	Lys	Arg
Specific activity	1	6 ± 1
Heat stability	+	–
Digestion by trypsin (native enzyme)	–	weak
Digestion of high molecular weight substrates	+	+
Digestion of low molecular weight substrates	+	weak $\rightarrow$ –
Inhibition by GlcNAc or (GlcNAc)$_2$	+	–

[a] Isolated by ion-exchange chromatography on Amberlite CG-50.
[b] From A.-C. Dianoux and P. Jollès, *Bull. Soc. Chim. Biol.* **51**, 1559 (1969).

Of these, the only one which showed any marked bacteriolytic effect was the turnip. Since Fleming's observations, few studies have been devoted to plant lysozymes and these all deal with the enzyme contained in latex. We recently purified turnip lysozyme by electrofocusing or by chromatographic procedures (32). Its molecular weight was around 25,000 and its pH optimum for the lysis of *M. lysodeikticus* cells was around 5 (0.066 *M* phosphate

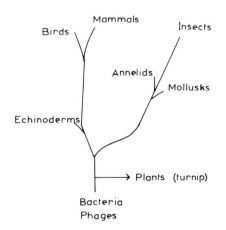

Fig. 2. Sources from which lysozymes were isolated.

buffer) instead of 7 for hen lysozyme (Fig. 3). At pH 5, it was more active between 45°–55°C, than at 20° or 37°C (Fig. 4). Thus the "classic" hen lysozyme and turnip lysozyme both hydrolyze β (1 → 4) glycosidic bonds (see below) in *M. lysodeikticus* cells. However, the results presented here show that the two lytic enzymes differ in many important properties. Under certain conditions, the turnip lysozyme seems to act more as a chitinase than as a lysozyme and thus resembles papaya latex lysozyme (33) (Fig. 5). These enzymes constitute an additional class of lysozymes.

In conclusion, phylogenetic studies point out the existence of different classes of lysozymes which will be discussed at the end of the section dealing with kinetic data.

Kinetic Studies

Definition of a Lysozyme

Previous studies with hen or duck egg-white as well as with human lysozymes led us to attribute the following characteristics to these enzymes (15): (a) basic protein; (b) low molecular weight (around 15,000); (c) stable at acidic pH values at higher temperatures; (d) lability at alkaline pH values;

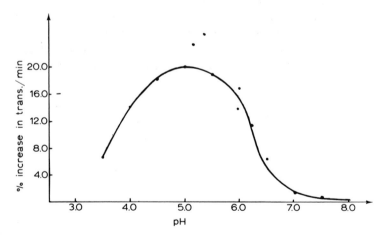

Fig. 3. pH activity profile for turnip lysozyme. Suspensions of acetone dried *M. lysodeikticus* cells were made in 0.066 *M* phosphate buffer (25 mg/100 ml) at the following pH's: 3.5, 4.0, 4.5, 5.0, 5.5, 6.0, 6.2, 6.5, 7.0, 7.5. Each suspension contained 0.1% NaCl and had an initial transmission of 20–25% at 650 nm. To 3 ml substrate at 25°C, 0.3 ml of enzyme (40 μg) was added and the increase in transmission was measured continuously. The percentage increase in transmission per minute was calculated from the steepest part of the curve.

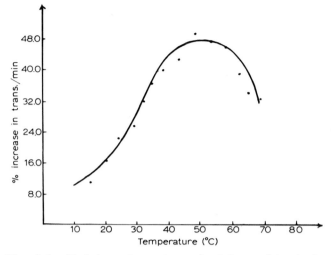

Fig. 4. The relationship between temperature and activity was determined as indicated in the legend to Fig. 3, using a suspension of cells at pH 5.0.

(e) lyse suspensions of *M. lysodeikticus*; and (f) their action on an appropriate substrate liberates compounds which can be detected by reagents for reducing sugars or amino sugars. Recent research in this field has shown that this definition of a lysozyme must be partially modified, as lysozymes have been described which are heat labile or show a higher molecular weight (see above) or present a different specificity (this section).

The common characteristic of all lysozymes is hydrolysis of glycosidic linkages in bacterial cell walls because they are primarily muramidases and only possess weak chitinase activity. Our kinetic studies have been performed with both substrates: the bacterial substrate (cells or cell walls) and the soluble derivatives of chitin, the most representative being short polymers of GlcNAc.

The Bacterial Substrate: Cells of M. lysodeikticus

Apparent Affinity Constants

Previous experiments indicated that human lysozymes reacted more slowly with the bacterial substrate than the enzymes from avian sources; goose lysozyme had the highest initial velocity, but its action was very short. In the course of our comparative studies, we wanted to express the differences observed from one enzyme to the other by the determination, for each of them, of an affinity constant for the bacterial substrate. The insolubility of the substrate presented a difficulty in the resolution of this

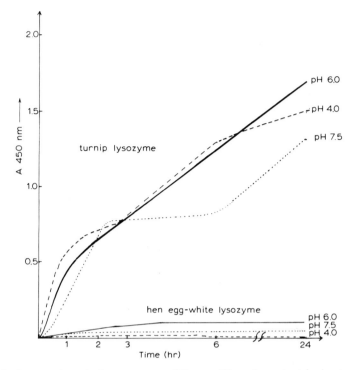

Fig. 5. Increase in reducing groups at different pH's and constant ionic strength (0.1) during the action at 20°C of turnip and hen egg-white lysozymes (40 μg) on 0.8 mg colloidal chitin determined by the method of Dygert *et al.* (33a).

problem. However, the type of kinetics observed seemed to follow the classic pattern for enzymes. The determination of an apparent affinity constant (K_a app.) became possible by expressing the data in units of substrate and not by the usual molarity system (34).

The various lysozymes can be ranged into two groups: a group including the human lysozymes which have very similar affinity constants, and a group including the bird egg-white lysozymes where important differences between the constants are noted from one enzyme to another (Table IV).

Influence of pH and Ionic Strength (I) on the Lysis of M. lysodeikticus Cells. Representation by Three-Dimensional Models

In observations by Meyer *et al.* (35) and others, the lytic activity vs. pH curves for hen lysozyme exhibited a unique maximum situated between pH 5.8 and 7: others have not confirmed this result. The different interpretations may be attributed to the different experimental conditions, especially variations in ionic strength. This latter factor plays an important

TABLE IV

Apparent Affinity Constants of Lysozymes from Different Origins for *Micrococcus lysodeikticus* Cells

Lysozyme	Number of assays	K_a app. (mg/liter)
Human		
Normal leukocytes	10	90 ± 10
Normal plasma	8	100 ± 5
Milk	13	110 ± 10
Avian		
Hen egg white	10	115 ± 10
Duck egg white II	8	150 ± 10
Duck egg white III	17	200 ± 20
Goose egg white	9	400 ± 100

role, as demonstrated by Davies *et al.* (36), in studies devoted to hen lysozyme. We examined the behavior of six human and four avian lysozymes on cell suspensions of *M. lysodeikticus* as a function of pH and I and built three-dimensional models [initial velocity, pH, I (Fig. 6)] (25). Alkaline pH improved the rate of lysis at low I. As an example, for human lysozymes,

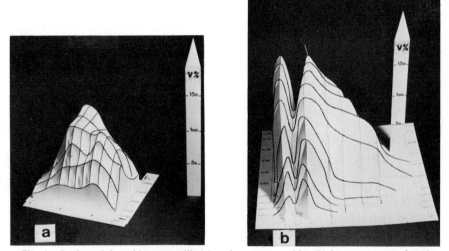

Fig. 6. Lytic activity of human milk (a) and goose egg-white (b) lysozymes as a function of pH and I. Activity is recorded as a percentage of the velocity at pH 6.2, $I = 0.107$. pH 4–9 and $I = 0.0125–0.100$ for (a); pH 3–11 and $I = 0.0125–0.190$ for (b).

TABLE V

K_a app. of Various Lysozymes for Bacterial Substrate at Different pH's nad I's

Lysozyme	pH	I	K_a, app. (mg/liter)
Human origin			
Normal plasma	7.5	0.025	107–103–105
	7.5	0.100	110–100
Normal leukocytes	7.5	0.025	108–105–111
	7.3	0.100	105–110–100
CML leukocytes, peak I	8.5	0.025	250–243–227
	8.6	0.118	200–250
AML urine			
Peak α	8.5	0.025	250–234
	8.5	0.100	250
Peak β	8.5	0.0375	303–307
	8.5	0.100	300
Milk	6.2	0.075	120–112
	6.0	0.118	110
Avian egg whites			
Hen	8.6	0.025	330–400
Guinea hen	7.9	0.075	312–310–300
	8.0	0.118	310–320
	8.6	0.0375	420–342–400
	8.5	0.100	400–357–416
Duck			
Lysozyme II	8.6	0.0375	328–360–333
	8.6	0.118	330–360–250
Lysozyme III	8.0	0.0375	320–350–280
	8.0	0.118	285–300
Goose	3.5	0.050	900
	3.8	0.125	1100
	5.2	0.125	300–370
	5.0	0.100	400–310

the pH optimum was 8–8.2 for $I = 0.025$–0.0375. Goose lysozyme, however, exhibited different behavior, its maximum activity occurring at acidic pH values (3.8 and 5.25) and also at a higher ionic strength (0.125–0.250) than for other lysozymes.

Influence of pH and I on the Apparent Affinity Constants (K_a app.). The Catalytic Mechanism

The affinity of the various human and bird lysozymes for the substrate is independent of the ionic strength (Table V). The variations of pK_a app.

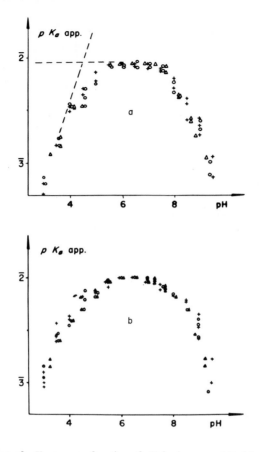

Fig. 7. Variation of $p K_a$ app. as a function of pH for hen egg-white (a) and normal human leukocyte (b) lysozymes.

as a function of pH suggest that analogous chemical groups are involved in the catalytic mechanism. The profiles of the curves are very similar; they all bend, first between pH 3.5 and 4 and again between pH 8 and 9 (including goose lysozyme) (Fig. 7). We assume that the first bend corresponds to the ionization of the carboxyl groups implicated in the catalytic mechanism. Our experiments also allow us to conclude that only one ionizable group is detected and that it probably belongs to the free carboxyl group of Glu 35 of hen lysozyme since this residue was found in identical or very similar sequences in several lysozymes (Table I). This Glu residue was replaced by a His residue in bovine α-lactalbumin which has a structure closely related to hen lysozyme but possesses no lytic activity.

Inhibition of Lysozymes by N-Acetylglucosamine and Its Short Polymers

The percentage of inhibition of all the bird and human lysozymes by GlcNAc is a function of pH, but not of I. The inhibition is particularly marked at alkaline pH where the rate of lysis is highest. However, goose lysozyme is not inhibited by this amino sugar. The inhibition by GlcNAc is competitive for human lysozymes (normal leukocytes, milk, tears) and for several bird enzymes (hen, guinea hen, duck II, duck III). The inhibition constants are nearly the same for the human lysozymes and increase for the bird lysozymes in the order mentioned above (28). For goose egg-white lysozyme an apparently uncompetitive inhibition is observed at several pHs (28). Turnip lysozyme is inhibited by GlcNAc at pH 6.2 (0.066 M phosphate buffer) and the inhibition is of the same order as that described for the duck lysozymes. The inhibition constants for chitobiose, chitotriose, and chitotetraose are indicated in Table VI (37).

Chitin as Substrate: The Way from Lysozymes to Chitinases

Lysozymes are not only muramidases, but also exhibit chitinase activity as observed for the first time by Berger and Weiser (38). Partially hydrolyzed chitin or polymers of GlcNAc such as chitotetraose or chitopentaose are usually employed as substrates.

In the course of our comparative studies of lysozymes of various origins, we tried to determine the influence of pH and I on the digestion of chitopentaose (39). The degradation products obtained after reaction with three bird egg-white, human milk, and turnip lysozymes under optimum conditions (pH 4.7; $I - 0.05$) were separated by gel filtration on BioGel P4 and

TABLE VI

Inhibition Constants K_i for Various Lysozymes by Short Polymers of GlcNAc at pH 6.2, $I = 0.217$

Lysozyme	Chitobiose $(10^{-4} M)$	Chitotriose $(10^{-5} M)$	Chitotetraose $(10^{-5} M)$
Human origin			
Milk	10	3	0.9
Leukemia urine	22	3.8	1.8
Egg white			
Hen	6	2.5	1.5
Duck II	10	3.4	2.6
Duck III	30	6.5	4.0
Guinea hen	9	3	2.6
Goose		no inhibition	

quantitatively determined. These five lysozymes attacked chitopentaose in three different ways. The question of a possible transfer reaction (transglycosylation products) is discussed in each case.

1. Goose lysozyme produced minimal digestion of the substrate in the first hour. After 16 hr, 67% of chitopentaose remained undigested; only chitobiose and chitotriose were found among the reaction products. No transfer reaction was catalyzed by this enzyme.

2. Hen and human milk lysozymes attacked chitopentaose in a similar manner; after 1 hr, 46% of the substrate remained undigested and this was reduced to 12% after 16 hr; chitotetraose, chitotriose, chitobiose, and GlcNAc were obtained. The transglycosylation products characterized after 30 min of action of hen lysozyme corresponded best to 12% of the total sugars (Fig. 8). They were ultimately hydrolyzed, but considering their low proportion, the final result was not very different from that obtained when only the hydrolysis products of chitopentaose were considered. With human milk lysozyme, transglycosylation compounds were barely detectable under similar experimental conditions.

Duck II lysozyme differed from hen and human lysozymes in having a faster rate and in yielding after 16 hr, mainly chitobiose and chitotriose. These latter polymers appear to be derived from chitopentaose. Duck lysozyme, like human milk lysozyme, did not produce an appreciable transfer reaction; after 30 min the higher polymers represented less than 5% of the total sugars.

3. Chitopentaose was rapidly and completely digested by turnip lysozyme at a concentration 100 times less than that employed for the enzymes from vertebrates. Chitobiose and chitotriose in equal amounts were characterized as the reaction products. Transfer reactions could not be demonstrated. Thus turnip lysozyme resembles papaya lysozyme. As reported by Dahlquist *et al.* (40) the hydrolysis of chitotriose by papaya lysozyme was more rapid than with the hen and human lysozymes, but did not result in any detectable transglycosylation.

Existence of Different Classes of Lysozymes. Their Common Feature

Our comparative studies of lysozymes of various origins allowed us to distinguish different classes.

1. Lysozymes which are primarily muramidases with only a weak chitinase activity; included in this class are the "classic" hen as well as several other bird egg-white lysozymes, and all the studied human and some invertebrate lysozymes.

2. Lysozymes which are only muramidases: goose egg lysozyme is included in this class.

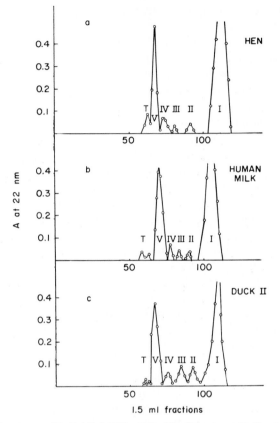

Fig. 8. Gel filtration on BioGel P 4 (250 × 0.9 cm) of the enzymic digest obtained by the action of hen (a), human milk (b), and duck II (c) lysozymes on chitopentaose (E/S:1/10; 0.5 h at 37°C; pH 4.7; $I = 0.05$). I = mixture of ammonium acetate and GlcNac; II to V = (GlcNAc)$_2$ to (GlcNAc)$_5$. T = transglycosylation products.

3. Lysozymes which are primarily chitinases but also possess a muramidase activity; plant lysozymes can be placed in the category.

4. The phage lysozymes studied by other investigators constitute a fourth class.

5. Finally, in addition to lysozymes, we have characterized from *Specia officinalis* some chitinases devoid of any muramidase activity under our experimental conditions.

We suggest that only the enzymes in the first two classes mentioned above might represent the "lysozyme" discovered by Fleming. The common factor shared by the enzymes within the different classes, however, is their capacity to hydrolyze glycosidic β $(1 \rightarrow 4)$ linkages in bacterial cell walls.

Immunologic Studies with Lysozymes in the Native State

Immunologic studies have been performed with several lysozymes purified in our laboratory and employed in their native state. Cross reactivities were noted between hen, guinea hen, duck II, and duck III egg-white lysozymes. but not between hen and goose egg-white or human milk lysozymes (41).

A large degree of homology in amino acid sequence between hen lysozyme and bovine α-lactalbumin has been described (26). However, no immunologic reactivity could be observed between any of the above mentioned lysozymes and an anti-bovine α-lactalbumin antiserum. Human α-lactalbumin also failed to react with anti-lysozyme antisera (42) (Table VII).

The degree of similarity between the primary structures of bovine α-lactalbumin and hen lysozyme, including their antigenic areas (43, 44), are of the same order. A comparable similarity—but often from different areas—exists with human milk lysozyme. However, the structural similarities are isolated and dispersed along the polypeptide chain. Rarely longer identical fragments have been observed and these latter may be necessary to form common antigenic sites.

These observations suggest two explanations: a common immunologic reactivity does not exist between these different structures in the native state because the regions of homology are not reactive, or the three-dimensional structures of the proteins are different. The fact that no lysozyme reacted with anti-α-lactalbumin antiserum suggests the key role of tertiary structure. It is also possible that antibodies directed against dominant three-dimensional configurations could not demonstrate the presence of minor common antigenic structures.

Sela's sensitive technique of "chemically modified bacteriophage" (45) was applied to some of our lysozymes in an effort to obtain selective information about changes in the unique "loop" region of lysozyme. By this method slight differences could be demonstrated between duck lysozymes II and III and two human leukemia enzymes (18).

Crystallographic Studies: High Temperature Crystallization of Lysozymes, An Example of Phase Transition (46)

The crystallization of proteins is unfortunately very often almost empirical. If we could define and control those external factors which are the easiest to manage, more general methods could perhaps be established. We have attempted to show the importance of constant temperatures (up to 55°C) for crystallization of enzymes. Experiments performed in this

TABLE VII

Reactivity of Six Native Lysozymes and of Bovine and Human α-Lactalbumins in the Presence of Three Different Anti-lysozyme Antisera and of an Anti-bovine α-Lactalbumin Antiserum

Lysozymes or α-Lactalbumins	Anti-hen egg white #1		Anti-hen egg white #2		Anti-human milk		Anti-bovine α-lactalbumin	
	Id[a]	Iha Titer[a]	Id	Iha Titer	Id	Iha Titer	Id	Iha Titer
Hen egg-white	+++ (complete identity)	512	+++	2048	0 (absence of reaction)	<1	0	<1
Guinea hen egg-white	+++	128	+++	256	0	<1	0	<1
Duck egg-white II	+++± (partial identity)	4	+++±	4	0	<1	0	<1
Duck egg-white III	+++±	2	+++±	2	0	<1		
Goose egg-white	0	<1	0	<1	0	<1		
Human milk	0	<1	0	<1	+++	512	0	<1
Bovine α-lactalbumin	0	<1	0	<1	0	<1	+++	1024
Human α-lactalbumin	0	<1	0		0	<1	0	<1

[a] Id = immunodiffusion; Iha = inhibition of hemagglutination. Titer 1 = 2 mg protein/ml. Titer <1 = >2 mg/ml.

manner with hen egg-white lysozyme allowed characterization of a phase transition.

The great variety of crystalline forms obtained from a given enzyme is a well-known phenomenon. It depends mainly on the following internal factors: (a) Different representatives of an enzyme family crystallize in quite different systems: hen (12), duck (14), and human (13) lysozyme crystals, for example, are all different. (b) When only one representative is considered (hen lysozyme) several crystalline forms have been described, depending on the ion bound to the enzyme: Cl^-, Br^-, NO_3^-, SO_4^{2-} (47), etc. (c) Furthermore, if we consider one form of hen lysozyme, for example, the hydrochloride derivative, we can obtain various types of crystals. Their occurrence depends on factors such as the protein concentration, the

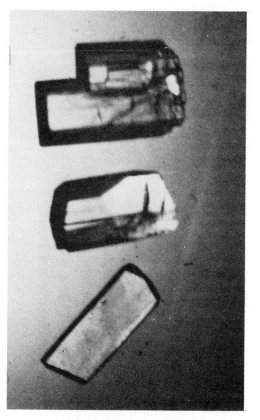

Fig. 9. Crystals of hen egg-white lysozyme obtained at pH 4.7 and 40°C (orthorhombic form) instead of the "classic" tetragonal form obtained at 18°C (× 5).

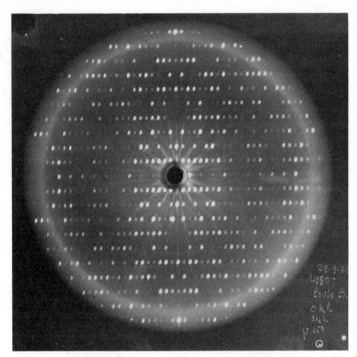

Fig. 10. X-ray diffraction pattern of the orthorhombic crystals of hen egg-white lysozyme obtained at pH 4.7 and 40°C.

buffer, the pH, the precipitating salt, the presence of small amounts of heavy atoms, ions, dyes, and traces of other materials (48).

In the case of inorganic substances, one particular structure can only exist within a definite temperature range, and if the temperature is altered there may be a rapid rearrangement. An extreme example is NH_4NO_3 which exists in five different crystalline forms, each of which changes to another at a definite temperature.

In this regard hen lysozyme behaves like an inorganic substance. It exhibits a polymorphism dependent upon one external physical parameter: variations in temperature can produce an A or a B crystal form with the first seeming more metastable.

In a first series of experiments, the crystallization of hen lysozyme was achieved at 18° and 37°C. At 18°C the classic tetragonal form (12, 49), called A, was obtained ($P4_32_12$, $a = 79.1$, $c = 37.9$, $\rho = 1.25$, content of water: 33.5%, MW = 14,700). At 37°C, an orthorhombic form, called B, appeared ($P2_12_12_1$, $a = 56.3$, $b = 73.8$, $c = 30.4$, $\rho = 1.24$, content of water: 36%, MW = 15,000) (Figs. 9 and 10). It was different from the crystalline form described by Palmer *et al.* (49). The question thus arose as to what happened between 18° and 37°C and at higher temperatures (46).

The transition point was around 25°–30°C. At this temperature, both forms were present. If pregrown A crystals were put at 33°C, they dissolved from the walls of the tubes and a transformation of the A into the B form was observed. The B crystals obtained at 33°C were of fine optical quality. From 37° to 55°C only the B form was observed. These crystals were also of excellent optical quality and reached several millimeters in size. At temperatures above 55°C, lysozyme precipitated, but even at 45°C a pre-cipitate was observed as a function of time after the appearance of the crystals. To preserve these crystals, it was necessary to remove them quickly; at 37°C they seemed stable, and the reversibility of the phenomenon was verified.

These findings establish for the first time to our knowledge that high temperatures (up to 60°C) are compatible with crystallization of the enzyme, hen lysozyme. This observation has been verified in the case of other lysozymes. It also permitted characterization of a phase transition dependent upon temperature (46). The crystals grown under these conditions were frequently rather large and of a rare optical quality, probably suitable for studies by other physical methods such as neutron diffraction. Other external physical parameters need to be more intensively studied, to answer the question: Might a protein have more than one three-dimensional structure?

References

1. Fleming, A., *Proc. R. Soc., Lond.* [*Biol.*] **93**, 306 (1922).
2. Fleming, A., *Proc. R. Soc. Med.* **26**, 71 (1932).
3. Jollès, J., Jauregui-Adell, J., Bernier, I., and Jollès, P., *Biochim. Biophys. Acta* **78**, 668 (1963).
3a. Rees, A. R., and Offord, R. E., *Biochem. J.* **130**, 965 (1972).
4. Canfield, R. E., *J. Biol. Chem.* **238**, 2698 (1963).
5. Phillips, D. C., *Sci. Am.* **215**, 78 (1966).
6. Hermann, J., and Jollès, J., *Biochim. Biophys. Acta* **200**, 178 (1970).
7. Hermann, J., Jollès, J., and Jollès, P., *Eur. J. Biochem.* **24**, 12 (1971).
8. Jollès, J., van Leemputten, E., Mouton, A., and Jollès, P., *Biochim. Biophys. Acta* **257**, 497 (1972).
9. Jollès, J., and Jollès, P., *Helv. Chim. Acta* **54**, 2668 (1971).
10. Jollès, J., and Jollès, P., *FEBS Lett.* **22**, 31 (1972).
11. Hermann, J., Jollès, J., Buss, G., and Jollès, P., *J. Mol. Biol.* **79**, 587 (1973).
12. Blake, C. C. F., Mair, G. A., North, A. C. T., Phillips, D. C., and Sarima, V. R., *Proc. R. Soc., Lond.* [*Biol.*] **167**, 365 (1967).
13. Blake, C. C. F., and Swan, I. D. A., *Nature* (*Lond.*), *New Biol.* **232**, 12 (1971).
14. Berthou, J., Laurent, A., and Jollès, P., *J. Mol. Biol.* **71**, 815 (1972).
15. Jollès, P., *Angew. Chem.* [*Engl.*] **8**, 227 (1969).

16. Jollès, J., Sportono, G., and Jollès, P., *Nature (Lond.)* **208**, 1204 (1965).
17. Mouton, A., and Jollès, J., *FEBS Lett.* **4**, 337 (1969).
18. Maron, E., Arnon, R., Sela, M., Périn, J.-P., and Jollès, P., *Biochim. Biophys. Acta* **214**, 222 (1970).
19. Jollès, P., Saint-Blancard, J., Charlemagne, D., Dianoux, A.-C., Jollès, J., and Le Baron, J. L., *Biochim. Biophys. Acta* **151**, 532 (1968).
20. Prager, E. M., and Wilson, A. C., *J. Biol. Chem.* **246**, 523 (1971).
21. Arnheim, N., and Steller, R., *Arch. Biochem. Biophys.* **141**, 656 (1970).
22. Périn, J.-P., and Jollès, P., *Clin. Chim. Acta* **42**, 77 (1972).
23. Robinson, A. B., McKerrow, J. H., and Cary, P., *Proc. Natl. Acad. Sci. U.S.A.* **66**, 753 (1970).
24. Canfield, R. E., Kammerman, S., Sobel, J. H., and Morgan, F. J., *Nature (Lond.), New Biol.* **232**, 16 (1971).
24a. La Rue, J. N., and Speck, J. C., Jr., *J. Biol. Chem.* **245**, 1985 (1970).
24b. Kaneda, M., Kato, I., Tominaga, N., Titani, K., and Narita, K., *J. Biochem. (Tokyo)* **66**, 747 (1969).
25. Saint-Blancard, J., Chuzel, P., Mathieu, Y., Perrot, J., and Jollès, P., *Biochim. Biophys. Acta* **220**, 300 (1970).
26. Brew, K., Castellino, F. J., Vanaman, T. C., and Hill, R. L., *J. Biol. Chem.* **245**, 4570 (1970).
27. Dianoux, A.-C., and Jollès, P., *Bull. Soc. Chim. Biol.* **51**, 1559 (1969).
28. Saint-Blancard, J., Hénique, J., Ducassé, D., Dianoux, A.-C., and Jollès, P., *Bull. Soc. Chim. Biol.* **50**, 1783 (1968).
29. Landureau, J. C., and Jollès, P., *Nature (Lond.)* **225**, 968 (1970).
30. Jollès, P., and Jollès, J., unpublished data.
31. Périn, J.-P., and Jollès, P., *Biochim. Biophys. Acta* **263**, 683 (1972).
32. Bernier, I., van Leemputten, E., Horisberger, M., Bush, D. A., and Jollès, P., *FEBS Lett.* **14**, 100 (1971).
33. Howard, J. B., and Glazer, A. N., *J. Biol. Chem.* **244**, 1399 (1969).
33a. Dygert, S., Li, L. H., Florida, D., and Thoma, J. A., *Anal. Biochem.* **13**, 367 (1965).
34. Locquet, J.-P., Saint-Blancard, J., and Jollès, P., *Biochim. Biophys. Acta* **167**, 150 (1968).
35. Meyer, K., Palmer, J. W., Thompson, R., and Khorazo, D., *J. Biol. Chem.* **113**, 479 (1936).
36. Davies, R. C., Neuberger, A., and Wilson, B. M., *Biochim. Biophys. Acta* **178**, 294 (1969).
37. Charlemagne, D., and Jollès, P., *C. R. Acad. Sci. [D] Paris* **270**, 2721 (1970).
38. Berger, L. R., and Weiser, R. S., *Biochim. Biophys. Acta* **26**, 517 (1957).
39. Charlemagne, D., and Jollès, P., *FEBS Lett.* **23**, 275 (1972).
40. Dahlquist, F. W., Borders, C. L., Jr., Jacobson, G., and Raftery, M. A., *Biochemistry* **8**, 694 (1969).
41. Faure, A., and Jollès, P., *FEBS Lett.* **10**, 237 (1970).
42. Faure, A., and Jollès, P., *C. R. Acad. Sci. [D] Paris* **271**, 1916 (1970).
43 Arnon, A., *Eur. J. Biochem.* **5**, 583 (1968).
44. Fujio, H., Imanishi, M., Nishioka, K., and Amano, T., *Biken J.* **11**, 219 (1968).
45. Haimovich, J., Hurwitz, E., Novick, N., and Sela, M., *Biochim. Biophys. Acta* **207**, 115 (1970).
46. Jollès, P., and Berthou, J., *FEBS Lett.* **23**, 21 (1972).
47. Steinrauf, L. K., *Acta Crystallogr. [Kbh.]* **12**, 77 (1959).
48. King, M. V., Magdoff, B. S., Adelman, M. B., and Harker, D., *Acta Crystallogr. [Kbh.]* **9**, 460 (1956).
49. Palmer, K. J., Ballantyre, M., and Galvin, J. A., *J. Am. Chem. Soc.* **70**, 906 (1948).

Comparison of the Structures of Alpha-Lactalbumin and Lysozyme*

ROBERT L. HILL, HOWARD M. STEINMAN, AND KEITH BREW

Although structural similarities between α-lactalbumin and egg-white lysozyme were noted earlier (1, 2), it was not until 1967 (3) that Brew, Vanaman, and Hill showed that the amino acid sequences of these two proteins are very similar. The subsequent determination of the complete amino acid sequence of bovine (4), human (5), and guinea pig (6) α-lactalbumin established beyond question that the sequences of these proteins are very similar to those of chicken (7, 8) and human (9) lysozymes. In view of these close similarities in primary structure, it was suggested (10) that the three-dimensional structure of α-lactalbumin may also be very similar to that of chicken egg-white lysozyme (11). Determination of the three-dimensional structure of α-lactalbumin by X-ray crystallographic studies would reveal the extent of similarity between the two structures, but unfortunately such studies have not progressed sufficiently to provide such information. In the absence of an X-ray determined three-dimensional structure, considerable effort has been expended with a variety of indirect methods to assess the extent of secondary and tertiary structural similarities between the α-lactalbumins and the lysozymes.

*The studies reported here, from the laboratory of Robert L. Hill, were supported by the National Heart and Lung Institute, National Institutes of Health (Grant HL-06400), and the National Science Foundation.

55

It is the purpose of this paper to compare briefly the primary structure of α-lactalbumin and lysozyme, based upon recent sequence analyses of α-lactalbumin from different species, and then to assess the extent to which the three-dimensional structures of the two proteins may be similar based upon present information. A complete discussion of a possible structure of α-lactalbumin has been given earlier (10) and need not be repeated here. Thus, only aspects of α-lactalbumin structure brought to light by recent sequence analyses will be considered.

The available sequence data on four α-lactalbumins, hen egg-white, and human lysozymes are shown in Table I.

Table II summarizes the numbers of identical residues found among the sequences of α-lactalbumins and lysozymes from different organisms. The sequences have been aligned and gaps introduced to achieve maximum homology. Clearly, all of the α-lactalbumins are more similar in sequence to one another than to either of the two lysozymes. For example, 89 (72%) of the 123 residues are identical on comparing human to bovine α-lactalbumin and 79 (64%) residues are identical in guinea pig and bovine α-lactalbumin. This is not an unexpected degree of homology on comparing sequences from different animal species. In contrast, 48 residues (39%) are identical in human α-lactalbumin and human lysozyme whereas 49 residues are identical in bovine α-lactalbumin and chicken lysozyme. Although this is about half the extent of identity as found on comparing any two α-lactalbumins or the two lysozymes, it appears to be sufficiently extensive to support the view that the lysozymes and α-lactalbumins have a common ancestral gene (3).

A comparable degree of similarity between the lysozymes and the α-lactalbumins is found by statistical analyses based upon the mRNA-amino acid code (12). Thus, additional sequence information supports the view proposed earlier that the structural gene for α-lactalbumin arose during evolution of premammals by duplication of a structural gene for lysozyme. Through subsequent evolution, the α-lactalbumin sequence came to be controlled by only one of the duplicate genes. Although α-lactalbumin assumed a function quite different from lysozyme, it retained similarities in its three-dimensional structure.

Each of the α-lactalbumin sequences can easily be accommodated into a three-dimensional structure similar to that of lysozyme. This was shown first for bovine α-lactalbumin on the basis of model building (11). Twenty-one residues (16%) or about 1 in 6 residues in the two lysozyme molecules and the three α-lactalbumin molecules are identical, including the 8 disulfide bonds. Despite this small number of identical residues, almost all of the replacements appear to be conservative and do not require that major features of the lysozyme conformation be altered. Since bovine α-lactalbumin has 7 fewer residues than the lysozymes, it was necessary to delete these residues in building a three-dimensional model of α-lactalbumin, as discussed earlier (11). These deletions involved the residues numbered 14 and 15, 46 and 47, 101, and 125–127 in the lysozyme sequence. It is of interest that kangaroo α-lactalbumin would not be expected to have a deletion corresponding to residues 14 and 15 (Arg-His) since its sequence from residues 14–18 is almost identical to that in the lysozymes but quite dissimilar to corresponding sequences in the α-lactalbumins. The deletion in the kangaroo α-lactalbumin sequence has therefore been placed in a position corresponding to residues 21 and 22 in the lysozyme sequences. This location preserves the homologies in the sequence between residues 14 and 18, and also takes into account the consequences of the deletion upon the proposed conformational similarity between α-lactalbumins and lysozymes. Residues 21 and 22 in the lysozyme sequence are near the surface of the molecule, and are contained in a loop of the polypeptide chain, about 12 residues in length, which bridges two helical segments. It appears that only minor changes in the overall conformation of the protein would be required, if residues 21 and 22 were deleted, and a peptide linkage formed between residues 20 and 23, to rejoin the polypeptide chain. As a further point regarding the conformational compatibility of this deletion, it is of some interest to note the location of a prolyl residue at position 24, replacing the seryl, threonyl, or alanyl residues which appear at this position in the lysozymes, or the other α-lactalbumins. Model building suggests that the sequence of residues 24–26, Pro-Leu-Pro, is capable of affecting a sharp turn in the polypeptide chain, which might be essential for accommodating the deletions of residues 21 and 22.

There are 61 (50%) residues in the α-lactalbumins which are in identical positions in each molecule. Although this is about 1 of every 2 residues, it is interesting to note that only 3 of the first 20 residues (1 in 7) or 7 in the first 30 residues (1 in 4) are identical. Thus, the amino terminal portion of α-lactalbumin is quite variable and much more so than the remainder of the molecule. Assuming that the gene for α-lactalbumin arose rather recently in vertebrate evolution, it is evident that α-lactalbumin has undergone a rather high rate of evolutionary change. Estimates of the rate suggest that it is 5 to 6 times that of cytochrome c and twice that of lysozyme.

As noted above, 8 of the 21 residues which are at identical positions in the α-lactalbumins and the two lysozymes are half-cystines. If the α-lactalbumins have a lysozyme-like conformation it is obvious that conservation of the half-cystines and the disulfide bridges would be essential. It is not at all obvious, however, why all of the other residues are conserved. In the α-lactalbumin sequences, residues 51 and 100 are glycine. The three-dimensional structure of lysozyme reveals that at the corresponding position

TABLE I

Comparison of α-Lactalbumin and Lysozyme Sequences[a]

Residues 1–20

α-Lactalbumins

	1										10[b]									20
Bovine	Glu –Gln –Leu –Thr –*Lys* –CYS–Glu –Val –Phe –Arg –*Glu* –*Leu* –*Lys* –Asp –Leu –Lys –Gly –Tyr –Gly –Gly
Guinea pig	Lys –Gln –Leu –Thr –*Lys* –CYS–Ala –Leu –Ser –His –*Glu* –*Leu* –*Asn* –Asp –Leu –Ala –Gly –Tyr –Arg –Asp
Human	Lys –Gln –Phe –Thr –*Lys* –CYS–Glu –Leu –Ser –Gln –*Leu* –*Leu* –*Lys* –Asp –Ile –Asp –Gly –Tyr –Gly –Gly
Kangaroo[c]	Ile –Asp –Tyr –Arg –*Lys* –CYS–Gln –Ala –Ser –Gln –Ile –*Leu* –*Lys* –Glu –His –Gly –Met –Asp –Lys –Val –

Lysozymes

	1										10									20
Chicken	*Lys* –*Phe* –Gly –Arg –CYS–*Glu* –*Leu* –*Ala* –*Ala* –Ala –Met –*Lys* –*Arg* –*His* –*Gly* –Leu –Asp –Asn –Tyr –Arg –Gly
Human	*Lys* –*Phe* –Glu –Arg –CYS–*Glu* –*Leu* –*Ala* –*Arg* –Thr –Leu –*Lys* –*Arg* –*Leu* –*Gly* –Met –Asp –Gly –Tyr –Arg –Gly

Residues 21–40

α-Lactalbumins

							30										40		
Bovine	Val –Ser –*LEU*–*Pro*–*Glu* –Trp –Val –CYS–Thr–Thr –*Phe* –*His* –Thr–SER–*Gly* –Tyr –Asp –Thr –Glu –ALA–Ile –*Val*
Guinea pig	Ile –Thr –*LEU*–*Pro*–*Glu* –Trp –Leu –CYS–Ile –Ile –*Phe* –*His* –Ile –SER–*Gly* –Tyr –Asp –Thr –Gln –ALA–Ile –*Val*
Human	Ile –Ala –*LEU*–*Pro*–*Glu* –Leu –Ile –CYS–Thr–Met –*Phe* –*His* –Thr–SER–*Gly* –Tyr –Asp –Thr –Gln –ALA–Ile –*Val*
Kangaroo	Ile –Pro –*LEU*–*Pro*–*Glu* –Leu –Val –CYS–Thr–Met –*Phe* –*His* –Ile –SER–*Gly* –Leu –Ser –Pro –Gln –ALA–Glu –*Val*

Lysozymes

							30										40		
Chicken	Tyr –*Ser* –*LEU*–Gly –Asn –Trp –Val –CYS–Ala –Ala –*Lys* –*Phe* –*Glu* –SER–Asn –Phe –Asn –*Thr* –Gln –ALA–*Thr* –Asn
Human	Ile –*Ser* –*LEU*–Ala –Asn –Trp –Met –CYS–Leu –Ala –*Lys* –*Trp* –*Glu* –SER–Gly –Tyr –Asn –*Thr* –Arg –ALA–*Thr* –Asn

Residues 41–60

α-Lactalbumins

						50										60		
Bovine	Glu –ASN–	Asn –Gln –Ser –Thr –Asp–TYR–GLY–Leu –Phe –GLN–ILE–Asn –*Asn* –*Lys* –Ile –Trp –CYS–Lys –Asn
Guinea pig	Lys –ASN–	Ser –Asn –His –Lys –Glu–TYR–GLY–Leu –Phe–GLN–ILE–Asn –*Asn* –*Lys* –Asp –Phe –CYS–Glu –Ser
Human	Glu –ASN–	Asn –Gln –Ser –Thr –Glu–TYR–GLY–Leu –Phe–GLN–ILE–Ser –*Asn* –*Lys* –Leu –Trp –CYS–Lys –Ser

Lysozymes

						50										60		
Chicken	Arg –*ASN*–Thr –	Asp –Gly –Ser –Thr –Asp–TYR–GLY–Ile –Leu–GLN–ILE–Asn –*Ser* –Arg –Trp –Trp –CYS–Asn –Asp
Human	Tyr –*ASN*–Ala –	Gly –Asp –Ser –Thr –Asp–TYR–GLY–Ile –Phe–GLN–ILE–Asn –*Ser* –Arg –Tyr –Trp –CYS–Asn –Asp

α-Lactalbumins

                      70

Bovine   Asp-Gln -Asp -Pro-His -*Ser* -Ser -Asn -*Ile* -CYS-Asn -Ile -Ser -CYS-Asp-Lys -Phe -LEU-*Asn*-Asn -Asp -Leu

Guinea pig Ser -Thr -Thr -Val -Gln -*Ser* -Arg -Asp -*Ile* -CYS-Asp -Ile -Ser -CYS-*Asp*-*Lys* -Leu -LEU-*Asn*-Asp -Asn -Leu

Human   Ser -Gln -Val -Pro-Gln -*Ser* -Arg -Asn -*Ile* -CYS-Asp -Ile -Ser -CYS-*Asp*-*Lys* -Phe -LEU-*Asn*-Asp -Asn -Ile

Lysozymes

                      70

Chicken   Gly -Arg -*Thr* -Pro-Gly -Ser -Arg -*Asn* -Leu-CYS-Asn -Ile -Pro-CYS-Ser -Ala -Leu -LEU-Ser -Ser -Asp -Ile

Human   Gly -Lys -*Thr* -Pro-Gly -Ala -Val -*Asn* -Ala -CYS-His -Leu-Ser -CYS-Ser -Ala -Leu -LEU-Gln-Asp -Asn -Ile

α-Lactalbumins

                 90   100

Bovine   Thr -*Asn* -*Asn* -Ile -*Met* -CYS-Val -LYS-Lys-Ile -Leu - Asp-Lys -Val -GLY-Ile -*Asn* -*Tyr* -TRP-Leu -ALA

Guinea pig Thr -*Asn* -*Asn* -Ile -*Met* -CYS-Val -LYS-Lys-Ile -Leu - Asp-Ile -Lys -GLY-Ile -*Asn* -*Tyr* -TRP-Leu -ALA

Human   Thr -*Asn* -*Asn* -Ile -*Met* -CYS-Ala -LYS-Lys-Ile -Leu - Asp-Ile -Lys -GLY-Ile -*Asn* -*Tyr* -TRP-Leu -ALA

Lysozymes

     90   100   110

Chicken   Thr-Ala -Ser -*Val* -Asn -CYS-Ala -LYS-Lys-Ile -Val -Ser -Asn-Gly -Asp-GLY-Met -Asn -*Ala* -TRP-*Val* -ALA

Human   Ala -Asp -Ala -*Val* -Ala -CYS-Ala -LYS-Arg-Val -Val -Arg -Asp-Pro -Gln-GLY-Ile -Arg -*Ala* -TRP-*Val* -ALA

α-Lactalbumins

             110   120

Bovine   His -Lys -Ala -Leu -CYS-Ser -Glu -Lys -Leu-Asp -Gln -Trp -Leu- CYS-Glu -Lys-Leu

Guinea pig His -Lys -Pro -Leu -CYS-Ser -Asp-Lys -Leu -Glu -Gln -Trp -Tyr- CYS-Glu -Ala-Gln

Human   His -Lys -Ala -Leu -CYS-Thr -Glu -Lys -Leu -Glu -Gln -Trp -Leu- CYS-Glu -Lys-Leu

Lysozymes

                 120   123

Chicken   Trp -*Arg* -*Asn* -Arg -CYS-Lys -Gly -Thr -Asp-*Val* -Gln -Ala -Trp-Ile -Arg-Gly -CYS- Arg-Leu

Human   Trp -*Arg* -*Asn* -Arg -CYS-Glu -Asn-Arg -Asp-*Val* -Arg -Gln -Tyr -Val -Gln-Gly -CYS- Gly-Val

[a] Residues printed in italics indicate identities between at least two α-lactalbumin sequences or identities between the two lysozyme sequences. Residues printed in capital letters indicate identities in all six sequences shown.

[b] Residues 10 and 17 in goat α-lactalbumin are lysine and aspartic acid, respectively. All other residues from 1–30 appear to be in a sequence identical to bovine α-lactalbumin.

[c] Brew, K., Steinman, H. M., and Hill, R. L. *J. Biol. Chem.* **248**, 4739 (1973).

TABLE II

Numbers of Identical Amino Acid Residues between Pairs of α-Lactalbumins and Lysozymes

	Bovine α-lactalbumin	Human α-lactalbumin	Guinea pig α-lactalbumin	Chicken egg-white lysozyme
Bovine α-lactalbumin	—	—	—	—
Human α-lactalbumin	89 (72%)	—	—	—
Guinea pig α-lactalbumin	79 (64%)	86 (70%)	—	—
Chicken egg-white lysozyme	49 (40%)	48 (39%)	41 (33%)	—
Human lysozyme	45 (37%)	48 (39%)	41 (33%)	77 (60%)

in the sequence, glycine is the only amino acid permitted by the adjacent conformation. Perhaps this is also true in the α-lactalbumins. Even less can be said with confidence about the other conserved residues. However, Asn 44, Gln 54, Ile 55, Trp 104, and Ala 106 are on either side of the substrate binding cleft and may be essential if the cleft is to be retained in α-lactalbumin.

There appear to be five lengths of sequence, containing 11–13 residues, which are identical, or nearly so, in all of the α-lactalbumins. These are as follows:

(a) *Residues 23–35.* 8 identical residues, 5 conservative replacements.

(b) *Residues 50–61.* 9 identical residues, 5 of which are conserved in the lysozymes as well; 2 of 3 nonidentical residues are conservative replacements.

(c) *Residues 72–82.* 9 identical residues, 2 conservative replacements.

(d) *Residues 86–97.* 11 identical residues, 1 conservative replacement.

(e) *Residues 100–111.* 11 identical residues, 1 conservative replacement. Three of the five stretches contain residues which line the cleft (23–35, 50–61, 100–111) and they may be conserved for this reason. It is unclear why the other two stretches (72–82, 86–97) would be vital, although they do contain residues which participate in organized structures, such as hydrophobic regions.

From the foregoing considerations one may conclude that the α-lactalbumin molecule should have a size and shape very similar to that of lysozyme and that the conformation of its polypeptide chain is probably not identical but very similar to that of lysozyme. These similarities would preserve in α-lactalbumin many major structural features, such as helices, the single stretch of antiparallel pleated sheet, and the two wings separated by a cleft (11). The exact environment of specific side chains is much more difficult to predict with assurance, although on model building there is no difficulty in finding highly plausible positions for every side chain. In the absence of an absolute three-dimensional structure derived from crystallographic studies many other methods have been used to assess the structural similarities of the two molecules. It is unnecessary to discuss the results of these studies in detail here since the consensus is that lysozyme and α-lactalbumin are very similar, regardless of what method is used for comparison. For example, low angle X-ray studies (13) originally suggested that the two molecules differed considerably, but reinvestigation of the two proteins by this technique (14) indicated that they were essentially indistinguishable in size and shape, and differed only by about 4% in degree of hydration. From circular dichroism and optical rotatory dispersion studies (15–17), α-lactalbumin and lysozyme are very similar structurally after compensating for the different content of aromatic chromophores. In addition, the nuclear magnetic resonance spectrum of α-lactalbumin (18) is also consistent with the proposed α-lactalbumin structure.

The reactivity of specific side chains in α-lactalbumin and lysozyme to chemical reagents has also been used to assess structural similarities. In this way the reactivity of histidine (19), methionine (19), tryptophan (20), tyrosine (21), and the carboxyl groups (22) has been investigated. The results of these studies suggest structural similarities as well as differences, but they are very difficult to evaluate. The microenvironment of a specific side chain cannot be easily inferred from its reactivity, and reactivities of corresponding side chains in the two molecules could differ considerably, even though their overall conformations were similar. Thus, measurement of the rate and extent of reaction of specific side chains gives at best only limited information as to whether or not the two proteins have similar conformations.

Finally, it should be noted that marked differences in the antigenic characteristics of lysozyme and α-lactalbumin have been reported (23, 24). Antisera to bovine α-lactalbumin do not cross react with egg-white lysozyme, nor do antisera to lysozyme cross react with α-lactalbumin, This has been interpreted as evidence that the conformations of the two proteins are different (24). However, this conclusion is likely erroneous since antisera to α-lactalbumins from different species, which very probably have similar three-dimensional structures, do not cross react. Finally, sequence similarities between lysozyme and α-lactalbumin are reflected by the cross-reactivity of the proteins with antisera prepared against the reduced and carboxymethylated proteins (23).

In conclusion, the weight of available evidence supports the view that the α-lactalbumins are closely related to the lysozymes, such as those in chicken egg-white, both with respect to their evolution as well as to their three-dimensional structures. It is not clear at present, however, why this relationship exists, since there remains no compelling explanation for the structural and evolutionary relatedness in terms of the function of the two molecules. Lysozyme catalyzes the cleavage of β (1 → 4) glycosides while α-lactalbumin aids in regulating the synthesis of a β (1 → 4) glycoside, lactose. Here the similarities in function end; neither molecule can replace the other functionally. In this light, one expects that further understanding of these two proteins must await further elucidation of their functional properties.

References

1. Yasunobu, K. T., and Wilcox, P. E., *J. Biol. Chem.* **231,** 309 (1958).
2. Brew, K., and Campbell, P. N., *Biochem. J.* **102,** 258 (1967).
3. Brew, K., Vanaman, T. C., and Hill, R. L., *J. Biol. Chem.* **242,** 3747 (1967).
4. Brew, K., Castellino, F. J., Vanaman, T. C., and Hill, R. L., *J. Biol. Chem.* **245,** 4570 (1970).
5. Findlay, J. B. C., and Brew, K., *Eur. J. Biochem.* **27,** 65 (1972).
6. Brew, K., *Eur. J. Biochem.* **27,** 341 (1972).
7. Canfield, R. E., and Liu, A. K., *J. Biol. Chem.* **240,** 1997 (1965).
8. Jollès, P., *Proc. R. Soc., Lond.* [*Biol.*] **167,** 350 (1967).
9. Canfield, R. E., Kammerman, S., Sobel, J. M., and Morgan, F. J., *Nature (Lond.), New Biol.* **232,** 16 (1971).
10. Browne, W. J., North, A. C. T., Phillips, D. C., Brew, K., Vanaman, T. C., and Hill, R. L., *J. Mol. Biol.* **167,** 365 (1967).
11. Blake, C. C. F., Mair, G. A., North, A. C. T., Phillips, D. C., and Sarma, V. R., *Proc. R. Soc., Lond.* [*Biol.*] **167,** 365 (1967).
12. Dayhoff, M. O., ed., "Atlas of Protein Sequence and Structure," Vol. 5. National Biomedical Research Foundation, Silver Spring, Maryland, 1972.
13. Krigbaum, W. R., and Kugler, F. R., *Biochemistry* **9,** 1216 (1970).
14. Pessen, H., Kumosinski, T. F., and Timasheff, S. N., *J. Agric. Food Chem.* **19,** 698 (1971).
15. Robbins, F. M., and Holmes, L. G., *Biochim. Biophys. Acta* **221,** 234 (1970).
16. Kronman, M. J., Holmes, L. G., and Robbins, F. M., *J. Biol. Chem.* **246,** 1909 (1971).
17. Cowburn, D. A., Brew, K., and Gratzer, W. B., *Biochemistry* **11,** 1228 (1972).
18. Cowburn, D. A., Bradbury, E. M., Crane-Robinson, C., and Gratzer, W. B., *Eur. J. Biochem.* **14,** 83 (1970).
19. Castellino, F. J., and Hill, R. L., *J. Biol. Chem.* **245,** 417 (1970).
20. Barman, T. E., *J. Mol. Biol.* **52,** 391 (1970).
21. Denton, W. L., and Ebner, K. E., *J. Biol. Chem.* **246,** 4053 (1971).
22. Lin, T. Y., *Biochemistry* **9,** 984 (1970).
23. Arnon, R., and Maron, E., *J. Mol. Biol.* **61,** 225 (1971).
24. Atassi, M. Z., Habeeb, A. F. S. A., and Rydstedt, L., *Biochim. Biophys. Acta* **200,** 184 (1970).

Human Leukemia Lysozyme*

ROBERT E. CANFIELD, JANNA C. COLLINS, AND JOAN H. SOBEL

This communication contains a brief resume of our observations with human leukemia lysozyme (HLL). In 1967, shortly after Elliott Osserman told us of his finding that patients with chronic monocytic leukemia often excrete large quantities of lysozyme in their urine, we had the opportunity to participate in the care of a patient who suffered from this disorder and who was willing to collect substantial quantities of his urine to serve as a source of lysozyme. This individual excreted approximately 500 mg of urinary lysozyme daily, permitting the eventual purification of over 50 gm of this protein. The level of lysozyme in his plasma was approximately 100 μg/ml, i.e., 1–2% of total protein. Since most of the protein in the urine was lysozyme, his kidneys had already accomplished a nearly 100-fold initial purification step.

Method of Purification

The initial description of the purification of human leukemia lysozyme involved adsorption to bentonite and subsequent crystallization of the enzyme (1, 2). In our laboratory, however, a somewhat different approach

* This research was supported by NIH Research Grants GM 11246 and AM 09579.

† Human leukemia lysozyme and hen egg-white lysozyme are abbreviated HLL and HEWL, respectively, in the text.

to purification was employed and because the method has proved to be simple and yields an extremely pure product, the details are included below.

Step 1

Pooled urine specimens, stored in the frozen state, are thawed just prior to purification, and all procedures are carried out at 4°C. Following adjustment of the urine to pH 6.3 with acetic acid, the sediment is removed by filtration. A volume of 30 liters is then filtered, at a rate of 300 ml/hr, through a 4 × 30 cm column of BioRex 70 (100–200 mesh) equilibrated with 0.05 M NaH_2PO_4 adjusted to pH 6.3 with NaOH. The effluent from the column should be periodically assayed for lysozyme activity to be certain that all the enzyme is adsorbed to the ion exchanger. Following passage of the entire batch of urine over the column, the BioRex is poured from the column and washed twice as a slurry with cold distilled water to remove urinary chromagens. The resin is then suspended in enough distilled water to permit stirring, after which sufficient $(NH_4)_2CO_3$ is added to make the liquid portion 0.05 M. Following titration of the slurry to pH 7.8 with NH_4OH, it is poured into the original chromatography column and the protein is eluted at room temperature with 0.4 M $(NH_4)_2CO_3$. (The adjustment in pH of the slurry prior to repouring the column is necessary to prevent the evolution of carbon dioxide bubbles in the column.) The eluate is collected in large fractions of approximately 100 ml which are assayed to locate the lysozyme-containing fractions and these are then lyophilized.

Step 2

Two gm batches of the lyophilized powder described above are dissolved in 150 ml distilled water and the solution is titrated to pH 7.0 with HCl and applied to a 2.5 × 20 cm column of CM-32 cellulose, equilibrated in 0.05 M $NH HCO_3$. Generally all the material becomes soluble after pH adjustment; if not, it is centrifuged and the supernatant is applied to the column. The column is developed with a three-chambered varigrad (500 ml each) in which the first chamber contains 0.05 M NH_4HCO_3, the second 0.10 M NH_4HCO_3 and the third 0.5 M $(NH_4)_2CO_3$. The column is developed at a rate of 75 ml/hr and 10-cm^3 fractions are collected. Figure 1 illustrates a typical result for this step. The fractions containing lysozyme are pooled and lyophilized. Amino acid analysis, NH_2-terminal end-group determinations, and polyacrylamide disc-gel electrophoresis of material from this peak indicate that it is a highly purified preparation of human leukemia lysozyme.

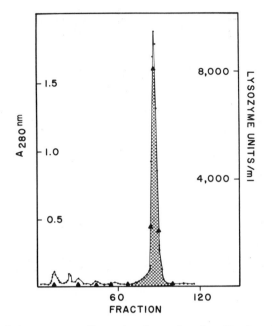

Fig. 1. Typical chromatogram illustrating the results of purification of 2 gm of human leukemia lysozyme on a 2.5 × 20 cm column of carboxymethyl cellulose (CM)-32, as described in the text. In addition to 280 nm determinations (small circles), fractions were assayed for lysozyme activity (triangles and cross-hatched area).

Step 3

To desalt the lysozyme the material may be passed over a column of G-50 Sephadex equilibrated with 0.1 N acetic acid, or it may be dissolved in and dialyzed against 0.1 N acetic acid and then lyophilized. The enzyme stored as a powder in the cold appears to be quite stable.

This procedure for the purification of lysozyme has been employed with samples obtained from six different patients.

Primary Structure

We have previously reported the complete amino acid sequence of human leukemia lysozyme that was isolated by this method from the single patient (3). These data are reproduced in Fig. 2 (4, 5). The amino acid sequences of hen egg-white lysozyme (HEWL) and of bovine α-lactalbumin (BAL) are included for comparison. This figure includes one revision from our

	1	2	3	4	5	6	7	8	9	10	11	12	13	14	15	16	17	18	19
HLL	LYS	VAL	PHE	GLU	ARG	CYS	GLU	LEU	ALA	ARG	THR	LEU	LYS	ARG	LEU	GLY	MET	ASP	GLY
HEWL	LYS	VAL	PHE	GLY	ARG	CYS	GLU	LEU	ALA	ALA	ALA	MET	LYS	ARG	HIS	GLY	LEU	ASP	GLY
BAL	GLU	GLN	LEU	THR	LYS	CYS	GLU	VAL	PHE	ARG	GLU	LEU	...	...	...	ASP	LEU	LYS	GLY

	20	21	22	23	24	25	26	27	28	29	30	31	32	33	34	35		36	37
HLL	TYR	ARG	GLY	ILE	SER	LEU	ALA	ASN	TRP	MET	CYS	LEU	ALA	LYS	TRP	GLU	...	SER	GLY
HEWL	TYR	ARG	GLY	TYR	SER	LEU	GLY	ASN	TRP	VAL	CYS	ALA	ALA	LYS	PHE	GLU	...	SER	ASN
BAL	TYR	GLY	GLY	VAL	SER	LEU	PRO	GLU	TRP	VAL	CYS	THR	THR	...	PHE	HIS	THR	SER	GLY

	38	39	40	41	42	43	44	45	46	47	48	49	50	51	52	53	54	55	56
HLL	TYR	ASN	THR	ARG	ALA	THR	ASN	TYR	ASN	ALA	GLY	ASP	ARG	SER	THR	ASP	TYR	GLY	ILE
HEWL	PHE	ASN	THR	GLN	ALA	THR	ASN	ARG	ASN	THR	...	ASP	GLY	SER	THR	ASP	TYR	GLY	ILE
BAL	TYR	ASP	THR	GLU	ALA	ILE	VAL	GLU	ASN	...	...	ASN	GLN	SER	THR	ASP	TYR	GLY	LEU

	57	58	59	60	61	62	63	64	65	66	67	68	69	70	71	72	73	74	75
HLL	PHE	GLN	ILE	ASN	SER	ARG	TYR	TRP	CYS	ASN	ASP	GLY	LYS	THR	PRO	GLY	ALA	VAL	ASN
HEWL	LEU	GLN	ILE	ASN	SER	ARG	TRP	TRP	CYS	ASN	ASP	GLY	ARG	THR	PRO	GLY	SER	ARG	ASN
BAL	PHE	GLN	ILE	ASN	ASN	LYS	ILE	TRP	CYS	LYS	ASN	ASP	GLN	ASP	PRO	HIS	SER	SER	ASN

	76	77	78	79	80	81	82	83	84	85	86	87	88	89	90	91	92	93	94
HLL	ALA	CYS	HIS	LEU	SER	CYS	SER	ALA	LEU	LEU	GLN	ASP	ASN	ILE	ALA	ASP	ALA	VAL	ALA
HEWL	LEU	CYS	ASN	ILE	PRO	CYS	SER	ALA	LEU	LEU	SER	SER	ASP	ILE	THR	ALA	SER	VAL	ASN
BAL	ILE	CYS	ASN	ILE	SER	CYS	ASP	LYS	PHE	LEU	ASN	ASN	ASP	LEU	THR	ASN	ASN	ILE	MET

	95	96	97	98	99	100	101	102	103	104	105	106	107	108	109	110	111	112	113
HLL	CYS	ALA	LYS	ARG	VAL	VAL	ARG	ASP	PRO	GLN	GLY	ILE	ARG	ALA	TRP	VAL	ALA	TRP	ARG
HEWL	CYS	ALA	LYS	LYS	ILE	VAL	SER	ASP	GLY	ASP	GLY	MET	ASN	ALA	TRP	VAL	ALA	TRP	ARG
BAL	CYS	VAL	LYS	LYS	ILE	LEU	...	ASP	LYS	VAL	GLY	ILE	ASN	TYR	TRP	LEU	ALA	HIS	LYS

	114	115	116	117	118	119	120	121	122	123	124	125	126	127	128		129	130
HLL	ASN	ARG	CYS	GLN	ASN	ARG	ASP	VAL	ARG	GLN	TYR	VAL	GLN	GLY	CYS	...	GLY	VAL
HEWL	ASN	ARG	CYS	LYS	GLY	THR	ASP	VAL	GLN	ALA	TRP	ILE	ARG	GLY	CYS	...	ARG	LEU
BAL	ALA	LEU	CYS	SER	GLU	LYS	LEU	ASP	GLN	...	TRP	LEU	...	...	CYS	GLU	LYS	LEU

Fig. 2.   The primary structures of human leukemia lysozyme (HLL), hen egg-white lysozyme (HEWL) (4), and bovine α-lactalbumin (BAL) (5) are shown. Residues have been aligned, with arbitrary deletions made, so as to maximize the homology among the three proteins.

original report which contained an error brought to our attention by Dr. J. Thomsen (6). The tryptic peptide assigned to positions 99–101 has the sequence Val-Val-Arg. The Val-Val bond is resistant to acid hydrolysis and when this tryptic peptide was first analyzed by us, the results were most consistent with a composition of Val_1, Arg_1. Following the recent communication from Dr. Thomsen, we repeated the hydrolysis for a longer period of time and the composition was indeed found to be Val_2, Arg_1.

As noted earlier, the primary structure of human leukemia lysozyme is remarkably homologous with that of hen egg-white lysozyme (3). The extra amino acid residue in HLL appears to be a glycine inserted in the region of the hinge of the β-pleated sheet at position 48. Thus the aspartic acid residue at position 53 in HLL is in the same location as the aspartic acid at 52 in HEWL.

While we continue to designate our material as a "leukemia" lysozyme to identify its source, it is apparent from amino acid sequence studies by Jollès and colleagues that the leukemia protein is identical to the lysozyme in human breast milk (7). Whether or not there is another form of human lysozyme remains an unsettled issue.

Extinction Coefficient

We have noted that it is the practice in many clinical laboratories to prepare standard assay solutions of lysozyme by weighing the preparation and then dissolving it in a known volume of buffer. Since the powdered material inevitably contains approximately 10% bound water and an additional unknown quantity of salt, the estimate of protein content in the standard solution will be in error. The only appropriate way to estimate the concentration of a solution of purified lysozyme is to measure its absorbance at 280 nm. We have determined the extinction coefficient for a solution of purified human leukemia lysozyme at a concentration of 1 mg/ml at neutral pH to be 2.55 when the absorbancy is measured in a 1-cm light path cell at 280 nm.

Activity Relative to Hen Egg-White Lysozyme

The relative specific enzyme activity of various lysozymes is a matter of interest to those who wish to evaluate evolutionary changes affecting the active site as well as to those concerned with measuring lysozyme activity in biological fluids. Hen egg-white lysozyme, for example, was employed for many years as the standard in estimating lysozyme concentrations in various human tissues, plasma, and urine. When purified human lysozyme became available, it was possible to compare the activity of these two enzymes, so that a correction factor could be applied to the values in the older literature.

Unfortunately, great variations exist among laboratories in the methods used to assess lysozyme activity, and standardization is urgently needed. In our own laboratory we employ an assay utilizing a suspension of *M. lysodeikticus* in a 0.066 *M* sodium phosphate buffer adjusted to pH 6.2. A Waring blender suspension of *M. lysodeikticus* cells (10 ml of a solution with an optical density of 1.0 measured in 4-cm light path cells at 650 nm) is placed in both cells of a double beam recording spectrophotometer (Beckman DB) and the lysozyme is added to the reference cell. As the bacterial walls are lysed by the enzyme, a positive slope is registered on the recorder and activity is expressed in arbitrary units as a function of the slope. Figure 3 illustrates the results that we obtained for three different lysozyme preparations assayed at different concentrations. When the slopes of these lines (calculated by the least squares method) are compared, human leukemia lysozyme is 2.5 times as active as hen egg-white lysozyme and the White Embden goose egg-white lysozyme is 9.3 times as active as that of the hen under the conditions of this assay.

Immunochemical Determinations

Several years ago, in collaboration with Dr. Norman Arnheim, we attempted to examine the immunochemical resemblance between HLL and HEWL and their reduced carboxymethyl derivatives (8). In agreement with others, we found that antibodies to HLL generally did not cross react with HEWL despite great similarity in the primary structures of these enzymes. When antisera were raised against the carboxymethyl derivatives of these lysozymes, we could show a structural resemblance between the two proteins by the criteria of cross-reactivity in the radioimmunoassay or by microcomplement fixation. Figure 4 illustrates the type of result that we observed.

It is of interest that antibodies raised against mouse lysozyme did not react with either human or hen egg-white lysozyme (9). In view of our results, cited above, this failure to demonstrate immunologic cross-reactivity does not necessarily rule out significant structural similarity, i.e., homology in primary structure. A similar situation exists with the cross-reactivity between bovine α-lactalbumin and hen egg-white lysozyme, i.e., antibodies

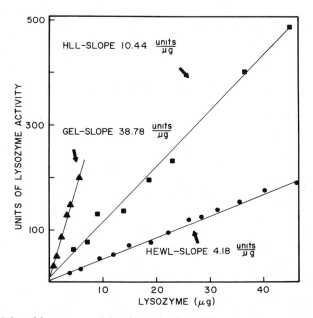

Fig. 3. Units of lysozyme activity determined at different enzyme concentrations. Hen egg-white lysozyme (HEWL) (circles) is least active, goose egg-white lysozyme (GEL) (triangles) is most active, and human leukemia lysozyme (HLL) (squares) is intermediate in its activity toward *M. lysodeikticus* cell walls (see text).

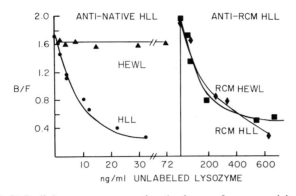

Fig. 4. (Left) Radioimmunoassay assessing the degree of cross-reactivity between HEWL and HLL. Iodine-labeled HLL was bound to anti-HLL and increasing amounts of "cold" HLL (closed circles) or "cold" HEWL (closed triangles) were each examined for their ability to displace the antibody-bound, labeled HLL. It is apparent that native HEWL does not cross-react with native HLL in the concentrations noted. (Right) Radioimmunoassay assessing the degree of cross-reactivity between reduced carboxymethyl (RCM) lysozyme derivatives. Iodine-labeled RCM-HLL was bound to anti-RCM HLL and increasing amounts of "cold" RCM HEWL (diamonds) or "cold" RCM HLL (squares) were each examined for their ability to displace the antibody-bound, labeled RCM HLL. The results illustrate a substantial degree of cross-reactivity (see Ref. 8).

raised against the native protein (BAL) do not cross react with native HEWL while antibodies to reduce carboxymethyl BAL do cross react with reduced, carboxymethyl HEWL (10).

On the basis of these observations we would suggest that when immunologic comparisons are made for the purpose of assessing evolutionary changes in the primary structure of proteins, reduced, carboxymethyl, or performic acid oxidized derivatives as well as the native forms of the proteins should be subjected to study.

The HLL curve at the left of Fig. 4 is typical of the type of radioimmunoassay that is employed in clinical medicine to measure the level of protein hormones in plasma. By changing several parameters in this lysozyme immunoassay, we were able to increase the sensitivity of the system so that it could be employed to quantitate the level of human lysozyme in the serum of 35 normal blood donors. An average level of 0.79 ± 0.35 μg/ml was found.

We have attempted to correlate the immunoassay results in plasma with estimates of enzyme activity for the same specimens. When these comparisons are made with leukemic serum or with serum from patients with febrile infectious diseases, the agreement is good. However, in normal plasma samples taken from blood donors, the enzyme activity estimates are less precise (perhaps due to the insensitivity of our assay) and appear to be higher than the immunoassay results. Thus the question remains whether all the lysozyme activity in plasma exists in the form of the protein depicted in Fig. 2 or whether there may be more than one form of human lysozyme. We believe that this approach of correlating immunoassay with enzyme activity measurements should be pursued further in an attempt to answer this question.

References

1. Osserman, E. F., and Lawlor, D. P., *J. Exp. Med.* **124,** 921 (1966).
2. Osserman, E. F., *Science* **155,** 1536 (1967).
3. Canfield, R. E., Kammerman, S., Sobel, J. H., and Morgan, F. J., *Nature (Lond.)* **232,** 16 (1971).
4. Canfield, R. E., *J. Biol. Chem.* **238,** 2698 (1963).
5. Brew, K., Vanaman, T. C., and Hill, R. L., *J. Biol. Chem.* **242,** 3747 (1967).
6. Thomsen, J., Lund, E. H., Kristiansen, K., Brunfeldt, K., and Malmquist, Jr., *FEBS Lett.* **22,** 34 (1972).
7. Jollès, J., and Jollès, P., *FEBS Lett.* **22,** 31 (1972).
8. Arnheim, N., Sobel, J., and Canfield, R., *J. Mol. Biol.* **61,** 237 (1971).
9. Riblet, R. J., and Herzenberg, L. A., *Science* **168,** 1595 (1970).
10. Arnon, R., and Maron, E., *J. Mol. Biol.* **61,** 225 (1971).

The High Resolution X-Ray Study of Human Lysozyme: A Preliminary Analysis

S. H. BANYARD, C. C. F. BLAKE, AND I. D. A. SWAN

Introduction

The earlier X-ray analysis of human lysozyme (4) resulted in an electron density map of the enzyme at 6 Å resolution. This clearly showed that human lysozyme was structurally homologous with hen egg-white lysozyme (2, 3). The primary structure of human lysozyme has been determined independently by Jollès and Jollès (7) and Canfield and his colleagues (6). The X-ray analysis has now been extended to 2.5 Å resolution and it has proved possible to produce a preliminary interpretation of the electron density map in terms of sequence and to define some of the similarities and differences between the structures of the human and hen egg-white molecules.

X-Ray Methods

Human lysozyme was isolated from the urine of a leukemic patient by Dr. Elliott F. Osserman who has generously supplied us with large quantities of the enzyme. The crystals were grown from 7 M ammonium nitrate at pH 4.7 as previously described (10). While large crystals could be readily obtained, they nearly always grew as pseudo-parallel aggregates that gave multiple X-ray patterns of little use for structural work. A wide variety of crystallizing conditions has been tried in an attempt to eliminate this effect, but without success. The only useful crystals that have been observed appeared occasionally in batches that took a year or more to grow and the availability of five true single crystals grown in this way was sufficient for high-resolution analysis.

The crystals of human lysozyme are orthorhombic, space group $P2_12_12_1$ with cell dimensions $a = 57.1$ Å, $b = 61.0$ Å, and $c = 33.0$ Å, and contain a single molecule in the asymmetric unit. In view of the small number of crystals available for this study, we decided to use only two derivatives and to restrict the resolution limit to 2.5 Å. The derivatives selected were the best used in the 6 Å work, mercuric acetate and uranyl nitrate.

The bulk of the intensity data was collected on a Linear Diffractometer modified to collect five reflections at a time in the flat-cone setting (11). The reflections in the inaccessible blind region in this geometry and data to scale the five counters were subsequently collected on a Hilger and Watts four-circle diffractometer from the same crystal in the same orientation. Using this system it was possible to obtain three complete and self-consistent sets of 2.5 Å data from three crystals without having to remount them. The 10,000 reflections in each set of data included independent measurements of the Bijvoet pairs hkl and $hk\bar{l}$ in order to make use of the anomalous scattering from the heavy atoms in the phase determination (8).

Each set of data was corrected for absorption by the method of North, Phillips, and Mathews (9) and for Lorenz and polarization factors using standard laboratory programs. The various levels of data were scaled together by the method of Hamilton, Rollett, and Sparks (6a) and then merged together. The scaling of each of the derivative sets of data to the native set was carried out using J. Kraut's method (1).

Three-dimensional Patterson maps using $(F_{ph}-F_p)^2$ as terms were calculated for both derivatives. Both maps contained a single peak in each of the Harker sections, indicating one dominant site in each derivative. The coordinates for these sites were in excellent agreement with those obtained in the previous low-resolution study. An initial refinement of the heavy atom parameters using only zonal data followed by the calculation of difference-Fourier maps revealed the presence of a minor site in the mercuric acetate derivative. Two minor sites were found in the map of uranyl nitrate and a clear indication that the main site was either undergoing marked anisotropic thermal vibrations or that it was composed of two close statistically occupied sites. The latter model was used for subsequent refinement. This refinement revealed no further minor sites and on its completion the refined heavy atom parameters and error values listed in Table I were used as input to the phase determination. The mean "figure of merit" for the 4277 reflections was 0.60.

TABLE I

Heavy Atom Parameters

	Site	X	Y	Z	O[a]	B[b]	E[c]	N[d]	R[e]
$UO_2(NO_3)_2$	I	0.4153	0.4710	0.7261	10.8	13.4	17.5	892	0.59
	II	0.4295	0.4567	0.7508	11.2	15.5	—	—	—
	III	0.1183	0.4917	0.4295	4.9	21.7	—	—	—
	IV	0.3856	0.4885	0.8060	3.6	12.6	—	—	—
$Hg(OOC \cdot CH_3)_2$	I	0.7398	0.3628	0.1343	18.4	12.5	18.2	889	0.63
	II	0.2132	0.2563	0.4163	3.9	11.9	—	—	—

[a] Occupancy in electrons, on a relative scale.
[b] The temperature factor in $Å^2$.
[c] The root mean square error in the centric data.
[d] The number of reflections used in refinement.
[e] The reliability index for the heavy atoms.

The Fourier Map

The best electron density distribution (5) was calculated from the phases, weighting each term by its figure-of-merit. The density was calculated at eightieths of the cell edge along *a* and *b* and set out in 40 sections perpendicular to the *c*-axis. A trial set of sections was prepared that contained the expected helix segment, residues 24–34, that was clearly visible in the 6 Å map. These sections are shown in Fig. 1. It can be seen that the helix itself and the disulfide bridge 30–115 appear as well-defined features. This trial was sufficiently encouraging for us to complete the map and attempt an interpretation of its features.

Although the quality of the map is high enough to warrant serious interpretation in a Richards' box (12), it does not compare very favorably with the high-resolution map of the hen egg-white enzyme (2). The reasons for this are clear enough and are directly related to the shortage of usable crystals. The reduction in resolution from 2 Å to 2.5 Å, though allowing us to collect only half as many data, inevitably results in poorer definition of molecular features. The use of only two derivatives as opposed to the three used in hen lysozyme structure again reduces the amount of data to be collected but also leads to poorer phase determination and less definition. Finally, only one measurement could be made of each *hkl* and *hk̄l* reflection, as compared with two or four made during the study of the hen enzyme, reducing the quality of the phase determination even further.

In these circumstances it is remarkable that the map is interpretable and it is not surprising that it suffers from some defects that did not affect the map of the hen egg-white lysozyme. It is particularly noticeable that in

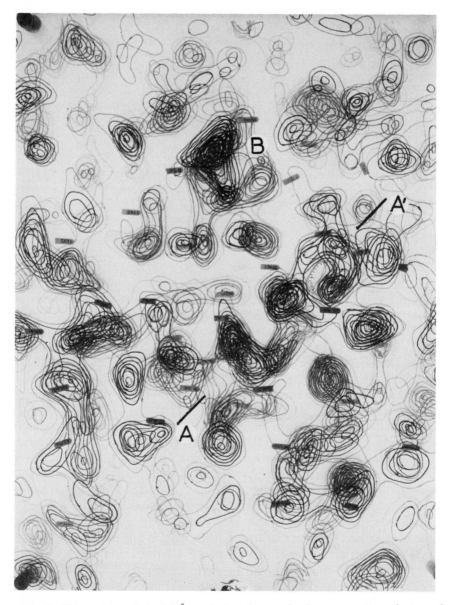

Fig. 1. Ten sections of the 2.5 Å resolution electron density map, contoured at equal but arbitrary intervals. AA′ shows the course of α-helix, 24–34, and B shows helix 5–15. Compare with Fig. 3 of Blake *et al.* (2).

many of the larger side chains the β-carbons are appreciably weaker than either the main chain density or the density in the body of the side chain itself. In a few places the main chain density drops to rather low values and some of the side chains are less well defined than expected. In the absence of any more usable crystals, except those set aside for substrate binding studies, it is not clear how the map may be improved, but studies are under way to see if refinement of the map is possible.

Interpretation of the Fourier Map

As yet no serious interpretation of the map has been made in a Richards' box and the preliminary interpretation described here has been made by simple inspection of the map and comparisons with the detailed model of hen egg-white lysozyme. The conclusions given here must therefore await confirmation from a more objective study of the map before it can be ascertained whether the proposed conformations are reasonable or not. The difficulties encountered in interpreting an electron density map by inspection have been noted by Blake *et al.* (2) in relation to hen egg-white lysozyme. In particular, it is easier to recognize the density corresponding to the larger side chains than to follow the main chain directly. In what follows, differences in main chain conformations have been inferred from observed differences in side chain orientation.

It is clear that the main chain of human lysozyme follows the same general conformation as that found in hen egg-white lysozyme and that no gross differences have been observed. The one slight exception is caused by the additional glycine residue that appears to loosen the tight hairpin loop that occurs at the end of the β-sheet structure. This glycine (47a) has been placed in the sequence (6) between residues 47 and 48. However, while the map suggests that this is a plausible position, in the absence of any side chain density this conclusion must await confirmation. In the region of residues 98–100, where the sequence has been revised to eliminate the deletion originally proposed, the map shows clearly that residues 98 and 99 are both small internal side chains consistent with their current identification as valines.

In many cases of the 53 sequence differences between hen and human lysozymes the map shows the change clearly. This is especially true of the internal residues which is fortunate since substitutions in the internal core appear to have more interesting effects than those on the exterior. Of course most of the 53 substitutions occur at residues that are on the surface of the molecule; at this stage of the interpretation, however, these changes appear to take place with little effect on the structure of the molecule as a whole.

The surface residues that are of most interest are those in the active site of lysozyme. The side chains of Asp 101, Glu 35, Asp 52, and Trp 63 are well defined and appear to be in much the same orientation as they are in the hen molecule. Tyr 62, like the equivalent tryptophan in hen lysozyme, is not particularly well defined and it seems probable that in both molecules the lack of definition can be ascribed to the side chain undergoing large thermal vibrations. Trp 108 is also ill defined in human lysozyme and this probably results from a disturbance around the mercury atom of the mercuric acetate derivative which binds to this side chain.

At this stage of the interpretation three substitutions are of particular interest because they result in surface residues becoming internal. These are

Residue 15, leucine for histidine
Residue 23, isoleucine for tyrosine
Residue 85, glutamine for serine

No substitutions that result in internal residues becoming external have been observed.

The environment of residue 15 is shown in Fig. 2 as observed in hen egg-white lysozyme. In this enzyme, the histidine in position 15 occurs at the carboxyl end of helix 5–15 and is located on the surface of the enzyme, forming a hydrogen bond with Thr 89. In human lysozyme the histidine

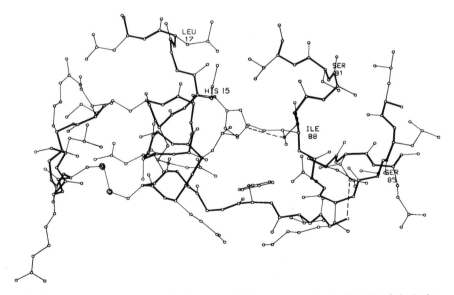

Fig. 2. A computer drawing of the hen egg-white lysozyme molecule. Ser 85 and the hydrophilic part of the core are shown on the right together with the amino terminus hydrogen bonded to Thr 40. His 15 is shown in the upper center.

is replaced by leucine, the threonine by alanine, and the hydrogen bond is lost. However, Leu 15 in human lysozyme is not found on the surface but by a rotation around its $\beta-\gamma$ bond, that appears to leave the main chain conformation essentially unchanged, largely buries itself in the hydrophobic core. The process appears to be aided by the replacement of the bulky forked side chain of Leu 17 by a methionine residue. This change is one of a number of examples of "coupled change," i.e., a number of changes in a relatively small volume that appear to allow a significant rearrangement in the packing of those residues.

The replacement of Ser 85 by a glutamine is interesting because it takes place in a small hydrophilic section of the otherwise hydrophobic core. Figure 2 shows this section as it appears in hen lysozyme. Ser 85 is located in the surface of the molecule, while behind it is a pocket lined with a number of main chain carbonyl residues, ending with the internal side chain of Ser 91. The pocket contains two water molecules hydrogen bonded to the carbonyls and the hydroxyl of Ser 91. Human lysozyme has Ser 85 replaced by glutamine and Ser 91 by alanine. The map clearly shows that Gln 85 is turned inward, replacing the water molecules. This change appears to be accomplished by a local alteration of the main chain conformation which apparently differs somewhat from that observed in the hen molecule. There appears to be a further change associated with this process involving the N-terminus. In human lysozyme this terminus extends beyond Thr 40, which hydrogen bonds the N-terminus in hen lysozyme, as shown in Fig. 2, and comes close to the main chain of residue 85. This may indicate that the main chain alteration required to allow Gln 85 to become internal also permits a main chain carbonyl to interact with the N-terminal amino group. However, we cannot at present rule out the possibility that the electron density at the amino terminus is extended by the presence of additional atoms at the terminus belonging to a blocking group or an ion.

The most extensive alteration to the structure of human lysozyme, as compared to the hen molecule, also involves a coupled change, Ile for Tyr 23 and Ile for Met 105. The structure of this region of the hen molecule is shown in Fig. 3, where Tyr 23 is found lying in the surface of the enzyme. It forms one side of the so-called hydrophobic box (2), in which the sulfur of Met 105 is surrounded by four aromatic side chains whose other members are the tryptophans 28, 108, and 111. The isoleucine that replaces the tyrosine in human lysozyme is clearly entirely internal, occupying part of the space that was taken by Met 105. However Ile 105 does not occupy the remaining space. Instead, Trp 111 appears to rotate to a considerable extent so that its indole ring comes into contact with Trp 28 and Ile 23. Ile 105 is then accommodated above Trp 111 in such a position that its main chain conformation must be considerably altered. This is probably also true of

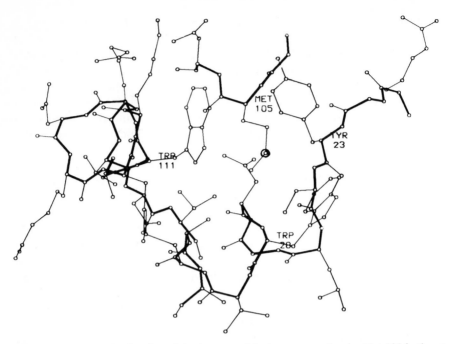

Fig. 3.   A computer drawing of the hen egg-white lysozyme molecule. Met 105 is shown in the center, its sulfur marked by the double ring. Three of the four aromatic residues, Tyr 23, Trp 28, and Trp 111, that surround Met 105 and form the hydrophobic box are also shown.

Trp 111 and it is most unfortunate that the main chain density in this region is particularly difficult to interpret. It is interesting to note that the substitution of a proline residue at position 102 must cause an alteration of the main chain in its immediate vicinity, because in hen lysozyme the NH bond is pointing inward while Pro 102, in which this bond becomes a NC bond, in human lysozyme is pointing outward. It is tempting to suppose that these three changes in a short stretch of chain are linked in some way.

We can conclude from this discussion that the core of human lysozyme is altered to a much greater extent than expected simply from a consideration of the substitution of those residues present in the core of hen lysozyme. In a discussion of this latter point (4) it was estimated that the homology between hen and human lysozymes is about 74% for internal residues. While it is difficult to estimate in percentage what the effect of the additional changes in the core regions described above amounts to, they clearly reduce the homology by a significant factor. Since even a 74% homology for internal residues is low in relation to the values obtained from somewhat similar studies of chymotrypsin, elastase (13), and the subtilisins (14), it is not clear, in the lysozymes at least, that the internal core plays such a dominating role in the maintenance of structure and function as has been thought, or that it is subject to a greater degree of conservation than the exterior of the molecule. It is hoped that the detailed interpretation of the electron density map of human lysozyme that is now under way will shed more light on this surprising, though tentative, conclusion.

Acknowledgments

We wish to thank Dr. Elliott F. Osserman for the gift of human lysozyme, Dr. A. C. T. North, and Mr. L. O. Ford for the computer drawings of hen egg-white lysozyme. We are grateful to the Medical Research Council for financial support and for providing research studentships to S. H. Banyard and I. D. A. Swan.

References

1. Arnone, A., Bier, C. J., Cotton, F. A., Day, V. W., Hazen, E. E., Richardson, D. C., Richardson, J. S., and Yonath, A., *J. Biol. Chem.* **246,** 2302 (1971).
2. Blake, C. C. F., Koenig, D. F., Mair, G. A., North, A. C. T., Phillips, D. C., and Sarma, V. R., *Nature* (*Lond.*) **206,** 757 (1965).
3. Blake, C. C. F., Mair, G. A., North, A. C. T., Phillips, D. C., and Sarma, V. R., *Proc. R. Soc., Lond.* [*Biol.*] **167,** 365 (1967).
4. Blake, C. C. F., and Swan, I. D. A., *Nature* (*Lond.*), *New Biol.* **232,** 12 (1971).
5. Blow, D. M., and Crick, F. H. C., *Acta Crystallogr.* [*Kbh.*] **12,** 794 (1959).
6. Canfield, R. E., Kammerman, S., Sobel, J. H., and Morgan, F. J., *Nature* (*Lond.*), *New Biol.* **232,** 16 (1971).
6a. Hamilton, W. C., Rollett, J. S., and Sparks, R. A., *Acta Crystallogr.* [*Kbh.*] **18,** 129 (1965).
7. Jollès, J., and Jollès, P., *Helv. Chim. Acta* **54,** 2668 (1971).
8. North, A. C. T., *Acta Crystallogr.* [*Kbh.*] **18,** 212 (1965).
9. North, A. C. T., Phillips, D. C., and Mathews, F. S., *Acta Crystallogr.* [*A*] [*Kbh.*] **24,** 351 (1968).
10. Osserman, E. F., Cole, S. J., Swan, I. D. A., and Blake, C. C. F., *J. Mol. Biol.* **46,** 211 (1969).
11. Phillips, D. C., *J. Sci. Instrum.* **41,** 123 (1964).
12. Richards, F. M., *J. Mol. Biol.* **37,** 225 (1968).
13. Shotton, D. M., and Watson, H. C., *Nature* (*Lond.*) **225,** 811 (1970).
14. Wright, C. S., Alden, R. A., and Kraut, J., *Nature* (*Lond.*) **221,** 235 (1969).

Preliminary Biochemical Studies of the Lysozymes of the Black Swan, *Cygnus atratus**

FRANK J. MORGAN AND NORMAN ARNHEIM

Introduction

Initial studies of the lysozyme from White Embden geese led to the interesting finding that this lysozyme differed significantly from that of chicken egg lysozyme and other homologous avian lysozymes in enzymatic activities (1–3), immunologic reactivity (4), and chemical characteristics (5, 6). The most striking confirmation of these phenomena was the complete absence of any similarity in an extended length of NH_2-terminal amino acid sequence (7).

Thus there seemed to exist in waterfowl at least two quite distinct types of lysozyme molecules. Further immunologic studies (4) indicated that both types of lysozymes coexisted in the egg white of the black swan, *Cygnus atratus*. The present paper describes the isolation, purification, and preliminary structural studies on both types of lysozyme present in the black swan. A reinvestigation of the properties of the goose-type lysozyme indicates that its molecular weight is significantly greater than previously believed.

*Supported by research grants from the National Institutes of Health (AM 09579) and the National Science Foundation (GB 17351).

81

Purification of Lysozymes from Black Swan Egg White

The initial purification of the lysozymes from black swan egg white followed the scheme described for goose lysozyme (5) except that in the chromatography step CM-Sephadex C-50 was substituted for CM-cellulose. The lysozyme activity emerged as a single peak from the CM-Sephadex column. This active fraction was then chromatographed on BioRex 70 and equilibrated with 0.2 M Na phosphate, pH 7.18 (8). Two major lysozyme-containing peaks were obtained. The first was eluted with the equilibration buffer; the second major peak of lysozyme activity was then eluted by a stepwise change to 0.2 M Na phosphate, pH 8.3. The peaks were provisionally called peak A and peak B in order of their emergence from the column.

Immunologic Reactivity of Black Swan Lysozymes

The reaction of both black swan lysozyme peaks was tested by gel diffusion with antisera to chicken egg lysozyme and to goose egg lysozyme. Antibodies to goose egg lysozyme do not cross react with those of chicken egg lysozyme (4).

Lysozyme from peak A was shown to form a precipitin line with anti-goose egg lysozyme serum, but did not react at all with anti-chicken egg lysozyme serum. On the other hand, lysozyme from peak B gave a precipitin line with anti-chicken egg lysozyme serum but did not react at all with anti-goose egg lysozyme serum. These immunologic findings thus suggested we had separated the two antigenic species earlier observed in black swan egg white (4). In the remainder of this paper we shall refer to peak A lysozyme as the black swan (goose-type) lysozyme and peak B lysozyme as the black swan (chicken-type) lysozyme.

Enzymatic Activity of Black Swan Egg Lysozymes

Both types of black swan egg lysozyme were tested for activity in a procedure based on lysis of *Micrococcus lysodeikticus* (9). Black swan (goose-type) lysozyme had an activity approximately fivefold that of black swan (chicken-type) lysozyme on a weight basis. Black swan (chicken-type) lysozyme activity was equal to that of chicken egg lysozyme. This is consistent with the finding that goose egg lysozyme is more active than chicken egg lysozyme in the *M. lysodeickticus* assay (1, 5).

Chemical Studies of the Lysozymes of Black Swan Egg White

Black Swan (Goose-Type) Lysozyme

The protein was reduced and alkylated with iodoacetic acid (10) using dithiothreitol in 6 M guanidine HCl–0.2 M Tris Cl, pH 8.2, as the reducing medium. A single NH_2-terminal amino acid, arginine, was found by phenylisothiocyanate degradation (11). The monomer molecular weight calculated from the amount of terminal amino acid released was between 19,500 and 20,500. On SDS-gel electrophoresis in 10% acrylamide gels, the S-carboxymethyl lysozyme migrated as a single band, and the molecular weight calculated from the migration distance as compared to known standards was between 19,500 and 21,500. Studies of White Embden goose lysozyme by SDS-gel electrophoresis and by quantitative NH_2-terminal amino acid determination also indicate that its molecular weight is between 19,500 and 21,500 (11a); this figure is somewhat higher than earlier estimates (1, 5).

Total amino acid analysis was performed after acid hydrolysis (6 N HCl; 110°C; 24, 48, and 72 hr); half-cystine was estimated as S-carboxymethylcysteine and the tryptophan content by the method of Liu and Chang (12). The amino acid composition to the nearest integer is shown in Table I. This is based on 20 glycine residues, which gives the closest correspondence to the observed molecular weight. On the basis of this assumption, the molecule possesses 181 amino acid residues. There are four half-cystine residues, half the number present in chicken egg lysozyme.

The amino acid composition of a sample of White Embden goose lysozyme (kindly provided by Dr. R. E. Canfield) is shown for comparison; this composition corresponds closely to that previously published for goose lysozyme (5) if the absolute values are calculated on the basis of 20 glycine residues. The compositions of black swan (goose-type) lysozyme and of Embden goose lysozyme are strikingly similar, and very different from chicken and similar lysozymes (see Table III). There are nine differences in composition between the two proteins, suggesting a minimum of 4 sequence differences in the 180 residues.*

The similarity is confirmed by a study of the NH_2-terminal sequence of

---

*Because of the low absorbance of the reaction product of proline with the ninhydrin reagent at pH 5.5, quantitation of this imino acid is not as accurate as that of the amino acids. The difference in proline values between the two proteins could therefore be due to an error in the analyses and this residue, which provides important structural constraints, may be present in equal amounts in both proteins. In this case, both proteins have equal numbers of amino acid residues, and the compositional difference between them is reduced to eight residues.

black swan (goose-type) lysozyme. A partial NH$_2$-terminal amino acid sequence was performed by automated Edman degradation (13) and the results are shown in Table II. The corresponding region of goose lysozyme (7) is shown for comparison. Both are identical in the region shown. This sequence bears no relationship to the amino acid sequence of chicken egg-white lysozyme (14, 15) or to the sequences of the homologous avian lysozymes (see Ref. 16).

Black Swan (Chicken-Type) Lysozyme

This lysozyme was totally reduced and alkylated with iodoacetic acid. The single NH$_2$-terminal amino acid was lysine: the molecular weight calculated from the amount of terminal amino acid released was approximately 13,000. On SDS-gel electrophoresis in 10% acrylamide gels, this lysozyme had a migration identical to that of chicken egg lysozyme, giving a molecular weight of 14,000–15,000.

TABLE I

Amino Acid Composition of Goose-Type Lysozymes[a]

Amino acid	Goose (White Embden)	Black swan (goose type)
Trp	3	3
Lys	18	17
His	5	5
Arg	11	9
Asp	20	20
Thr	13	12
Ser	9	9
Glu	15	15
Pro[b]	5	4
Gly	20	20
Ala	15	15
Half-cys	4	4
Val	10	12
Met	3	3
Ile	11	13
Leu	7	7
Tyr	9	9
Phe	3	3
Total[b]	180	181

[a] These values represent the average analysis of duplicate hydrolysates for 24, 48, and 72 hr. Appropriate corrections have been made to allow for changes in composition with time of hydrolysis.

[b] See footnote on p. 83.

TABLE II

Partial Amino Acid Sequence of Black Swan (Goose-Type) Lysozyme Compared to White Embden Goose Lysozyme[a]

					5				
Swan	Arg	Thr	Asp	Cys	Tyr	Gly	Asn	Val	Asn
Goose	Arg	Thr	Asp	Cys	Tyr	Gly	Asn	Val	Asn

	10					15			
Swan	Arg	Ile	Asp	Thr	Thr	Gly	Ala	Ser	Cys
Goose	Arg	Ile	Asp	Thr	Thr	Gly	Ala	Ser	Cys

	20	
Swan	Lys	Thr
Goose	Lys	Thr

[a] Sequence data for White Embden goose lysozyme is from Ref. 7.

The integral values for the amino acid composition, calculated from an amino acid analysis performed as described in the preceding section, are given in Table III. Values for other avian lysozymes are given for comparison. It can be seen that this enzyme corresponds closely to the lysozyme type of which chicken egg is the archetype. It possesses 129 residues, 8 half-cystine residues, and a high content of basic amino acids. There is an especially close resemblance in composition to the duck lysozymes, which also possess no histidine residues (17–19).

A partial NH$_2$-terminal amino acid sequence was determined by automated Edman degradation (Table IV). The relationship to the duck lysozymes is confirmed; the histidine residue 15 of chicken egg is replaced in all cases by a leucine. All half-cystines so far determined have identical positions.

Comments

These findings confirm the existence of at least two distinct lysozyme genes in waterfowl; the major features of each type are summarized in Tables V and VI. They represent the second demonstration of a goose-type lysozyme, and the first instance in which both forms have been isolated from one species. Although the studies were performed on pooled black swan egg whites, the original immunologic finding made with individual egg whites (4) excludes the possibility that we are examining two separate populations of swan. The several chicken-type lysozymes found in the duck, however, are clearly related to the presence of multiple alleles (18, 19).

One type of enzyme, represented by the well-studied chicken egg lysozyme, contains 129 amino acids, 4 disulfide bridges, and has a molecular

TABLE III

Amino Acid Compisition of Chicken-Type Lysozymes[a]

	Black swan (chicken type)	Duck II	Duck III	Chicken
Trp	6	6	6	6
Lys	11	6	7	6
His	0	0	0	1
Arg	10	13	13	11
Asp	17	19	19	21
Thr	8	7	7	7
Ser	8	11	9	10
Glu	8	5	6	5
Pro	2	2	2	2
Gly	11	12	11	12
Ala	11	11	12	12
Half-Cys	8	8	8	8
Val	6	7	7	6
Met	2	2	2	2
Ile	6	6	6	6
Leu	9	8	8	8
Tyr	5	5	5	3
Phe	1	1	1	3
Total	129	129	129	129

[a] The analysis of swan (chicken-type) lysozyme was performed as described for the goose-type lysozymes in Table I. Duck compositions from Ref. 18; chicken composition from Ref. 14.

weight of 14,000–15,000. Its enzymatic activity is that of a muramidase, cleaving between N-acetylglucosamine and N-acetylmuramic acid residues of the bacterial cell wall. The second type, represented most clearly at present by the White Embden goose lysozyme, has approximately 180 amino acid residues, 4 half-cystine residues and presumably 2 disulfide bridges, and a molecular weight of 19,500–21,500. It appears to act as a muramidase, but may have selective affinity for those N-acetylmuramic acid residues which have a peptide moiety attached (20).

It seems most likely that these enzymes are synthesized under the direction of two separate genes, which may be present in all birds. Synthesis of oviduct proteins (lysozymes, ovalbumin, conalbumin, and ovomucoid) is under the control of steroid hormones and recent studies have shown that the relative proportion of estrogens and progestins used to stimulate the oviduct determine the relative rate of synthesis of the different proteins produced (21). The existence in a bird egg of mainly one or the other lysozyme, or of both, presumably reflects some such control mechanism.

TABLE IV

Partial NH$_2$-Terminal Amino Acid Sequence of Black Swan (Chicken-Type) Lysozyme Compared to Other Bird Lysozymes[a]

				5						
Swan	Lys	Val	Tyr	Glu	Arg	Cys	Glu	Leu	Ala	
Duck III										
Duck II				Ser						
Hen				Gly						

	10					15			
Swan	Ala	Ala	Met	Lys	Arg	Leu	Gly	Leu	Asp
Duck III									
Duck II									
Hen						His			

		20
Swan	Asn	Tyr
Duck III		
Duck II		
Hen		

[a] Only the residues which differ from swan are shown. Duck lysozymes from Ref. 18; chicken lysozyme from Ref. 14.

There is at present no evidence of the existence of the goose-type gene in birds other than the waterfowl, although it would not be surprising if a careful search demonstrated this gene in other bird orders (Arnheim, Chap. 14, this volume).

Mammalian lysozymes from various species have been studied, including man (7, 22–24), baboon (25), and mouse (Riblet, Chapter 8, this volume). All these lysozymes clearly belong to the chicken egg type. Whether the gene for the second type exists in mammals is as yet not known.

TABLE V

Physiochemical Characteristics of Goose-Type Lysozymes

	Black swan (goose type)	White Embden goose
Molecular weight	19,500–21,500	19,500–21,500
NH$_2$-terminus	Arg	Arg
Half-cystine	4	4
Tryptophan	3	3
Total residues[a]	181	180

[a] See footnote on p. 83.

TABLE VI

Physicochemical Characteristics of Chicken-Type Lysozymes

	Black swan (chicken type)	Chicken
Molecular weight	14–15,000	14–15,000
NH_2-terminus	Lys	Lys
Half-cystine	8	8
Tryptophan	6	6
Total residues	129	129

References

1. Dianoux, A. C., and Jollès, P., *Biochim. Biophys. Acta* **133**, 472 (1967).
2. Jollès, P., Blancard, J. S., Charlemagne, D., Dianoux, A. C., Jollès, J., and LeBaron, J., *Biochim. Biophys. Acta* **151**, 532 (1968).
3. McKelvy, J. F., Eshdat, Y., and Sharon, N., *Fed. Proc.* **29**, 532 (1970) (abstr.).
4. Arnheim, N., and Steller, R., *Arch. Biochem. Biophys.* **141**, 656 (1970).
5. Canfield, R. E., and McMurry, S., *Biochem. Biophys. Res. Commun.* **26**, 38 (1967).
6. Dianoux, A. C., and Jollès, P., *Bull. Soc. Chim. Biol. (Paris)* **51**, 1559 (1969).
7. Canfield, R. E., Kammerman, S., Sobel, J. H., and Morgan, F. J., *Nature (Lond.), New Biol.* **232**, 16 (1971).
8. Tallan, H. H., and Stein, W. H., *J. Biol. Chem.* **200**, 507 (1953).
9. Parry, R. M., Jr., Chandan, R. C., and Shahani, K. M., *Proc. Soc. Exp. Biol. Med.* **119**, 384 (1965).
10. Crestfield, A. M., Moore, S., and Stein, W. H., *J. Biol. Chem.* **238**, 622 (1963).
11. Edman, P., *Acta Chem. Scand.* **4**, 283 (1950).
11a. Arnheim, N., and Morgan, F. J., unpublished results.
12. Liu, T.-Y., and Chang, Y. H., *J. Biol. Chem.* **246**, 2842 (1971).
13. Edman, P., and Begg, G., *Eur. J. Biochem.* **1**, 80 (1967).
14. Canfield, R. E., *J. Biol. Chem.* **238**, 2698 (1963).
15. Jollès, J., Jauregui, A. J., Bernier, I., and Jollès, P., *Biochim. Biophys. Acta* **78**, 668 (1963).
16. Dayhoff, M. O., ed., "Atlas of Protein Sequence and Structure," Vol. 5. National Biomedical Research Foundation, Silver Springs, Md., 1972.
17. Jollès, J., Hermann, J., Niemann, B., and Jollès, P., *Eur. J. Biochem.* **1**, 344 (1967).
18. Hermann, J., Jollès, J., and Jollès, P., *Eur. J. Biochem.* **24**, 12 (1971).
19. Prager, E., and Wilson, A. C., *J. Biol. Chem.* **246**, 523 (1971).
20. Arnheim, N., Inouye, M., Law, L., and Laudin, A., *J. Biol. Chem.* **248**, 233 (1973).
21. Palmiter, R. D., *J. Biol. Chem.* **247**, 6450 (1972).
22. Blake, C. C. F., and Swan, I. D. A., *Nature (Lond.), New Biol.* **232**, 12 (1971).
23. Jollès, J., and Jollès, P., *Helv. Chim. Acta* **54**, 2668 (1971).
24. Jollès, J., and Jollès, P., *FEBS Lett.* **22**, 31 (1972).
25. Buss, D., *Biochim. Biophys. Acta* **236**, 587 (1971).

Sequence Studies of Mouse Lysozyme[*]

ROY J. RIBLET

While lysozymes can be readily obtained from the milk of large mammals the study of lysozymes from small animals requires a more convenien source. Following the discovery by Osserman and Lawlor (1) of lysozyme in the urine of monocytic leukemia patients, lysozymuria was found asso-ciated with transplantable monocytic malignancies in laboratory rodents. Lysozyme is produced by monocytic leukemias in the rat (2) and BALB/c mouse (3) and by a reticulum cell sarcoma in the (BALB/c × NZB) F_1 mouse (4). These lysozymes will be useful, of course, both for the corre-lation of structural variations with changes in enzyme function and for the analysis of molecular evolution at the species level. They may also help refine the study of molecular evolution to variability within individual species since monocyte tumors are well known in other inbred strains of mice, e.g., the spontaneous reticulum cell sarcomas of SJL (5) and the inducible ones in DBA/2 (6). Examination of these and others may yield variant lysozymes available in sequenceable amounts. This report dis-cusses the first partial sequence of a single mouse lysozyme.

Mouse lysozyme was purified from the urine of (BALB/c × NZB) F_1 mice bearing GPC-11, a transplantable mineral oil induced reticulum cell

*This work was supported by a Damon Runyon Cancer Research Fellowship to R. J. Riblet and N.I.A.I.D. Research Grant No. AI05875 and N.I.A.I.D. Training Grant No. AI00430 to Melvin Cohn.

sarcoma. Mouse lysozyme was shown to be similar in physical, chemical, and enzymologic properties to human and chicken lysozymes (4), and it has now been sequenced to residue 60 on a Beckman Sequencer (7, 8). The partial sequence of mouse lysozyme is shown in Table I along with comparable sequences of human and various bird lysozymes (7–13). It is clear that mouse lysozyme is a member of this family of proteins related by evolution. It is also clear that the mouse and human sequences are more similar to each other than to the bird sequences, and that the bird sequences are a nearly identical subset. Over the first 60 residues, mouse and human lysozymes show 78% identity while the mouse–chicken and human–chicken comparisons give 60 and 68% identity, respectively. At 13 positions in the first 60 both mouse and human lysozyme share a "mammal-specific"

TABLE I

Comparison of Lysozyme Partial Sequences[a]

	1	2	3	4	5	6	7	8	9	10	11	12	13	14	15	16	17
M[b]			Tyr·	Glu			Glx·	Phe		Arg·	Thr·	Leu			Asx		Met
H				Glu						Arg·	Thr·	Leu			Leu		Met
C	Lys·	Val·	Phe·	Gly·	Arg·	Cys·	Glu·	Leu·	Ala·	Ala·	Ala·	Met·	Lys·	Arg·	His·	Gly·	Leu·
T			Tyr												Leu		
DII			Tyr·	Ser											Leu		
DIII			Tyr·	Gln											Leu		

	18	19	20	21	22	23	24	25	26	27	28	29	30	31	32	33	34
M	Ala·	Gly		Tyr		Val			Ala·	Asp				Leu		Glx·	His
H		Gly				Ile			Ala			Met		Leu			Trp
C	Asp·	Asn·	Tyr·	Arg·	Gly·	Tyr·	Ser·	Leu·	Gly·	Asn·	Trp·	Val·	Cys·	Ala·	Ala·	Lys·	Phe
T																	
DII																Asn·	Tyr
DIII																Asn·	Tyr

	35	36	37	38	39	40	41	42	43	44	45	46	47	48[c]	49	50	51
M			Asx·	Tyr·	Asx		Arg			Asx·	Tyr·	Asx·	Arg·	Gly·	Asx·	Glx	
H			Gly·	Tyr			Arg				Tyr·		Ala·	Gly		Arg	
C	Glu·	Ser·	Asn·	Phe·	Asn·	Thr·	Gln·	Ala·	Thr·	Asn·	Arg·	Asn·	Thr·	— ·	Asp·	Gly·	Ser·
T							His										
DII			Ser														
DIII			Gly														

	52	53	54	55	56	57	58	59	60	61
M		Asx				Phe·	Glx		?	
H						Phe				
C	Thr·	Asp·	Tyr·	Gly·	Ile·	Leu·	Gln·	Ile·	Asn·	Ser·
T										
DII							Glu			
DIII			?	?	?	?	?	?	?	

[a] The sequence of chicken lysozyme from residues 1 to 61 is listed in full; where the lysozyme of another species differs the variant residue is shown, otherwise the sequences are identical to that of chicken.

[b] M: Mouse lysozyme, (7, 8); H: human lysozyme, (9); C: chicken lysozyme, (10); T: turkey lysozyme, (11); DII: duck lysozyme II, (12); DIII: duck lysozyme III, (13).

[c] Position 48 is a gap in the bird lysozymes.

TABLE II

Mouse Lysozyme Tryptic Peptides

	60	61	62	63	64	65	66	67	68	69	70	71	72	73	74	75	76	77	78	79
													Arg							
Mouse Human	Asn ·	Ser ·	Arg ·	Tyr ·	Trp ·	Cys ·	Asn ·	Asp ·	Gly ·	Lys ·	Thr ·	Pro ·	Gly ·	Ala ·	Val ·	Asn ·	Ala ·	Cys ·	His ·	Leu

	80	81	82	83	84	85	86	87	88	89	90	91	92	93	94	95	96	97	98	99
Mouse Human	Ser ·	Cys ·	Ser ·	Ala ·	Leu ·	Leu ·	Gln ·	Asp ·	Asn ·	Ile ·	Ala ·	Asp ·	Ala ·	Val ·	Ala ·	Cys ·	Ala ·	Lys ·	Arg ·	Val ·

	100	101	102	103	104	105	106	107	108	109	110	111	112	113	114	115	116	117	118	119
															Ala	His				
Mouse Human	Val ·	Arg ·	Asp ·	Pro ·	Gln ·	Gly ·	Ile ·	Arg ·	Ala ·	Trp ·	Val ·	Ala ·	Trp ·	Arg ·	Asn ·	Arg ·	Cys ·	Gln ·	Asn ·	Arg

	120	121	122	123	124	125	126	127	128	129	130
	Leu					Asp					
Mouse Human	Asp ·	Val ·	Arg ·	Gln ·	Tyr ·	Val ·	Gln ·	Gly ·	Cys ·	Gly ·	Val ·

substitution not found in the bird sequences. One of these changes is the insertion of glycine at position 48. In addition the mouse sequence shows 8 or more independent substitutions, while the human sequence has 5, and the bird sequences 1 to 5 each. Hence the mouse sequence appears to be accumulating mutations somewhat more rapidly than the others, that is, it is evolving faster. The comparison between mouse and human proteins can be made in two other cases, hemoglobin α-chain, which shows 88% identity, and the constant region of κ-immunoglobulin light chains, which shows 61% identity. Thus lysozyme in these species appears to be evolving at a rate intermediate to hemoglobin and immunoglobulin.

None of the substitutions in mouse lysozyme would appear to alter the tertiary structure in a major way from that of human and chicken lysozymes. Hydrophobic residues which in chicken lysozyme were shown (14) to be internal or extensively interacting with other hydrophobic residues, for example positions 3, 8, 12, or 17, show only conservative hydrophobic substitutions. On the other hand positions which are external, with side chains directed into the solvent, have accumulated a variety of substitutions, e.g., Gly to Glu at position 4 or Ala to Arg at 10. A very interesting issue is that of coordinated changes within the groups of residues which interact with each other. Without model building only one example can be clearly seen in this data. In the chicken sequence the conformation of the main peptide chain requires a glycine at position 49 (14), but the insertion of a glycine residue at 48 alters the backbone structure sufficiently to allow replacement of glycine (now 50) by Arg in the human and Glx in the mouse sequence.

In current studies tryptic peptides of carboxymethylated and amino-ethylated mouse lysozymes have been eluted from peptide maps and their amino acid compositions analyzed. These peptides were aligned by homology to the human sequence after position 60 and the available data is shown in Table II. There the sequence of human lysozyme is listed and, for the mouse sequence, a solid line is shown where a mouse peptide of appropriate composition has been found. Additional peptides have been found which must correspond to the large vacant region, 78–95, but their compositions indicate extensive changes in this region. Although these data for the second half of the mouse lysozyme sequence are fragmentary, they do yield an interesting indication. The sequence substitutions seen in the first half of mouse lysozyme were frequent and rather uniformly spread along the molecule. The second half however shows two long regions of nearly invariant sequence and hints at an intervening region of high variability.

References

1. Osserman, E. F., and Lawlor, D. P., *J. Exp. Med.* **124,** 921 (1966).
2. Osserman, E. F., and Azar, H. A., *Fed. Proc.* **28,** 619 (1969).
3. Warner, N. L., Moore, M. A. S., and Metcalf, D., *J. Natl. Cancer Inst.* **43,** 963 (1969).
4. Riblet, R. J., and Herzenberg, L. A., *Science* **168,** 1595 (1970).
5. Wanebo, H. J., Gallmeier, W. M., Boyse, E. A., and Old, L. J., *Science* **154,** 901 (1966).
6. Rask-Nielsen, R., and Ebbesen, P., *J. Natl. Cancer Inst.* **35,** 83 (1965).
7. Riblet, R. J., Ph.D. thesis, Stanford University, 1972.
8. Riblet, R. J., Wang, A. C., Fudenberg, H. H., and Herzenberg, L. A., in preparation.
9. Canfield, R. E., Kammerman, S., Sobel, J. H., and Morgan, F. J., *Nature (Lond.), New Biol.* **232,** 16 (1971).
10. Canfield, R. E., and Liu, A. K., *J. Biol. Chem.* **240,** 1997 (1965).
11. Larue, J. N., and Speck, J. C., *J. Biol. Chem.* **245,** 1985 (1970).
12. Hermann, J., and Jollès, J., *Biochim. Biophys. Acta* **200,** 178 (1970).
13. Niemann, B., Hermann, J., and Jollès, J., *Bull. Soc. Chim. Biol. (Paris)* **50,** 923 (1968).
14. Browne, W. J., North, A. C. T., Phillips, D. C., Brew, K., Vanaman, T. C., and Hill, R. L., *J. Mol. Biol.* **42,** 65 (1969).

Lysozyme *Chalaropsis*

JOHN H. HASH

Lysozyme *Chalaropsis* is an extracellular bacteriolytic enzyme produced by a fungus that has been provisionally classified as *Chalaropsis* species. Produced concomitantly with this lysozyme is another extracellular enzyme, a ribonuclease that is specific for 3′-guanylic acid, and purification procedures have been directed toward the simultaneous isolation of both enzymes. This fungal lysozyme was first reported in 1963 (1) as a staphylolytic enzyme and the ribonuclease was found later to be a contaminant in the preparation (2).

Chalaropsis sp. produces both the lysozyme and ribonuclease in submerged cultures in a medium with a high carbon-to-nitrogen ratio (3). The elaboration of the lysozyme is constitutive and the levels of enzyme attained are unaffected by growth on bacterial cell walls. Procedures for the isolation of both enzymes are relatively simple as illustrated in the flow diagram shown in Fig. 1. Amberlite CG-50-H$^+$ quantitatively absorbs both the lysozyme and ribonuclease from crude culture filtrates. Both enzymes are displaced from this resin with ammonium acetate and then are concentrated by precipitation with ammonium sulfate. At this stage the preparation contains large amounts of contaminating peptones and colored impurities from the medium. Most of these are effectively removed by passage through a column of DEAE-cellulose at pH 5.0. The colored impurities are adsorbed while both ribonuclease and lysozyme pass through

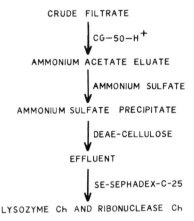

CRUDE FILTRATE

⬇ CG—50—H⁺

AMMONIUM ACETATE ELUATE

⬇ AMMONIUM SULFATE

AMMONIUM SULFATE PRECIPITATE

⬇ DEAE-CELLULOSE

EFFLUENT

⬇ SE-SEPHADEX-C-25

LYSOZYME Ch AND RIBONUCLEASE Ch

Fig. 1. Flow diagram for purification of lysozyme Ch and ribonuclease Ch.

the column. Final purification is by chromatography on sulfoethyl-Sephadex C-25 (pH 5.0) at which time both enzymes are completely separated (Fig. 2). Rechromatography (Fig. 3) yields lysozyme of high purity and it is easily crystallized (Fig. 4). The crystals are needles and to date all efforts to obtain crystals in other habits for X-ray crystallography have been unsuccessful.

In the original purification (1), contaminating peptones were removed on Amberlite CG-50 (pH 6.5) and two fractions, designated A and B, were obtained. The A enzyme was a small fraction of the total that was not

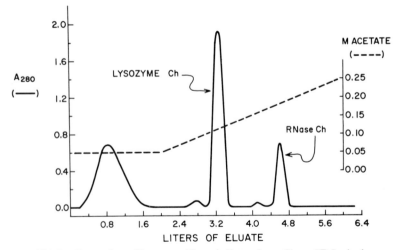

Fig. 2. Separation of lysozyme Ch and ribonuclease Ch on SE-Sephadex.

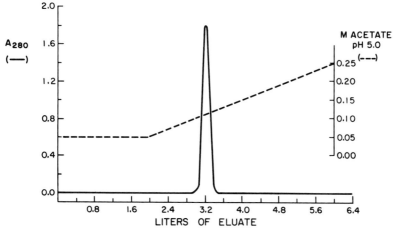

Fig. 3. Rechromatography of lysozyme Ch on SE-Sephadex.

absorbed to the resin at pH 6.5, whereas that designated B was adsorbed at this pH. When the chromatography was repeated at pH 5.0 (3) no activity corresponding to A was observed, and when material obtained as A at pH 6.5 was rechromatographed at pH 5.0, it behaved as B enzyme. The interpretation was made that small amounts of enzyme were complexed with medium constituents at the higher pH and that they dissociated at the lower pH. Only one molecular species of lysozyme is considered to be

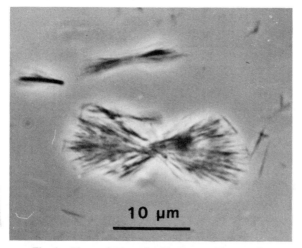

Fig. 4. Photomicrograph of lysozyme Ch crystals.

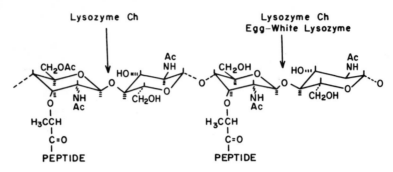

Fig. 5. Specificities of egg-white and *Chalaropsis* lysozymes.

produced by *Chalaropsis* sp. The early designations A and B were aban-
doned (3) and the trivial name lysozyme Ch was adopted (4).

The initial determination of specificity was made by Tipper *et al.* (5) on
some of the original preparation and by Hash and Rothlauf (3) on the
crystalline enzyme. The enzyme proved to be an *N*-acetylmuramidase (Fig.
5). In this figure the cell wall murein of *S. aureus* is drawn with conformational formulas, chair conformations assumed, and the site of hydrolysis
of egg-white lysozyme is included for comparison. Egg-white lysozyme is
a β-1,4-*N*-acetylmuramidase and lysozyme Ch has both β-1,4-*N*-acetyl-
muramidase and β-1,4-*N*,6-*O*-diacetylmuramidase activity. Staphylococci have been known to be insensitive to egg-white lysozyme for a long
time (6) and one of the reasons for this insensitivity is that staphylococci
contain large quantities of *N*,6-*O*-diacetylmuramic acid in their cell wall
murein (7). Egg-white lysozyme is inactive on walls that contain *N*,6-*O*-
diacetylmuramic acid because the 6-*O*-acetyl group is too large to fit into
the cleft of the molecule (8). Part of the interest in lysozyme Ch stems from
the fact that the cleft of this enzyme, assuming it has a cleft, must be quite
different from that of egg-white lysozyme. Yet both enzymes catalyze the
hydrolysis of the same bond.

There are other differences between egg-white lysozyme and lysozyme
Ch in their specificity. Egg-white lysozyme is a weak chitinase, catalyzing
the hydrolysis of chitin to soluble components (9). Therefore, 3-*O* substitution of *N*-acetylglucosamine, as in *N*-acetylmuramic acid, is not an
absolute requirement for egg-white lysozyme. Lysozyme Ch is totally inactive against chitin (3) and the requirements for 3-*O* substitution may be
more stringent than for egg-white lysozyme. Also, egg-white lysozyme
hydrolyzes and carries out transglycosylation reactions with soluble chitin
derivatives (10) and cell wall oligosaccharides (11). These soluble derivatives and the reactions catalyzed have been instrumental in elucidating the
mechanism of action of egg-white lysozyme. In contrast, and in accord
with the failure to hydrolyze chitin, lysozyme Ch neither hydrolyzes nor

transglycosylates chitin oligosaccharides (11a). Consequently there have been no soluble substrates available for studies on the mechanism of action of Ch. The cell wall disaccharide, *N*-acetylmuramyl-*N*-acetylglucosamine, is not hydrolyzed by Ch (12) and, in fact, linear murein strands are less susceptible to hydrolysis than the intact cell wall (13). The available evidence seems to suggest that substituted *N*-acetylmuramic residues are desirable for the action of lysozyme Ch. The lack of suitable soluble substrates has greatly hampered studies on the mechanism of action of this enzyme.

The physical properties of lysozyme Ch were determined by Mitchell and Hash (4) and the amino acid composition was established by Shih and Hash (14). The properties of lysozyme Ch and egg-white lysozyme are contrasted in Table I and the amino acid compositions of several lysozymes are compared in Table II (14–18).

Lysozyme Ch is a much larger molecule than egg-white lysozyme (23,400 vs. 14,300). Both are single chain peptides, neither contains carbohydrate, and both have similar extinction coefficients and partial specific volumes. As reflected by their isoelectric points, egg-white lysozyme is a much more basic protein than lysozyme Ch. The positive b_0 value for lysozyme Ch from optical rotatory dispersion data indicates a large amount of β-structure in lysozyme Ch and contrasts sharply with the negative b_0 value for egg-white lysozyme. In accord with the failure of lysozyme Ch to hydrolyze chitin oligosaccharides, *N*-acetylglucosamine is not an inhibitor of lysozyme Ch.

TABLE I

Comparative Properties of Lysozyme Ch and Egg-White Lysozyme[a]

	Lysozyme Ch	Egg-white lysozyme
MW (centrifugation)	$21,800 \pm 1350$	$14,400 \pm 100$
MW (amino acid analysis)	23,385	14,307
Sedimentation coefficient ($s_{20,w}$)	2.27 ± 0.14	1.91
Partial specific volume ($\bar{v}$)	0.726 ml/gm	0.703 ml/gm
Isoelectric point	7.53	11.1
Extinction coefficient ($E_{1\,cm}^{1\%}$)	24.8 ± 0.4	26.35 ± 0.18
Optical rotary dispersion		
Cotton minima	227 nm	233 nm
λ_c (Drude, one term)	218 nm	245 nm
b_0 (Moffitt)	$+195$	-145
λ_0 (best fit of data in Moffitt plot)	220 nm	212 nm
Disulfide bonds	1	4
NAG inhibition	No	Yes

[a] From Ref. 4.

TABLE II

Amino Acid Compositions of Lysozymes

	Ch (14)	Hen egg white (15)	Papaya (16)	T4 Phage (17)	Human leukemic (18)
Lysine	4	6	10	13	5
Histidine	4	1	3	1	1
Arginine	5	11	13	13	14
Aspartic acid	23	21	22	22	18
Threonine	18	7	13	11	5
Serine	21	10	16	6	6
Glutamic acid	11	5	11	13	9
Proline	12	2	18	3	2
Glycine	29	12	26	11	11
Alanine	27	12	21	15	14
Half-cystine	2	8	8	2	8
Valine	8	6	8	9	9^a
Methionine	3	2	4	5	2
Isoleucine	12	6	11	10	5
Leucine	10	8	12	16	8
Tyrosine	14	3	13	6	6
Phenylalanine	11	3	12	5	2
Tryptophan	7	6	7	3	5
Ammonia	23	16		16	16
Number of residues	221	129	228	164	130
MW	23,385	14,307	24,700	18,635	14,695
NH$_2$-terminus	Thr	Lys	Gly	Met	Lys
COOH-terminus	Gly	Leu	Gly	Leu	Val

[a] Corrected by Dr. R. E. Canfield, at the conference, from 8 residues of valine.

From the amino acid compositions of several lysozymes it is quite evident that there are wide differences with respect to amino acid compositions, terminal residues, and molecular weights. Some generalizations such as the relative scarcity of histidyl and methionyl residues and the relative abundance of aspartyl (or asparaginyl) residues can be made but differences outweigh similarities. With respect to size, lysozyme Ch is similar to that of papaya lysozyme but papaya lysozyme contains 4 disulfide bonds as compared to a single disulfide bond in lysozyme Ch. There is a marked difference between the two in terms of catalytic activity. Papaya lysozyme displays a chitinase activity toward chitin that is 10 times that of egg-white lysozyme and an activity toward chitotetraose that is 400 times that of

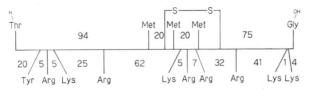

Fig. 6. Schematic diagram of the lysozyme Ch molecule. Cyanogen bromide peptides are shown on the top and tryptic peptides are shown on the bottom. The number of residues in each peptide is indicated.

egg-white lysozyme (19). Thus papaya lysozyme may be preferentially a chitinase. Lysozyme Ch is inactive on chitin and chitin oligosaccharides.

Phage T4 lysozyme has 2 half-cystine residues, as does lysozyme Ch. However, they are present as free sulfhydryl groups rather than a disulfide bond (17).

Lysozyme Ch has 3 methionyl residues and a total of 9 lysyl and arginyl residues—residues that are susceptible to chemical and proteolytic cleavage. We have elected to determine the primary sequence of lysozyme Ch by fragmentation of the molecule with cyanogen bromide and trypsin at these residues. Cyanogen bromide cleaved the molecule at the 3 methionyl residues as expected and the 4 resulting peptides were separated and ordered (14). One internal 20 residue peptide was completely sequenced by frag-

–PRO–SER–VAL–ASN–PHE–ALA–GLY–ALA–TYR–SER–ALA–GLY–ALA–ARG

PHE–VAL–ILE–ILE–LYS–ALA–THR

GLY–GLY–TYR–HIS–PHE–ALA

HIS–PRO–GLY–GLU–THR–THR–GLY–ALA–ALA–GLN–ALA–ASP–TYR–PHE–ILE–ALA–HIS–GLY–GLY–GLY–TRP–SER–GLY–ASP–GLY

ILE–THR–LEU–PRO–GLY–MET–LEU–ASP–LEU–GLU–SER–GLU–GLY–SER–ASN–PRO–ALA–CYS–TRP–GLY–LEU–SER–ALA–ALA–SER

MET–VAL–ALA–TRP–ILE–LYS–ALA–PHE–SER–ASP–ARG–TYR–HIS–ALA–VAL–THR–GLY–ARG–TYR–PRO–MET–LEU–TYR–THR–ASN

PRO–SER–TRP–TRP–SER–SER–CYS–THR–GLY–ASN–SER–ASN–ALA–PHE–VAL–ASN–THR–ASN–PRO–LEU–VAL–LEU–ALA–ASN–ARG

TYR–ALA–SER–ALA–PRO–GLY–THR–ILE–PRO–GLY–GLY–TRP–PRO–TYR–GLN–THR–ILE–TRP–GLN–ASN–SER–ASP–ALA–TYR–ALA

TYR–GLY–GLY–SER–ASN–ASN–PHE–ILE–ASN–GLY–SER–ILE–ASP–ASN–LEU–LYS–LYS–LEU–ALA–THR–GLY

Fig. 7. Partial sequence of lysozyme Ch.

menting the molecule with trypsin and by sequencing the tryptic peptides by subtractive Edman degradation. Tryptic digestion of lysozyme Ch was accomplished in 2 *M* urea and the expected peptides plus a peptide due to an atypical tryptic cleavage have been separated by a combination of gel filtration and ion-exchange chromatography (20). A schematic diagram of the cyanogen bromide and tryptic peptides is shown in Fig. 6, with the number of residues in each peptide shown. Placement of the tryptic peptides within the cyanogen bromide structure was relatively straightforward with the exception of 3 peptides in the amino terminal portion of the molecule. These have been placed tentatively in the order shown with reduced carboxymethylated protein in the automated Edman sequenator.

Using these cyanogen bromide and tryptic fragments and automated Edman degradation, and carboxy and aminopeptidases, the sequence of this molecule has been partially completed (21). The completed portions are shown in Fig. 7.

A major portion of the enzyme has been sequenced. There is an ambiguity in the sequence of the amino terminal peptide that is as yet unresolved and only the portion established with carboxypeptidase is shown. The largest task remaining is the sequencing of an arginine peptide that analyzes for 25–27 amino acids. Tryptic and cyanogen bromide peptides, together with carboxypeptidase and occasionally chymotryptic digestion of a larger peptide, have been sufficient to provide all the necessary overlaps.

The single disulfide loop of 40 residues is located in the center of the molecule. No apparent sequence homologies exist between lysozyme Ch and hen egg white, human leukemic, or T4 phage lysozymes. Thus the *Chalaropsis* lysozyme appears to be completely different from all other lysozymes, not only in the nature of its substrate specificity but also in its primary sequence.

Acknowledgments

Assistance from United States Public Health Service Grant Number AI-06712 and Career Development Award AI-38631 is gratefully acknowledged.

References

1. Hash, J. H., *Arch. Biochem. Biophys.* **102,** 379 (1963).
2. Hash, J. H., and Robinson, J. P., *Fed. Proc.* **25,** 742 (1966).
3. Hash, J. H., and Rothlauf, M. V., *J. Biol. Chem.* **242,** 5586 (1967).
4. Mitchell, W. M., and Hash, J. H., *J. Biol. Chem.* **244,** 17 (1969).
5. Tipper, D. J., Strominger, J. L., and Ghuysen, J.-M., *Science* **146,** 781 (1964).

6. Thompson, R., and Khorazo, D., *Proc. Soc. Exp. Biol. Chem.* **33,** 299 (1935).
7. Ghuysen, J.-M., and Strominger, J. L., *Biochemistry* **2,** 1119 (1963).
8. Phillips, D. C., *Proc. Natl. Acad. Sci. U.S.A.* **57,** 484 (1967).
9. Berger, L. R., and Wiser, R. S., *Biochim. Biophys. Acta* **26,** 517 (1957).
10. Rupley, J. A., and Gates, V., *Proc. Natl. Acad. Sci. U.S.A.* **57,** 496 (1967).
11. Chipman, D. M., Pollock, J. J., and Sharon, N., *J. Biol. Chem.* **243,** 487 (1968).
11a. Felch, J. W., and Hash, J. H., unpublished data.
12. Tipper, D. J., and Strominger, J. L., *Biochem. Biophys. Res. Commun.* **22,** 48 (1966).
13. Tipper, D. J., Strominger, J. L., and Ensign, J. C., *Biochemistry* **6,** 906 (1967).
14. Shih, J. W.-K., and Hash, J. H., *J. Biol. Chem.* **246,** 994 (1971).
15. Canfield, R. E., *J. Biol. Chem.* **238,** 2698 (1963).
16. Smith, E. L., Kimmel, J. R., Brown, D. M., and Thompson, E. O. P., *J. Biol. Chem.* **215,** 67 (1955).
17. Tsugita, A., and Inouye, M., *J. Mol. Biol.* **37,** 201 (1968).
18. Canfield, R. E., Kammerman, S., Sobel, J. H., and Morgan, F. J., *Nature (Lond.), New Biol.* **232,** 16 (1971).
19. Howard, J. B., and Glazer, A. N., *J. Biol. Chem.* **242,** 5715 (1967).
20. Wahba, N., and Hash, J. H., *Fed. Proc.* **30,** 1296 (1971).
21. Felch, J. W., and Hash, J. H., *164th Meet. Am. Chem. Soc.* Biology 20 (1972).

Structural Aspects of Lysozyme — From the Viewpoint of an Antibody

RUTH ARNON

Antibodies elicited by proteins may be directed against groupings which involve various structural features of the antigen. Thus it has been demonstrated that antibodies recognize antigenic determinants dictated by the primary structure—*sequential determinants*—or by *conformational determinants,* those molded by the secondary, tertiary, and even quaternary structures (1, 2). The specific antibodies differentiate and discriminate between those various structural characteristics of the protein molecule. For this reason an immunologic approach has often been used in the investigation of structural aspects of proteins, particularly in the elucidation of structure–function correlation of enzyme molecules (3), and for following conformational changes (e.g., 4, 5).

In the present paper several examples of recent studies from our laboratory are described which illustrate the manner in which the study of specific interactions with antibodies was employed in the elucidation of some structural aspects concerning the lysozyme molecule.

The immunologic specificity of lysozyme is almost entirely dependent on its three-dimensional native conformation, as has been implied by several observations: First, there is no cross-reactivity between the native enzyme and its unfolded polypeptide chain. Thus, the reduced carboxymethylated derivative (RCM) does not react with antibodies to native lysozyme, and the antibodies it elicits do not react at all with the native enzyme (6, 7). On the other hand, if the native globular protein is subjected to mild enzymatic cleavage it yields, among others, two fragments which retain immunologic activity and encompass two independent antigenic determinants, of specificity similar to that of the intact protein. One of these immunologically active fragments consists of two peptides, derived from the NH_2-terminus (residues 1–27) and the COOH-terminus (residues 122–129) of lysozyme, linked together by a disulfide bond. This fragment binds to antilysozyme antibodies with an average affinity constant of 1.75×10^5, and the percentage of antibodies directed toward it was evaluated at 47% (8).

The second immunologically active component isolated was a large fragment derived from the region located between residues 57 and 107 of the lysozyme sequence (9–12). This peptide, which contains two disulfide bridges, was also capable of binding to anti-lysozyme antibodies and in that way to interfere both with their precipitation with lysozyme and with their inhibition of its catalytic activity. The antibody fraction specific toward this fragment (amounting to 30% of the total antibodies) inhibited lysozyme activity (11). Each of these two immunologically active fragments is quite large, and probably embodies more than a single antigenic determinant.

Antibodies to a Unique Region of Lysozyme— Properties and Specificity

Three years ago we showed that the latter of the two fragments mentioned above can yield a smaller peptide which still retains immunologic activity (13). This fragment, consisting of the amino acid sequence 60–83 and containing one intrachain disulfide bond, was denoted "loop." The location of this region in the three-dimensional structure of lysozyme is shown in Fig. 1. Antibodies specific to this region only have been prepared by two alternate procedures. One of these utilized a specific immunoadsorbent, containing the loop peptide attached to a solid support, for selective isolation of antibodies from anti-lysozyme serum. The alternate procedure consisted of the immunization of rabbits and goats with a conjugate containing the loop peptide bound to a synthetic carrier (multi-poly-DL-alanyl-poly-L-lysine, abbreviated A—L), which elicited the formation of antibodies reactive with lysozyme. By adsorption on, and elution from, a lysozyme immunoadsorbent, a preparation of antibodies was obtained, with a specificity directed against a unique region in native lysozyme. The anti-loop antibodies obtained by either of these two procedures showed,

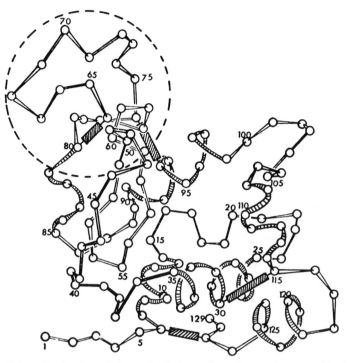

Fig. 1. Schematic drawing of the main chain conformation of hen egg-white lysozyme [from C. C. F. Blake, D. F. Koenig, G. A. Mair, A. C. T. North, D. C. Phillips, and V. R. Sarma, *Nature (London)* **206**, 757 (1965)]. The area encompassing the loop peptide is encircled.

as expected, less structural heterogeneity than the totality of the anti-lysozyme antibody population. This was manifested both in the acrylamide electrophoresis of the respective light chains and in the isoelectric focusing of the intact antibodies (14).

The detailed specificity of anti-loop antibodies was investigated by three different sensitive techniques—antigen binding capacity using radioactively labeled antigens; inactivation of modified bacteriophage (15) preparations coated chemically with either lysozyme or the loop peptide; and a fluorometric method using a loop derivative to which a fluorescent chromophore was attached—with the following results:

The capacity of the antibodies to bind radioactively labeled antigen was the same for lysozyme and for the isolated loop fragment—in high antibody concentration both antigens were bound to an extent of 90%. On the other hand, the open-chain loop peptide, in which the disulfide bond had been disrupted by performic acid oxidation, showed only very limited binding (14). The experiments with the modified bacteriophage (15) yielded similar results. Bacteriophage preparations, conjugated with either lysozyme or the loop peptide, were efficiently inactivated by the anti-loop antibodies. This inactivation was in turn inhibited by the lysozyme, as well as by the loop peptide. But the open-chain peptide, obtained by reduction and carboxymethylation of the loop, was a much weaker inhibitor (14).

The fluorometric method was developed in our laboratory specifically for use on the loop system. For this purpose the loop peptide was labeled with the dansyl (1-dimethylaminonaphthalene-5-sulfonyl) group as an external fluorophore (16). The resultant fluorescent derivative served as an environmental probe for the interaction of the loop with the antibody. Indeed, binding of anti-loop antibodies to this peptide derivative led to a specific excitation energy transfer from the antibody to the dansyl group, and was manifested as an apparent fluorescent enhancement, which increased as a function of the amount of antibodies. This enhancement was, in turn, competitively inhibited by the addition of unlabeled loop peptide or by equimolar amounts of intact lysozyme (Fig. 2). The loop peptide was a very efficient inhibitor and brought about 50% inhibition when added in a ratio of 2:1 to the dansylated loop. The open-chain loop peptide, obtained by reduction and carboxymethylation did not have any effect on the fluorescence of the dansyl-loop-antibody complex.

The data accumulated with the aid of all three techniques are evidence of one and the same phenomenon: the immunologic reactivity of the loop is drastically reduced as a result of unfolding of the peptide chain, indicating the decisive role played by spatial conformation in the antigenic specificity of this unique region in the lysozyme molecule. Further corroborating evidence for this conclusion will be presented in a later section describing synthetic analogs of the loop peptide.

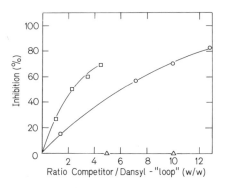

Fig. 2. Inhibition of enhanced fluorescence of the mixture of dansyl-loop (4×10^{-6} M) and anti-loop antibodies (0.6×10^{-6} M) by varying concentrations of the loop peptide (square), hen egg-white lysozyme (circle), and the open-chain peptide obtained by reduction and carboxymethylation of the loop peptide (triangle).

Affinity and Homogeneity of Anti-loop Antibodies

It was mentioned earlier that electrophoresis and electrofocusing experiments indicated that the anti-loop antibodies showed restricted structural heterogeneity (14). Using physicochemical methods for the measurement of the binding parameters, their functional homogeneity was also demonstrated. Two methods were used for this purpose: (1) Equilibrium dialysis, using ^{14}C-labeled loop peptide (by attaching a radioactive carboxymethyl group to the free sulfhydryl group, cysteine residue 76). The results, shown by the Sips plot in Fig. 3(a), permitted the calculation of the association constant of the antibodies ($3.0 \times 10^6 M^{-1}$) and their homogeneity index— 1.02 ± 0.05 (the slope). (2) Fluorometric measurements, utilizing the interaction between the anti-loop antibodies and the dansylated loop derivative described above. By using the data obtained from the "titration" curve of increase in fluorescence as a function of antibody concentration, and inserting the extrapolated value F_∞ of the fluorescence expected with fully bound loop peptide, the association constant and the homogeneity index of the antibodies were calculated as shown in Fig. 3(b). The values obtained ($3.2 \times 10^6 M^{-1}$) for the association constant, and 0.98 ± 0.05 for the homogeneity index, are in very close agreement with the results obtained in the equilibrium dialysis experiments. Both show a homogeneity index close to unity which is indicative of a high degree of functional homogeneity.

The high sensitivity of the fluorometric method using the specific excitation energy transfer reaction made it possible to compare small samples of purified antibody isolated from different individual bleedings. We followed the binding parameters of anti-loop antibodies produced by a single goat over a course of immunization of 20 months in which the goat received several boosting injections (17). The results, illustrated in Fig. 4, show that the homogeneity index of the antibodies had a value of 1.0 (reflecting the uniformity of the binding process), and remained unchanged during the course of the experiment. On the other hand, the association constants of the various antibody samples showed a marked increase following each of the repeated immunizations of the animal (indicated by arrows). These data lead to a very interesting conclusion: The immune response to a unique antigenic region gives rise, indeed, to antibodies of a uniformly high degree of homogeneity which are produced, probably, by a very limited number of clones. However, these clones do not stay invariant during the course of immunization. With the amplification of the immune response achieved by boosting injections, more efficient clones are produced, synthesizing antibodies of higher affinity and concomitantly the older less efficient clones are eliminated from the system. These findings introduce a new concept into theories of antibody biosynthesis which should be tested in other suitable systems.

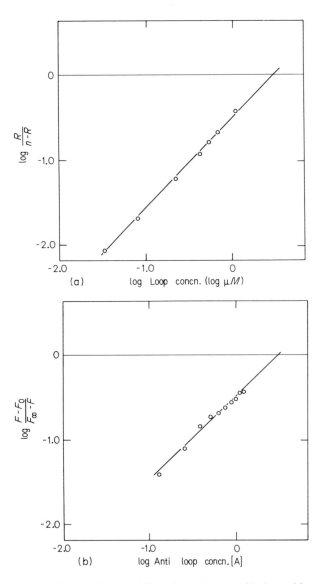

Fig. 3. (a) Sips plot for the binding of ^{14}C-labeled loop peptide by anti-loop antibodies. Loop concentrations expressed in millimicromoles per ml (or 10^{-6} M). Antibody concentration used in the experiment was 0.5×10^{-6} M. (b) A plot of the fluorescence data as a function of anti-loop antibodies concentration, for calculation of the reaction parameters, F_0, the fluorescence intensity of the dansyl-loop alone; F, the fluorescence intensity of the dansyl-loop in the presence of concentration [A] of the antibodies; F_∞, the extrapolated value of the fluorescence for the dansyl-loop saturated with the antibodies. The heterogeneity index was evaluated from the slope of the curve, and the association constant is indicated by the value of [A] when $F - F_0 = F_\infty - F$, or $\log[(F - F_0)/F_\infty - F)] = 0$ (intersection point).

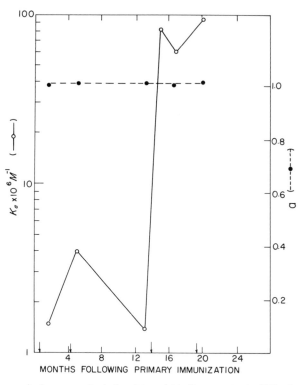

Fig. 4. Change in homogeneity index (*a*) and binding constants (K_a) of antibodies to a specific antigenic region of lysozyme as a function of time lapse following primary immunization.

Genetic Control of the Immune Response to the Lysozyme Loop

The genetic control of the immune response has in recent years been extensively studied (18). The problem was approached using mainly synthetic polypeptides which form immunogens of relatively simple structure. These studies led to the conclusion that the antibody responses of inbred mouse strains are quantitative traits, under a dominant, determinant-specific type of genetic control (19, 20). Such studies were unapproachable by the use of natural multiple determinant antigens, such as proteins, since the simultaneous production of antibodies of differing specificities complicated the study of the processes involved. One may very well envisage a situation where the specificity of antibodies formed in various inbred strains against a particular protein may vary, being due to different antigenic determinants, but this effect could escape notice since the overall amount of antibodies against the whole protein molecule might still be similar. In effect this phenomenon was observed even in the case of a simpler synthetic antigen (multichain polyproline to which peptides of phenylalanine and glutamic acid were attached). Two inbred strains immunized with this antigen responded equally well to the whole polypeptide, but one of them produced antibodies directed mainly toward the side chain peptide portion of the molecule, whereas the antisera of the second strain reacted primarily with the backbone moiety of the immunogen (21). It appears, therefore, that two different genetic controls are operating for two determinants on the same molecule. With multideterminant protein antigen the situation is expected to be much more complex.

Using the lysozyme–loop system we took advantage of the feasibility of eliciting antibodies reactive specifically with a unique region of this native protein. It was thus possible to analyze the immune response of various inbred mouse strains to the loop region, as compared to their response to other portions of the molecule (22). The results shown in Table I demonstrate that most of the mouse strains tested responded well when immunized with lysozyme, but only some strains produced antibodies when injected with the loop-A—L conjugate. The strains which did not respond to this conjugate were also unable to elicit antibodies specific toward the loop region when injected with intact lysozyme. Analysis of the immune response in the F1 hybrid between low and high responders (showing intermediate

TABLE I

Immune Response of Different Inbred Mouse Strains to Hen Egg-White Lysozyme and Its Loop Region[a]

Mouse strain	Immunization with loop-A–L: Assay with lysozyme	Immunization with lysozyme	
		Assay with lysozyme	Assay with loop
C3H · SW	1:8 (10)[b]	1:16 (5)	–
C57BL/6	1:8–1:16 (10)	1:8 (5)	–
BALB/c	1:8 (10)	1:64 (5)	–
DBA/2	1:32 (20)	1:32–1:64 (5)	1:16 (6)
AKR	1:8–1:16 (10)	–	–
C3H/He	1:16 (20)	1:16–1:32 (5)	–
DBA/1	1:64 (15)	1:32–1:64 (10)	1:16–1:32 (10)
SJL	0 (25)	1:32 (10)	0 (6)
(DBA/1 X SJL)F$_1$	1:8–1:16 (6)	1:32 (5)	1:8 (5)
C3HeB/Fe	1:8 (10)	1:32 (5)	0 (5)
SWR	1:64 (30)	1:64 (15)	–

[a] The mice were immunized with either lysozyme or loop-A–L, and the sera were tested by the passive hemagglutination tests using, respectively, lysozyme-coated or loop-coated erythrocytes.

[b] The numbers in parentheses indicate the number of animals tested in each experiment.

response) and their backcrosses with the parental strains demonstrated that the antibody production to the loop is indeed genetically controlled by a unigenic dominant trait which is not linked to the major histocompatibility locus *H-2*.

Evolution of Lysozymes and Possible Relationship to Bovine α-Lactalbumin

The immunologic approach has often been used in the study of biochemical evolution, especially in cases of enzymes, where specific antibodies can be employed in the search for biological pathways which disappeared in the course of evolution, or to detect the extent of similarity between enzymes that persisted through the ages. In cases where the primary sequence of the enzyme is known, the elucidation of the immunologic behavior is feasible in precise molecular terms and may provide a sensitive probe of the surface conformation. This type of investigation is probably best exemplified by the extensive study of Margoliash and his collaborators on cytochrome c (23).

Lysozymes of many species were also compared by immunological cross-reactivities in order to follow the phylogenetic distances between them (24, 25). This topic is covered in detail in Dr. Wilson's chapter. In studies in our laboratory (26) we have also investigated several lysozymes, by measuring their capacity to inhibit the reaction of (anti hen egg-white lysozyme) antibodies either with the intact hen egg-white lysozyme or with the loop region alone [the sensitive technique of chemically modified bacteriophage (15) was used for this purpose]. In that manner we were able to obtain selective information about differences in a defined region of lysozyme. The results yielded interesting information concerning the relatedness of several lysozymes. For example, guinea hen lysozyme proved to be very close to the hen lysozyme, particularly in the loop region. Two fractions of duck lysozyme were tested; they had very similar reactivities in interfering with lysozyme–anti-lysozyme reaction, but only one of them (duck II) was inhibitory in the interaction with the loop. This immunologic difference may be explained by the known replacement within the loop region of glycine residue 71 of hen or duck II lysozyme with an arginine residue in duck III. Thus, this system offers the advantage of high sensitivity as well as a selective capacity to investigate and compare different regions of an immunogenic macromolecule.

In connection with studies on the evolution of lysozyme, it was of interest to study the relationship between hen egg-white lysozyme and bovine α-lactalbumin. These two proteins show a striking similarity in their amino acid sequence (49 out of 123 amino acid residues are identical) including an identity in the position of their disulfide bridges (27). On this basis it has been suggested that they might be functionally related (28) and might also exhibit structural similarities (29). Measurements of various physical properties of the two proteins did not lead to a unified concept concerning the similarity in their structure. We employed an immunologic approach as a probe for the detection of a conformational relationship. Animals were immunized with either lysozyme or α-lactalbumin and the antisera were tested for immunologic interactions with the two antigens by a variety of sensitive assays, such as antigen-binding capacity, passive cutaneous anaphylaxis, microcomplement fixation, passive hemagglutination, and phage inactivation. The antisera toward each of the proteins, raised either in goats or in rabbits, revealed strong positive reaction with the homologous protein antigen, but showed no cross-reaction whatsoever with the heterologous antigen (30). Immunologic cross-reaction between the two native proteins could be detected only on the cellular level of the immune response, using either *in vivo* or *in vitro* measurements (31).

On the other hand, when the unfolded peptide chains of lysozyme and lactalbumin (obtained by complete reduction and carboxymethylation) were used for immunization (32), each of the resultant antisera showed an appreciable extent of immunologic cross-reaction which could be measured by various techniques, including the precipitin test and antigen-binding capacities (Fig. 5). The above findings indicate that hen's egg-white lysozyme and bovine α-lactalbumin probably share structural features, although they might differ in most areas which are exposed to the surrounding medium. This conclusion is in accord with the assumption that the structural genes for these two proteins might have been derived from a common ancestor gene (28).

Antibodies to Lysozyme Induced with a Completely Synthetic Antigen

In the studies described in the preceding pages concerning the loop region of lysozyme and the specific antibodies against it, use was made of the loop fragment as prepared from native lysozyme. In more recent studies we have synthesized, using the solid-phase technique (33), a looplike peptide in which the only difference from the natural loop of lysozyme is the replacement of cysteine in position 76 with alanine, to avoid ambiguous disulfide bond formation. The synthesized loop, corresponding to the sequence 64–82 in the amino acid sequence of lysozyme, was identical in its properties and immunologic reactivity to the natural loop peptide. The

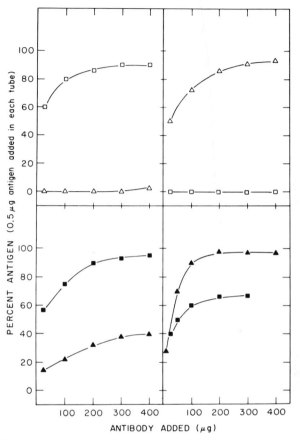

Fig. 5. Antigen-binding capacity of native and open-chain hen egg-white lysozyme and bovine α-lactalbumin by isolated antibodies against native lysozyme (upper left), native lactalbumin (upper right), reduced carboxymethylated lysozyme (lower left), and reduced carboxymethylated lactalbumin (lower right). The antigens are: open square, native lysozyme; triangle, native lactalbumin; solid square, reduced carboxymethylated lysozyme; solid triangle, reduced carboxymethylated lactalbumin.

synthetic loop was attached to multichain poly-DL-alanine carrier, and the resultant completely synthetic macromolecule was used for immunization of both rabbits and goats (34). The antibodies it elicited were similar in every respect to the antibodies obtained upon immunization with conjugates of the natural loop fragment. They were reactive with lysozyme and their specificity was directed toward a conformation-dependent determinant. Thus, the antibodies reacted very efficiently with either intact lysozyme or the closed loop fragment, but did not react at all with the unfolded loop peptide.

One of the advantages offered by a synthetic approach in biochemical and biological studies is that, once the biological properties of one synthetic material have been unequivocally demonstrated, many analogs of it may be prepared and tested. Since the chemistry of these analogs is known and controlled, they can lead to an understanding of the role played by different parameters of the molecule in conferring the specific biological activity. In the case of the lysozyme loop system, we have recently prepared (35), by the solid-phase technique, several analog derivatives in which one or two amino acid residues were replaced by alanine (Fig. 6), for the purpose of testing the extent of involvement of proline, arginine, leucine, or isoleucine in the immunologic reactivity of this antigenic determinant. The results are shown in Fig. 7, where the capacity of anti-loop antibodies to bind the various loop analogs is demonstrated. The derivatives in which either

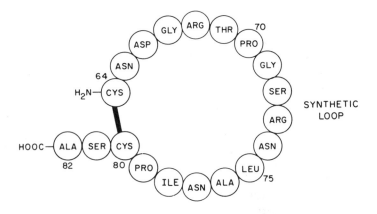

Fig. 6. Schematic description of the synthetic loop derivatives. In the lower part of the figure the intact synthetic loop is depicted. The upper part of the figure contains the linear amino acid sequence of the various analogs prepared (all analogs were subsequently oxidized to yield the loop form). The black positions represent those amino acids which were replaced by alanine.

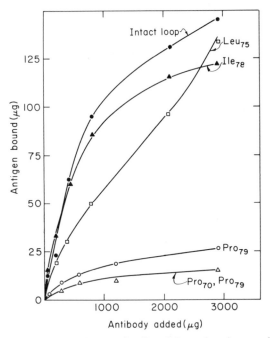

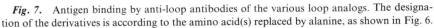

Fig. 7. Antigen binding by anti-loop antibodies of the various loop analogs. The designation of the derivatives is according to the amino acid(s) replaced by alanine, as shown in Fig. 6.

leucine (residue 75) or isoleucine (residue 78) were replaced by alanine were almost indistinguishable from the intact synthetic loop. On the other hand, replacing the proline residues (70 and 79) or even one proline only (residue 70) brought about a drastic decrease in the antigenic reactivity.

Another analog derivative in which the arginine residues (68 and 73) were replaced also lacked completely the ability to bind to the antibodies against the original loop peptide (Fig. 8). By replacement of the arginines two factors have been introduced: the depletion of positive charges and the removal of the bulky arginyl side chains. Each of these factors may have an effect on the immunologic properties of the derivative. However, as also shown in Fig. 8, a loop derivative in which the arginyl residues were modified by biacetyl reagent (thereby blocking their charge) rather than replaced by alanine, was just as reactive with the antibodies as the unmodified intact loop. This indicates that it is the bulky side chain of the arginine residues which makes a contribution to the antigenic specificity of the loop rather than their positive charge. It is of interest that when the linear peptides of these loop analogs were tested for their binding to antibodies to unfolded reduced carboxymethylated lysozyme, they all showed a similar reduced binding as compared to the linear peptide of the intact loop.

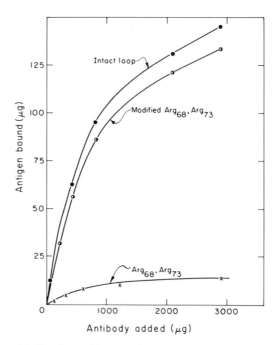

Fig. 8. Antigen binding by anti-loop antibodies of synthetic loop derivatives in which arginine was either replaced by alanine or modified with biacetyl reagent.

Our findings, therefore, provide evidence of the crucial effect of conformation in the antigenic specificity of native proteins. Whereas in the unfolded polypeptide chain each individual amino acid partakes in the immunologic activity to a similar extent, in the system of the conformation-dependent antigenic determinant (the loop), only those residues which have a decisive role in determining the overall shape of the molecule affect its antigenic properties and reactivity.

Conclusions

The common denominator in the studies discussed herein is the use of antibodies and the immunologic reactivity in order to contemplate biological properties and structural features of an intact native protein molecule. With a protein such as lysozyme, whose three-dimensional structure has been elucidated in detail, the immunochemical analysis has, on the one hand, corroborated findings which have been obtained by other methods, but, on the other hand, in some cases it has offered a solution to problems to which ambiguous answers had been obtained by different procedures.

The main message of this investigation is probably the finding that a fragment of the molecule stabilized by an intrachain disulfide bond maintains, in the isolated form, the same conformation it assumes in the intact molecule. Thus, antibodies elicited by a synthetic conjugate of such a peptide are specific to the conformation-dependent determinant present in the native protein. Changes in those positions which are crucial for the spatial conformation destroy the antigenic reactivity as well.

The availability of this system made possible the study, on a molecular level, of several biological aspects such as: the genetic control of the immune response elicited by a multideterminant protein molecule, and the biochemical evolution both of isofunctional enzymes (phylogenetic relationships among various lysozymes) and of chemically homologous proteins not sharing a biological function (lysozyme and lactalbumin). The high sensitivity of the immunologic techniques and the selective specificity of the immunologic reactions establish the antibodies as a powerful tool in protein research.

References

1. Sela, M., Schechter, B., Schechter, I., and Borek, F., *Symp. Quant. Biol.* **32**, 537 (1967).
2. Benjamini, E., Michaeli, D., and Young, J. D., *Curr. Top. Microbiol. Immunol.* **58**, 85 (1972).
3. Arnon, R., *Curr. Top. Microbiol. Immunol.* **54**, 47 (1971).
4. Crumpton, M. J., and Wilkinson, J. M., *Biochem. J.* **100**, 223 (1965).
5. Neumann, H., Steinberg, I. Z., Brown, J. R., Goldberger, R. F., and Sela, M., *Eur. J. Biochem.* **3**, 171 (1967).
6. Gerwing, J., and Thompson, K., *Biochemistry* **7**, 3892 (1968).
7. Young, J. D., and Leung, C. Y., *Biochemistry* **9**, 2755 (1970).
8. Fujio, H., Imanishi, M., Nishioka, K., and Amano, T., *Biken J.* **11**, 207 (1968).
9. Canfield, R. E., and Liu, A. K., *J. Biol. Chem.* **240**, 1997 (1965).
10. Shinka, S., Imanishi, M., Miyagawa, N., Amano, T., Inouye, M., and Tsugita, A., *Biken J.* **10**, 89 (1967).
11. Arnon, R., *Eur. J. Biochem.* **5**, 583 (1968).
12. Fujio, H., Imanishi, M., Nishioka, K., and Amano, T., *Biken J.* **11**, 219 (1968).
13. Arnon, R., and Sela, M., *Proc. Natl. Acad. Sci. U.S.A.* **62**, 163 (1969).
14. Maron, E., Shiozawa, C., Arnon, R., and Sela, M., *Biochemistry* **10**, 763 (1971).
15. Haimovich, J., Hurwitz, E., Novik, N., and Sela, M., *Biochim. Biophys. Acta* **207**, 115 and 125 (1970).
16. Pecht, I., Maron, E., Arnon, R., and Sela, M., *Eur. J. Biochem.* **19**, 368 (1971).
17. Maron, E., Arnon, R., and Sela, M., *Isr. J. Med. Sci.* **8**, 635 (1972).
18. McDevitt, H. O., and Benacerraf, B., *Adv. Immunol.* **11**, 31 (1969).
19. Pinchuk, P., and Maurer, P. H., *in* "Regulation of the Antibody Response," p. 97. Thomas, Springfield, Illinois, 1968.
20. McDevitt, H. O., and Sela, M., *J. Exp. Med.* **122**, 517 (1965).
21. Mozes, E., McDevitt, H. O., Jaton, J.-C., and Sela, M., *J. Exp. Med.* **130**, 1263 (1969).
22. Mozes, E., Maron, E., Arnon, R., and Sela, M., *J. Immunol.* **106**, 862 (1971).

23. Margoliash, E., Reichlin, M., and Nisonoff, A., *in* "Conformation of Biopolymers" (G. N. Ramachandran, ed.), Vol. 2, p. 253. Academic Press, New York, 1967.
24. Arnheim, N., Jr., and Wilson, A. C., *J. Biol. Chem.* **242,** 3951 (1967).
25. Prager, E. M., and Wilson, A. C., *J. Biol. Chem.* **246,** 7010 (1971).
26. Maron, E., Arnon, R., Sela, M., Perin, J.-P., and Jollès, P., *Biochim. Biophys. Acta* **214,** 222 (1970).
27. Brew, K., Castellino, F. J., Vanaman, T. C., and Hill, R., *J. Biol. Chem.* **245,** 4570 (1970).
28. Hill, R. L., Brew, K., Vanaman, T. C., Trayer, I. P., and Mattock, P., *Brookhaven Symp. Biol.* **1,** No. 21, 139 (1969).
29. Brew, K., Vanaman, T. C., and Hill, R., *Proc. Natl. Acad. Sci. U.S.A.* **59,** 491 (1968).
30. Arnon, R., and Maron, E., *J. Mol. Biol.* **51,** 703 (1970).
31. Maron, R., Webb, C., Teitelbaum, D., and Arnon, R., *Eur. J. Immunol.* **2,** 294 (1972).
32. Arnon, R., and Maron, E., *J. Mol. Biol.* **61,** 225 (1971).
33. Merrifield, R. B., *Science* **150,** 178 (1965).
34. Arnon, R., Maron, E., Sela, M., and Anfinsen, C. B., *Proc. Natl. Acad. Sci. U.S.A.* **68,** 1450 (1971).
35. Teicher, E., Maron, E., and Arnon, R., *Immunochemistry* **10,** 265 (1973).

Immunologic Comparison of the Active Site and the Loop Regions of Bird and Human Lysozymes

ELCHANAN MARON*

Hen egg-white lysozyme (HEWL) is a protein of special interest for comparative studies, since both its primary and three-dimensional structures as well as its catalytic functions and the location of the antigenic determinants have been clearly elucidated. On the other hand, no such complete data are available for any one of the lysozymes obtained from other sources.

The use of specific antibodies provides a sensitive tool for the study of structural relationship among these molecules. Various bird and human lysozymes have been compared by immunologic techniques (1, 2). This approach provided evidence concerning the overall homology and phylogeny of the different lysozymes which corroborated the conclusions arrived at by the elucidation of their chemical structure. However, in order to obtain information regarding selected and defined regions of the molecules, another approach has been applied, namely, the use of more defined interaction between the antibodies and the tested protein. Examples for the latter approach are given below.

Some of the recent studies of the antigenic determinants of HEWL have been concentrated on a conformation-dependent determinant within a unique region denoted the loop region (sequence 60–83 containing an intra-

*Deceased August 8, 1973.

molecular disulfide bridge) (3, 4). A peptide consisting of the loop region has been isolated and was shown by chemical as well as by immunologic tests to preserve the structure it possesses in the intact molecule (4).

The sensitive technique known as the "chemically modified bacteriophage" (5), in which either the whole lysozyme molecule or only the isolated loop fragment were attached to the bacteriophage for the purpose of studying the specific immunologic reactions, was applied. Both lysozyme-phage and loop-phage preparations were inactivated efficiently with goat anti-HEWL antibodies. This inactivation, caused by specific antibodies, may be prevented or inhibited by means of related antigens or fragments. The efficiency of such inhibition may reflect the chemical similarity of the inhibitor to the specific determinants on the immunogen.

Several lysozymes have thus been investigated by measuring their capacity to inhibit the reaction of anti-HEWL antibodies, either with the intact molecule or with the isolated loop fragment (6). It was feasible, in the latter case, to obtain selective information regarding changes in a unique region. The results are summarized in Table I.

Guinea hen egg-white lysozyme (GHL) was only 4 times less efficient as an inhibitor than HEWL in the lysozyme-phage system, whereas both molecules had the same inhibitory capacity in the loop-phage system. This implies that any differences in the loop region of GHL and HEWL are not reflected immunologically. The two duck egg-white lysozymes (D II, D III) are indistinguishable in the lysozyme-phage system. On the other hand, a very striking different behavior in the loop-phage system was

TABLE I

Inhibition by Various Lysozymes of the Inactivation of Modified Bacteriophages with Anti-Hen Egg-White Lysozyme[a]

	Modified bacteriophage preparation	
Lysozyme source	Lysozyme-T4	Loop-T4
Hen egg-white	$1.5 \cdot 10^{-6}$	$1.7 \cdot 10^{-5}$
Guinea hen egg-white	$6.3 \cdot 10^{-6}$	$2.3 \cdot 10^{-5}$
Duck egg-white II	$5.7 \cdot 10^{-4}$	$1.4 \cdot 10^{-2}$
Duck egg-white III	$5.7 \cdot 10^{-4}$	—[b]
Human milk	$2.1 \cdot 10^{-3}$	—[b]
Human leukemic urine	—[b]	—[b]
Bovine α-lactalbumin	—[b]	—[b]

[a]The numbers in the table indicate the concentration (mg per sample) of each lysozyme which brought about 50% inhibition of the bacteriophage inactivation by the antibodies.

[b]No inhibition was observed with an inhibitor concentration as high as 0.15 mg per sample.

observed, as only D II was capable of inhibition. This immunologic difference may be due to the known replacement of Gly71 in the amino acid sequence of D II with an arginine residue in D III (7).

Human milk lysozyme had a definite and reproducible, though weak, capacity to inhibit the lysozyme-phage system, whereas no such inhibition could be detected in the loop-phage system. The lysozyme derived from human leukemic urine (HL), had even less in common with HEWL than human milk lysozyme had. Bovine α-lactalbumin did not cross-react with anti-HEWL antibodies either (8), due to differences in their three-dimensional folding of their peptide chains, or to differences in their exposed side chains.

A comparison of the active site regions of different lysozyme molecules has also been conducted using another approach (9). Antibodies directed solely against the active site region of HEWL have been isolated from a goat antiserum. Their reaction with lysozyme was inhibited by oligosaccharides of *N*-acetyl-D-glucosamine (10). These antibodies inactivate the lysozyme-bacteriophage conjugate in a very efficient manner, but they do not interact with the loop region of HEWL as indicated by the lack of inactivation of the loop-bacteriophage conjugate.

The reversible complex formed between HEWL and a large excess of $(GlcNAc)_2$ (40:1 molar excess) was compared to the free enzyme molecule. When antibodies to the whole enzyme were applied, both forms of the enzyme had an identical inhibitory capacity. However, when the "active site directed" antibodies were used for the assay, the enzyme-saccharide complex had lost its inhibitory capacity by 150 fold as compared to the native enzyme (Fig. 1). Thus the latter assay system could serve for the examination of the degree of immunologic relatedness between the active site of HEWL and other molecules.

The following pattern of reactivity was observed when the different lysozymes were assayed with antibodies against the whole hen egg-white molecule. Turkey egg-white lysozyme (TL) was eightfold less efficient as an inhibitor compared to HEWL, whereas ring-necked pheasant egg-white lysozyme (RNPL) was 50 times less, and D II was 320 fold less efficient (Fig. 1).

The different proteins tested show a great similarity in their enzymatic activities and amino acid sequences, particularly in the amino acids which correspond to the residues that make contact with the substrate in the HEWL-substrate complex. These enzymes differed however in the extent of interaction with the "active site-directed" antibodies. Thus, D II was 25-fold less active as inhibitor than HEWL, and RNPL was 70-fold less active; TL had even much weaker inhibitory capacity, namely 460-fold less than HEWL (Fig. 1).

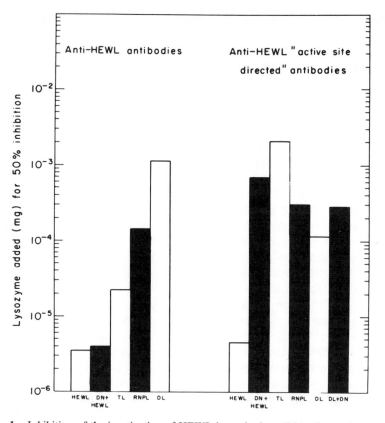

Fig. 1. Inhibition of the inactivation of HEWL-bacteriophage T4 conjugate by goat anti-HEWL antibodies and by anti-HEWL "active site-directed" antibodies, caused by different lysozymes. The following abbreviations are used: HEWL: hen egg-white lysozymes; DN: di(N-acetyl-D-glucosamine) added at a 40 times molar excess; TL: turkey egg-white lysozyme; RNPL: ring-necked pheasant lysozyme; DL: duck egg-white lysozyme (II).

When a large excess of $(GlcNAc)_2$ was added to duck lysozyme, its inhibitory activity in the latter system was reduced only by a factor of two. Therefore, it may be assumed that the "active site-directed" antibodies are specific not only for contact amino acids, but for an active site region, which includes the amino acids lining the lysozyme cleft, as well as neighboring surface amino acids. In addition, the conformations of the various enzymes may differ, and such differences may also affect their immunologic cross reactions.

Taking into consideration those amino acids which occupy, in the different molecules, positions corresponding to the residues which are within 4 Å from substrate atoms in the HEWL-substrate complex ("contact" residues) (11), a possible explanation can be derived. Although the proportion of the changes in the active site region and in the whole TL molecule are about the same, the single change in the active site is a very marked one: replacement of the highly exposed Asp 101 of HEWL by a glycyl residue in TL (12). The weakening of the interaction of TL with the active site-directed antibodies, compared to its interaction with the anti-HEWL antibodies, can probably be attributed to this change.

A different relationship exists between D II and HEWL. The single change in the active site region (out of the total 22 differences) involves the replacement of Gln 57, a deeply buried residue in HEWL, by a glutamic acid residue (7). It is not surprising, therefore, that the interaction of D II and the anti-HEWL active site-directed antibodies is stronger than with the whole anti-HEWL antibody population. RNPL interacts to about the same extent with both types of antibody. Since the primary structure of RNPL is not yet known, it can be assumed that the seven differences found between the amino acid composition of RNPL and HEWL (1) are distributed equally in the active site region as well as in other parts of the molecule.

In summary, the use of the modified bacteriophage to which either the whole antigenic molecule or only a part of it were attached, and the use of highly specific antibodies directed against a defined region on the molecule for the purpose of elucidating the specific immunologic reactions have been described. Such a system has the advantages of great sensitivity as well as a selective capacity to investigate and compare different regions of intact molecules belonging to a homologous series.

References

1. Arnheim, N., Prager, E. M., and Wilson, A. C., *J. Biol. Chem.* **244**, 2085 (1969).
2. Prager, E. M., and Wilson, A. C., *J. Biol. Chem.* **246**, 5978 (1971).
3. Arnon, R., and Sela, M., *Proc. Natl. Acad. Sci. U.S.A.* **62**, 163 (1969).
4. Maron, E., Shiozawa, C., Arnon, R., and Sela, M., *Biochemistry* **10**, 763 (1971).
5. Sela, M., and Haimovich, J., *Protides Biol. Fluids, Proc. Colloq.* p. 391 (1971).
6. Maron, E., Arnon, R., Sela, M., Perin, J. P., and Jollès, P., *Biochim. Biophys. Acta* **214**, 222 (1970).
7. Hermann, J., Jollès, J., and Jollès, P., *Eur. J. Biochem.* **24**, 12 (1971).
8. Arnon, R., and Maron, E., *J. Mol. Biol.* **51**, 703 (1970).
9. Maron, E., Eshdat, Y., and Sharon, N., *Biochim. Biophys. Acta* **278**, 243 (1972).
10. Imanishi, M., Miyagawa, N., Fujio, H., and Amano, T., *Biken. J.* **12**, 85 (1969).
11. Hess, G. P., and Rupley, J. A., *Ann. Rev. Biochem.* **40**, 1013 (1971).
12. LaRue, J. N., and Speck, J. C., *J. Biol. Chem.* **245**, 1985 (1970).

Antigenic Comparison of Animal Lysozymes

A. C. WILSON AND E. M. PRAGER

As Fleming (21) and others in the succeeding 50 years found, most animal species as well as many plant and microbial species produce lysozyme. More recently we learned that lysozyme is a rather rapidly evolving protein; even closely related species seldom have identical lysozymes. As the number of species living on earth today is probably at least 10 million (43), the number of different lysozymes in existence is similarly large. This estimate, of course, does not take into account the many rare allelic variants of lysozyme that must occur within each species.

To estimate the relatedness of the lysozymes from different species, individuals, or cell types, it is most valuable to determine their amino acid sequences and three-dimensional structures. However, it is not always desirable to resort to such tedious and expensive methods. Certain immunologic* methods can, far more easily, provide estimates of the approximate degree of sequence resemblance and conformational similarity among related lysozymes. The aim of this article is to summarize the evidence for a correlation between immunologic resemblance and sequence resemblance among bird lysozymes. The article also presents the results of a comparative immunologic study of lysozyme from 19 species of primates, including man.

*The authors of this chapter have consistently used the term "immunological," not "immunologic," in reports from their laboratory. However, for the sake of uniformity with the rest of the volume, they have consented to the use of "immunologic" here.

The Sequence-Immunology Correlation

Amino Acid Sequences

We have compared immunologically all seven of the bird lysozymes whose amino acid sequences are known. The species are listed in Table I, which also gives the number of sequence differences between each pair of lysozymes. All of the sequence differences result from amino acid substitutions: no deletions, additions, inversions, or translocations seem to have occurred since the divergence of bird lysozymes from a common ancestor.

Figure 1 depicts possible phylogenetic relationships among these lysozymes and human lysozyme.

TABLE I

The Number of Differences in Amino Acid Sequence among Animal Lysozymes[a]

	Chicken	Bobwhite quail	Japanese quail	Turkey	Guinea fowl	Duck II	Duck III	Human
Chicken	0	4	6	7	10	22	23	51
Bobwhite	4	0	10	11	8	23	24	52
Japanese quail	7	11	0	10	15	25	26	53
Turkey	7	11	10	0	16	22	23	53
Guinea fowl	10	8	16	16	0	29	30	56
Duck II	27	28	29	27	33	0	6	52
Duck III	27	28	29	27	33	7	0	51
Human	70	71	72	72	75	69	67	0

[a] Minimal mutation distances are given on the lower left. In the comparison of bird lysozymes with human lysozyme, we ignored the deletion of one amino acid in bird lysozymes.

Micro-complement Fixation Tests

Antisera were made by immunizing rabbits with lysozymes purified from the egg whites of the chicken, turkey, Japanese quail, bobwhite quail, and duck.* Each antiserum was tested for reactivity with each of the lysozymes listed in Table I. The reactions were detected by micro-complement fixation, which is superior to other immunologic techniques in its capacity to discriminate between closely related lysozymes (3, 41, 53). The differences in reactivity are expressed in units of immunologic distance (see Fig. 2). As

*The antisera to duck lysozyme were made with duck A lysozyme, which is extremely similar in amino acid composition, electrophoretic mobility and immunologic properties to duck II lysozyme, whose sequence is known (39, 42).

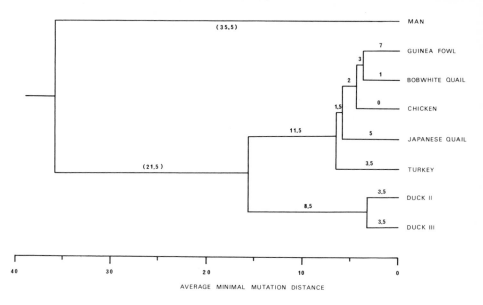

Fig. 1. Possible phylogenetic relationships among animal lysozymes of known sequence. This tree was constructed by the Farris (20) method from the matrix of minimal mutation distances in Table I.

Fig. 2 illustrates, there is a correlation between the degree of antigenic difference and the degree of sequence difference among bird lysozymes.

Antiserum Variability

To obtain the above correlation, several precautions must be taken. Cross-reactivity measurements are affected by several parameters of the immunologic response of the animals being immunized. As shown in Fig. 3, the length of the immunization period influences the ability of antisera to chicken lysozyme to discriminate between chicken lysozyme and the lysozymes of other species, such as turkey, duck, and chachalaca. Only after 4 months of immunization does the ability to discriminate become constant with time. The importance of the time factor was also demonstrated in an analogous study with alkaline phosphatases (12). Individual rabbits also vary significantly in the specificity of the antiserum they produce; therefore, at least four rabbits should be immunized with the same antigen, and pools should be prepared by mixing the antisera in inverse proportion to their titers, so that each rabbit contributes an equal concentration of complement-fixing antibodies (40). The results depicted in Fig. 2 were obtained by following the above recommendations.

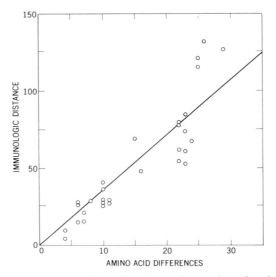

Fig. 2. Dependence of immunologic distance on the number of amino acid sequence differences among bird lysozymes. The seven lysozyme sequences were obtained from the following sources: chicken (8, 27), Japanese quail (31), turkey (33), duck II (25), duck III (26), guinea fowl (30), and bobwhite quail (38). The immunologic distance values were obtained with five antiserum pools from rabbits immunized for 6 months with the lysozymes of chicken, Japanese quail, turkey, duck (see footnote on p. 128), and bobwhite quail; the values are taken from Prager and Wilson (40) and Prager (37a).

The problem of antiserum variability can be further reduced if one averages the results of reciprocal tests (i.e., the results of an anti-X vs. Y test and an anti-Y vs. X test). A stronger correlation between sequence resemblance and immunologic resemblance than that shown in Fig. 2 is obtained if one plots the average immunologic distance obtained in reciprocal tests against the number of amino acid differences (40). Nevertheless, the correlation is still not perfect.

Other Sources of Error

The imperfect nature of the sequence-immunology correlation undoubtedly stems from additional factors, besides the variability of cross-reactivity measurements. Some amino acid replacements may produce bigger antigenic effects than others, a possibility that is discussed further below. Another possibility is that of errors in the determination of amino acid sequences. When two laboratories independently determined the amino acid sequence of chicken lysozyme they obtained slightly different answers (8, 27) and when they determined the sequence of human lysozyme both groups made an error (29, 52). No other lysozyme sequences have yet been

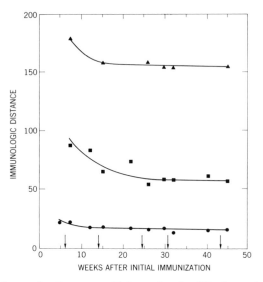

Fig. 3. Dependence of antiserum specificity on length of the immunization period. Four rabbits were immunized with pure chicken lysozyme and bleedings were taken at fairly regular intervals for almost a year. The arrows indicate the times of intravenous booster injections. The antisera obtained on a given date were titered with chicken lysozyme and then mixed in inverse proportion to their titers. Each antiserum pool was then tested for reactivity with the following lysozymes: chachalaca (triangle), duck A (square), and turkey (circle). The results of the cross-reaction tests are expressed in units of immunologic distance. Adapted from Prager and Wilson (40).

reinvestigated. Sequencing errors may thus contribute to the scatter of points in Fig. 2.

Rationale for the Correlation

It is important to discuss the rationale for the existence of the relationship we have observed between immunologic resemblance and sequence resemblance. This relationship was not predicted by immunochemists, who had several reasons for believing that some amino acid substitutions are bound to produce much larger immunologic effects than others.

Immunochemists were captivated by many dramatic experiments showing that the antigenic determinants of globular proteins are usually "conformational" determinants (48). If one unfolds such a determinant, its antigenic properties change dramatically. Accordingly, those amino acid replacements which preserve the conformation* of a determinant produce little alteration in antigenicity whereas those that alter the conformation produce

---

*The conformation of a polypeptide determinant is used here to refer to the conformation of the polypeptide backbone and the orientation of the side chains.

gross antigenic changes (51; Arnon, this volume, Chapter 10). As the sequence-immunology correlation shows, however, the immunologic effects of those amino acid replacements which have survived in naturally occurring lysozymes are usually small, roughly equal, and, to a first approximation, additive.

To resolve this apparent paradox, it is helpful to consider the findings of Blake and Swan (4), who compared the three-dimensional structures of human and chicken lysozymes by X-ray methods. In spite of 51 amino acid replacements, these two lysozymes are strikingly alike in both conformation of the polypeptide backbone and orientation of the amino acid side chains. Evidently natural selection has not often permitted survival of those amino acid replacements causing gross alterations in conformation.

Another reason for not anticipating a sequence-immunology relationship came from the idea that the number of antigenic determinants on proteins is small. Antisera to chicken lysozyme, for example, normally consist predominantly of antibodies to two determinant regions, the loop region and the region embracing both the amino and carboxyl termini of lysozyme (22, 35, 49). This finding makes it seem plausible that a large fraction of the amino acid replacements occurring in lysozyme would be in nonantigenic regions and hence immunologically silent. Conversely, amino acid substitutions in one of the major determinants would be expected to produce large immunologic effects. It is possible, however, that strongly immunogenic proteins like lysozyme have many minor antigenic determinants, eliciting production of only small amounts of antibody, yet collectively embracing almost the entire outer surface of lysozyme. These minor classes of antibody could be active in micro-complement fixation, but this remains to be demonstrated.

A third reason for overlooking the sequence-immunology relationship was the suggestion that internal substitutions would not be detectable immunologically. This is almost certainly true because interactions between antibodies and globular proteins are a surface phenomenon (48), but we now think that lysozyme evolution is also mainly a surface phenomenon (4, 38, 40). As indicated in Table II, only 13% of the sites at which amino acid replacements have occurred among bird lysozymes are fully buried in the interior of lysozyme. Eighty-seven percent of the sites at which replacements have occurred are partly or completely exposed and, in principle, accessible to antibody molecules. For this reason, immunologic techniques are well suited for the detection of a large fraction of the amino acid substitutions occurring in lysozymes. Furthermore, it is noteworthy that if one plots immunologic distance among lysozymes against the number of amino acid replacements at exposed sites, thereby eliminating the "hidden"

TABLE II

Location of Amino Acid Replacements among Seven Bird Lysozymes[a]

Location of amino acid side chain	All sites	Sites having amino acid replacements
External	55	22
Surface	47	11
Internal	27	5
	——	——
	129	38

[a] The information concerning accessibility of the amino acid side chains in chicken lysozyme is taken from Browne *et al.* (6) and Lee and Richards (34). It is assumed that the lysozymes of the bird species (listed in Table 1) have the same conformation as chicken lysozyme and that the amino acid side chains replacing those in the chicken have the same degree of accessibility as the side chains they replaced.

replacements from consideration, the sequence-immunology correlation is enhanced (Fig. 4).

In conclusion, the results of X-ray studies help us to understand why cross-reactivity is a function of sequence resemblance, but a full understand-

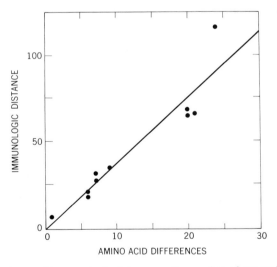

Fig. 4. Dependence of immunologic distance on the number of exposed amino acid substitutions among bird lysozymes. The immunologic distance values are the averages of reciprocal tests. The number of sequence differences plotted for each pair of lysozymes compared is the sum of the external plus surface substitutions; fully internal substitutions are not included. Adapted from Prager *et al.* (38).

ing will have to await further elucidation of the antigenic structure of lysozyme.

Advantages and Disadvantages

Despite its approximate nature, the sequence-immunology correlation provides a justification for the use of an immunologic approach for estimation of the relatedness of lysozymes from different biological sources. The immunologic approach has some important advantages over conventional sequence determination. It is easy to do, requires little material, and, if one dispenses with doing reciprocal tests (which would be acceptable when an approximate answer would suffice), permits a survey of a large number of related proteins without purification. There are of course disadvantages to this approach: internal substitutions may not be registered, the precise location and nature of the substitutions cannot be determined, only an approximate value for the number of substitutions is obtained, and distantly related proteins are beyond the range of this method.*

In summary, if proper precautions are taken in preparing the antisera and executing the cross-reactivity tests, micro-complement fixation can be a useful technique for obtaining an approximate measure of the degree of sequence difference among related lysozymes.

Primate Lysozymes

Introduction

Although the amino acid sequence of human lysozyme has been determined (9, 28, 29, 52), little is known about the lysozymes of other primates, except that tears and milk are rich in this enzyme (1, 7, 19). We therefore undertook a comparative micro-complement fixation study of milk lysozymes in primates.

It is important to compare human macromolecules with those of other species, especially other primates. Such comparisons contribute quantitative information concerning the degree of phenotypic and genetic relatedness of man to other species. Such information will be valuable not only

---

*Proteins that differ by more than 40% in sequence generally appear to be unrelated in tests requiring the antigens to be multivalent (e.g., complement fixation and precipitin tests). However, immunologic resemblance between more distantly related proteins can often be detected by tests requiring only that the antigen be univalent (e.g., radioimmunoassay, neutralization, phage inactivation, or hemagglutination tests) or by the use of antibodies to the unfolded proteins (2; Arnon, this volume, Chapter 10).

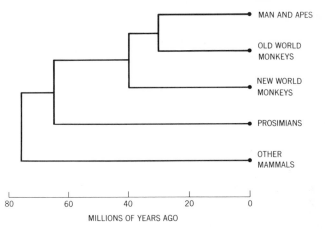

Fig. 5. Phylogenetic relationships of four groups of primates. The times of divergence of the lineages are based on consideration of both fossil evidence (50) and macromolecular evidence (45). From Hanke *et al.* (24).

to those who use animals or animal cells in medical research but also to those who want to view man's similarity to other species objectively, those who are interested in human evolution, and those who would like eventually to discern the molecular basis for the characteristics that are unique to man.

The species whose lysozymes were studied are divisible into the following four taxonomic groups: Hominoidea (apes and man), Cercopithecoidea (Old World monkeys), Ceboidea (New World monkeys), and Prosimii (prosimians). The phylogenetic relationships among these groups, which have been worked out primarily by classical morphological methods, with support from molecular methods, are depicted in Fig. 5. This shows, for example, that the two most closely related of these groups are the Hominoidea and Cercopithecoidea and that they diverged from a common ancestor about 30 million years ago, whereas the common ancestor of all four groups lived about 65 million years ago.

Results

The antisera used in this investigation were produced essentially as described by Prager and Wilson (40) by immunizing rabbits for 6 months with human lysozyme, purified from the urine of humans with monocytic and monomyelocytic leukemia (37), and baboon lysozyme, purified from milk (7). The two resulting antiserum pools were each tested for reactivity with the purified lysozymes of man, baboon, rat, and chicken as well as with the unpurified lysozymes present in the milk of 17 other primate species, whose names are listed in Table III.

TABLE III

Immunologic Differences among Primate Lysozymes[a]

| | Immunologic distance | |
| | Measured with anti-human lysozyme | Measured with anti-baboon lysozyme |
Species		
Man and apes		
Man, *Homo sapiens*	0	66
Chimpanzee, *Pan troglodytes*	0	67
Orangutang, *Pongo pygmaeus*	1	76
Gorilla, *Gorilla gorilla*	32	38
Old World monkeys		
Green monkey, *Cercopithecus aethiops*	93	6
Talapoin, *C. talapoin*	114	3
Rhesus, *Macaca mulatta*	122[b]	1
Stump-tailed macaque, *M. speciosa*	124	2
Baboon, *Papio cynocephalus*	127	0
Crab-eating macaque, *M. fascicularis*	130	2
Bonnet macaque, *M. radiata*	131	2
New World monkeys		
Squirrel monkey, *Saimiri sciureus*	127	36
Tamarin, *Saguinus oedipus*	134	37
Marmoset, *Callithrix jacchus*	137	36
Prosimians		
Ring-tailed lemur, *Lemur catta*	151	203
Greater bushbaby, *Galago crassicaudatus*	175	>247
Black lemur, *L. macaco*	202	239
Brown lemur, *L. fulvus fulvus*	212	235
Lemur hybrid, *L. f. fulvus* × *L. f. rufus*	>226	not tested
Nonprimates		
Rat, *Rattus norvegicus*	>230	218
Domestic fowl, *Gallus gallus*	>230	>250

[a] Taken from Hanke *et al.* (24).
[b] A similar value was obtained by Arnheim (1a) using rhesus tears as the lysozyme source.

As indicated in Table III, the milk lysozyme of only one species (chimpanzee) was indistinguishable from human leukemic lysozyme; the latter is known to have the same amino acid sequence as human milk lysozyme (28, 29). Orangutang lysozyme may differ slightly from that of man and chimpanzee. The only other ape tested, the gorilla, had a milk lysozyme that was clearly different from human lysozyme. The lysozymes of all the Old World and New World monkeys tested reacted less well than gorilla lysozyme with the antiserum pool made against human lysozyme. Still weaker reactions were obtained with the lysozymes of prosimians. Neither rat nor chicken lysozyme reacted at the highest concentrations of anti-human lysozyme tested.

The antiserum pool to baboon lysozyme reacted very strongly with the lysozymes of other Old World monkeys, less strongly with those of New World monkeys, man, and apes, very weakly with prosimian and rat lysozymes, and not at all with chicken lysozyme.

Discussion

These results are only in rough agreement with expectations based on our knowledge of the phylogenetic relationships among primates (see Fig. 5). The major surprises are the anomalous antigenic properties of gorilla lysozyme and the fact that the New World monkey lysozymes are not as different from those of Old World monkeys as expected.

The great antigenic difference between gorilla and human lysozyme is noteworthy. The African apes (gorillas and chimpanzee) are man's closest relatives at the morphological as well as the molecular level. As a general rule, the degree of sequence difference between human proteins and those of African apes is no more than 1%, as indicated in Table IV. If we assume that the lysozyme difference is due only to amino acid substitutions and that the relationship between immunologic resemblance and sequence resemblance is the same for primates as for bird lysozymes,* then gorilla lysozyme may differ from that of man by 8 amino acid substitutions, a 6% sequence difference. It is important to investigate this possibility more thoroughly because it may turn out to be the first case of a relatively large sequence difference between human and African ape proteins.

The immunologic resemblance between New World and Old World monkey lysozymes was expected to be less than that between the lysozymes of hominoids and Old World monkeys because the latter two groups are the more closely related phylogenetically, according to morphological evidence as well as evidence obtained in studies with other macromolecules (23, 32, 45, 46). This discrepancy could be real or else an artifact arising from the imperfect correlation between sequence resemblance and immunologic resemblance. It would be unwise to conclude that New World monkey lysozymes are anomalously close in sequence to those of Old World monkeys until reciprocal tests are made (with an antiserum pool prepared against lysozyme purified from a New World monkey).

We already know that reciprocal tests do not always agree closely in the lysozyme immune system (40) and primate lysozymes are no exception. Thus the average immunologic distance between baboon and human lysozyme is 97 but the distance measured with anti-human lysozyme is 66,

---

*The latter assumption is sound because there is increasingly good evidence that the sequence-immunology relation found for bird lysozymes applies to other proteins as well (10, 41, 44).

TABLE IV

Comparison of Human Proteins with those of African Apes[a]

	Percentage difference in amino acid sequence			
Protein	Chimpanzee vs. man	Gorilla vs. man	Method[b]	Reference
Hemoglobin chains				
α	0	0.7	Sequencing	(14)
β	0	0.7	Sequencing	(14)
γ	0	nd[c]	Sequencing	(5, 15, 16)
δ	0.7	0	Sequencing	(5, 15, 16)
Myoglobin	0.7	nd	Sequencing	(44a)
Fibrinopeptides A and B	0	0	Sequencing	(18)
Cytochrome c	0	nd	Sequencing	(14)
Carbonic anhydrase	1	nd	Micro-complement fixation	(36)
Serum albumin	1	1	Micro-complement fixation	(47)
Transferrin	1.3	0.5	Micro-complement fixation	(13, 54)
Lysozyme	0	6	Micro-complement fixation	(24)

[a] Taken from Hanke *et al.* (24).

[b] To convert microcomplement fixation results into percentage differences in amino acid sequence (x), use was made of the equation $y = 5x$, where y is immunologic distance. It is thus assumed that the relationship found for lysozymes applies also to other proteins, an assumption for which there is increasingly good evidence (10, 41, 44).

[c] Not done.

whereas that measured with anti-baboon lysozyme is 127. From this we estimate that the number of sequence differences between these two lysozymes is 25 ± 8. The amino acid composition data reported for baboon lysozyme are also consistent with a large number of sequence differences between baboon and human lysozyme. As the Jollès laboratory is currently working out the complete amino acid sequence of amino acids in baboon lysozyme, the exact number of sequence differences will hopefully be available soon.

The data in Table III may be used as shown in Table V to calculate an approximate value for the average rate of lysozyme evolution in primates. On the average it has taken about 2 million years for a 1% sequence difference to accumulate between the lysozymes of any two primate species. This means that primate lysozymes appear to be among the most rapidly evolving proteins and that lysozyme has been evolving more rapidly in primates than in birds, where the average period required for a 1% sequence difference to develop is about 6 million years (38). It is noteworthy that primate lysozymes have evolved about as fast as the related protein, lactalbumin (cf. Ref. 14).

Summary

Rabbit antisera were produced against highly purified human and baboon lysozymes and tested for reactivity with the lysozymes of 19 primate species. Micro-complement fixation tests revealed no antigenic differences between the lysozymes of man and the chimpanzee. Major antigenic differences were detected between human lysozyme and the lysozymes of the gorilla, Old World monkeys, New World monkeys, and prosimians. The calibration established earlier between degree of antigenic difference and degree of sequence difference among bird lysozymes allows estimation of the probable number of amino acid sequence differences between human and baboon lysozymes and the lysozymes of the other species tested. Though the results do not permit construction of an accurate phylogenetic tree for primates,

TABLE V

Estimation of the Average Rate of Lysozyme Evolution in Primates[a]

Taxa compared	Mean immunologic distance	Percent change, n	Corrected percent change, m	Divergence time (millions of years)	Unit evolutionary period (millions of years)
Hominoidea vs. Cercopithecoidea	91	18.2	20.0	30	1.5
Catarrhini[b] vs. Ceboidea	85	17.0	18.6	40	2.2
Catarrhini[b] vs. Prosimii	202	40.4	51.8	65	1.3
					Mean 1.6

[a] The mean immunologic distances between taxonomic groups are calculated from the data in Table III and then converted to percent differences in amino acid sequence (n) by means of the calibration curve in Fig. 2. To correct for the possibility of repeated change at the same site, the "corrected percent change" (m) is calculated from the formula

$$m/100 = -\ln[1 - (n/100)]$$

The divergence times used are those given in Fig. 5. One then obtains the unit evolutionary period, i.e., the time required for a 1% sequence difference to arise between two lineages, by dividing m into the divergence time (17). Taken from Hanke et al. (24).

[b] Catarrhini comprise Hominoidea plus Cercopithecoidea, i.e., man, apes, and Old World monkeys.

they do permit estimation of the average rate of lysozyme evolution. Lyso-zyme evolution appears to have proceeded very rapidly in primates.

Acknowledgments

This article is based extensively on research by Prager and Wilson (40), Prager, Arnheim, Mross, and Wilson (38), Hanke, Prager, and Wilson (24), and Champion, Prager, Wachter, and Wilson (11) and on unpublished work by Prager (37a). Financial support for this work came in part from the National Science Foundation, the National Institutes of Health, and the Guggenheim Memorial Foundation. We thank those who kindly supplied purified lysozyme or milk samples.

References

1. Allensmith, M. R., Drell, D., Anderson, R. P., and Newman, L., *Am. J. Ophthalmol.* **71,** 525–529 (1971).
1a. Arnheim, N., personal communication.
2. Arnheim, N., Sobel, J., and Canfield, R. E., *J. Mol. Biol.* **61,** 237–250 (1971).
3. Arnheim, N., and Wilson, A. C., *J. Biol. Chem.* **242,** 3951–3956 (1967).
4. Blake, C. C. F., and Swan, I. D. A., *Nature (Lond.), New Biol.* **232,** 12–15 (1971).
5. Boyer, S. H., Crosby, E. F., Noyes, A. N., Fuller, G. F., Leslie, S. E., Donaldson, L. J., Vrablik, G. R., Schaefer, E. W., and Thurmon, T. F., *Biochem. Genet.* **5,** 405–488 (1971).
6. Browne, W. J., North, A. C. T., Phillips, D. C., Brew, K., Vanaman, T. C., and Hill, R. L., *J. Mol. Biol.* **42,** 65–86 (1969).
7. Buss, D. H., *Biochim. Biophys. Acta* **236,** 587–592 (1971).
8. Canfield, R. E., *J. Biol. Chem.* **238,** 2698–2707 (1963).
9. Canfield, R. E., Kammerman, S., Sobel, J. H., and Morgan, F. J., *Nature (Lond.), New Biol.* **232,** 16–17 (1971).
10. Champion, A. B., Ambler, R. P., Doudoroff, M., and Wilson, A. C., unpublished work.
11. Champion, A. B., Prager, E. M., Wachter, D., and Wilson, A. C., *in* "Biochemical and Immunological Taxonomy of Animals" (C. A. Wright, ed.), pp. 397–416. Academic Press, New York, 1974.
12. Cocks, G. T., and Wilson, A. C., *J. Bacteriol.* **110,** 793–802 (1972).
13. Cronin, J. E., and Sarich, V. M., unpublished work.
14. Dayhoff, M., "Atlas of Protein Sequence and Structure," Vol. 5. National Biomedical Research Foundation, Silver Spring, Maryland, 1972.
15. De Jong, W. W. W., *Nature (Lond.), New Biol.* **234,** 176–177 (1971).
16. De Jong, W. W. W., *Biochim. Biophys. Acta* **251,** 217–226 (1971).
17. Dickerson, R. E., *J. Mol. Evol.* **1,** 26–45 (1971).
18. Doolittle, R. F., Wooding, G. L., Lin, Y., and Riley, M., *J. Mol. Evol.* **1,** 74–83 (1971).
19. Erickson, O. F., Feeney, L., and McEwen, W. K., *AMA Arch. Ophthalmol.* **55,** 800–806 (1956).
20. Farris, J. S., *Am. Natur.* **106,** 645–668 (1972).
21. Fleming, A., *Proc. R. Soc., Lond. [Biol.]* **93,** 306–317 (1922).
22. Fujio, H., Imanishi, M., Nishioka, K., and Amano, T., *Biken J.* **11,** 207–218 (1968).
23. Goodman, M., *in* "Taxonomic Biochemistry and Serology" (C. A. Leone, ed.), pp. 467–486. Ronald Press, New York, 1964.

24. Hanke, N., Prager, E. M., and Wilson, A. C., *J. Biol. Chem.* **248**, 2824–2828 (1973).
25. Hermann, J., and Jollès, J., *Biochim. Biophys. Acta* **200**, 178–179 (1970).
26. Hermann, J., Jollès, J., and Jollès, P., *Eur. J. Biochem.* **24**, 12–17 (1971).
27. Jollès, J., Jauregui-Adell, J., Bernier, I., and Jollès, P., *Biochim. Biophys. Acta* **78**, 668–689 (1963).
28. Jollès, J., and Jollès, P., *Helv. Chim. Acta* **54**, 2668–2675 (1971).
29. Jollès, J., and Jollès, P., *FEBS Lett.* **22**, 31–33 (1972).
30. Jollès, J., Van Leemputten, E., Mouton, A., and Jollès, P., *Biochim. Biophys. Acta* **257**, 497–510 (1972).
31. Kaneda, M., Kato, I., Tominaga, N., Titani, K., and Narita, K., *J. Biochem.* (*Tokyo*) **66**, 747–749 (1969).
32. Kohne, D. E., *Q. Rev. Biophys.* **3**, 327–375 (1970).
33. LaRue, J. N., and Speck, J. C., Jr., *J. Biol. Chem.* **245**, 1985–1991 (1970).
34. Lee, B., and Richards, F. M., *J. Mol. Biol.* **55**, 379–400 (1971).
35. Maron, E., Arnon, R., and Bonavida, B., *Eur. J. Immunol.* **2**, 181–185 (1971).
36. Nonno, L., Herschman, H., and Levine, L., *Arch. Biochem. Biophys.* **136**, 361–367 (1970).
37. Osserman, E. F., and Lawlor, D. P., *J. Exp. Med.* **124**, 921–952 (1966).
37a. Prager, E. M., unpublished work.
38. Prager, E. M., Arnheim, N., Mross, G. A., and Wilson, A. C., *J. Biol. Chem.* **247**, 2905–2916 (1972).
39. Prager, E. M., and Wilson, A. C., *J. Biol. Chem.* **246**, 523–530 (1971).
40. Prager, E. M., and Wilson, A. C., *J. Biol. Chem.* **246**, 5978–5989 (1971).
41. Prager, E. M., and Wilson, A. C., *J. Biol. Chem.* **246**, 7010–7017 (1971).
42. Prager, E. M., and Wilson, A. C., *Biochem. Genet.* **7**, 269–272 (1972).
43. Raven, P. H., Berlin, B., and Breedlove, D. E., *Science* **174**, 1210–1213 (1971).
44. Rocha, V., Crawford, I. P., and Mills, S. E., *J. Bacteriol.* **111**, 163–168 (1972).
44a. Romero Herrera, A. E., and Lehmann, H., *Biochim. Biophys. Acta* **278**, 62–67 (1972).
45. Sarich, V. M., *in* "Old World Monkeys" (J. R. Napier and P. H. Napier, eds.), pp. 175–226. Academic Press, New York, 1970.
46. Sarich, V. M., and Wilson, A. C., *Science* **154**, 1563–1566 (1966).
47. Sarich, V. M., and Wilson, A. C., *Science* **158**, 1200–1203 (1967).
48. Sela, M., *Science* **166**, 1365–1374 (1969).
49. Shinka, S., Imanishi, M., Miyagawa, N., Amano, T., Inouye, M., and Tsugita, A., *Biken J.* **10**, 89–107 (1967).
50. Simons, E. L., "Primate Evolution." Macmillan, New York, 1972.
51. Teicher, E., Maron, E., and Arnon, R., *Immunochemistry* **10**, 265–271 (1973).
52. Thomsen, J., Lund, E. H., Kristiansen, K., Brunfeldt, K., and Malinquist, J., *FEBS Lett.* **22**, 34–36 (1972).
53. Wilson, A. C., unpublished work.
54. Wilson, A. C., and Sarich, V. M., *Proc. Natl. Acad. Sci. U.S.A.* **63**, 1088–1093 (1969).

Immune Response to Lysozyme:
Limited Heterogeneity Caused by Restricted T Cells

ELI SERCARZ, BENJAMIN BONAVIDA, ALEXANDER MILLER,
ROBERT J. SCIBIENSKI, AND JOAN A. STRATTON

Introduction

We have made extensive use of the gallinaceous lysozymes in our studies of the regulation of the immune response. We felt that since two or more different cells take part in processes leading to the formation of antibodies, it might be beneficial to have a structurally mapped molecule in which the spatial relationships of given antigenic determinants could be plotted. The choice of the lysozyme molecule has proven to be fortunate, since it possesses a variety of antigenic determinants (epitopes) and we were encouraged to focus on a vexing question of prime importance to immunologists, the nature of immunogenicity. Why are certain determinants "immuno-potent" and others "immuno-silent"? What are the factors controlling the engagement of particular sets of specific immunocytes in the response?

How Heterogeneous is the Anti-lysozyme Response?

It has long been axiomatic for the immunochemist to emphasize the heterogeneity of the molecular species of antibody that are raised in response to a given antigen (1). Thus, if there are n epitopes (unique antigenic determinants) on a given molecule, it might be imagined that all the animal's cells with the appropriate n' specificities (unique immunoglobulins) complementary to the n epitopes would be turned on (where the prime refers to the fact that not only ideally fitting receptors but also some more loosely fitting receptor-bearing cells would be activated). If the species can elaborate n' receptors of this specificity, what proportion of this total is found in any one individual? We have suggestive evidence that at least a dozen different epitopes can be recognized by the mouse or rabbit on the hen egg-white lysozyme molecule (HEWL) (2). It, therefore, seemed reasonable to assume that a heterogeneous response would be found. In fact, however, many different experiments in our laboratory lead to the conclusion that quite a restricted anti-lysozyme response is found in any one animal.

T Lymphocytes and B Lymphocytes

It has been established in recent years that there are at least two distinct lymphoid cell types which are involved in the events leading from an antigenic challenge to the induction of antibody synthesis (3). One of these cell types, termed the B lymphocyte, is destined to become an antibody-forming cell after it has completed an unknown number of differentiative and inductive steps. The other, "T" lymphocyte, is known to be essential for the induction of antibody synthesis to many antigens, particularly those which are monomeric, but the mechanism by which it exerts its role is unknown. It is thought to somehow assist in presenting the antigen to B cells in some preferred manner, either through direct cell-to-cell contact with antigen as a bridge, or through the mediation of soluble factors produced by the T cells (4).

Evidence has accumulated suggesting that the T cell has a different order of specificity than the B cell. For example, although humoral cross-reactivity between HEWL and carboxymethyl-HEWL is virtually absent, at the T-cell level native or reduced lysozyme induce memory or unresponsiveness which appears to be totally cross-reactive (5). Moreover, it is thought that most self-antigens induce T-cell unresponsiveness but not B-cell unresponsiveness (6). Therefore, it would be expected that the remaining T cells would have a narrower spectrum of reactivity than B cells. It can be asked whether this limited spectrum of T specificities will place any restriction on expression in the B-cell line.

The Restricted Presentation Model

In accord with current notions that a matrix of identical epitopes may be needed to activate the B cell (7), it would be expected that an ordering template would have to exist which would expose the desired facet of the antigen at each point in the matrix. Although the T cell provides the ideal ordering template, surrogates such as macrophage surfaces, antigen–antibody complexes, or antigens with repeating subunits might also suffice. Actually, the T cell-independent antigens that have been described all contain repeating epitopes (8).

We propose that a stringent stereospecificity underlies the presentation of antigen, which will place severe restrictions upon which B cells will be activated during the immune response. Only a very few of the "carrier" epitopes on a small antigen such as lysozyme would be in the appropriate relationship on an ordering template (e.g., opposite) to the available epitope designated to combine with the B-cell receptors. If we represent the molecule as in Fig. 1, a T cell with specificity for 1 will be able to present 4 or 6, but *not* 3 or 7. It is postulated that if the animal is ever going to make anti-7

Fig. 1. Model of the lysozyme molecule obtained by X-ray analysis at 6 Å resolution. The numbers represent epitopes on the molecule whose interrelations are described in the text.

antibody, there must exist a carrier (T) epitope 3 that is in exactly the right orientation vis-à-vis 7 to present it properly. The result would be that restrictions in the T-cell specificities available will show up in restricted antibody specificities.

The alternative to this model is one in which a given epitope is capable of presenting most, if not all, of the other epitopes on the molecule to appropriate B cells. Such an alternative model would minimize any contribution by T cells to the specificity of antibody produced by B cells. Restriction in the specificity of T cells would be of little moment.

In order to decide between these two models, it is necessary to examine situations in which limited T-cell recognition exists. Otherwise, in view of the wide variety of possible epitopes on even an antigen the size of lysozyme, (epitope composition ABCDMNOP), one would expect to induce many specificities of antibodies regardless of which model is correct. Thus, if antibodies M, N, O, P are produced, it would be difficult to know whether A presented M; B, N; C, O; and D, P; or, alternatively, whether A presented M, N, O, and P! With the gallinaceous lysozyme system, we have been able to fulfill our analytical requirements and can cite several lines of evidence which we feel support the restricted presentation model.

Results

Limitations in T-Cell Specificity as a Result of Unresponsiveness

Mice and rabbits can be made immunologically unresponsive to HEWL in such a way that their T cells are suppressed but their B cells can still be activated, provided that a new carrier epitope can be found. Lysozymes that are even very closely related might be expected to have a small number of unique epitopes which could serve this carrier function. Table I shows cross-reactivity at the antibody level (the B-cell receptor).

In T-cell unresponsive mice, the antibody titers obtained after challenge with related gallinaceous lysozymes are always subnormal (9). This is what would be expected if only a limited number of B cells were being turned on by the few new carrier epitopes. Furthermore, in HEWL-unresponsive rabbits, of four lysozymes only turkey (TEL) will induce a response at all, and the response is TEL-specific. The assumption we must make is that only on TEL is there the correct relationship between unique turkey T-cell epitopes and turkey B-cell epitopes. Clearly, it is not a tenable concept that one carrier epitope can present all other epitopes on the molecule to ambient B cells.

TABLE I

**Cross-Reactivity of Gallinaceous Lysozymes with Rabbit and Mouse
Anti-HEWL (Average Percent Inhibition)**

Lysozymes	Rabbit sera	Mouse sera
Hen	100	100
Bobwhite	90	95
Peahen	70	89
Turkey	74	75
Ringed-neck pheasant	52	62
Japanese quail	49	50

Deficiencies in Carrier Epitopes on T Cells

It might be supposed that responses restricted in specificity could arise solely from a limited library of B-cell specificities. We would like to set forth evidence, however, showing that in certain cases, deficiency in an appropriate carrier epitope can be held responsible for the existence of some "immuno-silent" epitopes.

Bobwhite lysozyme, BEL, when used as an immunogen in the rabbit, elicits antibodies which will bind to I*-HEWL (^{125}I-HEWL). Some of these cross-reactive antibodies are directed against epitopes found on peptide 106–129 of HEWL, as detected by inhibition studies. It is fascinating that when HEWL is used as immunogen, no antibodies with this specificity are ever raised. It can be seen in Table II that the cyanogen bromide-treated HEWL, as with BEL, the epitopes are immunogenic (10).

Our explanation has been that whereas rabbits possess B cells capable of producing the antibodies in question, no carrier epitope with the correct steric relationship exists on HEWL! The presence of other nonimmunogenic epitopes can be deduced quite often when they are sought (11), and we are again led to believe that the right carrier epitope is an essential requirement for most responses.

Restricted Responses

We have shown that the existence of an epitope on HEWL does not guarantee that it will be translated by the immune machinery into a complementary antibody to that epitope. We also have other evidence which indicates that antibody responses to lysozyme are quite often highly restricted.

TABLE II

Inhibition of Binding between ^{125}I-HEWL and Anti-lysozymes by Cyanogen Bromide Peptide of HEWL

Antisera vs.[a]	Percent inhibition by peptide 106–129 (HEWL)
HEWL	nil
BEL	28
HEWL(CB)	32

[a] HEWL, hen egg-white lysozyme; BEL, bobwhite egg lysozyme; HEWL(CB) hen egg-white lysozyme (cyanogen bromide).

An Epitope Found on Human and Turkey Lysozyme

Most published reports indicate that hen lysozyme and human lysozyme are totally noncross-reactive despite considerable structural homology. We have found, however, that in 3 of 8 rabbits the antihuman lysozyme *was* cross-reactive with HEWL, albeit poorly, and highly cross-reactive with turkey lysozyme (TEL) (Table III). It is of interest that about half of the antibodies reacting with human lysozyme in these rabbits were cross-reactive with turkey lysozyme, strongly suggesting that these rabbits were responding against few epitopes (12).

This was borne out by antigen-binding studies indicating that all three rabbit sera reacted with an epitope present on both turkey and human lysozymes which is not present in HEWL or Japanese quail lysozyme (see Table III). Only at one position is there unique homology between turkey and human lysozymes (15 = leucine, whereas 15 = histidine for HEWL or JEL), so, on first approximation, we can suppose that the antibody may be directed at an epitope including leucine 15.

TABLE III

Average Binding Capacity of Three Rabbit Anti-human Lysozyme Sera for Gallinaceous Lysozymes

Lysozyme	Antigen-binding capacity[a]
Human	321
Turkey	148
Hen	34
Japanese quail	28

[a] Micrograms antigen bound per ml serum.

Such highly homogeneous antibody responses might be explained by a lack of B cells directed at other determinants. An equally likely explanation, we believe, is that the T-cell specificities are restricted in most rabbits such that the epitope involving residue 15 on human lysozyme is only presented rarely.

The Response of Normal Mice and Rabbits to HEWL

The anti-HEWL response of each inbred mouse or normal rabbit bears its own specificity imprint as assessed by differing patterns of cross-reactivity with other gallinaceous lysozymes. As shown in Table IV, rabbit A serum can be distinguished by its high cross-reactivity with TEL, rabbit B by its poor reactivity with JEL, mouse C by the fact that it cross reacts better with JEL than with TEL, etc. Moreover, if small pieces of mouse spleen are studied, localized evidence of restricted numbers of clones is found (12a). The evidence seems to support the notion that only a limited response is made in any one animal. Such different patterns of cross-reactivity and restricted responses are not consistent with the "one-presents-all" model.

A/J Mouse Response to Turkey Lysozyme

A striking example of a restricted response is evident from examination of A/J mouse anti-TEL sera. These sera gave a much higher (sometimes 10-fold) ABC with I*-HEWL than with I*-TEL (Table V). Noniodinated TEL, however, is as effective an inhibitor of the binding of I*-HEWL by mouse anti-TEL as is HEWL. This effect is not seen with rabbit anti-TEL, indicating that iodination does not denature the TEL molecule in a general

TABLE IV

Pattern of Inhibition in Individual Anti-HEWL Sera (Percent Inhibition)[a]

	Rabbit			Mouse		
Inhibitor	A	B	C	A	B	C
Hen	100	100	100	100	100	100
Turkey	91	64	68	80	66	73
Bobwhite quail	89	89	91	94	90	93
Peahen	82	57	72	95	92	97
Japanese quail	66	30	50	70	69	84
Ringed-neck pheasant	55	48	55	78	62	75

[a] 100-fold excess of unlabeled inhibitor.

TABLE V

Antigen-Binding Capacity of Mouse Anti-TEL for ^{125}I-TEL and ^{125}I-HEWL

Antigen-binding capacity		Ratio
I*-HEWL	I*-TEL	HEWL/TEL
232	115	2.0
395	79	5.0
505	44	11.5
420	90	4.7
350	37	9.5

way. It seems more likely that the A/J mouse simply directs most of its anti-TEL antibody versus the N-terminal–C-terminal part of the molecule, and that under conditions where TEL is iodinated uniquely at position 3 (13), the iodine molecule obscures the determinant.

Discussion

We have presented evidence that argues for restriction in the specificities of anti-lysozyme antibodies which arise in a single animal. Whereas this might have been expected with the repeating unideterminant microbial polysaccharides, proteins such as the gallinaceous lysozymes, against which at least a dozen antibody specificities can be raised (2), would be predicted to elicit heterogeneous responses. The reasons for the restrictiveness of the responses are not fully understood. We might imagine, in fact, that all antibody specificities eventually appear, but only sequentially (14–16).

It seems more likely that there is a limitation on the expression of the available spectrum of unique immunoglobulins in any one animal. It may be assumed that failure to make a particular antibody molecule is due to absence of the right V genes in the available B-cell pool. We propose that in addition, the limited spectrum of epitopes recognized by T cells will sterically restrict the variety of specificities suited for effective bridging to B cells. It is generally accepted that an antigen-mediated interaction between T and B cells is necessary for B-cell stimulation.

This would predict that untriggered B cells of certain specificities could be found. It was possible to show that BEL or HEWL (CB), when used as immunogens, induced some peptide-reactive antibodies which could not be induced by HEWL, although the B-cell population was present and the epitopes were available for reaction on HEWL. This was a strong indication that nonimmunogenic epitopes may be quite common; they are "epitopes in search of a carrier epitope." Such a search would hardly seem necessary if *any* epitope on the molecule sufficed to present any other! This formulation, by the way, would also have predicted the common finding that there is not always reciprocal cross-reactivity to two antigens, A and B. (Some cross-reactive epitopes on A may never elicit a response for want of an appropriately placed carrier epitope.)

Although our explanations are logical, we are now attempting to test them more critically. We are examining the specificity and isoelectric-focusing spectrum of sera arising in T-cell unresponsive animals when challenged by closely related lysozymes (with few unique carrier epitopes) to see how restricted they really are. Furthermore, we are raising T-cell memory against peptides of lysozyme and asking whether the resulting antibody is restricted after boosting with intact lysozyme.

Summary

Several gallinaceous lysozymes (hen, turkey, bobwhite quail, Japanese quail, ringed-neck pheasant, peahen), because of their known structure, relatedness, and small size, have been studied in immune induction experiments. Anti-lysozyme responses were most often restricted to certain areas on the molecule. In analyzing the reasons for limited heterogeneity of the antibodies, we proposed that restriction at the level of antigen recognition by the T cell will result in restriction in the specificities produced by B cells, owing to the marked stereospecificity of antigen presentation.

References

1. Kabat, E. A., "Structural Concepts in Immunology and Immunochemistry," Chapter 9. Holt, New York, 1968.
2. Scibienski, R. J., Stratton, J. A., Miller, A., and Sercarz, E., unpublished.
3. "Antigen-Sensitive Cells, Their Source and Differentiation," *Transplant. Rev.* **1**, (1969).
4. Dutton, R. W., Falkoff, R., Hirst J. A., Hoffman, M., Kappler, J. W., Kettman, J. R., Lesley, J. F., and Vann, D., *Prog. Immunol.* **1**, 355 (1971).
5. Thompson, K., Harris, M., Benjamini, E., Mitchell, G. F., and Nobel, M., *Nature (Lond.), New Biol.* **238**, 20 (1972).
6. Chiller, J. M., Habicht, G. S., and Weigle, W. O., *Science* **171**, 813 (1971).
7. Diener, E., and Feldmann, M., *Transplant. Rev.* **8**, 76 (1971).
8. Howard, J. G., Christie, G. H., Courtenay, B. M., Leuchars, E., and Davies, A. J. S., *Cell Immunol.* **2**, 614 (1971).
9. Scibienski, R. J., and Sercarz, E., *J. Immunol.* **110**, 540 (1973).
10. Bonavida, B., and Sercarz, E., *Eur. J. Immunol.* **1**, 166 (1971).

11. Benjamini, E., Scibienski, R., and Thompson, K., *Contemp. Top. Immunochem.* **1,** 1 (1972).
12. Miller, A., Bonavida, B., Stratton, J. A., and Sercarz, E., *Biochim. Biophys. Acta* **243,** 520 (1971).
12a. Stratton, J. A., Ph.D. thesis, UCLA, 1970.
13. Stratton, J. A., and Miller, A., *Fed. Proc.* **29,** 700 (1970).
14. Young, J. D., and Leung, C. Y., *Biochemistry* **9,** 2755 (1970).
15. Maron, E., Arnon, R., and Bonavida, B., *Eur. J. Immunol.* **2,** 181 (1972).
16. Bonavida, B., Miller, A., and Sercarz, E., unpublished.

14

Multiple Genes for Lysozyme*

NORMAN ARNHEIM

Chemical data have suggested that the lysozymes from diverse species of vertebrates have significant structural similarities. Amino acid sequence data on seven avian egg-white lysozymes (3, 13, 16, 19, 20, 22, 26) and two mammalian lysozymes (4, 5, 17, and Riblet, Chapter 8, this volume) reveal sequence homologies between the lysozymes of these vertebrate groups approximate 60%.

However, the lysozyme found in Embden goose egg white stands in marked contrast to these other vertebrate lysozymes on the basis of amino acid composition (6, 10) immunologic cross-reactivity (2), and partial amino acid sequence studies (5). In order to learn whether the lysozyme in Embden goose egg white was unique, an investigation of the egg-white lysozymes present in other species of birds was carried out in our laboratory. These immunologic studies used antibodies directed against either purified chicken egg-white lysozyme or purified Embden goose egg-white lysozyme (2). The antiserum directed against the chicken enzyme does not cross react with Embden goose lysozyme; the antiserum directed against the Embden goose enzyme does not react with chicken lysozyme. Thus, these two antiserums could be used to determine unambiguously whether other species had a "chicken-type" or an "Embden goose-type" lysozyme.

A totally unexpected observation was made during studies on individual egg whites from three black swans (*Cygnus atratus*). It appeared

*Supported by a research grant from the National Science Foundation (GB 17351).

153

that this species contained not one but two lysozymes in its egg white (2). One of them was immunologically an Embden goose-type, whereas the other was immunologically a chicken-type. To support the idea that two distinct molecules were present in each egg white of this species rather than a single enzyme with two distinct antigenic specificities, further studies using the lysoplate inhibition test (25) were carried out.

An aliquot of black swan egg white was absorbed with antibodies directed against chicken lysozyme and the precipitate that formed was removed by centrifugation. The supernatant was placed in the well of a lysoplate, an agar plate containing *M. lysodeikticus*. Incubation of the plate at room temperature was followed by a clearing of the agar around the well that formed because of the lysis of the bacteria contained in the agar (Fig. 1). This lysis was a consequence of the diffusion of a molecule with lysozyme activity from the well into the surrounding agar. That this activity was not

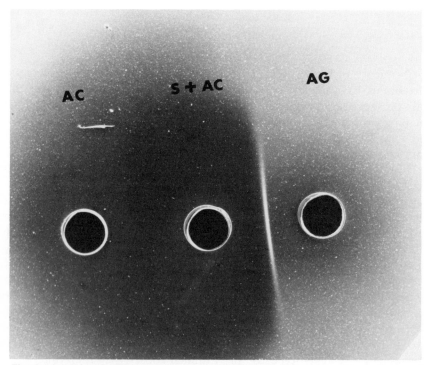

Fig. 1. Lysoplate inhibition experiment with black swan egg white. The wells contain, from left to right, antibodies directed against chicken egg-white lysozyme (AC), black swan egg white absorbed with antibodies directed against chicken egg-white lysozyme (S + AC), and antibodies directed against Embden goose egg-white lysozyme (AG).

the result of any residual chicken-type lysozyme left as a consequence of incomplete absorption of the swan egg white was demonstrated by the failure of antichicken egg-white lysozyme to inhibit the diffusion of this enzyme when placed in an adjacent well. On the contrary, antibodies directed against goose egg-white lysozyme did inhibit the diffusion of this remaining lysozyme. Absorption studies using black swan egg white and antibodies directed against Embden goose lysozyme also supported the conclusion that each egg white of *Cygnus atratus* examined contained both a chicken-type and an Embden goose-type lysozyme.

The fact that two distinct lysozymes are found in individual egg whites of the black swan can be explained in either of two ways. These enzymes may be different forms of the same structural gene, i.e., allelic variants of a single genetic locus. Alternatively, each enzyme may be the product of a different genetic locus. The following evidence that is available supports the latter possibility: (a) No immunologic resemblance can be detected between the Embden goose-type and the chicken-type enzymes of the black swan, whereas the multiple allelic forms of Peking duck egg-white lysozyme (chicken-type enzymes) show strong antigenic similarities to each other (27). (b) Chemical analysis of the purified black swan lysozymes shows that these proteins differ radically in amino acid composition and amino acid sequence of their amino-terminal regions as well as in molecular size (Morgan, F. J., and Arnheim, N., Chapter 7, this volume). (c) Studies on sixteen individual eggs from another species (Canada Goose) containing both an Embden goose-type lysozyme and a chicken-type lysozyme showed that both enzymes were present in every egg white (1). If these lysozymes were allelic products of the same gene, one would have expected some egg whites to contain only the chicken-type enzyme, others only the Embden goose-type lysozyme, and only some of the egg whites to contain both. All of the above observations therefore support the idea that these two lysozymes are the products of two distinct genetic loci.

In contrast to the black swan and Canada goose, other species of birds appear to contain only a chicken-type lysozyme in their egg white (chicken, Peking duck) while still others appear to contain only an Embden goose-type lysozyme (Embden goose). This distribution of lysozymes in the egg whites of these species may be a consequence of either of two distinct phenomena: (a) The absence of a particular lysozyme in the egg white could mean that the structural gene for the enzyme is missing from the genome of the species. Considering the phylogenetic relationships among the bird species mentioned above, one would have to argue further that several independent losses of one or the other gene loci occurred during evolution. (b) A more likely explanation of this distribution is that the presence or absence of either enzyme in egg white is not an accurate reflection of whether or not the structural genes for these lysozymes are present in the genome of the species. It would not be unreasonable to argue that all of these birds contain both a chicken-type and an Embden goose-type lysozyme locus but that the expression of each of these genes in the cells of the oviduct responsible for the production of egg-white lysozyme varies from species to species.

If the latter suggestion were true, one might expect to find, in those birds who appear to express only one lysozyme in their egg white, the other form of lysozyme in another tissue. To explore this possibility, we have recently carried out a preliminary tissue survey on a 7-week-old duck. This species apparently produces only a chicken-type lysozyme in its egg white. Our tissue survey was based upon the immunologic approach described above and showed, as we expected, that several tissues of the duck do in fact have an Embden goose-type lysozyme. The results of a lysoplate inhibition test are shown in Fig. 2. The absorption of an extract of the duck's bursa of Fabricius with antibodies directed against chicken lysozyme revealed that an additional lysozyme was present in this organ. Antibodies directed against Embden goose lysozyme placed in an adjacent well of the lysoplate were found to inhibit the diffusion of this residual enzyme. The purification of the duck's Embden goose-type lysozyme is currently under way in our laboratory as is the search for a chicken-type lysozyme in tissues of the Embden goose.

These data suggest that many other waterfowl may have two distinct lysozyme genes, but that not all species express both loci to the same extent in the oviduct. This observation leads to a variety of interesting questions concerning the control of lysozyme synthesis in the oviduct. Studies on the

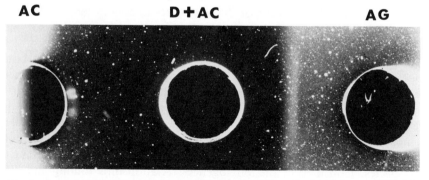

Fig. 2. Lysoplate inhibition experiment with an extract from the duck's bursa of Fabricius. The wells contain, from left to right, antibodies directed against chicken egg-white lysozyme (AC), an extract of the duck's bursa absorbed with antibodies directed against chicken egg-white lysozyme (D + AC), and antibodies directed against Embden goose egg-white lysozyme (AG).

mechanisms which result in the apparently exclusive production of either the chicken-type or the Embden goose-type lysozyme in the egg whites of some species or the production of both enzymes in others may provide valuable information on the hormonal control of protein synthesis.

Aside from the problem of how the synthesis of the two lysozymes is controlled in the tissues of different species, the fact that two distinct lysozymes exist in waterfowl and perhaps in other birds (21) and in mammals as well (30) raises the question as to what is the functional significance of two lysozymes in each individual. While lysozyme may have more than one function (29), only its ability to carry out the lysis of bacterial cell suspensions is known in great detail (9). Comparative studies of chicken egg-white lysozyme and Embden goose egg-white lysozyme, therefore, might reveal catalytic differences which could help to explain the presence of two distinct lysozyme genes.

Chicken egg-white lysozyme acts on the peptidoglycan of bacterial cell walls. The peptidoglycan is composed of alternating *N*-acetylglucosamine (NAG) and *N*-acetylmuramic acid (NAM) residues joined together by a $\beta(1,4)$ glycosidic linkage. The lactyl group of NAM can be substituted with a peptide moiety, the amino acid composition of which varies among bacterial groups (24). The frequency with which the NAM residues are joined to a peptide constituent also varies among bacterial species.

Chicken egg-white lysozyme is known to catalyze the hydrolysis of the $\beta(1,4)$ linkage between NAM and NAG. Until recently the exact nature of the catalytic cleavage carried out by goose egg-white lysozyme was unknown. On the other hand, a number of other enzymatic studies had been carried out on the Embden goose enzyme. Table I summarizes the conclusions of several of these experiments. Inspection of Table I shows that the chicken and Embden goose egg-white enzymes differ substantially in a number of important enzymatic properties. Since a variety of enzymes are known in other organisms to hydrolyze the peptidoglycan of bacterial cell walls not as muramidases but as glucosaminidases and peptidases (12), we decided to examine specifically whether or not Embden goose egg-white lysozyme was in fact a muramidase (1a).

The tests on the enzymatic specificity of Embden goose egg-white lysozyme were carried out using purified peptidoglycans isolated from the cell walls of *M. lysodeikticus* and *E. coli*. If goose lysozyme has muramidase activity on these substrates the $\beta(1,4)$ linkage between NAM and NAG will be hydrolyzed. After incubating the substrate with lysozyme, sodium borohydride ($NaBH_4$) reduction is carried out. The NAM residues that are involved in a glycosidic bond left unhydrolyzed by lysozyme remain unmodified by the $NaBH_4$ treatment and are detected after acid hydrolysis and amino acid analysis as muramic acid. The NAM residues whose glyco-

TABLE I

Summary of Differences in Enzymatic Properties between Chicken Egg-White and Embden Goose Egg-White Lysozymes

	Activity on *M. lysodeikticus* cells (Ref. 11)	Hydrolysis of the purified tetrasaccharide, NAG–NAM–NAG–NAM	Chitinase activity (Ref. 7)	Inhibition of *M. lysodeikticus* lysis by the mono- or tetrasaccharide of NAG (Ref. 18)	Reaction with the 2′,3′-epoxypropyl β-glycoside of the dimer of NAG (Ref. 23)
Chicken egg-white lysozyme	Lysis readily achieved and accompanied by solubilization of NAG–NAM and NAG–NAM–NAG–NAM from the peptidoglycan	Yes (Ref. 28)	Yes	Yes	Yes
Embden goose egg-white lysozyme	Lysis readily achieved but with little if any solubilization of NAG–NAM and NAG–NAM–NAG–NAM	No (Ref. 15)	Little if any	No	No

sidic bond is cleaved by lysozyme are modified by the $NaBH_4$ treatment and are detected after acid hydrolysis and amino acid analysis as muramicitol. The higher the ratio of muramicitol to muramic acid + muramicitol recovered after lysozyme treatment of the peptidoglycan, the greater the extent of hydrolysis of the $\beta(1,4)$ linkage between NAM and NAG (muramidase activity).

Table II summarizes the results of our experiments on the enzymatic activity of Embden goose egg-white lysozyme. At several pHs ranging from 3.5 to 7.1 the goose enzyme was found to have significant muramidase activity on the peptidoglycans of *E. coli* and *M. lysodeikticus* (1a).

The observation that Embden goose egg-white lysozyme has muramidase activity must be reconciled with the enzymatic differences between it and chicken egg-white lysozyme as summarized in Table I. The inability of the goose enzyme to hydrolyze polymers of NAG or to be inhibited in its lysis of *M. lysodeikticus* by the small molecular weight NAG inhibitors might be due to significant architectural differences in the active site of the two enzymes, at least with respect to the binding of NAG. On the other hand, the inability of the goose enzyme to solubilize peptide-free NAG–NAM or NAG–NAM–NAG–NAM from *M. lysodeikticus* cells or to hydrolyze purified peptide-free NAG–NAM–NAG–NAM is somewhat more difficult to explain, since the goose enzyme is a muramidase. One suggestion that has been put forward to explain these observations is that Embden goose egg-white lysozyme is specific for those NAM residues whose lactyl group are linked to a peptide constituent (1a). If this were the case the solubilization of peptide-free saccharides from *M. lysodeikticus* cell walls would not be expected nor would the hydrolysis of purified peptide-free NAG–NAM–NAG–NAM. That the peptide portion of the peptidoglycan may be important in lysozyme activity is suggested by comparative studies of the binding of NAG–NAM–NAG–NAM with and without a peptide

TABLE II

Summary of Experimental Data Demonstrating the Lysozyme Activity of Embden Goose Egg-White Lysozyme

	Ratio of muramicitol to muramic acid + muramicitol observed in experiments with *E. coli* peptidoglycan		Ratio of muramicitol to muramic acid + muramicitol observed in experiments with *M. lysodeikticus* peptidoglycan goose lysozyme
	Chicken lysozyme	Goose lysozyme	
pH 3.5	0.38	0.93	0.14
pH 5.3	0.36	0.92	0.36
pH 7.1	0.24	0.90	0.28

constituent (8) to the chicken egg-white enzyme. Although it is clear that chicken lysozyme has no absolute requirement for the presence of a peptide moiety, the behavior of the goose enzyme would be consistent with a very strong preference for substituted NAM residues. It is obvious that further chemical studies are needed before other alternative explanations can be ruled out.

Although it may be premature to draw any conclusions about the significance of the possible difference in specificity among these two lysozymes, such a difference could be related to their antimicrobial function since different species of bacteria vary in the extent to which NAM is substituted with a peptide (24).

In conclusion, two lysozymes with distinct immunologic and chemical properties have been found in birds. The egg whites of some species appear to contain only the "chicken-type" enzyme, some contain only the "Embden goose-type," and still others contain both lysozymes. Analysis of the lysozymes in individual egg whites, in addition to chemical and immunologic data, suggest that these enzymes are the products of two distinct genetic loci. Although all species apparently do not contain both lysozymes in their egg white, evidence from a tissue survey of the duck suggests that the genes for both proteins may be present in the genome.

Attempts have been made to understand the functional significance of the presence of two lysozyme genes. A study of the enzymatic properties of Embden goose egg-white lysozyme suggests that, in contrast to the chicken enzyme, it may have a strong preference for cleaving the glycosidic bond of N-acetylmuramic residues substituted with a peptide constituent. Further studies on the cellular distribution of these lysozymes and the hormonal control of their synthesis in the oviduct may increase our understanding of their biological role or roles.

References

1. Arnheim, N., unpublished studies.
1a. Arnheim, N., Inouye, M., Law, L., and Laudin, A., *J. Biol. Chem.* **248**, 233 (1973).
2. Arnheim, N., and Steller, R., *Arch. Biochem. Biophys.* **141**, 656 (1970).
3. Canfield, R. E., *J. Biol. Chem.* **238**, 2698 (1963).
4. Canfield, R. E., cited in Johnson *et al.* (14).
5. Canfield, R. E., Kammermann, S., Sobel, J. H., and Morgan, F. M., *Nature (Lond.)*, *New Biol.* **232**, 16 (1971).
6. Canfield, R. E., and McMurry, S., *Biochim. Biophys. Res. Commun.* **26**, 38 (1967).
7. Charlemagne, D., and Jollès, P., *Bull. Soc. Chim. Biol. (Paris)* **49**, 1103 (1967).
8. Chipman, D. M., Grisaro, V., and Sharon, N., *J. Biol. Chem.* **242**, 4388 (1967).
9. Chipman, D. M., and Sharon, N., *Science* **165**, 454 (1969).

10. Dianoux, A. C., and Jollès, P., *Biochim. Biophys. Acta* **133,** 472 (1967).
11. Dianoux, A. C., and Jollès, P., *Helv. Chim. Acta* **52,** 611 (1969).
12. Ghuysen, J. M., *Bacteriol. Rev.* **32,** 425 (1968).
13. Hermann, J., Jollès, J., and Jollès, P., *Eur. J. Biochem.* **24,** 12 (1971).
14. Johnson, L. N., Phillips, D. C., and Rupley, J. A., *Brookhaven Symp. Biol.* **21,** 136 (1968).
15. Jollès, P., *in* "Homologous Enzymes and Biochemical Evolution" (N. T. Thoai and J. Roche, eds.), pp. 341–358. Gordon & Breach, New York, 1968.
16. Jollès, J., Jauregui-Adell, J., Bernier, I., and Jollès, P., *Biochim. Biophys. Acta* **78,** 668 (1968).
17. Jollès, J., and Jollès, P., *Helv. Chim. Acta* **54,** 2668 (1971).
18. Jollès, P., Saint Blancard, J., Charlemagne, D., Dianoux, A. C., Jollès, J., and Le Baron, J. L., *Biochim. Biophys. Acta* **151,** 532 (1968).
19. Jollès, J., Van Leemputten, E., Mouton, A., and Jollès, P., *Biochim. Biophys. Acta* **257,** 497 (1972).
20. Kaneda, M., Kato, I., Tominaga, N., Titani, K., and Narita, K., *J. Biochem. (Tokyo)* **66,** 747 (1969).
21. Kuettner, K. E., Eisenstein, R., Soble, L. W., and Arsenis, C., *J. Cell Biol.* **49,** 450 (1971).
22. La Rue, J. N., and Speck, J. C., Jr., *J. Biol. Chem.* **245,** 1985 (1970).
23. McKelvy, J. F., Eshdat, Y., and Sharon, N., *Fed. Proc.* **29,** 532 (1970).
24. Osborn, M. J., *Annu. Rev. Biochem.* **38,** 501 (1969).
25. Osserman, E., and Lawlor, D. P., *J. Exp. Med.* **124,** 921 (1966).
26. Prager, E., Arnheim, N., Mross, G., and Wilson, A. C., *J. Biol. Chem.* **247,** 2905 (1972).
27. Prager, E. M., and Wilson, A. C., *Biochem. Genet.* **1,** 269 (1972).
28. Sharon, N., Jollès, J., and Jollès, P., *Bull. Soc. Chim. Biol. (Paris)* **48,** 731 (1966).
29. Sorgente, N., Hascall, V. C., and Kuettner, K. E., *Biochim. Biophys. Acta* **284,** 441 (1972).
30. Wolinsky, I., and Cohn, D. V., *Nature (Lond.)* **210,** 413 (1966).

SECTION

2

STRUCTURAL BASIS OF ENZYMATIC AND NONENZYMATIC BEHAVIOR

Introduction

SHERMAN BEYCHOK

Introduction

The papers of this section deal with two broad areas in lysozyme research, namely: (1) the mechanism of enzymatic activity and (2) the solution characteristics and behavior of hen egg-white and human lysozyme.

Lysozyme was first recognized and is still most widely known and understood as an enzyme that catalyzes a reaction leading to the lysis of certain bacterial cells. The first area represents an attempt to probe this function in molecular detail. To go further still and attempt to explain the whole range of biological phenomena in which vertebrate lysozyme is increasingly implicated requires the fullest possible inquiry into the behavior of the molecule in solution and its interactions with a variety of tissue and cellular constituents.

The quest for insight into an enzyme mechanism properly begins with a chemical description of substrate and products. In the case of lysozyme, this appeared a formidable—even hopeless—task not too many years ago, for the substrate was nothing less than the bacterial cell wall itself. Strominger and Tipper describe in the first paper the experimental tour de force which culminated in the complete chemical structure of the giant macromolecule which the cell wall comprises. They detail the experiments which led ultimately to the identification of the actual bond attacked and the configurations about the substrate atoms involved. This leads them finally to a discussion of the basis of susceptibility and resistance to lysozyme of different bacteria in terms of substrate configuration, three-dimensional conformation, and composition.

Ghuysen further explores the specificity of lysozyme and examines this specificity from the point of view of effect of substitution of O-acetyl groups or of peptides on the N-acetylmuramic acid residues. An important conclusion drawn by Ghuysen is that lytic glycosidases exhibit relaxed specificity when exposed to nonpeptidoglycan substrates.

Paralleling the characterization of the peptidoglycan substrate have been intensive studies from many laboratories on smaller substrates and inhibitors. The availability of these smaller molecules made possible the crystalline enzyme-inhibitor complex and, ultimately, led to the proposal of a mechanism of enzymatic activity by Phillips and his colleagues. That proposal triggered a great deal of detailed research into the role of every active site residue in catalytic function and has been shown to be correct in its essential predictions. However, many features of the chemical reactivity of atoms in the active site still require further elucidation, especially if the catalysis by lysozymes other than the hen egg-white enzyme, with altered sequences in the active site, is to be understood. In their paper Sharon and Eshdat describe the design and preparation of a homologous series of "affinity label" irreversible inhibitors. They characterize most thoroughly a particular one, $2', 3'$-epoxypropyl β-glycoside, and demonstrate its reaction with Asp(52) in hen egg-white lysozyme. The inhibitor is shown to block catalysis in human lysozyme also, as well as in six different bird lysozymes.

Three papers follow on nuclear magnetic resonance spectroscopy of hen egg-white lysozyme, each utilizing the most recent advances in nmr spectroscopy. Thus, these measurements are utilized not only to provide greater detail on the precise arrangement of atoms in the active site and of atoms of the bound sugars, but they serve also to calibrate new applications and developments against a protein of known structure and reactivity so that these techniques may be expanded to other proteins whose structures are being explored. Lysozyme has, of course, served well in this role many times before, as Dr. Canfield has pointed out, and as each of the papers amply demonstrate.

Nmr spectra are especially revealing of distance and geometric relationships between the atoms which generate the bands. In the first paper of the three mentioned above, Campbell, Dobson, Williams, and Xavier use several sensitivity-enhancing techniques that have not been previously applied to proteins. Together with these, they employ so-called chemical shift reagents that selectively perturb certain proton resonances. Lanthanide ions are used for this purpose. The binding site in the crystal of Gd (III), for example, is known and from the spectra the identification and location of the nearest groups can be deduced.

Patt, Dolphin and Sykes make use of the coupling constant between the anomeric proton on carbon-1 of the sugar ring and the axial proton on carbon-2 to observe directly the distortion of the ring in subsite D. That distortion is a central feature of the Phillips mechanism and it is here demonstrated using the tetrasaccharide (NAG-NAM)$_2$.

Brewer, Marcus, Grollman, and Sternlicht do a comparative study of ^{13}C magnetic resonance of lysozyme and concanavalin A, a lectin which, in common with HEWL, has a high specificity for binding of certain sugars. By studying and identifying the resonances of each carbon atom in sugars that are enriched with ^{13}C, the orientation within the site of the sugar rings is determined.

Rupley, Banerjee, Kregar, Lapanjee, Shrake, and Turk present the first paper dealing with solutions properties and behavior of lysozyme. They begin with an analysis of the pH dependence of catalysis of HEWL and determine the pK_a's of Glu (35) and Asp (52). Enthalpy measurements by calorimetry secure the assignments. They then present a novel and powerful calculation of the exposure of atoms and groups in lysozyme to solvent which may be used to describe the character of interactions between the protein and ligands and between groups within the folded protein. Glycine is identified as among the most exposed residues in proteins generally and as a prominent sidechain in β-bends. Finally, the exposure to solvent of atoms at ends and inside of helices is investigated.

Wetlaufer, Johnson, and Clauss continue with questions of protein folding, generally, and lysozyme folding, particularly. They present results of studies of nonenzymatic reoxidation and regeneration of reduced human lysozyme. These studies are of great physiological significance in that they examine processes related to the control of *in vivo* folding. Moreover, very basic questions in molecular biology are probed—namely, whether a native protein is in a state of conformational equilibrium and how cellular constituents may influence conformational states and reactivity.

In the last paper of this section, Mulvey, Gualtieri, and Beychok present a comparative study of HEWL and human lysozyme with respect to acid base behavior, fluorescence, and difference and circular dichroism spectroscopy, especially in regard to differences these reveal in binding constants of various inhibitors to the two proteins. Such differences are analyzed with reference to a particular amino acid substitution in the active site and to other sequence replacements in an effort to identify those features of the sequence which lead to common enzymatic and folding properties and those which modify these properties. The sequence differences are also analyzed in terms of the maintenance of a common secondary structure.

There are a number of properties of lysozyme which are now so well known that they are sometimes not given attention. Among the most important are the following: (1) vertebrate lysozyme, as a class of proteins of very similar structure and enzymatic specificity, is widely distributed both as regards the number of animal species in which lysozyme is found and the variety of tissues and secretions in which it occurs within any animal. (2) Lysozyme is unusual in its distinctive cationic character. With isoionic points near 11, it possesses net positive charge over most of the entire accessible pH range in aqueous media. (3) The native structure withstands sharp change in pH, temperature, and salt concentration. Binding of anions with resultant charge alterations are thus not likely to cause disruption in structure. Indeed, the native protein reversibly combines with a great number of anions of different mass and valence. (4) Substrate analogs are commonly distributed on or protrude from surfaces of plasma and other membranes. In considering the papers of this section, and in anticipation of the biological and clinical studies of the next section, it is well to bear these properties in mind because they provide much of the logical basis for a fuller understanding of the diverse roles which lysozyme plays in nature.

Structure of Bacterial Cell Walls:
The Lysozyme Substrate*

JACK L. STROMINGER and DONALD J. TIPPER

The natural substrate for lysozyme is the bacterial cell wall. Investigations of the lysozyme substrate began in the late 1930's and early 1940's. Karl Meyer was one of the pioneers in this field. He and his collaborators (1) and Epstein and Chain (2) were the first to show that a bacterial polysaccharide of some kind was the lysozyme substrate, and that this polysaccharide contained N-acetyl-D-glucosamine. Bacterial cell walls were first isolated in the late 1940's and early 1950's by a number of investigators among whom Milton Salton was very prominent (3). Salton showed that isolated cell walls were substrates for the enzyme lysozyme (4).

The cell wall is a gigantic macromolecule, a three-dimensional network which envelops the microbial cell. It is a highly water-insoluble polymer and it is this material, or more exactly one of its constituents, the peptidoglycan (a polymer containing two amino sugars, N-acetylglucosamine and N-acetylmuramic acid, and amino acids) which is the substrate for the enzyme lysozyme.

The physiological importance of the cell wall was soon emphasized by Weibull (5) who demonstrated that if the treatment with lysozyme was

* The work described has been supported by research grants from the U.S. Public Health Service (AM-13230, AI-09152, HD-06518) and National Science Foundation (GB-29747X).

carried out not in ordinary buffer but in hypertonic sucrose solution, bacterial cells were not lysed, but they were converted to spheroplasts, or protoplasts. The sucrose is needed for stability in the absence of the cell wall to counterbalance the high internal osmotic pressure of microbial cells, estimated at 5–30 atmospheres. This, then defined the cell wall as the rigid outer layer of microorganisms which determined its shape and served to protect it from such deleterious influences as osmotic shock.

Not long after this development it was shown by Berger and Weiser (6) that chitin was also a lysozyme substrate. Karl Meyer (7) then showed that chitodextrins were also substrates. This was the first evidence that lysozyme was a carbohydrase with a possible specificity for $\beta(1, 4)$ linkages between sugars.

Structure of Bacterial Cell Walls

Beginning in the early 1960's, a great deal of work commenced on the structure of the gigantic macromolecule, the cell wall. Salton and Ghuysen (8) and Perkins (9) were the first to use lysozyme to isolate soluble fragments from cell walls of *Micrococcus lysodeikticus*. Among these they obtained in low yield a disaccharide, *N*-acetylglucosaminyl-*N*-acetylmuramic acid. Shortly afterward Ghuysen and Strominger (10) isolated from cell walls of *Staphylococcus aureus* the same disaccharide in high yield and a second disaccharide having an *O*-acetyl substituent, pointing out another feature of the structure of the glycan in cell walls; viz., some of the acetyl-muramic acid residues in some cell walls bear *O*-acetyl substituents. The isolation of this disaccharide was possible because another lysozyme from *Streptomyces,* called the "32 enzyme" which had been isolated by this time (11), can catalyze the cleavage of glycosidic linkages involving *N,O*-dia-cetylmuramic acid residues [and because of the discovery of another enzyme, the *N*-acetylmuramyl-L-alanine amidase (see below)]. The Chaloropsis B enzyme, also a lysozyme, has a similar specificity (12, 13) but egg-white lysozyme is unable to cleave linkages involving *N,O*-diacetylmuramic acid.

All of these early workers obtained data which led them to believe that the disaccharides which were isolated were $\beta(1, 6)$ linked structures. The clue to the correct structure was provided by synthetic studies of Jeanloz, Sharon, and Flowers (14) which led them and us (15) to the conclusion that the natural disaccharide was not $\beta(1, 6)$ linked, but was $\beta(1, 4)$ linked. However, the structure of these disaccharides gave little information as to the nature of the lysozyme substrate, because the bond which is cleaved by lysozyme is obviously missing in these materials.

Now, at about that time attention was also being paid to the peptide substituent on the acetylmuramic acid residues of the glycan. Virtually all of these residues in most bacterial cell walls are substituted by a tetrapeptide. In 1961, Weidel and Pelzer (16) showed that there were peptide-linked dimers in the cell wall of *Escherichia coli,* i.e., they isolated from among the products of lysozyme hydrolysis two dissacharide–tetrapeptide units linked together by a direct peptide cross-link. At the same time, Mandelstam and Strominger (17) observed that in cell walls of *S. aureus* there appeared to be intervening amino acids between the two tetrapeptides on the acetyl-glucosaminyl–acetylmuramic acid, so that a small peptide cross bridge could exist between disaccharide–tetrapeptide units. Subsequent analyses have amply confirmed this important general feature of the structure of bacterial cell walls. It is a structural feature which was of crucial importance in understanding the mechanism by which penicillin (another bacteriolytic substance discovered by Fleming) kills bacterial cells (18). The work on cell wall structure was greatly catalyzed by the discovery that penicillins and other antibacterial agents are specific inhibitors of bacterial cell wall synthesis [reviewed in Strominger (18)].

After a great deal more work, a general outline for the structure of bacterial cell walls could be put down on paper, illustrated for the cell wall of *S. aureus* (Fig. 1). The glycan strands contain two sugars X, acetylglucos-amine, and Y, the 3-*O*-D-lactic acid ether of acetylglucosame, called acetyl-muramic acid. Four glycan strands in which these two sugars strictly alternate are represented, but to account for the dimensions of the wall many more strands would be needed. In *S. aureus* all of the acetylmuramic acid residues are substituted by a tetrapeptide (represented by the four open circles), the sequence of which is L-alanyl-D-isoglutaminyl-L-lysyl-D-alanine (19, 20); all of the acetylmuramic acid residues are so substituted.

These glycan strands substituted by the tetrapeptide are called peptido-glycan strands, and they are then cross-linked to each other by means of an interpeptide bridge which is to some extent a genus-specific character. In the genus *S. aureus* the interpeptide bridge is a pentaglycine chain (represented by the five black circles) which interconnect the terminal carboxyl group of the D-alanine moiety in the tetrapeptide with the ε-amino group of lysine in another tetrapeptide. Lysine is thus the focus for the formation of a branched polypeptide which is the polymer going in the second direction of the network. Little or nothing is known about the three-dimensional structure, because present physical methods are inadequate to investigate the structure of highly ordered, noncrystalline networks of this kind. One can imagine that the third dimension of the structure might be built up by interpeptide bridges which extend not in the plane of the figure, but forward or backward to other sheets of peptidoglycan strands.

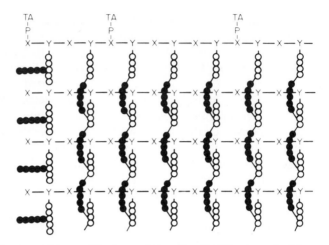

Fig. 1. Structure of the peptidoglycan of the cell wall of *S. aureus*. In this representation, X (acetylglucosamine) and Y (acetylmuramic acid) are the two sugars in the peptidoglycan. Open circles represent the four amino acids of the tetrapeptide, L-alanyl-D-isoglutaminyl-L-lysyl-D-alanine. Closed circles are pentaglycine bridges which interconnect peptidoglycan strands. The nascent peptidoglycan units bearing open pentaglycine chains are shown at the left of each strand. TA-P is the teichoic acid antigen of the organism which is attached to the polysaccharide through a phosphodiester linkage. From J. L. Strominger, *Harvey Lect.* **64,** 179 (1970).

The Interpeptide Cross Bridges of Bacterial Cell Walls

Analyses of the peptidoglycan structures from a number of bacterial species have allowed the recognitition of four classes of peptidoglycans, differing in the nature of the interpeptide bridge [reviewed in Ghuysen (21)]. All evidence suggests that with minor modifications the glycan component of the peptidoglycan is the same in all of these structures. In the first class there are no intervening amino acids, and the terminal carboxyl group of the tetrapeptide, a D-alanine residue, cross-links directly to the dibasic amino acid, *meso*-diaminopimelic acid, in another tetrapeptide. *Escherichia coli* is a good example of this class. In the second class are a large number of organisms in which a small peptide, varying in size from one to five amino acids, interconnects the terminal carboxyl group of one tetrapeptide with the dibasic amino acid of another. *Staphylococcus aureus* is a good example of this type.

In the third class, represented by *M. lysodeikticus,* a very unusual situation occurs. Some of the acetylmuramic acid residues are naked, i.e., they bear no tetrapeptide substituent. The tetrapeptide has been "moved" from one acetylmuramic acid residue into the bridge position, so that the bridge is composed of the same tetrapeptide as is ordinarily linked to muramic acid, leaving many acetylmuramic acid residues without any peptide substituent. It is for this reason that disaccharides free of peptide could be isolated by early workers from the walls of *M. lysodeikticus.*

The fourth type of cross bridge is represented by an organism, e.g., *Corynebacterium poinsettiae,* which has no dibasic amino acid in the fourth position of the tetrapeptide. A dibasic amino acid in the cross bridge links the terminal carboxyl group of the tetrapeptide to the carboxyl of the second amino acid, D-glutamic acid, in another tetrapeptide. In this type of organism the bridge is not between D-alanine and L-lysine, but is between D-alanine and D-glutamic acid.

So far little has been said about the two- and three-dimensional organization of the peptidoglycan. In fact very little information is available on this subject. The thickness of the peptidoglycan in *E. coli* is sufficient to accommodate 1–3 sheets of cross-linked glycan and peptide strands, whereas that in *S. aureus* is at least 10 times as thick. Each disaccharide peptide unit of the peptidoglycan can be linked to a maximum of four others through the glycan or through peptide cross-links and the structure obviously has potential for three-dimensional extension. Several models for the structure of this polymer have been proposed, most of which are two-dimensional in character. Little data are available to compare with these models since present physical methods are inadequate for the structural analysis of noncrystalline polymers such as peptidoglycan. A two-dimensional representation of the peptidoglycan of *S. aureus* (Fig. 2) can be compared with a similar representation of the peptidoglycan of *E. coli* (Fig. 3). The most obvious difference between these structures is the much greater extent of cross-linking in *S. aureus,* which involves 75% of the peptide subunits as compared with 33% of the units in *E. coli.* This means that the peptidoglycan structure of *S. aureus* is a relatively tightly woven network while that in *E. coli* is much more open and presumably considerably more flexible and hydrated. Similarly, the peptidoglycan of *M. lysodeikticus,* with its stretches of peptide-free glycan and its very long peptide cross bridges, is even more open than that of *E. coli.* It seems very probably that the susceptibility of these three peptidoglycan structures to the lytic action of lysozyme is related to the openness of their structures.

Use of Bacteriolytic Enzymes in the Analysis of Cell Wall Structure

How was the structure of bacterial cell walls worked out? It became obvious very early that chemical methods of degradation of a complex polymer of this kind would yield such a large variety of fragments that it would be impossible to work out the structures of all of them, to separate

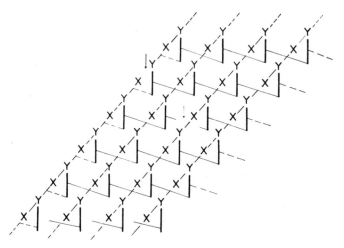

Fig. 2. Schematic representation of the structures of the peptidoglycans of *S. aureus* and *M. roseus*. These are tight networks. Note the large size of the polypeptide network. X, acetylglucosamine, and Y, acetylmuramic acid, are the two sugars in the polysaccharide. Short lines are chemical bonds; long light lines represent the cross-bridges (pentaglycine in *S. aureus* or tri-L-alanyl-L-threonine in *M. roseus*); heavy lines represent the tetrapeptide linked to acetylmuramic acid. From J. L. Strominger and J.-M. Ghuysen, *Science* **156**, 213 (1967), copyright 1967 by the American Association for the Advancement of Science.

them, and to finally put back the jigsaw puzzle; and therefore enzymes—i.e., bacteriolytic enzymes—with high specificities for cleavage of particular bonds were employed in these studies. At that time all of these enzymes were called lysozymes, and it was believed that, like egg-white lysozyme, they all cleaved the same type of linkage in the cell wall. It very quickly became obvious that that was not the case.

These enzymes come from a variety of animal sources and from bacterial extracellular enzymes (lysins), enzymes which are built into the cell wall (autolysins), and enzymes which are induced by bacteriophages (virolysins). Methods for determining what type of linkage was cleaved by different bacteriolytic enzymes were developed (22). Of great importance in these studies were the *Streptomyces* enzymes discovered in Liège and studied successively by Gratia and Dath (23), Welsch (24), and Ghuysen (25). Gratia's discovery was contemporary to the discovery of lysozyme by Fleming. These enzymes were brought to St. Louis and studied there by Ghuysen, Tipper, and Strominger (10, 15, 22) and reviewed in Strominger and Ghuysen (26).

It became clear quickly that there were two kinds of carbohydrases, the acetylmuramidases (or lysozymes), and another type of carbohydrase, *endo*-acetylglucosamidases which split the alternate link in the bacterial cell wall—not the one between acetylmuramic and acetylglucosamine but the

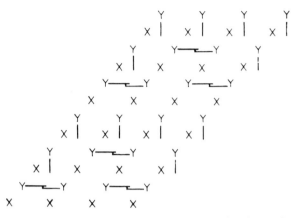

Fig. 3. Schematic representation of the structure of the peptidoglycan of *E. coli*. This is a loose network in which all of the acetylmuramic residues are substituted either by tetrapeptide monomers or by peptide-linked dimers. For symbols, see legend to Fig. 2. From J. L. Strominger and J.-M. Ghuysen, *Science* **156**, 213 (1967), copyright 1967 by the American Association for the Advancement of Science.

one between acetylglucosamine and acetylmuramic acid.

There were also a variety of peptidases. They could be divided into two major groups, interpeptide bridge splitting enzymes and acetylmuramyl-L-alanine amidases, which cleave the linkage between acetylmuramic acid and L-alanine, the first amino acid in the tetrapeptide. The specificity of some of these enzymes and their sources is given in Table I (26). All of these enzymes had originally been called lysozymes, but it is very clear now that the term "lysozyme" should be reserved for the first group of these enzymes, those which are acetylmuramidases.

In summary, a representation of a small segment of the walls of *S. aureus* and *E. coli* is shown in Figs. 4 and 5 with an indication of which bond is

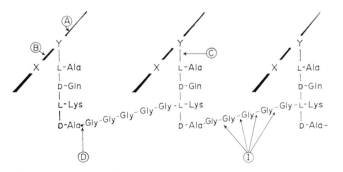

Fig. 4. A fragment of the peptidoglycan of *S. aureus* showing points of cleavage by various enzymes (see Table I). From J. L. Strominger and J.-M. Ghuysen, *Science* **156**, 213 (1967), copyright 1967 by the American Association for the Advancement of Science.

TABLE I

Enzymes Degrading Peptidoglycans of Bacterial Cell Walls[a]

Enzymes	Type of linkage split
Carbohydrases[b]	
Endoacetylmuramidases	N-acetylmuramyl-N-glucosamine (A)[c]
Plant and animal lysozymes (such as egg white)	
32 enzyme from *Streptomyces albus* G	
F_1 enzyme from *Streptomyces albus* G	
B enzyme from *Chalaropsis*	
T2 phage lysozyme	
Autolysin from *Streptococcus faecalis*	
Endoacetylglucosaminidases	N-acetylglucosaminyl-N-acetylmuramic acid (B)
Streptococcal muralysin	
Glycosidase in lysostaphin	
Autolysin from *Staphylococcus aureus*	
Exoacetylglucosaminidases	
Enzyme from pig epididymis	
Enzyme from *Escherichia coli* autolytic complex	
Acetylmuramyl-L-alanine amidases[d]	
Enzyme from *Streptomyces albus* G	N-acetylmuramyl-L-alanine (C)
Autolysin from *Bacillus subtilis*	
Enzyme from *Escherichia coli* autolytic complex	
Autolysin from *Listeria monocytogenes*	
Autolysin from *Staphylococcus aureus*	
L_{11}, L_3 Lysostaphin and enzymes from *Myxobacterium* and *Sorangium*[e]	

Endopeptidases (bridge-splitting enzymes)

SA endopeptidase from *Streptomyces albus* G — D-alanylglycine (D); and D-alanyl-L-alanine (E)

ML endopeptidase from *Streptomyces albus* G — D-alanyl-N^γ-L-lysine (F)

MR endopeptidase from *Streptomyces albus* G — L-alanyl-L-threonine (G) or L-alanyl-L-alanine

Endopeptidase from *Escherichia coli* autolytic complex — D-alanyl-*meso*-DAP (H)

L_3 enzyme from *Streptomyces* — D-alanyl-*meso*-DAP (H)

Lysostaphin — Glycylglycine (I)

Nonlytic endopeptidase from *Streptomyces albus* G — Glycylglycine (I)

L_{11} enzyme from *Flavobacterium* — Glycylglycine (I) and D-alanylglycine (D)

Enzyme from *Myxobacterium* — Glycylglycine (I) and D-alanylglycine (D)

Enzyme from *Sorangium* — D-alanyl-N^γ-L-lysine (F)

Exopeptidases

L-alanine and glycine aminopeptidase and D-alanine and glycine
carboxypeptidases from *Streptomyces albus* G

D-alanine carboxypeptidase from *Escherichia coli* autolytic complex

[a] The endoenzymes listed are bacteriolytic with the exception of the glycosidase in lysostaphin, the acetylmuramyl-L-alanine amidase from *Streptomyces albus* G, and the nonlytic endopeptidase from *Streptomyces albus* G, which, when acting on walls of *Staphylococcus aureus*, require peptidoglycan solubilized by prior treatment of the walls with lytic enzymes. The endopeptidase and the acetylmuramyl-L-alanine amidase from *Escherichia coli* cells have not been obtained free of endoacetylmuramidase, and therefore it is not known whether they are lytic by themselves. The exoenzymes are also not bacteriolytic, and they remove terminal groups that are exposed as the consequence of action of endoenzymes.

[b] In addition to the enzymes indicated, many other enzymes which hydrolyze the carbohydrate polymer in bacterial cell walls have been found. However, these enzymes were discovered before methods were available for defining which of the glycosidic linkages was hydrolyzed.

[c] Letters in parentheses refer to bonds hydrolyzed as identified in Figs. 4 and 5. The methods employed in our laboratories in elucidating the specificities of bacteriolytic enzymes have been described (22).

[d] A number of other enzymes have been identified which may lyse cell walls through hydrolysis of peptide linkages. Available data suggest that they are bridge-splitting enzymes and acetylmuramyl-L-alanine amidases, although the nature of the split linkage has not been completely defined.

[e] Many of the bridge-splitting enzymes also hydrolyze the acetylmuramyl-L-alanine linkage. This hydrolytic activity is slower and is not complete until long after solubilization of cell walls. Therefore, the bridge-splitting activity is probably responsible for bacteriolysis and cell wall solubilization.

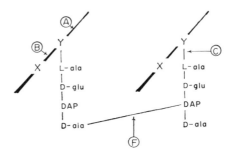

Fig. 5. A fragment of the peptidoglycan of *E. coli* showing points of cleavage by various enzymes (see Table I). From J. L. Strominger and J.-M. Ghuysen, *Science* **156**, 213 (1967), copyright 1967 by the American Association for the Advancemen, of Science.

split by which kind of enzyme. Lysozymes, or acetylmuramidases, split bond A between acetylmuramic acid and acetylglucosamine. The *endo*-acetylglucosaminidases split bond B, the alternate bond between acetyl-glucosamine and acetylmuramic acid. The amidases split bond C, between acetylmuramic acid and the first amino acid. And the bridge-splitting enzymes split either at the beginning of the bridge or within the interpeptide bridge (bonds D, H, and I). It is of some interest that enzymes which split within the tetrapeptide itself, a peptide containing alternating L- and D-amino acids, are comparatively rare. Only two examples are known, a phage lysin from *B. stearothermophilus* which splits between L-alanine and D-glutamic acid residues (27) and an enzyme from *B. sphaericus* which splits between D-glutamic acid and *meso*-diaminopimelic acid (28). Paren-thetically another phage lysin, the phage lambda "lysozyme," has re-cently been shown not to be a lysozyme at all. It is a peptidase which splits within the interpeptide bridges of *E. coli* (29).

Isolation of the Disaccharide, Acetylmuramyl–Acetylglucosamine, and the Elucidation of the Structure of the Lysozyme Substrate

Two degradation sequences will illustrate how small fragments whose structures could be elucidated were obtained from the cell wall of *S. aureus*. The two degradation sequences which have been selected end up in com-pounds which are of special interest in terms of the reaction catalyzed by egg-white lysozyme.

The first degradation sequence made use of the enzyme mixture, lyso-staphin, a lysin excreted by a strain of *Staphylococcus* (30). This mixture has in it two enzymes, one of which is an *endo*-acetylglucosaminidase, and catalyzes cleavage of the linkage between acetylglucosamine and acetyl-muramic acid. The second is a peptidase or peptidases which catalyzes cleavages at both the acetylmuramyl-L-alanine linkage and within the interpeptide bridges (31).

The consequence of this lysis of the cell wall of *S. aureus* with lysostaphin is that the disaccharide, acetylmuramyl–acetylglucosamine (in which the linkage which is split by egg-white lysozyme remains intact) is released and the peptide falls apart into small peptide fragments. It was then relatively easy, using various chromatographic techniques, to obtain the disaccharide and the peptide subunits in high yield and in relatively pure form (32, 33).

With this disaccharide in hand the nature of the linkage actually split by lysozyme could be characterized. It turned out to be a $\beta(1, 4)$ linkage as in chitobiose, as had been guessed earlier—but this was the first time that this had actually been established. The disaccharide itself was resistant to periodate oxidation but, after reduction with $NaBH_4$, it consumed 1 mole of periodate with production of 1 mole of formaldehyde (Fig. 6). These data are compatible only with a 1,4 linkage between the two sugars.

The nmr spectrum of the unreduced disaccharide in D_2O (33) is consistent with the disaccharide structure drawn above it (Fig. 7). If it is assumed to contain 6 *N*-acetyl protons in the 2 barely separated peaks of equal size at $\tau 8.01$ and 8.03, it contains 3 protons at $\tau 8.75$ which correspond to the lactic acid methyl groups, and 12 ring protons in the τ range 6–6.7. Then only the 2 anomeric protons and the proton on the lactyl group need to be specified. α-Anomeric protons are known to absorb in the region of $\tau 4.9$ with a low J of about 3 CPS while β-anomeric protons absorb near $\tau 5.6$ with a higher J of about 7 CPS. Free *N*-acetylglucosamine in D_2O solution at equilibrium exists mostly in the α-configuration and this would also probably be true for the disaccharide, so that the one proton doublet of $\tau 4.88$, J 2.5 was presumably due to the α-proton of the reducing GlcNAc moiety, of the disaccharide. The molecular environment of the β-proton of the lactyl moiety would be very similar to that of a β-anomeric proton at carbon atom 1 and these 2 protons presumably correspond to the 2 doublets in the region of $\tau 5.5$–5.8. Irradiation at the frequency of the

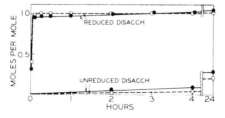

Fig. 6. Periodate oxidation of reduced and unreduced disaccharides. (—) MurNAc-GlcNAc and (- - -) Mur-*N,O*-diAcOGlcNac. From D. J. Tipper, M. Tomoeda, and J. L. Strominger, *Biochemistry* **10**, 4683 (1971).

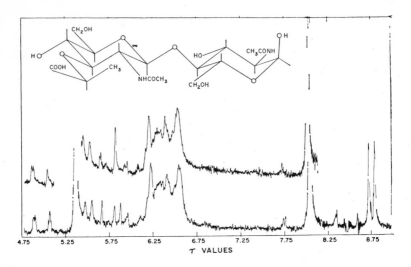

Fig. 7. nmr spectrum of MurNAc-GlcNAc. The structure of the disaccharide with $\beta(1,4)$ linkage and both sugars in the C-1 conformation is shown at the top. The upper tracing is for the sample irradiated at $\tau 8.75$ while the lower tracing is the unirradiated sample. The peak at $\tau 5.36$ is due to HDO, with side bands of $\tau 5.05$ and 5.66. From D. J. Tipper, M. Tomoeda, and J. L. Strominger, *Biochemistry* **19**, 4683 (1971).

lactyl methyl group decoupled the doublet at $\tau 5.8$ demonstrating that this doublet corresponded to the lactyl proton. Conversely, irradiation at the frequency of $\tau 5.8$ decoupled the lactyl methyl doublet, identifying these as the lactyl protons.

The nmr spectra of the reduced disaccharide (Fig. 8) revealed that the doublet in the α-anomeric region had disappeared, that the number of protons in the $\tau 6$–6.7 region had increased from 12 to 14 as expected, and that the doublet at $\tau 5.47$, J 7.5 persisted. Again irradiation at the average frequency of the doublet at 5.83 decoupled the 3 proton doublet at $\tau 7.5$ and vice versa, identifying these as the lactyl protons. The single proton peak at $\tau 5.7$ must thus correspond to the anomeric proton of the glycosidic linkage in the β-configuration, and its quantitation shows that at least 90% of these linkages had the β-configuration. Therefore, it has been conclusively established that the bond cleaved by the enzyme, lysozyme, is a $\beta(1,4)$ linkage between acetylmuramic acid and acetylglucosamine.

Isolation of the Intact Cell Wall Polysaccharide and its Susceptibility to Various Lysozymes

A second degradation sequence which is of interest made use of an enzyme which is a pure peptidase and contained no carbohydrase activity

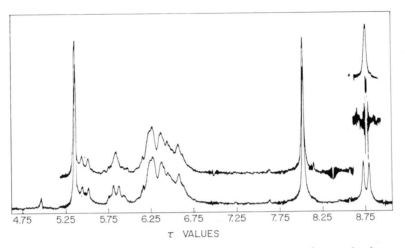

Fig. 8. nmr spectrum of reduced MurNAc-GlcNAc. The upper tracings are for the sample irradiated at $\tau 8.75$ and 5.83, and the lower tracing is for the unirradiated sample. The HDO peak and its side bands are $\tau 5.36$, 4.96, and 5.78. From D. J. Tipper, M. Tomoeda, and J. L. Strominger, *Biochemistry* **10**, 4683 (1971).

whatever—the *Myxobacterium* enzyme (34, 35). It splits all of the linkages between acetylmuramic acid and L-alanine and carries out several splits in the interpeptide bridge. Since it has no carbohydrase activity, its use permitted the isolation of the intact polysaccharide, free of peptide (35). The yields obtained were exceedingly high, and the polysaccharide was obtained in 90% yield. The polysaccharide was polydisperse with chain lengths varying from 12 to 90 hexosamine units, with a number average of about 20 hexosamines.

An unexpected and interesting feature of this polysaccharide product is its resistance to hydrolysis by egg-white lysozyme and by other *endo-N*-acetylmuramidases. The kinetics of this hydrolysis are shown in Fig. 9. The substrate was the unfractioned *S. aureus* polysaccharide, incubated in the appropriate buffers with the enzymes indicated at 37°C. At the extremely high enzyme-to-substrate ratio (by weight) of 0.46, the "32" enzyme from *Streptomyces* was totally incapable of hydrolyzing this substrate, the *Chalaropsis* and F_1 enzymes gave 45 to 60% hydrolysis after several hours and then stopped. Lysozyme, the most effective of these enzymes in hydrolyzing this substrate, gave maximally 80% hydrolysis at 4 hr. By contrast, the hydrolysis by an *endo*-acetylglucosaminidase ("π" enzyme) was total and was complete within the same time at an enzyme-to-substrate ratio of 0.006. This enzyme was an impure fraction of lysostaphin enriched for the *endo-N*-acetylglucosaminidase activity. The relative resistance of the peptide-free polysaccharide to the action of lysozyme demonstrates that this enzyme has specificity for the amino acid substituents on the carboxyl

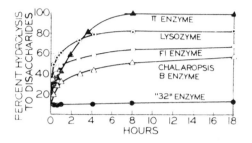

Fig. 9. Kinetics of hydrolysis of *S. aureus* polysaccharide by *endo*-*N*-acetylhexosaminidases. SP 2 (41 μg) was incubated at 37°C with lysozyme (19 μg) in 38.4 μl of ammonium acetate buffer, pH 6.8, and with Fl enzyme (19 μg), *Chalaropsis* B enzyme (19 μg), or 32 enzyme (15 μg) at pH 4.5 in 38.4 μl of 0.01 *M* sodium acetate buffer. SP 2 (71.3 μg) was incubated with 0.46 μg of π enzyme in 61.5 μl of 0.01 *M* phosphate, pH 7.5. From D. J. Tipper, J. L. Strominger, and J. C. Ensign, *Biochemistry* **6,** 906 (1967).

group of the acetylmuramic acid. The simplest interpretation of these data may be that the existence of the unsubstituted charged COOH group on the lactic acid moieties of the peptide-free glycan may impede fit into the cleft of the enzyme (perhaps electrostatically) and so impede the rate of hydrolysis. However, it is also possible that the intact peptidoglycan of lysozyme-sensitive cell walls may present the glycan moiety of its structure to the enzyme in a conformation which is better suited to fit into the enzyme cleft. In any case this points out a feature of lysozyme catalysis which has not so far been taken into account in the mechanisms proposed, viz., there is some specificity involved in the peptide substituent on the polysaccharide. The polysaccharide itself is a very bad substrate for egg-white and other lysozymes.

Basis of Susceptibility or Resistance to Lysozyme

Finally, a few concluding remarks about the basis of susceptibility or resistance to egg-white lysozyme are in order. In addition to the peptide substituent, it is obvious that a microorganism will be resistant to this enzyme if the linkage between acetylmuramic acid and acetylglucosamine is anything other than a $\beta(1,4)$ linkage. No other type of linkage has so far been demonstrated in peptidoglycans but the question has also not been very extensively investigated, and it would not be a great surprise to discover that some microorganism contained, for example, an $\alpha(1,4)$ linkage as the basis for its resistance to lysozyme. 6-*O*-Acetyl groups on acetylmuramic acid also provide a basis of lysozyme resistance (10, 13, 36, 37). The absence of *N*-acetyl groups on muramic acid, such as occurs in *B. cereus*

(38), and the absence of *N*-acetyl, and the presence of muramic lactam, as occurs in spore walls (39), are other bases of lysozyme resistance. *N*-Glycolyl residues, instead of *N*-acetyl residues on muramic acid (as occurs in *Mycobacteria*) (40), may also alter the susceptibility to lysozyme.

In addition to those detailed features, there are a number of gross features of cell wall structure which are involved in lysozyme resistance. For example, the approach to the network which is the substrate may be hindered by other polymers. Teichoic acid is a highly negatively charged polymer present in the cell walls of *S. aureus* and many other bacteria. It certainly can bind a highly basic protein, such as egg-white lysozyme and probably provides some basis for resistance. The lipopolysaccharides in the gram-negative bacteria are another example of substances which may prevent approach.

Now, if the enzyme can approach, then the tightness of the net may affect the susceptibility to lysozyme. Is there any difference in susceptibility of a cell wall which is very highly cross-linked and that of a cell wall which is very loosely cross-linked, and which enzyme molecules can presumably get into much more readily?

Finally, there is one problem which has been difficult to focus on. In the studies of these bacteriolytic enzymes, it became obvious that some of them are not truly bacteriolytic, i.e., that they cleave linkages in the cell wall, but only after the wall has been solubilized through the action of another enzyme. Others of these enzymes, like egg-white lysozyme, are able to act on highly water-insoluble substrates. How does that happen? How does an enzyme approach a substrate which is not in solution and catalyze the hydrolysis of linkages within it? Some of these enzymes are obviously unable to do that, and can catalyze the hydrolysis of linkages in the wall only after the whole polymer has been solubilized through the action of another enzyme. Microorganisms are primarily responsible for the biodegradation of carbohydrate polymers within the biosphere, and this depends upon the production by these microorganisms of enzymes capable of hydrolyzing substrates in essentially biphasic environments. It might be worth contemplating what characteristics are required by such enzymes in the performance of this function.

It seems obvious that some of these problems will provide the basis for still further work on Fleming's lysozyme.

References

1. Meyer, K., Palmer, J. W., Thompson, R., and Khorazo, D., *J. Biol. Chem.* **113**, 479 (1936).
2. Epstein, L. A., and Chain, E., *Br. J. Exp. Pathol.* **21**, 339 (1940).
3. Salton, M. R. J., "The Bacterial Cell Wall." Elsevier, Amsterdam, 1964.
4. Salton, M. R. J., *Biochim. Biophys. Acta* **8**, 510 (1952).

5. Weibull, C., *J. Bacteriol.* **66,** 688 (1953).
6. Berger, L. R., and Weiser, R. S., *Biochim. Biophys. Acta* **26,** 517 (1957).
7. Meyer, K., *in* "Polysaccharides in Biology" (G. Springer, ed.), p. 78. Josiah Macy, Jr. Found., New York, 1959.
8. Salton, M. R. J., and Ghuysen, J.-M., *Biochim. Biophys. Acta* **36,** 552 (1959).
9. Perkins, H. R., *Biochem. J.* **73,** 33 (1959).
10. Ghuysen, J. M., and Strominger, J. L., *Biochemistry* **2,** 110 (1963).
11. Ghuysen, J. M., Leyh-Bouille, M., and Dierickx, L. *Biochim. Biophys. Acta* **63,** 286 (1962).
12. Hash, J., *Arch. Biochem. Biophys.* **102,** 379 (1963).
13. Tipper, D. J., Strominger, J. L., and Ghuysen, J.-M., *Science* **146,** 781 (1964).
14. Jeanloz, R. W., Sharon, N., and Flowers, H. M., *Biochem. Biophys. Res. Commun.* **13,** 20 (1963).
15. Tipper, D. J., Ghuysen, J. -M., and Strominger, J. L., *Biochemistry* **4,** 468 (1965).
16. Weidel, W., and Pelzer, H., *Adv. Enzymol.* **26,** 193 (1964).
17. Mandelstam, M. H., and Strominger, J. L., *Biochem. Biophys. Res. Commun.* **5,** 466 (1961).
18. Strominger, J. L., *Harvey Lect.* **64,** 179 (1970).
19. Ghuysen, J.-M., Tipper, D. J., Birge, C. H., and Strominger, J. L., *Biochemistry* **4,** 2245 (1965).
20. Jarvis, D., and Strominger, J. L., *Biochemistry* **6,** 2591 (1967).
21. Ghuysen, J.-M., *Bacteriol. Rev.* **32,** 425 (1968).
22. Ghuysen, J.-M., Tipper, D. J., and Strominger, J. L., *in* "Methods in Enzymology" (E. F. Neufeld, and V. Ginsburg, eds.), Vol. 8, p. 685. Academic Press, New York, 1966.
23. Gratia, A., and Dath, S., *C. R. Soc. Biol.* (*Paris*) **91,** 1442 (1924).
24. Welsch, M., *Rev. Belge Pathol. Med. Exp.* **28,** Suppl. 2, 1 (1947).
25. Ghuysen, J. L., *Biochim. Biophys. Acta* **40,** (1960).
26. Strominger, J. L., and Ghuysen, J.-M., *Science* **156,** 213 (1967).
27. Welker, N. E., *J. Bacteriol.* **107,** 697 (1971).
28. Guinand, M., and Tipper, D. J., unpublished observations.
29. Taylor, A., *Nature* (*Lond.*), *New Biol.* **234,** 144 (1971).
30. Schindler, C. A., and Schuhardt, V. T., *Proc. Natl. Acad. Sci. U.S.A.* **51,** 414 (1964).
31. Browder, H. P., Zygmunt, W. A., Young, J. R., and Tavormina, P. A., *Biochem. Biophys. Res. Commun.* **19,** 383 (1965).
32. Tipper, D. J., and Strominger, J. L., *Biochem. Biophys. Res. Commun.* **22,** 48 (1966).
33. Tipper, D. J., Tomoeda, M., and Strominger, J. L., *Biochemistry* **10,** 4683 (1971).
34. Ensign, J. C., and Wolfe, R. S., *J. Bacteriol.* **90,** 395 (1965).
35. Tipper, D. J., Strominger, J. L., and Ensign, J., *Biochemistry* **6,** 906 (1967).
36. Brumfitt, W., Wardlaw, A. C., and Park, J. T., *Nature* (*Lond.*) **181,** 1783 (1958).
37. Brumfitt, W., *Br. J. Exp. Pathol.* **40,** 441 (1959).
38. Araki, Y., Fukuoka, S., Oba, S., and Ito, E., *Biochem. Biophys. Res. Commun.* **45,** 751 (1971).
39. Warth, A. D., and Strominger, J. L., *Proc. Natl. Acad. Sci. U.S.A.* **64,** 528 (1969).
40. Adam, A., Petit, J. F., Wietzerbin-Falszpan, J., Sinay, P., Thomas, D. W., and Lederer, E., *FEBS Lett.* **4,** 87 (1969).

Substrate Requirements of Glycosidases for Lytic Activity on Bacterial Walls

JEAN-MARIE GHUYSEN

Following Fleming's discovery in 1922 (1) that hen egg-white lysozyme caused the dissolution of living cells of *Micrococcus lysodeikticus* and the investigations of Meyer *et al.* (2), Epstein and Chain (3), and Meyer and Hahnel (4) on the degradation of a "soluble mucopolysaccharide" by lysozyme, modern studies on the mode of action of this enzyme and other bacteriolytic and antibiotic agents started in the 1950's when Salton succeeded in isolating the cell walls from mechanically disrupted bacteria and showed that the isolated wall of *M. lysodeikticus* could be used as "the substrate" for lysozyme (5). The isolation of the protoplasts of *Bacillus megaterium* by Weibull (6) beautifully confirmed the idea that the action of lysozyme was solely restricted to the wall structure of the cell, and the formation of spheroplasts of *Escherichia coli* under the action of either lysozyme (7) or penicillin (8) pointed to the identity of the cell target (i.e., the wall) impaired by these two antimicrobial agents (for complete references, see 9).

Structure of the Walls of *Micrococcus lysodeikticus*

It took about 15 years of work in several laboratories to establish the exact primary structure of the polymer which forms the solid matrix of the wall of *M. lysodeikticus* and is susceptible to solubilization by lysozyme. This polymer is a peptidoglycan network in which linear glycan strands are interconnected by peptide chains.* The glycan strands essentially consist of alternating $\beta(1,4)$ linked pyranoside *N*-acetylglucosamine and *N*-acetylmuramic acid residues, i.e., a chitin-like structure in which each alternate *N*-acetylglucosamine residue is ether-linked at C-3 to a lactyl group which has the D configuration. Small amounts of mannomuramic acid were recently reported to occur along with glucomuramic acid (11). Less than 50% of the *N*-acetylmuramic acid residues have their D-lactyl groups substituted by pentapeptide units having the sequence N^{α}-/[L-alanyl-γ-(α-D-glutamyl-glycine)]-L-lysyl-D-alanine. Most of these pentapeptide units are in turn interconnected through "bridges" which extend from the C-terminal D-alanine residue of one pentapeptide unit to the ϵ-amino group of L-lysine of another peptide unit. A few of these interpeptide bridges are extremely short and consist of direct N^{ϵ}-(D-alanyl)-L-lysine amide bonds between two pentapeptide units. Most of the interpeptide bridges, however, are much longer and composed of one or several peptides, each having the same sequence as the pentapeptide units which substitute the *N*-acetylmuramic acid residues. In this type of peptidoglycan both D-alanyl-L-alanine peptide linkages and N^{ϵ}-(D-alanyl)-L-lysine amide bonds are involved in peptide cross-linking and the hydrolysis of either of these linkages causes the dissolution of the wall and the lysis of the cell. The D-alanyl-L-alanine linkages are sensitive to the *Myxobacter* Al$_1$ enzyme whereas the N^{ϵ}-(D-alanyl)-L-lysine linkages are sensitive to the *Streptomyces* ML enzyme.

The total number of pentapeptide units of the wall peptidoglycan is roughly equivalent to the number of disaccharide units of the glycan strands. It is known, however, that UDP-*N*-acetylglucosamine and UDP-*N*-acetyl-muramyl-L-alanyl-γ-D-glutamyl-L-lysyl-D-alanyl-D-alanine are the cytoplasmic nucleotide precursors and that the glycine residue is added to the α-carboxyl group of D-glutamic acid at a later stage of the biosynthesis (for references, see 12). Evidently, the observed one-to-one molar ratio between the peptidoglycan constituents and the accommodation of unsubstituted glycan fragments must result from the migration at a certain stage of the biosynthesis, of some pentapeptide units from the *N*-acetylmuramic acid residues into a bridging position with the formation of oligomeric units. The mechanism of these migrations and polymerizations is still entirely unknown (10).

The observed yields of peptidoglycan fragments obtained by various degradation procedures established that both the peptide and glycan moieties of the peptidoglycan are polydisperse systems. These yields are consistent with a structure in which 5% of the pentapeptide units occur as monomers,

---

*For details and illustrations, see Ref. 10.

4% as dimers, 15% as trimers, and 60% as hexamers. The hexamer, of course, represents a statistical average among oligomers of various sizes. Similarly, the glycan strands appear to be composed, on the average, of about 15–20 disaccharide units. It is known that at most 50% of the *N*-acetyl-muramic acid residues are peptide-substituted but the exact distribution of the peptide-free disaccharide units along the glycan chains is still a matter of speculation. It seems likely, however, that large unsubstituted glycan segments up to the size of octasaccharides may occur.

The tightness of a peptidoglycan network depends upon the frequency with which the glycan chains are substituted by peptide units and upon the frequency with which these peptide units are, in turn, interlinked. The low order of cross-linking both at the junction between the glycan and the peptide constituents and in the peptide moiety itself makes the *M. lyso-deikticus* peptidoglycan a rather loose network and explains why its dissolution can be brought about by the hydrolysis of only a few bonds located either in the glycan or in the peptide part of the peptidoglycan. For example, the exposure of about 120 neq of reducing groups (expressed as *N*-acetyl-glucosamine) through the cleavage of glycosidic bonds by lysozyme, or, alternatively, the hydrolysis of about 150–200 nEq of N^{ε}-(D-alanyl)-L-lysine linkages by the ML endopeptidase are sufficient to cause the dissolution of 1 mg of walls (containing about 600 neq of disaccharide-pentapeptide units). The looseness of the *M. lysodeikticus* peptidoglycan is also reflected by the high rate of degradation observed at low lysozyme concentrations. Complete solubilization of the walls occurs in about 30 min at a ratio wall/enzyme of 100/1 (w/w).

In addition to the peptidoglycan, the walls of *M. lysodeikticus* contain another polymer composed of equimolar amounts of glucose and 2-acet-amido-2-deoxymannuronic acid residues (13). This polymer is devoid of insolubility and rigidity and is anchored into the wall peptidoglycan matrix via phosphodiester bonds extending to the C-6 position of some of the *N*-acetylmuramic acid residues. On the average, one such phosphodiester bond would occur for each glycan strand of 15–20 disaccharides in length (14).

Action of Hen Egg-White Lysozyme on Walls of *M. lysodeikticus*

From the studies of Salton on the products found in the digested walls, it was clear that lysozyme attacked glycosidic bonds in endoposition in the glycan strands (9). Later on, it was established that lysozyme specifically hydrolyzed $\beta(1–4)$ linkages between *N*-acetylmuramic acid and *N*-acetyl-glucosamine so that *N*-acetylmuramic acid was always at the reducing end of the degraded fragments. Among these fragments, peptide-free disaccharide β-1,4-*N*-acetylglucosaminyl-*N*-acetylmuramic acid, tetrasaccharide and higher oligosaccharides were isolated and characterized (10). The choice of *M. lysodeikticus* for these early studies was very fortunate. Indeed the peptidoglycans of all bacteria, with the exception of *M. lysodeikticus* and some related *Micrococcaceae* (10, 14), contain peptide substituents on all of their *N*-acetylmuramic acid residues so that the action of a glycosidase alone cannot bring about the release of unsubstituted di- and oligosaccharides. Kinetics of the lysis of *M. lysodeikticus* cell walls and the release of glycan fragments by lysozyme are a complex phenomenon. This complexity is due to a number of factors which are discussed elsewhere in this volume. Lysozyme performs a transglycosidase activity and its lytic activity is product-inhibited, thus precluding complete hydrolysis of all sensitive bonds. At completion of the reaction of lysozyme on walls of *M. lysodeikticus,* only 30–40% of the peptidoglycan *N*-acetylhexosamine residues are found in the form of disaccharides (either free or peptide substituted), i.e., the end product of the degradation. However, through the combined use of lysozyme and a mixture of *Streptomyces exo-N*-acetylhexosaminidases (which further degrade those di- and oligosaccharide fragments liberated by lysozyme treatment), about 70% of the glycan chains were found to have undergone complete conversion into monomeric *N*-acetylhexosamine residues (15).

Structure of Bacterial Walls Other than Those of *M. lysodeikticus*

Although there is a remarkable consistency of the structure of the wall peptidoglycans throughout the bacterial world, variations occur which have been used to divide the bacterial species into different chemotypes (10). The tetrapeptide units have the general sequence L-alanyl-γ-D-glutamyl-L-R_3-D-alanine where the L-R_3 residue may be a neutral amino acid (L-alanine or L-homoserine), or a dicarboxylic amino acid (L-glutamic acid) or a diamino acid (L-ornithine, L-lysine, LL- or *meso*-diaminopimelic acid). The α-carboxyl group of D-glutamic acid can be either free, amidated, or substituted by a glycine amide or a C-terminal glycine residue (as found in *M. lysodeikticus*). Similarly, the carboxyl group of diaminopimelic acid not engaged in peptide bond may be substituted by an amide. In almost all bacteria, the interpeptide bridges extend from the C-terminal D-alanine of one peptide to the ω-amino group of the diamino acid at the L-R_3 position of another peptide. However, the situation where these interpeptide bridges are made up of one or several peptides each having the same amino acid sequence as the peptide units that substitute the *N*-acetylmuramic acid residues is exceptional and restricted to *M. lysodeikticus* and some related

Micrococcaceae (chemotype III). Usually, the bridges either consist of direct N^ω-(D-alanyl)-L-R_3 peptide bonds (chemotype I) or are mediated via one or several additional amino acids (chemotype II). Finally, the peptide cross linking of a few bacterial species (plant pathogenic *Corynebacteria*, *Butyribacterium rettgeri*) extends from the C-terminal D-alanine residue of one peptide unit to the α-carboxyl group of D-glutamic acid of another peptide unit (chemotype IV). This type of bridging links two carboxyl groups and, hence, necessarily involves a diamino acid residue or a diamino acid-containing peptide.

Variations also occur in the glycan strands. They include (*i*) the *O*-acylation on C-6 of some of the *N*-acetylmuramic acid residues; (*ii*) the replacement of *N*-acetylmuramic acid by *N*-glycolylmuramic acid (in *Nocardia* and *Mycobacterium* sp.) (16); (*iii*) the occurrence of a high proportion of the muramic acid residues in the form of the lactam derivative (in the spore peptidoglycan of *Bacillus*) (17); (*iv*) the presence of small amounts of mannomuramic acid instead of glucomuramic acid (11) (in *M. lysodeikticus*, see above).

Early ideas about the organization of the peptidoglycan were that the glycan strands completely extended around the cell (18). Accurate analyses, however, showed that many terminal groups were present in both the glycan and the peptide moieties of all bacterial peptidoglycans. The average length of the glycan chains vary from 20 to 100 *N*-acetylhexosamine residues. In all bacteria, with the exception of *M. lysodeikticus* (see above) and the spores of *Bacillus*, virtually all of the *N*-acetylmuramine acid residues are peptide substituted. In *E. coli* about 50% of the peptide units occur as uncross-linked monomers, and the other units as peptide dimers. In *S. aureus,* which has one of the most cross-linked peptidoglycans, the average size of the peptide moiety does not exceed 10 cross-linked peptide units (10).

In addition to the peptidoglycan layer, the envelope of gram-negative bacteria contains an outer membrane very similar in appearance in thin sections to the inner plasma membrane. This outer membrane has different permeability properties and contains lipopolysaccharides and lipoproteins. In *E. coli,* the lipoproteins have been shown to be covalently linked to the peptidoglycan (about 1 lipoprotein molecule for every 10th repeating peptide unit) through lysl-arginine dipeptides extending to some *meso*-diaminopimelic acid residues (19). The envelope of gram-positive bacteria lacks the outer membrane. However, an almost endless variety of polysaccharides that are frequently negatively charged, and of polyol-phosphates that are collectively called teichoic acids, are covalently linked to the peptidoglycan. One such covalent link, but probably not the only one between the peptidoglycan and these other wall polymers, is via phosphodiester bridges to the C-6 position of *N*-acetylmuramic acid (10).

Substrate Requirements of Hen Egg-White Lysozyme for Lytic Activity on Bacterial Peptidoglycans

Following the discovery of O-acetyl groups in walls of *Streptococcus faecalis* by Abrams in 1958 (20), the importance of these substituents in relation to sensitivity to lysozyme was demonstrated by Brumfitt *et al.* (21). Resistance to lysozyme exhibited by mutants of *M. lysodeikticus* was shown to be due to the substitution of the glycan strands by O-acetyl groups (in fact on C-6 of N-acetylmuramic acid). Cells and walls could be made resistant to lysozyme by chemical O-acetylation with acetic anhydride. Similarly, resistance to lysozyme of mutants of *B. megaterium* was shown to be related to acetylation of wall hydroxyl groups [for references, see Salton (9) and Ghuysen (10)]. We now know from the X-ray crystallographic studies of Phillips and co-workers that subsite D of hen egg-white lysozyme could not accommodate an acetyl group at the C-6 position.

The frequency with which the glycan strands are substituted by peptides is also of importance for lysozyme action. This idea rests upon the following observations: (*i*) Intact glycan strands from which the peptide substituents were enzymatically removed (through the action of *Myxobacter* Al_1 N-acetylmuramyl-L-alanine amidase) were obtained from several types of lysozyme-sensitive peptidoglycans (10). The rate of hydrolysis of these peptide-free glycan chains by lysozyme was only a few percent of the rate of hydrolysis of the peptidoglycans themselves and much higher enzyme concentrations were needed for the degradation. (*ii*) The peptidoglycan of *E. coli* in which all of the N-acetylmuramic acid residues are peptide substituted could be degraded (after removal of the lipoproteins which are covalently linked to it) almost quantitatively into disaccharide peptide monomers and *bis*-disaccharide peptide dimers by lysozyme (18, 22). Only very small amounts of peptide-substituted tetrasaccharide fragments were found among the degradation products. Such a complete degradation contrasts with the incomplete degradation of the walls of *M. lysodeikticus* (see above) in which case long unsubstituted glycan segments occur in the native peptidoglycan. These observations suggest that the presence of a substituted N-acetylmuramyl carboxyl group facilitates the hydrolysis of the neighboring glycosidic linkage.

A great number of peptidoglycans that are not O-acetylated and in which all the N-acetylmuramic acid residues are peptide substituted are totally or at least highly resistant to hen egg-white lysozyme. Elimination of the O-acetyl groups of the glycan chains and removal of the teichoic acids do not render such peptidoglycans, for examples those of *Staphylococcus aureus* and *Lactobacillus acidophilus,* significantly more sensitive to lysozyme than the native cell walls. There are many examples of this type which all point to the fact that some intrinsic property of the peptidoglycan must be involved in the resistance (or sensitivity) to lysozyme. From a survey of the sensitivity (or resistance) of peptidoglycans of different chemotypes to lysozyme, it appears, however, that the primary structure of the peptide moieties is probably not in itself a feature that is important for lysozyme action. Both the *M. lysodeikticus* and *E. coli* peptidoglycans are very sensitive to lysozyme and yet the peptide moieties of these two peptidoglycans differ in many respects. The α-carboxyl group of D-glutamic acid is free in *E. coli* and substituted by a glycine residue in *M. lysodeikticus*; the L-R_3 residue is *meso*-diaminopimelic acid in the former and L-lysine in the latter; the interpeptide bridges in these two organisms are very different both in length and in amino acid composition. From the foregoing, one can hypothesize that the overall macromolecular structure of the peptidoglycan which is imparted by its primary structure—perhaps the tightness of the network—is one of the main factors involved in lysozyme sensitivity. Several molecular models of peptidoglycans have been constructed which suggest possible conformations (23–25). All of them involve extensive hydrogen bonding within the network. To date, however, there is no physical evidence in support of any of these, or other, molecular models. The three-dimensional organization of the peptidoglycan is still completely unknown.

Lysozyme-Like Enzymes

Hen egg-white lysozyme has been a tool of great value for the isolation and purification of membranes and other cellular organelles from a wide variety of bacteria via the transformation of the bacteria into osmotically fragile protoplasts and spheroplasts. In this process, not all the glycosidic bonds of the peptidoglycans have to be hydrolyzed to sufficiently weaken the tensile strength of the wall and to prevent it from playing its role as a cell supporting structure. By contrast and because of its incomplete action on many types of walls, hen egg-white lysozyme has been of very limited use for the determination of the chemical structure of a great number of bacterial peptidoglycans (10). Indeed, interpreting the structure of a peptidoglycan on the basis of the fragments liberated through the action of a lytic enzyme is an exceedingly difficult task if only a fraction of the bonds, which should be hydrolyzed owing to the intrinsic specificity of the enzyme for the chemical bonds hydrolyzed, are actually split. Therefore, a search for other lytic agents has been undertaken which has resulted in the discovery of other lysozyme-like glycosidases as well as of two other classes of lytic enzymes, *N*-acetylmuramyl-L-alanine amidases and endopeptidases.

The use of these enzymes in the elucidation of the primary structures of the bacterial peptidoglycans has been reviewed (10).

Both the F1 enzyme isolated from *Streptomyces albus* G and the enzyme B from *Chalaropsis* (26) are of great interest. Among the many glycosidases so far studied, these two *endo-N*-acetylmuramidases have the broadest lytic spectrum, lysing cells or solubilizing walls of virtually all gram-positive bacteria and degrading the isolated peptidoglycans from gram-negative bacteria. With these enzymes, the glycan strands are usually quantitatively degraded into peptide-substituted disaccharide units, irrespective of the possible presence of *O*-acetyl groups and the chemotypes of the peptidoglycans. Thus far, the only requirement is that the *N*-acetylmuramic acid residues must be substituted by peptides. Both enzymes hydrolyze the peptidoglycan of *S. aureus* into disaccharide-peptide and *O*-acetyl disaccharide peptide units. However, by acting on *M. lysodeikticus,* they release a mixture of substituted disaccharide units and unsubstituted di-, tetra-, and octasaccharides. They have no or very little action on intact peptide-free glycan chains.

Of equal interest has been the discovery of the streptococcal muralysin and the staphylococcal glycosidase which act as *endo-N*-acetylglucosaminidases, i.e., they hydrolyze the glycosidic linkages between *N*-acetylglucosamine and *N*-acetylmuramic acid so that *N*-acetylglucosamine is at the reducing end of the degradation fragments [for references, see Ghuysen (10) and Wadstrom (27)]. These two enzymes are not readily available and hence their lytic spectra have not been thoroughly investigated. In contrast to hen egg-white lysozyme and other *endo-N*-acetylmuramidases, however, the staphylococcal *endo-N*-acetylglucosaminidase seems to act preferentially on glycan chains that are free of peptide or on unsubstituted segments of them. *Micrococcus lysodeikticus,* which has many unsubstituted *N*-acetylmuramic acid residues, is one of the most susceptible organisms. The enzyme has very little if any action on the *S. aureus* peptidoglycan but completely degrades the intact peptide-free polysaccharides obtained by hydrolysis of the walls with the *Myxobacter* Al$_1$ enzyme.

The foregoing observations emphasize the strict specificity of the lytic glycosidases which act either as *endo-N*-acetylmuramidases (lysozymes; glycoside hydrolases 3.2.1.17) or *endo-N*-acetylglucosaminidases (chitinases; glycoside hydrolases 3.2.1.14) with regard to the bonds they attack in the wall peptidoglycans. However, hen egg-white lysozyme degrades purified chitin (28) and soluble oligosaccharides isolated from chitin (29). Tetra-*N*-acetylchetotetraose yields chitobiose and higher oligosaccharides after incubation with lysozyme (9, 30). The *Streptomyces* F1 "*endo-N*-acetylmuramidase" was also found to be able to degrade the above tetrasaccharide into chitobiose (9). Turnip lysozyme, which is an excellent chitinase, was actually isolated by following its lytic action on *M. lyso-deikticus* (31). Hence, when exposed to appropriate nonpeptidoglycan substrates, some of the lytic glycosidases appear not to have the seemingly strict lysozyme- or chitinase-type of specificity that is suggested by their activity on the isolated wall peptidoglycans.

References

1. Fleming, A., *Proc. R. Soc., Lond.* [*Biol.*] **93,** 306 (1922).
2. Meyer, K., Palmer, J. W., Thompson, R., and Khorazo, D., *J. Biol. Chem.* **113,** 479 (1936).
3. Epstein, L. A., and Chain, E., *Br. J. Exp. Pathol.* **21,** 339 (1940).
4. Meyer, K., and Hahnel, E., *J. Biol. Chem.* **163,** 723 (1946).
5. Salton, M. R. J., *Biochim. Biophys. Acta* **10,** 512 (1953).
6. Weibull, C., *J. Bacteriol.* **66,** 688 (1953).
7. Zinder, N. D., and Arndt, W. F., *Proc. Natl. Acad. Sci. U.S.A.* **42,** 586 (1956).
8. Lederberg, J., *J. Bacteriol.* **73,** 144 (1957).
9. Salton, M. R. J., *in* "The Bacterial Cell Wall." Elsevier, Amsterdam, 1964.
10. Ghuysen, J. M., *Bacteriol. Rev.* **32,** 425 (1968).
11. Hoshino, O., Zehavi, U., Sinay, P., and Jeanloz, R. W., *J. Biol. Chem.* **247,** 381 (1972).
12. Strominger, J. L., *Harvey Lect.* **64,** 179 (1970).
13. Perkins, H. R., *Biochem. J.* **86,** 475 (1962).
14. Campbell, J. N., Leyh-Bouille, M., and Ghuysen, J. M., *Biochemistry* **8,** 193 (1969).
15. Ghuysen, J. M., Bricas, E., Lache, M., and Leyh-Bouille, M., *Biochemistry* **7,** 1450 (1968).
16. Azuma, I., Thomas, D. W., Adam, A., Ghuysen, J. M., Bonaly, R., Petit, J. F., and Lederer, E., *Biochim. Biophys. Acta* **208,** 444 (1970).
17. Warth, A. D., and Strominger, J. L., *Proc. Natl. Acad. Sci. U.S.A.* **64,** 528 (1969).
18. Weidel, W., and Pelzer, H., *Adv. Enzymol.,* **26,** 193 (1964).
19. Braun, V., and Sieglin, V., *Eur. J. Biochem.* **13,** 336 (1970).
20. Abrams, A., *J. Biol. Chem.* **230,** 949 (1958).
21. Brumfitt, W., Wardlaw, A. C., and Park, J. T., *Nature* (*Lond.*) **181,** 1783 (1958).
22. Van Heijenoort, J., Elbaz, L., Dézélée, P., Petit, J. F., Bricas, E., and Ghuysen, J. M., *Biochemistry* **8,** 207 (1969).
23. Tipper, D. J., *J. Syst. Bacteriol.* **20,** 361 (1970).
24. Kelemen, M. L., and Rogers, H. J., *Proc. Natl. Acad. Sci. U.S.A.* **68,** 992.(1971).
25. Higgins, M. L., and Shockman, G. D., *CRC Crit. Rev. Microbiol.* **1,** 29 (1971).
26. Hash, J. H., and Rothlauf, M. V., *J. Biol. Chem.* **242,** 5586 (1967).
27. Wadstrom, T., *Biochem. J.* **120,** 745 (1970).
28. Berger, L. R., and Weiser, R. S., *Biochim. Biophys. Acta* **26,** 517 (1957).
29. Meyer, K., *in* "Polysaccharides in Biology" (G. Springer, ed.), p. 84. Josiah Macy, Jr. Found., New York, 1959.
30. Kravchenko, A., and Maksimov, V. I., *Izv. Akad. Nauk SSSR* [*Khim.*] p. 584 (1964).
31. Bernier, I., Van Leemputten, E., Horisberger, M., Bush, D. A., and Jollès, P., *FEBS Lett.* **14,** 100 (1971).

Affinity Labeling of Lysozyme

NATHAN SHARON and YUVAL ESHDAT

Introduction

Hen egg-white lysozyme (HEWL) is undoubtedly one of the best characterized enzymes. Its three-dimensional structure (3, 4), as well as the structure of the lysozyme saccharide complexes (2, 23) that have been known
since 1965 have provided the foundation for the eventual understanding of
the mechanism of action of the enzyme in terms of the precise arrangement
of its atoms in space (13). There is, however, still insufficient chemical
evidence on the amino acids in the active site of HEWL and on the different
factors contributing to the catalytic power of the enzyme. Further studies of
HEWL are also important because they can serve as an excellent model
for the study of enzymes whose structure has not been elucidated and in
particular of lysozymes derived from sources other than hen egg white.

Following the initial finding that lysozyme can catalyze transglycosylation in addition to hydrolysis (43, 44, 46), our group at the Weizmann
Institute has been interested in the active site of this enzyme. From binding
studies with different saccharide substrates and inhibitors, we obtained
evidence on the interactions of these saccharides with subsites A, B, C, and
D in the active site of HEWL, and have concluded that association at subsite D induces strain in the bound sugar ring (11). These results are in ex-

cellent agreement with the conclusions made by Phillips and his co-workers from their X-ray crystallographic studies.

Detailed investigations of the acceptor specificity in the lysozyme-catalyzed transglycosylation reaction (12, 37) have provided chemical evidence on the nature of subsites E and F. These studies were of special interest, since binding to subsites E and F has not been observed by X-ray crystallography and the existence of these subsites has been inferred from model building. More recently, we have been engaged in an investigation of HEWL and related enzymes, using the affinity labeling technique. The results of this investigation are the subject of the present communication.

The affinity labeling reagents were designed by us on the basis of the knowledge of the structure of the active site of the enzyme. It is well established that subsites A, B, and C in the active site of HEWL bind N-acetyl-D-glucosamine (GlcNAc) residues strongly and reversibly, with C being the strongest and A the weakest of the three; binding is always with the reducing end at subsite C (13). Thus, GlcNAc binds in subsite C; (GlcNAc)$_2$ in B C; and (GlcNAc)$_3$ in A B C, and in this way they act as reversible, competitive inhibitors of the enzyme. In order to convert these oligosaccharides to irreversible inhibitors, it was necessary to link them to a chemically reactive group. An epoxide function seemed to us to be admirably suited for this purpose because of its small size and chemical reactivity under mild conditions. Such a group was successfully used in the past for the affinity labeling of α-chymotrypsin (48) and of β-glucosidase (27), and more recently for the affinity labeling of pepsin (17).

Accordingly, the 2',3'-epoxypropyl β-glycosides of the above oligomers [(GlcNAc)$_n$-Ep] were synthesized (50) (Fig. 1). It could be expected that incubation of such glycosides with HEWL would result in their rapid reversible binding to the active site of the enzyme, leaving the epoxide group intact in the region of subsite D (Fig. 2). The epoxide group, being a strong alkylating agent, would then attack electrophilic centers located on amino

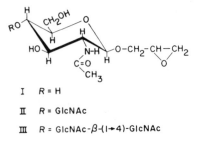

I   R = H

II   R = GlcNAc

III   R = GlcNAc-β-(1→4)-GlcNAc

Fig. 1.   2',3'-Epoxypropyl β-glycosides of GlcNAc oligomers.

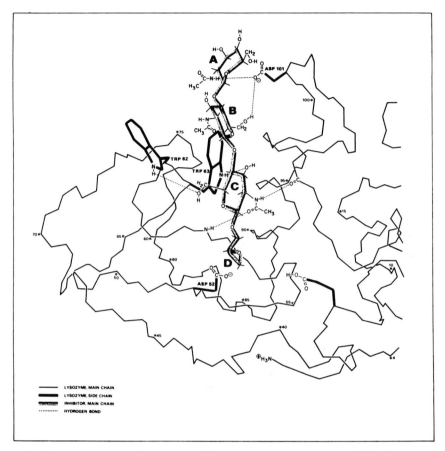

Fig. 2. Presumed reversible binding of 2′,3′-epoxypropyl β-glycoside of (GlcNAc)$_3$ in the active site of hen egg-white lysozyme.

acid side chains in this region, forming a covalent bond between the inhibitor and the enzyme. Two such side chains are the carboxylates of Glu 35 and Asp 52, believed to be involved in the catalytic mechanism of HEWL (2). At pH 5–6, in which Asp 52 (pK ~ 4.5) is mainly ionized while Glu 35 (pK ~ 6.0–6.5) is not (20, 35), it was most likely that the β-carboxyl group of Asp 52 could be reactive toward electrophilic reagents while Glu 35 would remain inert. A mechanism for a specific chemical reaction between Asp 52 and the affinity label was therefore suggested (Fig. 3).

When tested, the epoxypropyl β-glycosides fulfilled all the expectations and behaved as typical active site-directed irreversible inhibitors (51). Thus,

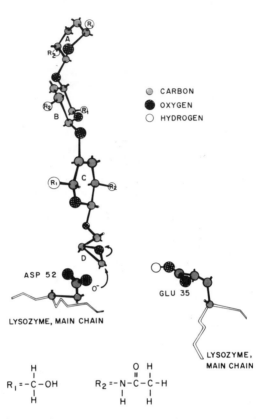

inactivation was stoichiometric, time dependent, and was markedly affected by the chain length of inhibitor, with (GlcNAc)$_3$-Ep being most effective, and GlcNAc-Ep very poor. Oligosaccharides of GlcNAc afforded effective protection of the enzyme against inactivation. Also, simple epoxides, such as propylene oxide, were very poor inhibitors, if at all. By peptide mapping it was shown that the β-carboxyl group of Asp 52 is the only amino acid side chain which interacts with the affinity label, forming an ester bond between the two. Immunologic studies (31) and X-ray crystallographic studies (33) provided evidence that the sugar moiety in the enzyme inactivated by (GlcNAc)$_2$-Ep occupies subsites B and C of the active site of the enzyme. Finally, the epoxypropyl β-glycosides inactivated irreversibly a number of lysozymes closely related to the hen egg-white enzyme, such as

human leukemic lysozyme and several bird lysozymes. They did not, however, inhibit the lysozymes from goose egg white, phage T, and papaya
latex, which differ in many other properties from the hen enzyme (30).

These studies are of interest not only in that they have provided further
evidence on the location of a particular amino acid side chain in the active
site of HEWL; they also open the way for the direct examination of subsites
E and F, since in the inactive enzyme subsites A, B, C, and D can be preferentially blocked. Displacement by suitable reagents of the affinity label
bound to the enzyme may be used for the preparation of derivatives of
lysozyme in which only the carboxyl side chain Asp 52 has undergone
specific chemical modification. Studies of the affinity labeling of other
lysozymes may provide information about the nature of their active site
and the amino acids located in this region. Last, but not least, we wish to
point out the usefulness of the techniques used in our work for affinity
labeling studies of specific sites of other proteins.

Synthesis of the Affinity Labeling Reagent

The 2',3'-epoxypropyl β-glycosides of $(GlcNAc)_{1-3}$ were synthesized from
the corresponding acetylated chitin oligosaccharides according to the
method of Thomas (50) (Fig. 4). When radioactive compounds were re-

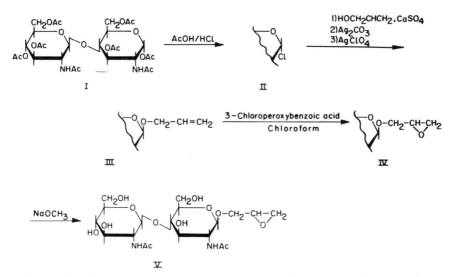

Fig. 4. Reaction scheme for the preparation of the 2',3'-epoxypropyl β-glycoside of
$(GlcNAc)_2$.

quired, additional steps were included in order to introduce an acetyl group labeled either with ^{14}C or ^{3}H in the sugar residue which is nearest to the aglycon (Fig. 5). Recently we modified this method and synthesized the 2',3'-epoxypropyl β-glycoside of GlcNAc by direct condensation of 1,2-epoxypropanol with the acetochloro derivative of *N*-acetyl-D-glucosamine (13b). This modification may be useful for the preparation of related glycosides from readily available epoxy alcohols.

Affinity Labeling of Hen Egg-White Lysozyme

Characterization of the Inactivation Reaction

Unless otherwise noted, the affinity labeling of HEWL was studied with the 2',3'-epoxypropyl β-glycoside of di(*N*-acetyl-D-glucosamine) [(GlcNAc)$_2$-Ep]. Activity of the enzyme was routinely assayed using cells of *M. lysodeikticus* as substrate (47).

Incubation of HEWL (1.25×10^{-4} *M*) with (GlcNAc)$_2$-Ep (1.0×10^{-3} *M*) at 37°C in water at pH 5.5 resulted in progressive loss of the lytic activity of the enzyme [Fig. 6(A)]. The inactivation was exponential with time, and under those conditions the enzyme lost 50% of its initial activity in 3 hr.

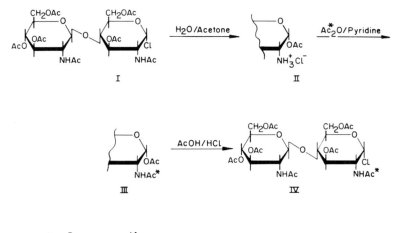

$$Ac^* = C^3H_3CO \quad \text{or} \quad ^{14}CH_3CO$$

Fig. 5. Reaction scheme for the introduction of a radioactively labeled *N*-acetyl group at the reducing moiety of (GlcNAc)$_2$.

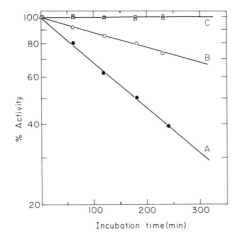

Fig. 6. (A) Rate of inactivation of hen egg-white lysozyme (1.25×10^{-4} M) by (GlcNAc)$_2$-Ep (1.0×10^{-3} M) at 37°C in water at pH 5.5. (B) Same as (A) but in the presence of (GlcNAc)$_2$ (4.1×10^{-2} M). (C) Control experiments: HEWL (1.25×10^{-4} M) only (open square) and with propylene oxide (0.5 M) (open triangle).

The rate of inactivation was markedly decreased when (GlcNAc)$_2$ (4.1×10^{-2} M) was added to the incubation mixture [Fig. 6(B)]. In the absence of (GlcNAc)$_2$-Ep, the enzymatic activity of HEWL remained constant under the same incubation conditions. Incubation of HEWL with an epoxide without a sugar moiety, such as propylene oxide (up to 0.5 M) did not affect the activity of HEWL under the experimental conditions [Fig. 6(C)].

Correlation was found between the rates of inactivation and the oligosaccharide chain length of the affinity labels. The inactivation rates of HEWL by GlcNAc-Ep, (GlcNAc)$_2$-Ep, and (GlcNAc)$_3$-Ep were directly proportional to the association constants of their saccharide moieties with the enzyme (51). At pH 3.6, where the carboxyl groups of the aspartic and glutamic acid side chains in the enzyme are mainly un-ionized, no inactivation of the enzyme by the epoxypropyl β-glycoside occurred.

For the quantitative examination of the extent of incorporation with time of the affinity label into HEWL, the enzyme was incubated with radioactive [^{3}H]-labeled (GlcNAc)$_2$*-Ep, and the excess of reagent separated from the protein by thin layer chromatography on silica gel G in methanol-acetone (1:1 v/v). The silica layer at the origin, containing the protein only, was transferred to a scintillation vial; water and Bray's scintillation solution (6) were added, and the sample counted in a scintillation counter. A direct correlation was found between the molar incorporation of the radioactive affinity label and the inactivation of the enzyme (Fig. 7).

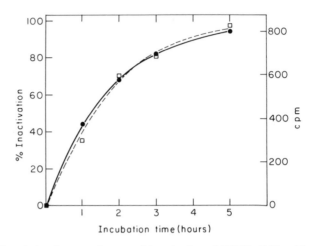

Fig. 7. Correlation between the rate of inactivation of HEWL ($1.38 \times 10^{-4} M$) by [³H]-labeled (GlcNAc)$_2$-Ep ($8.3 \times 10^{-3} M$) at 37°C in water at pH 5.5 (closed circle) and the incorporation of radioactivity into the enzyme (open square). Radioactivity measured on samples containing 0.5 mg protein. The completely inactivated enzyme (24 hr incubation) had activity of 850 cpm/0.5 mg.

Isolation of the Affinity Labeled Hen Egg-White Lysozyme

HEWL completely inactivated by (GlcNAc)$_2$*-Ep was obtained by incubation of the enzyme with a large excess of the radioactive affinity label at 37°C for 24 hr. The affinity labeled enzyme, (GlcNAc)$_2$*-Pr-HEWL, was isolated by dialysis against water, followed by gel filtration on Sephadex G-25. Only one radioactive peak which showed absorbance at 280 nm and had constant specific radioactivity corresponding to 1 mole of (GlcNAc)$_2$* per mole of enzyme, was obtained (Fig. 8). The fractions containing the protein were combined and dialyzed against water. After lyophilization, the isolated protein was homogeneous by polyacrylamide gel electrophoresis at pH 4.5. Its enzymatic activity when assayed with cells of *M. lysodeikticus* (47) was found to be less than 2% of that of an equal weight of HEWL. Analysis of the acid hydrolysate (HCl 6 *N*, 110°C, 22 hr) of (GlcNAc)$_2$*-Pr-HEWL on an amino acid analyzer revealed, in addition to the expected amino acids, the presence of glucosamine in a ratio of 1.9 moles per mole of enzyme.

Identification of the Site of Attachment

Proteolytic digestion was chosen as a suitable method to obtain small peptide fragments from the affinity labeled enzyme. Inactivated radioactive labeled enzyme, isolated as described above, was reduced with dithio-threitol and carboxymethylated by iodoacetic acid. All the radioactivity of (GlcNAc)$_2$*-Pr-HEWL was recovered in the reduced carboxymethylated product [RCM-(GlcNAc)$_2$*-Pr-HEWL].

Different methods and enzymes were used for proteolytic digestion (Table I). In one type of experiment, RCM-(GlcNAc)$_2$*-Pr-HEWL (5.35 mg, 40600 cpm) was subjected to total enzymatic digestion (34) with sub-tilisin and aminopeptidase M. The incubation mixture was then separated by high voltage paper electrophoresis at pH 1.9. Two radioactive peaks were found by scanning; one of these contained about 6% of the total radioactivity and was located in the area where uncharged compounds migrate at this pH. The second peak migrated as a positively charged com-pound and contained about 91% of the total radioactivity. The radioactive materials under the two peaks were recovered in high yield by elution from the paper with water (14). Analysis on the amino acid analyzer of an acid hydrolysate (6 N HCl, 22 hr, 110°C) of the neutral peak revealed the pres-ence of glucosamine only. This glucosamine is most probably derived from

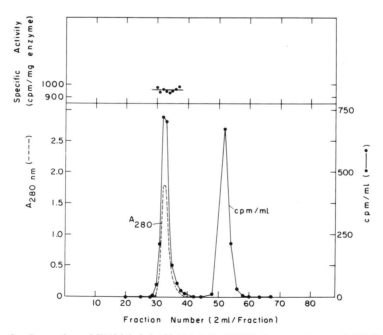

Fig. 8. Separation of [^{14}C]-labeled (GlcNAc)$_2$-Pr-HEWL from an excess of (GlcNAc)$_2$*-Ep by gel filtration chromatography on Sephadex G-25 (fine) column (1.5 × 80 cm) using 0.1 M ammonium acetate as eluent. (---) absorbance at 280 nm; (closed circle, bottom) radioactivity; (closed circle, top) specific radioactivity (cpm/mg protein).

TABLE I

Peptides Derived from Proteolytic Digests of the Reduced Carboxymethylated Affinity Labeled HEWL

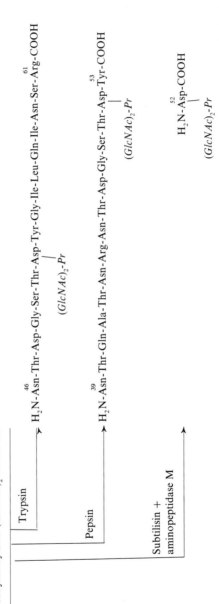

the 2,3-propanediol β-glycoside of $(GlcNAc)_2$ which has been released from the affinity labeled enzyme during the proteolytic digestion. The major radioactive product was further purified by gel filtration on a column of Sephadex G-25 (fine). A single radioactive peak, which contained all the radioactivity applied to the column, was eluted with Ve similar to that of $(GlcNAc)_2$. Amino acid analysis of the acid hydrolysate of this material revealed the presence of aspartic acid and of glucosamine (GlcN) in a molar ratio of 1.0 to 2.1. This finding shows that in the inactivated enzyme the affinity label is bound solely to aspartic acid.

To establish the position of this aspartic acid residue in the polypeptide chain of the enzyme, the reduced and carboxymethylated $(GlcNAc)_2$*-Pr-HEWL was digested for 16 hr with pepsin in 5% formic acid and the digest was subjected to preparative high voltage electrophoresis at pH 3.5. Approximately 80% of the radioactivity applied to the paper was recovered in the area of positively charged peptides. The radioactive material was eluted and fractionated on a column of Sephadex G-25 (fine). Several peptide peaks (absorbance at 220 nm) were collected, only one of which was radioactive. This radioactive peptic peptide was homogeneous on electrophoresis at pH 1.9 and 6.5, and its composition, established by analysis of the acid hydrolysate on an amino acid analyzer, was $Asp_{5.0}Thr_{3.9}Ser_{1.0}Gly_{1.3}Ala_{1.1}$-$Tyr_{0.9}Arg_{1.0}Glu_{1.0}GlcN_{1.8}$, corresponding to the sequence of amino acid residues 39–53 in HEWL (8) (Table I). This conclusion was substantiated by the finding that carboxypeptidase A released tyrosine from the peptide, showing it to be the C-terminal amino acid, and that the N-terminal amino acid was found, by the dansyl chloride method, to be aspartic acid. The peptic peptide, which contained a single arginine residue and no lysine, was cleaved by trypsin to give two peptides having the amino acid composition expected to result from cleavage at Arg 45; the radioactivity and the glucosamine were associated only with the peptide possessing the composition corresponding to that of residues 46–53 (Table II). The digestion of the peptic peptide with pronase gave another radioactive peptide containing glucosamine, whose amino acid composition corresponded to that of the peptide located between Asp 48 and Asp 52 in HEWL.

Mild treatment on paper of the radioactive peptic peptide with triethylamine resulted in 90% loss of radioactivity. The labeled peptide, which was neutral at pH 6.5, was converted by this treatment to a negatively charged peptide. Thus, the attachment of $(GlcNAc)_2$*-Ep to HEWL is via an ester bond to Asp 52 or Asp 48.

To identify unequivocally the aspartic acid to which the affinity label is linked, the peptic peptide was digested with an aminopeptidase obtained from *Clostridium histolyticum* (25). The incubation mixture was fractionated on a column of Sephadex G-25 (fine) using 0.05 M acetic acid as eluent.

TABLE II

Peptides Derived from Proteolytic Digests of the Labeled Peptic Peptide

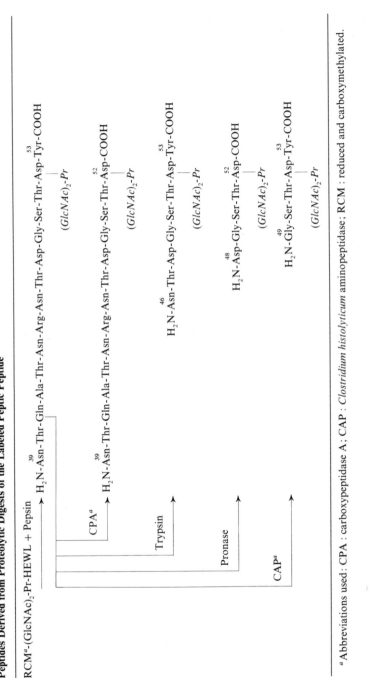

[a] Abbreviations used: CPA: carboxypeptidase A; CAP: *Clostridium histolyticum* aminopeptidase; RCM: reduced and carboxymethylated.

Only one radioactive peptide peak was obtained, which proved to be homogeneous on paper electrophoresis at pH 1.9 and 6.5. Amino acid analysis of its acid hydrolysate gave the following composition: $Asp_{1.0}$-$Thr_{1.1}Ser_{1.0}Gly_{1.0}Tyr_{0.9}GlcN_{1.9}$. The C-terminal amino acid of this peptide was shown by carboxypeptidase A digestion to be tyrosine, and glycine was found to be the N-terminal by the dansyl chloride method. Upon treatment of the radioactive pentapeptide with triethylamine, a ninhydrin-positive nonradioactive peptide which was negatively charged on paper electrophoresis at pH 6.5 was obtained. The uncharged radioactive spot was now ninhydrin-negative. Both compounds were eluted from the paper electrophoretogram. The nonradioactive peptide had the following composition: $Asp_{1.0}Thr_{0.9}Ser_{0.9}Gly_{1.0}Tyr_{0.8}$. The radioactive material was found to migrate on silica gel G plates (acetone, methanol, 2:1) as the 2,3-propanediol ν-glycoside of $(GlcNAc)_2$; analysis on an amino acid analyzer after acid hydrolysis revealed only the presence of glucosamine.

These experiments establish that digestion of the peptic peptide with the aminopeptidase derived from *C. histolyticum* affords a pentapeptide, whose composition, partial structure, and other properties correspond to those of the peptide located between Gly 49 and Tyr 53 in HEWL. To this pentapeptide, the affinity label is covalently bound via an ester bond to a carboxyl group. Therefore, the amino acid to which the affinity label is linked in $(GlcNAc)_2$*-Pr-HEWL is Asp 52.

Physical and Chemical Characterization of the Affinity Labeled Enzyme

As mentioned previously, $(GlcNAc)_2$*-Pr-HEWL was homogeneous when examined by gel filtration on Sephadex G-25, and by polyacrylamide gel electrophoresis at pH 4.5. In the latter system it migrated to the cathode slightly faster than the native enzyme. Similar results were obtained on electrophoresis of the native and inactivated enzyme on cellulose acetate at pH 6.8.

Reactivation experiments, now in progress, show that the affinity label remains attached to the inactivated enzyme even after prolonged incubation (24–48 hr) at 37°C, between pH 2 and 10. However, in an aqueous solution of triethylamine at pH 11, very slow reactivation was observed.

Potentiometric titrations in the pH range 2.8–8.4, in 0.15 M KCl, showed that the proton uptake of the affinity labeled lysozyme was 8.84 protons/mole, as compared to 9.88 protons/mole of the native enzyme. Thus, the affinity labeling results in the disappearance of one acid-titratable group. As shown by Sakakibara and Hamaguchi (42), the groups titrated in the native enzyme in this pH range and salt concentration are eight of the eleven carboxyls, the single histidine, and part of the N-terminal amino group.

Even without the data obtained from studies of peptide mapping, it is very unlikely that the epoxide of (GlcNAc)$_2$-Ep reacted with the histidine or the N-terminal amino group, since this would have led to the formation of acid-stable derivatives of the corresponding amino acids. However, no difference in the amino acid composition between the native and affinity labeled enzyme was detected. This, together with our finding on the location of the affinity label in the polypeptide chain of HEWL, clearly shows that the change in proton uptake is the result of the blocking of the β-carboxyl group of Asp 52. Indeed, the difference titration curve (native vs. affinity labeled HEWL, Fig. 9) is very similar to that obtained by Parsons and Raftery (35) for the β-ethyl ester derivative of the Asp 52 residue of HEWL. Analysis of such curves has been carried out by Parsons and Raftery. They have shown that a curve of this type is a result of the blocking of Asp 52 (pK ~4.4) and the perturbation of the pK of the nearby Glu 35 from about 6.1 to 5.2, which is caused by the blocking of Asp 52.

Spectrofluorimetric titrations of (GlcNAc)$_2$-Pr-HEWL and of (GlcNAc)$_3$-Pr-HEWL revealed the presence of an ionizable group with a pK of about 5 (Fig. 10) (14a), instead of the group with pK 6.0–6.5 (ascribed to Glu 35) found in similar titrations of the reversible complexes formed between HEWL and (GlcNAc)$_2$ or (GlcNAc)$_3$ (28). It should be noted that both (GlcNAc)$_2$-Pr-HEWL and (GlcNAc)$_3$-Pr-HEWL gave identical spectrofluorimetric titration curves, whereas the corresponding curves for the reversible (GlcNAc)$_2$-HEWL and (GlcNAc)$_3$-HEWL complexes differ from

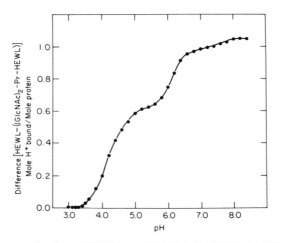

Fig. 9. Difference titration of HEWL vs. (GlcNAc)$_2$-Pr-HEWL in 0.15 M KCl at 25°C. Protein concentration 2.42 × 10^{-4} M.

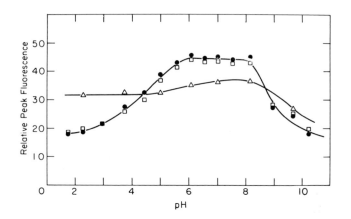

Fig. 10. Fluorescence plotted against pH: (open triangle) HEWL; (open square) (GlcNAc)$_2$-Pr-HEWL; (closed circle) (GlcNAc)$_3$-Pr-HEWL. Temperature = 23°C; excitation at 285 nm; enzyme emission measured at 340 nm, 1 cm path length; protein concentration, 2.6×10^{-6} *M*.

each other. This may be attributed to the difference in the nature of binding in the two types of complex. In particular, in the affinity labeled derivatives, it may be assumed that (GlcNAc)$_2$ and (GlcNAc)$_3$ are bound with the same strength and are constantly located in the active site region. In the reversible complexes, a dynamic equilibrium exists between the saccharide and the enzyme and, moreover, (GlcNAc)$_3$ binds considerably stronger than (GlcNAc)$_2$.

X-Ray Crystallography

The location of the (GlcNAc)$_2$ in the affinity labeled enzyme was unequivocally demonstrated by the use of X-ray crystallography (33). For this purpose, HEWL in crystalline form was inactivated by (GlcNAc)$_2$-Ep. The diffraction patterns obtained from the modified crystals were analyzed by the technique used for the analysis of the patterns of the crystals of the reversible complex formed between HEWL and the β-phenyl glycoside of (GlcNAc)$_2$ (32).

In these experiments, tetragonal crystals of HEWL (1) were employed. This is the same crystalline form used by Phillips and his co-workers (2–4, 23) in the elucidation of the three-dimensional structure of HEWL and of the lysozyme-saccharide complexes. The crystals were incubated for several days in mother liquor containing (GlcNAc)$_2$-Ep (0.06 *M*). This resulted in irreversible inactivation to the extent of 80% and in incorporation of 1.67 moles of glucosamine per mole of lysozyme. After removal of the

excess of inhibitor, precession photographs of the inhibited crystals, with
$\mu = 18°$ (minimum spacing of 2.5 Å) of the *hko, okl,* and *hhl* zones, were
taken. The crystals were isomorphous with those of native lysozyme, and
showed only small changes in intensities. The diffraction patterns were
analyzed essentially as has been done by Phillips *et al.* in the study of lysozyme-saccharide complexes. Difference-Fourier projections were calculated using as amplitudes the differences between the amplitudes of native
HEWL and of (GlcNAc)$_2$-Pr-HEWL, weighted with the native figures of
merit and native phases. For comparison, the same difference projections
were calculated using the structure factors extracted from the three-dimensional data of the reversible complex formed between HEWL and the
β-phenyl glycoside of (GlcNAc)$_2$. The three-dimensional structure of this
complex is known (32).

A comparison of the difference maps obtained for the affinity labeled
enzyme and for the complex of HEWL with the β-phenyl glycoside of
(GlcNAc)$_2$ shows the same set of peaks representing the known binding
site for (GlcNAc)$_2$ (2). In the *hko* projection, two peaks—A and B— related by the fourfold crystallographic axis, appear in both maps (Fig. 11).
These peaks represent one binding site per molecule. Similar results were
obtained from the examination of the *okl* and *hhl* projections (33).

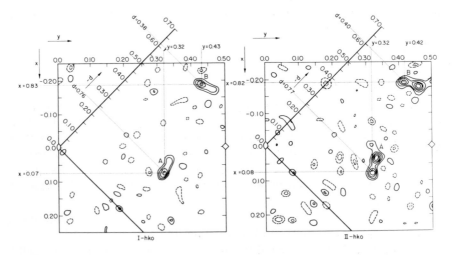

Fig. 11. Difference-Fourier projection maps (*hko* projection) of hen egg-white lysozyme
irreversibly inhibited by the 2′,3′-epoxypropyl β-glycoside of (GlcNAc)$_2$ (I) and of the reversible
complex formed between hen egg-white lysozyme and the β-phenyl glycoside of (GlcNAc)$_2$
(II). The coordinates of two crystallographically related binding sites, A and B, each one
representing subsites B and C in the active site of lysozyme, are indicated. Reproduced from
J. Moult, Y. Eshdat, and N. Sharon, *J. Mol. Biol.* **75,** 1 (1973).

The maps of both crystals show other peaks, some of which are of the same order of magnitude. The presence of these peaks is to be expected since the three-dimensional map of the complex of HEWL with the β-phenyl glycoside of (GlcNAc)$_2$ (32) contains a number of small features, some of which may be identifiable as minor conformational changes ($<$ 0.75 Å), and in projection these may sum up at particular points to give large peaks. None of these peaks is present in more than one projection, however, and in the *hhl* and *okl* projections none are related by the required translation of $\frac{1}{4}$ of the unit cell in the z direction.

The fact that the peaks in the two sets of maps coincide shows clearly that the sugar rings in the affinity labeled enzyme occupy the same positions on the enzyme surface as those of the β-phenyl glycoside, i.e., subsites B and C in the active site of HEWL. This is in excellent agreement with the results obtained from the sequence studies which show that the affinity label is bound covalently to Asp 52. Moreover, the maps suggest that the conformational changes induced by the binding of the saccharides in the two complexes examined are small.

In addition to the specific information obtained from this study, HEWL served here, for the first time, as a model for characterization of an affinity labeled enzyme with the aid of X-ray diffraction studies. It is worthwhile to point out that such relatively simple crystallographic techniques may be of use as a routine biochemical tool for studies of affinity labeled enzymes whose three-dimensional structure is known.

Immunochemical Studies of Affinity Labeled Hen Egg-White Lysozyme

Another approach, which may prove most helpful for the identification of the site of attachment of an irreversible inhibitor to an active site of an enzyme, is exemplified by the following immunochemical study, in which "active site-directed" antibodies were employed (31).

Antibodies directed against the active site region of HEWL were prepared from the whole population of anti-HEWL antibodies, according to the method of Imanishi *et al.* (21). In this method, antibodies to HEWL are reacted with the enzyme in the presence of an excess of GlcNAc oligomers. Under these conditions, the active site-directed antibodies do not combine with HEWL and can be readily isolated. The reaction of the anti-HEWL antibodies and of the active site-directed antibodies with HEWL and with (GlcNAc)$_2$-Pr-HEWL was examined using the "chemically modified bacteriophage assay" (15, 16). It was found that the anti-HEWL antibodies reacted equally well with HEWL, with HEWL in the presence of large molar excess of (GlcNAc)$_2$, and with the affinity enzyme [Fig. 12(A)]. In contrast, the active site-directed antibodies reacted with the affinity labeled enzyme

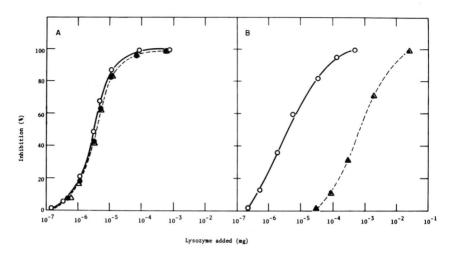

Fig. 12. Inhibition of the inactivation of HEWL-bacteriophage T4 conjugate by goat anti-HEWL antibodies (A) and by anti-HEWL "active site-directed" antibodies (B) in the presence of HEWL, $(GlcNAc)_2$-Pr-HEWL, and HEWL-$(GlcNAc)_2$ reversible complex. The following symbols are used: (open circle) hen egg-white lysozyme; (closed circle) $(GlcNAc)_2$-Pr-HEWL; (open triangle) HEWL-$(GlcNAc)_2$ (1:40 molar ratio). [Reprinted with permission from Ref. (31).]

to a much lesser extent than with HEWL [Fig. 12(B)]. The reversible complex formed between HEWL and $(GlcNAc)_2$ behaved in the assay with the active site-directed antibodies in a manner identical to the affinity labeled enzyme, strongly suggesting that in both cases the $(GlcNAc)_2$ occupies the same site on the enzyme, i.e., subsites B and C. This finding is also in agreement with those described before, and especially with results of the X-ray crystallographic studies.

Study of Subsites E and F

The affinity labeled HEWL provided an opportunity for the study of the interaction of saccharides with subsites E and F, which are unoccupied in the irreversibly inhibited enzyme.

As shown by Teichberg and Shinitzky (49), the fluorescence polarization of HEWL is markedly affected by the addition of GlcNAc oligomers. Thus, at low concentrations of $(GlcNAc)_4$ (up to $5 \times 10^{-5}\ M$) the fluorescence polarization of the enzyme is enhanced. It then remains constant up to a concentration of $(GlcNAc)_4$ of about $10^{-3}\ M$, but at higher concentrations of $(GlcNAc)_4$ the fluorescence polarization decreases. From analysis of the bell shaped curve obtained, Teichberg and Shinitzky have concluded that at the low $(GlcNAc)_4$ concentrations, subsites A, B, and C are occupied, with an association constant of $8 \times 10^4 \ M$; this value is close to that obtained by several other techniques (13). At high saccharide concentrations, occupation of subsites E and F takes place, and the association constant to these subsites can be calculated. This constant was found for several GlcNAc oligomers to be about 2×10^2, which is close to the value found, for example, from transglycosylation studies (13, 37).

Evidence for the simultaneous binding of two saccharide molecules to HEWL was obtained by a study of the fluorescence polarization of (GlcNAc)$_2$-Pr-HEWL and its complexes with saccharides. The fluorescence polarization spectrum of this derivative was found to be very similar to that of HEWL in the presence of $1.1 \times 10^{-4} \ M$ (GlcNAc)$_4$, with a maximal polarization value of $p = 0.29$ when excited at 303 nm. Since the reversible binding of the first (GlcNAc)$_4$ molecule to HEWL caused an enhancement of the fluorescence polarization up to a maximal value of $p = 0.29$, which is identical to that observed in the affinity labeled enzyme, it was concluded that HEWL is effectively saturated at a saccharide concentration of $10^{-4} \ M$ and that the saccharide is bound to the same sites as those occupied by the disaccharide covalently linked to the enzyme. Since (GlcNAc)$_4$ binds to subsites A, B, and C with one sugar moiety outside the enzyme cleft (40), the enhancement of fluorescence polarization originates from the occupation of subsites A, B, and C (49).

Using (GlcNAc)$_2$-Pr-HEWL, the association constants of a series of disaccharides to this lysozyme derivative were measured. Except for cellobiose, which had only a small effect on the fluorescence polarization of (GlcNAc)$_2$-Pr-HEWL, all other disaccharides tested produced, at high concentrations ($\sim 10^{-2} \ M$), a decrease in the fluorescence polarization similar to that observed with HEWL. The association constant of (GlcNAc)$_2$ to subsites E and F found by this method ($K = 2.4 \times 10^2 \ M^{-1}$) was in agreement with that found by the same method, using the native enzyme.

Affinity Labeling of Lysozymes From Different Sources

The affinity labeling method, using the reagents described above, has been applied by us to obtain information on the active sites of lysozymes isolated from different organisms. To the extent that they have been investigated, many of these lysozymes exhibit close similarity in their amino acid sequences (9, 18, 24, 26, 38), flourescence spectra (41, 45a), and enzymatic properties (29, 39). The latter include not only the ability to digest cell walls of *M. lysodeikticus,* but also to catalyze transglycosylation reactions with a variety of saccharide acceptors, when the cell wall tetrasaccharide or chitin oligosaccharides are used as substrates (30a, 45).

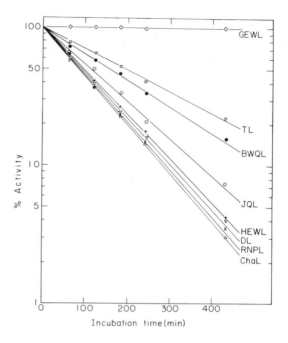

Fig. 13. Rate of inactivation of different bird egg-white lysozymes (1.25×10^{-4} *M*) by (GlcNAc)$_2$-Ep (3.8×10^{-3} *M*) at 37°C in water at pH 5.5. (Open diamond) goose (GEWL); (open square) turkey (TL); (+) hen (HEWL); (closed circle) bobwhite quail (BWQL); (open circle) Japanese quail (JQL); (∇) duck II (DL); (×) ring-necked pheasant (RNPL); (open triangle) chachalaca (ChaL).

As can be seen from Fig. 13, all the bird lysozymes examined, except the goose egg-white lysozyme, were inactivated by (GlcNAc)$_2$-Ep at pH 5.5, 37°C. The enzymes which interacted with the affinity label were, in addition to HEWL, the lysozymes from the egg white of the turkey, the bobwhite quail, the Japanese quail, duck II, ring-necked pheasant, and chachalaca. The rate of inactivation differed with the different enzymes, but in all cases decreased significantly upon the addition of an excess of (GlcNAc)$_2$. The fact that these bird lysozymes, except that of the goose, were all affinity labeled by the same reagent, implies that the general architecture and the orientation of the amino acid side chains of the active site region of these enzymes are similar.

The goose egg-white lysozyme was not irreversibly inhibited by (GlcNAc)$_2$-Ep even at pH 4.7 at which it attacks *N*-acetylglucosamine oligosaccharides better than at higher pH (10). The active site of this enzyme is, therefore, markedly different from those of the other bird lysozymes tested. This conclusion is in accord with the results of the study of the primary structure of that enzyme (9) which showed an amino terminal peptide of 30 residues completely different from those of the HEWL.

The finding that the turkey lysozyme is inactivated by the affinity label seems of particular interest. Though the primary structures of TL and of HEWL are very similar, there is one important difference: Asp 101, assumed to participate in saccharide binding in subsites A and B of the active site of HEWL, is absent in turkey lysozyme where it is replaced by Gly 101 (26). The finding that the two enzymes are inactivated by (GlcNAc)$_2$-Ep casts doubt on the significance of Asp 101 for the binding of a GlcNAc moiety in subsite B. Further kinetic studies with (GlcNAc)$_2$-Ep as well as with (GlcNAc)$_3$-Ep are required to support this conclusion, and to elucidate the role of Asp 101 in saccharide binding at subsite A.

It should be pointed out that the overall rate of inactivation is a function both of the binding to the active site and the rate of chemical reaction of the enzyme with the epoxide. Since binding constants can be readily evaluated by a variety of techniques, separate information on the rate of reaction with the epoxide function may be obtained. In this way, some properties of the presumed reactive carboxyl group in the active site of the different lysozymes may be evaluated.

Lysozymes derived from sources other than bird egg whites were also tested for their susceptibility to inactivation by the epoxypropyl β-glycosides. Among these, human leukemic urine lysozyme was inactivated by (GlcNAc)$_3$-Ep at pH 5.5 at a lower rate than HEWL (Fig. 14) while the lysozymes from papaya latex and the bacteriophage T4 were not inactivated at all. Similar results were obtained with (GlcNAc)$_2$-Ep at pH 5.5 and 4.6 [at the latter pH, the papaya lysozyme acts on chitin most efficiently (19)]. These results show that the active site of the human and the hen lysozymes are similar, whereas the active sites of the T4 phage and the papaya lysozymes are different. Indeed, HEWL and the human lysozyme are closely similar in their amino acid sequences (9) and three-dimensional structures (5) while the papaya and the T4 phage lysozymes differ from the HEWL in their amino acid composition, their molecular weight, and substrate specificity (19, 22, 52).

A protein which was found to be closely related in its primary structure to HEWL is α-lactalbumin. α-Lactalbumin serves as a modifier in the lactose synthetase system of the enzyme known as the A protein, and inhibits the formation of N-acetyllactosamine by this enzyme (7). We have tested the effect of GlcNAc-Ep on α-lactalbumin. Incubation of α-lactalbumin with GlcNAc-Ep at pH 5.5 at 37°C for 17 hr resulted in irreversible binding of 1 mole of the reagent per mole protein (13a). Preliminary experiments have shown that the incorporation of the reagent into α-lactalbumin markedly decreased its ability to enhance the synthesis of lactose by the A protein and

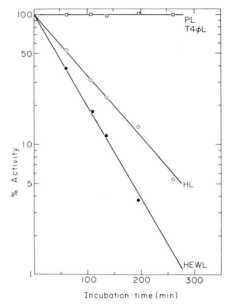

Fig. 14. Rate of inactivation of lysozymes from different sources (1.38×10^{-4} M) by $(GlcNAc)_3$-Ep (2.76×10^{-3} M) at 37°C in water at pH 5.5. (Open square) papaya latex (PL) and bacteriophage T4 (T4ΦL) lysozymes; (open circle) human leukemic urine lysozyme (HL); (closed circle) hen egg-white lysozyme (HEWL).

to inhibit synthesis of *N*-acetyllactosamine by this protein. Also, the modified α-lactalbumin competed with native α-lactalbumin in its effect on the A protein in both reactions. We therefore assume that the labeled α-lactalbumin preserves its ability to bind the A protein, thus competing with native α-lactalbumin, but its ability to modify the enzyme specificity of the A protein is decreased.

However, neither *N*-acetyl-D-glucosamine nor D-glucose at high concentrations protected α-lactalbumin in the reaction with GlcNAc-Ep. Moreover, propylene oxide, in the same concentration at GlcNAc-Ep (7.2×10^{-2} M) exerted the same effects on α-lactalbumin as the β-epoxypropyl glycoside of GlcNAc. This finding suggests that though α-lactalbumin is not affinity labeled by GlcNAc-Ep, it contains reactive groups which interact with epoxides thus affecting its biological activity.

Concluding Remarks

In this paper we have described the rational design and preparation of an homologous series of active site-directed irreversible inhibitors for hen egg-white lysozyme, and their use in the study of related enzymes. The reaction of one of the inhibitors, the $2'.3'$-epoxypropyl β-glycoside of GlcNAc $\beta(1 \rightarrow 4)$GlcNAc, with HEWL has been thoroughly investigated. As expected, this glycoside reacted with Asp 52 in the active site of the enzyme, forming an ester bond between the two. Although it has not been established which carbon atom of the epoxide was attacked by the β-carboxyl group of Asp 52, the result is in line with the mechanism proposed for the reaction between the inhibitor and the enzyme (Fig. 3). Though our findings clearly show that Asp 52 forms part of the active site of HEWL, they do not give evidence on the proposed catalytic role of this group.

The affinity labeled enzyme has been fully characterized by a variety of methods, among which were included X-ray crystallographic and immunologic studies. Binding to subsites B and C of the GlcNAc $\beta(1 \rightarrow 4)$GlcNAc moiety of the inhibitor was thus unequivocally established. The modified enzyme has also been used in the study of the binding of saccharides to subsites E and F.

The epoxypropyl β-glycosides have been shown to inhibit human lysozyme and six different bird lysozymes, in a manner analogous to their inhibition of the hen enzyme. In addition to providing further evidence on the great similarity of these enzymes, our work opens the way for the detailed characterization of their active sites, as well as those of related enzymes.

Acknowledgments

We thank Drs. A. K. Allen, R. E. Canfield, P. Jollès, E. F. Osserman, Ellen Prager, A. Tsugita, and A. C. Wilson for their generous gifts of the various lysozymes used in this study. Thanks are also due to Miss Efrat Kessler and Dr. A. Yaron from the Department of Biophysics, Weizmann Institute, for the aminopeptidase of *C. histolyticum*.

This work was supported in part by grant GM-19143 from the National Institutes of Health, U.S. Public Health Service.

References

1. Alderton, G., and Fevold, J., *J. Biol. Chem.* **164,** 1 (1946).
2. Blake, C. C. F., Johnson, L. N., Mair, G. A., North, A. C. T., Phillips, D. C., and Sarma, V. R., *Proc. Roy. Soc. Lond.* [*Biol.*] **167,** 378 (1967).
3. Blake, C. C. F., Koenig, D. F., Mair, G. A., North, A. C. T., Phillips, D. C., and Sarma, V. R., *Nature* (*Lond.*) **206,** 757 (1965).
4. Blake, C. C. F., Mair, G. A., North, A. C. T., Phillips, D. C., and Sarma, V. R., *Proc. Roy. Soc., Lond.* [*Biol.*] **167,** 365 (1967).
5. Blake, C. C. F., and Swan, I. D. A., *Nature, (Lond.) New Biol.* **232,** 12 (1971).
6. Bray, G. A., *Anal. Biochem.* **1,** 279 (1960).
7. Brew, K., *Essays Biochem.* **6,** 93 (1970).
8. Canfield, R. E., *J. Biol. Chem.* **238,** 2698 (1963).
9. Canfield, R. E., Kammerman, S., Sobel, J. H., and Morgan, F. J., *Nature* (*Lond.*) *New Biol.* **232,** 16 (1971).

10. Charlemagne, D., and Jollès, P., *Bull. Soc. Chim. Biol.* **49**, 1103 (1967).
11. Chipman, D. M., Grisaro, V., and Sharon, N., *J. Biol. Chem.* **242**, 4388 (1967).
12. Chipman, D. M., Pollock, J. J., and Sharon, N., *J. Biol. Chem.* **243**, 487 (1968).
13. Chipman, D. M., and Sharon, N., *Science* **165**, 454 (1969).
13a. Eshdat, Y., Bernstein, Y., Thomas, E., and Sharon, N., unpublished data.
13b. Eshdat, Y., Flowers, H., and Sharon, N., in preparation.
14. Eshdat, Y., and Mirelman, D., *J. Chromat.* **65**, 458 (1972).
14a. Eshdat, Y., Teichberg, V. I., and Sharon, N., in preparation.
15. Haimovich, J., Hurwitz, E., Novik, N., and Sela, M., *Biochim. Biophys. Acta* **207**, 115 (1970).
16. Haimovich, J., Hurwitz, E., Novik, N., and Sela, M., *Biochim. Biophys. Acta* **207**, 125 (1970).
17. Hartsuck, J. A., and Tang, J., *J. Biol. Chem.* **247**, 2575 (1972).
18. Hermann, J., and Jollès, J., *Biochim. Biophys. Acta* **200**, 178 (1970).
19. Howard, J. B., and Glazer, A. N., *J. Biol. Chem.* **244**, 1399 (1969).
20. Hess, G. P., and Rupley, J. A., *Annu. Rev. Biochem.* **40**, 1013 (1971).
21. Imanishi, M., Miyagawa, N., Fujio, H., and Amano, T., *Biken J.* **12**, 85 (1969).
22. Inouye, M., Imada, M., and Tsugita, A., *J. Biol. Chem.* **245**, 3479 (1970).
23. Johnson, L. N., and Phillips, D. C., *Nature (Lond.)* **206**, 761 (1965).
24. Kaneda, M., Kato, I., Tominaga, N., Titani, K., and Narita, K., *J. Biochem. (Tokyo)* **66**, 747 (1969).
25. Kessler, E., and Yaron, A., *Biochem. Biophys. Res. Commun.* **50**, 405 (1973).
26. LaRue, J. N., and Speck, J. C., *J. Biol. Chem.* **245**, 1985 (1970).
27. Legler, G., *Biochim. Biophys. Acta* **151**, 728 (1968).
28. Lehrer, S. S., and Fasman, G. D., *J. Biol. Chem.* **242**, 4644 (1967).
29. Locquet, J. P., Saint-Blancard, J., and Jollès, P., *Biochim. Biophys. Acta* **167**, 150 (1968).
30. McKelvy, J. F., Eshdat, Y., and Sharon, N., *Fed. Proc.* **29**, 532 (1970).
30a. Maoz, I., and Sharon, N., unpublished data.
31. Maron, E., Eshdat, Y., and Sharon, N., *Biochim. Biophys. Acta* **278**, 243 (1972).
32. Moult, J., D.Phil. thesis, University of Oxford, 1970.
33. Moult, J., Eshdat, Y., and Sharon, N., *J. Mol. Biol.* **75**, 1 (1973).
34. Parsons, S. M., and Raftery, M. A., *Biochemistry* **8**, 4199 (1969).
35. Parsons, S. M., and Raftery, M. A., *Biochemistry* **11**, 1623 (1972).
36. Phillips, D. C., *Sci. Amer.* **215**, 78 (1966).
37. Pollock, J. J., and Sharon, N., *Biochemistry* **9**, 3913 (1970).
38. Prager, E. M., Arnheim, N., Mross, G. A., and Wilson, A., *J. Biol. Chem.* **247**, 2905 (1972).
39. Prager, E. M., and Wilson, A. C., *J. Biol. Chem.* **246**, 5978 (1971).
40. Rupley, J. A., *Proc. Roy. Soc., Lond. [Biol.]* **166**, 416 (1967).
41. Saint-Blancard, J., Capbern, A., Ducassé, D., and Jollès, P., *C. R. Acad. Sci (Paris)* **269**, 858 (1969).
42. Sakakibara, R., and Hamaguchi, K., *J. Biochem. (Tokyo)* **64**, 613 (1968).
43. Sharon, N., *Proc. Int. Symp. Fleming's Lysozyme, 3rd, 1964* p. 44/T (1900).
44. Sharon, N., *Proc. Roy. Soc., Lond. [Biol.]* **167**, 402 (1967).
45. Sharon, N., Jollès, J., and Jollès, P., *Bull. Soc. Chim. Biol.* **48**, 731 (1966).
45a. Sharon, N., Prager, E. M., and Wilson, A. C., unpublished data.
46. Sharon, N., and Seifter, S., *J. Biol. Chem.* **239**, 2398 (1964).
47. Shugar, D., *Biochim. Biophys. Acta* **8**, 302 (1952).
48. Stevenson, K. J., and Smillie, L. B., *J. Mol. Biol.* **12**, 937 (1965).
49. Teichberg, V. I., and Shinitzky, M., *J. Mol. Biol.* **74**, 519 (1973).
50. Thomas, E. W., *Carbohyd. Res.* **13**, 225 (1970).
51. Thomas, E. W., McKelvy, J. F., and Sharon, N., *Nature (Lond.)* **222**, 485 (1969).
52. Tsugita, A., *in* "The Enzymes" (P. D. Boyer, ed.), 3rd ed., Vol. 5, p. 361. Academic Press, New York, 1971.

New nmr Techniques for the Quantitative Determination of the Structure of Proteins in Solution Applied to Hen Egg-White Lysozyme (HEWL)

I. D. CAMPBELL, C. M. DOBSON, R. J. P. WILLIAMS,
and A. V. XAVIER

In this paper several new developments in the study of proteins by proton nuclear magnetic resonance (pmr) are described. We believe that these will ultimately allow the detailed determination of the structures of such molecules and their bound substrates and inhibitors in solution. Figure 1(A) shows the resolution which we are able to achieve in a conventional Fourier transform pmr spectrum using a 270 MHz spectrometer for a part of the high field region of HEWL. The first of our requirements is a spectrometer of such resolution or better.

McDonald and Phillips (1) have discussed the type of information which high-resolution protein spectra can provide. Assignment of only a few resonances was able to be achieved, and these required a knowledge of the crystal structure as determined by X-ray diffraction (2).

In order to determine the structure of lysozyme or any other protein in detail we clearly need far more information than such spectra can provide. The first step is to enhance the resolution in the pmr spectrum. We have found that by using adaptations of the Fourier transform experiment, which we call (3) CDS* techniques, new line shapes can be produced

*Convolution difference spectroscopy.

219

giving better effective resolution by a factor of two or more. The effect of performing one of these procedures on lysozyme is shown in Fig. 1(B). As we shall prove later, multiplet structure is revealed and singlets, doublets, and triplets can often be distinguished, in itself a considerable advance as a means of assignment. Separation of the resonances using shift probes reveals this structure more clearly.

This and other methods to improve resolution that are subsequently described allow use of spin decoupling techniques which were previously inapplicable to proteins. These techniques make it possible to assign many resonances to types of amino acids, particularly in the methyl and aromatic regions of the spectrum. In order to determine a structure we need the geometric information which defines it, and at some stage must assign resonances to particular amino acids.

In the case of small molecules, for example nucleotides, we have shown (4) that the selective perturbations of resonances in the pmr spectrum induced by the series of paramagnetic lanthanide ions can provide this geometric information. The paramagnetic lanthanide ions are of two types (Fig. 2). Those, for example Gd(III), with long electron relaxation times

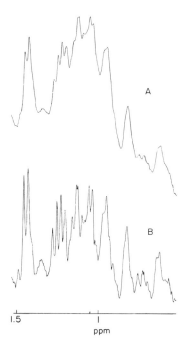

Fig. 1. Resolution enhancement in part of the high field region of HEWL. (A) Conventional spectrum. (B) CD spectrum.

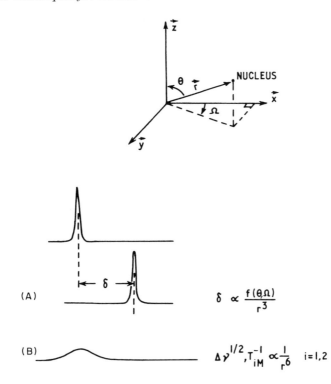

Fig. 2. Perturbation of nmr parameters by lanthanide ions. (A) Effect of shift probe. (B) Effect of relaxation probe.

relax or broaden the pmr resonances of nuclei close to them. The broadening falls off as $1/r^6$ and this can be of use in two ways: to simplify the spectrum and to enable distances from the metal ion to individual resonances to be found.

In the crystal, Gd(III) has been shown to bind between the carboxyl groups of residues Asp 52 and Glu 35 in the active site of lysozyme. From this structure, the positions of the amino acids relative to the metal ion can be represented on a two-dimensional projection (Fig. 3) which shows that the resonances of Ala 110 (CH_3), Val 109 (2 CH_3), and Trp 108 (C_2-H) are nearer than any others.

As we are in the situation of "fast exchange" (5) the observed pmr spectrum is a weighted average of the spectra of the bound and unbound species. Increasing the concentration of Gd(III) in the solution increases the broadening observed on any given resonance. The value of this is illustrated in Fig. 4 as difference spectra can be used to reveal separate groups of resonances at varying distances from the metal ion, i.e., resolution has

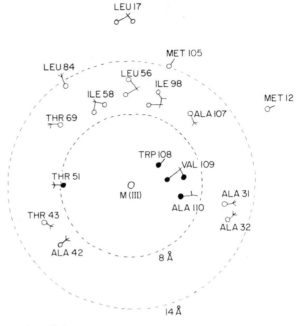

Fig. 3. Schematic radial distribution diagram of part of the active site of HEWL as revealed by X-ray diffraction, the metal position being represented in the center of the diagram. (We thank D. C. Phillips for providing the crystallographic information and for useful discussions.)

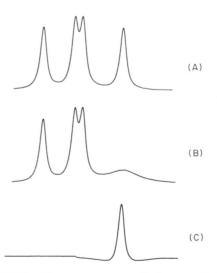

Fig. 4. Schematic Gd(III) difference spectrum. (A) Sample. (B) Sample + Gd(III). (C) Difference (A − B).

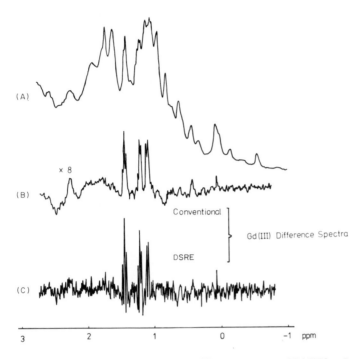

Fig. 5. (A) The methyl region of HEWL. (B) Difference spectrum $[Gd(III)] = 5 \times 10^{-5}\ M$, $[Lys] = 5 \times 10^{-3}\ M$. (C) CD spectrum of B.

been improved by reduction of the number of peaks observed in the spectrum. In Fig. 4 the addition of a small concentration of Gd(III) is assumed to greatly broaden the resonance nearest to the bound Gd (III) and the further resonances, because of the $1/r^6$ dependence of broadening, are almost unaffected. Thus, subtraction of the two spectra allows observation of this one resonance alone. Addition of a higher concentration of Gd(III) enables further resonances to be observed separately.

Figure 5 shows an experiment of this type performed on lysozyme. Three main peaks are observed in the methyl region and one single peak in the aromatic region. Figure 4 indicated from the crystal structure that there are indeed three methyl groups and one aromatic proton nearer than the rest. Thus the methyls of Val 109 (2 doublets), Ala 110 (1 doublet), and the C_2-H of Trp 108 (a singlet) can be confidently assigned. Similarly, more distant resonances can be observed, for example a resonance to very high field can be assigned to the methyl of Thr 51 whose doublet resonance is expected to be shifted to high field by Tyr 53 (1, 6, 7). These few examples, shown in Fig. 6, illustrate precisely one way of assignment in lysozyme, and how to obtain distance information.

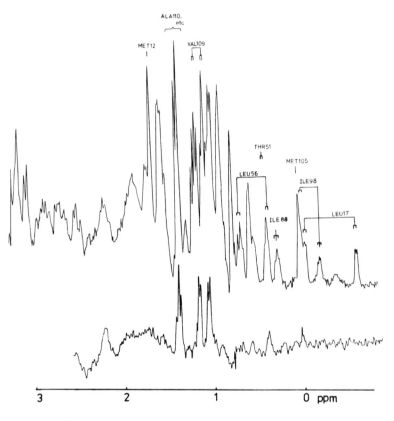

Fig. 6. Some assignments in the methyl region of HEWL.

To place an atom in space we need angular as well as distance information, which brings us to the second class of lanthanides (Fig. 2). The paramagnetic lanthanides other than Gd(III) have fast electron relaxation times [e.g., Pr(III)] and shift the resonances of the protons, in accordance with the distance and two angle functions (8):

$$\delta = \frac{D}{r^3}(3 \cos^2 \theta - 1) + \frac{D'}{r^3}(\sin^2 \theta \cos 2\Omega)$$

where D and D' are constants. D and D' can vary for different lanthanide ions, thus allowing information on both θ and Ω to be determined by comparing the shifts induced by several lanthanide ions. This is most readily illustrated by use of difference spectra, as shown in Fig. 7. A titration curve of the first fifteen high field resonances with Pr(III) is shown in Fig. 8.

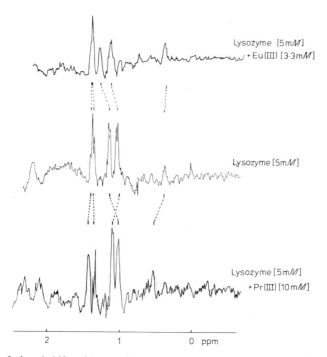

Gd (III) Difference Spectra

Lysozyme [5mM]
+ Eu(III) [3·3mM]

Lysozyme [5mM]

Lysozyme [5mM]
+ Pr(III) [10mM]

2   1   0 ppm

Fig. 7. Induced shifts with Eu(III) and Pr(III) observed using difference spectra.

This information can be obtained throughout the spectrum, and more than half of the residues in lysozyme (those containing methyl and aromatic protons) can be investigated readily. The results will be discussed in more detail in a subsequent paper. So far we have relied heavily on the crystal structure. However, computer programs developed for small molecules (4) are being adapted to the needs of larger molecules. Using these, conformations consistent with the broadening and all the shift data can be sought.

The conclusions, therefore, are that these nmr procedures, in conjunction with the lanthanide ion probes, have enabled data which define the structure of proteins (in solution) to be obtained for the first time. In the case of lysozyme the crystal structure has greatly helped the development of procedures although this is not envisaged as a necessity. Another feature of the CDS method is that sharp peaks can be readily observed (3) thus enabling detection and study of resonances of bound inhibitors and substrate even in the presence of the large numbers of protein resonances. We have previously shown that the resonances of bound inhibitors can be

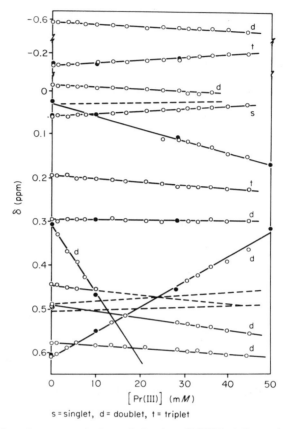

Fig. 8. Titration of resonances in the methyl region of HEWL. A linear plot is obtained by keeping the total concentration of lanthanide constant using the diamagnetic La(III) such that [La(III)] + [Pr(III)] = 50 m*M*.

both relaxed and shifted (9), and so the structure of an inhibitor or substrate lysozyme complex can be determined.

Acknowledgments

C. M. Dobson and A. V. Xavier acknowledge financial support from Merton College and the Calouste Gulbenkian Foundation, respectively. This paper is a contribution from the Oxford Enzyme Group which is supported by the Science Research Council. The work was also supported by the Medical Research Council.

References

1. McDonald, C. C., and Phillips, W. D., *in* "Fine Structure of Proteins and Nucleic Acids" (G. D. Fasman, and S. N. Timasheff, eds.), Vol. 4, p. 1. Marcel Dekker, New York, 1970.
2. Blake, C. C. F., Johnson, L. N., Mair, G. A., North, A. C. T., Phillips, D. C., and Sarma, V. R., *Proc. Roy. Soc., Lond.* [*Biol.*] **167,** 378 (1967).
3. Campbell, I. D., Dobson, C. M., Williams, R. J. P., and Xavier, A. V., *J. Mag. Res.* **11,** 172 (1973).
4. Barry, C. D., North, A. C. T., Glasel, J. A., Williams, R. J. P., and Xavier, A. V., *Nature* (*Lond.*) **232,** 236 (1971).
5. Dwek, R. A., Richards, R. E., Morallee, K. G., Nieboer, E., Williams, R. J. P., and Xavier, A. V., *Eur. J. Biochem.* **21,** 204 (1971).
6. Sternlicht, H., and Wilson, D., *Biochemistry* **6,** 2881 (1967).
7. Cowburn, D. A., Bradbury, E. M., Crane-Robinson, C., and Gratzer, W. B., *Eur. J. Biochem.* **14,** 83 (1970).
8. Lamar, G. N., Horrocks, W. de W., and Allen, L. C., *J. Chem. Phys.* **41,** 2125 (1954).
9. Morallee, K. G., Nieboer, E., Rossotti, F. J. C., Williams, R. J. P., Xavier, A. V., and Dwek, R. A., *Chem. Commun.* p. 1132 (1970).

The nmr Detection of the Distorted Intermediate in the Lysozyme Catalyzed Cleavage of Cell Wall Tetrasaccharide

STEVEN L. PATT,* DAVID DOLPHIN, and BRIAN D. SYKES†

The proposal by Phillips and co-workers (1–4) that the cleavage of oligosaccharides by lysozyme involves the distortion of the ring bound in subsite D toward a half-chair conformation has been discussed in several recent investigations (5–18). The major approach to the problem has been to measure or estimate the contribution to the free energy of binding of oligosaccharides from the ring bound in subsite D. These studies have shown that $(NAG)_3$, $(NAG)_4$, $(NAG)_5$, and $(NAG)_6$‡ have the same association constant to lysozyme ($K_A = 1.0 \times 10^5 \ M^{-1}$) (5, 10, 15); that (NAG–NAM)$_2$ is bound much less strongly than NAG-NAM-NAG ($K_A = 2.1 \times 10^3 \ M^{-1}$ vs. $K_A = 2.8 \times 10^5 \ M^{-1}$) (5); and that $(NAG)_3$–NAL ($K_A = 3.6 \times 10^6 \ M^{-1}$) (17, 18) and $(NAG)_3$–NAX ($K_A = 5.5 \times 10^6 \ M^{-1}$) (8) are bound much more strongly than $(NAG)_4$. The results imply that $(NAG)_4$,

* National Science Foundation Predoctoral Fellow 1970–1973.

† Alfred P. Sloan Fellow 1971–1973.

‡ Abbreviations used: $(NAG)_n$, $\beta(1 \rightarrow 4)$ linked oligomer of 2-acetamido-2-deoxy-D-glucose; NAM, 2-acetamido-2-deoxy-D-muramic acid; NAL, δ-lactone of NAG; NAX, 2-acetamido-2-deoxy-D-xylose.

$(NAG)_5$, and $(NAG)_6$ avoid unfavorable interactions in subsite D by binding only to subsites A, B, and C; that the interaction of NAM with subsite D is unfavorable by 2.9 kcal mole^{-1}; and that the interaction of NAL (which is already in a half-chair conformation) and NAX [which lacks the CH_2OH group of NAG that makes contact with lysozyme and has been suggested to be the cause of the distortion (3, 4)] with subsite D are favorable by 2.1–2.2 kcal mole^{-1}. A second approach has focused on a concomitant feature of the proposed mechanism, the carbonium ion character of the transition state in the reaction (10, 11, 13, 16). We have taken a third approach and report here the direct observation of a distorted intermediate in the lysozyme catalyzed cleavage of $(NAG-NAM)_2$ using nmr techniques. Preliminary reports of these studies have been published (19, 20).

Method

The conformation of the reducing ring of $(NAG-NAM)_2$ when this substrate is bound to lysozyme in subsites A–D, its primary mode of binding (7), can be determined in part by measuring the coupling constant between the anomeric proton H_1 and the ring proton H_2 (Fig. 1) (19, 20). The dihedral angle dependence of such coupling constants is well known (21–24), and the angle so obtained provides information about the conformation of the glucopyranose ring. The relationship between the coupling constant and the dihedral angle θ between H_1 and H_2 is given in general by

$$J_{H_1, H_2}(\theta) = A \cos^2 \theta + B \cos \theta + C \tag{1}$$

with the $\cos^2 \theta$ term dominant. Thus if the anomeric hydroxyl group is axial and the ring is in the chair conformation, $\theta \simeq 60°$ and $J_{H_1, H_2} \simeq 2$–3 Hz. As the ring is distorted toward the half-chair conformation, θ decreases toward $0°$ and J_{H_1, H_2} increases toward 7–9 Hz (Fig. 2).

Several factors are involved in the determination of the coupling constant for the bound substrate. First, the spectra of both the free and the bound

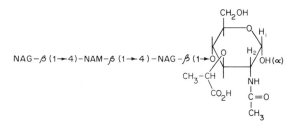

NAG$-\beta$ (1→4)$-$NAM$-\beta$ (1→4)$-$NAG$-\beta$ (1→O

Fig. 1. The α-anomer of cell wall tetrasaccharide.

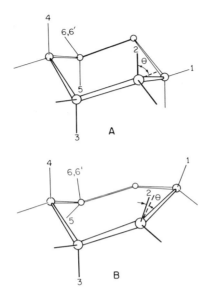

Fig. 2. (A) Chair conformation of the 2-acetamido-2-deoxy-D-glucopyranose ring. The numbers indicate the position of hydrogens. The dotted line marks the intersection between the C-2–C-1–H-1 plane and the H-2–C-2–N plane. The dihedral angle θ is the angle between the H-2–C-2–C-1 plane and the C-2–C-1–H-1 plane. (B) Half-chair conformation of the 2-acetamido-2-deoxy-D-glucopyranose ring. With the C-5–O–C-1 confined to a plane, the equatorial groups attached to C-1 and C-5 rotate to nearly axial positions, thus eliminating the substrate–enzyme interation at C-6 and closing the dihedral angle θ toward $0°$.

substrate are not necessarily simple. Fortunately the α-anomeric proton of the reducing N-acetylmuramic acid is well removed from the remaining protons in the $(NAG-NAM)_2$ spectrum (Fig. 3) so that the spectrum of the reducing ring protons is essentially ABCDEFX. The two strongest lines in the X (α-anomeric) portion of such a spectrum are centered at ν_X and separated by $J_{AX} + J_{BX} + J_{CX} + \cdots$ (25). Since all of these couplings with the exception of J_{H_1, H_2} are approximately zero for N-acetylmuramic acid, the observed doublet reflects only that coupling constant. The line is broadened, however, by the presence of other smaller, unresolved splittings.

The second factor is the presence of exchange of the substrate between free solution and the active site of lysozyme which in turn exchanges the anomeric proton of the reducing N-acetylmuramic acid between free solution (with chemical shift δ^S and coupling constant $J^S_{H_1, H_2}$) and subsite D in lysozyme (with chemical shift δ^{ES} and coupling constant $J^{ES}_{H_1, H_2}$). If the exchange is fast enough to satisfy the condition (26)

$$\left| \frac{(\delta^S - \delta^{ES}) \pm \frac{1}{2}(J^S_{H_1, H_2} - J^{ES}_{H_1, H_2})}{2/\tau} \right| \ll 1 \qquad (2)$$

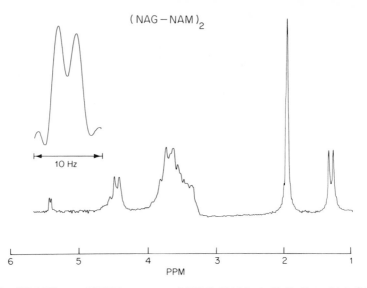

Fig. 3. 100 MHz nmr WEFT spectrum of (NAG–NAM)$_2$ in D$_2$O. From high field, the resonances are the CH$_3$ protons of the lactyl side chains, the acetyl protons, the ring protons, the β-anomeric protons, and the α-anomeric protons (shown also at 10 × horizontal expansion).

where τ is the lifetime of the bound substrate, then the observed coupling constant is the weighted average of that for the free and bound substrate:

$$J_{\text{OBS}} = P_S J_{\text{H}_1,\text{H}_2}^{\text{S}} + P_{\text{ES}} J_{\text{H}_1,\text{H}_2}^{\text{ES}} \tag{3}$$

where $P_S = [\text{S}]/[\text{S}]_{\text{TOTAL}}$ and $P_{\text{ES}} = [\text{ES}]/[\text{S}]_{\text{TOTAL}}$. $J_{\text{H}_1,\text{H}_2}^{\text{ES}}$ can then be determined given the known initial concentrations of E and S, the known association constant, the coupling constant for the free substrate, and the observed coupling constant in a mixture of substrate and lysozyme.

The third factor that must be considered is the cleavage of the substrate into dimer, which at the concentration levels used proceeds to approximately 50% cleavage, presumably according to the transglycosylation mechanism discussed by Chipman (7). As the tetramer is cleaved, the fraction of the remaining tetramer bound to lysozyme increases and thus the coupling constant and line width for the α-anomeric proton will increase during the course of the reaction. However, the α-anomeric proton of the weakly bound dimer which is formed in the reaction has a chemical shift identical to that of the α-anomeric proton of the tetramer. The observed doublet becomes, therefore, increasingly dominated by this resonance, causing the observed coupling constant to decrease to that of the free dimer during the course of the reaction.

Since the time course of the reaction is relatively rapid, spectra have to be taken as quickly as possible. Using Fourier transform techniques spectra can be accumulated at a rate of the order of the inverse of the resolution desired in the final spectrum (27). The sample is subjected to a series of rf pulses and the free induction decay resultant from each pulse is digitized and stored in the memory of the computer. Periodically the accumulated free induction decays are transferred to a storage device, where they are stored sequentially for retrieval and inspection following the completion of the reaction. Thus one may observe a reaction of the type

$$E + S \rightleftharpoons ES \rightleftharpoons ES' \rightleftharpoons E + P$$

if the steady-state concentration of the intermediate of interest lasts longer than the time required to acquire and store several free induction decays.

Experimental

Cell wall tetrasaccharide was prepared by the method of Hoshino *et al.* (28), and checked for purity by thin-layer chromatography using the solvent system 2-butanone, acetic acid, and water (14:6:5). R_f values obtained were NAG = 0.53, NAG-NAM = 0.41, and $(NAG-NAM)_2$ = 0.27. Three times crystallized lysozyme was purchased from P-L Biochemicals and used after repeated dialysis against or lyophilization from D_2O. The D_2O was 99.8% D as obtained from Stohler Isotope Chemicals. 100 MHz nmr spectra were taken on either a Varian HA-100 spectrometer equipped with Fourier transform capabilities, or a Varian XL-100-FT spectrometer. A Computer Operations tape drive was used as a storage device.

In a typical experiment, $1.2 \times 10^{-2}\ M$ $(NAG-NAM)_2$ and $3 \times 10^{-3}\ M$ lysozyme (both in $0.1\ M$ CD_3COOD buffer, pD* = 4.90, equilibrated in a temperature bath at 25.0°C) were mixed in 1:1 proportions directly in an nmr tube, which was quickly inserted into the nmr probe previously equilibrated at 25.0°C. A series of rf pulses was then applied to the sample for 15 sec (30 pulses) and the accumulated free induction decay rapidly (~ 2.5 sec) transferred to magnetic tape, the process being repeated for the course of the reaction.

For the non-nmr kinetic experiments, the reaction was initiated as above and then 5λ aliquots were extracted periodically and quenched into 9 ml of boiling Park–Johnson reagent (29) (3 ml H_2O, 3 ml carbonate-cyanide solution, 3 ml ferricyanide solution). After boiling for 15 min, a 3 ml aliquot

* Meter reading +0.4.

of this solution was added to 5 ml of ferric iron solution and the absorbance
of the final solution was read at 690 nm after 30 min color development.

Results and Discussion

The 100 MHz nmr spectrum of (NAG-NAM)$_2$ in D$_2$O solution is shown
in Fig. 3, with WEFT (30) having been used to eliminate the residual HDO
resonance at $\sim$4.6 ppm from the spectrum. The α-anomeric doublet is seen
to lowest field, with the expanded spectrum revealing more clearly its
coupling constant to H$_2$ of 2.5 $\pm$ 0.1 Hz. After rapid mixing of (NAG-NAM)$_2$
with lysozyme, the coupling constant is observed to be initially larger,
3.0 Hz being the largest value observed to date. The coupling constant decreases toward the value for the free NAG-NAM as seen in Fig. 4(A), with
a time constant of 90($\pm$30) sec. Extrapolation of the observed coupling
constant to zero time is difficult because of the complex reaction mechanism (7), but the estimated magnitude and the direction of the change are
consistent with a distortion of the dihedral angle between H$_1$ and H$_2$ toward
0°, and thus provides direct evidence for the presence of the distorted
species postulated by Phillips and co-workers.

Two methods were used to provide an independent check on the observed
time course of the reaction. It is well known (31–33) that the acetyl protons
of the ring bound in subsite C of the enzyme experience a large ($\sim$ 50 Hz)
upfield shift due to proximity to tryptophan 108. If the exchange between
the tetramer free in solution and bound to the enzyme is rapid (as is presumed by comparison with the known fast exchange of NAG-NAG with
its similar association constant) (33), then when the tetramer is first mixed
with the enzyme the main acetyl peak in the spectrum should include only
75% of the intensity observed in the absence of enzyme. As the tetramer is
cleaved, dimer is formed whose acetyl peaks are largely unshifted from
their normal position (34), and the integrated intensity of the acetyl resonance
will increase to the extent that the tetramer is cleaved. This experiment, seen
in Fig. 4(B), shows a time constant of 112 ($\pm$14) sec, and a magnitude of
change which indicates the cleavage of 55 ($\pm$10)% of the tetramer to dimer.

The observed magnitude of change in the unshifted acetyl intensity also
serves to confirm the fast exchange situation, since in the case of slow
exchange, only the actual number of acetyl groups bound in subsite C would
be shifted upfield. Since the enzyme remains essentially saturated even at
the end of the reaction, the number of acetyl groups bound in the subsite C
is largely unchanged during the course of the reaction, and thus in the slow
exchange limit an increase in the unshifted acetyl intensity would not be
observed. Since the changes between the free and bound spectrum are
much smaller for the α-anomeric doublet ($\delta_\text{S} - \delta_\text{ES} \simeq 0, |J^\text{S}_{\text{H}_1, \text{H}_2} - J^\text{ES}_{\text{H}_1, \text{H}_2}| \simeq$

6 Hz) than for the acetyl peak, it is clear that the α-anomeric doublet must also be in the fast exchange limit, as was assumed above.

A second check on the time course of the reaction was performed as described in the Experimental section by testing for the reducing power in the solution following rapid quenching of the reaction. The data, presented in Fig. 4C after correction for the reducing power of the enzyme, show a time constant of 73 ($\pm$ 22) sec and indicate, after correction for the different reducing powers of the tetramer and dimer (35), 51 ($\pm$ 10)% cleavage of the tetramer. Thin-layer chromatography showed no detectable quantities of higher oligomers present at the end of 5 min and essentially no change in the reaction mixture was observed from that time until at least 24 hr later. The extent of cleavage was again found to be ~ 50%.

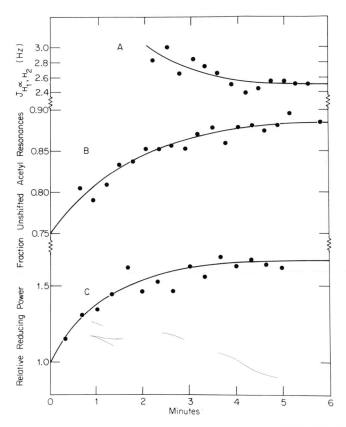

Fig. 4. Time course of the reaction of $1.2 \times 10^{-2} M$ (NAG–NAM)$_2$ with $3 \times 10^{-3} M$ lysozyme: (A) $J^{2}_{H_1,H_2}$; (B) integrated intensity of the unshifted acetyl resonances (see text); (C) reducing power of the saccharides in the reaction mixture estimated by the Park–Johnson test.

The nmr procedure described above was also repeated on mixtures of NAG-NAM and lysozyme, $(NAG)_3$ and lysozyme, or previously reacted mixtures of $(NAG-NAM)_2$ and lysozyme. In all cases no detectable change in the coupling constant was observed.

The reaction of 1.2×10^{-2} M $(NAG-NAM)_2$ and 3×10^{-3} M lysozyme at pD = 4.9 and 25°C has been studied. Three independent checks confirm that the reaction goes to completion with a cleavage of $\sim 50\%$ of the substrate, and three methods provide an approximate time constant for this reaction of 90 sec. The observed change of $J_{H_1^\alpha, H_2}$ proceeds at the same rate as the cleavage of tetramer to dimer, and the evidence strongly suggests that the observed coupling constant is a weighted average of the coupling constant for the free species with that for the species bound to the enzyme. Since no other phenomenon is expected under the present experimental conditions to induce changes in the coupling constant of this magnitude and direction except changes in the dihedral angle, this observation is then a clear indication that the reducing ring of $(NAG-NAM)_2$ is substantially distorted toward a half-chair conformation when bound to the enzyme. In conclusion, it should be noted that only the α-anomer has been observed in this experiment, which by nmr integration forms $\sim 60\%$ of the total tetramer. Since the group proposed to be the cause of the distortion is common to both anomers, and since all previous experiments have been performed with equilibrium mixtures of the two anomers, we have no reason to believe that the conclusion we have reached is not valid for both anomeric forms.

Acknowledgments

The authors would like to thank Prof. J. Strominger for a gift of tetrasaccharide; Deborah D. Jamison and Emilia Line Sciammas for assistance in the preparation of tetrasaccharide; and the National Institutes of Health [grant GM-17190 (BDS) and AM-14343 (DHD)], the National Science Foundation (Predoctoral Fellowship to S. L. Patt), and the Alfred P. Sloan Foundation (Fellowship to B. D. Sykes) for financial support.

References

1. Blake, C. C. F., Mair, G. A., North, A. G. T., Phillips, D. C., and Sarma, V. R., *Proc. Roy. Soc., Lond. [Biol.]* **167**, 365 (1967).
2. Blake, C. C. F., Johnson, L. N., Mair, G. A., North, A. C. T., Phillips, D. C., and Sarma, V. R. *Proc. Roy. Soc., Lond. [Biol.]* **167**, 378 (1967).
3. Phillips, D. C., *Sci. Am.* **215**, 78 (1966).
4. Phillips, D. C., *Proc. Natl. Acad. Sci. U.S.A.* **57**, 484 (1967).
5. Chipman, D. M., Grisaro, V., and Sharon, N., *J. Biol. Chem.* **242**, 4388 (1967).
6. Chipman, D. M., and Sharon, N., *Science* **165**, 454 (1969).

7. Chipman, D. M., *Biochemistry* **10,** 1714 (1971).
8. Van Eikeron, P., and Chipman, D. M., *J. Am. Chem. Soc.* **94,** 4788 (1972).
9. Dahlquist, F. W., Jao, L., and Raftery, M., *Proc. Natl. Acad. Sci. U.S.A.* **56,** 26 (1966).
10. Dahlquist, F. W., Rand-Meir, T., and Raftery, M., *Proc. Natl. Acad. Sci. U.S.A.* **61,** 1194 (1968).
11. Dahlquist, F. W., Rand-Meir, T., and Raftery, M. A., *Biochemistry* **8,** 4214 (1969).
12. Johnson, L. N., Phillips, D. C., and Rupley, J. A., *Brookhaven Symp. Biol.* **21,** 120 (1969).
13. Piskierwicz, D., and Bruice, T. C., *J. Am. Chem. Soc.* **90,** 5844 (1968).
14. Rupley, J. A., and Gates, V., *Proc. Natl. Acad. Sci. U.S.A.* **57,** 496 (1967).
15. Rupley, J. A., Butler, L., Gerring, M., Hartegen, F. J., and Pecoraro, R., *Proc. Natl. Acad. Sci. U.S.A.* **57,** 1088 (1967).
16. Rupley, J. A., Gates, V., and Bilbrey, R., *J. Am. Chem. Soc.* **90,** 5633 (1968).
17. Secemski, I. I., and Lienhard, G. E., *J. Am. Chem. Soc.* **93,** 3549 (1971).
18. Secemski, I. I., Lehrer, S. S., and Lienhard, G. E., *J. Biol. Chem.* **247,** 4740 (1972).
19. Sykes, B. D., and Dolphin, D., *Nature (Lond.)* **233,** 421 (1971).
20. Sykes, B. D., Patt, S. L., and Dolphin, D., *Symp. Quant. Biol.* **36,** 29 (1971).
21. Barfield, M., and Karplus, M., *J. Am. Chem. Soc.* **91,** 1 (1969).
22. Karplus, M., *J. Chem. Phys.* **30,** 11 (1959).
23. Karplus, M., *J. Chem. Phys.* **33,** 1842 (1960).
24. Karplus, M., *J. Am. Chem. Soc.* **85,** 2870 (1963).
25. Emsley, J. W., Feeney, J., and Sutcliffe, L. H., "High Resolution Nuclear Magnetic Resonance Spectroscopy," p. 424. Pergamon, Oxford, 1967.
26. Alexander, J., *J. Chem. Phys.* **37,** 967 (1962).
27. Ernst, R. R., and Anderson, W. A., *Rev. Sci. Instrum.* **37,** 93 (1966).
28. Hoshino, O., Zehavi, U., Sinay, P., and Jeanloz, R. W., *J. Biol. Chem.* **247,** 381 (1972).
29. Park, J. T., and Johnson, M. J., *J. Biol. Chem.* **181,** 149 (1949).
30. Patt, S. L., and Sykes, B. D., *J. Chem. Phys.* **56,** 3182 (1972).
31. Dahlquist, F. W., and Raftery, M. A., *Biochemistry* **7,** 3269 (1968).
32. Dahlquist, F. W., and Raftery, M. A., *Biochemistry* **8,** 713 (1969).
33. Sykes, B. D., and Parravano, C., *J. Biol. Chem.* **244,** 3900 (1969).
34. Raftery, M. A., Dahlquist, F. W., Parsons, S. M., and Wilcott, R. G., *Proc. Natl. Acad. Sci. U.S.A.* **62,** 44 (1969).
35. Ghuysen, J.-M., Tipper, D. J., and Strominger, J. L., *in* "Methods in Enzymology" (E. F. Neufeld and V. Ginsburg, eds.), Vol. 8, p. 685. Academic Press, New York, 1966.

Kinetics and Mechanism of Binding of α- and β-Methyl-D-Glucopyranosides to Concanavalin A as Studied by ¹³C-Carbon Magnetic Resonance: A Comparison with Lysozyme*

C. F. BREWER, D. MARCUS, A. P. GROLLMAN,
and H. STERNLICHT

Concanavalin A (Con A) and the lysozymes presently are the only sugar-binding proteins whose X-ray structures are known. Con A, however, is not an enzyme. Thus, an understanding of the mechanism of sugar binding to Con A may help separate the binding features from the catalytic features of lysozyme.

Con A, a hemagglutinin protein (1a) isolated from the jack bean (*Canavalia ensiformis*), interacts with biopolymers containing appropriate carbohydrate residues in a manner similar to an antigen–antibody reaction (2). The protein binds monosaccharides with the D-glucopyranosyl configuration at the 3-, 4-, and 6-hydroxy group positions, and interacts preferentially with the α-anomers of these sugars (3). The biological properties of Con A include selective agglutination of cells transformed by oncogenic viruses (4) and inhibition of tumor growth in experimental animals (5).

* Part of this work has been presented elsewhere (see reference 1).

The protomeric unit of Con A has a molecular weight of 27,000 (6, 7). The protein is a dimer between pH 3.5 and 5.6, and forms tetramers and other polymeric species at higher pH. Each protomeric unit has one sugar-binding site and two binding sites for metal ions: a site for transition metal ions and a specific calcium site. The protein binds sugars only when both metal ion sites are occupied (8, 9). Con A derivatives containing Zn^{2+}, Co^{2+}, or Mn^{2+} in the transition metal ion site possess equal sugar-binding activity. The primary sequence and electron density map of the protein at 2 Å resolution have recently been published by Edelman and co-workers (10).

We have studied the mechanism of sugar binding to Con A at pH 5.6 by carbon magnetic resonance relaxation measurements of uniformly enriched 14% $[^{13}C]$ α- and β-methyl-D-glucopyranosides (α- and β-MDG) in the presence of Zn^{2+}, Co^{2+}, and Mn^{2+} transition metal derivatives of the protein. Unlike earlier fluorine or proton nmr studies of saccharide binding to lysozyme (11, 12) which were experimentally restricted to measurements of a few resolvable protons or specifically labeled fluorine sites on the sugars, ^{13}C-nmr studies, as exemplified in our study, provides dynamic and structural information from essentially all of the carbons of the sugar when bound to the protein. We have been able to obtain information on (1) the binding constants of α- and β-MDG, (2) kinetics of binding, (3) distance and orientation of the bound sugars relative to the transition metal site in the protein, and (4) the possible site of sugar binding on the protein. Comparisons to lysozyme will be made.

Figure 1 shows the proton-decoupled carbon nmr spectrum of ^{13}C-enriched α-MDG at pH 5.6 in (A) the absence of protein, (B) the presence of Zn-Con A, and (C) the presence of Zn-Con A and α-methyl-D-mannopyranoside (α-MDM). Each of the carbon atoms in the sugar has a separate resonance line. The carbon assignments have been previously made in the literature (13, 14). No change in the chemical shifts of the sugar carbons is observed upon addition of protein. In the presence of Zn-Con A, α-MDG is observed to have broadened resonance lines (spectrum B). In spectrum C, addition of excess α-MDM, a competitive binding sugar, displaces the ^{13}C sugar from the protein and reduces the line broadening of the α-MDG resonances. These data are consistent with the observed line broadening of the ^{13}C sugar being due to specific binding to Con A.

The carbon linewidths of α- and β-MDG in the presence of "native" Con A at pH 5.60 are shown in Fig. 2 for the temperature range of 4° to 50°C. C-1 refers to the anomeric carbon and C-6 refers to the hydroxymethyl carbon. All of the sugar ring carbons give essentially the same results as the C-1 carbon. f is the ratio of protein bound to unbound sugar and $\Delta v - \Delta v_0$ is the change in linewidth upon adding protein. The ordinate is equal to $T_{2m} + \tau_m$ (15). T_{2m} is the transverse relaxation time of the protein-

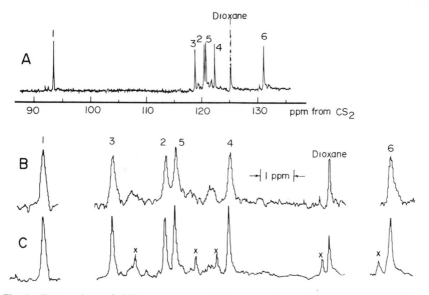

Fig. 1. Proton decoupled ^{13}C nmr spectra of ^{13}C enriched α-methyl-D-glucopyranoside at pH 5.6 and 25°C in (A) absence of protein; (B) presence of Zn-Con A; and (C) presence of Zn-Con A and α-methyl-D-mannopyranoside. The concentrations used were: 1.25×10^{-2} M ^{13}C enriched α-methyl-D-glucopyranoside, 8.7×10^{-4} M Zn-Con A, 4×10^{-2} M α-methyl-D-mannopyranoside. Spectra (B) and (C) are expanded scale presentations. The small peaks (X) in spectrum (C) are those of the unenriched α-methyl-D-mannopyranoside. Reprinted with permission from C. F. Brewer, H. Sternlicht, D. M. Marcus, and A. P. Grollman, *Biochemistry* **12**, 4448 (1973). Copyright by the American Chemical Society.

bound sugar, and is determined by the intramolecular dipole–dipole interaction between the carbon spin and its directly bonded protons. The C-6 carbon with two directly bonded protons, therefore, relaxes with a T_{2m} that is one-half that of the C-1 carbon which has one bonded proton (Fig. 2). τ_m is the residence time of the sugar on the protein which decreases by a factor of 2 per 10°C increase in temperature for both sugars. The τ_m of the β-anomer is approximately a factor of 10 shorter than that of the α-anomer at the same temperature.

Binding constants and the kinetics of sugar binding were obtained from measurements of α- and β-MDG linewidths as a function of temperature. These results are summarized in Table I. The affinity constant, K_a, of both anomers decreases by approximately a factor of 2 per 10°C increase in temperature. The off-rate, k_{-1}, which is equal to τ_m^{-1}, increases by approximately a factor of 2 per 10°C increase in temperature, and has a 13 kcal/mole activation energy. The on-rate, k_1, is essentially constant with increasing temperature and has a near zero activation energy. The two anomers have

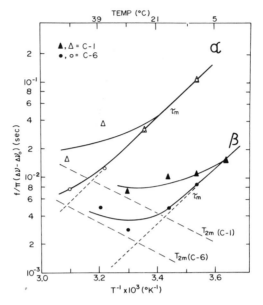

Fig. 2. Plot of the line broadening of the C-1 and C-6 resonances of ^{13}C enriched α- and β-methyl-D-glucopyranoside in the presence of "native" Con A at pH 5.6 as a function of temperature. The line broadening is expressed as $f/\pi(\Delta v - \Delta v_0)$ where $\Delta v - \Delta v_0$ denotes the change in linewidth in the presence of protein at the appropriate temperature, and f denotes the ratio of bound to unbound sugar. The total sugar concentration varied from $6.1 \times 10^{-3} M$ to $5 \times 10^{-2} M$; the Con A concentration varied from $4.8 \times 10^{-4} M$ to $8.7 \times 10^{-4} M$. The measurements at 50° were complicated by the instability of the protein at this temperature. Reprinted with permission from C. F. Brewer, H. Sternlicht, D. M. Marcus, and A. P. Grollman, *Biochemistry* **12**, 4448 (1973). Copyright by the American Chemical Society.

k_1 rate constants that differ by a factor of 2. The factor of 25 difference in the affinity constants for the two anomers is therefore primarily due to the large difference in k_{-1}.

A comparison of kinetic results for inhibitors binding to lysozyme with α- and β-MDG binding to Con A is shown in Table II. The activation energy for k_1 and k_{-1} of chitobiose binding to lysozyme (16) is essentially the same as we have found for α- and β-MDG binding to Con A. The chitobiose results have been interpreted (16) as indicative of two or more steps in a binding mechanism which involves a slow enzyme-inhibitor isomerization to give the final complex. The kinetics of α- and β-methyl-N-acetylglucosamine binding to lysozyme (17) also suggests multistep binding (Table II). We propose that the kinetics of binding of α- and β-MDG to Con A also involves a multistep binding mechanism in which isomerization of the protein–sugar complex occurs to give the final complex. This proposed isomerization is consistent with circular dichroism (18) measurements on Con A which indicated a protein conformational change upon sugar binding.

In an effort to learn more about the structural features of α- and β-MDG binding to Con A, the spin-lattice (i.e., longitudinal relaxation times) T_1 were measured. The τ_m contribution to T_1 was small relative to its contribution to T_2. Figure 3 shows the magnetization recovery of $[^{13}C]$ α-MDG in the presence of Zn- and Mn-Con A versus the pulse interval, Δt, in the $180°$-Δt-$90°$ pulse sequence used for the T_1 measurements (19). This is a technique in which the carbon spins are perturbed, and the signal intensities are measured as a function of Δt as the system relaxes back to equilibrium. M is the nonequilibrium magnetization value of a given carbon along the external magnetic field (represented by the observed signal intensity) as measured after a time interval, Δt, following the $180°$ pulse. M_0 is the equilibrium magnetization value. The ordinate is equal to exponential e, 2,718, when $\Delta t = T_1$. In the presence of Zn-Con A [Fig. 3(A)], T_1 of all the ring carbons are shortened to the same degree. The factor of 2 difference between T_1 of the ring carbons and the 6-carbon is consistent with the primary relaxation mechanism of the sugar bound to Zn-Con A being an intramolecular dipole–dipole interaction between the ^{13}C nuclei and its directly bonded hydrogen(s). The ^{13}C relaxation differences between the free and bound sugars are related to the differences in their molecular rotational

TABLE I

Affinity Constant and Rate Constants as a Function of Temperature for α- and β-MDG + Con A (pH 5.6)[a]

	$2°$	$10°$	$18°$	$25°$	$30°$	Arrhenius activation energy (kcal/mole)
β-MDG[b]						
K_a	210 (176)[c]	140	75 ± 15	70 ± 15	35 ± 6	
$k_1(\times 10^{-4})$	1.4	1.6	1.8 ± 0.4	2.8 ± 0.6	2.1 ± 0.4	$2.5 \left\{ {+0.5 \atop -2} \right\}$
k_{-1}	6.4×10	1.2×10^2	2.4×10^2	4.0×10^2	6×10^2	
α-MDG						
K_a	$4.9 \times 10^{3\ c}$	—	—	$1.7(\pm 0.4) \times 10^3$	—	
$k_1(\times 10^{-4})$	2.5	—	—	5 ± 1.2	—	$5 \left\{ {+1.4 \atop -2.2} \right\}$
k_{-1}	5 est.	0	1.7×10	3.1×10	4.8×10	13.2

[a] Portions of these data will be published elsewhere (C. F. Brewer, H. Sternlicht, D. M. Marcus, and A. P. Grollman, manuscript in preparation).

[b] K_a in liter mole^{-1}; k_1 in liter mole^{-1} sec^{-1}; k_{-1} in sec^{-1}.

$$\text{E} + \text{S} \underset{k_{-1}}{\overset{k_1}{\rightleftharpoons}} \text{ES}; \quad K_a = k_1/k_{-1} = [\text{ES}]/[\text{E}][\text{S}]$$

[c] L. L. So and I. J. Goldstein, *Biochim. Biophys. Acta* **165**, 398 (1968).

TABLE II

Comparison of the Kinetics of Inhibitors Binding to Lysozyme with Sugar Binding to Con A[a]

	Methyl-D-glucopyranoside + Con A (25°C)		Chitobiose + LYS[b] (30°C)	N-Acetylglucosamine + LYS (33°C)[c]	
	β-	α-		β-	α-
K_a (liter mole^{-1})	70 ± 15	1.7×10^3	5×10^3	35	25
$k_1 \times 10^{-4}$ (liter mole^{-1} sec^{-1})	2.8 ± 0.6	5 ± 1.2	380	16	14
$k_{-1} \times 10^{-2}$ (sec^{-1})	4.0	0.3	7.7	45	55
Activation energy (kcal/mole)					
k_1	$2.5\begin{Bmatrix} +0.5 \\ -2 \end{Bmatrix}$	$5\begin{Bmatrix} +1.4 \\ -2.2 \end{Bmatrix}$	~0	−2.3	—
k_{-1}	13.2	13.2	13	5	—

[a] Portions of these data will be published elsewhere (C. F. Brewer, H. Sternlicht, D. M. Marcus, and A. P. Grollman, manuscript in preparation).

[b] F. W. Dahlquist and M. A. Raftery, private communication.

[c] B. D. Sykes, *Biochemistry* **8**, 1110 (1969).

correlation times. In the case of the bound sugar, the slower tumbling time of the sugar–protein complex facilitates spin relaxation (20). A rotational correlation time, τ_r, can be calculated (20) for the protein–sugar complex using the Zn-Con A T_1 data and T_{2m} data from Fig. 2. τ_r at 25°C has a value of 8×10^{-8} sec and suggests that the sugar is firmly bound to the protein.

In the presence of Mn-Con A [Fig. 3(B)], the carbons of the α-anomer experience differing degrees of relaxation. Similar results were obtained for the β-anomer. These data indicate that the unpaired electron spins of the Mn^{2+} ion provides an additional mechanism of relaxation for the carbons of the bound sugar.

The observed T_1 of the sugar carbons in the presence of protein is related to T_{1m}, the relaxation times of the protein-bound sugar, and T_{1o}, the relaxation times in the absence of protein by

$$(T_1)^{-1} = (T_{1o})^{-1} + f/(T_{1m} + \tau_m) \tag{1}$$

where f < 1.

Table III lists T_{1m} for the carbons of α-MDG when bound to Zn- and Mn-Con A. T_{1p} denotes the contribution of the paramagnetic Mn^{2+} ion to the spin–lattice relaxation times of the bound sugar carbons, and is given by

$$(T_{1p})^{-1} = (T_{1m})^{-1}_{Mn\text{-Con A-sugar}} - (T_{1m})^{-1}_{Zn\text{-Con A-sugar}} \tag{2}$$

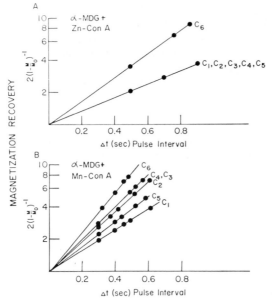

Fig. 3. Plot of the ^{13}C magnetization recovery versus pulse interval, Δt, for $2.5 \times 10^{-2}\ M$ [^{13}C] α-methyl-D-glucopyranoside in the presence of (A) $8.7 \times 10^{-4}\ M$ Zn-Con A and (B) $8.7 \times 10^{-4}\ M$ Mn-Con A. $2(1 - M/M_0)^{-1}$ is equal to 2.718 when $\Delta t = T_1$. Reprinted with permission from C. F. Brewer, H. Sternlicht, D. M. Marcus, and A. P. Grollman, *Biochemistry* **12**, 4448 (1973). Copyright by the American Chemical Society.

Somewhat different T_{1p} values were obtained for the β-anomer. The sugar carbons have different T_{1p} values because of their different distances to the Mn^{2+} ion. This is expressed quantitatively as (20, 21):

$$(T_{1p})^{-1} = \left(\frac{2}{15}\right) g^2 \beta^2 \gamma_I^2 S(S + 1) R_{CM}^{-6} \tau_C (1 + \omega_I^2 \tau_C^2)^{-1} \tag{3}$$

where g is the electronic "g" factor; β, the Bohr magneton; S, the electron spin quantum number of the paramagnetic ion; R_{CM}, the metal ion–carbon internuclear distance; ω_I and γ_I are the carbon Larmor frequency at 24 kgauss and the carbon gyromagnetic constant, respectively. τ_C is the correlation time for the magnetic perturbation at the carbons produced by the unpaired electrons of Mn^{2+}. τ_C was determined by Koenig *et al.* (22) from T_1 measurements of solvent water in the presence of Mn-Con A. At 24 kgauss, τ_C was shown to be equal to τ_r, the rotational correlation time which was a value of 8×10^{-8} sec obtained for the sugars bound to Zn-Con A.

Table IV lists the $Mn^{2+} - {}^{13}C$ distances for the α- and β-anomers calculated from the T_{1p} data. Each carbon of both sugars has a unique distance to the Mn^{2+} ion. $Mn^{2+} - {}^{13}C$ distances for the 2 anomers are somewhat different, suggesting that the two sugars bind in different orientations.

The mean distance of the Mn^{2+} to the sugars is 10 Å–11 Å. To determine the geometric consistency of the distance measurements determined from T_{1p} data, a computer program was written which used the X-ray crystal coordinates (23) of the α-anomer in the C-1 chair conformation for the

TABLE III

Spin-lattice Relaxation Time of [${}^{13}C$] α-Methyl-D-Glucopyranoside in the Presence of the Zinc and Manganese Derivatives of Con A at 25°C ($\tau_m = 0.030 \pm 0.005$ sec)[a,b]

| Con A transition metal derivative | Carbon | T_1 (sec)[c] | | | T_{1m} (sec) | $T_{1p}{}^{f,g}$ (sec) |
		Free sugar	$f^d = 0.071$	$f^e = 0.14$		
Free sugar	1	1.09				
(no protein)	2	1.06				
	3	1.03				
	4	1.03				
	5	1.07				
	6	0.60				
Zn-Con A	1		0.86	0.71	0.26	
	2		0.82	0.72	0.26	
	3		0.82	0.72	0.29 0.28[f]	
	4		0.84	0.72	0.30	
	5		0.83	0.75	0.28	
	6		0.45	0.40	0.12	
Mn-Con A	1		0.56	0.45	0.069	0.091
	2		0.46	0.32	0.036	0.040
	3		0.40	0.29	0.026	0.028
	4		0.40	0.29	0.026	0.028
	5		0.49	0.38	0.048	0.057
	6		0.33	0.24	0.029	0.038

[a] Reprinted with permission from C. F. Brewer, H. Sternlicht, D. M. Marcus, and A. P. Grollman, *Biochemistry* **12**, 4448 (1973). Copyright by the American Chemical Society.

[b] Concentration of protein was 8.7×10^{-4} M.

[c] T_1 was determined accurately to ± 0.03 sec.

[d] Total sugar concentration was 2.5×10^{-2} M.

[e] Total sugar concentration was 1.25×10^{-2} M.

[f] An average T_{1m} of ca. 0.28 sec for the ring carbons of the Zn-Con A bound sugars were used in calculating T_{1p}. The differences in the individual T_{1m} values reflect experimental errors in determining T_1 values.

[g] The estimated error in T_{1p} is ca. $\pm 15\%$ for the ring carbons and ca. $\pm 25\%$ for the C-6 carbon.

TABLE IV

Experimental (nmr) and Computer Fitted (c.f.) Distances between the Manganese Ion and Carbons of α- and β-Methyl-D-Glucopyranoside[a]

Sugar	Carbon	T_{1p} (sec)	R(Å)exp.[b]	R(Å)c.f.
[^{13}C]α-Methyl-D-glucopyranoside	1	0.091	11.7	11.5
	2	0.040	10.2	10.4
	3	0.028	9.6	9.8
	4	0.028	9.6	9.4
	5	0.057	10.8	10.6
	6	0.038	10.1	10.4
[^{13}C]β-Methyl-D-glucopyranoside	1	0.11	12.0	12.0
	2	0.075	11.2	11.4
	3[b]	0.054[c]	10.6	10.0
	4	0.034	9.9	10.2
	5[b]	0.054[c]	10.6	10.9
	6	0.10	11.8	11.5

[a] Reprinted with permission from C. F. Brewer, H. Sternlicht, D. M. Marcus, and A. P. Grollman, *Biochemistry* **12**, 4448 (1973). Copyright by the American Chemical Society.

[b] τ_r was set equal to 8×10^{-8} sec at 25°C. The estimated error in R(Å) for this τ_r value is ± 0.3 Å.

[c] These resonances are unresolved at 24 kgauss. T_{1p} is an apparent average value for the relaxation times of these two carbons.

internal bond angles and bond distances for the bound sugar. The program attempted to find an orientation(s) of the sugar which would satisfy all of the $Mn^{2+} - {}^{13}C$ distances determined from the T_{1p} data and the internal bond angles and distances of the sugar. One unique orientation was found for the α-anomer; two orientations were found for the β. Only one of the orientations for β-MDG will be discussed in that it explains the stereochemical factors involved in binding. The computer-calculated distances for α- and β-MDG are shown in the last column in Table IV and are in excellent agreement with the distances calculated from the T_{1p} data. This agreement rapidly vanishes if the absolute distance of the sugar carbons to the Mn^{2+} is varied more than ± 0.5 Å.

The computer-determined orientations of the α- and β-anomers are shown in Fig. 4. The rotamer orientations of the methoxy and hydroxyl groups for both sugars are not known from this study and are arbitrarily positioned. In the proposed orientations, the 2-, 3-, and 4-hydroxyl groups of the β-anomer bind at positions occupied by the α-anomer's 6-, 4-, and 3-hydroxyl groups, respectively. The two sugar ring conformations are, however, the same. In this connection, it is interesting to note that studies of lysozyme-catalyzed transglycosylation reactions (24) suggest relative

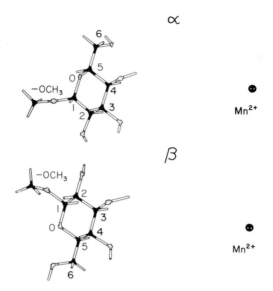

Fig. 4. Binding orientation of α- and β-methyl-D-glucopyranoside relative to the transition ion site in Con A. The rotamer orientations of the –OCH₃ and –OH groups for both sugars are not known from this study. Reprinted with permission from C. F. Brewer, H. Sternlicht, D. M. Marcus, and A. P. Grollman, *Biochemistry* **12**, 4448 (1973). Copyright by the American Chemical Society.

rotations of sugar rings in the E subsite of lysozyme. These rotations also maintain superimposable ring conformations.

The position of the 6-hydroxymethyl group of β-MDG is located at the position occupied by the 2-hydroxyl group of the α-anomer (Fig. 4). There appears to be sufficient steric tolerance in this binding region to accommodate the 6-hydroxymethyl group of the β-anomer based on binding studies by Goldstein of a series of 2-substituted α-D-glucopyranoside derivatives to Con A (25). The proposed unfavorable steric interaction of the β-anomer's equatorial methoxy group with the protein, if the sugar were to bind in an orientation similar to the α-anomer's orientation, is overcome in the proposed binding orientation since the C-1 carbon and the methoxy group are rotated to a position over the ring oxygen of the α-anomeric binding orientation. The 3-, 4-, and 6-hydroxy groups of the α-anomer have been implicated as the primary determinants for binding to Con A (3). The difference in the binding orientations of the two sugars shown in Fig. 4 suggests that the difference in binding constants between the two sugars, which favors the α-anomer by a factor of 25 at 25°C, can be explained by the lack of efficient binding of the β-anomer's 2-hydroxyl group with the protein at the position occupied by the 6-hydroxymethyl group of the α-anomer. This suggestion is consistent with the observation by So and

Goldstein (26) that α- and β-methyl-D-xylopyranosides bind with essentially the same affinity constant. These sugars lack the 6-hydroxymethyl group.

Our data indicate that the mean distance separating the α- and β-MDG sugars from the Mn^{2+} site is approximately 10–11 Å. Edelman and co-workers (10) have reported that the iodine atom of β-(o-iodophenyl)-p-glucopyranoside is located in the major cleft in the protein about 20 Å from the Mn^{2+} ion. The glucopyranoside ring was not observed, but its position was inferred to be in the cleft above the iodine atom. In this position the nonreducing end of the sugar moiety faces away from the Mn^{2+} ion and is directed toward the surface of the protein, approximately 20 Å from the Mn^{2+} ion. Our data, which were obtained from solution measurements, suggest that if binding occurs in the major cleft as defined in the crystal structure, then α- and β-MDG and probably polysaccharides bind at the bottom of the cleft 10–11 Å from the manganese with their nonreducing end facing the Mn^{2+}. The data do not rule out the possibility that the carbohydrate binding site may be located elsewhere in the protein.

In conclusion, the techniques used in this study provide a unique method for determining the orientation of sugars bound to proteins *in solution*. These techniques may be applied to a variety of small molecules enriched in ^{13}C that bind to proteins containing a paramagnetic ion binding site(s). When X-ray crystallographic data are available for the metalloprotein, distance calculations derived from similar ^{13}C nmr measurements should help define the binding points of the protein to the small molecule. The orientation of small molecules bound to macromolecules as determined by these nmr techniques can explain the differences in their affinity constants as well as predict structure–activity relationships. The techniques in this study could be applied to the study of sugar binding to lysozyme.

Note Added in Proof

We have synthesized β-(o-iodophenyl)-D-glucopyranoside, uniformly labeled with 14% ^{13}C in the sugar moiety, and used the ^{13}C nmr techniques described in this paper to demonstrate that a fraction of this aryl glycoside binds to Mn-Con A with the nonreducing end of the sugar moiety directed toward the Mn^{2+} ion at a mean distance of 10 Å. Additional experiments indicate that some of the remaining β-(o-iodophenyl)-D-glucopyranoside binds at a second site in the protein which may relate to the site discussed by Edelman *et al.* (10) for the crystalline complex. In addition, C. F. Brewer and K. D. Hardman [*Biochemistry* **12**, 4442–4448 (1973)] have demonstrated that β-(o-iodophenyl)-D-galactopyranoside, an analog of β-(o-iodophenyl)-D-glucopyranoside that does not inhibit hemagglutination or dextran precipitation by Con A, forms a crystalline complex with Con A in which the iodine atom of β-(iodophenyl)-D-galactopyranoside occupies the same site in the protein as that of β-(o-iodophenyl)-D-glucopyranoside. Our results suggest that the iodophenyl ring rather than the saccharide moiety is principally responsible for the binding of β-(o-iodophenyl)-D-glucopyranoside and β-(o-iodophenyl)-D-galactopyranoside to crystalline Con A.

Acknowledgments

The authors wish to thank Dr. Allen Tonelli of Bell Laboratories, Murray Hill, New Jersey, for his excellent assistance in writing the computer program used in this study.

C. F. Brewer is the recipient of a National Institute of Health Postdoctoral Traineeship (USPHS GM 00065) in the Department of Pharmacology, Albert Einstein College of Medicine.

References

1. *Abstr. Pap., 164th Nat. Meet., Amer. Chem. Soc.* Pap. No. 241, Biol. Chem. Sect. (1972).
1a. Summers, J. B., and Howell, S. F., *J. Bacteriol.* **32**, 227–237 (1936).
2. Goldstein, I. J., Hollerman, C. E., and Merrick, J. M., *Biochim. Biophys. Acta* **97**, 68–76 (1965).
3. Goldstein, I. J., Hollerman, C. E., and Smith, E. E., *Biochemistry* **4**, 876–883 (1965).
4. Inbar, M., and Sachs, L., *Proc. Natl. Acad. Sci. U.S.A.* **68**, 1418–1425 (1969).
5. Shoham, J., Inbar, M., and Sachs, L., *Nature* (*Lond.*) **227**, 1244–1246 (1970).
6. Kalb, A. J., and Lustig, A., *Biochim. Biophys. Acta* **168**, 366–367 (1968).
7. McKenzie, G. H., Sawyer, W. H., and Nichol, L. W., *Biochim. Biophys. Acta* **263**, 283–293 (1972).
8. Kalb, A. J., and Levitzki, A., *Biochem. J.* **109**, 669–672 (1968).
9. Agrawal, B. B. L., and Goldstein, I. J., *Can. J. Biochem.* **46**, 1147–1150 (1968).
10. Edelman, G. M., Cunningham, B. A., Reeke, G. N., Jr., Beeker, J. W., Waxdal, M. J., and Wang, J. L., *Proc. Natl. Acad. Sci. U.S.A.* **69**, 2580–2584 (1972).
11. Dahlquist, F. W., and Raftery, M. A., *Biochemistry* **8**, 713–725 (1969).
12. Butchard, C. G., Dwek, R. A., Kent, P. W., Williams, R. J. P., and Xvvier, A. V., *Eur. J. Biochem.* **27**, 548–553 (1972).
13. Perlin, A. S., Casu, B., and Koch, H. J., *Can. J. Chem.* **48**, 2596–2609 (1970).
14. Dorman, D. E., and Roberts, J. D., *J. Am. Chem. Soc.* **92**, 1355–1361 (1970).
15. Swift, T. J., and Connick, R. E., *J. Chem. Phys.* **37**, 307–320 (1962).
16. Dahlquist, F. W., and Raftery, M. A., private communication (1972).
17. Sykes, B. D., *Biochemistry* **8**, 1110–1116 (1969).
18. Pflumm, M. N., Wang, J. L., and Edelman, G. M., *J. Biol. Chem.* **246**, 4369–4370 (1971).
19. Vold, R. L., Waugh, J. S., Klein, M. P., and Phelps, D. E., *J. Chem. Phys.* **48**, 3831–3832 (1968).
20. Abragram, A., "Principles of Nuclear Magnetism," Chapter 8, pp. 264–322. Oxford Univ. Press, London and New York, 1961.
21. Bloembergen, N., *J. Chem. Phys.* **27**, 572–573 (1957).
22. Koenig, S., Brown, R., and Brewer, C. F., *Proc. Natl. Acad. Sci U.S.* **70**, 475–479 (1973).
23. Berman, H. M., and Kim, S. H. *Acta Crystallogr.* [*B*] **24**, 897–904 (1968).
24. Pollock, J. J., and Sharon, N., *Biochemistry* **9**, 3913–   (1970).
25. Poretz, R. D., and Goldstein, I. J., *Biochemistry* **9**, 2890–2896 (1970).
26. So, L. L., and Goldstein, I. J., *Carbohyd. Res.* **10**, 231–244 (1969).

Structure and Chemistry of Lysozyme: pH-Rate Profile, Calorimetric Studies, and Computations on Exposure to Solvent*

J. A. RUPLEY, S. K. BANERJEE, I. KREGAR, S. LAPANJE,
A. F. SHRAKE,[†] and V. TURK

Introduction

High-resolution crystallographic analyses of lysozyme and several lyso-zyme inhibitor complexes were completed in 1966 (5, 6). Thus the detailed chemical studies that have been carried out since then have had the crystal structure as a guide to experimental design and interpretation, which un-doubtedly has reduced confusion and controversy. It is satisfying that pre-dictions about mechanism made by Phillips and his colleagues and based largely on the crystallographic information have in the last five years been shown to be essentially correct through subsequent chemical and crystal-lographic experiments. These studies have given depth to the understanding of mechanism which could not have come from the crystallography alone. There remain points of importance with regard to function that are not clear, however, and some of these are considered in the work described following.

*This work was supported by the American Cancer Society, the National Institutes of Health, and the National Science Foundation.
†National Institutes of Health Postdoctoral Fellow.

Ionizations Important for Catalysis

The ten carboxyl groups of hen egg-white lysozyme constitute a particularly interesting class of side chains. Six are strongly affected by folding and have the p*K* presumably perturbed (15). Four groups have the environment altered through interaction with substrate or inhibitors: Glu 35, Asp 52, and Asp 101 contact saccharide, and Asp 66 is buried within a wing of the cleft that moves slightly when saccharide binds.

A variety of experiments have given the following information on the p*K* of these groups in the free protein and in protein–ligand complexes [summarized by Imoto *et al.* (15)]. Glu 35 is largely withdrawn from solvent in the free protein. Its environment is relatively apolar and the carboxyl does not participate in hydrogen bonds. A considerable body of data indicates that the p*K* of Glu 35 is near 6 in the free protein and is increased through binding of saccharides. It is of interest that the ionization of Glu 35 in the free protein was precisely measured a decade ago by Donovan *et al.* (11), who studied the pH dependence of the UV spectrum. Asp 101 is in the free protein essentially fully exposed to solvent, and in the enzyme–$(GlcNAc)_3$

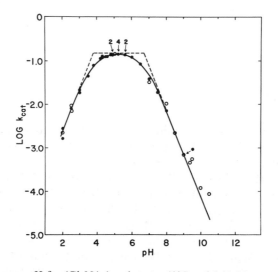

Fig. 1. Log k_{cat} vs. pH for $(GlcNAc)_6$ substrate. 40°C and 0.01 *M* acetate, phosphate, or carbonate buffer, with NaCl to ionic strength 0.1. Overlaps of pH ranges for the buffers showed no buffer effect. Measurements (open circles) of extent of hydrolysis using charcoal columns to separate the products of glycosyl transfer reaction (saturating concentration ca. 10^{-2} *M* $(GlcNAc)_6$ and 0.1 *M* GlcNAc) were used so that the low rates at very high pH could be measured. Other data (closed circles) are from measurements of increase in reducing sugar concentration (automated ferricyanide assay), with extraction of kinetic parameters by weighted least squares fit of Lineweaver–Burk equation to the data. From Ref. 3a.

complex forms two hydrogen bonds and is largely withdrawn from solvent. The free energy of formation of the enzyme-(GlcNAc)$_3$ complex shows pH dependence in the range pH 2 to 5 that indicates an ionization with pK ca. 4.2 in the free protein and pK about one unit lower in the complex. This ionization is almost certainly that of Asp 101. This is concluded from the observations that the pH dependence of binding of mono- and trisaccharide differs significantly only over the pH range 2–5, and crystallographic analysis shows Asp 101 to be the only ionizable group having a significantly different environment in the two complexes. A group of pK 1.5–2 is seen in saccharide binding (2) and in measurements of the pH dependence of lysozyme denaturation (1, 25). Asp 66 is completely withdrawn from solvent and is believed to be the only ionization of lysozyme with pK so abnormally low. The structural changes about Asp 66 presumably produce the small pK shift associated with saccharide binding. The pK of Asp 52 has been reported as ca. 4.5 (19). Comparison of the titration of native lysozyme and lysozyme specifically esterified at Asp 52 indicates that a group of pK 4.5 in the native protein is not present in the ester. Similar conclusions were reached by Sharon and his colleagues (this volume) from their studies of epoxide derivatives of inhibitors that also specifically esterify Asp 52.

In the mechanism proposed by Phillips and his colleagues Glu 35 and Asp 52 participate in the catalysis of bond rearrangement. Glu 35 is understood to be a general acid catalyst and in the active enzyme form this side chain should be protonated. Asp 52 is understood to act as an electrostatic catalyst and in the active form it should be deprotonated. If the mechanistic proposal is correct, these two ionizations should appear in the pH-rate profile for lysozyme.

Figures 1 and 2 give, respectively, the pH dependence of k_{cat} and of k_{cat}/K_m. The measurements were made using (GlcNAc)$_6$. Only a single bond in this substrate, that between the 4th and 5th units from the nonreducing end, is hydrolyzed in the enzyme-catalyzed reaction (22). The time course of the reaction was determined by the increase in reducing sugar associated with hydrolysis of the glycosidic bond. It is interesting that the progress curve follows first-order kinetics. This is to be expected [eq. (1)] for an enzymatic reaction in which the product competitively inhibits, with inhibition constant K_p equal to K_m (13).

$$v = \frac{k_{cat}(E)(S)}{K_m\left(1 + \dfrac{(P)}{K_p}\right) + S}$$

$$= \frac{k_{cat}(E)(S)}{K_m + S_0} \text{ (for } K_m = K_p\text{)} \tag{1}$$

$$= k_1(S)$$

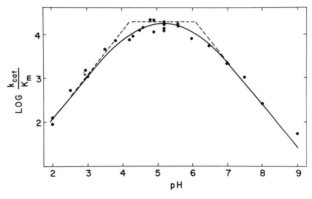

Fig. 2.   Log k_{cat}/K_m vs. pH. Conditions as in Fig. 1.

The products of $(GlcNAc)_6$ hydrolysis are $(GlcNAc)_2$ and $(GlcNAc)_4$. The tetrasaccharide shows the same free energy of binding to lysozyme as found for the hexasaccharide (3). $(GlcNAc)_2$ binds almost two orders of magnitude more weakly than hexasaccharide and competition between this product and substrate can be ignored.

The pH-rate profiles of Figs. 1 and 2 are remarkably simple in that the data closely follow theory for dependence of rate on two ionizations, in agreement with the mechanistic proposal involving Asp 52 and Glu 35. Studies with bacterial cell walls (9) have shown strong ionic strength dependence and a pH optimum ca. 9 which depends on ionic strength and is changed through lysyl modification. Hydrolysis of the nitrophenyl glycoside of GlcNAc-Glc shows (21) complex pH dependence of the kinetic constants in the acid region. These difficulties found for both very large and very small substrates were not observed with the hexasaccharide. This is perhaps understandable in view of the crystallographic suggestion that the hexasaccharide is the smallest substrate that can fill the active site. As noted $(GlcNAc)_6$ undergoes catalyzed hydrolysis at a unique position, and smaller oligomers of N-acetylglucosamine do not. The hexasaccharide of N-acetylglucosamine and the corresponding cell wall hexasaccharide are the most effective of the known substrates for lysozyme. These compounds show the smallest values for K_s and K_m (ca. 10^{-5} M) and the largest values for k_{cat} (0.1–0.5 sec^{-1}), which are four orders of magnitude greater than k_{cat} for small saccharides such as $(GlcNAc)_3$ or the nitrophenyl glycoside of GlcNAc-Glc.

Analysis of the pH-rate profiles of Figs. 1 and 2 gives kinetic apparent pK values of 3.8 and 6.7 for k_{cat} and 4.2 and 6.1 for k_{cat}/K_m. According to the simplest mechanistic model, pK values estimated from a k_{cat}/K_m profile apply to the free enzyme and those estimated from a k_{cat} vs. pH profile apply to the enzyme-substrate complex. The alkaline limb of the curve agrees well with expectation for Glu 35. As noted, in equilibrium studies a pK near 6 for the free protein is found shifted to higher pK through interaction with saccharides that resemble substrate, which should be compared with kinetic pK of 6.1 and 6.7 for free protein and complex. The acid limb of the pH-rate profile is not easily understood in terms of the pK near 4.5 reported for Asp 52 in free protein and enzyme-substrate complex (19). We feel that the accuracy of the kinetic data is sufficient to preclude a fit of either the k_{cat} or k_{cat}/K_m profiles using pK 4.5. The effect of binding of the substrate on ionizations other than Asp 52 must be taken into account in interpretation of k_{cat}/K_m. There is a strong effect of pH on binding of trisaccharide and tetrasaccharide in the pH range 2 to 5 (2–4) which is ascribed to perturbation of Asp 101. A comparable effect should obtain for hexamer. pK_m and pK_s for the $(GlcNAc)_6$ substrate show the same pH dependence within experimental error (0.1 log unit). If the pH dependence reflects Asp 101, then the pK of Asp 52 in the free protein is substantially lower than the value of 4.2 estimated on the simple assumption above. The data shown in Fig. 2 can be fit well using pK 3.8 for Asp 52 in the free protein and the pH dependence measured for K_s and K_m. This modified interpretation of the k_{cat}/K_m data suggests that the pK of Asp 52 in the free enzyme is the same as that in the enzyme-substrate complex and is 3.8.

The kinetic apparent pK of 3.8 which according to the mechanism should be assigned to Asp 52 is in conflict with the estimate of 4.5 for the pK of this side chain determined from the comparison of native lysozyme with lysozyme esters. This can be resolved in two ways: (i) kinetic parameters can be complex functions of pH and protein ionization constants. The discussion given above was according to the simplest formulation of the mechanism of an enzymatic reaction, and the reaction for lysozyme is more complex. Substrates bind to lysozyme in several modes, only one of which is productive and passes over into products. At least one nonproductive mode of binding of hexamer has been established (14). Equation (2) gives the simplest mechanism that can obtain for lysozyme:

$$E + S \quad \begin{array}{c} \overset{K_p}{\nearrow} \quad ES_p \overset{k_p}{\longrightarrow} E + P \\ \Big\Updownarrow R \\ \underset{K_{np}}{\searrow} \quad ES_{np} \end{array} \qquad (2)$$

Subscripts p and np denote productive and nonproductive complexes of enzyme with substrate. K_p and K_{np} are association constants. Equation (3) gives for this mechanism the dependence of $\log k_{cat}$ upon the extents of ionization (α) of protein groups, which in turn are functions of pH and pK.

$$k_{cat} = \frac{K_p}{K_p + K_{np}} \cdot k_p$$

$$k_{cat} = \frac{K_p}{K_p + K_{np}} \cdot k_{max} \cdot \alpha^1_{ES,p} \cdot (1 - \alpha^2_{ES,p})$$

$$\log k_{cat} = \log(k_{max} \cdot R \cdot K^1_{ES,p}/K^1_{ES,np}) + \log \alpha^1_{ES,np} + \log(1 - \alpha^2_{ES,np}) \quad (3)$$

$$- \log\left\{\left[1 + R \prod_{i=1}^{n} \frac{(1 - \alpha^i_{ES,np})}{(1 - \alpha^i_{ES,p})}\right]\left[\prod_{i=3}^{n} \frac{1 - \alpha^i_{ES,p}}{1 - \alpha^i_{ES,np}}\right]\right\}$$

$$R = \frac{K_{p,ref}}{K_{np,ref}}$$

Superscripts 1 and 2 denote the ionizable groups understood important in the mechanism, i.e., group 1 must be deprotonated and group 2 protonated for catalysis. Groups 3 to n include all other ionizations that might be affected by substrate binding, but are not directly involved in the catalysis. Two points are to be made. First, the observed kinetic apparent pK for groups 1 and 2 can depend entirely upon ionizations within nonproductive complexes, if productive complexes are present in small amounts relative to nonproductive (a small value of the constant R). Second, and more important for explanation of the kinetic pK, the last term of Eq. (3) includes ionizations that are not involved in bond rearrangement but are affected differently through productive and nonproductive binding of substrate. It is possible that contributions of this kind can rationalize the kinetic apparent pK of 3.8 and the pK of 4.5 determined for Asp 52 through comparison of native and modified proteins.

(ii) Alternatively, it is possible that esterification of Asp 52 results in structural changes that are reflected in the ionizations of other side chains and that must be taken into account in explanation of the difference titration. Asp 52 participates in hydrogen bonding interactions with other elements of the wing that forms part of the lysozyme active site. Furthermore, there is no reason to expect that the last term of Eq. (3) should be large, because groups other than Asp 52 and Glu 35 should not be perturbed differently in productive and nonproductive enzyme-substrate complexes. The only group that is not catalytic that could be expected to contribute in the pH region 2 to 5 is Asp 101, and the contacts of this side chain with substrate should be essentially insensitive to changes in mode of substrate binding in the lower part of the active site and would not be responsible for a shift of 0.7 pK between productive and nonproductive complexes. The enthalpy measurements described below suggest that indeed the kinetic apparent pK may be close to the equilibrium pK of Asp 52.

Enthalpy Measurements

Figure 3 shows the pH dependence of the heat of reaction of lysozyme with the β-methyl glycoside of GlcNAc. The curve of Fig. 3 is calculated for pK 3.8. Because binding of β-methyl-GlcNAc is independent of pH in the range 2 to 5 (8), the pK observed in the enthalpy measurements can be simply interpreted as reflecting ionization of a group with the same pK in both free enzyme and enzyme–saccharide complex. It is difficult to explain these data except in terms of ionization of Asp 52. Other groups shown by the crystallography to have their environment changed through substrate binding have pK considerably different from the value 3.8 seen in the enthalpy and kinetic measurements. The agreement between the pK found in the thermodynamic measurement of heat of reaction and the kinetic apparent pK indicates that the latter reflects rather directly Asp 52 in accordance with the Phillips' mechanism. The additional complexity described in the preceding section [the last term of Eq. (3)] need not be brought in to explain the kinetics.

The 1.6 kcal change in enthalpy of reaction without significant change in free energy is striking. Enthalpy–entropy compensation of this sort may be common where protein ionizations are involved. Substantially larger effects of pH change on enthalpy compared with free energy have been

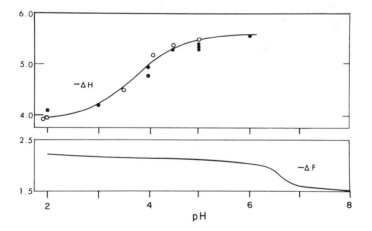

Fig. 3. pH dependence of the heat of binding of the β-methyl glycoside of N-acetylglucosamine. Calorimetric determinations were carried out at ionic strength 0.1 and 30°C, using an LKB batch reaction calorimeter. Generally 2 ml of 10 mg/ml protein were mixed with 4 ml saccharide solution, to give a final saccharide concentration that produced ca. 75% saturation. Molar heats of reaction (kcal) were calculated from observed reaction heats after correction for heats of dilution and using the equilibrium constant for the reaction. Open circles are for 0.1 M NaCl, closed circles for 0.09 M NaCl with 0.01 M NaOAc.

observed for reactions of chymotrypsin (10, 24), ribonuclease (12), and other proteins. In the binding of $(GlcNAc)_3$ to lysozyme, a 6 kcal change in ΔH is to be compared with a 1 kcal change in ΔF over the pH range 2 to 5 (2, 4). Substantial compensating effects upon enthalpy and entropy are probably associated with solvent participation. It is tempting to interpret the enthalpy difference between solvent-accessible and solvent-shielded ionizations in terms of organization of the aqueous solvent about the accessible carboxylate. Coordination of water about carboxylate ions results in a *net* decrease in ordered water structure, because of second sphere effects, and this leads to large positive enthalpy and entropy contributions that would not obtain for ionization of a carboxylate shielded from solvent, for which there would be no change in the number of hydrogen bonds or in structure of the environment.

The sensitivity of heat of reaction to environment and changes in environment makes it particularly useful as a probe for protein reactions. This approach has been used with great success by Sturtevant and his coworkers. Recent advances in design of batch and flow reaction calorimeters (17, 27) have led to instruments that make the measurement of reaction heat a standard laboratory operation.

Two other examples of calorimetric measurements that bear upon lysozyme chemistry are given in Table I and Fig. 4. Table I gives the enthalpy of reduction of disulfide bonds of lysozyme and oxidized glutathione by dithiothreitol. The heat of reduction of glutathione at pH 8 is near 0 in water (ΔH, -0.5 kcal/mole) and in 8 M guanidine hydrochloride (-0.4 kcal/mole). The enthalpy of reduction of the disulfides of disorganized

TABLE I

Enthalpy of Reduction of Disulfides of Lysozyme and Glutathione by Dithiothreitol at 30°

		Concentration		
pH	Solvent	Protein or GSSG (mg/ml)	Dithiothreitol (M)	ΔH (kcal/mole SS)
A. Native lysozyme				
8.0	6 M GuHCl	4.2–38.9	0.03–0.05	-1.51 ± 0.06
8.0	8 M GuHCl	9.0–16.6	0.03–0.04	-1.45 ± 0.15
6.0	6 M GuHCl	17.8–34.4	0.04–0.05	-1.17 ± 0.07
B. 16-hr peptic digest of lysozyme				
8.0	6 M GuHCl	20.3–21.5	0.03–0.04	-1.09 ± 0.07
C. Glutathione				
8.0	6 M GuHCl	9–13	0.06	-0.43 ± 0.10
8.0	water	15	0.05	-0.51 ± 0.05

[a] From Ref. 15a.

native lysozyme (6 or 8 M guanidine, pH 6 or 8) is only slightly more negative, -1.2 to -1.5 kcal/mole disulfide bond. Lysozyme extensively digested with pepsin shows in the same solvent a heat of reaction -1.1 kcal/mole disulfide. The heat of breaking an intrachain disulfide bridge thus is only ca. 0.5 kcal more negative than that for breaking an interchain disulfide bond. These data show that the several kcal negative conformational free energy estimated to be associated with breaking of a disulfide bridge (23) would be nearly a pure entropy effect. Experimental determination of the conformational free energy and entropy of breaking of disulfide bridges in a disorganized protein is not possible. It is therefore particularly satisfying that the expected near zero value for the enthalpy contribution is observed. Restrictions imposed by disulfide bridge formation do not restrict the chain to energetically unfavorable states, although they clearly must restrict the number of configurations of the chain.

Figure 4 shows the enthalpy of dilution of lysozyme as a function of lysozyme concentration. Measurements obtained for several concentration ranges were adjusted to bring the data onto the same scale. The solid curve is calculated for a monomer-dimer equilibrium constant that accords with the literature values [reviewed by Imoto *et al.* (15)] and an enthalpy of dimerization of -7 kcal/mole. It is noteworthy that reaction microcalorimetry can accommodate very high concentrations of protein, in this case near 100 mg/ml.

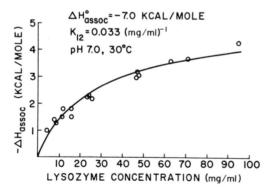

Fig. 4. Heat of dilution of lysozyme as a function of lysozyme concentration. Calorimetric measurements at 30°C, 0.01 M Na phosphate with NaCl to ionic strength 0.1, pH 7. 0.5–2 ml of 10–100 mg/ml deionized protein stock solution was diluted with 4 ml buffer. Measurements for different dilution series (different protein stock concentrations) were normalized by addition of a calculated heat of infinite dilution that was fit to the data using a nonlinear least squares analysis, to establish a common reference point for all experiments. Thus the observed heats of reaction are differences between the data points shown on the graph. The calculations used the value 0.033 mg^{-1}/ml^{-1} for the equilibrium constant for dimerization, and estimated the heat of dimerization as -7 kcal/mole.

Exposure Calculations

High-resolution crystallographic structures are available for more than 30 proteins. For several reasons a method of quantitatively describing the environments of protein atoms appears to be of value. Use of a crystal structure with a view to understanding the chemistry usually has required careful examination of a model. It is not generally possible for a laboratory to construct a model unless it is deeply involved with a particular system. Also, buried within the tremendous amount of crystallographic data there should be generalizations that bear on the chemistry of protein folding and protein reactivity, and some of these can perhaps best be brought forward through comparison of a number of high-resolution structures; for any single structure individual peculiarities should obscure more general aspects. Comparisons require summary listings of structural features.

The effect of protein folding on the chemistry of a group is expected to depend upon (i) hydrogen bonding in which it is involved, (ii) neighboring charges, and (iii) the extent of the exposure of the group to solvent. The first two points are listable in straightforward fashion from the crystal structure data. Qualitative or semiquantitative descriptions of exposure have been based upon visual examination of the structure [e.g., Browne *et al.* (7)]. Lee and Richards (16) have made quantitative calculations of exposure based upon atomic coordinates. They represented the protein by a set of van der Waals spheres. Cross sections through this representation gave a set of two-dimensional images analogous to the Fourier sections derived from the crystallographic data. The program used to generate the cross-section plots also generated information on what fraction of a particular atom sphere was not included within the similar spheres drawn about other atoms. If the radius of each sphere is the sum of the van der Waals radii of protein atom and solvent, then if two neighboring spheres touch or intersect, it is impossible to bring a solvent up to the occluded surfaces. The surface not occluded is considered exposed to solvent.

Several years ago we began similar computations in collaboration with Prof. D. C. Phillips.* The program used differs slightly from that of Lee and Richards (16). The surface of the sphere is represented by a set of 92 uniformly distributed points. The program retains for those surface elements which are included in other spheres the identity of the occluding atom which is closest to the surface element. Information of this kind can be summed to give the extent of contact between atoms or groups of atoms, and to give a semiquantitative idea of the nonpolar/polar nature of the environment about a particular side chain. The van der Waals atom radii used in these computations are those of Pauling (20); the solvent radius was 1.4 Å.

*This work was begun during the tenure at Oxford of a National Institute of Health Special Fellowship (1970) held by J. A. Rupley.

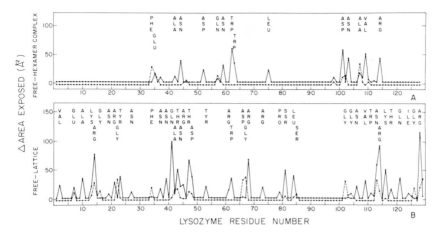

Fig. 5. (A) Changes in area exposed to solvent that follow formation of the lysozyme-(GlcNAc)$_6$ complex. The changes in area are given for the backbone atoms (dashed curve) and side chain atoms (solid curve) of each residue. (B) Changes in area exposed to solvent that develop when lysozyme is transferred into the tetragonal crystal lattice. From Ref. 24a.

Figures 5(A) and (B) give, respectively, the changes in area exposed to solvent when (GlcNAc)$_6$ binds to lysozyme and when lysozyme is incorporated into the crystal lattice. The hexasaccharide when bound covers about 10% of the lysozyme surface. Neighboring molecules in the crystal lattice cover 25–30% of the lysozyme surface. At pH 5 and in dilute salt, saccharide binding has an equilibrium constant ca. 10^3 more favorable than crystallization. Clearly the specificity of the interaction is of greater importance than the extent of surface covered. The difference between substrate binding and crystal formation is not surprising, considering evolution of lysozyme presumably has been toward more favorable interaction with substrate and has not been concerned with crystal lattice formation. The contacts between residues of the enzyme and elements of the hexasaccharide are described in more detail in Fig. 6. Contacts less than 10 Å^2 are of questionable importance. The contact information accords well with the chemistry of binding. Considering the interaction with (GlcNAc)$_3$ at sites ABC, it is seen from the relative contact areas that Trp 62 and Asp 101 are the residues most strongly affected, and chemical studies have shown these to be strongly perturbed in the (GlcNAc)$_3$-lysozyme complex. Similarly, binding at site C does not perturb Asp 101 and Fig. 6 shows there is no significant interaction between this residue and saccharide bound at site C.

Ooi and his collaborators (18) have put forward a simple representation of the secondary and tertiary structure of proteins. This is a two-dimensional map showing distances between C$_\alpha$-atoms for each pair of residues. Contouring this map at 15 Å intervals gives patterns in which certain

	33 LYS	34 PHE	35 GLU	36 SER	37 ASN	42 ALA	43 THR	44 ASN	46 ASN	52 ASP	56 LEU	57 GLN	58 ILE	59 ASN	62 TRP	63 TRP	75 LEU	98 ILE	101 ASP	102 GLY	103 ASN	107 ALA	108 TRP	109 VAL	110 ALA	112 ARG	114 ARG
α-MONOMER NAG C'			36/43				3/2	35/35	69/98	1/2	48/74	44/37	82/97	0/3	58/51		54/34					62/88	120/85	92/122			
HEXAMER NAG A															20/18	10/9	49/27	143/185	38/46	50/58							
NAG B															112/93	81/70		18/19	50/59		37/57	11/20					3/3
NAG C										5/8	21/22	68/46	56/96	68/89	68/78		52/56				90/134	95/58					
NAG D			31/32				110/118	93/120	8/7	76/85	5/3	53/63									9/17	50/39	57/63				
NAG E	3/3	100/139	24/13	2/0	28/19	12/8	139/132	5/8	33/30					83/93											75/95	29/32	
NAG F	5/4	64/94	34/26	16/7	24/14																		3/0				74/99

LYSOZYME

Fig. 6. Contacts between elements and protein residues in the active site. The lower part of the figure gives contacts between protein residues and the 6 saccharide elements of $(GlcNAc)_6$. The top part of the figure gives the contacts with the α-anomer of GlcNAc which binds differently than the corresponding unit (C) of the hexasaccharide. The top number of each pair gives the area in $Å^2$ occluded on the abscissa residue by the ordinate saccharide element, and the lower number gives the reverse. From Ref. 24a.

structural features stand out. Contact information can be similarly represented, as in Fig. 7(A). The extent of the area occluded by an abscissa residue on an ordinate residue is indicated by the letter. The strong contacts along the diagonal reflect near-neighbor interactions. As Ooi and his collaborators noted, the broad regions of contact along the diagonal represent α-helical segments, within which a residue contacts neighbors 4 and 5 residues distant along the chain. The bands orthogonal to the helix axis represent β-bends, where the chain folds back upon itself and the regions of contact describe lines of unit slope. The poor quality of the helix 5 near the C-terminus of the molecule is evident compared with the regularity of helices 1, 2, and 4. Interactions between distant elements of the molecule are indicated by the off-diagonal regions of contact. It is clear that folding

Fig. 7. (A) Ooi plot of contacts between residues of the folded lysozyme molecule. The letters indicate the extent to which abscissa residues occlude surface of ordinate residues. Each increase in alphabet stands for 15 $Å^2$. The numbers on the right give the surface area exposed to solvent in units of 30 $Å^2$. The contacts between atoms within a residue and with backbone atoms of the two neighboring residues are not included in the sums (thus the diagonal is blank). Secondary structure is indicated by the bars along the diagonal. The labels "h" and "β" stand for helical and β structure, respectively (15). Helix h3 is 3_{10}, and other helices are α. (B) Ooi plot showing contacts between side chain atoms. The characters designate the type of contact: O, nonpolar side chain to nonpolar side chain; X, polar to polar; *, polar to nonpolar. The diagonals delimit a band of 21 residues width centered on the main chain. Contacts outside this band are considered distant.

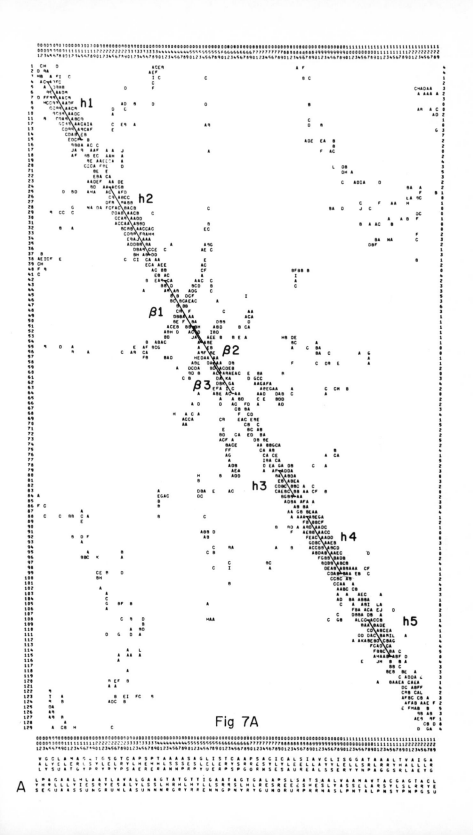

Fig 7A

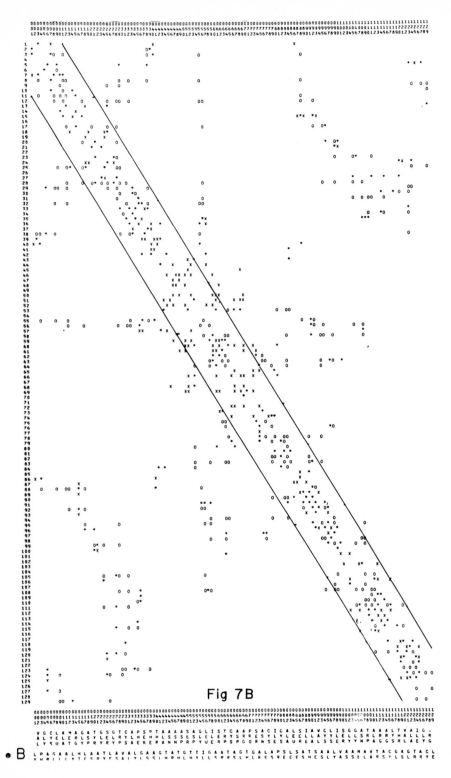

Fig 7B

brings residues near the N- and C-termini into contact and that residues 25–40 constitute a core with which several other regions of the chain have contact.

Representations of this sort can be used to describe the extent of contact between particular elements of residues, e.g., backbone vs. side chain, and can be used to show the nature of the contact, as between nonpolar side chains and polar side chains. Figure 7(B) shows that the off-diagonal elements (distant interactions) are dominated by contacts between nonpolar side chains. Nonpolar–nonpolar, nonpolar–polar, and polar–polar contacts are in the ratio 55-30-15, contrasted with essentially equal frequency of the three types of contacts along the diagonal. This dominance of non-polar contacts in the off-diagonal region is expected from the importance placed upon hydrophobic interactions in determining protein folding and the stability of folded structures. It is interesting that the substantial effect of folding on the polar side chains apparently reflects local structure and involvement of backbone. It should be noted, moreover, that the surface of the protein molecule has a considerable proportion of nonpolar atoms (16), and not all nonpolar side chains are in the interior of the molecule.

A principal value of exposure calculations is that they give numbers which can be used to compare protein structures. Figure 8 gives, for each type of amino acid residue, the fractional exposure summed over all back-bone atoms vs. the fractional exposure summed over all side chain atoms. These values are averages over 11 structures. Several points can be made. There is good correlation between exposure of backbone and exposure of side chain elements (correlation coefficient 0.90). The bold-printed residues are nonpolar, and as expected these constitute the most buried class of residues, with the exception of proline. This confirms the observation of Lee and Richards who from their calculations on ribonuclease, lysozyme, and myoglobin, concluded that proline was abnormally exposed. They attributed this to the probability of proline being in bends of the chain and thus at the surface of a molecule. It is also of interest that glycine is abnormally exposed. One would expect this residue, with no side chain, to have equal probability of being at the surface or inside the protein. Venkatachalam (26) has proposed several β-bends that are commonly found in proteins and that require glycine as an element.

The carbonyl oxygen is the main chain atom which in the unfolded model has the greatest surface area and which shows the largest changes associated with folding of the chain. It was found for lysozyme that each of the five helices has the carbonyl oxygen of the first helical residue completely withdrawn from solvent, and four of the five helices have the carbonyl oxygen of the last helical residue exposed to solvent. This correlation was borne out through observations of 44 helices in other proteins, of which 39

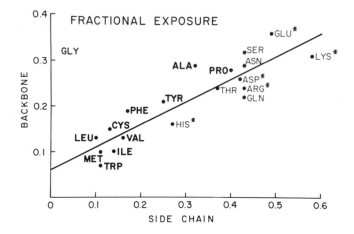

Fig. 8. Fractional exposure of backbone atoms   side chain atoms. Fractional exposure (exposure in folded molecule/exposure in unfolded model) was averaged over 11 proteins for all residues of each type. Nonpolar residues are shown in bold print and charged residues are indicated by asterisks. The least-squares line shows a correlation coefficient of 0.90.

had the carbonyl oxygen of the first helical residue buried, and 37 had the carbonyl oxygen of the last residue exposed. This should be compared with the 42% and 52% of the main chain oxygen atoms found to be exposed, averaging over all helical and all nonhelical residues, respectively. It is not clear why helices should tend to begin within the folded structure. However, it is to be hoped that theories of secondary structure formation in proteins can accommodate observations of this kind.

Summary

The k_{cat}-pH-rate profile for lysozyme shows kinetic apparent pK of 3.8 and 6.7. Consideration of the mechanism proposed by Phillips and his colleagues suggests that these pK reflect Asp 52 and Glu 35. The kinetic pK value for Glu 35 (6.7) accords with a considerable literature on the pK of this residue. Measurements of the pH dependence of the enthalpy of binding of the β-methyl glycoside of N-acetylglucosamine suggest that the kinetic pK of 3.8 reflects ionization of Asp 52. Other measurements of heats of reaction for lysozyme are described. The results of calculations on the exposure of atoms and groups of atoms of a protein to solvent can be used to concisely describe the character of the interaction between protein and ligands, and between the elements within a folded protein. Glycine is among the most exposed residues in protein structures, in agreement with the involvement of this residue in β-bends. The first residue of protein helices has the carbonyl oxygen completely withdrawn from solvent, which stands against the approximately 0.5 probability for exposure of both helical and nonhelical carbonyl oxygens at other positions in the chain.

References

1. Aune, K. C., and Tanford, C., *Biochemistry* **8**, 4579 (1969).
2. Banerjee, S. K., and Rupley, J. A., *J. Biol. Chem.* **248**, 2117 (1973).
3. Banerjee, S. K., and Rupley, J. A., *Arch. Biochem. Biophys.* **155**, 19 (1973).
3a. Banerjee, S. K., Kregar, I., Turk, V., and Rupley, J. A., *J. Biol. Chem.* **248**, 4786 (1973).
4. Bjurulf, C., and Wadsö, I., *Eur. J. Biochem.* **31**, 95 (1972).
5. Blake, C. C. F., Johnson, L. N., Mair, G. A., North, A. C. T., Phillips, D. C., and Sarma, V. R., *Proc. Roy. Soc., Lond. [Biol.]* **167**, 378 (1967).
6. Blake, C. C. F., Koenig, D. F., Mair, G. A., North, A. C. T., Phillips, D. C., and Sarma, V. R., *Nature (Lond.)* **206**, 757 (1965).
7. Browne, W. J., North, A. C. T., Phillips, D. C., Brew, K., Thomas, C., and Hill, R. C., *J. Mol. Biol.* **42**, 65 (1969).
8. Dahlquist, F. W., and Raftery, M. A., *Biochemistry* **7**, 3277 (1968).
9. Davies, R. C., Neuberger, A., and Wilson, B. M., *Biochim. Biophys. Acta* **178**, 294 (1969).
10. Doherty, D. G., and Vaslow, F., *J. Am. Chem. Soc.* **74**, 931 (1952).
11. Donovan, J. W., Laskowski, M., Jr., and Scheraga, H. A., *J. Am. Chem. Soc.* **83**, 2686 (1961).
12. Hammes, G. G., and Schimmel, P. R., *J. Am. Chem. Soc.* **87**, 4665 (1965).
13. Harmon, K., and Niemann, C., *J. Biol. Chem.* **178**, 743 (1949).
14. Holler, E., Rupley, J. A., and Hess, G. P., *Biochem. Biophys. Res. Commun.* **40**, 166 (1970).
15. Imoto, T., Johnson, L. N., North, A. C. T., Phillips, D. C., and Rupley, J. A., *in* "The Enzymes" (P. D. Boyer, ed.), 3rd ed., Vol. 7, p. 665. Academic Press, New York, 1972.
15a. Lapanje, S., and Rupley, J. A., *Biochemistry* **12**, 2370 (1973).
16. Lee, B., and Richards, F. M., *J. Mol. Biol.* **55**, 379 (1971).
17. Monk, P., and Wadsö, I., *Acta Chem. Scand.* **22**, 1842 (1968).
18. Nishikawa, K., Ooi, T., Isogai, Y., and Nobuhiko, S., *J. Phys. Soc. Jap.* **32**, 1331 (1972).
19. Parsons, S. M., and Raftery, M. A., *Biochemistry* **11**, 1623, 1630, and 1633 (1972).
20. Pauling, L., "Nature of the Chemical Bond," 3rd ed. Cornell Univ. Press, Ithaca, New York, 1960.
21. Rand-Meir, T., Dahlquist, F. W., and Raftery, M. A., *Biochemistry* **8**, 4206 (1969).
22. Rupley, J. A., and Gates, V., *Proc. Natl. Acad. Sci. U.S.* **57**, 496 (1967).
23. Scheraga, H., *in* "The Proteins" (H. Neurath, ed.), 2nd ed., Vol. 1, p. 477. Academic Press, New York, 1963.
24. Shiao, D. D. F., and Sturtevant, J. M., *Biochemistry* **8**, 4910 (1969).
24a. Shrake, A. F., and Rupley, J. A., *J. Mol. Biol.* **79**, 351 (1973).
25. Sophianopoulos, A. J., and Weiss, B. J., *Biochemistry* **3**, 1920 (1964).
26. Venkatachalam, C. M., *Biopolymers* **6**, 1425 (1968).
27. Wadsö, I., *Acta Chem. Scand.* **22**, 927 (1968).

Rapid Nonenzymic Regeneration
of Reduced Human Leukemia Lysozyme

DONALD B. WETLAUFER, ERIC R. JOHNSON,
and LORRAINE M. CLAUSS

Introduction

In an earlier study from this laboratory, it was demonstrated that the oxidative regeneration of lysozyme activity from reduced, disorganized hen egg-white lysozyme (HEWL) could be achieved rapidly and in high yield by mixtures of oxidized glutathione (GSSG) and reduced glutathione (GSH), or other low molecular weight thiol, disulfide pairs (8). Since it has been claimed by other investigators that an enzymic system carries out this function in protein biosynthesis, it was of interest to test our non-enzymic system in attempts to regenerate native protein from reduced, disordered human leukemia lysozyme (HLL). These attempts have proved successful. Characterization of the process is reported here.

Materials and Methods

Hen egg-white lysozyme was obtained from Worthington Biochemical Corp. (lot LYSF OCC, activity 10,900 units/mg). Human leukemia lysozyme was the gift of Dr. R. E. Canfield of Columbia University. These materials were purified chromatographically (see below) before use. Reduced glutathione, oxidized glutathione, and dithiothreitol were obtained from Sigma Chemical Co. Dried *Micrococcus lysodeikticus* cells and "Tris" (free base, Trizma quality) were also obtained from Sigma Chemical Co. Urea was of "Ultra-pure" quality from Schwarz-Mann. EDTA was an AR grade product of the Mallinckrodt Chemical Works. Other reagents used were of Reagent Grade. Water was deionized, followed by glass distillation from 0.2 N H_2SO_4.

Ion-exchange chromatography of hen egg-white lysozyme was performed using the method described by Saxena and Wetlaufer (8). The same method was used for the fractionation of human leukemia lysozyme. Under the experimental conditions detailed in the legend for Fig. 1, the HLL protein was resolved into one major peak, which comprised over 90% of the original sample, and two minor peaks. The major peak was desalted on a 4.7 cm × 50 cm Sephadex G-25 (medium) column equilibrated with and eluted with 0.1 N acetic acid. This material was lyophilized, reduced, and used in the regeneration experiments.

The preparation of both reduced HEWL and reduced HLL used in this study was similar to that employed by Saxena and Wetlaufer (8) with the exception that 2-mercaptoethanol was replaced with dithiothreitol. The hen egg-white lysozyme was reduced in 8 M urea which contained 0.1 M Tris acetate buffer, 1 mg dithiothreitol/mg protein, and was adjusted to an

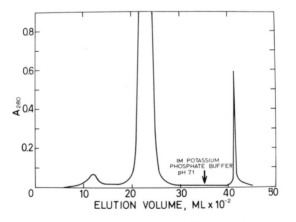

Fig. 1. Preparative Biorex-70 chromatography of HLL on a 2.6 cm × 170 cm column equilibrated and eluted with 0.2 M potassium phosphate buffer, pH 7.14 at 25°C. The column was loaded with 0.7 gm of lysozyme in 10 ml of the buffer. The flow rate was 1 ml/min. The material used for the reduction and regeneration experiments here reported was taken from the large peak at 2250 ml.

apparent pH of 8.6. After 2 hours incubation at room temperature, the solution was acidified to pH 3 with glacial acetic acid and passed through a 2.5 cm × 70 cm Sephadex G-25 (coarse) column equilibrated with and eluted with 0.1 N acetic acid. The protein was immediately frozen and lyophilized. Reduced human leukemia lysozyme was prepared using the above procedure, with 10 M urea instead of 8 M. Analysis by Ellman's method (2) showed the reduced hen egg-white lysozyme to have a sulfhydryl content of 7.7–7.9 Eq/mole. Reduced human leukemia lysozyme was found to have 7.6–7.9 sulfhydryl Eq/mole. These values were found to be constant over a period of 4 weeks when stored desiccated at $-20°C$.

All regeneration experiments were performed at a reduced lysozyme concentration of 1.0 mg/100 ml. Standard solutions containing 1.0 mg lysozyme/100 ml were used for comparison with the regenerated solutions to determine the percentage of native lysozyme activity attained during the regeneration experiments. The concentrations of the lysozyme solutions were determined spectrophotometrically using $A_{280}^{1\%} = 26.3$ for native HEWL, $A_{280}^{1\%} = 23.7$ for reduced HEWL, $A_{280}^{1\%} = 25.5$ for native HLL, and $A_{280}^{1\%} = 22.4$ for reduced HLL. These values were determined from absorbance measurements of samples dissolved in 0.1 M acetic acid, and dry weight determinations on the same solutions.

Regeneration media containing 0.10 M Tris acetate, pH 8.0, 10^{-4} M EDTA, and the desired amounts of GSH and GSSG were prepared only 1 or 2 min before beginning the regeneration in order to minimize air oxidation of GSH. It should be noted that glutathione ratios and total concentrations are based on equivalents of glutathione. In the regeneration system containing 6.0×10^{-3} M GSH and 6.0×10^{-4} M GSSG, the ratio [GSSG]/[GSH] is $\frac{1}{5}$ since 1 mole of GSSG contains two equivalents of the tripeptide.

The regeneration process was started by adding 1.00 ml of reduced lysozyme (1.0 mg/ml 0.1 N acetic acid) to 99.0 ml of regeneration media. At specified time intervals, 1.00 ml of the regenerating lysozyme solution was added to 1.00 ml of 0.2 N acetic acid to quench the regeneration and bring the pH to 4.0. This method of stopping the regeneration reaction was found to be effective for at least 90 min, as shown in Fig. 2. This successful acid quenching of the regeneration samples allowed us to obtain many data points during the first few minutes of rapid reactivation process.

The lysozyme assay used was a modification of the method of Jollès (4). Dried *M. lysodeikticus* cells were prepared in a suspension at a concentration of 10 mg/100 ml in 0.2 M sodium acetate buffer, pH 5.5. The 2.00 ml quenched regeneration sample was added to 3.00 ml of the cell suspension. Both solutions had previously been incubated at 37°C. The resulting suspension was mixed well and transferred to a 1.00 cm pathlength cuvet. The

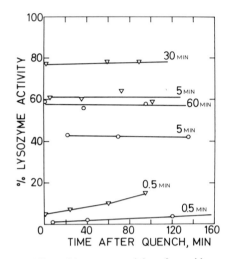

Fig. 2. Tests of the stability of lysozyme activity after acid quench. After the indicated time period of regeneration, 1.00 ml of 0.2 N acetic acid was added to 1.00 ml of regeneration mixture containing either HEWL (circle) or HLL (triangle). Regeneration conditions: 0.1 M Tris acetate, pH 8.0, [GSSG]/[GSH] = 1/5 at 37°C with a total glutathione concentration of 7.2×10^{-3} eq/l.

decrease in apparent absorbance at 350 nm was measured from 20 to 120 sec after mixing, in a Beckman DU spectrophotometer modified by a Gilford power supply and detector unit, with an attached chart recorder. Under these conditions a linear relationship of activity vs. lysozyme concentration was routinely obtained.

Since we observed a slight decrease in apparent activity of the native lysozyme standard solutions at times approaching 90 min after the start of the regeneration, the apparent activity of a 1.00 mg/100 ml standard lysozyme solution was redetermined every third assay. Thus any change in the system, particularly in the cell suspension, is accounted for during the regeneration. We found that neither EDTA nor glutathione had an effect on the lysozyme assay at concentrations resulting from the regeneration experiments.

Results

In general, the pattern of regeneration of HLL is very similar to that of HEWL, the human enzyme being different mainly in showing a slightly more rapid regeneration. The more rapid regeneration of reduced HLL is seen in an exaggerated form in the comparison (Fig. 2) of the time-stability of the acid-quench of the 0.5 min regeneration solutions. HLL shows 6%

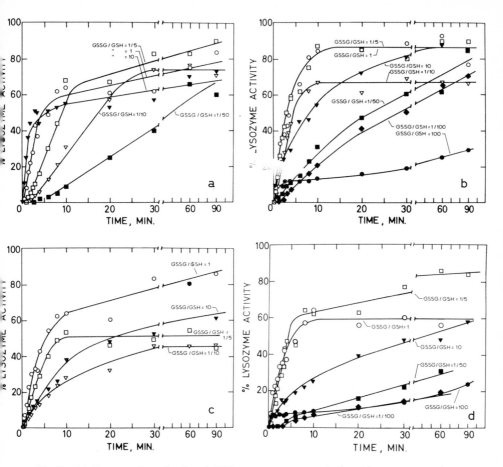

Fig. 3. (a) Regeneration of reduced HLL at a constant total glutathione concentration of 7.2×10^{-4} Eq/liter and varying [GSSG]/[GSH] ratios. Conditions: 0.1 M Tris acetate, pH 8.0, at 37°C with 1.0 mg lysozyme/100 ml buffer. (Solid square) [GSSG]/[GSH] = 1/50; (open triangle) [GSSG]/[GSH] = 1/10; (open square) [GSSG]/[GSH] = 1/5; (open circle) [GSSG]/[GSH] = 1; (solid triangle) [GSSG]/[GSH] = 10. (b) Regeneration of reduced HLL at a constant total glutathione concentration of 7.2×10^{-3} Eq/liter and varying [GSSG]/[GSH] ratios. Conditions: 0.1 M Tris acetate, pH 8.0, at 37°C with 1.0 mg lysozyme/100 ml buffer. (Solid diamond) [GSSG]/[GSH] = 1/100; (solid square) [GSSG]/[GSH] = 1/50; (open triangle) [GSSG]/[GSH] = 1/10; (open square) [GSSG]/[GSH] = 1/5; (open circle) [GSSG]/[GSH] = 1; (solid triangle) [GSSG]/[GSH] = 10; (solid circle) [GSSG]/[GSH] = 100. (c) Regeneration of reduced HLL at a constant total glutathione concentration of 2.0×10^{-2} Eq/liter and varying [GSSG]/[GSH] ratios. Conditions: 0.1 M Tris acetate, pH 8.0, at 37°C with 1.0 mg lysozyme/100 ml buffer. (Open triangle) [GSSH]/[GSH] = 1/10; (open square) [GSSG]/[GSH] = 1/5; (open circle) [GSSG]/[GSH] = 1; (closed triangle) [GSSG]/[GSH] = 10. (d) Regeneration of reduced hen egg-white lysozyme at a constant total glutathione concentration of 7.2×10^{-3} Eq/liter and varying [GSSG]/[GSH] ratios. Conditions: 0.1 M Tris acetate, pH 8.0, at 37°C, with 1.0 mg lysozyme/100 ml buffer. (Closed diamond) [GSSG]/[GSH] = 1/100; (closed square) [GSSG]/[GSH] = 1/50; (open square) [GSSG]/[GSH] = 1/5; (open circle) [GSSG]/[GSH] = 1; (closed triangle) [GSSG]/[GSH] = 10; (closed circle) [GSSG]/[GSH] = 100.

regain of activity when HEWL has less than 2% of its activity, and at this very short regeneration time HLL has a distinctly more pronounced increase of activity with time than does HEWL. We must point out, however, that the time-dependence of the aliquots taken after very short regenerations was no practical problem, since the assays for $\frac{1}{2}$ and 1 min regeneration times were completed within 10 min of the quench time, and the rate of increase of activity in quenched solutions was slow on this time scale. We made many more observations of activity at short periods of regeneration than can be plotted on our figures, due to physical space limitations. In many rapid regenerations we found no detectable activity at times of 30 sec or less. In other words, there was a lag period in the regenerations. This can be actually seen on the time scale of Figs. 3(a), 3(d), and 7(b) for several regenerations taking place under very unfavorable conditions (proceeding at low rates).

Figure 3 shows the results of regeneration experiments using a glutathione system with different ratios of [GSSG]/[GSH], but other conditions invariant (ΣG, temperature, red. lysozyme concentration, pH, etc.). Figure 3(a), (b), and (c) show regenerations of HLL at $\Sigma G = 7.2 \times 10^{-4}\ M$, 7.2 $\times\ 10^{-3}\ M$, and $2.0 \times 10^{-2}\ M$, respectively. Figure 3(d) shows the regeneration of HEWL at $\Sigma G = 7.2 \times 10^{-3}\ M$, and is therefore directly comparable with Fig. 3(b) above it. In these regeneration curves, as in those to be shown below, the experimental scatter of points is large enough to be observable but small enough to be of no real consequence for present considerations.

To summarize the foregoing group of regeneration experiments, we have plotted the two parameters: lysozyme activity after 5 min regeneration, and lysozyme activity after 60 min regeneration, as a function of [GSSG]/[GSH]. It can be seen by inspection that the 5 min activity provides approximate measure of the regeneration rate, while the 60 min activity provides a similar measure of the final yield of lysozyme activity. Figure 4 shows these summary plots for the data of Fig. 3, Fig. 3(a) being summarized by Fig. 4(a), etc. It can be seen that, of the conditions sampled, regeneration is most rapid and yields are highest for $\Sigma G = 7 \times 10^{-3}\ M$, and [GSSG]/[GSH] = 1/10 and 1 [Fig. 4(b)]. Under substantially different conditions of ΣG, the pattern of rate and yield is perturbed from that of Fig. 4(b), but regenerations do proceed with substantial rates and yield. From this first set of data, it appears that both ΣG and [GSSG]/[GSH] can be varied over a broad range without drastically reducing regeneration rate or yield. The similarity between regeneration of the hen and the human enzymes is readily apparent in the strong similarities of pattern between Fig. 4(b) (HLL) and 4(d) (HEWL). The similarity is probably closer than Fig. 4(d) shows, since repeat of two of the regenerations of Fig. 4(d) with another batch of reduced

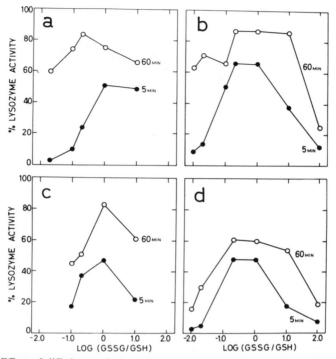

Fig. 4. Effect of differing total glutathione concentrations and [GSSG]/[GSH] ratios on the rate and final yield of the regeneration of reduced HLL and HEWL. Regeneration was carried out in 0.1 M Tris acetate, pH 8.0, at 37°C, with 1.0 mg lysozyme/100 ml buffer. Five minute (closed circle) and 60 minute (open circle) yields are given for the regeneration at each different [GSSG]/[GSH] and total glutathione concentration. (a) HLL, [GSSG] + [GSH] = 7.2×10^{-4} Eq/liter. (b) HLL, [GSSG] + [GSH] = 7.2×10^{-3} Eq/liter. (c) HLL, [GSSG] + [GSH] = 2.0×10^{-2} Eq/liter. (d) HEWL, [GSSG] + [GSH] = 7.2×10^{-3} Eq/liter.

HEWL gave definitely higher 60 min activities ($\sim 80\%$) than those shown in Fig. 4(d). Since for the present we are more interested in pattern than in particular results, we retain comparison of Fig. 4(b) and (d) as drawn.

We carried out a systematic exploration of the effects of the variation of ΣG at constant [GSSG]/[GSH], with the regeneration results partially shown in Fig. 5. We again prepared a summary figure (Fig. 6) showing the 5 min and 60 min activities as a function of ΣG. Figure 6 shows a maximum in 5 min yield (rate) from 4 to 12×10^{-3} M, and a maximum in 60 min yield from 1 to 10×10^{-3} M, similar in breadth but slightly displaced from the rate maximum. These results confirm our conclusion from the earlier experiments: The regeneration of HLL varies with ΣG and [GSSG]/[GSH], but maximal rates and yields are obtained over a fairly broad range of concentration of GSH and GSSG.

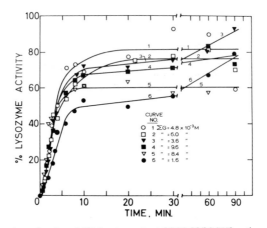

Fig. 5. Regeneration of reduced HLL at constant [GSSG]/[GSH] ratio, constant tempera-ture, and varying total glutathione concentrations. The regeneration experiments were carried out in 0.1 M Tris acetate, pH 8.0, at 37°C with 1.0 mg lysozyme/100 ml buffer and a [GSSG]/[GSH] ratio of 1/5. The total glutathione concentrations are indicated in the figure.

As a further characterization of the regeneration process, we examined the temperature-dependence of the process for both HLL and HEWL. We chose $\Sigma G = 7.2 \times 10^{-3} M$ and [GSSG]/[GSH] $= 1/5$, conditions shown by the foregoing experiments to be approximately centered in the broad optima at 37°C. Results of reactivation experiments at several temperatures are shown in Fig. 7(a) for HLL, in Fig. 7(b) for HEWL, and in the summary Fig. 8(a) and 8(b). The temperature-dependence of the human and the hen enzyme form a remarkably similar pattern, both in rate and in yield. The HLL regeneration is somewhat faster than that of HEWL, but the tem-perature optima are clearly the same (37°C) for regeneration of the two enzymes.

Discussion

The pattern of nonenzymic regeneration of HLL by an oxidoreduction buffer composed of oxidized and reduced glutathione is overall very similar to that of HEWL, as shown in these and earlier studies from this labora-tory (8). Thus, not only are the enzymic activity and three-dimensional structure very similar for the two enzymes (Phillips, this volume), but the self-assembly mechanisms also. In addition, nonenzymic regeneration of activity in high yield has been achieved for bovine pancreatic RNase (1, 9), and for the Bowman–Birk trypsin inhibitor (5) with essentially the same regeneration system. These findings appear to be extensive enough that we

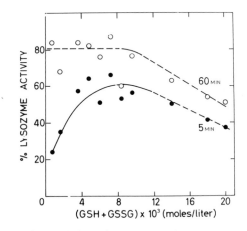

Fig. 6. Effects of varying total glutathione concentration at constant [GSSG]/[GSH] ratio and temperature on the regeneration of reduced HLL. Regeneration experiments were carried out in 0.1 M Tris acetate, pH 8.0, at 37°C with 1.0 mg lysozyme/100 ml buffer and a [GSSG]/[GSH] ratio of 1/5. Five minute (closed circle) and 60 minute (open circle) yields are given for each glutathione concentration.

can now claim that this regeneration system has generality. Physiological feasibility for the system was earlier demonstrated (8). We believe that the evidence now at hand is sufficient to claim that our nonenzymic system (or something very similar) is more plausible as a general physiological mechanism to facilitate assembly of disulfide proteins than the "shuffle-ase enzyme" of Givol *et al.* (3). Our nonenzymic system has been shown to function in the regeneration of four different reduced proteins, the same number as the shuffle-ase enzyme. However, the demonstrated distribution of the shuffle-ase enzyme is very limited compared with ubiquitous glutathione. On present evidence, we find our nonenzymic system a moᵿ plausible general agency for promoting *in vivo* oxidation and assembly of reduced disulfide proteins.

Direct evidence of intermediates in the Cu^{2+}-catalyzed regeneration of reduced HEWL has been presented (6). The finding of a lag period in the regenerations studied here gives further support to a mechanism which produces inactive intermediates. Thus a "two-state" analysis can not be applied here.

Much interest attends the mechanism of three-dimensional structure formation in proteins. In spite of a common belief that such a structure is thermodynamically determined, recent substantial evidence has appeared in support of kinetic determinism of protein architecture (6, 7, 10). Reversible denaturation has been mistakenly interpreted by many as compelling evidence for thermodynamic determinism of structure. It is nothing of the

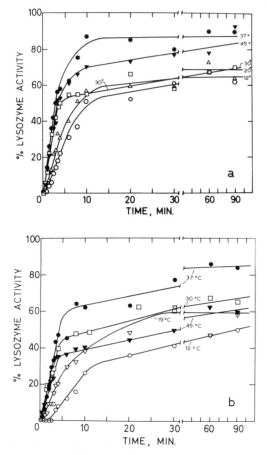

Fig. 7. (a) Regeneration of reduced HLL at a constant total glutathione concentration of 7.2×10^{-3} Eq/liter constant [GSSG]/[GSH] ratio of 1/5, and varying temperatures. The regeneration experiments were carried out in 0.1 *M* Tris acetate pH 8.0, with 1.0 mg lysozyme/100 ml buffer. (Open circle) 12°C; (open triangle) 20°C; (open square) 30°C; (closed circle) 37°C; (closed triangle) 45°C. (b) Regeneration of reduced HEWL at a constant glutathione concentration of 7.2×10^{-3} Eq/liter constant [GSSG]/[GSH] ratio of 1/5, and varying temperatures. The regeneration experiments were performed in 0.1 *M* Tris acetate, pH 8.0, with 1.0 mg lysozyme/100 ml buffer. (Open circle) 12°C; (open triangle) 19°C; (open square) 30°C; (closed circle) 37°C; (closed triangle) 45°C.

sort. Consider the analogy of a two-pan rough balance on a laboratory bench top. By changing the position of a small weight, the balance can be reversibly repeatedly perturbed so that the mass in the left pan is greater than that in the right pan, and then vice versa. However, if the balance is knocked onto the floor by the activation energy of a carelessly wielded

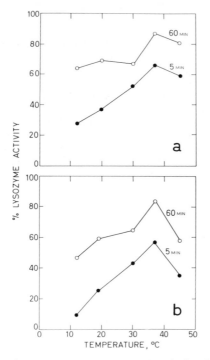

Fig. 8. Effects of varying temperature at constant total glutathione concentration and [GSSG]/[GSH] ratio on the regeneration of reduced HLL (a) and HEWL (b). The regeneration conditions were 0.1 *M* Tris acetate, pH 8.0, with 1.0 mg lysozyme/100 ml buffer. The [GSSG]/[GSH] ratio was 1/5 at a total glutathione concentration of 7.2×10^{-3} Eq/liter.

mop handle, we obtain a new, much lower free energy state for the balance than either of the two states on the bench top. Moral: Ready reversibility does not prove access to the global free energy minimum. The present study presents a kind of reversible denaturation, but this should not be understood as meaning that the native state is the thermodynamically most stable state. Rather, it is, under regeneration conditions, the most stable of the kinetically accessible states.

Acknowledgments

We thank Jeff Gabe for skilled and enthusiastic technical assistance, and Dr. R. E. Canfield for the gift of HLL. This work was supported by USPHS grants GM-10900 and GM 18814-01 and by a grant from the American Cancer Society (IN-13L). E. R. Johnson was a trainee under USPHS Training Grant 5 T01 GM 00157. L. M. Clauss held a postdoctoral fellowship of USPHS while participating in this work.

References

1. Ahmed, A. K., Ph.D. thesis, University of Minnesota, Minneapolis, 1969.
2. Ellman, G., *Arch. Biochem. Biophys.* **82,** 70 (1959).
3. Givol, D., de Lorenzo, F., Goldberger, R. F., and Anfinsen, C. B., *Proc. Natl. Acad. Sci. U.S.* **53,** 676 (1965).
4. Jollès, P., *in* "Methods in Enzymology" (S. P. Colowick and N. O. Kaplan, eds.) Vol. 5, p. 117. Academic Press, New York, 1962.
5. Liener, I. E., and Hogle, J., *Canad. J. Biochem.* **51,** 1014 (1973).
6. Ristow, S., and Wetlaufer, D. B., *Biochem. Biophys. Res. Commun.* **50,** 544 (1973).
7. Ristow, S., and Wetlaufer, D. B., *Annu. Rev. Biochem.* **42,** 135 (1973).
8. Saxena, V. P., and Wetlaufer, D. B., *Biochemistry* **9,** 5015 (1970).
9. Schaffer, S. W., Ph.D. thesis, University of Minnesota, Minneapolis, 1970.
10. Wetlaufer, D. B., *Proc. Natl. Acad. Sci. U.S.* **70,** 697 (1973).

Comparative Studies of the Solution
Behavior of Hen Egg-White and Human Lysozyme*

RODERICK S. MULVEY, RICHARD J. GUALTIERI, and
SHERMAN BEYCHOK

Our knowledge of how polypeptide chains fold into a native structure
and why a particular conformation results rather than another is still scant.
Some of the major questions were already thrown into relief with the discovery that sequences as different from one another as the α and β chains
of various hemoglobins, and the different myoglobins, somehow generate
structures of great similarity. Clearly, selection pressures are exerted on
function rather than structure, but this knowledge does not take us very
far. It does not seem surprising that different sequences generate different
structures, stunning though it may be to see the structural and functional
changes occasionally wrought by a single substitution. Indeed, it was so
essential to establish that only sequence had to be coded for, and that structure would follow speedily without additional coding, that great stress was
naturally placed on the sequence–structure implication: if this sequence,
then this structure. However, the converse is clearly not always true for
backbone structure. The usual explanation for the counter examples is that
the substitutions are conservative, i.e., hydrophobic for hydrophobic. In

*Supported in part by grants from the National Institutes of Health, 5 R01 CA 13014-01;
and the National Science Foundation, GB 29481x1.

the α and β chains of human hemoglobin, however, only 40 of the 75 differences are conservative (8).

Human and hen egg-white lysozyme (HEWL) differ in 40% of the 130 positions when the two sequences are aligned to maximize homology (5, 17). X-Ray crystallographic studies of HEWL at 2 Å resolution (2) and of human lysozyme at 6 Å resolution (4; see also paper by Phillips) suggest that the two enzymes have very similar secondary structures, as do studies of the far-UV circular dichroism (10, 14).

At least 12 and possibly 16 of the 50 sequence differences are nonconservative. These two lysozymes thus comprise the kind of system we are interested in, that is, one in which the sequences are substantially different, in which a sizable number of substitutions is nonconservative and in which the structures and functions are highly similar.

At least some of the sequence differences do, however, generate functional differences in human and hen egg-white lysozyme. Through a broad interval of pH, human lysozyme is more active in hydrolyzing the walls of killed *Micrococcus lysodeikticus* cells than is HEWL (19, 22). Corresponding structural differences are implied in the strikingly different near ultraviolet circular dichroism spectra (10), in the fluorescence spectra, and in the oligosaccharide binding constants, to all of which I shall return in a few moments.

Our initial efforts in comparing the two enzymes were stimulated by Dr. Canfield who, remembering our acid–base and denaturation studies on HEWL some years ago (1), suggested that it might be interesting, for reasons he detailed, to examine the acid–base properties of human lysozyme.

The carboxyl region of the titration curve of human lysozyme presented some technical difficulties, which we will not detail now, but it turned out to be very similar to that of HEWL, that is, highly anomalous by virtue of being very spread out. In HEWL, that is due in part to the atypically depressed pK of Asp 52 and the elevated pK of Glu 35 (11). Between pH 7 and 12, the curves are identical in terms of intrinsic pKs and interaction factors required for fitting. In the free human lysozyme, there are no carboxyl groups that remain inaccessible at high pH. The isoionic point, determined as the pH of a concentrated solution that has been deionized on a mixed bed resin, is between 10.8 and 10.9. This is the expected value after taking into account 3 abnormal tyrosine residues (19) and the net charge required for electroneutrality in the solution.

The anomalous carboxyl ionization is, of course, closely related to the mechanism of enzymatic activity of HEWL. In Fig. 1 the initial rates of turbidity clearing of killed *M. lysodeikticus* cells are compared. Two different batches of human lysozyme are shown, with good agreement between them. The activities are about four times as great as that of HEWL. The

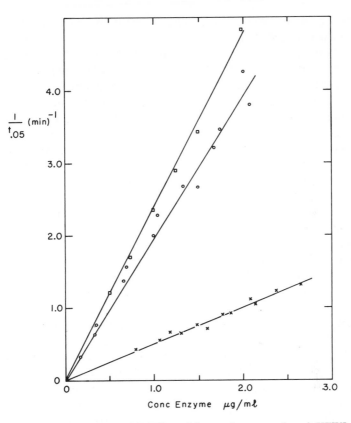

Fig. 1. Comparison of activities of HEWL and human lysozyme; (cross) HEWL, (circle and square) two different lots of human lysozyme, pH 6.2, phosphate buffer, 0.1 *M*. Room temperature.

dependence of these rates on pH was measured for human lysozyme by Kammerman (18a) and by Saint-Blancard *et al.* (24), the latter a very detailed study. Those reports showed that if the ionic strength is maintained constant within a certain range, then both HEWL and human lysozyme exhibit a plateau of activity rather than a more typical bell-shaped curve. We have confirmed these findings with very similar results. Both enzymes, then, are about as active at pH 9 as at 5, where the activities are customarily measured.

In view of what is believed about the mechanism of action, this finding has important implications about the energy of interaction with substrate. To begin with, Glu 35 has a p*K* near 6 in uncomplexed HEWL, and a slightly higher p*K* still when the enzyme binds oligosaccharide inhibitors (16). At 25°C, a free energy expenditure of 2050 cal/mole is required to maintain the protonated form of Glu 35, with an additional 400 cal/mole necessary when triNAG is bound.

Since both enzymes are fully, or almost fully, active at pH 9, Glu 35 remains protonated at that pH when the enzymes are bound to bacterial cell walls. This requires an additional 4000 or more calories of work. Most of this free energy change must come from the binding interactions, per se, rather than from conformational rearrangements since (at least for small substrates) the conformational adjustments are actually minor (3).

In view of the greater activity of human lysozyme, it appeared worthwhile to examine the binding constants of oligosaccharide inhibitors to determine whether the interactions were significantly different. To do so, we resorted to fluorescence spectra in the absence and presence of the inhibitors. Before discussing the results, I should say that Dr. Sharon recently showed us a paper describing some of the fluorescence properties of human lysozyme (26). Those studies parallel some of the experiments reported here and there is generally good agreement, particularly with regard to the binding constant of triNAG and human lysozyme.

Figure 2 shows the fluorescence emission spectra of human lysozyme and its complexes with oligosaccharides of N-acetyl-D-glucosamine. The emission spectrum of free HEWL is shown also for comparison. Except for a slightly shorter maximum wavelength, the latter spectrum agrees with previously reported measurements (15, 20, 27). There are two significant features of the uncomplexed human enzyme to note in comparison with HEWL: (i) Calculation shows that the tryptophan quantum efficiency of human lysozyme is significantly lower (ca. 40%) than that of HEWL (21) and (ii) The emission maximum for human lysozyme is 330 nm, which indicates that the average environment of fluorescent tryptophan residues is less polar than in the egg-white lysozyme.

Imoto et al. (15) have estimated that Trp 62 in HEWL is responsible for 35–38% of the fluorescence in the pH interval 2–8. Substitution of Trp 62 by Tyr 63 in the human enzyme then affords a quantitative explanation for the reduced quantum efficiency. Moreover, the fact that the emission maximum of human lysozyme is blue-shifted relative to that of HEWL also accords with the suggested contribution of Trp 62, since Trp 62 is the most exposed of the 6 tryptophan residues in HEWL and should emit at the longest wavelength. The replacement of Trp 62 would accordingly be expected to eliminate a long wavelength band with resultant blue-shifted spectrum. This argument implies comparable contributions by the homologous tryptophan residues (see below). Of these, Trp 108, the other major fluorophore of HEWL (15), is partially buried and emits at a shorter wavelength. By implication, then, Trp 108 dominates the fluorescence spectrum of human lysozyme.

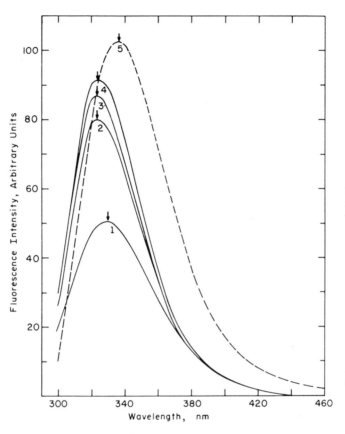

Fig. 2. Fluorescence spectra of (1) human lysozyme; (2) human lysozyme + diNAG, $9 \times 10^{-3}\ M$; (3) human lysozyme + triNAG, $1.2 \times 10^{-3}\ M$; (4) human lysozyme + tetraNAG or pentaNAG, $1.2 \times 10^{-3}\ M$; (5) HEWL. Protein concentration was 0.003% in phosphate buffer. pH = 7.5, μm = 0.1. Excitation at 280 nm. Room temperature.

Figure 2 also demonstrates the magnitudes of the dequenching at pH 7.5 brought about by the binding of di-, tri-, tetra-, and pentaNAG. This dequenching allows measurement of the respective binding constants.

In Fig. 3 is illustrated the pH dependence of the fluorescence intensity of HEWL, human lysozyme, and human lysozyme triNAG complex. For HEWL, the rise in emission between pH 5 and 7 is very slightly greater than that recorded by Lehrer and Fasman (20), but otherwise the titration is in excellent agreement with their data and those of other investigators. An unexpected finding is the contrasting behavior of the free enzymes in the pH region 5.5–7.5. HEWL fluorescence is dequenched as the pH increases in this interval, while that of human lysozyme is quenched, generat-

ing a shallow step in the curve. In HEWL, the rise in fluorescence intensity
has been attributed to the ionization of Glu 35. The reason for the opposite
effect in human lysozyme is unclear and intriguing, probably suggestive of
a subtle difference in the local conformation which relates Glu 35 to Trp
108 in the two enzymes. Protonation of Glu 35 almost certainly quenches
Trp 108. The increased quenching of the fluorescence of the human lysozyme triNAG complex as pH is lowered from about 7.5 to about 5.5 supports this proposal.

Still another important difference occurs in the acid region of this titration.
In HEWL, the fluorescence intensity is more or less constant below pH
5.5 (only partly shown in Fig. 3). This is observed also in human lysozyme.
However, inhibitor complexes of HEWL show a fall in intensity as pH is
lowered (not shown) (20) whereas human lysozyme–inhibitor complexes
are again constant, much as in the free enzyme. The additional quenching
of the HEWL complex in acid was interpreted by Lehrer and Fasman as
arising from protonation of Asp 52 or Asp 101, affecting the fluorescence
of Trp 63. However, Trp 63 is now believed to contribute only weakly to
HEWL fluorescence (15) and, as just noted, there is no corresponding

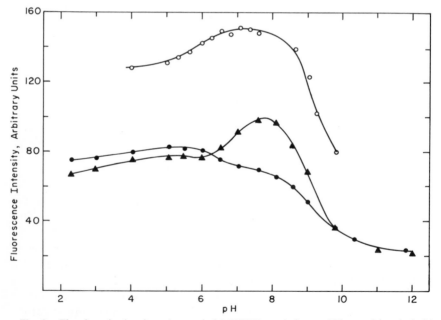

Fig. 3. Fluorimetric titration: (open circle) HEWL, emission at 337 nm; (closed circle)
human lysozyme, emission at 330 nm; (triangle) human lysozyme + triNAG, 5.0×10^{-4} M,
emission at 323 nm. Protein concentration, 0.004%; excitation wavelength: HEWL 280 nm,
human lysozyme 287 nm. Room temperature.

effect in human lysozyme (Trp 64 in the human enzyme is homologous with Trp 63 in HEWL). These various results are difficult to reconcile with the quenching of Trp 63 at low pH. Rather, they suggest that in the HEWL complex Trp 62, having moved closer to the binding-cleft, is susceptible to carboxyl quenching. The absence of such quenching in the free enzyme may then reflect the different orientation of Trp 62 in the binding site. Moreover, this hypothesis is consistent with the absence of additional acid quenching in both free and complexed human enzyme below pH 5, as is revealed in Fig. 3.

Figure 4 shows the calculated difference fluorescence spectra of the human enzyme–tetraNAG complex minus the free enzyme and confirms the absence of significant quenching at low pH. In addition, the difference spectrum at pH 7.5 reveals that fluorescence in the complex is markedly enhanced at short wavelengths (320 nm maximum) with a blue shift signaled by the small negative band at longer wavelength. At pH 5.5 and 2.5,

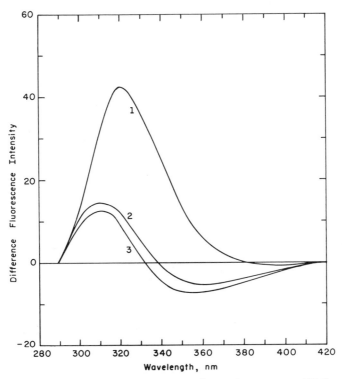

Fig. 4. Calculated difference fluorescence spectra, (human lysozyme-tetraNAG complex)—(human lysozyme) at (1) pH 7.5, (2) pH 5.5, (3) pH 2.9. Protein concentration, 0.003%; tetra-NAG 1.3×10^{-3} M. Excitation at 280 nm. Room temperature.

the spectra are almost identical and are mainly the result of the blue shift consequent to binding.

At pH 7.5, the enhancement of fluorescence due to inhibitor binding is maximal and the binding constants were evaluated at that pH. The results are presented as a modified Scatchard plot in Fig. 5. With the assumption of a single binding site for each inhibitor, the constants may be evaluated from the Y-intercepts. The binding constants will be summarized below, but I should like to call attention to the marked difference between the slopes and intercepts of triNAG and tetraNAG binding. This is in very sharp contrast to the result with HEWL. In common, however, with the behavior of HEWL, pentaNAG and tetraNAG exhibit the same binding to human lysozyme.

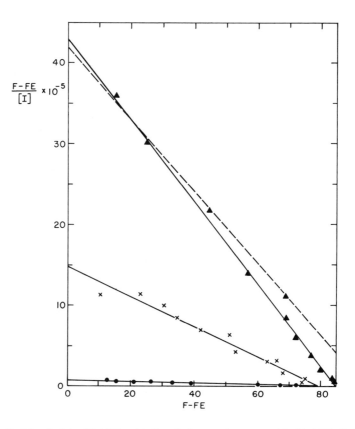

Fig. 5. Scatchard plot of inhibitor binding to human lysozyme: (circle) diNAG; (cross) triNAG; (triangle) tetraNAG; (dashed line) pentaNAG. Fluorescence emission measured at 323 nm in phosphate buffer, pH 7.5.

TABLE I

TABLE I

Fluorescence and Binding Data: Comparison of Human and Egg-White Lysozyme[a]

	λ_{max} (nm)		Q.E.		K_a (M^{-1})		$-\Delta F^\circ$ (kcal/mole)	
	HEWL	HL	HEWL	HL	HEWL[b]	HL	HEWL	HL
Enzyme	337	330	0.06	0.040				
diNAG complex		323		0.057		9.4×10^2		4.05
triNAG complex		323		0.060	7.0×10^4	1.9×10^4	6.60	5.80
tetraNAG complex		323		0.064		5.0×10^4		6.40

[a] Excitation at 280 nm. pH 7.5; phosphate buffer; $\mu = 0.1$; ambient temperature.
[b] Value of HEWL-triNAG K_a from S. S. Lehrer and G. D. Fasman [*Biochem. Biophys. Res. Commun.* **23**, 133 (1966)].

Table I summarizes several of the pertinent fluorescence and binding results presented thus far. The binding constants of triNAG to human lysozyme, 1.9×10^4 M^{-1}, is seen to be less than 30% of that to HEWL representing a difference of 800 cal/mole in the standard free energy of association. The binding constant for tetraNAG is 5×10^4 M^{-1}, more than 2.5 times that for triNAG but still lower than the constant for binding to HEWL.

I would like to turn now to absorption spectral measurements. To understand the comparative behavior we should first examine the differences in composition with respect to aromatic residues. Table II presents such a comparison.

Figure 6 shows the spectrophotometric titration of tyrosine residues as determined from extinction coefficient changes at 290.5 nm (19), at 245 nm, and by ionization difference spectroscopy. The three methods agree closely. There are six tyrosine residues in human lysozyme as compared to three in HEWL (Table II). From earlier results and these we have concluded that three of the six residues titrate normally, two have abnormally elevated pK_a values, and the sixth is inaccessible in the native protein. Its titration is accompanied by a time-dependent, irreversible denaturation.

Figure 7 illustrates the effect of inhibitor binding. High absorbance of tri- and tetraNAG interfered with measurements at 245 nm and the titration here is recorded at 290.5 nm. All of the points for the complexes lie above the titration curve of free enzyme, suggesting a lowering in pK_a of

TABLE II

Comparison of Aromatic Residues in Human and Egg-White Lysozyme

Tryptophan		Tyrosine		Phenylalanine	
HL	HEWL	HL	HEWL	HL	HEWL
Trp 28	Trp 28	Tyr 20	Tyr 20	Phe 3	Phe 3
Trp 34	Phe 34	Ile 23	Tyr 23	Trp 34	Phe 34
Tyr 63	Trp 62	Tyr 38	Phe 38	Tyr 38	Phe 38
Trp 64	Trp 63	Tyr 45	Arg 45	Phe 57	Leu 56
Trp 109	Trp 108	Tyr 54	Tyr 53		
Trp 112	Trp 111	Tyr 63	Trp 62		
Tyr 124	Trp 123	Tyr 124	Trp 123		

one or more tyrosine residues. However, the displacements are small and virtually within the experimental error at these protein concentrations.

To be more secure about the phenomenon, we measured difference spectra with tandem cells and these spectra are recorded in Fig. 8 at pH values of 6.6, 8.6, 9.6, and 10.1. The difference spectra are clearly pH-dependent, whereas such spectra are pH-independent in HEWL inhibitor difference spectra (21). Much of the spectrum at any pH is a tryptophan perturbation spectrum. To isolate the tyrosine portion, we calculated a "difference–difference" spectrum, that is $[\Delta\epsilon_\lambda, \mathrm{pH}\,10 - \Delta\epsilon_\lambda, \mathrm{pH}\,7.6]$, and show this in Fig. 9. From the value of the maximum near 250 we can estimate the different degree of tyrosine ionization in the complex and the free enzyme. A more direct measurement may be made now of the difference extinction at 250 nm as a function of pH (Fig. 10). The maximum in this curve is at pH 10 and represents very nearly 0.3 equivalent of tyrosine residues per mole at pH 10. This must be interpreted as arising from a pK_a difference(s) in the free and complexed enzymes. From these various results, we have been able to compute the pK_a of the perturbed residue in the free and complexed enzyme and we *assume* that this is Tyr 63 (21). In the free enzyme, pK_a is 10.55; in the complex, 10.07. The assignment to Tyr 63 is the one that makes sense, but it is not proved. However, the fact that a perturbed tyrosine has a *lowered* pK_a is demanded from the data.

The combined fluorescence, binding, and spectral titration results may now be interpreted a bit further and more conclusively. The proximity of Tyr 63 to the substrate binding cleft suggests that this is the tyrosine which is perturbed by inhibitor binding. The distance of others from the binding site makes them far less likely candidates. Since the pK_a is lowered, Tyr 63 is not a donor in a hydrogen bond to inhibitor. A lowered pK_a may be explained in a number of ways. For example, the ionized form of the residue might serve as an acceptor, with a sugar hydroxyl acting as donor, or an

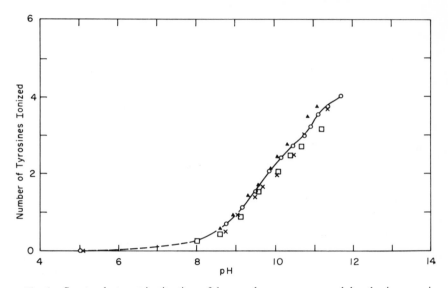

Fig. 6. Spectrophotometric titration of human lysozyme measured by the increase in molar extinction coefficient at 245 nm (cross); 290.5 nm (square), and 245 nm by ionization difference spectroscopy (triangle). The circles are taken from previously published data (19). Protein concentration was 0.013% for measurements at 245 nm, and 0.06% for measurements at 290.5 nm.

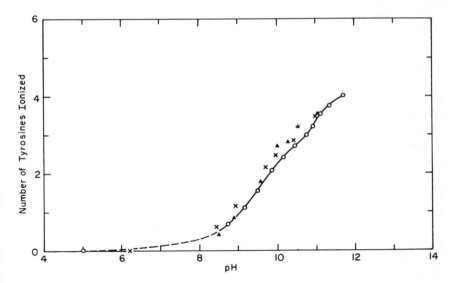

Fig. 7. Spectrophotometric titration of human lysozyme–inhibitor complex, measured by the increase in molar extinction coefficient at 290.5 nm. (Triangle) triNAG, 2.1×10^{-3} M; (cross) tetraNAG, 1.3×10^{-3} M; circles as in Fig. 1. Protein concentration was 0.04%.

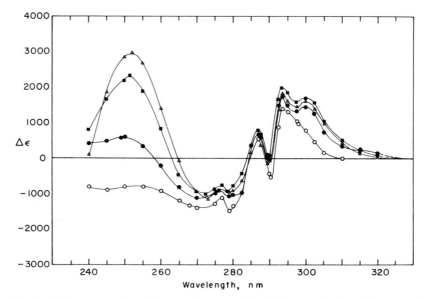

Fig. 8. Difference spectra of human lysozyme–tetraNAG complex (complex–enzyme). (Open circle) pH 6.6; (closed circle) pH 8.6; (square) pH 9.6; (triangle) pH 10.1. Protein concentration, 0.04%; tetraNAG concentration, 1.14×10^{-3} M.

existing hydrogen bond with tyrosine as donor in the free enzyme might be broken when inhibitor is bound.

The interactions between triNAG and the A, B, and C subsites of HEWL are detailed by Imoto *et al.* (16). There are 48 van der Waal's contacts and six hydrogen bonds, including the one formed by Trp 62 and the sugar C-6 hydroxyl in subsite C. The discussion above argues against a similar hydrogen bond involving Tyr 63 and inhibitor, below pH 10. It is likely that the absence of this bond accounts for the loss of 800 cal/mole in the free energy of binding.

The different association constants for tri- and tetraNAG also deserve additional comment, especially since the tri-, tetra-, penta-, and hexaNAG compounds bind to hen egg white with the same constants. (Dr. Rupley has pointed out, in a personal communication, that this statement is true for 25°C in HEWL but that the constants do differ at other temperatures.) That has been interpreted to mean that all these inhibitors bind in the same nonproductive mode (7, 23). Three pyranoside rings at the reducing end of the molecule occupy subsites A, B, and C, while additional rings extend out from the protein. The different constants of tri- and tetraNAG with the human enzyme would suggest, from this line of reasoning, different binding

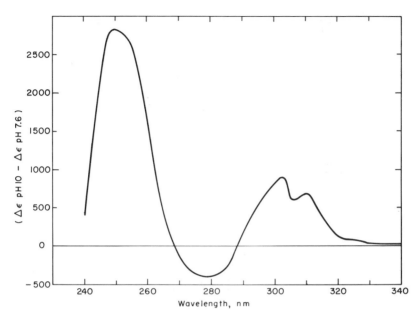

Fig. 9. Calculated "difference–difference" spectrum of human lysozyme–tetraNAG complex: [$\Delta\epsilon_\lambda$, pH 10 − $\Delta\epsilon_\lambda$, pH 7.6].

modes for the two inhibitors. It should be noted, though, that no other spectral features are significantly different.

The last kind of spectral measurement which I would like to touch on briefly is the CD spectra of human lysozyme and of its inhibitor complexes. Figure 11 shows the near-UV CD spectra of the free enzyme and the triNAG complex at pH 6. The main observation here has an exact parallel in HEWL, namely, that binding of inhibitor exerts a substantial effect on the near-UV CD bands (6, 9, 12, 13, 25). The spectrum of human lysozyme looks quite different, on first inspection, from free HEWL (not shown here), but curve resolution reveals very considerably similarity (10). For example, each of the longest wavelength bands (at neutral and acid pH) in HEWL is paralleled by bands of the same shape, position, and (comparable) intensity in the human enzyme. At neutral pH, the dominant difference is the much more intense negative band centered near 270 nm, which is due to ^1L_A tryptophan transitions. Total tryptophan intensity is greater in the human enzyme, notwithstanding the smaller number of residues.

Figures 12 and 13 show these spectral regions at pH values of 8.5 and 10, respectively. Inhibitor effects persist, apparently undiminished, at pH 10, confirming continued triNAG binding at that pH. At pH 10, a highly

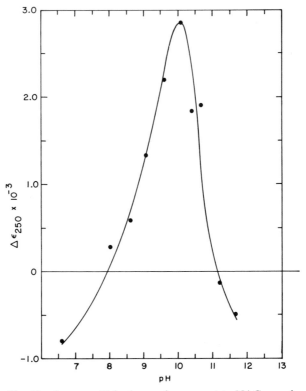

Fig. 10. $\Delta\epsilon_{250}$ vs. pH for human lysozyme–tetraNAG complex.

unusual CD band at 313 nm is seen in both free and complexed enzyme (10). This is the longest wavelength intrinsic chromophore band found in any protein to date and it must arise from an uncommon interaction. This band is not observed in HEWL.

Figure 14 shows the pH dependence of the band in the triNAG complex and Fig. 15 compares the titration of the band measured at its maximum in the free enzyme and in the tri- and tetraNAG complexes. The band appears to titrate as if it were a tyrosine residue, but its wavelength maximum, albeit anomalous, suggests a strongly perturbed tryptophan vibronic band. Coupled with its absence in HEWL, its pH behavior suggested to Halper *et al.*, (10) that it might arise from an interaction between Tyr 63 (in the ionized state) and Trp 64 or Trp 109.

It is very difficult, given experimental uncertainty in values near pH 8, to decide whether the midpoints in Fig. 15 are different for free and bound enzyme, though there does appear to be a shift of the entire curve to lower pH. This would be compatible with the difference spectral arguments presented above in connection with the depression in pK_a of Tyr 63 in complexed enzyme. The midpoints of the curves are somewhat lower than our calculated pK_a for Tyr 63 even after electrostatic interaction corrections, so the assignment is still indefinite. We are presently attempting to decide, also, whether the chromophore is really a tryptophan residue or not, and, if it is, which one.

The far-UV CD spectra of human and HEWL have been compared (10). They are similar enough to suggest very similar secondary structures, in

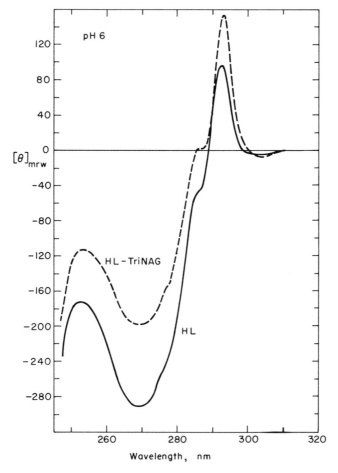

Fig. 11. Near ultraviolet CD spectra of human lysozyme and human lysozyme triNAG complex, pH 6. Lysozyme concentration, 0.46 mg/ml; triNAG, 1.3 mg/ml.

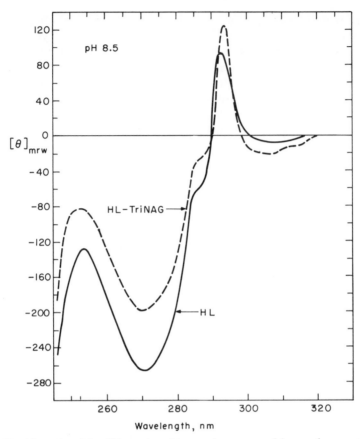

Fig. 12. Near-ultraviolet CD spectra of human lysozyme and human lysozyme triNAG complex, pH 8.5. Concentrations as in Fig. 11.

agreement with the X-ray diffraction results presented by Dr. Phillips in Chapter 2 and previously by Blake and Swan (4).

This leads us finally to a quite different kind of comparison related to the similar secondary structures generated by different sequences, in connection with our introductory comments. In Table III, we have listed the residues in α-helical segments in HEWL, as found by X-ray studies of the crystal (2). These segments are compared to segments predicted to be permissively helical by a method suggested by Wu and Kabat (18, 28). In their procedure, the sequences of proteins of known three-dimensional structure are analyzed in terms of nearest neighbor influence on the ϕ, ψ angles adopted by each residue in the sequence. Each residue is tabulated according to its occurrence in helical backbone conformation when surrounded by a particular pair of residues. When HEWL is analyzed in this way, the overall predicted helix content agrees well with the known helix content; moreover, the actual stretches are in reasonably good agreement. In addition, the β-pleated sheet residues do not occur in the predicted sequences. These results were good enough that we were encouraged to examine the sequence of human lysozyme by the same method. The results are shown in the last column of Table III. The predicted overall helix content is good and a comparison of predicted stretches of α-helix in the two enzymes reveal striking, though hardly perfect, similarity, notwithstanding the extensive sequence differences.

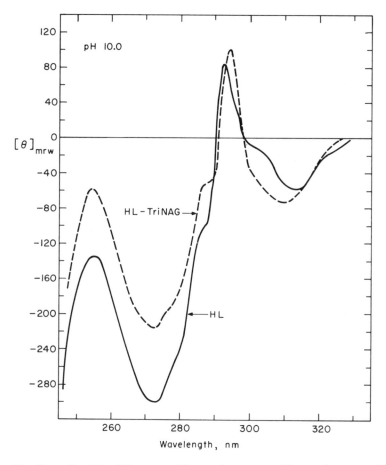

Fig. 13. Near-ultraviolet CD spectra of human lysozyme and human lysozyme triNAG complex, pH 10.0. Concentrations as in Fig. 11.

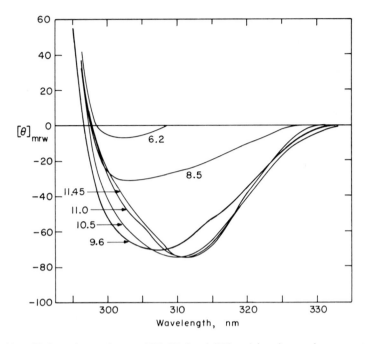

Fig. 14. pH dependence of a near-UV CD band (313 nm) in a human lysozyme–triNAG complex.

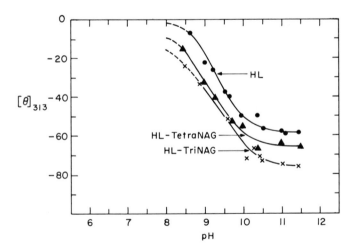

Fig. 15. Residue ellipticity change at 313 nm as a function of pH in human lysozyme, and its triNAG and tetraNAG complexes.

TABLE III

Calculation of Permissively Helical Residues

HEWL		HL
X-ray	Calculated	Calculated
4–15	2–14	2–10
	17–20	16–19
24–36	27–38	27–36
60–63		
		74–79
80–85		81–84
88–95	91–98	87–104
97–101		
108–115	104–115	108–117
119–125	117–122	119–122
45.7%	42.6%	50%

Acknowledgments

We are pleased to express our indebtedness to Dr. James P. Halper, Harvey Bernstein, and Norman Latovitzki for their contributions in the early phases of this work and to Dr. Peter Kahn and Duane Sears for many valuable discussions and suggestions.

References

1. Beychok, S., and Warner, R. C., *J. Am. Chem. Soc.* **81,** 1892 (1959).
2. Blake, C. C. F., Koenig, D. F., Mair, G. A., North, A. C. T., Phillips, D. C., and Sarma, V. R., *Nature (Lond.)* **206,** 757 (1965).
3. Blake, C. C. F., Mair, G. A., North, A. C. T., Phillips, D. C., and Sarma, V. R., *Proc. Roy. Soc., Lond. [Biol.]* **167,** 365 (1967).
4. Blake, C. C. F., and Swan, I. D. A., *Nature (Lond.), New Biol.* **232,** 12 (1971).
5. Canfield, R. E., Kammerman, S., Sobel, J. H., and Morgan, F. J., *Nature (Lond.), New Biol.* **232,** 16 (1971).
6. Cowburn, D. A., Bradbury, E. M., Crane-Robinson, C., and Gratzer, W. B., *Eur. J. Biochem.* **14,** 83 (1970).
7. Dahlquist, F. W., Jao, L., and Raftery, M. A., *Proc. Natl. Acad. Sci. U.S.A.* **56,** 26 (1966).
8. Dickerson, R. E., and Geis, I., "Structure and Action of Proteins." Harper, New York, 1969.
9. Glazer, A. N., and Simmons, N. S., *J. Am. Chem. Soc.* **88,** 2335 (1966).
10. Halper, J. P., Latovitzki, N., Bernstein, H., and Beychok, S., *Proc. Natl. Acad. Sci. U.S.A.* **68,** 517 (1971).
11. Hess, G. P., and Rupley, J. A., *Annu. Rev. Biochem.* **40,** 1013 (1971).
12. Ikeda, K., and Hamaguchi, K., *J. Biochem. (Tokyo)* **66,** 513 (1969).

13. Ikeda, K., Hamaguchi, K., Imanishi, M., and Amano, T., *J. Biochem.* (*Tokyo*) **62,** 315 (1967).
14. Ikeda, K., Hamsguchi, K., Miwa, S., and Nishina, T., *J. Biochem.* (*Tokyo*) **71,** 371 (1972).
15. Imoto, T., Forster, L. S., Rupley, J. A., and Tanaka, F., *Proc. Natl. Acad. Sci. U.S.A.* **69,** 1151 (1971).
16. Imoto, T., Johnson, L. N., North, A. C. T., Phillips, D. C., and Rupley, J. A., *in* "The Enzymes" (P. D. Boyer, ed.), Vol. 7, p. 665. Academic Press, New York, 1972.
17. Jollès, J., and Jollès, P., *FEBS Lett.* **22,** 31 (1972).
18. Kabat, E. A., and Wu, T. T., *Biopolymers* **12,** 751 (1973).
18a. Kammerman, S., personal communication (1969).
19. Latovitzki, N., Halper, J. P., and Beychok, S., *J. Biol. Chem.* **246,** 1457 (1971).
20. Lehrer, S. S., and Fasman, G. D., *Biochem. Biophys. Res. Commun.* **23,** 133 (1966).
21. Mulvery, R. S., Gualtieri, R. J., and Beychok, S., *Biochemistry* **12,** 2683 (1973).
22. Osserman, E. F., and Lawlor, D. P., *J. Exp. Med.* **124,** 921 (1966).
23. Rupley, J. A., and Gates, V., *Proc. Natl. Acad. Sci. U.S.A.* **57,** 496 (1967).
24. Saint-Blancard, J., Chuzel, P., Matthiew, Y., Perrot, J., and Jollès, P., *Biochim. Biophys. Acta* **220,** 300 (1970).
25. Teichberg, V. I., Kay, C. M., and Sharon, N., *Eur. J. Biochem.* **16,** 55 (1970).
26. Teichberg, V. I., Plasse, T., Sorell, S., and Sharon, N., *Biochim. Biophys. Acta* **278,** 250 (1972).
27. Teichberg, V. I., and Sharon, N., *FEBS Lett.* **7,** 171 (1970).
28. Wu, T. T., and Kabat, E. A., *Proc. Natl. Acad. Sci. U.S.A.* **68,** 1501 (1971).

SECTION

3

BIOLOGICAL AND CLINICAL STUDIES

Introduction

ELLIOTT F. OSSERMAN

Introduction

This section is devoted to a consideration of certain aspects of the biology of lysozyme in mammals, particularly man, and changes in lysozyme metabolism and turnover associated with various disease states, particularly the leukemias. As stated in several papers in the preceding sections, we still are not certain of lysozyme's functions in higher vertebrates other than its probable antibacterial role. The possibility has been repeatedly suggested that lysozyme may have other important physiological functions, some of which may be related to specific enzymatic activities on nonbacterial substrates and others associated with its marked cationic property. These provocative questions are considered in several of the papers in this section.

In the first paper, Hansen, Karle, and Andersen summarize their extensive studies of the turnover of lysozyme in man and in rats employing the radioactively labeled enzyme. These studies document the key role of granulocytes in the production of lysozyme and of the kidneys in the elimination of lysozyme.

Maack and Sigulem further extend consideration of the complex mechanisms involved in the renal handling of lysozyme, i.e., its filtration, reabsorption, and possible metabolism by the kidney. Their own observations have been made on intact animals, perfused kidneys, and isolated tubule preparations.

The association of abnormalities of lysozyme metabolism with various types of leukemias has been the subject of numerous studies in the past ten years. Finch and his associates at Yale were among the first to describe an increase in serum lysozyme levels in certain patients with monocytic and monomyelocytic leukemia. Their paper reviews their extensive experience in this area and summarizes the relationship between serum lysozyme levels and physiologic as well as neoplastic changes in granulocyte and monocyte turnover. They also compare the clinical patterns of two groups of adults with acute leukemia, those with high and those with low serum lysozyme levels. The patients in the high lysozyme category tend to be older, survive longer, and have a relatively lower incidence of serious bacterial infections than the patients with low lysozyme levels. They conclude that the increased serum lysozyme in certain cases of leukemia probably reflects the presence of a relatively large pool of circulating and tissue leukocytes which are still functional and may delay the onset of infectious complications.

Malmquist provides further documentation of the association of elevated serum and urine lysozyme levels in monocytic leukemias. He has employed the Laurell method of electroimmunoassay for the quantitation of lysozyme and has demonstrated that this method correlates well with enzymatic assay methods. The electroimmunoassay procedure is possibly more convenient for routine clinical laboratories already equipped for this type of procedure. Malmquist further explores the possible interrelationship of lysozyme with two other leukocyte proteins, myeloperoxidase and lactoferrin. In contrast to lysozyme, the serum concentrations of myeloperoxidase and lactoferrin are only moderately elevated in the monocytic and granulocytic leukemias.

Gilbert has determined the intracellular concentrations of lysozyme and unsaturated B_{12}-binding protein (UBBP) in the leukocytes of chronic myelocytic leukemia, polycythemia vera and other myeloproliferative conditions. These studies indicate that the chronic myelocytic leukemic leukocytes have a disproportionately high concentration of intracellular lysozyme, suggesting a major metabolic abnormality in these cells.

Perillie and Finch examine the question of the diagnostic usefulness of lysozyme determinations and their value for assessing chemotherapy in the leukemias. They also review the available data with respect to the concentration of lysozyme in granulocytic vs. monocytic leukemic cells. These data indicate that the intracellular concentration of lysozyme is greater in myelocytic than in monocytic cells, whereas the serum and urine lysozyme levels are greater in the monocytic than in the myelocytic leukemias. This suggests that lysozyme is released more rapidly from monocytes than from myelocytes and granulocytes.

Asamer, Schmalzl, and Braunsteiner review their studies of the intracellular localization of lysozyme in leukemic monocytes and myelocytes as determined by immunofluorescence techniques and also the changes in lysozyme levels associated with the treatment of monocytic and monomyelocytic leukemia.

Farhangi and Osserman describe studies from our laboratory which demonstrate the *de novo* synthesis of lysozyme by bone marrow but not the peripheral blood cells of a patient with monomyelocytic leukemia. These studies also provide evidence that lysozyme is very rapidly released from these cells.

A substantial body of evidence is now developing which appears to establish that lysozyme is, indeed, elaborated by cells of *both* the granulocytic and the monocytic series. There would seem, however, to be significant differences between the "dynamics" of lysozyme in the two cell types, viz., in the granulocytic–myelocytic series, lysozyme is apparently localized quite strictly to the lysosomes and very little, if any, of the enzyme is discharged by "leakage" except insofar as this involves lysosomal lysis. In the monocytic–histiocytic series, however, lysozyme is apparently continuously liberated from intact cells. The intracellular concentration of lysozyme in monocytes is, therefore, not unexpectedly lower than that of myelocytes, but it is quite possible that the total synthesis of lysozyme is greater in cells of the monocytic than the myelocytic series.

Rosenthal, Greenberger, and Moloney describe their observations on the Shay chloroleukemia in rats which elaborates lysozyme and, like human monomyelocytic leukemia, is associated with hyperkaluria. This tumor thus represents a very useful model of the human disease as well as an excellent source of rat lysozyme.

Pascual, Perillie, Gee, and Finch at Yale have clearly documented increased serum lysozyme levels in the majority of cases of sarcoidosis—a diffuse granulomatous disease resembling tuberculosis. Their studies demonstrate that serum lysozyme levels may provide a very useful index of disease activity and response to therapy in sarcoidosis.

The functional significance of the high concentrations of lysozyme in cartilage has puzzled many investigators from Fleming to the present day. Kuettner, Eisenstein, and Sorgente review the evidence that lysozyme is synthesized by chondrocytes and that cationic lysozyme forms complexes with the anionic proteoglycans of cartilage. They speculate that lysozyme may serve some function in the calcification process as a consequence of this property. Alternatively or additionally, it is also proposed that lysozyme may enzymatically cleave an as yet unidentified polysaccharide substrate in cartilage and other connective tissues. Josephson and Greenwald describe their studies of the effects of pH, salt concentration, and specific cations on the extractability of lysozyme from various types of human cartilage. They have also shown that the intravenous injection of papain into rabbits produces a simultaneous increase in serum levels of chondroitin sulfate and lysozyme, both apparently derived from altered cartilage. This interesting animal model deserves further investigation. Pruzanski, Ogryzlo, and Katz have demonstrated an increase in serum and synovial fluid lysozyme in rheumatoid arthritis, and they postulate that in addition to the damaged cartilage, synovial membrane cells may be a source of this lysozyme.

As described by Schumacher, lysozyme is present in human seminal plasma and cervical secretions, and its production is strongly influenced by sex hormones. Again, the precise biological functions of lysozyme in these locations are still obscure but it may have an important role in reproductive processes.

Hyslop, Kern, and Walker have carried out detailed serial studies of lysozyme and other proteins in human colostrum and milk. Once again the key question is whether lysozyme's role in these secretions is strictly antibacterial.

The last two papers present studies from our laboratory which are specifically concerned with determining what effects lysozyme may have on mammalian cells and their constituent organelles. Adinolfi, Loeb, and Osserman present our findings that mammalian mitochondria are agglutinated by lysozyme and that this agglutination is inhibited by concanavalin A. In the final paper, Osserman, Klockars, Halper, and Fischel present additional evidence that lysozyme has significant effects on the morphology and behavior of a variety of mammalian cells in culture. We postulate that these effects are due to an action of lysozyme on the membranes of these cells. The key question is whether certain mammalian cell polysaccharide(s) are susceptible to enzymatic cleavage by lysozyme or whether these *in vitro* effects are related to lysozyme's marked cationic property. The answer to this question will have to be provided by further studies, but the suggestion is made that lysozyme may possibly play an important role in certain mammalian cell functions.

Production and Elimination of Plasma Lysozyme

NIELS EBBE HANSEN, HANS KARLE, and VAGN ANDERSEN

This survey presents a quantitative description of lysozyme turnover in man and an attempt to evaluate factors of importance for plasma lysozyme production and elimination. Studies in the rat are included to the extent that such studies serve to illustrate conditions in man. Due to the design of the studies, problems pertaining to the elimination of lysozyme are treated prior to the discussion of factors influencing the rate of production.

Methods

Studies on lysozyme turnover and the organ distribution of lysozyme were carried out with purified human and rat lysozyme labeled with ^{125}Iodine. These purified lysozymes, obtained from the urine of human leukemic patients and tumor-bearing rats, were kindly provided by Dr. Osserman, who has demonstrated that these lysozymes are identical with naturally occurring lysozymes in man and rat, respectively (1, 2). The technique of the studies have been reported in detail elsewhere (3, 4).

In order to evaluate one factor of possible significance for the "production" of lysozyme, kinetic studies of radioactively labeled (DF^{32}P) neutrophilic granulocytes were carried out on 36 patients covering a wide range of neutrophil counts, and data from these studies compared with plasma lysozyme levels. The details of this study are reported elsewhere (5). Lysozyme activity was measured with the lysoplate method of Osserman and Lawlor (1) or with the turbidometric method of Litwack (6) as earlier described (7). In both methods human lysozyme was used as standard. Values with the lysoplate method are 3–4 times higher than values with the turbidometric method (7), which again are in good accordance with those obtained with a quantitative immunoelectrophoretic method (8). Reference to the method used in the individual experiments will be given below.

Elimination of Plasma Lysozyme

The technique used for the evaluation of lysozyme elimination from plasma was the conventional technique for measuring the turnover rate of plasma proteins, i.e., intravenous injection of radioactively labeled lysozyme; after injection, lysozyme radioactivity in plasma is followed. Calculations of kinetic data are based on the analyses of the plasma disappearance curve. The most important figure obtained with this method is the fractional catabolic rate (FCR) which gives the fraction of the total plasma lysozyme content eliminated per hour. This figure is the reciprocal of the area under the plasma disappearance curve followed to "infinity."

Figure 1 shows the ^{125}I-labeled lysozyme disappearance curve from nine persons without hematological or renal disease. It is seen that plasma lysozyme has a very rapid elimination rate; analysis of the disappearance curves showed a fractional catabolic rate of 0.76/hr (S.D., 0.09), which means that 76% of the total plasma lysozyme content is eliminated per hour. This is a very fast turnover rate compared with plasma albumin, which has a fractional catabolic rate of 0.10 per day but is quite comparable with the turnover rate of immunoglobulin light chains, which are of slightly higher molecular weight (9).

It is well known that kidney function is of great importance for plasma lysozyme concentration, although the quantitative relationship has not been extensively studied previously. This relationship is illustrated in Fig. 2 which shows the relation between plasma lysozyme concentration and the glomerular filtration rate (measured with the endogenous creatinine clearance method) in 36 patients. It is seen that a curvilinear relationship exists between these two variables with a statistically significant correlation coefficient between log creatinine clearance and plasma lysozyme concentration ($r = 0.86$, p less than 0.001).

In an attempt to examine this relationship further, lysozyme turnover studies were carried out on three bilaterally nephrectomized patients and four patients with varying degrees of reduced glomerular filtration rate. Figure 3 shows the results of these investigations. It is seen that lysozyme

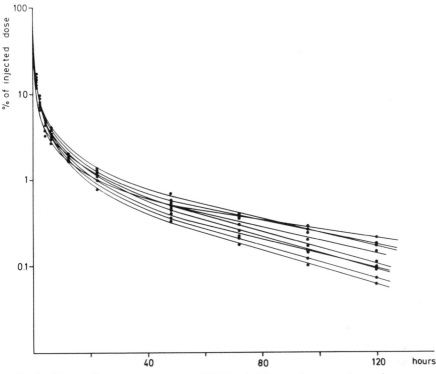

Fig. 1. Plasma disappearance curves of ^{125}I-labeled human lysozyme from nine persons without hematological or renal disease. The ordinate shows radioactivity in the total plasma pool as percent of injected dose. [Reprinted with permission from Ref. (4).]

was eliminated in anephric man, although only at a rate amounting to about 15% of the rate found in persons with intact kidneys (FCR = 0.11/hr); it is furthermore seen that the fractional catabolic rate decreases with the glomerular filtration rate. The importance of the kidneys for the rapid elimination of plasma lysozyme is also illustrated in Fig. 4 which shows the plasma disappearance curves during the first 180 min following injection in bilaterally nephrectomized patients and normals. It is seen that already at the first sampling after 5 min, plasma lysozyme radioactivity was lower in subjects with intact kidneys than in nephrectomized patients. The dependence of the fractional catabolic rate on glomerular filtration is demonstrated in Fig. 5, which shows results from 22 patients. The relationship between glomerular filtration rate and lysozyme fractional catabolic rate is statistically significant ($r = 0.90$, p less than 0.001).

We do not know from these investigations if extrarenal catabolization is a feature of the uremic state, or if it is also operative in normal man. However, it appears most likely that lysozyme, in addition to renal elimina-

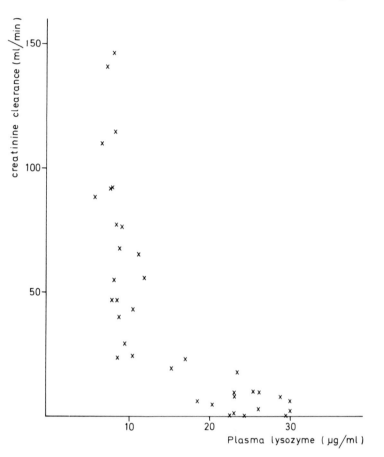

Fig. 2. The relationship between plasma lysozyme concentration and endogenous creatinine clearance in 36 patients without hematological disease. Plasma lysozyme concentrations were measured with the lysoplate method with human lysozyme as the reference standard. [Reprinted with permission from Ref. (4).]

tion, is subjected to the same mechanisms of degradation as other proteins in the blood (4). Assuming this to be the case, and provided there is no significant recirculation of lysozyme from the kidneys to the blood stream, it is possible to calculate the glomerular filtration rate of lysozyme by subtracting the fractional catabolic rate in anephric patients from the fractional catabolic rate of persons with normal glomerular filtration (4) In our nine persons without renal or hematological disease we found the lysozyme glomerular filtration rate to be 35% (S.D. 7) of that of endogenous creatinine. This agrees well with the clinical and experimental findings of others (10, 11).

The importance of the kidneys for lysozyme elimination is also evident from a comparison of plasma and kidney levels of radioactivity following intraperitoneal injection of labeled rat lysozyme in the rat. Figure 6 shows that kidney lysozyme rose concomitantly with a decrease in plasma radio-activity, and that the kidney activity stayed at a higher level than plasma activity for several days. Thus, even if the kidney uptake of lysozyme from plasma is a fast process, degradation within the kidneys seems to be rather slow.

Lysozyme Production

In conventional plasma protein turnover studies, the rate of synthesis of a protein may be calculated from its fractional catabolic rate (FCR)

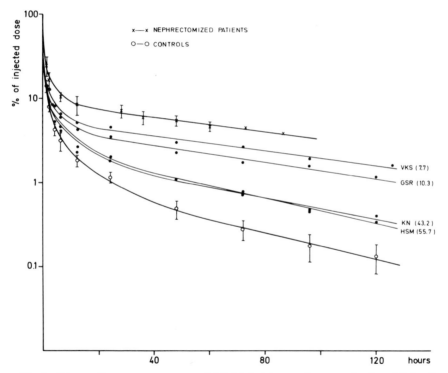

Fig. 3. Plasma disappearance curves of ^{125}I-labeled human lysozyme in three bilaterally nephrectomized patients (upper curve) and four patients with varying degrees of renal insufficiency; the endogenous creatinine clearance values (ml/min) are given in parentheses. The composite disappearance curve from nine persons with normal glomerular filtration is shown for comparison (lower curve). Vertical bars indicate S.D. The ordinate shows radio-activity in the total plasma pool as percent of injected dose. [Reprinted with permission from Ref. (4).]

provided steady-state conditions. This calculation may be carried out for plasma lysozyme:

lysozyme rate of synthesis (μg/kg/hr)

$$= \frac{\text{FCR/hr} \times \text{plasma volume (ml)} \times \text{plasma lysozyme (}\mu\text{g/ml)}}{\text{body weight (kg)}}$$

In nine normal persons we found a mean lysozyme rate of synthesis of 287 μg/kg/hr (S.D. 72) corresponding to a lysozyme production of about

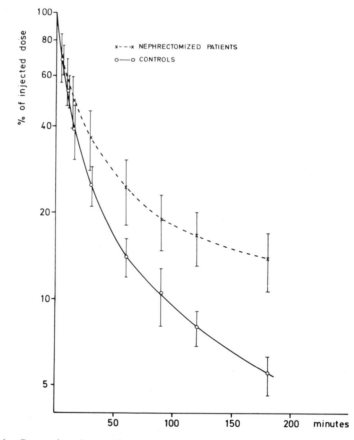

Fig. 4. Composite plasma disappearance curves of ^{125}I-labeled human lysozyme from three bilaterally nephrectomized patients (upper curve) and from nine normal persons (lower curve) during the first 180 min following intravenous injection of labeled lysozyme. Vertical bars indicate S.D. The ordinate shows radioactivity in the total plasma pool as percent of injected dose. [Reprinted with permission from Ref. (4).]

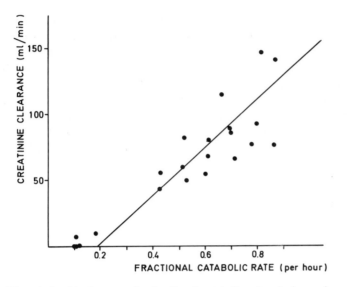

Fig. 5. The relationship between the fractional catabolic rate of plasma lysozyme and endogenous creatinine clearance in 22 patients. The correlation coefficient was statistically significant ($r = 0.90$, p less than 0.001).

½ gm per day in normal man (the mean plasma lysozyme concentration in these nine patients was 8.5 μg/ml as measured with the lysoplate method).* However, this only applies for lysozyme appearing in the plasma; as discussed below, the overall lysozyme rate of synthesis is presumably considerably higher.

The same calculations were carried out in six patients with myeloproliferative disorders (chronic myelocytic leukemia, myelofibrosis, acute myeloid leukemia), all of whom had elevated plasma lysozyme levels. The plasma disappearance curves in these patients were normal, i.e., the fractional catabolic rates as expected from these patients' creatinine clearance values; this means that the fractional catabolic rate does not vary with plasma lysozyme concentration. According to the above equation for the calculation of the rate of synthesis, the combination of a normal fractional catabolic rate and increased plasma lysozyme concentration means increased rate of synthesis; in these six patients with hematological disorders the rate of synthesis was 2–4 times the normal values. Calculations carried out in the uremic patients showed that in these patients the rate of synthesis was decreased, which was a quite unexpected finding.

---

*With the turbidometric and immunoelectrophoretic methods the daily lysozyme production would amount to about 150 mg per day.

The demonstration of increased rate of synthesis in the patients with myeloproliferative disorders supports the hypothesis that in such patients elevated plasma lysozyme levels are due to release of lysozyme from neutrophilic granulocytes, possibly because of an increased neutrophil turnover rate. If this assumption has general validity, the low rate of synthesis in the uremic patients should be explicable either by a decrease in the intraneutrophilic lysozyme activity and/or by a decreased neutrophil turnover rate. In a study of lysozyme activity in neutrophilic granulocytes (Fig. 7) we have in fact demonstrated a 25% reduction of the neutrophil lysozyme

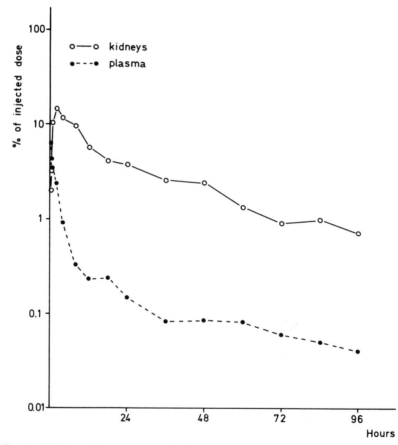

Fig. 6. ^{125}I-labeled lysozyme activity in plasma and kidneys following intraperitoneal injection of labeled rat lysozyme to the rat. Values are expressed as percent of injected dose for total plasma volume and for total amount of kidney tissue. Each point represents the mean value from two rats. [Reprinted with permission from Ref. (3).]

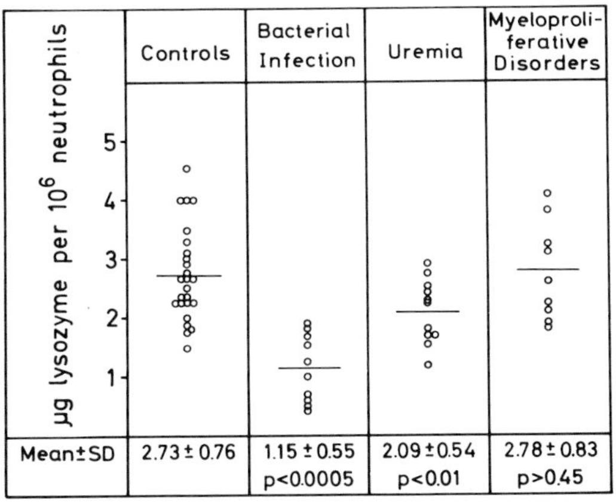

μg lysozyme per 10⁶ neutrophils	Controls	Bacterial Infection	Uremia	Myeloproliferative Disorders
Mean±SD	2.73±0.76	1.15±0.55 p<0.0005	2.09±0.54 p<0.01	2.78±0.83 p>0.45

Fig. 7. The intraneutrophilic lysozyme content in various groups of patients. Lysozyme activity was measured with the turbidometric method with human lysozyme as reference. The *p* values represent the statistical significance of difference from the values of the control group. [Reprinted with permission from Ref. (7).]

content in uremic patients (7). Figure 7 also shows that the intraneutrophilic lysozyme activity in patients with myeloproliferative disorders was normal; consequently, the reason for the high lysozyme production rate in these patients cannot be release of lysozyme from neutrophils with higher than normal neutrophil lysozyme activity.

So far, the results of our studies on the rate of lysozyme synthesis are quite compatible with the hypothesis that the neutrophilic granulocytes are major contributors to the plasma lysozyme activity. We have attempted to evaluate this hypothesis further in two ways.

First, we studied the variations of plasma lysozyme in patients with neutropenia induced with cytotoxic drugs (12). Figure 8 shows the results of such a study in one patient; it is seen that plasma lysozyme activity rather closely followed the variations in the neutrophil counts. Similar changes were found in five additional patients; in all cases a statistically significant relationship was found between plasma lysozyme concentrations and neutrophil counts in the blood (*p* less than 0.001). It is remarkable, however, that in no patient was the plasma lysozyme concentration suppressed to zero even if the neutrophil counts were very low. Part of the explanation, at least, must be "background" lysozyme, from nonneutrophilic sources, e.g., monocytes.

Second, kinetic studies with DF^{32}P-labeled neutrophils were carried out in 36 patients covering a wide range of neutrophil counts and plasma lysozyme levels (5). If it is true that plasma lysozyme mainly stems from dis-

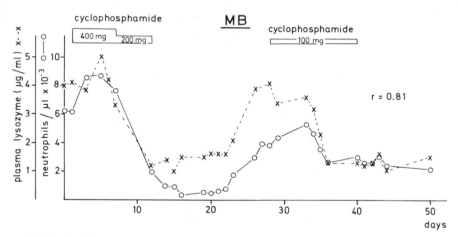

Fig. 8. Variations of plasma lysozyme and blood neutrophil counts in a patient with neutropenia induced with cytotoxic drugs. The *r* value is the correlation coefficient between plasma lysozyme concentration and neutrophil counts. Plasma lysozyme was measured with the turbidometric method with human lysozyme as reference. [Reprinted with permission from Ref. (12).]

integrating neutrophils, a linear relationship should be obtained between the granulocyte turnover rate (GTR) measured with the $DF^{32}P$ technique and the GTR calculated from plasma lysozyme levels. This lysozyme derived GTR (LZM-GTR) may be calculated:

LZM-GTR (neutrophils/kg/day)

$$= \frac{\text{plasma lysozyme } (\mu g/ml) \times \text{plasma volume} \times \text{lysozyme FCR/hr} \times 24}{\text{body weight (kg)} \times \text{neutrophilic lysozyme } (\mu g/cell)}$$

Figure 9 shows the relationship between the $DF^{32}P$-GTR and the LZM-GTR. It is seen that a linear relationship exists between these two variables. We consider this linear relationship between the $DF^{32}P$- and lysozyme-derived GTR strong support of the hypothesis that disintegrating neutrophils are the main contributors to the plasma lysozyme content. However, it also appears from the figure that not all neutrophils passing through the blood—and thereby measured with the $DF^{32}P$ technique—deliver their lysozyme content to the plasma. In patients with normal and increased neutrophil counts the underestimate of the GTR with the lysozyme method averaged 40% (5). Such a loss of cells is well known since it has been demonstrated that a considerable number of neutrophils (quantitative data are not available) are lost through egress to the alimentary tract including the oral cavity (13). Part of the discrepancy might in addition be explicable if some lysozyme from disintegrating neutrophils were catabolized locally in the tissues. Nonneutrophilic sources of plasma lysozyme would further increase the discrepancy between neutrophil turnover rate and neutrophil lysozyme release to plasma.

These findings have an important bearing on the estimate of the lysozyme rate of synthesis as measured with radioactively labeled lysozyme as described above. This method is based on the assumption that all lysozyme passes through the plasma compartment. Since, as demonstrated, at least 40% of the neutrophilic granulocytes do not deliver their lysozyme to the

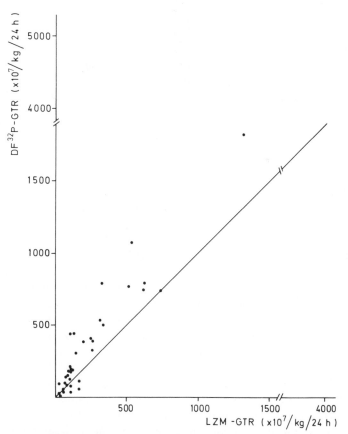

Fig. 9. The relationship between the granulocyte turnover rate calculated from plasma lysozyme values (LZM-GTR, see text) and the granulocyte turnover rate measured with DF^{32}P labeling (DF^{32}P-GTR). The line indicating identity between the two methods is shown in the figure. [Reprinted with permission from Ref. (5).]

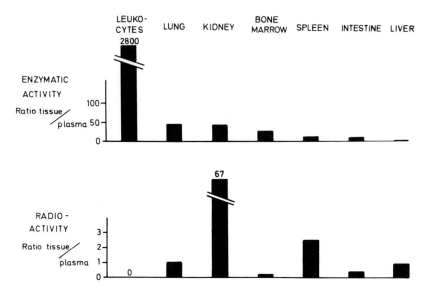

Fig. 10. Tissue/plasma ratios for lysozyme enzymatic activity (upper part) and lysozyme radioactivity (lower part) 3 h after intraperitoneal injection of ^{125}I-labeled rat lysozyme to the rat. Activities were calculated per gm tissue and per ml plasma.

plasma, the above figure of $\frac{1}{2}$ gm for the lysozyme rate of synthesis should be increased accordingly. This means that the rate of lysozyme synthesis amounts to about 1 gm per day in normal man.* This figure for the rate of lysozyme synthesis should be even further increased to the extent that lysozyme is produced and catabolized locally, as for instance in tissue macrophages, tears, nasal secretion, and cartilage.

We may thus conclude that the neutrophilic granulocytes are the main contributors to plasma lysozyme and that the kidneys are the most important sites of degradation. These conclusions are compatible with studies in the rat on the organ distribution of lysozyme enzymatic activity and radioactivity after injection of labeled rat lysozyme. High enzymatic activity in an organ may mean that the organ in question (a) contains lysozyme synthesizing cells, (b) is a site of storage and/or destruction of lysozyme containing cells, and/or (c) that the organ takes up lysozyme from the plasma. Injected radioactive lysozyme would presumably accumulate only in organs of group (c). The results of such a study are shown in Fig. 10, which shows the ratio tissue/plasma for enzymatic activity and radioactivity, respectively. Two "organs" in particular stand out clearly, viz., the leukocytes with their extremely high lysozyme enzymatic activity and zero uptake of

* With the turbidometric and immunoelectrophoretic methods the daily lysozyme production would amount to about 300 mg per day.

radioactivity, and the kidneys with their high enzymatic activity and their extremely high uptake of radioactivity (ratio tissue/plasma $= 67:1$). This pattern in the leukocytes, seen in conjunction with the low protein synthesis of these cells (14), is indirect evidence that neutrophil lysozyme originates in the neutrophil precursor cells in the bone marrow, and the kidney pattern clearly indicates the kidney to be the main organ of uptake and possibly degradation of lysozyme from the plasma.

Acknowledgments

We thank Dr. E. F. Osserman for making available to us purified human and rat lysozymes. Work reported in this survey was supported by grants from The Danish Medical Research Council, The Danish Foundation for the Advancement of Medical Science, and The Medical Research Foundation for Copenhagen, the Faroe Islands, and Greenland.

References

1. Osserman, E. F., and Lawlor, D. P., *J. Exp. Med.* **124,** 921 (1966).
2. Osserman, E. F., personal communication.
3. Hansen, N. E., Karle, H., and Andersen, V., *J. Clin. Invest.* **50,** 1473 (1971).
4. Hansen, N. E., Karle, H., Andersen V., and Ølgaard, K., *J. Clin. Invest.* **51,** 1146 (1972).
5. Hansen, N. E., *Br. J. Haematol.* **25,** 767 (1973).
6. Litwack, G., *Proc. Soc. Exp. Biol. Med.* **89,** 401 (1955).
7. Hansen, N. E., and Andersen, V., *Br. J. Haematol.* **24,** 607 (1973).
8. Johansson, B. G., and Malmquist, J., *Scand. J. Clin. Lab. Invest.* **27,** 255 (1971).
9. Jensen, K., *Scand. J. Clin. Invest.* **25,** 281 (1970).
10. Harrison, J. F., Lunt, G. S., Scott, P., and Blainey, J. D., *Lancet* **1,** 371 (1968).
11. Harrison, J. F., and Blainey, A. D., *Clin. Sci.* **38,** 533 (1970).
12. Hansen, N. E., Andersen, V., and Karle, H., *Br. J. Haematol.* **25,** 485 (1973).
13. Boggs, D. R., *Semin. Hematol.* **4,** 359 (1967).
14. Stjernholm, R. L., *Plen. Sess. Pap., Congr. Int. Soc. Hematol. 12th, 1968* p. 175 (1968).

Renal Handling of Lysozyme

THOMAS MAACK and DANIEL SIGULEM

Evidence gathered to the present points to the fact that the kidneys play an important role in the regulation of plasma levels of lysozyme. Indeed, it is now well established that the kidneys accumulate a large fraction of both administered heterologous and homologous lysozyme (5, 9, 11, 15). Furthermore, plasma levels of lysozyme are markedly increased after bilateral nephrectomy or in renal failure (6, 7). However, the basic parameters of the renal handling of lysozyme are not yet fully understood. In this respect, the following initial considerations are pertinent:

1. Essentially, there are two routes by which lysozyme can be handled by the kidneys: by glomerular filtration and subsequent tubular uptake (absorption) or by leakage from post-glomerular capillaries with subsequent physical adsorption to the renal tissue. The relative contribution of these two routes may be established by determining, on one hand, the degree of hindrance the glomerular capillaries present to the passage of lysozyme into the glomerular filtrate (sieving coefficient), and on the other hand, the independent measurement of the physical adsorption of lysozyme to the renal tissue. We made such determinations and will show in the present report that the filtration–absorption process is the predominant one.

2. In order to understand the role the kidneys play in the overall regulation of plasma levels of lysozyme, it is important to determine the charac-

teristics of the absorption process with regard to its capacity, threshold, and nature. We will show that the absorption process is characterized by a very high capacity, a relatively low threshold, and a constant fractional absorption over a wide range of plasma concentrations of lysozyme. We will also present initial results showing that the absorption process is either directly or indirectly dependent on aerobic metabolism.

3. The culmination of the renal processing of lysozyme resides in the final disposal of the absorbed protein. This problem is difficult to approach experimentally and definitive answers are not yet possible. We will show that the lysozyme absorbed by the renal cells is returned to the circulation either as intact molecules or degradation products or both. We will also present our latest attempts to clarify the relationship between intracellular catabolism and transport of intact lysozyme (10, 11).

Glomerular Filtration of Lysozyme

Proteins are hindered at the glomerular filter in proportion to their molecular dimensions. However, the relationship is not a simple one since molecular shape, charge, and binding to other plasma macromolecules have a complex influence on the degree of sieving (8). Fortunately, the problem can be approached experimentally in a relatively simple manner. Two methods were used in our laboratory to determine the degree of glomerular sieving of lysozyme:

The sieving coefficient of egg-white lysozyme was estimated in anesthetized dogs by measuring the clearance of lysozyme (C_{LY}) in relation to the clearance of the glomerular marker, creatinine (C_{Cr}) at plasma levels of lysozyme much above the saturation of the absorption process. At lysozyme plasma levels of approximately 0.5 mg/ml, which produced filtered loads at least 5 times above the maximal capacity of absorption (see Fig. 1), the C_{LY}/C_{Cr} ratio was 0.71 ± 0.11 S.D. $(n = 5)$.

In the second approach, tubular absorption of homologous rat lysozyme* was inhibited by KCN and the sieving coefficient measured at low concentrations of lysozyme. These experiments were done in the isolated rat kidney perfused in a closed-circuit system [see Bowman and Maack (2) for details of the technique]. The perfusate consisted of Krebs–Henseleit medium with 7.5 gm% albumin. ^{125}I-labeled homologous rat lysozyme, the glomerular

---

*Rat lysozyme kindly provided by Dr. Ray Sherman, Department of Medicine, Cornell University Medical College, New York Hospital. We wish to acknowledge the excellent technical help of Mrs. Mary Wagner in the preparation of the rat lysozyme. Rat lysozyme was iodinated in our laboratory by the chloramine-T procedure. The label was stable and did not change the lysozyme enzymatic activity.

marker, inulin (In), and KCN were added to the perfusate. Table I shows that with 3 mM KCN, which almost completely blocked the accumulation of lysozyme in the renal tissue, the C_{LY}/C_{In} ratio was maximal and equal to 0.83.

The results in either case indicate that the concentration of lysozyme in the glomerular filtrate is about 70–80% of that found in plasma water. This value is much higher than those of 30–40% previously reported in the literature [e.g., Marshall and Deutsch (12)]. The probable reason for this discrepancy is that, in the previous studies, the plasma levels of lysozyme used to determine the C_{LY}/C_{Cr} ratios were insufficient to saturate the absorption process. Hence, the ratios measured the combined effect of glomerular sieving and tubular absorption of lysozyme. It cannot be completely ruled out that the present experiments underestimated the degree of glomerular hindrance in physiological conditions, since it could be argued that very high plasma levels of lysozyme or KCN may increase the glomerular permeability to proteins. However, the value of 0.7–0.8 agrees well with the theoretically predicted value for the sieving coefficient of a 14,000 MW globular protein (8). More importantly, if the sieving coefficient of lysozyme is 0.4 instead of 0.8 it would be impossible to account for the high levels of renal accumulation of injected lysozyme solely on the basis of filtration and subsequent tubular absorption (5, 9, 11, 15; see also Fig. 4 and 5). With a

TABLE I

Effect of KCN on the Renal Handling of Homologous Lysozyme by the Isolated Perfused Rat Kidney [a]

Exp	KCN (mM)	U_{IN}/P_{IN}	C_{LY}/C_{IN}	LY "space" [d] %
A [b]	0	15.80	0.09	420
B [c]	1	2.04	0.17	–
	3	1.01	0.83	24

[a] Isolated perfused kidney prepared as previously described (2). Perfusate consisted of Krebs–Henseleit medium with 7.5 gm% albumin.

[b] ^{125}I-Lysozyme and ^{3}H-inulin were added at the beginning of the perfusion. U_{IN} and C_{LY}/C_{IN} are average values of four 15 min clearance periods. The lysozyme "space" was determined from the radioactivity in kidney tissue and perfusate at the end of the experiment.

[c] ^{3}H-Inulin and KCN, 1 mM, were added at the beginning of the perfusion; 10 min thereafter ^{125}I-lysozyme was added. After 5 min equilibration a 15 min clearance period was taken. Then KCN concentration was increased to 3 mM and after 15 min equilibration another 15 min clearance period was taken. Lysozyme "space" was measured at the end of the experiment.

[d] LY "space" $= \dfrac{\text{cpm/gm kidney}}{\text{cpm/ml perfusate}} \times 100$.

sieving coefficient of 0.4 it would be necessary to postulate that a large fraction of the accumulated lysozyme originated as leakage from peritubular capillaries with subsequent physical adsorption to the renal tissue. Table I provides additional evidence that this cannot be the case (10). When tubular absorption was inhibited by KCN, renal accumulation of lysozyme was only 5% of that observed without inhibition. This result indicates that the bulk of the lysozyme accumulated by the kidney originates from glomerular filtration and subsequent absorption by renal tubular cells.

Renal Tubular Absorption of Lysozyme

Previous morphological evidence showed that administered lysozyme is absorbed almost exclusively by the proximal tubular cells of the kidney (11). Once the sieving coefficient has been determined, it is now possible to study some of the characteristics of the absorption process quantitatively, by standard renal titration and renal clearance techniques.

Maximal Capacity and Threshold

Figure 1 shows a complete renal titration curve for egg-white lysozyme in the dog. As can be seen, there is a linear relationship between filtration

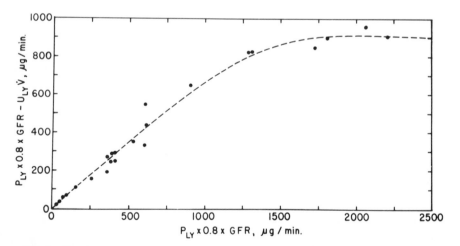

Fig. 1. Results of clearance experiments in 6 anesthetized dogs, with 4–6 20 min clearance periods in each dog. Each point represents 1 clearance period. Abscissa: filtered loads of lysozyme. P_{LY} = plasma concentration of lysozyme; GFR = creatinine clearance (C_{Cr}); 0.8 = sieving coefficient of lysozyme, estimated by the C_{LY}/C_{Cr} ratio at filtered loads above 10 mg/min (see text). Ordinate: lysozyme absorption calculated by the differences between the filtered load and urinary excretion rate of lysozyme ($U_{LY}V$). The dotted line was fitted by eye.

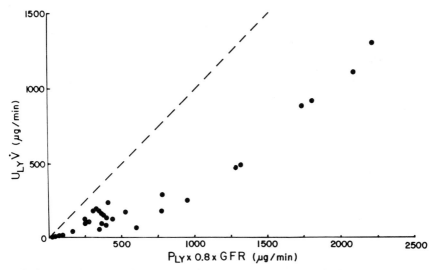

Fig. 2. Lysozyme excretion rate as a function of filtered load derived from the same experiments shown in Fig. 1. The dotted line is the line of identity for lysozyme filtration rates.

and absorption rates over a wide range of filtered loads of lysozyme. Saturation of the absorption process occurs only at the very high filtered loads of 1.5–2.0 mg/min, which corresponds to 20–50 times the normal filtered loads of endogenous lysozyme in the dog. Despite this very high absorption capacity, the renal threshold for lysozyme is relatively low. Figure 2 shows the corresponding excretion pattern in the renal titration curve; measurable lysozymuria is absent at the very low filtered loads but lysozyme appears in the urine much before the saturation of the absorption process has been reached. At filtered loads above the saturation point there is, as expected, a tendency for the lysozyme excretion rate to parallel the rate of lysozyme filtration.

These characteristics of high capacity and low threshold differentiate the lysozyme absorption process from the renal transport of simple organic compounds such as glucose. With glucose, at least in the dog, apparent transport maximum and threshold are reached at about the same plasma levels. However, the disparity between high maximal capacity and low threshold is by no means unique to lysozyme. The same pattern has been demonstrated for the renal transport of certain amino acids such as glycine (16) and other small molecular weight proteins such as insulin (unpublished results).

Nature of the Process of Lysozyme Absorption

Further insight into the nature of lysozyme absorption can be obtained by examining the C_{LY}/C_{Cr} ratios over a wide range of plasma concentrations of lysozyme below the saturation point. Figure 3 shows that, under these conditions, the C_{LY}/C_{Cr} ratio is relatively constant. In conjunction with the data shown in Fig. 1, this indicates that, over a wide range of plasma concentrations, the tubular cells remove lysozyme in a fixed proportion to the plasma concentrations of the protein. It should be pointed out that this relationship does not hold at very low plasma levels of lysozyme, where the C_{LY}/C_{Cr} ratio is close to zero, or at levels above the saturation point, where C_{LY}/C_{Cr} increases continuously up to the maximal values of 0.6–0.8. A relatively constant fractional absorption of lysozyme indicates that there is a definite, albeit imperfect, glomerulo–tubular balance for lysozyme. This finding suggests that conservation of lysozyme and/or its renal degradation products is of importance for the total body economy.

It has been previously demonstrated that administered lysozyme accumulates in the renal tissue (5, 9, 15). The data presented here demonstrate what has been previously suspected (1, 5, 9, 11, 15), i.e., that this accumulation is the result of lysozyme glomerular filtration with subsequent removal by the renal cells. Furthermore, from the clearance data it is possible to predict that, below the saturation point, lysozyme accumulation in renal tissue should be a constant fraction of the administered dose. Figures 4 and 5 demonstrate that this is indeed the case. They show that there is a linear relationship between dose and accumulation of lysozyme in the renal tissue and, hence, a constant fractional accumulation over a wide range of doses. We have previously demonstrated that the accumulation process, like the absorption process, is only saturated at very high doses of lysozyme

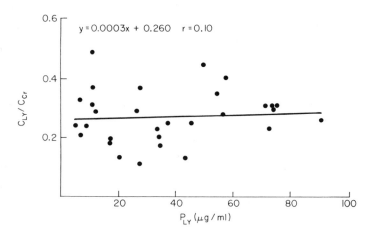

Fig. 3.   Relative lysozyme clearance (C_{LY}/C_{Cr}) as a function of plasma concentrations of lysozyme below the saturation point. Data derived from the same experiments shown in Fig. 1. The slope of the regression line is not significantly different from zero ($p > 0.10$).

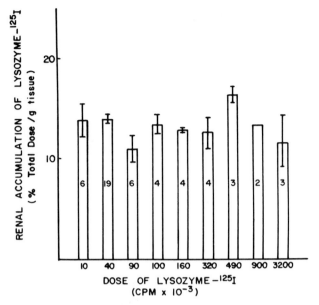

Fig. 4. Fractional renal accumulation of lysozyme as a function of dose. [125]I egg-white lysozyme was administered intravenously to rats and the kidneys removed 30 to 60 min after the injection. The height of the bars represents the mean fractional accumulation ($\pm$ SE) of the number of experiments shown for each dose. There were no significant differences in the fractional accumulations with different doses ($p > 0.10$). 1 μg lysozyme ~ 1000 CPM.

(9). Taken together, the results show that the kidneys, besides filtering and absorbing lysozyme, serve as a "sink" for the circulating plasma lysozyme.

Energy Dependence and Cellular Mechanisms of Lysozyme Absorption

Table I shows that cyanide, a powerful inhibitor of aerobic metabolism, inhibits lysozyme absorption and accumulation in renal tissue. The effect is maximal at a concentration of 3 mM but it is already significant at the concentration of 1 mM of KCN. It has been demonstrated that cyanide also hinders the formation of pinocytotic vesicles in macrophages (3). It is possible therefore that the effect of cyanide on the renal absorption of lysozyme is due to the inhibition of the formation of new membranes that are necessary for the transport of substances by pinocytosis. The cellular mechanism of renal lysozyme absorption remains to be clarified. Morphological and differential centrifugation studies (11) demonstrated that part of the absorbed lysozyme is accumulated within phago-lysosomes, indicating that the protein entered the cell by a process of pinocytosis. However, in the same study a large fraction of the absorbed lysozyme

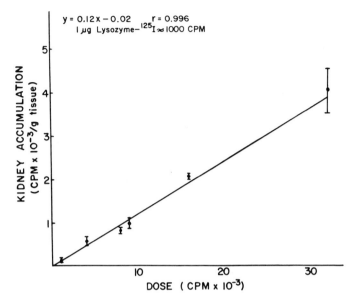

Fig. 5. Absolute renal accumulation of intravenously administered ^{125}I egg-white lysozyme as a function of dose. Data derived from the same experiments shown in Fig. 4. Points represent average $\pm$ SE of at least three experiments. The linear regression was highly significant ($p < 0.001$).

was also found in the tissue supernatant (cytoplasmic sap). The origin of this latter fraction is not entirely clear. It may have originated from the intracellular rupture of phago-lysosomes or from a direct passage of lysozyme across the cell membrane. In this regard it should be pointed out that, at least in macrophages, cationic molecules, and lysozyme in particular, are poor inducers of the formation of pinocytotic vesicles (4).

Table I also shows that, at the doses of cyanide tested, the decrease in lysozyme absorption corresponded to the decrease of fluid reabsorption. It is thus possible that the effect of cyanide upon lysozyme absorption was indirect and that lysozyme absorption is linked to fluid reabsorption. Further studies using other metabolic inhibitors and inhibitors of sodium transport are necessary to clarify this point. In the meantime, the present data demonstrate that lysozyme absorption by the kidney is either directly or indirectly dependent on aerobic metabolism.

The Fate of the Absorbed Lysozyme

The intracellular pathways of the renal transport of lysozyme have been previously studied and it has been demonstrated that the absorbed protein is distributed in two intracellular compartments: in phago-lysosomes and in the cytoplasmic sap (11). Direct evidence with regard to the final fate of the absorbed lysozyme is difficult to obtain. Theoretically, the absorbed protein may be returned to the circulation or excreted in the urine either as intact molecules or as degradation products such as amino acids and polypeptides. There is experimental evidence for both mechanisms. Transcellular transport of intact lysozyme has been postulated on the basis of findings in the flounder that injected lysozyme, which was accumulated by renal cells *in vivo,* was transported intact across the peritubular cell membrane when the flounder renal tubules were incubated *in vitro* (10). On the other hand, accumulation of lysozyme in phago-lysosomes and the finding of increased plasma levels of lysozyme after bilateral nephrectomy or in renal failure have been interpreted as indicating that at least part of the absorbed lysozyme is catabolized by the renal cells (6, 11, 17). However, the available evidence is not conclusive and a final answer will require the direct determination of the absorption products released from the renal cells.

We have attempted to elucidate this point with two experimental approaches. In one group of studies we injected rats with either cold or ^{125}I-labeled lysozyme. After renal accumulation of lysozyme, the right kidney was removed and perfused in a closed-circuit system. The products released to the perfusate and urine were quantitatively measured and characterized by enzymatic activity and/or gel chromatography on Sephadex G-50. Results obtained with this technique demonstrated that, after 1 hr of perfusion, 50% of the absorbed lysozyme was released to the perfusate either intact or as catabolic products or both. Insignificant amounts were released to the urine.

When high doses of lysozyme were administered to the intact rat and were accumulated by the kidney it was feasible to measure the lysozyme enzymatic activity released to the perfusate. Under these conditions it was possible to demonstrate that most of the released product was in the form of intact lysozyme molecules. Figure 6 shows the chromatogram of a typical experiment. With 78% recovery, all of the released product eluted together with lysozyme. These results confirm those obtained with isolated flounder tubules (10). They show that mammalian renal cells also have the capability of transporting intact lysozyme molecules. However, the doses of lysozyme used in these experiments produced plasma lysozyme concentrations much above those found in most pathological conditions (14). Unfortunately, because of the limited sensitivity of the method of measuring lysozyme activity, it was not possible to characterize the released product in this manner when more physiological doses of lysozyme were administered.

As an alternative procedure, we administered trace amounts of ^{125}I-labeled lysozyme and characterized the released products by gel chromatography,

TCA solubility, and radioactivity counting. Figure 7 shows a typical chromatogram of such an experiment. Here, very little of the radioactivity was associated with intact lysozyme molecules, the bulk of radioactivity being soluble in TCA and eluting after the elution region of lysozyme. Further characterization of the released product on Sephadex G-25 and Sephadex G-10 showed that the product eluted together with [125]I. In no instance were we able to detect radioactivity associated with amino acids or polypeptides. The results of these experiments cannot be interpreted in a simple fashion. Thus, it is possible that, at physiological doses, the kidney catabolized the absorbed labeled lysozyme and in the process liberated its incorporated label. However, it is also possible that lysozyme was transported in an intact form to the perfusate but that the attached label was set free by deiodinazing enzymes known to be present in kidney tissue. A definitive interpretation has to await the development of more sensitive methods of measuring lysozyme activity and/or the production of an internally labeled lysozyme. Both procedures are feasible.

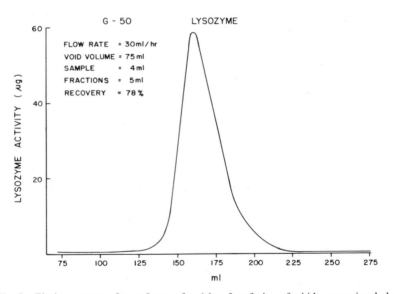

Fig. 6. Elution pattern of a perfusate after 1 hr of perfusion of a kidney previously loaded *in vivo* with lysozyme. 63 mg of egg-white lysozyme was administered intravenously in the rat; 30 min after the injection, the right kidney was removed and perfused as previously described (2). After 1 hr of perfusion, 4 ml of perfusate was layered on a 45 × 2.5 column of Sephadex G-50 and eluted with 0.9 gm% NaCl; 5 ml fractions were collected and lysozyme was measured as previously described (9). Lysozyme activity was detected only in the region of lysozyme elution previously determined with egg-white lysozyme on the same column (not shown).

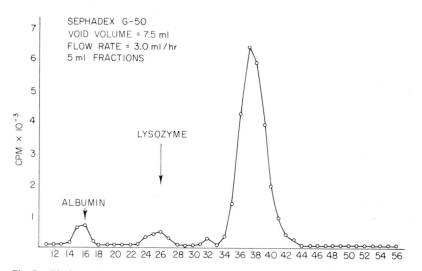

Fig. 7. Elution pattern of a perfusate after 1 hr of perfusion of a kidney which had accumulated ^{125}I-lysozyme *in vivo*. Procedure identical to that in Fig. 6 except that ^{125}I egg-white lysozyme was administered in amounts which did not change the plasma concentration of lysozyme significantly. The bulk of the radioactivity eluted after lysozyme and was soluble in TCA. Recovery $= 93\%$. Lysozyme elution region determined by passing ^{125}I-lysozyme through the same column (not shown).

At present, it is only possible to speculate on the physiological importance of the renal handling of lysozyme. As already stated, the kidneys have a very large capacity to handle lysozyme. From the clearance data it is possible to estimate that the kidneys filter and absorb a quantity of lysozyme a day which corresponds to at least 50 times the total extracellular pool of the protein. We also have shown that the absorption process is highly adaptable, the kidneys being able to absorb lysozyme in direct proportion to the plasma concentrations of the enzyme. Finally, we have demonstrated that the bulk of the absorbed lysozyme is returned to the circulation either as intact molecules or degradation products or both. Thus, the kidneys are in an ideal position to regulate the plasma levels of lysozyme. It is possible to consider the kidneys as a major regulator for the changes in plasma concentrations of lysozyme, and the products of absorption, which are released back to the circulation, as the feedback mechanism for the control of the rate of synthesis and/or release of lysozyme to the plasma. The finding of an increased synthesis of lysozyme after bilateral nephrectomy (7) fits this hypothetical homeostatic model for the control of plasma concentrations of lysozyme but this observation needs to be confirmed by other laboratories.

Pathophysiology of the Renal Handling of Lysozyme

The studies described in this report provide a physiological explanation of most of the described abnormalities in the renal handling of lysozyme. These abnormalities and their physiological interpretations may be summarized as follows:

1. Increased plasma levels of lysozyme in renal failure and in conditions in which the glomerular filtration rate is decreased (6, 7): The capacity of the kidneys to filter, absorb, and serve as a "sink" for circulating plasma lysozyme explains the increased plasma levels of lysozyme observed in these conditions. Thus, decrease in glomerular filtration rate leads to the accumulation in plasma of the fraction of lysozyme which is normally handled by the kidney. However, it is also possible that the increases in plasma levels of lysozyme under these conditions are partially due to the influence of the kidneys on the rate of synthesis and/or release of lysozyme by other organs (7).

2. Lysozymuria due to increased plasma levels of lysozyme (6, 14, 17): The relatively low threshold for lysozyme accounts for the fact that even small increments in plasma lysozyme levels may lead to lysozymuria. Due to the very high tubular absorptive capacity, uptake of lysozyme by the renal cells will continue to rise even while increased amounts of lysozyme are excreted in the urine. This process may eventually lead to a very large renal accumulation of lysozyme, with the consequent formation of "droplets." Since it has been shown that, at least in macrophages, high levels of lysozyme are cytotoxic (4), the large renal cell accumulation of lysozyme may lead to renal dysfunctions (13).

3. Lysozymuria due to tubular dysfunctions (1, 6, 17): Since under normal conditions, plasma lysozyme is extensively filtered and absorbed by the tubular epithelium, certain tubular dysfunctions can result in the excretion of large amounts of lysozyme. This lysozymuria may be unrelated to total proteinuria. This lack of relationship can be explained by considering, on one hand, the high glomerular permeability and tubular absorptive capacity for lysozyme and, on the other hand, the normally low glomerular permeability to most plasma proteins. Indeed, when the tubular absorptive mechanism is impaired, proteins appear in the urine in proportion to their filtered loads. Since the filtered load of lysozyme is high and that of most plasma proteins is low, massive lysozymuria may be present with relatively little excretion of other proteins (17). When increased glomerular permeability to plasma proteins is the predominant pathology, such as in nephrotic syndrome, the filtered load of lysozyme is relatively little affected while that of most plasma proteins is significantly increased.

As a result, massive proteinuria with only slight lysozymuria may be present in these conditions. Naturally, because of the varying degrees of glomerular and tubular involvement in different renal diseases, an entire spectrum of relationships between lysozymuria and proteinuria may be observed.

Acknowledgment

These studies were supported by the NIH grant AM 14241.

References

1. Balazs, T., and Roepke, R. R., *Proc. Soc. Exp. Biol. Med.* **123**, 380–385 (1966).
2. Bowman, R. H., and Maack, T., *Am. J. Physiol.* **222**, 1499–1504 (1972).
3. Cohn, Z. A., *J. Exp. Med.* **124**, 557–571 (1966).
4. Cohn, Z. A., and Parks, E., *J. Exp. Med.* **125**, 213–232 (1967).
5. Hansen, N. E., Karle, H., and Andersen, V., *J. Clin. Invest.* **50**, 1473–1477 (1971).
6. Hayslett, J. P., Perrilie, P. E., and Finch, S. C., *N. Engl. J. Med.* **279**, 506–512 (1968).
7. Keeler, R., *Can. J. Physiol. Pharmacol.* **48**, 131–138 (1970).
8. Landis, E. M., and Pappenheimer, J. R., *in* "Handbook of Physiology" Am. Physiol. Soc., (W. F. Hamilton, ed.), Sect. 2, Vol. II, p. 961. Williams & Wilkins, Baltimore, Maryland, 1963.
9. Maack, T., *J. Cell. Biol.* **35**, 268–273 (1967).
10. Maack, T., and Kinter, W. B., *Am. J. Physiol.* **216**, 1034–1043 (1969).
11. Maack, T., Mackensie, D. D. S., and Kinter, W. B., *Am. J. Physiol.* **221**, 1609–1616 (1971).
12. Marshall, M. E., and Deutsch, H. F., *Am. J. Physiol.* **163**, 461–467 (1950).
13. Muggia, F. M., Heinemann, H. O., Farhangi, M., and Osserman, E. F., *Am. J. Med.* **47**, 351–366 (1969).
14. Osserman, E. F., and Lawlor, D. P., *J. Exp. Med.* **124**, 921–951 (1966).
15. Perri, G. C., Faulk, M., Shapiro, E., and Money, W. L., *Proc. Soc. Exp. Biol. Med.* **115**, 189–192 (1964).
16. Pitts, R. F., *Am. J. Physiol.* **140**, 156–167 (1943).
17. Prockop, D. J., and Davidson, W. W., *N. Engl. J. Med.* **270**, 269–274 (1964).

Lysozyme in Leukopathic States*

S. C. FINCH, O.CASTRO, M. E. LIPPMAN,
J. A. DONADIO, and P. E. PERILLIE

Clinical and experimental studies indicate that from 70 to 80% of the total body lysozyme is either within or released from neutrophils and monocytes. This estimate is derived from several observations, which may be summarized as follows: (a) Serum and total body lysozyme activities in rodents are reduced from 60 to 80% following the administration of either total body irradiation or nitrogen mustard (Fig. 1 and 2) (1–3). (b) The nadir of total body, serum, and tissue lysozyme activities corresponds to the time of maximum leukopenia (Fig. 1 and 2). (c) Patients with severe agranulocytosis associated with bone marrow myeloid hypoplasia may have serum lysozyme activity reduced to 20–30% of normal (Fig. 3) (4–6). (d) Either acutely or chronically increased granulocyte destruction may increase serum lysozyme activity (Fig. 3) (2, 5). (e) Most patients with aplastic anemia, extremely blastic myeloid leukemia, acute lymphoblastic leukemia, leukemic reticuloendotheliosis, and most forms of aleukemic leukemia have low serum lysozyme activity.

Organ distribution of lysozyme in the rat (Fig. 4) varies in accord with the probable tissue concentrations of mature myeloid and monocytic cells, except for the kidney in which the high lysozyme concentration probably

*This investigation was supported by PHS Training Grant No. HL 05316-12 from the Heart and Lung Institute and Research Grant CA 11106-09.

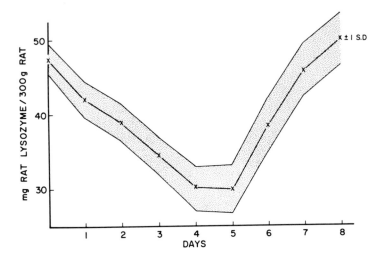

Fig. 1. Change in total rat lysozyme content following the administration of nitrogen mustard (HN$_2$). A total of 24 rats was used in the study. Each point represents the average of two or more rats.

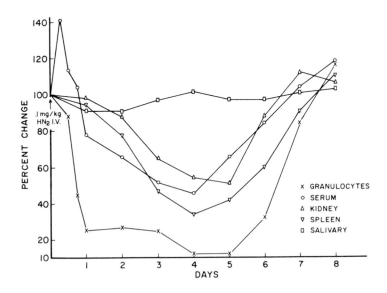

Fig. 2. Change in total granulocyte count and lysozyme activity in various organs of rats following the administration of nitrogen mustard (HN$_2$) in a dose of 0.1 mg/kg intravenously. The values shown are average percentages of pretreatment lysozyme concentrations. The constancy of the salivary gland lysozyme concentration would indicate autonomous enzyme production.

reflects enzyme entrapment in the proximal tubules (3, 7, 8). These observations permit construction of a model for lysozyme turnover in man (Fig. 5).

Reduced serum lysozyme activity is invariably associated with impaired granulopoiesis and a reduced total body granulocyte pool. Recent studies in our laboratory have demonstrated in both rabbits and man that reduced serum lysozyme activity may be a better indicator of reduced myelopoiesis than either the total B_{12} binding capacity or the αB_{12} binding capacity of the serum (Table I) (6). In acute drug-induced agranulocytosis, serum lysozyme activity will fall with the white count to extremely low levels whereas the B_{12} binding components will show some reduction, but rarely to levels below the control range. This may be a function of relatively slow plasma turnover for the B_{12} binding proteins in comparison to lysozyme. Not all patients with granulocytopenia, however, necessarily have reduced serum lysozyme activity. Thus, patients with granulocytopenia associated with a normocellular or hypercellular myeloid marrow will have normal or increased serum enzyme activity (Fig. 3) (4, 5). Ineffectual granulopoiesis due to defective cell lines or increased peripheral granulocyte destruction probably are responsible for the relatively high serum lysozyme in these granulocytopenia patients.

The major role of mature granulocytes and monocytes in lysozyme production is clearly seen in a great variety of myeloproliferative disorders associated with increased serum lysozyme activity (Fig. 6) (9). A physiological classification of the various clinical conditions which influence serum lysozyme is shown in Table II. Any condition which will increase proliferation of either mature granulocytes or monocytes is likely to in-

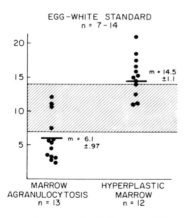

Fig. 3. Comparison of serum lysozyme levels in cases of bone marrow hypo- and hyperplasia. The horizontal bars represent the mean serum lysozyme values in μg/ml (egg-white standard) for the two groups of patients studied. The shaded area shows the control range.

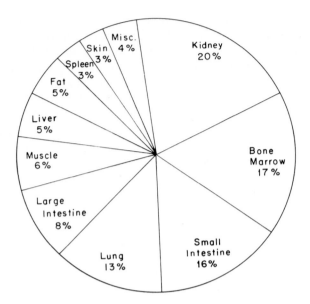

Fig. 4. Diagrammatic representation of the tissue distribution of lysozyme in normal Sprague Dawley rats. The results are based on average values from a pool of 36 rats assayed on by the lysoplate method using human lysozyme as the standard.

crease serum lysozyme activity (9–11). Since the concentrations of lysozyme in mature granulocytes and monocytes are comparable, the relatively high lysozyme activity in patients with monocytic disorders must be due to excessive monocyte secretion of enzyme, increased monocyte infiltration of body tissues, increased monocyte turnover, or decreased enzyme degrada-

TABLE I

Serum Lysozyme and B_{12} Binding Capacity in Acute Agranulocytosis

	Polys + Mono/mm³	Serum lysozyme μg/ml	Total U-B_{12}BC pg/ml	α-B_{12}BC pg/ml	β-B_{12}BC pg/ml
Human studies					
Patient A.B.	408	4.2	1645	313	1332
Patient E.G.	0	3.3	1305	377	928
Patient H.C.	266	4.5	507	180	327
Control range	—	7.0–14.0	800–2500	250–800	500–1600
Rabbit studies					
Nitrogen mustard					
treated ($\times$ 4)	127	4.5	34,000	7821	26,185
Controls ($\times$ 4)	4472	5.0–11.0	6000–15,000	2000–5000	4000–10,000

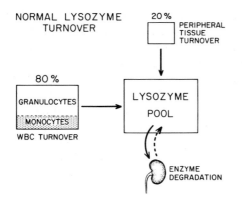

Fig. 5. Schematic representation of lysozyme turnover in man. The extent of reabsorption of lysozyme from the kidney is uncertain.

tion. Reduced serum lysozyme activity is associated with disorders which diminish myelopoiesis (Table II). Bone marrow replacement with fibrous tissue or foreign cells usually permits extramedullary hemopoiesis which is associated with increased serum enzyme activity. The fact that serum lysozyme activity is not related to the concentration of circulating monocytes in control subjects probably explains why we do not recognize a reduced lysozyme syndrome due to diminished monocytopoiesis (12).

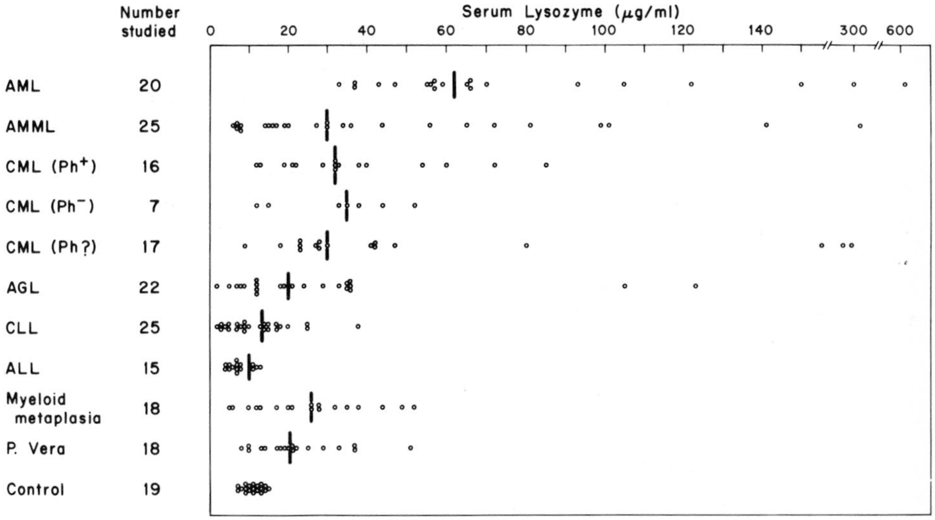

	Number studied	Serum Lysozyme (μg/ml)
AML	20	
AMML	25	
CML (Ph⁺)	16	
CML (Ph⁻)	7	
CML (Ph?)	17	
AGL	22	
CLL	25	
ALL	15	
Myeloid metaplasia	18	
P. Vera	18	
Control	19	

Fig. 6. Serum lysozyme values in μg/ml (egg-white standard) are shown for patients with leukemia and related disorders. All patients in this study had total leukocyte counts of greater than 5000/mm³. The median value for each group is shown as a vertical bar.

The significance of changes in serum lysozyme activity in patients with severe leukopathic states is unclear at this time, but there are some interesting clinical correlations. These are related to the enzymatic classification of adult acute leukemias. A study of 68 adult patients with acute leukemia showed that serum lysozyme was increased (high lysozyme leukemia, HLL) in 39 patients and was normal or reduced (low lysozyme leukemia, LLL) in 29 patients at the time of diagnosis (Fig. 7). In 78% of the patients with marked increase in serum lysozyme activity, the morphologic diagnosis was monocytic or myelomonocytic leukemia. All stem cell and promyelocytic leukemias were in the low lysozyme group. Auer rods were observed with the same frequency in both groups of patients. The median age of patients

TABLE II

Causes of Variation in Serum Lysozyme Activity

Increased activity	Reduced activity
Increased enzyme production	Decreased enzyme production
(1) Myeloid hyperproliferation (benign)	(1) Myeloid hypoproliferation (benign)
(a) Acute bacterial infection	(a) Aplastic anemia
(b) Leukemoid reaction	(b) Agranulocytosis with hypocellular myeloid marrow
(c) Accelerated myeloid destruction (i.e., PA, hypersplenism, etc.)	
(2) Myeloid hyperproliferation (malignant)	(2) Myeloid hypoproliferation (malignant)
(a) Acute granulocytic leukemia	(a) Leukemic reticuloendotheliosis (i.e., stem cell leukemia, hairy cell disease, etc.)
(b) Chronic granulocytic leukemia	(b) Acute lymphoblastic and lymphosarcoma leukemias
(c) Myeloid metaplasia	(c) Most aleukemic leukemias
(d) Polycythemia vera	
(3) Monocyte hyperproliferation (benign)	
(a) Sarcoid and tuberculosis	
(b) Monocytic leukemoid reaction	
(4) Monocyte hyperproliferation (malignant)	
(a) Acute monocytic leukemia	
(b) Chronic monocytic leukemia	
(c) Myelomonocytic leukemia	
Diminished enzyme degradation	Increased enzyme excretion
(1) Impaired renal glomerular filtration	(1) Proximal tubular renal disease

in the HLL group was 58 years in comparison to 50 years for those with LLL. Similar degrees of anemia, thrombocytopenia, and leukocyte count were found in each group.

The relationship between serum lysozyme and the number of peripheral blood granulocytes is shown in Fig. 8. Generally there was a tendency for higher serum lysozyme levels with an absolute increase in mature granulocytes. The relationship between serum lysozyme and circulating monocytes was less striking.

The relationship between the total leukocyte count and the number of circulating granulocytes for HLL and LLL patients is seen in Fig. 9. Thirteen of 20 patients with HLL who had total leukocyte counts of 20,000 or greater had a concomitant increase in mature granulocytes. In contrast, most patients with LLL presented with neutropenia regardless of the total leukocyte count.

Bacteremia documented by positive blood cultures occurred during life in only 20% of the patients with HLL in comparison to 50% of patients in the low lysozyme group. This difference was most striking in the patients who survived for only short periods following diagnosis (Table III). No positive blood cultures were found in the patients with high serum lysozyme who survived 30 days or less, while over half of the short survivors in the LLL

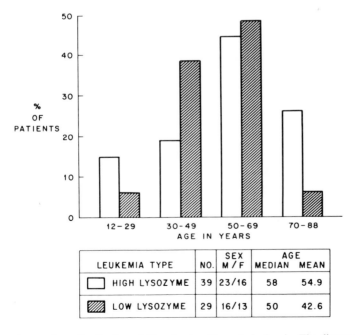

LEUKEMIA TYPE	NO.	SEX M/F	AGE MEDIAN	MEAN
☐ HIGH LYSOZYME	39	23/16	58	54.9
▨ LOW LYSOZYME	29	16/13	50	42.6

Fig. 7. Age and sex distribution of 68 patients with acute leukemia. The disease was classified according to the serum lysozyme activity at the time of diagnosis.

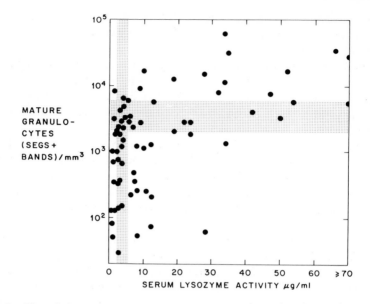

Fig. 8. The relationship of peripheral blood granulocytes to serum lysozyme in acute leukemia. Serum lysozyme assays were performed by a turbidimetric method using human lysozyme as the standard. Shaded areas show control ranges.

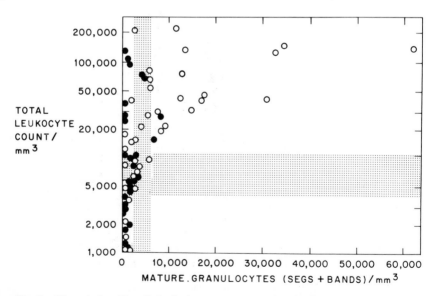

Fig. 9. The relationship of the leukocyte count to the absolute number of peripheral blood granulocytes in acute leukemia patients with high (open circle) and low (closed circle) serum lysozyme. Serum lysozyme assays were performed by a turbidimetric method using human lysozyme as the standard. Shaded areas show control ranges.

TABLE III

Bacteremias in Acute Leukemia

Leukemia type	All patients		Patients surviving one month or less		Patients surviving more than one month	
	Number	% Patients with bacteremia	Number	Patients with bacteremia	Number	Patients with bacteremia
High lysozyme	39	20	12	0 (0%)	24	8 (33%)
Low lysozyme	29	50	8	5 (62%)	17	9 (52%)

group had bacteremias confirmed during life. The number of days spent in the hospital and the number of blood cultures taken per patient were similar in both groups. *Pseudomonas aeruginosa* was by far the organism cultured most frequently from the blood in both types of leukemias.

Significant bacterial or fungal infection was present at the time of death in 50% of the patients with HLL (Table IV). One of these was the primary cause of death in 22%, all of whom were granulocytopenic at the time of death. In the LLL group, bacterial or fungal infection was a major cause of death in 80% and was the main cause of death in 49% despite the fact that about $\frac{1}{3}$ of the patients had adequate numbers of circulating granulocytes, terminally. The most striking findings were observed in the group of patients who survived less than 1 month. Infection was present at death in 33% of patients with HLL and in no case was it the main cause of death. These early deaths were more frequently caused by central nervous system hemorrhage and leukostasis. In the LLL group, infection was the most frequent cause of death regardless of length of patient survival. In both high and low lysozyme leukemia, the most frequent sites of terminal infection were the lung and gastrointestinal tract.

There were several major differences in the clinical characteristics of the acute leukemia patients in the HLL and LLL groups (Table V). Patients with HLL were older, and hemorrhage was a more important factor in early deaths than infection. The incidence of (a) absolute neutropenia at the time of diagnosis, (b) bacteremias during the course of the illness, and (c) infection as a cause of death was less frequent in this group of patients than in those with a low serum lysozyme. Despite older age and less frequent remissions, the median survival of the high lysozyme group was somewhat longer than the LLL group with 7 of 39 surviving for over 1 year. Survival of over 1 year in patients with LLL was relatively rare and it was only seen in those patients who had had a successful course of chemotherapy.

TABLE IV

Infection as Cause of Death in Acute Leukemia

Leukemia type	All patients			Patients surviving one month or less			Patients surviving more than one month		
	Number	Infection contributing to death	Infection main cause of death	Number	Infection contributing to death	Infection main cause of death	Number	Infection contributing to death	Infection main cause of death
High lysozyme	39	50%	22%	12	4 (33%)	0 (0%)	24	14 (58%)	8 (33%)
Low lysozyme	29	80%	49%	8	7 (87%)	5 (62%)	17	14 (82%)	7 (41%)

TABLE V

Differences in Clinical Characteristics of High and Low Lysozyme Acute Leukemia in Adults

	High serum lysozyme	Low serum lysozyme
Median age (yrs)	58	50
Number of peripheral blood graunlocytes	Frequently normal or increased	Neutropenia frequent
Bacteremias	Not very frequent	Frequent
Infection as cause of death	Frequent	Very frequent
Usual cause of early death	Hemorrhage	Infection
Median survival (days)	79	53
Survival of more than one year	Not rare	Rare
Role of remission induction in one year survival	Undetermined	Appears essential

Increased serum lysozyme activity in adult patients with acute leukemia probably reflects the presence of a relatively large pool of circulating and tissue leukocytes which are functionally competent. These leukocytes may delay the time of onset of infectious complications and, since infection is the major cause of death in acute leukemia, the relatively longer survival of patients with HLL may result.

References

1. Hook, W. A., Carey, W. F., and Muschel, L. H., *J. Immunol.* **84**, 569 (1960).
2. Fink, M. E., and Finch, S. C., *Proc. Soc. Exp. Biol. Med.* **127**, 365 (1968).
3. Lippman, M. E., and Finch, S. C., *Yale J. Biol. Med.* **45**, 463 (1972).
4. Vietzke, W. M., and Finch, S. C., *Clin. Res.* **16**, 543 (1968) (abstr.).
5. Vietzke, W. M., Perillie, P. E., and Finch, S. C., *Yale J. Biol. Med.* **45**, 457 (1972).
6. Donadio, J. A., and Finch, S. C., to be published.
7. Lippman, M. E., and Finch, S. C., unpublished oversvations.
8. Balazs, T., and Roepke, R. R., *Proc. Soc. Exp. Biol. Med.* **123**, 380 (1966).
9. Finch, S. C., Gnabasik, F. J., and Rogoway, W., *Proc. Int. Symp. Fleming's Lysozyme, 3rd, 1964*, pp. 1–5.
10. Perillie, P. E., Khan, K., and Finch, S. C., *Am. J. Med. Sci.* **265**, 297 (1973).
11. Pascual, R. S., Perillie, P. E., Sulavik, S., Donadio, J. A., Gee, J. B. L., and Finch, S. C., *Ann. Intern. Med.* **76**, 880 (1972) (abstr.).
12. Finch, S. C., Lamphere, J. P., and Jablon, S., *Yale J. Biol. Med.* **36**, 350 (1964).

Lysozyme in Hematology and Its Relation to Myeloperoxidase and Lactoferrin

JORGEN MALMQUIST

Lysozyme: Purification and Measurement

The bentonite adsorption method for lysozyme (LZM) purification was successfully employed by Osserman for the isolation of human lysozyme (1, 2) from the urine of patients with monocytic leukemia. In the present studies, the author has used a cation exchanger, carboxymethyl-Sephadex, for lysozyme purification (3).

For lysozyme quantitation, an immunochemical method, rather than enzymatic activity measurement, has been employed. Anti-lysozyme sera were produced in rabbits. The test samples are electrophoresed into antiserum-containing agarose gel (electroimmunoassay) as described by Laurell (4). The LZM concentration in the samples are obtained by measuring the heights of the rocket-shaped immunoprecipitates (Fig. 1). By this method normal serum LZM ranges between 1 μg/ml and 5 μg/ml. In most normal urine samples LZM is not detectable by this method; occasionally 1–2 μg/ml is present.

In laboratories where immunochemical methods for protein determination are in frequent use, this method is an attractive alternative to enzymatic

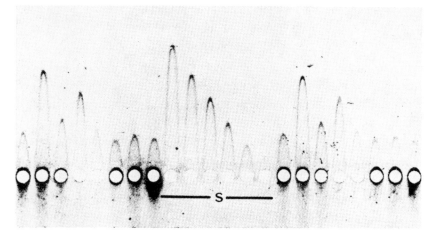

Fig. 1. Immunoprecipitates obtained by the determination of LZM by electroimmuno-assay. Standards (indicated by —S—) contain 8.5–0.7 μg/ml.

methods. The latter, however, are potentially more sensitive and may accordingly be used when measuring serum LZM in the subnormal range.

The results of immunochemical determinations of serum and urinary LZM in various leukemias have been reported (5). The highest serum LZM levels tended to occur in acute or subacute leukemias with a predominance of monocytoid cells in the marrow and peripheral blood. However, LZM values did not clearly separate this group from cases of acute leukemia lacking monocytoid features. Cases of "lysozyme nephropathy" (6) with pronounced renal potassium loss and persistently low serum potassium, together with slowly decreasing glomerular filtration rate, have been encountered.

It is quite clear that the serum LZM concentration is influenced by the glomerular filtration rate. Reduced filtration is associated with a retention of lysozyme (Fig. 2). This fact reduces the diagnostic value of minor elevations of serum LZM. Conversely, determination of LZM in urine alone should not be used in hematology since it is imperative to ascertain whether lysozymuria is caused by primary renal disease or whether it reflects an elevation of serum LZM.

Other Leukocytic Proteins

Although LZM is not exclusively localized in leukocytes and related cells it is reasonable to assume that the increased amounts of lysozyme encountered in leukemias and allied conditions originate in the involved leukocytes. It was therefore felt to be of interest to examine other leukocyte constituents that might possibly be released in measurable amounts to blood plasma. Table I is a list of some proteins occurring in the granulocytic and monocytic series. Myeloperoxidase and lactoferrin were selected for assay.

A detailed review of present knowledge on molecular properties and possible function of these two proteins is beyond the scope of this presentation. As for the subcellular localization and time of appearance during cell maturation, available data derive from light and electron microscopy, peroxidase cytochemistry, and immunofluorescence studies. In addition, valuable information stems from centrifugal separation of granule classes. Studies with the latter method, however, have been mainly restricted to granules from the rabbit heterophilic granulocyte, and a complete similarity to the human neutrophil cannot be assumed.

Myeloperoxidase is a heme enzyme present in neutrophils where it is localized in the primary cytoplasmic granules, i.e., the granules appearing at the transition from the myeloblast to the promyelocyte stage (7, 8). The peroxidase occurring in monocyte granules is indistinguishable from the

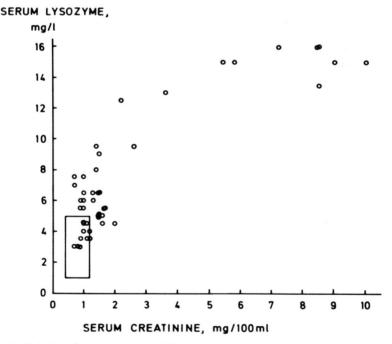

Fig. 2. Relationship between serum LZM and serum creatinine in 46 patients with various renal diseases. [Reprinted with permission from Ref. (5).]

TABLE I

Some Cytoplasmic Constituents of Neutrophilic Granulocytes and Monocytes

Neutrophils	Monocytes
Primary granules	Lysozyme
Lysozyme	Myeloperoxidase
Myeloperoxidase	
Acid hydrolases	
Cationic proteins	
Glycosaminoglycans	
Secondary granules	
Lactoferrin (? nuclei)	
Alkaline phosphatase	

neutrophil enzyme, whereas eosinophil peroxidase seems to be an entirely different enzyme (9, 10). Myeloperoxidase has never been demonstrated in cells of the lymphocytic series or in any nonhemic cells (11).

Lactoferrin is an iron-binding protein originally isolated from milk. In addition to its occurrence and probable synthesis in some exocrine glands, the detailed studies of Masson and associates (12–14) have demonstrated this protein in neutrophils and their precursors. It has not been detected in monocytes. Lactoferrin seems to be localized in the secondary (specific) cytoplasmic granules, i.e., the structures appearing at the early myelocyte stage. However, recent immunofluorescence studies (15) indicate the exclusive localization of lactoferrin in nuclei of human neutrophils, an interesting and unexplained difference from the results obtained by Masson *et al.* (13) with a similar method.

In the present context it is pertinent that LZM is known to be present in neutrophils as well as in monocytes and occurs in both primary and secondary granules of the rabbit heterophil (16, 17). Corresponding data for the human neutrophil are not available.

For the present studies, human lactoferrin was purified from milk (18) and myeloperoxidase from human neutrophils and myelocytes (19). Antisera to the purified proteins were prepared in rabbits. Attempts to measure the serum concentrations of these proteins by radial immunodiffusion or its modification, electroimmuno assay, were unsuccessful. To increase sensitivity, the method of radioactive radial immunodiffusion described by Rowe (20) was utilized (Fig. 3). By this method, myeloperoxidase and lactoferrin could be detected in some normal sera, the upper limit being about 2 μg/ml for myeloperoxidase and about 3.5 μg/ml for lactoferrin (19, 21). The lower limit of normal could not be defined. The Rowe method has the disadvantage of requiring about 2 weeks before results can be read. Therefore, attempts at quantitation by radioimmunoassay without gel diffusion are presently being made.

The limited observations made to date on serum myeloperoxidase and lactoferrin are presented in Fig. 4, where LZM values are also included. In the acute leukemia group, the frequent occurrence of myeloperoxidase elevation and the absence of measurable lactoferrin are features compatible with the data on the appearance of myeloperoxidase and lactoferrin in the granulocytopoietic maturation chain. There was no close parallelism between serum lysozyme and serum myeloperoxidase concentrations. However, none of the three cases of acute leukemia lacking detectable serum myeloperoxidase had a significant serum LZM elevation. These cases may therefore possibly represent lymphoblastic leukemias.

In the monocytic leukemia cases the myeloperoxidase detected in sera may have originated in the monocytoid cells, but a contribution from the more or less coinvolved granulocytic series cannot be excluded.

With myeloperoxidase and lactoferrin no counterpart to the massive LZM elevation in monocytic leukemias has been observed. Since, unlike LZM, myeloperoxidase and lactoferrin would not be expected to be cleared from the plasma by glomerular filtration, increased production and release of the latter two proteins would be expected to result in greatly elevated serum levels, unless other effective disposal mechanisms exist. Thus, at present, the apparent overproduction of lysozyme by monocytoid leukemic cells must be regarded as a unique phenomenon, as reported by Osserman and Lawlor (1).

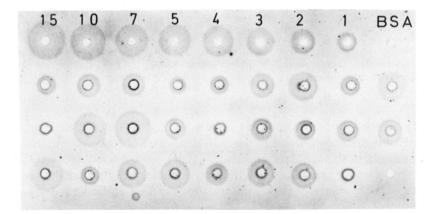

Fig. 3. Radioautographic film from a determination of lactoferrin by radioactive radial immunodiffusion. Top row contains lactoferrin standards with concentrations as indicated (μg/ml). [Reprinted with permission from Ref. (21).]

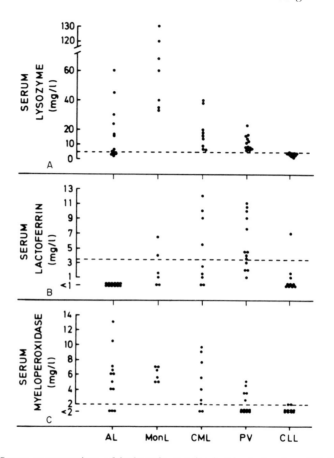

Fig. 4. Serum concentrations of leukocytic proteins in blood disorders. (A) LZM; (B) lactoferrin; (C) myeloperoxidase. The dashed lines indicate the upper limits of normal concentrations. AL, acute leukemia; MonL, monocytic leukemia; CML, chronic myelocytic leukemia; PV, polycythemia vera; CLL, chronic lymphocytic leukemia.

It is generally recognized that there is a need for additional clinically applicable methods for evaluating individual cases of leukemia in terms of cellular origin and the stage of disease. Lysozyme assays have proved to be of some value in this respect. Further observations will be needed to elucidate whether clinically useful information can be obtained by assays of additional components presumably released from leukemic cells.

References

1. Osserman, E. F., and Lawlor, D. P., *J. Exp. Med.* **124,** 921 (1966).
2. Osserman, E. F., *Science* **155,** 1536 (1967).
3. Johansson, B. G., and Malmquist, J., *Scand. J. Clin. Lab. Invest.* **27,** 255 (1971).
4. Laurell, C. -B., *Anal. Biochem.* **15,** 45 (1966).
5. Malmquist, J., *Scand. J. Haematol.* **9,** 258 (1972).
6. Muggia, F. M., Heinemann, H. O., Farhangi, M., Osserman, E. F., *Am. J. Med.* **47,** 351 (1969).
7. Dunn, W. B., Hardin, J. H., and Spicer, S. S., *Blood* **32,** 935 (1968).
8. Bainton, D. F., Ullyot, J. L., and Farquhar, M. G., *J. Exp. Med.* **134,** 907 (1971).
9. Salmon, S. E., Cline, M. J., Schultz, J., and Lehrer, R. J., *N. Engl. J. Med.* **282,** 250 (1970).
10. Yam, L. T., Li, C. Y., and Crosby, W. H., *Am. J. Clin. Pathol.* **55,** 283 (1971).
11. Schultz, J., and Berger, S. J., *in* "Biochemistry of the Phagocytic Process" (J. Schultz, ed.), pp. 115–129. North-Holland Publ., Amsterdam, 1970.
12. Masson, P. L., "La Lactoferrine. Protéine des sécrétions externes et des leucocytes neutrophiles." Editions Arscia, Brussels, 1970.
13. Masson, P. L., Heremans, J. F., and Schonne, E., *J. Exp. Med.* **130,** 643 (1969).
14. Baggiolini, M., de Duve, C., Masson, P. L., and Heremans, J. F., *J. Exp. Med.* **131,** 559 (1970).
15. Green, J., Kirkpatrick, C. H., and Dale, D. C., *Proc. Soc. Exp. Biol. Med.* **137,** 1311 (1971).
16. Baggiolini, M., Hirsch, J. G., and de Duve, C., *J. Cell Biol.* **40,** 529 (1969).
17. Baggiolini, M., Hirsch, J. G., and de Duve, C., *J. Cell Biol.* **45,** 586 (1970).
18. Malmquist, J., and Johansson, B. G., *Biochim. Biophys. Acta* **236,** 38 (1971).
19. Malmquist, J., *Scand. J. Haematol.* **9,** 311 (1972).
20. Rowe, D. S., *Bull. W. H. O.* **40,** 613 (1969).
21. Malmquist, J., *Scand. J. Haematol.* **9,** 305 (1972).

Leukocyte Lysozyme and Unsaturated Vitamin B_{12} Binding Protein Content in Chronic Myeloproliferative Disease: Demonstration of an Altered Ratio in Chronic Myelocytic Leukemia*

HARRIET S. GILBERT

Introduction

Elevated levels of lysozyme (LZM) and unsaturated vitamin B_{12} binding protein (UBBP) have been observed in the serum of patients with chronic myeloproliferative disorders and have been attributed to increased leukocyte turnover (1–4). The extent to which each of these abnormalities reflects alterations in the leukocyte population was investigated by measuring and comparing the content of LZM and UBBP in the leukocytes of normal subjects and those with various forms of myeloproliferative disease.

Materials and Methods

Leukocyte-rich plasma was obtained from heparinized whole blood by gravity sedimentation and rendered erythrocyte-free by hypotonic lysis. Where necessary, dextran or methyl cellulose was added to promote erythrocyte sedimentation. Leukocytes were washed three times with Tris-buffered

*Supported in part by the Sarah Chait Memorial Grant for Cancer Research from the American Cancer Society, and the Jack Martin Fund.

saline, pH 7.2, and disrupted by freeze–thaw lysis and sonication. LZM was measured by the lysoplate assay method using purified human LZM as the standard (5). The UBBP was measured by $^{57}CoB_{12}$ uptake (6). Values were expressed per milligram of leukocyte lysate protein, determined by the method of Lowry *et al.* (7).

Results

The findings in 11 normal subjects, 19 patients with active polycythemia vera, and 8 with chronic myelocytic leukemia are shown in Fig. 1. Levels of LZM and UBBP were similar in leukocytes of normal subjects and those with polycythemia vera. In contrast, chronic myelocytic leukemia leukocyte lysates contained significantly greater amounts of LZM ($p < 0.001$) and less UBBP ($p < 0.005$) than normal or polycythemia vera leukocytes. These differences were reflected in the ratio of leukocyte LZM ($\mu g/mg$ protein) to leukocyte UBBP (ng/mg protein), calculated for each patient and plotted according to clinical diagnosis in Fig. 2. The mean ratio of 11.1 ± 3.9 found in chronic myelocytic leukemia was significantly higher than that in normal subjects (4 ± 1.4) and patients with polycythemia vera (4.5 ± 3.3). Chromosome analysis performed in 7 of the 8 patients with the clinical diagnosis of chronic myelocytic leukemia revealed 4 with and 3 without the Philadelphia chromosome. No differences were found between Philadelphia chromosome-positive and chromosome-negative cases as regards content

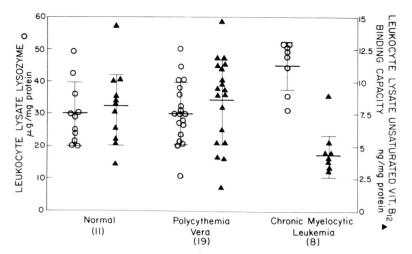

Fig. 1. Content of lysozyme (circle) and unsaturated vitamin B_{12} binding capacity (triangle) in leukocyte lysates prepared from normal subjects and patients with myeloproliferative disease. The number studied in each group appears below the diagnosis. The mean ± 1 S.D. is indicated by an I bar.

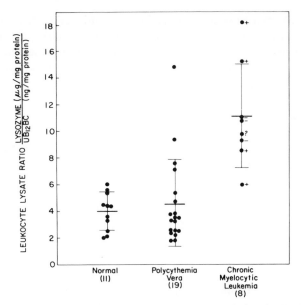

Fig. 2. Ratio of lysozyme (μg/mg protein) to unsaturated vitamin B_{12} binding capacity (UB$_{12}$BC, ng/mg protein) in leukocyte lysates prepared from normal subjects and patients with myeloproliferative disease. The number in each group appears below the diagnosis. The mean $\pm$ 1 S.D. is indicated by an I bar. The presence or absence of a Philadelphia chromosome is indicated to the right of each patient with the clinical diagnosis of chronic myelocytic leukemia (Philadelphia chromosome-negative = $-$, Philadelphia chromosome-positive = $+$, karyotype unknown = ?).

or ratio of LZM and UBBP (Fig. 2). There was no correlation of leukocyte LZM, leukocyte UBBP, or LZM:UBBP ratio with the peripheral blood leukocyte or granulocyte count in the 38 subjects studied.

Discussion

These findings suggest that more than one mechanism may be responsible for the elevated serum levels of LZM and UBBP seen in various myeloproliferative disorders and that derangements in leukocyte LZM and UBBP synthesis and/or secretion may occur in certain forms of myeloproliferative disease. The preservation of relatively normal leukocyte content of and ratio between LZM and UBBP in polycythemia vera would indicate that elevated serum levels result primarily from an increased turnover of leukocytes. In contrast, the finding in chronic myelocytic leukemia of a distorted LZM:UBBP ratio due to increased LZM and decreased UBBP content of the leukocyte population suggests that abnormal serum levels in chronic myelocytic leukemia may be a reflection not only of increased leukocyte turnover, but of a more profound disturbance in the leukocyte population, with a shift toward a cell type having synthetic and/or secretory characteristics that differ from normal and polycythemia vera leukocytes.

Summary

Measurement of LZM and unsaturated vitamin B_{12} binding protein (UBBP) content of leukocytes from normal subjects and patients with polycythemia vera and chronic myelocytic leukemia were performed to elucidate the mechanisms responsible for elevated serum levels of these proteins observed in myeloproliferative disease. In polycythemia vera, leukocyte LZM, UBBP content, and LZM:UBBP ratios did not differ significantly from normal. An elevated LZM:UBBP ratio was found in chronic myelocytic leukemia, resulting from a significantly higher LZM and lower UBBP content of the leukocyte lysates. These findings suggest that elevations of serum LZM and UBBP in polycythemia vera reflect primarily an increased turnover of leukocytes, whereas in chronic myelocytic leukemia they are due to a more profound disturbance in the synthetic and/or secretory functions of the leukocyte population.

Acknowledgment

The author gratefully acknowledges the technical assistance of Mrs. Olga Castre in the performance of the assays described herein.

References

1. Perillie, P. E., and Finch, S. C., *N. Engl. J. Med.* **283,** 456 (1970).
2. Binder, R. A., and Gilbert, H. S., *Blood* **36,** 228 (1970).
3. Miller, A., and Sullivan, J. F., *J. Clin. Invest.* **38,** 2135 (1959).
4. Gilbert, H. S., Krauss, S., Pasternack, B., Herbert, V., and Wasserman, L. R., *Ann. Intern. Med.* **71,** 719 (1969).
5. Osserman, E. F., and Lawlor, D. P., *J. Exp. Med.* **124,** 921 (1966).
6. Gottlieb, C. W., Lau, K. S., Wasserman, L. R., and Herbert, V., *Blood* **25,** 875 (1965).
7. Lowry, O. H., Rosebrough, N. J., Farr, A. L., and Randall, R. J., *J. Biol. Chem.* **193,** 265 (1951).

Lysozyme Measurements
in Acute Leukemia: Diagnostically Useful?

PASQUALE E. PERILLIE and STUART C. FINCH

Diagnostic Usefulness of Lysozyme in the Leukemias

Classification of the acute leukemias on the basis of cell morphology alone is frequently very difficult. Every clinical hematologist has faced the more than occasional diagnostic dilemma in which on the basis of hematologic and clinical features, he can establish the diagnosis of acute leukemia but has difficulty in accurately determining the basic cell type. In such situations he usually interprets clinical features such as age, presence of lymphadenopathy, gingival changes, etc., with morphological and histochemical observations to arrive at a probable cellular classification of the leukemia. Thus any additional laboratory or clinical observations which can assist the clinician in such situations would be useful.

In recent years, there have been several studies of the diagnostic significance of measurement of serum and urinary lysozyme in the leukemias (1–15). Investigation of the possible value of such measurements was based on earlier observations regarding lysozyme (LZM) and leukocytes. Following the original demonstrations by Fleming and Florey that LZM activity was demonstrable in many body tissues and fluids including blood, it was established that hematopoietic LZM was limited almost exclusively to the lysosomal and cytoplasmic fraction of mature granulocytes and mono-cytes (16–22). Additional experimental evidence indicated that most of the measurable LZM of serum or plasma was derived from the *in vivo* lysis of granulocytes (23–25).

Earlier studies of LZM in leukemic subjects were concerned with relating serum LZM levels to the various morphological types of acute and chronic leukemia (1–4). These studies utilized hen egg-white LZM as the standard and employed turbidimetric methods for the assay. The results of these earlier studies were both confirmatory and contradictory. All described increased levels of serum LZM in chronic granulocytic leukemia (CGL) and normal values in chronic lymphocytic leukemia (CLL). Finch *et al.* (3) and Jollès *et al.* (4) described moderately elevated levels in acute granulocytic leukemia (AGL) and reduced levels in acute lymphocytic leukemia (ALL). Inai and co-workers (1) reported reduced levels in AGL, perhaps related to the unusually low white counts of most of their patients, and the possibility that some of these patients were studied after treatment. They did not study patients classified as having acute monocytic leukemia (AML). Marcolongo and Conti (2) also reported low serum LZM activity in their patients with acute leukemia. However, they did not subclassify their patients into the various morphological forms of acute leukemia. Finch *et al.* (3) and Jollès *et al.* (4) described markedly elevated serum enzyme levels in most patients with acute myelomonocytic (AMML) and AML. Shortly thereafter, Osser-man and Lawlor (5), utilizing human LZM isolated from the urine of patients with AMML as a standard and a lysoplate technique for the enzyme assay, confirmed the findings of markedly elevated levels of serum LZM activity in AMML and AML, and in addition described the presence of increased LZM in the urine of such patients. Following Osserman and Lawlor's report, several investigators, utilizing either hen egg-white or human LZM as standards and turbidimetric or lysoplate assay methods, confirmed the finding of markedly elevated levels of serum and urinary LZM activity in most patients with AMML and AML (6–14) (Table I). Some of these later studies also reported the occasional demonstration of lysozymuria in CGL and AGL (6–9). In no situation, however, did the levels of serum or urine LZM activity approach the levels typical of most AMML and AML patients. All subsequent studies also confirmed the earlier observations of normal or reduced serum enzyme levels and absent urinary LZM activity in the lymphocytic forms of leukemia. The only striking exception to the results of most reported studies was that of Youman and associates (15). These authors reported finding elevated serum LZM levels in only 4 of 20 patients classified as AMML leukemia and found significant lysozymuria in only 3 of 16 patients with AMML. It is difficult to reconcile the results of this study with those of most other similar studies cited. As noted in Table II, of the 145 cases of AMML and AML reported in the literature, including the 20 patients of Youman and associates, 120 were shown to have elevated serum LZM values, and 89 of 112 were shown to have significant lysozymuria. If one excludes the patients studied by Youman *et al.*, then 116 of 125 cases had elevated LZM levels and 85 of 94 had lysozymuria. Certain features of the study of Youman *et al.* are puzzling. Included in the control group were subjects whose total leukocyte count ranged between 2900 and 16,000/mm³. One control subject was shown to have 30 mg of LZM in a 24 hr urine collection which would be considered increased on the basis of all other studies involving normal or control subjects (Table I).

It is apparent from the studies reported that a standard method for assaying LZM has not yet been decided upon. The range of serum LZM values in control groups utilizing hen egg-white LZM and the turbidimetric methods of assay has been quite wide ranging from 3.2 to 40 μg/ml. Similarly the range of values in the leukemia patients reported has been quite wide. Of interest is the closer relationship of serum enzyme values in series utilizing human LZM as the standard and the lysoplate technique for assay. This is perhaps attributed to the fact that most of these studies have utilized a human LZM provided by Dr. Elliott Osserman.

With the above reservations, results of most studies to the present would permit the following general conclusions in regard to the diagnostic usefulness of serum and urine LZM measurements in acute leukemia: (a) In the absence of evidence of renal dysfunction,* markedly elevated levels of both serum and urinary LZM activity would permit concluding that the patient has either AMML or AML; (b) In the absence of evidence of renal dysfunction, elevated levels of serum and/or urine LZM activity would exclude the diagnosis of ALL; (c) Reduced levels of serum LZM activity have no differential diagnostic significance.

Usefulness of Lysozyme Measurements for Evaluating Therapy in Leukemia

In addition to evaluating the diagnostic usefulness of serum and urine LZM activity in acute leukemia, recent studies have demonstrated the value of these measurements in the management of certain patients with leukemia. Serial measurements of serum and urine LZM activity have been performed

---

* Serum LZM levels have been shown to vary directly with the level of serum urea nitrogen and creatinine. A less constant relationship has been demonstrated between azotemia and lysozymuria (26, 27).

TABLE I

Serum and Urine Lysozyme Values in Leukemia Patients

| Ref. | Method | Standard | Controls | Serum or plasma (μg/ml) | | | | |
				CGL	CLL	AGL	AMML	AML
6	Turbidi-metric	Hen egg white	8.2±0.2[a]	38.6±6.2	10.3±2.2	32.1±7.5	61.3±13	133.9±3.2
8	Turbidi-metric	Hen egg white	(6–15)[b]	(15–38)		(3.4–38)		
9	Turbidi-metric	Hen egg white	16.4±1 (9.8–19.8)	51.6±12.9 (22–98)	9.1±1.1 (7.8–12.3)	20.5±1.6 (11.8–32)	64±5.3 (41–84)	133.6±4.6 (108–154)
4	Turbidi-metric	Hen egg white	13.2 (7–20)	32.8	13.1		33.2	122
26	Turbidi-metric	Hen egg white	(3.2–7.4)	20.0	9.0		(2–1320)[c]	
15	Turbidi-metric	Hen egg white	(5–18)					
12	Turbidi-metric	Hen egg white	27.7±5.6 (16–40)	105.3		31.5 (17–43)	182.5 (55–475)	404.3 (12.5–800)
7	Lysoplate	Human	7±1.3 (4–9)	17.5±5.7 (4.5–26)	4.0	10±6.3 (2.7–29)	47.8±17.3[c] (25–75)	
11	Lysoplate	Human	10.1±3.6 (4.1–17.6)	20.4±12.4 (6.2–50)	13.7±5.1 (6.5–23)		43.3±18.7[c]	
10	Lysoplate	Human	9.69±1.8 (6.6–13.8)		7.18±2.7 (2.7–11.8)		49.7±34.8[c] (5–230)	
13	Lysoplate	Hen egg white	57.3±13 (45–85)			116.0 (64–185)	151.6[c] (135–180)	
5	Lysoplate	Human	7.0				(40–150 μg/ml)	

[a] Standard deviation.
[b] Parentheses indicate range.
[c] In these studies a single value was given for patients with either AMML or AML.

in acute leukemia patients undergoing active chemotherapy (7, 9, 12, 14). In all the studies reported to date it has been concluded that serum and urine LZM determinations were of value in determining the efficacy of anti-leukemic therapy. In the first of these studies (14) serial serum and urine LZM measurements were obtained in a group of 35 patients with acute leukemia and the results were correlated with qualitative and quantitative changes in the peripheral blood and marrow. In 5 patients (2 AGL and 3 AML) serum LZM levels remained elevated despite drug-induced leuko-penias of less than 1000 cells/mm^3. In each of these cases bone marrow

			Urine				
ALL	Controls	CGL	CLL	AGL	AMML	AML	ALL
4.2±0.4	0	0.9 (gm/day)	0	0.4	1.0	1.5	0
	—	—					
7.9±1.2	—	14±21.2		32	27.0±10.2	128±22.8	
(2–12.3)	—	μg/ml			(10–87)	(62–211)	
4.0	—	—					
(2–13)	—	—					
	<10 mg/day				0–6700 mg/day^c		4–6
	0	—		1.2 mg/day	1444	11,908	
2.9±1.2	0	0.6 (0–1.9 mg/day)	0	0.1 (0–1.2)	802^c (160–3910)		0
	1.57±0.8 0.4–2.9	1.6±1.8 (0–6.17)	0.69±0.59 (0.22–2.0)		235±403^c (0.95–1320)		
6.45±2.5 (4.1–14.3)	0.73 (0–2.04 μg/ml)	1.2 (0.2–5.6)			223.6^c (0.48–1600)		
36 (25–45)	0			28 (0–100 μg/ml)	1965^c (65–5250)		0
	—				(25–420 μg/ml)		

examination revealed persistent leukemic infiltration. In 3 patients (2 AGL and 1 AML) serum LZM levels decreased to normal levels with concomitant leukopenia and marrow hypoplasia. In 5 of 7 patients with ALL who experienced complete remission, reduced pretreatment serum LZM activity returned to normal. Two additional patients were of interest. One patient with AMML (Fig. 1) experienced a complete remission and pretreatment elevated serum LZM activity returned to normal levels. Later in the course, when the peripheral blood still indicated a remission, there was an abrupt rise in serum and urine LZM activity. This was followed shortly by morphological evidence of relapse. Another patient with AMML (Fig. 2) although not experiencing a complete hematological remission, demonstrated a similar phenomenon of rising serum enzyme activity preceding evidence of

TABLE II

Total of Reported Studies of Serum and Urine Lysozyme in Acute Leukemia

	Serum			Urine	
Diagnosis	No. of patients	Increased	Reduced	No. of patients	Increased
AGL	121	54	9	65	4
AMML	73	56	3	52	31
AML	72	64	1	60	58
ALL	85	1	52	46	0
	351	175	65	223	93

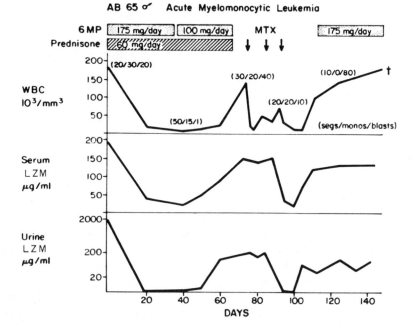

Fig. 1. Patient A. B.—acute myelomonocytic leukemia. Serum and urine LZM, total white cell count and differential count of mature neutrophils, monocytes, and blasts during course of antileukemic therapy. The serum LZM returned to normal on the 32nd day of therapy when marrow indicated a complete remission. On the 45th day serum and urine LZM began to rise followed shortly by peripheral blood and marrow evidence of leukemic relapse.

hematological relapse. In this study, urine LZM in general disappeared when serum LZM levels fell to approximately 45 μg/ml, thus suggesting a possible threshold phenomenon.

Wiernik and Serpick (9) in similar serial studies also found that in ALL, reduced pretreatment levels of serum LZM returned to normal with remission while in AMML and AML elevated levels returned to normal with remission or remained elevated in therapeutic failures even in the presence of leukopenia. In 2 patients (1 ALL and 1 AMML) changes in serum LZM values indicated relapse 1 week and 4 days, respectively, earlier than marrow examinations. They concluded that serial measurements of serum LZM activity provided as much information as serial marrow examinations. Catovsky and associates (12) also found serial LZM determinations helpful in evaluating therapeutic efficacy in patients with AMML and AML. They performed serial measurements of serum and urine LZM activity in 7 patients with AML and AMML during therapy. In 5 patients, elevated pretreatment LZM levels fell to normal when a remission was achieved. In one patient a

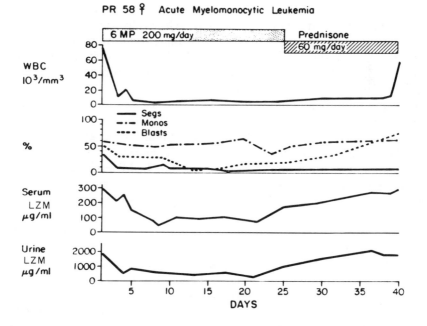

Fig. 2. Patient P. R.—acute myelomonocytic leukemia. Same parameters as Fig. 1, demonstrating a rise of serum and urine LZM prior to peripheral blood evidence of leukemic relapse.

subsequent morphological exacerbation was preceded by a rise in serum LZM activity. Ohta and Nagase (7) studied 10 patients with AML. In all cases in which a complete remission was induced, elevated serum LZM levels returned to normal and urine LZM activity disappeared. Therapeutic failures were characterized by persistent elevation of serum and urine enzyme activity.

It has not yet been established whether the increase of serum LZM in monocytic leukemia is the result of greater proliferation of the body pool of leukemic monocytes, of an increased cell turnover, or an increased production or secretion of LZM by monocytes. Cytobacterial techniques, i.e., using *Micrococcus lysodeikticus* on blood and marrow smears, have demonstrated LZM activity in monoblasts but not in myeloblasts (22) (Fig. 3). Only a few studies have attempted to quantitate leukocyte LZM activity in the leukemias (Table III) (6, 8, 28). Values obtained have shown some variation in part due to extraction of the enzyme by different methods and in part due to different assay techniques. In some of these studies serial measurements of leukocyte LZM activity were performed and the effects of treatment could not be determined (8). In two studies (6, 7) in which the values in CGL and AML were measured prior to treatment, the average leukocyte LZM activity expressed as $\mu g/10^8$ white cells were higher in CGL than in AML even though serum levels of the enzyme were higher in the AML patients studied. These findings would suggest that the increased serum LZM in AMML and AML may result in part from increased secretion or cell permeability to LZM in monocytes.

Attempts have been made to utilize serum LZM activity to evaluate prognosis in the acute leukemias. Castro *et al.* (29) found that patients with higher serum LZM activities had fewer bacterial infections and lived longer than those with reduced levels. Noble and Fudenberg (8) also speculated on the possible relationship between low serum LZM levels and infection in AML patients. Wiernick and Serpick (9) on the other hand found that AML patients with increased levels of serum LZM had the shortest survival in their series of patients.

The finding of lysozymuria in the leukemias (5) has proven perhaps to be of greater diagnostic and prognostic value than elevated serum values. As indicated above, in the acute leukemias, with the exception of the occasional case of AGL, marked lysozymuria in the absence of renal dysfunction is virtually pathognomonic of AMML or AML.

The mechanism(s) leading to lysozymuria in the leukemias has not been firmly established. Knowledge of the manner in which the kidney normally handles LZM is incomplete. Currently it is postulated that LZM, because of its low molecular weight (14,000–15,000), probably is freely filtered by the glomerulus. Although some studies have reported that the clearance of

TABLE III

Leukocyte Lysozyme in Leukemia ($\mu g/10^8$ cells)

Reference	Control	CGL	CLL	AGL	AML
8	390[a]	440	—	4.7	—
	(330–720)[b]	(350–590)		(0.9–7.1)	
7	240	160	—	—	140
	(190–290)	(80–300)			(34–280)
6	220	600	—	—	225
	(20–300)	(250–900)			(20–700)
26	300	450	150	—	—
	(203–363)	(150–600)	(50–350)		

[a] Mean value.
[b] Values in parentheses indicate range.

LZM is not as great as might be anticipated by its molecular weight (30–32). Since lysozymuria is not demonstrable in normal individuals it is further postulated that it is maximally reabsorbed by renal tubules and that normally tubular secretion of LZM does not occur. Studies in humans and animals (31) suggest that altered LZM is maximally absorbed by the proximal renal tubules. Lysozymuria can be produced in rats by administration of mercuric chloride, a known tubule toxin. Sodium chromate, an agent thought to damage specifically the first portion of the proximal convoluted tubule, also produces marked lysozymuria in rats (32, 33). Immunofluorescence studies of rat kidneys have demonstrated LZM in the proximal renal tubule cells (31). Human studies have demonstrated a good correlation between proximal renal tubular dysfunction and the presence of lysozymuria (26, 33).

Osserman and co-workers (34) have postulated that lysozymuria in acute monocytic leukemia results from a toxic effect of reabsorbed LZM on proximal tubular function. He and others have detected other features of proximal tubular dysfunction in acute monocytic leukemic patients demonstrating lysozymuria. Studies in our laboratory corroborated in part by other reports have suggested that lysozymuria in such patients may result from the massive load of filtered LZM exceeding the reabsorptive capacity of the tubular cell (9, 11, 33, 35, 36). It seems most likely that both factors, tubular dysfunction and the threshold phenomenon, may produce lysozymuria in AMML and AML. It is also possible that additional mechanisms may function in the production of lysozymuria in the monocytic leukemias. In experimental animals, kidney LZM concentrations have been shown to rise disproportionately to all other tissues following BCG immunization or transplantation of nonisogeneric renal tumors (27, 37, 38). In most of these

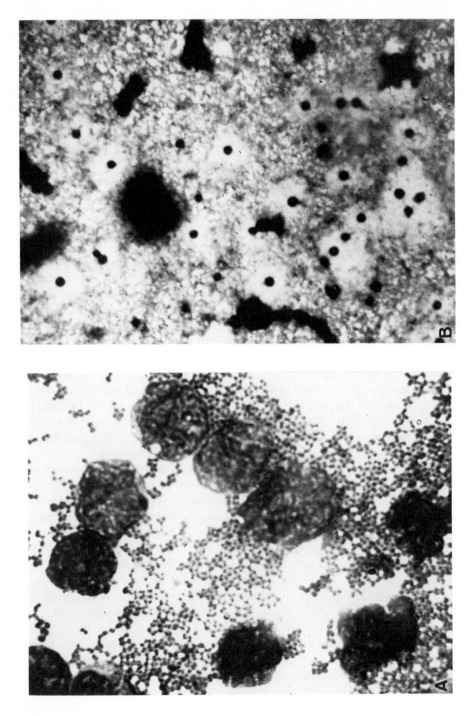

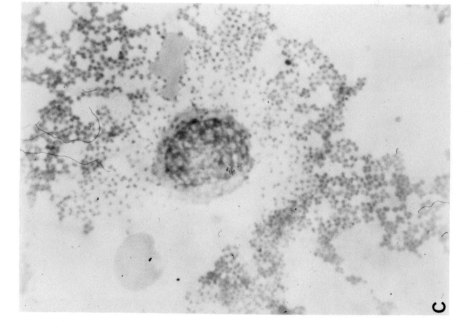

Fig. 3. Cytobacterial demonstration of leukocyte LZM activity. LZM activity is demonstrated by lysis and clearing of the LZM-sensitive organism (*M. lysodeikticus*) around each cell (Wright–Giemsa stain). (A) Myeloblasts showing no evidence of LZM activity (× 1200). (B) Immature monocytes with strong LZM effect (× 100). (C) Immature monocyte with evidence of LZM activity (× 1200). [Reproduction from P. E. Perillie, S. S. Kaplan, E. Lefkowitz, W. Rogaway, and S. C. Finch, *J. Am. Med. Ass.* **203**, 317 (1968), with permission of the publisher].

TABLE IV

Organ Assay for Lysozyme Activity[a]

Patient	Diagnosis	Serum lysozyme[b] μg/ml	Urine lysozyme[b] μg/ml	Kidney	Spleen	Lung	Liver	Heart	Thyroid	Adrenal	Skeleton muscle	Pancreas
	Control[c]	—	—	140 ± 52	430 ± 75	190 ± 37	160 ± 30	80 ± 21	75 ± 16	100 ± 36	40 ± 12	34 ± 6
1	AMML	88	150	2730	1160	483	52.5	69	60.8	0	90	—
2	AMML	—	—	330	364	89	65	57	—	—	—	40.2
3	AMML	111.5	1880	2600	1355	424	633	243	310	489	—	—
4	AGL	17	120	476	97.6	100	50	18	46.8	50	20.6	10.8
5	AGL	8.9	0	56	155	71	30	21	—	—	—	0
6	AGL	56	676	448	846	178	197	170	206	333	—	—
7	AGL	18.9	0	91	1240	322	85	54	—	90	—	—
8	AGL	33	0	348	618	351	331	—	63	—	—	—
9	ALL	7.5	0	23	56	6.6	38	17	—	—	—	7.1

[a] μg/gm of dry weight. Egg-white LZM standard.
[b] Last determination before death.
[c] Patients dying without primary hematologic disorders (represents average of 10 patients).

situations, tissue macrophages accumulate in the renal tissue and it is possible that the LZM released from these cells may contribute to the lysozymuria. A similar mechanism may result from renal infiltration by monocytes in AMML and AML. Table IV depicts the measurable levels of tissue LZM activity detected in a group of patients with acute leukemia. It is to be noted that the highest levels were detected in the kidneys of patients with monocytic leukemia demonstrating lysozymuria.

Summary

Measurement of serum and urinary LZM activity is useful in the classification of the acute leukemias. Furthermore, serial measurement of serum LZM in such patients during chemotherapy may prove useful to evaluate the efficacy of therapy.

References

1. Inai, S., Hirao, F., Kishimoto, S., and Takahash, H., *Med. J. Osaka Univ.* **9,** 33 (1958).
2. Marcolongo, R., and Contu, L., *Hematol. Latina* (*Milan*) **5,** 141 (1962).
3. Finch, S. C., Gnabasik, F. J., and Rogaway, W., *Bull. Med. Hyg.* **22,** 972 (1964).
4. Jollès, P., Sternberg, M., and Mathe, G., *Isr. J. Med. Sci.* **1,** 445 (1965).
5. Osserman, E. F., and Lawlor, D. P., *J. Exp. Med.* **124,** 921 (1966).
6. Perillie, P. E., Kaplan, S. S., Lefkowitz, E., Rogaway, W., and Finch, S. C., *J. Am. Med. Ass.* **203,** 317 (1968).
7. Ohta, H., and Nagase, H., *Acta Haematol. Jap.* **46,** 257 (1971).
8. Noble, R. E., and Fudenberg, H. H., *Blood* **30,** 465 (1967).
9. Wiernik, P. H., and Serpick, A. A., *Am. J. Med.* **46,** 330 (1969).
10. Pruzanski, W., and Saito, S. G., *Am. J. Med. Sci.* **258,** 406 (1969).
11. Pruzanski, W., and Platts, M. E., *J. Clin. Invest.* **49,** 1694 (1970).
12. Catovsky, D., Galton, D. A. G., and Griffin, C., *Br. J. Haematol.* **21,** 565 (1971).
13. Rudders, R. A., and Bloch, K. J., *Am. J. Med. Sci.* **262,** 79, (1971).
14. Perillie, P. E., and Finch, S. C., *Abstr. 10th Annu. Meet. Am. Soc. Hematol.*, *1967.*
15. Youman, J. D., III, Saarni, M. I., and Linman, J. W., *Mayo Clin. Proc.* **45,** 219 (1970).
16. Fleming, A., *Proc. R. Soc. Med.* **93,** 306 (1922).
17. Florey, H., *Br. J. Exp. Pathol.* **11,** 251 (1930).
18. Barnes, J. M., *Br. J. Exp. Pathol.* **21,** 264 (1940).
19. Flanagan, P., and Lionetti, F., *Blood* **10,** 497 (1955).
20. Jollès, P., *in* "Enzymes" (P. D. Boyer, H. Lardy, and K. Myrbäck, eds.), 2nd ed., Vol 4, p. 431. Academic Press, New York, 1960.
21. Cohn, Z. A., and Hirsch, J. G., *J. Exp. Med.* **112,** 983 (1960).
22. Briggs, R. S., Perillie, P. E., and Finch, S. C., *J. Histochem. Cytochem.* **14,** 167 (1966).
23. Ribble, J. C., and Bennett, I. L., Jr., *Clin. Res.* **7,** 41 (1959).
24. Fink, M. E., and Finch, S. C., *Proc. Soc. Exp. Biol. Med.* **127,** 365 (1968).
25. Perillie, P. E., Kaplan, S. S., and Finch, S. C., *N. Engl. J. Med.* **277,** 10 (1967).
26. Harrison, J. F., Lunt, G. S., Scott, P., and Blainey, J. D., *Lancet* **1,** 371 (1968).

27. Perri, G. C., Cappuccino, J. G., Faulk, M., Mellors, J., and Stock, C. C., *Cancer Res.* **23,** 431 (1963).
28. Crowder, J. G., and White, A., *Am. J. Med. Sci.* **255,** 327 (1968).
29. Castro, O., Perillie, P. E., and Finch, S. C., *Abstr., Int. Congr. Hematol., 13th, 1970* p. 231 (1970).
30. Marshall, M. E., and Deutsch, H. F., *Am. J. Physiol.* **163,** 461 (1950).
31. Glynn, A. A., *Expos. Annu. Biochim. Med.* **27,** 111 (1966).
32. Sussman, M., Asscher, A. W., and Jenkins, J. A. S., *Invest. Urol.* **6,** 148 (1968).
33. Prockop, D. J., and Davidson, W. D., *N. Engl. J. Med.* (1964).
34. Muggia, F. M., Heinemann, H. O., Farhangi, M., and Osserman, E. F., *Am. J. Med.* **47,** 351 (1969).
35. Perri, G. C., Faulk, M., and Money, W. L., *Proc. Soc. Exp. Biol. Med.* **115,** 185 (1964).
36. Pepitone, V., *et al., Atti Symp. Int. Lisozima Fleming, 1st, 1959* p. 182.
37. Cappuccino, J. G., Winston, S., and Perri, G. C., *Proc. Soc. Exp. Biol. Med.* **116,** 869 (1964).
38. Hayslett, J. P., Perillie, P. E., and Finch, S. C., *N. Engl. J. Med.* **279,** 506 (1968).

Immunocytological and Biochemical Demonstration of Lysozyme in Leukemic Cells

H. ASAMER, F. SCHMALZL, and H. BRAUNSTEINER

In normal persons the serum lysozyme (LZM) level seems to be mainly related to the degradation rate of granulocytes (5). The difficulties in finding a correlation between serum LZM activity and the absolute number of the peripheral blood cells in many patients with leukemia obviously came from insufficient knowledge of the LZM in the leukemic cells. Some observations suggest that the production of enzyme in these cells starts with the appearance of specific granules and increases according to the degree of cell maturation (3, 4). Noble and Fudenberg found in some cases with granulocytic leukemia that the leukocytes had a low LZM although no immature cells were found by differential counting (6). The present study was undertaken to evaluate the leukocyte LZM in different forms of leukemia by two different methods, an immunologic one and a biochemical one, and to relate the results to the serum and urinary LZM levels of these patients.

Methods

The immunofluorescent method was used for the demonstration of intracytoplasmic LZM in leukemic cells (2). These results were compared to the enzyme activity estimated by the quantitative lysoplate method (7). The different forms of leukemia were diagnosed according to cytochemical criteria (1).

Results

Table I shows the biochemically assayed LZM in leukocytes, serum, and urine compared to the immunocytological findings. The normal values are from 5 to 10 μg/ml for serum, 0 to 3 μg/ml for urine, and 1 to 2.7 μg/10^6 cells. In three cases with histochemically diagnosed mature monocytic leukemia, LZM was immunocytologically demonstrable in the cytoplasm of the leukemic monocytes and the neutrophils. Correspondingly, the serum and urinary levels of LZM were high. In Patient No. 4 the leukocyte LZM increased within 6 weeks from 2 to 24 μg/10^6 cells although the absolute peripheral monocytic cell count was constant, indicating that in this case a real increase of LZM in the leukemic cells occurred. During this time the serum and urinary levels also increased from 34 to 70 and from

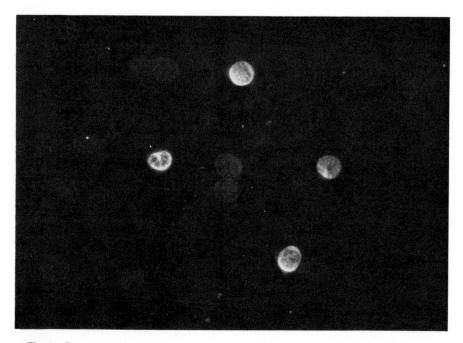

Fig. 1. Immunocytological demonstration of LZM in peripheral leukemic promyelocytes showing absence of enzyme in immature leukemic monocytes from a patient with two distinct leukemic cell populations ($\times$ 400).

TABLE I

Biochemically Assayed Serum and Urinary Lysozyme Activity (LZM) Compared to the Immunocytologically Demonstrated Lysozyme Content in Leukemic Cells[a]

Diagnosis	Serum-LZM (5–10 μg/ml)[a]	Urine-LZM (0–3 μg/ml)[a]	Leukocytes-LZM (1.0–2.7 μg/10^6)[a]	Immunocytologic demonstrated LZM in leukemic cells	
				Peripheral	Bone marrow
Monocytic leukemia					
1. Pi	n.d.	140	n.d.	+ +	n.d.
2. Le	32	230	26.0	+ +	n.d.
3. Ste	65	750	n.d.	+ +	+ +
4. Gru	34–70[b]	250–420[b]	2.0–24.0[b]	+ +	n.d.
5. Hu	12–16[b]	0	0.6–1.0[b]	–	–
Promyelocytic leukemia					
6. Tou	n.d.	n.d.	n.d.	n.d.	+ +
7. Ko	14–60[b]	4–50[b]	6.0	+ +	n.d.
8. Ca	18	33	n.d.	n.d.	+
9. Ta	3	0	trace	n.d.	–
Monocytic promyelocytic leukemia					
10. Ha	16	2	n.d.	only promyelocytes +	
Chronic granulocytic leukemia	17	0	3.0	+	+
Blast cell leukemia	4	0	trace	–	–

[a] Purified human lysozyme, kindly supplied by E. F. Osserman, was used as standard. Numbers in parentheses indicate normal range.
[b] Values of serial determinations.

150 to 420 µg/ml. Patient No. 5 with rather immature leukemic cells had no immunocytologically demonstrable LZM and exhibited only moderately elevated serum and urinary LZM. Three patients with promyelocytic leukemia showed very low to moderately elevated serum and urinary LZM, and in these cases the leukemic cells were positive for LZM.

Patient No. 10 is of special interest, displaying two distinct leukemic cell populations—one monocytic and the other promyelocytic. With the immunofluorescent method, LZM was only demonstrable in the promyelocytic cells and neutrophils but not in the immature monocytes (Fig. 1). In this case the serum LZM was consistent with the immunocytological finding and was rather low.

The mean value of leukocyte LZM from 8 patients with chronic granulocytic leukemia was slightly elevated. There was positive staining of most peripheral granulocytic cells.

Serial LZM determinations during the course of disease progression have been made in several cases. In two patients with undifferentiated blast cell leukemia, a decrease of the leukocyte LZM was observed 2 and 4 months before death. In these cases a decrease of the LZM of the neutrophils apparently occurred since no blast cells were seen in the peripheral blood at this time. The same occurred in the terminal stage of 4 patients with chronic

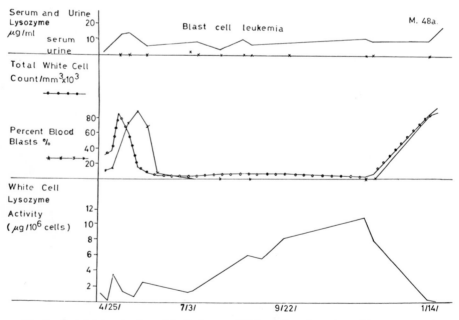

Fig. 2. Serial determination of the leukocyte LZM activity of a patient with acute granulocytic leukemia.

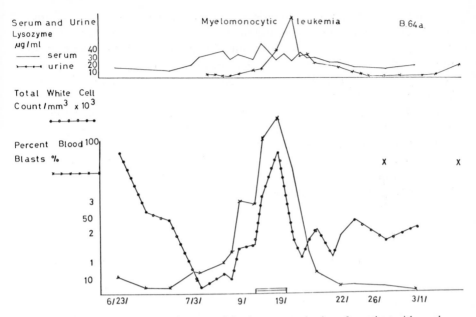

Fig. 3. Serial determination of LZM activity in serum and urine of a patient with myelomonocytic leukemia.

granulocytic leukemia. A very different pattern was observed in another patient with acute granulocytic leukemia as shown in Fig. 2. In this case a steady rise of the leukocyte LZM was observed during chemotherapy. Again, since no blast cells were seen in the peripheral blood, this change was apparently due to the LZM of the granulocytes.

Figure 3 shows a case of myelomonocytic leukemia in which there was a close correlation between the blast cell count and urinary LZM during 2 months of observation. In this patient the highest blast cell count was followed 2 days later by the peak of LZM excretion. It is of interest that the leukocyte LZM in this patient was within the normal range during the entire period of observation. We would postulate that the increased LZM excretion in this case was secondary to an increase in the leukemic cell population rather than a change in LZM concentration in individual leukemic cells.

Summary

We observed a close correlation between intracytoplasmic LZM content as detected by immunofluorescence and the biochemically assayed values on cell lysates. With these two independent methods the enzyme was found in mature leukemic monocytes and in granulocytes of patients with chronic myelocytic leukemia. A variable immunofluorescent staining was observed in promyelocytes of three patients with acute promyelocytic leukemia. No staining was obtained in immature monocytic cells, myeloblasts, lymphoblasts, or undifferentiated blast cells.

In the biochemical assay of the leukocyte LZM a considerable variation was observed which is possibly related to variable enzyme content of the neutrophils during the course of the disease.

The immunofluorescent method used allows an estimate of the LZM content of the individual cell and offers additional information on the functional stage of the leukemic cell.

References

1. Abbrederis, K., Schmalzl, F., and Braunsteiner, H., *Schweiz. Med. Wochenschr.* **99,** 1425 (1969).
2. Asamer, H., Schmalzl, F., and Braunsteiner, H., *Acta Haematol.* (*Basel*) **41,** 49 (1969).
3. Asamer, H., Schmalzl, F., and Braunsteiner, H., *Br. J. Haematol.* **20,** 571 (1971).
4. Briggs, R. S., Perillie, P. E., and Finch, S. C., *J. Histochem. Cytochem.* **14,** 167 (1966).
5. Fink, M. E., and Finch, S. C., *Proc. Soc. Exp. Biol. Med.* **127,** 365 (1968).
6. Noble, R. E., and Fudenberg, H. H., *Blood* **30,** 465 (1967).
7. Osserman, E. F., and Lawlor, D. P., *J. Exp. Med.* **124,** 921 (1966).

De Novo Synthesis of Lysozyme by Bone Marrow Cells of a Patient with Monomyelocytic Leukemia*

MEHDI FARHANGI and ELLIOTT F. OSSERMAN

Although the presence of lysozyme (LZM) in granulocytes and monocytes has been well documented (1–7), its *de novo* synthesis has not been previously demonstrated. In an attempt to elucidate this point, we have cultured the peripheral blood and bone marrow cells of a patient with monomyelocytic leukemia in the presence of $[1\text{-}^{14}C]$leucine and determined the extent of *in vitro* labeling of LZM by radioimmunoelectrophoresis. By this method LZM was found to be synthesized by the bone marrow but not by the peripheral blood cells.

Materials and Methods

Peripheral blood and bone marrow samples were obtained from a 75-year-old male diagnosed as having monomyelocytic leukemia (Naegeli-type) two years before this study.

* Study supported by Grants CA-02332 and CRTY-5011 of the National Cancer Institute.

At the time of the study, the patient had enlargement of the liver (8 cm) and spleen (6 cm), marked pallor, scattered petechial hemorrhages and ecchymoses. Hemoglobin was 7.6 gm %; white blood cell count, 77,300 per mm³. Differential count, 52% granulocytes (myeloblasts 1, myelocytes 3, bands 17, and segmented polys 31); eosinophils 1%, monocytes 31%, and lymphocytes 15%; platelets 20,000 per mm³. The bone marrow was markedly hypercellular, and approximately 90% of the nucleated cells were granulocytic and monocytic precursors. The remainder were erythroid elements, lymphocytes, and plasma cells. Cytogenetic studies of the bone marrow showed a normal karyotype. Serum LZM was 110 µg/ml (lysoplate method; purified human LZM standard). The urine LZM concentration was in the range of 600 µg/ml with a total 24-hr LZM excretion of approximately 1.5 gm. The serum vitamin B_{12} and B_{12} binding capacity were within normal limits. Serum protein electrophoresis pattern showed a diffuse hypergammaglobulinemia. Immunoelectrophoresis and Immunoplate analyses confirmed the elevations of IgG, IgA, and IgM. There was no monoclonal type γ-globulin abnormality.

Cell Cultures

Peripheral blood buffy coat and bone marrow leukocytes were separated from red cells by allowing heparinized samples to sediment for 30 min at 37°C. The cells were washed twice and suspended at a cell density of 8×10^6 per ml in Eagle's minimal essential medium (MEM) containing 2.5% calf serum and 2.5 µCi of [1-¹⁴C]leucine per ml. Cultures were carried out in glass tissue culture dishes at 37°C in a 5% CO_2: air atmosphere. As indicated by trypan blue dye exclusion, 98% of the cells were viable at the start of culturing. After 72 hr, most of the cells were firmly adherent to the surface and were harvested by scraping with a scalpel blade. Seventy percent of these cells excluded trypan blue dye. The cell-free culture medium and the cells were separated by centrifugation. The cells were washed twice with MEM prior to disruption by 3 cycles of freeze-thawing. The cell extracts and culture media were then exhaustively dialyzed against normal saline to eliminate unincorporated [1-¹⁴C]leucine, and lyophilized.

Radioimmunoelectrophoresis

To demonstrate and identify newly synthesized proteins, we used the technique of radioimmunoelectrophoresis described by Hochwald, *et al.* (8). Purified human LZM (2 mg/ml) and the patient's serum (approx. 1:1 ratio) were added to all samples as carriers. Immunoelectrophoresis was carried out in 1% Agarose with barbital buffer (pH 8.6), ionic strength 0.075. The relatively high ionic strength is necessary to permit LZM to migrate in the agar medium (9). The immunoelectrophoretic patterns were developed with rabbit antisera to human LZM (9) and IgG and a horse antiserum to κ-type Bence Jones protein (#HABJκ #117 DOlO: Kallestad Laboratories, Los Angeles).

After the immunoprecipitin arcs were fully developed, the plates were washed exhaustively with buffered saline, covered with filter paper, and air-dried. The plates were then stained with Ponceau 3R, and autoradiography was carried out by placing the plates in contact with conventional X-ray film for 14 days.

Results

In two separate experiments no incorporation of [1-^{14}C] leucine into LZM could be shown in peripheral blood cell cultures of one to five days' duration. As seen in Fig. 1, however, there was unequivocal labeling of LZM in the bone marrow culture *medium* (BM/TC medium) but no labeled LZM was demonstrable in the marrow *cell* lysate. This presumably reflects a rapid rate of release of the newly synthesized LZM into the medium resulting in a very low intracellular concentration. It is noteworthy, as shown in Fig. 1, that the marrow cell lysate did contain a labeled protein precipitated by anti-BJκ and most likely representing free κ-chains. For unexplained reasons, this precipitin arc was not stained by the Ponceau dye. It does not correspond in position to the gamma arc of the immunoelectrophoretic (IEP) pattern which probably represents IgG with κ-chains in the carrier serum. If this protein does represent κ-chains, it presumably arose from plasma cells or lymphocytes of the marrow rather than from monocytes or myelocytes, but this point needs further study.

Discussion

These results clearly demonstrate the *de novo* synthesis of LZM by marrow cells but not the peripheral blood cells of a patient with monomyelocytic leukemia. The most probable marrow elements responsible for the enzyme synthesis are the leukemic cells. However, sources other than leukemic cells (e.g., normal granulocytes, monocytes) present in the bone marrow cannot be ruled out.

The *absence* of labeled LZM in the marrow cell lysate, but its *presence* in the culture medium, indicates that the LZM is probably rapidly released from these cells after its synthesis. In contrast, another newly synthesized protein, possibly representing free κ-chains, was detected in the cell lysate only and not in the medium. We presume that this immunoglobulin constituent was probably synthesized by plasma cells or lymphocytes included among the marrow cells, but its synthesis by leukemic cells (monocytes?) cannot be excluded.

In previous studies (10, 11) it has been found that the intracellular concentration of LZM in peripheral blood monocytes from patients with monocytic and monomyelocytic leukemia with high serum and urine LZM levels is generally *lower* than its intracellular concentration in peripheral blood myelocytes and granulocytes of patients with chronic granulocytic

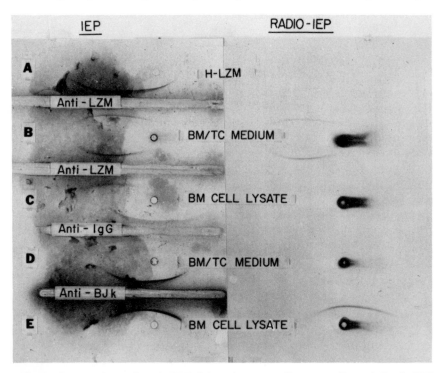

Fig. 1. Immunoelectrophoresis (IEP), left, and corresponding autoradiograph (Radio-IEP) of the following samples: (A) Purified human lysozyme (H-LZM), unlabeled; (B) and (D), bone marrow tissue culture medium (BM/TC medium); (C) and (E), bone marrow cell lysate after freeze-thawing (× 3). Aliquots of human LZM and the patient's serum were added as carrier proteins to antigen wells (B) through (E). The antisera used were rabbit anti-human lysozyme (Anti-LZM), rabbit anti-human IgG (Anti-IgG) and horse anti-Bence Jones kappa (Anti-BJκ). Radio-IEP (B) clearly demonstrates labeled LZM in the BM/TC medium, but the corresponding are in (C) (BM Cell lysate) shows no labeling. IEP's (D) (BM/TC medium) and (E) (BM Cell lysate) both show an arc of mid-γ mobility precipitated with Anti-BJκ and presumably representing IgG with κ light chains in the carrier serum. Radio-IEP (E) (BM Cell lysate) shows labeling of one arc of fast γ mobility. This arc *does not* correspond to the mid-γ arc of the IEP of this section. Note the relative positions of these arcs with respect to the antigen well.

leukemia with relatively normal or only slightly elevated serum LZM levels and no LZM uria. We believe that our present studies provide a possible explanation for this seeming paradox, namely, that monocytes, either normal or leukemic, apparently rapidly release their LZM after synthesis, whereas granulocytes, again either normal or leukemic, retain LZM, probably within lysosomes, and possibly release it only in the course of lysosomal disruption. This would imply that there may be two different cell populations (i.e., cells of the monocyte–macrophage series and those of the granulocytic series) which handle and presumably utilize the same enzyme (LZM) in two distinctly different ways. This also suggests that there may be different specific biological functions of LZM, as mediated by these different cell types.

References

1. Glynn, A. A., and Parkman, R., *Immunology* **7,** 724 (1964).
2. Barnes, J. M., *Br. J. Exp. Pathol.* **21,** 264 (1940).
3. Flanagan, P., and Lionetti, F., *Blood* **10,** 497 (1955).
4. Brumfitt, W., and Glynn, A. A., *Br. J. Exp. Pathol.* **42,** 408 (1961).
5. Briggs, R. S., Perillie, P. E., and Finch, S. C., *J. Histochem. Cytochem.* **14,** 167 (1966).
6. Speece, A. J., *J. Histochem. Cytochem.* **12,** 384 (1964).
7. Asamer, H., Schmalzl, F., and Braunsteiner, H., *Br. J. Haematol.* **20,** 571 (1971).
8. Hochwald, G. M., Thorbecke, G. J., and Asofsky, R., *J. Exp. Med.* **114,** 459 (1961).
9. Osserman, E. F., and Lawlor, D. P., *J. Exp. Med.* **124,** 921 (1966).
10. Perillie, P. E., Kaplan, S. S., Lefkowitz, E., Rogaway, W., and Finch, S. C., *J. Am. Med. Ass.* **203,** 317 (1968).
11. Ohta, H., and Osserman, E. F., *Tohoku J. Exp. Med.* **107,** 229 (1972).

Studies of Kidney Lysozyme in Human Acute Monomyelocytic Leukemia and in the Rat Chloroma*

DAVID S. ROSENTHAL, JOEL S. GREENBERGER, and
WILLIAM C. MOLONEY

Our interest in lysozyme (LZM) first stemmed from a patient who initially had myeloma with Bence Jones (κ) proteinuria but in the later phase of his disease developed the clinical and hematologic picture of acute leukemia. Associated with the leukemia was a different and unidentifiable proteinuria, and the patient and his urines were referred to Dr. Osserman. The rest of the story is well known, and this was the first case of acute monomyelocytic leukemia (AMML) in which lysozymuria was documented (1).

Since that time our group has gathered and reported studies of LZM in 168 patients with leukemia and a variety of myeloproliferative disorders carried out over the past 4 years. Of particular interest were serial follow-up studies of 8 patients with myeloproliferative disorders who terminated in an acute leukemic process (2). All the cases of AMML had striking elevations of LZM levels heralding the leukemic transformation, and one-half of those studied had very significant hypokalemia, as low as 2.1 mEq K/liter The microscopic findings in the kidneys of these patients were not uniform, e.g., acute tubular necrosis, leukemic infiltrates, or focal pyelonephritis.

*Supported by USPHS Grants CA 05691-05, C 6516, and CA 13589-01S1, and the American Cancer Society (Massachusetts Division) Grant #1410-Cl.

The association of lysozymuria and hypokalemia in acute monomyelocytic leukemia has been previously reported (3). The lysozymuria has been ascribed to either a simple overloading of the renal threshold or renal tubular damage produced by the LZM, per se. In human subjects with leukemia and suspected renal damage due to LZM, studies have generally been retrospective, and renal pathology examined only at autopsy. Prospective studies in acutely ill leukemic patients are inherently difficult with many variables. Therefore, in an effort to elucidate the metabolic abnormalities and consequences of elevated LZM levels on kidney function, we turned our attention to an animal model.

Since 1958, our laboratory has been passaging the Shay chloroleukemia in Sprague–Dawley (SD) rats. This tumor was initially induced with methylcholanthrene (4, 5). After subcutaneous transplantation in SD pups, local tumors (chloromas) develop within 10–14 days, and many animals also develop leukemia. The chloroma cells are primarily myeloblasts and promyelocytes and stain positively for cytoplasmic esterase and myeloperoxidase. At autopsy, green "chloromatous" tumors are found at the site of injection or in the peritoneal cavity; other organs often show the same greenish hue and may also demonstrate a bright pink fluorescence when exposed to ultraviolet light. Histochemically, these cells are almost identical to immature human myeloid cells.

Studies of serum and urinary LZM were carried out pre- and posttransplantation of the chloroleukemia and were compared to other rat leukemia models (6) (Table I). Striking increases of serum and urine LZM activity were found only in the chloroleukemic animals. In addition, there were significant elevations of LZM activity in the tumor ascites and kidney homogenates.

Because of the histochemical and biochemical similarities between the rat chloroleukemia and human AMML leukemia, it seemed possible that studies of the rat tumor could help answer some of the questions regarding the handling and excretion of lysozyme in normal and leukemic conditions. Rats developing chloroma and chloroleukemia following transplantation again showed significant increase in serum LZM after 10–14 days and subsequently developed lysozymuria. Chloroma-bearing animals had a mean serum LZM level of 75.1 + 8.2 μg/ml (range of 23–125) with a control of 11.6 + 0.7 (range of 9–16) (Table II). No LZM was detected in the urine of control animals, but in the leukemic rats, levels ranged from 3.3 to 20,250 μg/24 hr with a mean of 1,762 μg/24 hr. The mean serum potassium concentration in 17 rats with chloroma was 6.17 mEq, compared to 6.83 mEq in control animals. The difference was not statistically significant. However, the mean urinary potassium excretion in 16 leukemic rats was 0.86 $\pm$ 0.06 mEq/24 hr compared with 0.50 $\pm$ 0.06 in controls. This difference was statistically significant ($p\,0.005$). Although there was no direct

TABLE I

Lysozyme Activity in Rat Serum and Urine[a]

Breed	No.	Serum[b]			Urine[c]	
		Av.	Med	Range	No.	Range
		Normal values				
Sprague–Dawley	57	17.4	17.0	10–22	6	0
Wistar–Furth	14	17.7	17.5	12–22	4	0
Fischer	9	17.2	15.0	12–25	6	< 100
		Chloroleukemia				
Sprague–Dawley	27	280	290	70–410	9	2.5×10^5 8.0×10^5
		Mononuclear cell leukemia				
Wistar–Furth	16	18.9	16.0	10–42	—	—
Fischer	8	41.1	28.5	13–110	—	—

[a] From Ref. 6.

[b] Units = $\Delta OD_{540}/90$ sec/0.2 ml. Chloroma cell homogenate—102 units/mg tissue. Chloroleukemia ascites—480 units/0.2 ml.

[c] Units = $\Delta OD_{540}/90$ sec/24 hr.

correlation between the amounts of LZM and potassium excreted in the urine over a 24 hr period ($r = 0.4$), three of four rats followed serially showed concomitant elevations in urine LZM and potassium (7). Figure 1 shows the data from one of these animals.

The plasma disappearance and clearance of LZM in normal and leukemic rats have also been studied in our laboratory (8). These studies are summarized in Fig. 2. After collection of control serum and urine samples, 1 mg of purified LZM in a 2 ml volume was injected cephalad into the surgically exposed jugular vein of adult rats. In the control group [Fig. 2(a)], peak serum LZM levels, ranging from 50–330 μg/ml, were reached within 2 min of injection and fell to within 20% of preinjection levels by 15 min. These results are similar to those of Hansen and his co-workers (9). Urine LZM was not detectable in controls after the LZM injection.

After LZM injection in nephrectomized rats [Fig. 2(b)], peak serum LZM levels were reached within 5 min and fell more slowly than in the controls, reaching 40% of the peak level at 15 min post-injection. Thereafter, serum LZM levels rose continuously to 126% at 1 hr and 415% at 24 hr (420 μg/ml in one animal).

In chloroleukemic rats [Fig. 2(c)], clearance studies were done 6–7 days after intravenous tumor injection, at which time serum LZM levels were beginning to increase, but there was no detectable LZM-uria. After LZM injection, serum LZM levels returned to 20% of peak levels by 15 min,

TABLE II

Lysozymuria and Hyperkaluria in Chloromatous Rats[a]

	Serum LZM μg/ml	Urine LZM μg/24 hr	Urine K mEq/24 hr
Chloromatous rats	75.1 ± 8.2 (11)[b]	1762.4 (23)	0.862 ± 0.057 (16)
Control	11.6 ± 0.7 (7)	0 (14)	0.501 ± 0.059 (11)

[a] From Ref. 7.
[b] Numbers in parentheses indicate number of rats studied.

quite similar to the time pattern in control animals. LZM-uria was detectable in some animals 15 to 30 min after injection, but disappeared within 60 min.

In the chloromatous group [Fig. 2(d)], i.e., rats bearing solid tumors for 3 to 4 weeks with no leukemia, serum LZM levels prior to injection were significantly higher (110 ± 60 μg/ml) than controls, and there was marked LZM-uria. The rate of clearance of a similarly injected bolus of LZM in these animals was delayed as compared with the other groups, reaching 20% of the peak value in 26 min compared to 15 in the chloroleukemic and 14.7 in the normal rats, respectively.

These clearance studies suggest that there is not a simple threshold mechanism for renal handling of LZM, i.e., transient serum levels of over 300 μg/ml were rapidly cleared in control animals with little or no detectable enzyme activity in the urine. The finding of elevated serum LZM levels with

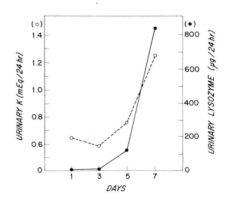

Fig. 1. Association of increased urinary potassium and lysozymuria in a rat bearing the Shay chloroleukemia (7).

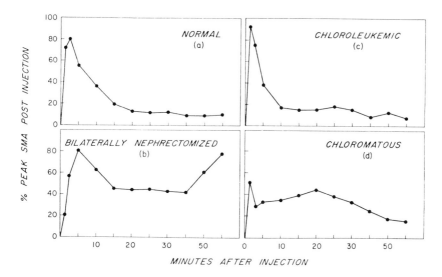

Fig. 2. LZM clearance studies after intravenous injection of purified rat LZM in (a) normal (b) bilaterally nephrectomized, (c) chloroleukemia, and (d) chloromatous Sprague–Dawley rats (8).

little or no LZM-uria in the acutely chloroleukemic rats compared to the marked LZM-uria of solid chloroma-bearing animals with lower serum levels also indicates a mechanism other than a simple filtration threshold. The data suggest that the kidneys have a finite capacity to absorb and store (or degrade) LZM and that this capacity becomes saturated in the chloroma-bearing rats by autologous LZM. With a subsequent injection, the enzyme cannot be reabsorbed after it passes from the glomerular filtrate and is thus lost in the urine. Therefore, the chromatous rat with chronic production of LZM demonstrates a state of overloading and interference with further renal handling of LZM.

With respect to the specific site of the renal lesion associated with LZM-uria, Osserman and Azar (10) have demonstrated abnormalities in the proximal convoluted tubules by electron and light microscopy and Glynn has demonstrated LZM in the proximal convoluted tubules by immunofluorescence (11). Indeed, the associated hyperkaluria and hypokalemia are consistent with a renal tubule defect. We have examined the kidneys of both human and animal leukemics by light microscopy and immunofluorescence. The antibodies to human and rat lysozyme were prepared with the help of Drs. Parkman and Greenberger. Light microscopy showed little in the way of specific renal abnormalities in either chloromatous rats or normal rats given injections of the LZM. Likewise, in studies of the kidneys of a patient with AMML with urinary LZM levels of greater than 1,000 μg/ml and persistent hypokalemia and hyperkaluria, no microscopic changes were noted with conventional (H and E) staining, nor were there any leukemic infiltrates within the kidneys. Immunofluorescent studies, however, showed significant localization of LZM in tubular cells and glomeruli.

Immunofluorescent studies of LZM in the rat leukemia model have been very informative. Normal rat kidney shows little or no localization of LZM as compared with the chloromatous rat kidney in which LZM can be visualized not only in tubules but also in glomeruli and around blood vessels. We are currently comparing normal and chloroma kidney slices with the immunofluorescent technique at varying intervals after LZM infusions.

In summary, we have demonstrated the rat chloroma and chloroleukemia to be morphologically, histochemically, and functionally similar to human AMML with respect to the increased serum and urine LZM levels and the development of marked hyperkaluria. The finding of marked LZM-uria in chloromatous rats with chronically elevated serum LZM levels and delayed LZM clearance indicates that chronic, but not acute elevations of LZM can overload the renal mechanism for LZM storage and/or degradation. Finally, immunofluorescence studies have provided evidence that the renal tubules are probably the major site of increased LZM activity and, although the proximal tubules may not be histologically abnormal, they may be functionally defective.

References

1. Osserman, E. F., and Lawlor, D. P., *J. Exp. Med.* **124,** 921 (1966).
2. Skarin, A. T., Matsuo, Y., and Moloney, W. C., *Cancer* **27,** 1336 (1972).
3. Muggia, F. M., Heinemann, H. O., Farhangi, M., and Osserman, E. F., *Am. J. Med.* **47,** 351 (1969).
4. Shay, H., Gruenstern, M., Mary, H. E., and Glazer, L., *Cancer Res.* **11,** 29 (1951).
5. Moloney, W. C., Dorr, A. D., Dowd, G., and Boschetti, A. E., *Blood* **19,** 45 (1962).
6. Rosenthal, D. S., and Moloney, W. C., *Proc. Soc. Exp. Biol. Med.* **126,** 682 (1967).
7. Rosenthal, D. S., Maglio, R., and Moloney, W. C., *Proc. Soc. Exp. Biol. Med.* **141,** 499 (1972).
8. Greenberger, J. S., Rosenthal, D. S., and Moloney, W. C., *J. Lab. Clin. Med.* **81,** 116 (1973).
9. Hansen, N. E., Karle, H., and Andersen, V., *J. Clin. Invest.* **50,** 1473 (1971).
10. Osserman, E. F., and Azar, H. A., *Fed. Proc.* **28,** 619 (1969).
11. Glynn, A. A., *Sci. Basis Med. Annu. Rev.* pp. 31–52 (1968).

Serum and Urine Lysozyme in Sarcoidosis

R. S. PASCUAL, P. E. PERILLIE,
J. B. L. GEE, and S. C. FINCH

The clinical usefulness of serum and urinary lysozyme activities in the differential diagnosis of leukemia is now well recognized (1, 2). It is also suggested that increased urinary lysozyme activity is a good index of proximal tubular damage (3, 4). There are several theoretical reasons suggesting that serum lysozyme levels might also be utilized to study the activity and pool size of mononuclear phagocytes, particularly in such granulomatous disorders as tuberculosis, certain fungal infections, and sarcoidosis. Both monocytes and fixed tissue macrophages are known to contain high lysozyme activity. Alveolar macrophages from rabbit lungs made granulomatous by the injection of heat-killed tubercle bacilli showed elevated lysozyme levels (5). Among the granulomatous diseases mentioned, tuberculosis is known to cause increased serum lysozyme activity (6). However, apart from the initial report of Osserman and Lawlor indicating increased serum and urine lysozyme activity in a few patients with sarcoid (1), little is known about lysozyme activity in sarcoidosis. We have investigated the relationship between disease activity and changes in serum and urine lysozyme.

The clinical sample consisted of 19 male and 26 female patients (or a total of 45 patients) with proved sarcoid, as shown in Table I. The majority of these subjects are black and had liver biopsies for histological diagnosis. Routine blood counts, blood urea nitrogen (BUN), and serum calcium

TABLE I

Subjects in Study with Proved Sarcoid (Total: 45)

Sex	Male	19
	Female	26
Ethnic groups	Black	35
	White	8
	P.R.	2
Stage	I	16
	II	21
	III	8
Tissue diagnosis	Liver	23
	Multiple organs	9
	Scalene node	8
	Others	5

levels were obtained on most patients. In no patient was there hypercalcemia and in only one patient was there borderline azotemia. Serum and urine lysozyme activities were determined on 35 and 42 patients, respectively, by either the lysoplate (1) or turbidimetric (7) technique, using either a human or egg-white lysozyme standard. For the purpose of this presentation, values are expressed in percent of normal rather than the usual μg/ml,

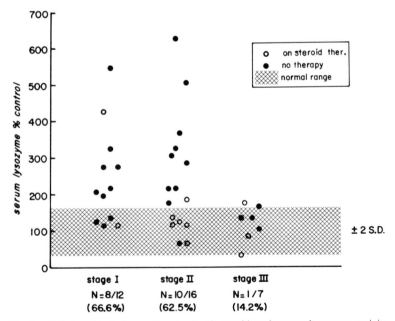

Fig. 1. Relationship between the stages of sarcoid and serum lysozyme activity.

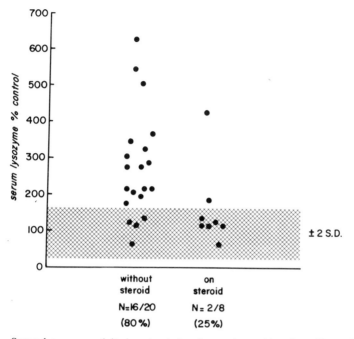

Fig. 2. Serum lysozyme activity in untreated and treated sarcoid patients (Stages I and II).

since different standards were used at various times. Each patient was staged according to his chest radiographic findings. Stage I patients have bilateral hilar lymph adenopathy (BHL) with or without right paratracheal node enlargement; Stage II consists of BHL with pulmonary infiltrates, and in Stage III, there are pulmonary infiltrates only, with fibrosis and with or without bullous changes.

The relationship between the clinical staging of sarcoid and serum lysozyme activity is shown in Fig. 1. Eight of the 12 patients (or 66%) with Stage I sarcoid showed increased serum lysozyme activity. The results were similar with Stage II disease, where 10 of 16 (or 62%) showed increased enzyme activity. By contrast, only 1 of 7 (or 14%) of Stage III sarcoid patients were significantly elevated. In addition to the clear difference in serum lysozyme levels in the three stages of the disease, it is apparent that steroids affect the lysozyme levels. This point is emphasized in Fig. 2, which compares the lysozyme levels in patients before and during steroid therapy. Serum lysozyme activity is increased in 80% of untreated patients and 25% of those receiving steroids.

Figure 3 demonstrates the relationship between the stage of sarcoid and urine lysozyme activity. Lysozymuria was present in 5 of 14 patients (or 35%) in Stage I and in 11 of 21 (or 52%) in Stage II. None of the Stage III

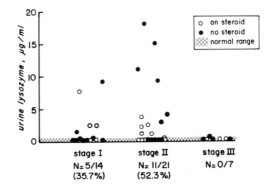

Fig. 3. Relationship between the stages of sarcoid and urine lysozyme activity.

patients had lysozymuria. Only 2 of the 42 patients studied had evidence of significant renal disease. Both had proven renal calculi and both had lysozymuria.

There is poor correlation between serum and urine lysozyme activity, as shown in Fig. 4. Lysozymuria is associated with increased serum lysozyme in Stages I and II and not in Stage III. Presumably this is due to glomerular filtration in excess of maximum tubular reabsorption, with the unabsorbed

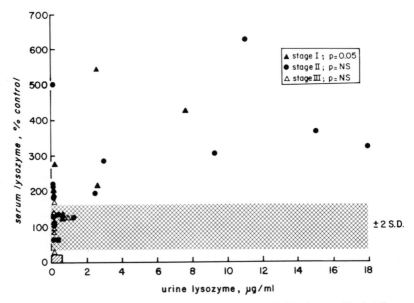

Fig. 4. Relationship between serum and urine lysozyme activities in sarcoidosis (all stages).

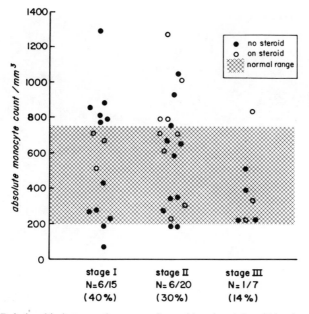

Fig. 5. Relationship between the stages of sarcoid and peripheral blood monocytes.

enzyme appearing in the urine. However, there are 2 patients with increased serum lysozyme approximately between 3 and 5 times the control level without lysozymuria. Whether this represents variation in renal threshold is purely speculative.

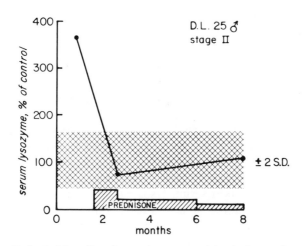

Fig. 6. Patient D. L. Serial studies of serum lysozyme activity during corticosteroid therapy.

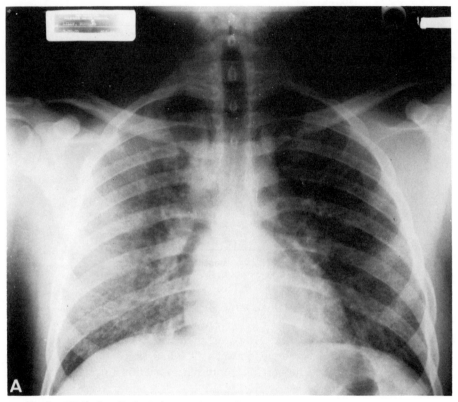

Fig. 7. (A) Patient D. L. Before corticosteroid therapy showing bilateral hilar lymphade-nopathy and pulmonary infiltrates. (B) Patient D. L. Improvement with decrease in the size of hilar nodes and pulmonary infiltrates.

The possibility that serum lysozyme may reflect peripheral monocytosis is considered. There was monocytosis in about one-third of the sarcoid patients, as is shown in Fig. 5. There is little relationship between the num-bers of blood monocytes and serum lysozyme activity. These findings strongly suggest that the increased body pool of lysozyme in patients with active sarcoid is derived from the turnover of increased numbers of tissue monocytes, probably located in the numerous granulomatous lesions in pulmonary and other body tissues. This contention is further supported by the finding of increased serum lysozyme in all 6 patients in this series with sarcoid involving the spleen. Presumably, the more extensive the tissue sarcoid involvement, the greater the amount of enzyme released to the blood plasma.

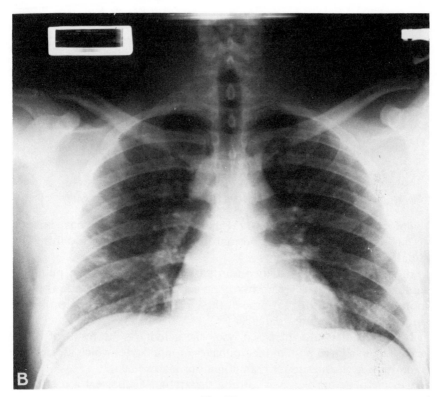

Fig. 7B

Serum lysozyme activity was followed serially in 7 patients during corticosteroid therapy. In 6, serum lysozyme activity returned to the normal range concomitant with clinical and chest radiographic improvement. Figures 6–8 demonstrate the effect of corticosteroid therapy in 2 patients. The serum lysozyme value of patient D. L. (Fig. 6) declined to the normal range weeks after he was started on 40 mg of prednisone and remained normal in the next 6 months of therapy. Figure 7(A) and (B) demonstrate concomitant improvement in his chest radiograph. Patient E. S. (Fig. 8) showed a similar response to corticosteroid therapy. The response to corticosteroid therapy in these patients suggests two possibilities: (a) that there is a reduction in the total number of tissue macrophages, and/or (b) reduction in the amount of lysozyme released from individual cells due to lysosomal stabilizing effect.

Studies of serum lysozyme activity in patients with lymphoma and other interstitial lung diseases are currently under way in our laboratory. Studies

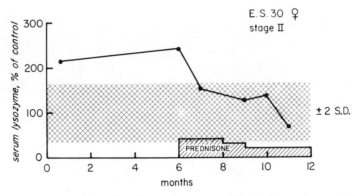

Fig. 8. Patient E. S. Serial studies of serum lysozyme activity during corticosteroid therapy.

to date have shown that only 3 of 31 patients with lymphoma have increased serum lysozyme activity and normal in 4 of 4 patients with diffuse interstitial lung disease.

In summary, more than two-thirds of patients with active pulmonary sarcoid have increased serum lysozyme activity and lysozymuria is present in more than one-third. In some of these patients, lysozymuria probably is evidence of renal tubular damage.

It is likely that this increased lysozyme activity is the result of enzyme release from a large pool of macrophages or monocytes which are degraded at multiple systemic sites of granulomatous involvement.

Change in serum lysozyme during therapy may represent a useful laboratory parameter for determination of clinical progress in response to therapy.

Serum lysozyme may be of value in the differential diagnosis of sarcoidosis, especially with regard to lymphoma or lymphocytic leukemia, since enzyme activity is normal or reduced in the latter conditions. Preliminary studies in a limited number of patients suggests that this enzyme may prove useful in differentiating sarcoid from other forms of interstitial lung disease.

References

1. Osserman, E. F., and Lawlor, D. P., *J. Exp. Med.* **124,** 921 (1966).
2. Perillie, P. E., Kaplan, S. S., Lefkowitz, E., Rogaway, W., and Finch, S. C., *J. Am. Med. Ass.* **203,** 317 (1968).
3. Hayslett, J. P., Perillie, P. E., and Finch, F. C., *N. Eng. J. Med.* **279,** 506 (1968).
4. Prockop, D. J., and Davidson, W. D., *N. Eng. J. Med.* **270,** 269 (1964).
5. Myrvik, Q. N., Leake, E. S., and Oshima, S., *J. Immunol.* **89,** 745 (1962).
6. Kerby, G. P., and Chaudhuri, S. N., *J. Lab. Clin. Med.* **41,** 632 (1953).
7. Litwack, G., *Proc. Soc. Exp. Biol. Med.* **89,** 401 (1955).

Lysozyme of Cartilage and
Other Connective Tissues (A Speculative Review)

KLAUS E. KUETTNER, REUBEN EISENSTEIN, and
NINO SORGENTE

It was approximately half a century between Fleming's discovery of cartilage lysozyme (1) and our description of the topographic distribution of lysozyme in the epiphyseal plate of the calf scapula (2). We found the highest concentration of diffusible lysozyme in that portion of the epiphyseal cartilage undergoing rapid transformation into bone. (The adjective "diffusible" is used with regard to lysozyme purposely and will be discussed later in more detail.) There it was several times higher than in cartilage more distant from the cartilage–bone junction, newly formed bone, or cartilage not destined to ossify. It was present in much higher concentration in the cartilage of young growing dogs than in that of older more slowly growing animals. We therefore hypothesized that lysozyme might have some function in cartilage related either to cartilage transformation or calcification.

As a preliminary approach to this problem we studied the effects of adding lysozyme to organ cultures of cartilage (3–5). The morphologic effects were dramatic in cartilage derived either from puppies or embryonic mice, humans or chickens. There was a striking opacification of the cartilaginous portion of the tissue exposed to high concentration of hen egg-white lysozyme. The distribution of the opacification was remarkable in that it spared the hypertrophic zone of the epiphyseal growth plate, precisely the locus of highest concentration of diffusible lysozyme in the native tissue. These effects were observed only in living cartilage containing an epiphyseal growth plate and, therefore, destined to be transformed into bone. They were not seen in nonossifying cartilage such as the nasal septum, or embryonic cartilage too young to possess a plate. Hematoxylin and eosin stained sections demonstrated an increased acidophilia of those areas of the matrix which were opacified when exposed to lysozyme (6). Some cationic proteins such as protamine or lysozyme altered by irradiation with UV light also induced changes in the staining properties of the matrix, but in different and less specific locations (7). These effects were not produced by the addition of more neutral proteins such as albumin (6).

Studies using fluorescein-labeled lysozyme clearly showed that these histological changes were due, at least in part, to the formation of complexes between lysozyme and the extracellular matrix (6). In electron micrographs of tissue exposed to high concentrations of egg-white lysozyme, electron-dense deposits were found in areas corresponding to the sites where fluorescein-labeled lysozyme was seen by light microscopy. The higher resolution afforded by this instrument showed the deposits, presumably complexes between lysozyme and some component of the cartilage matrix, to be largely restricted to two sites. These were the lacunae of the resting and columnar zones and the collagen fibrils. On collagen, they were arrayed 640 Å apart, an observation of some interest since collagen in these areas lacks a 640 Å axial period. Like the gross and light microscopic changes they spared the hypertrophic zones (8). Histochemical evidence suggested that these changes were due to interactions between lysozyme and proteoglycans (3).

These observations demonstrated a very precise anatomic distribution of sites to which exogenous lysozyme was bound in cartilage matrix. They also provided some new insights into the organization of cartilage matrix, and a new histochemical method for studying morphology (9).

We then investigated whether the sites to which exogenous lysozyme was bound corresponded to the sites where lysozyme is normally distributed in the tissue. Immunocytochemical studies using labeled antisera to purified avian and mammalian lysozymes showed that this was, indeed, the case (10) (Fig. 1).

Lysozyme is known to be localized in lysosomes or lysosome-like organelles in some cells (11, 12). To determine whether this was true of cartilage embryonic chicken chondrocytes were isolated, their subcellular fractions separated by the methods of de Duve (13), and the fractions assayed for lysozyme activity. Surprisingly, it was found that these chondrocytes contained almost no diffusible lysozyme and the minute amount present was found in the microsomal fraction (14).

It therefore appears that the lysozyme in cartilage is virtually restricted to the extracellular matrix, where it has a very precise distribution. Three key questions are then presented: (1) What is lysozyme bound to? (2) What is the source of cartilage lysozyme? (3) What is the function of lysozyme in cartilage metabolism? The second question is now answered, the first partially so, and the third remains an enigma. We will discuss them in order.

1. What is lysozyme bound to? Connective tissues and cartilage, in particular, are rich in polyanionic sulfated polysaccharides. These molecules are covalently bound to proteins to form proteoglycans (15). Lysozyme does not degrade these molecules. Schubert and Franklin (16) studied the interaction of lysozyme with the proteoglycans as well as with purified chondroitin sulfate of cartilage. A series of salt-like compounds form and their composition is expressed by the equivalent ratio of polyanion equivalent to lysozyme. These studies show that chondroitin sulfate and its proteoglycans show the typical polyelectrolyte property of binding counter ions behaving as weak electrolytes. Because of the high positive charge of lysozyme, its binding with polyanions would be expected to be favored over the binding of univalent counter ions. In the presence of a sufficiently high concentration of small counter ions the competitive effect displaces lysozyme from association with a polyanion. It therefore appears that lysozyme-anion compounds are simple salts since they are dissociated rapidly by 0.15 M KCl or carboxylmethyl-cellulose (16).

Recent studies of Sadjera and Hascall (15, 17) and Rosenberg *et al.* (18) have shown that the proteoglycans of cartilage are present in two forms: as subunit proteoglycan molecules (PGS) or as large molecular weight aggregates of these subunits (PGC). The aggregation of subunits into large complexes is mediated by glycoproteins, termed "link glycoproteins" (GP-L). We speculated that a role of lysozyme in cartilage might be to mediate the formation of PPC, but we have been unable to demonstrate that lysozyme acts as a linking component in the aggregation of proteoglycans (19).

Since lysozyme is a basic molecule and the proteoglycans are highly acidic, other types of interactions are possible between these two components of cartilage. We have demonstrated that lysozyme is extracted from cartilage by low concentrations of guanidinium chloride (GuCl) within a narrow range while proteoglycans require much higher molarities and are extracted over a wider range (19). We have interpreted this as evidence that lysozyme is bound specifically to some other component or components within the tissue matrix. Treatment of cartilage slices with chondroitinase,

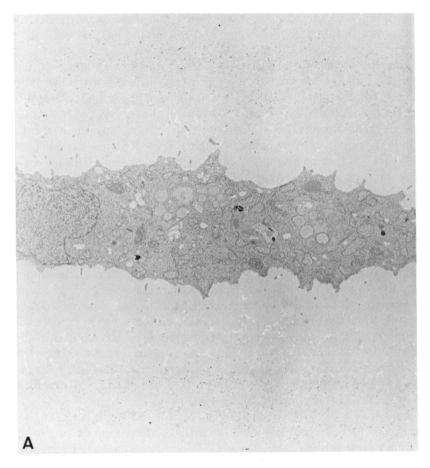

A

Fig. 1. (A) Calf scapula columnar zone. With routine methods of tissue preparation, the extracellular matrix appears to be relatively empty optically. From the *J. Cell Biol.* **46,** 627 (1970). Reproduced with their permission. (B) The same tissue stained with egg-white lysozyme as a vital stain (6). Electron-dense lysozyme–matrix complexes fill much of the lacunae and are also seen as fine droplets outside the lacunae. At higher resolution (8) the extracellular deposits are arrayed on the surface of collagen fibrils in a 640 Å repeating sequence. From the *J. Cell. Biol.* **46,** 627 (1970). Reproduced with their permission. (C) Chick cartilage exposed to purified anti-egg-white lysozyme. A soluble peroxidase–antiperoxidase method was used to localize the antigen. The darkly stained areas represent sites of peroxidase activity and thus the locus of lysozyme. Note that these sites are in the lacunar and extralacunar areas and closely correspond to sites of binding of exogenous lysozyme.

which digests chondroitin sulfate, released 85% of the uronic acid but no lysozyme. Trypsin digestion resulted in a release of about 90% of the proteoglycan and about 75% of the lysozyme. These data indicate that lysozyme is probably not bound directly to chondroitin sulfate (19). It also

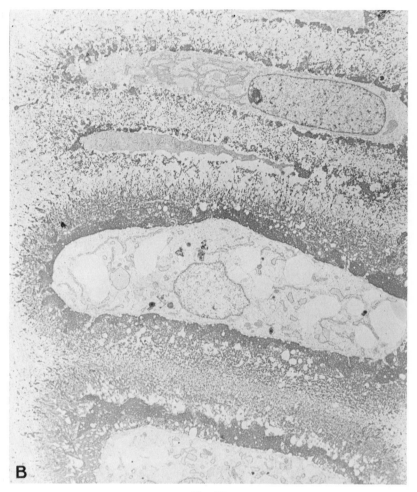

Fig. 1B

appears that the negatively charged polysaccharides do not act as an ionic barrier to the diffusion of lysozyme (19).

Other extraction studies led to similar conclusions. When cartilage is extracted with water or physiological salt, only a small amount of lysozyme can be extracted regardless of the extraction period (20). When 3 or 4 M GuCl is used, all the lysozyme is extracted from bovine nasal septum within 48 hr (19). During this period 90% of all cartilage proteoglycans are also solubilized (17). In articular cartilage which is richer in collagen and more resistant to extraction, 95% of the proteoglycans and all lysozyme can also be extracted within 48 hr. When 3 M $MgCl_2$ is used for extraction, the same results are obtained with hyaline cartilage (nasal septum). However, when

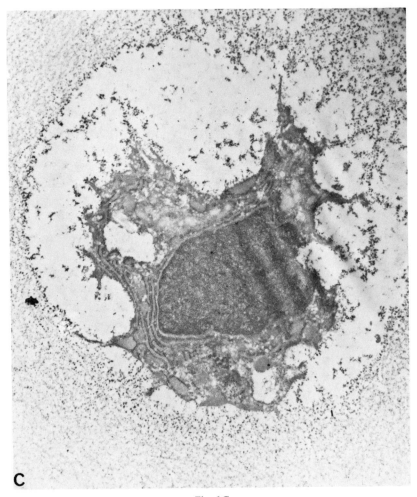

Fig. 1C

articular cartilage is extracted with 3 M $MgCl_2$, only about 50% of the proteoglycans are extracted (21). About 50% of the lysozyme also remains in the tissue and can be subsequently extracted with 3 or 4 M GuCl (22).

The observation that, in every tissue tested thus far, some lysozyme diffuses out in isotonic salt while another, larger fraction requires high salt concentration for extraction raises the possibility that there are two lysozyme pools in tissue: one bound and the other not. The diffusible and nondiffusible lysozymes may also not be identical. Data from animal experiments in which vitamin D (23) or parathyroid extract (24) were injected suggest that bone and aorta may have more than one lysozyme.

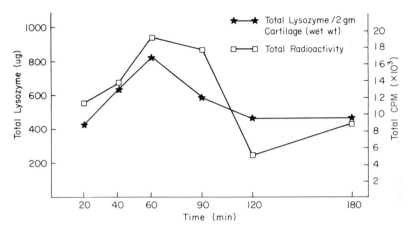

Fig. 2. 2.0 gm of fresh cartilage from 16-day-old chick embryos was incubated in 10 ml of Krebs–Ringer buffer (pH 7.4) containing $5\mu Ci$ [^{14}C]lysine and 3 mg/ml glucose. At the times indicated, the cartilage was homogenized in 0.15 M KCl. The homogenates and the media were chromatographed on CM-Sephadex C-50. The lysozyme peaks were pooled and assayed for lysozyme and radioactivity. The maximum incorporation occurred at 60 min, decreasing from 60–120 min and then leveling off paralleling the total amount of lysozyme.

Preliminary electrophoretic analyses of purified calf cartilage lysozyme indicate that there are several bands with lysozyme activity (22). Similar data are available for other lysozymes (25). If this is so, it may be of physiological significance. Since the "bound" lysozyme is not retained in the tissue by proteoglycans, we speculate that it is bound to something else, perhaps a glycoprotein. The immunocytochemical studies of binding of exogenous lysozyme, and some observations described below, suggest it may be associated with collagen.

It has been found that hen egg-white lysozyme is capable of binding to neutral sugars and oligosaccharides without hydrolyzing them (26). The oligosaccharides are apparently bound to the active site of the enzyme in a manner similar to that of native substrates. It is possible, therefore, that certain functions of lysozyme in the tissue may be related to its ability to interact with oligosaccharides built into the structure of the matrix. The binding loci could be oligosaccharides such as have been reported to occur in collagen or of other glycoproteins (27). This hypothesis might explain why the active site of cartilage lysozyme has been preserved through evolution even though no known substrates for the enzymatic activity of lysozyme are found in cartilage. It would further suggest that the biological role of lysozyme in cartilage is an important one (28–30).

2. What is the source of cartilage lysozyme? To determine whether chondrocytes synthesize lysozyme we have used several experimental approaches. Chick limb cartilage slices from 16-day-old embryos were incu-

TABLE I

Sedimentation of ^{14}C-labeled Nondialyzable Macromolecules in a CsCl Gradient[a]

	cpm/50 liter	Lysozyme (mg%)
Unfractionated sample	6800	2.20
Gradient fraction		
1	30,000	3.64
2	10,600	1.10
3	1600	0.00
4	650[b]	0.00
5	2100[b]	0.00

[a] 2.0 gm fresh nasal septum cartilage incubated for 3 hr at 37°C in 10 ml Krebs–Ringer bicarbonate buffer (pH 7.4) containing 50 μCi of [14C]lysine and 3 mg/ml glucose. The slices were extracted with 15 volumes of 3 M GuCl. The extract was dialyzed, CsCl added, and a density gradient established (19). The gradient was cut into five fractions and assayed for lysozyme activity and radioactivity. Most of the radioactivity and all the lysozyme were found in the top two fractions.

[b] 90% of the uronic acid was recovered in the bottom two fractions.

bated in Krebs–Ringer buffer (pH 7.4) containing [14C]lysine (31). The slices were then homogenized in 0.15 M KCl. The supernatant and the incubation media were chromatographed on a cation exchanger, CM-Sephadex C-50. The lysozyme-containing peaks were combined and assayed for enzymatic activity and for radioactivity. Figure 2 shows the time course of incorporation of [14C]lysine into the lysozyme-containing fractions. Both radioactivity and lysozyme activity reach a maximum at 60 min and then decrease.

In another experiment bovine nasal septum cartilage slices were incubated for 3 hr in Krebs–Ringer bicarbonate buffer (pH 7.4) containing [14C]lysine and then extracted with 3 M GuCl. The extract was dialyzed and a CsCl density gradient established (19). The two uppermost fractions contained most of the radioactivity and all the lysozyme (Table I). Aliquots of the dialyzed extract and of the upper two fractions of the gradient were electrophoresed on a cellulose acetate membrane. An autoradiograph was prepared from the membrane and the location of lysozyme activity on the strip was then determined by the sandwich procedure (2). Figure 3 shows a tracing of the results. The autoradiograph indicated that most of the incorporated isotope was in a cationic molecule and corresponded to the position of the marker hen egg-white lysozyme.

When chick embryo femora are cultured in the presence of specific anti-lysozyme antibody and [14C]lysine, the immunoprecipitin lines formed by lysozyme diffusing from the explant and the antibody in the medium in a living radial immunodiffusion preparation are radioactive, again indicating that lysozyme is synthesized by cartilage and bone (32).

Finally, to prove conclusively that chondrocytes are responsible for lysozyme synthesis, chondrocytes from embryo chick cartilage and from the epiphyseal plate of young calves were isolated and cultured in a protein-free media. In both cases lysozyme was found in the media clearly proving that the chondrocytes do synthesize lysozyme (22).

3. What is the function of lysozyme in cartilage? In some of our earlier studies, we found that antibodies to egg-white lysozyme had no discernible effect on organ cultures of chick limbs (32). Egg-white lysozyme is immunologically identical to chick cartilage lysozyme, and does not cross react immunologically with mammalian lysozymes.

The possibility that lysozyme may be involved in the calcification process must be considered. The high concentration of lysozyme near the cartilage–bone junction (2) and the ontogony of lysozyme accumulation in the growth plate suggest that this may be true since diffusible lysozyme appears in limb buds only after the growth plate has formed (4). The fact that parathyroid extract injection (24) and rickets decrease bone lysozyme while vitamin D increases it in both cartilage and aorta (23) further suggest an association between lysozyme and the calcification process.

More direct but still inconclusive evidence has been obtained from *in vitro* studies in two other laboratories. First, Termine and Posner (33) have

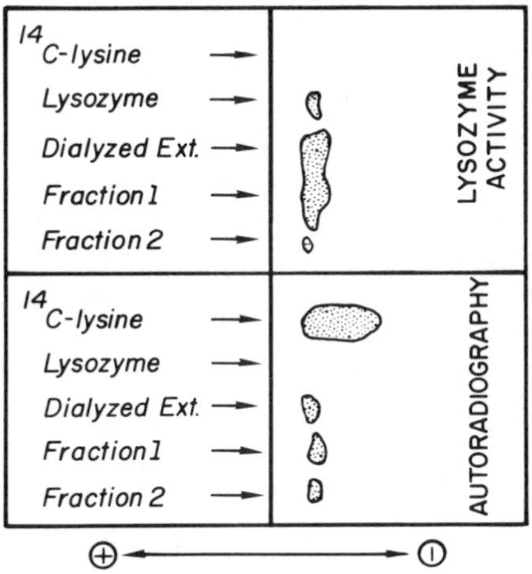

Fig. 3. Aliquots of the gradient fractions and untreated dialyzed extract were electrophoresed on cellulose acetate membranes with veronal buffer (pH 8.4) for 20 min. Autoradiographs of the membrane were prepared. Lysozyme activity was determined from the same membrane with the sandwich procedure (2). Note that the migration of the radioactivity paralleled migration of lysozyme activity.

reported that lysozyme accelerates the precipitation of amorphous calcium phosphate from metastable solutions. Second, Howell (34), using elegant micropuncture techniques, has aspirated extracellular material from the hypertrophic cell zone of calcifying cartilage and has found that this material inhibits mineral accretion *in vitro*. After ultracentrifugation the inhibitory activity was found in the sedimentable fraction and was destroyed by incubation with trypsin or hyaluronidase. If the aspirate is mixed with purified cartilage lysozyme and incubated for 6 hr at 37°C, its ability to inhibit mineral accretion is considerably diminished. The extent of inhibition approximates that of hyaluronidase (22). Pita *et al.* (35) have speculated that a specific proteoglycan in the extracellular fluid of cartilage which is susceptible to trypsin or hyaluronidase acts as a calcification inhibitor. Lysozyme does not degrade glycosaminoglycans. Preliminary ultracentrifugation studies indicate that cartilage lysozyme decreases the sedimentation coefficient of high molecular weight proteoglycan aggregates (22). We may, therefore, speculate that one function of lysozyme could be the regulation of calcification locally, perhaps by inhibiting a macromolecular complex from inhibiting calcification. Some preliminary tissue culture experiments in which dog cartilage lysozyme was added to cultures of dog scapula growth plate support this notion. In these experiments, more extensive calcification was seen than in control samples (22).

Apart from any role in calcification, lysozyme is one of the few nonnuclear cationic proteins. It is, therefore, theoretically capable of chemically interacting with the glycoproteins and strongly acidic proteoglycans of the extracellular matrix. Although it is generally thought that there is no substrate in mammalian tissues that lysozyme can catalytically degrade, this possibility cannot be excluded. In connective tissues there are many as yet unidentified glycoproteins and preliminary studies have indicated that some of these glycoproteins may interact with lysozyme.

In dentin, an acellular calcified tissue, a significant amount of lysozyme can be extracted by EDTA. The majority of this extractable lysozyme seems to follow the chromatographic behavior of a phosphoprotein isolated by Veis *et al.* (36). Dentin lysozyme not extracted with EDTA can be solubilized from the organic matrix with 3 M GuCl (22). In compact bone fractions, kindly supplied to us by G. M. Herring, lysozyme has also been found to be divided into fractions which are or are not extractable with EDTA. The extractable lysozyme is all found in Herring's G-I glycoprotein fraction (22, 37).

We have found that virtually every connective tissue contains some lysozyme. We have also begun some fractionation studies and have made the observation that if skin collagen is purified according to the method of Gross and Kirk (38) lysozyme follows the collagen during purification and can be extracted from the final preparation with 3 *M* GuCl (22). Lysozyme can be demonstrated by similar methods in an aortic structural glyco-protein isolated by Robert *et al.* (22, 39) and in the RNA-containing pre-dentin matrix vesicles isolated by Slavkin *et al.* (22, 40).

Our working hypothesis remains that the biological role of lysozyme in connective tissues involves an interaction between this cationic protein and structural proteins or glycoproteins and that these interactions may pos-sibly alter or regulate the biological function of these substances. Whether this interaction is primarily related to charge and molecular dimension or to an as yet unidentified catalytic function of lysozyme remains to be de-termined.

Acknowledgments

The part of the work performed in our laboratory and discussed in this review has been supported by Grant No. AM-09132 from the United States Public Health Service, by the Illinois Arthritis Foundation and in part by Grant Nos. AM-16020 and HL-14968 from the United States Public Health Service and the Hulbert Fund, Presbyterian-St. Luke's Hospital, Chicago, Illinois.

The stimulating collaboration of several collaborators listed as co-authors in our publi-cation is gratefully acknowledged as well as the secretarial assistance of Mrs. D. Cacorovski and Mrs. M. Gilbert.

References

1. Fleming, A., *Proc. Roy. Soc., Lond.* [*Biol.*] **93**, 306 (1922).
2. Kuettner, K., Guenther, H., Ray, R. D., and Schumacher, G., *Calcif. Tissue Res.* **1**, 298 (1968).
3. Kuettner, K., Soble, L., Eisenstein, R., and Yeager, J. A., *Calcif. Tissue Res.* **2**, 93 (1968).
4. Schrodt, M. J., Eisenstein, R., Ray, R. D., and Kuettner, K. E., *Surg. Forum* **19**, 461 (1968).
5. Kuettner, K. E., Soble, L. W., Ray, R. D., Croxen, R. L., Passavoy, M., and Eisenstein, R., *J. Cell Biol.* **44**, 329 (1970).
6. Kuettner, K. E., Soble, L. W., Guenther, H. L., Croxen, R. L., and Eisenstein, R., *Calcif. Tissue Res.* **5**, 56 (1970).
7. Eisenstein, R., Soble, L. W., and Kuettner, K. E., *Am. J. Pathol.* **60**, 43 (1970).
8. Eisenstein, R., Arsenis, C., and Kuettner, K. E., *J. Cell Biol.* **46**, 626 (1970).
9. Eisenstein, R., and Kuettner, K. E., *Calcif. Tissue Res.* **4**, Suppl., 137 (1970).
10. Kuettner, K. E., Eisenstein, R., Soble, L. W., and Arsenis, C., *J. Cell Biol.* **49**, 450 (1971).
11. Cohn, Z. A., and Hirsch, J. G., *J. Exp. Med.* **112**, 983 (1960).
12. Weissman, G., *N. Engl. J. Med.* **273**, 1084 (1965).
13. de Duve, C., Pressman, B. C., Gianetto, R., Wattiaux, R., and Appelmans, F., *Biochem. J.* **60**, 604 (1955).
14. Arsenis, C., Eisenstein, R., Soble, L. W., and Kuettner, K. E., *J. Cell Biol.* **49**, 459 (1971).
15. Sajdera, S. W., and Hascall, V. C., *J. Biol. Chem.* **244**, 77 (1969).
16. Schubert, M., and Franklin, E. C., *J. Am. Chem. Soc.* **83**, 2920 (1966).

17. Hascall, V. C., and Sajdera, S. W., *J. Biol. Chem.* **244,** 2384 (1969).
18. Rosenberg, L., Pal, S., Beale, R., and Schubert, M., *J. Biol. Chem.* **245,** 4112 (1970).
19. Sorgente, N., Hascall, V. C., and Kuettner, K. E., *Biochim. Biophys. Acta* **284,** 441 (1972).
20. Economou, J. S., *J. Surg. Oncol.* **3,** 89 (1971).
21. Rosenberg, L., personal communications.
22. Kuettner, K. E., unpublished observations.
23. Eisenstein, R., Sorgente, N., Arsenis, C., and Kuettner, K. E., *Arch. Pathol.* **94,** 479 (1972).
24. Wolinsky, I., and Cohn, D. V., *Nature (Lond.)* **210,** 413 (1966).
25. Jolles, P., *Angew. Chem. [Engl.],* **8,** 227 (1969).
26. Rupley, J. A., Butler, L., Gerring, M., Hartdegen, F. J., and Pecoraro, R., *Proc. Natl. Acad. Sci. U.S.A.* **57,** 1088 (1967).
27. Spiro, R. G., *J. Biol. Chem.* **244,** 602 (1969).
28. Eisenstein, R., Sorgente, N., and Kuettner, K. E., *Am. J. Pathol.* **65,** 515 (1971).
29. Kuettner, K. E., Sorgente, N., Arsenis, C., and Eisenstein, R., *Isr. J. Med. Sci.* **7,** 407 (1971).
30. Eisenstein, R., Arsenis, C., and Kuettner, K. E., *Isr. J. Med. Sci.* **7,** 415 (1971).
31. Sorgente, N., and Guenther, H. L., *Fed. Proc.* **29,** 932 (1970).
32. Kuettner, K. E., Wezeman, F. H., Simmons, D. J., Lisk, P. Y., Croxen, R. L., Soble, L. W., and Eisenstein, R., *Lab. Invest.* **27,** 324 (1972).
33. Termine, J. D., and Posner, A. S., *Arch. Biochem. Biophys.* **140,** 307 (1970).
34. Howell, D. S., *J. Bone Joint Surg. [Am.]* **53,** 250 (1971).
35. Pita, J. C., Cuervo, L. A., Madruga, J. E., Muller, F. J., and Howell, D. S., *J. Clin. Invest.* **49,** 2188 (1970).
36. Veis, A., Spector, A. R., and Zamoscianyk, H., *Biochim. Biophys. Acta* **257,** 404 (1972).
37. Herring, G. M., *in* "The Biochemistry and Physiology of Bone" (G. H. Bourne, ed.), 2nd ed., Vol 1, pp. 128–184. Academic Press, New York, 1972.
38. Gross, J., and Kirk, D., *J. Biol. Chem.* **233,** 355 (1958).
39. Robert, A. M., Robert, B., and Robert, L., *in* "Chemistry and Molecular Biology of the Intercellular Matrix" (E. A. Balazs, ed.), Vol. 1, p. 237. Academic Press, New York, 1970.
40. Slavkin, H. C., Croissant, R., and Bringas, P., Jr., *J. Cell Biol.* **53,** 841 (1972).

Lysozyme as a Component of Human Cartilage

ALAN S. JOSEPHSON* and ROBERT A. GREENWALD

Cartilage is a grossly homogeneous, pauci-cellular tissue, It is the major constituent of the skeletons of lower vertebrates and acts as the embryological precursor of the bony skeleton of higher vertebrates. The articular surfaces of diathrodial joints of vertebrates are composed of hyaline cartilage and the proper function of these joints is dependent on the integrity of the cartilage surfaces (1).

Hyaline cartilage contains collagen fibrils but its unique physical properties depend on a complex and seemingly ordered arrangement of polyanionic glycosaminoglycans. The proteoglycans are formed by the covalent linkage of acid mucopolysaccharides to a polypeptide core (2).

Fleming noted the presence of lysozyme in the joint (3). Meyer and Hahnel (4) found lysozyme in human and guinea pig costal cartilage. Kuettner and co-workers (5–8) have systematically examined cartilage lysozyme and have demonstrated its presence in cartilage formed in tissue culture, effectively eliminating the hypothesis that blood is a source of cartilage lysozyme.

This report details quantitative data on lysozyme content of human cartilage and the conditions of extraction of lysozyme from cartilage.

*A portion of this work was performed under Dr. Josephson's tenure as a Career Scientist of the Health Research Council of the City of New York under Contract I 315. Figures and tables reproduced courtesy of the *Journal of Clinical Investigation.*

Methods

Human costal cartilage was obtained at autopsy and articular cartilage was obtained from orthopedic surgical specimens. The cartilage was scraped free of bone and extraneous soft tissue, minced, frozen, lyophilized, and powdered.

Extraction of lysozyme was undertaken with electrolyte solutions of varying pH or varying concentrations of specific ions. Incubation of the cartilage powder with the buffer was carried out for 24 hr at 37°C with agitation. The suspension was then centrifuged and the powder discarded. All extraction mixtures 0.2 M or greater in concentrations of any electrolyte were dialyzed against 0.15 M sodium chloride, prior to lysozyme determination. Incubation of standard lysozyme solutions under identical conditions ensured that no loss of lysozyme activity occurred under the conditions of extraction. Lysozyme was assayed by an automated technique (9) in which the decrease in optical density caused by a timed incubation of a suspension of *Micrococcus lysodeikticus* with a lysozyme solution was compared with the decrease in optical density achieved by a series of standard solutions of purified human lysozyme (kindly provided by Dr. E. F. Osserman of Columbia University).

Results

Incubation of costal cartilage with universal buffer titrated to various pH values (Fig. 1) indicated that lysozyme was released when the suspending solution was above pH 9.0 and increased up to pH 10.5.

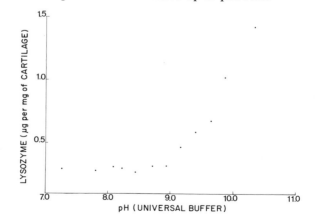

Fig. 1. Extraction of lysozyme from costal cartilage as a function of pH.

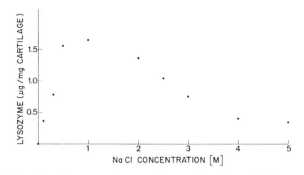

Fig. 2. Extraction of lysozyme from cartilage as a function of Na⁺ concentration.

When increasing concentrations of sodium ion, as sodium chloride, were added to a neutral incubation mixture, lysozyme extraction was also achieved (Fig. 2). Maximum extraction was obtained with 1 *M* sodium chloride solution, and less lysozyme was extracted at higher salt concentrations. Similar results were obtained with potassium. Although concentrations of sodium and potassium chloride above 1.5 *M* failed to extract the maximum lysozyme from cartilage, the same concentrations of these ions had little effect on the enzymatic activity of standard solutions of lysozyme. Lithium chloride had an extraction maximum of 0.75 *M* but higher concentrations of this salt irreversibly decreased the enzymatic activity of lysozyme. Bivalent ions (Fig. 3) such as calcium were effective in lower concentrations than monovalent cations and no decrease in lysozyme extraction was noted at salt concentrations as high as 1.6 *M*. The concentration maximum for extraction (Table I) was 1 *M* for K⁺ and Na⁺. The divalent ions showing a lower and more variable value. Thus Ca^{2+} was more effective than Mg^{2+} while Ba^{2+} achieved maximal extraction at a

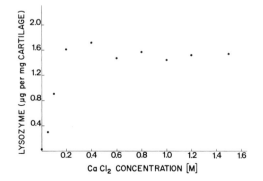

Fig. 3. Extraction of lysozyme from cartilage as a function of Ca^{2+} concentration.

TABLE I

Maximal Extraction of Lysozyme from Human Hyaline Cartilage by Various Cations

Cation	Lowest molarity for maximal extraction	Lysozyme yield (μg/mg)
K^+	1.0	2.13
Na^+	1.0	1.43
Mg^{2+}	0.40	1.57
Ca^{2+}	0.20	1.72
Ba^{2+}	0.12	2.07

lower concentration than either. This same order has been described as the order of affinity of these cations to chondroitin sulfate (2), a constituent of hyaline cartilage.

The content of lysozyme of a variety of cartilage specimens was determined by repeated extraction of a given specimen with 1 M sodium chloride solutions, the resulting extracts being combined prior to lysozyme determination. The average value for five costal cartilage specimens (Table II) was 2.40 μg of lysozyme per mg of dry weight of cartilage; the average for five specimens of articular origin was 1.67 μg/ml.

The lack of significant amounts of lysozyme in the articular fibrocartilage may be related to the high collagen and low glycosamineoglycan content of this tissue (2).

Confirmation that the lytic principal extracted from cartilage was indeed lysozyme was obtained by boiling representative samples at pH 3.7 or pH 10.5 for 90 sec. Eighty percent of the lytic activity was lost after boiling at alkaline pH while boiling at acid pH had no effect. The property of heat stability in acid, but not alkaline media, is a characteristic of animal lysozymes (10). Furthermore, sequential addition of anti-lysozyme antisera to extracts caused sequential loss of lytic activity and the electrophoretic mobility of the extracted lytic principal was comparable to that of purified standard lysozyme.

Hexuronic acid release, a measure of acid mucopolysaccharide release, was determined in 10 incubation supernates from 10 different cartilage specimens under mild extraction conditions. The correlation of lysozyme and hexuronic acid release (Fig. 4) was linear and statistically highly significant ($p = 0.003$).

Serum and synovial fluid lysozyme are elevated in some cases of rheumatoid arthritis (11). It was postulated that the degradation of joint cartilage with the release of lysozyme could be at least partially responsible for this phenomenon. To test if cartilage breakdown can contribute to alterations in serum lysozyme, use was made of a well-studied animal model. Intravenous injection of the proteolytic enzyme papain causes the substantial release of chondroitin sulfate from the ear cartilage of immature rabbits resulting in the reversible loss of rigidity of this appendage (12). We injected papain intravenously into 6 rabbits and measured chondroitin sulfate release by the hexamminecobaltic chloride reaction (13). Serum lysozyme was measured by the automated technique. At 24 hr the serum of the injected rabbits contained large amounts of chondroitin sulfate and also contained increased amounts of lysozyme compared with control serums from animals injected with saline.

Discussion

The extraction studies here presented suggest that lysozyme acts as at least one of the cationic counter ions to the polyanionic glycosamino-glycans of cartilage. Kuettner (8) has obtained evidence by tissue culture

TABLE II

Lysozyme Extraction (μg/mg) from Cartilage

Tissue	Age	Sex	NaCl[a]	Buffer[b]	Total extractable lysozyme
Costal cartilage	25	M	1.43	0.68	2.06
	48	F	0.89	0.40	1.45
	21	M	1.67	1.28	2.52
	22	M	2.53	1.64	3.36
	45	F	1.52	0.44	2.62
Mean values for five costal cartilages			1.61	0.89	2.40
Articular hyaline cartilage					
Hip	65	F	0.69	0.33	1.38
Hip	85	M	1.45	0.65	3.03
Knee	65	M	0.81	0.30	1.31
Knee	43	M	0.21	0.17	0.80
Knee	25	F	0.99	0.43	1.83
Mean values for five hyaline articular cartilages			0.83	0.38	1.67
Articular fibrocartilage					
Knee	34	M	Trace	Trace	—
Knee	30	F	Trace	Trace	—
Knee	25	F	Trace	Trace	—

[a] Results of single incubation in 1.0 M NaCl.
[b] Results of single incubation in universal buffer, pH 10.5.

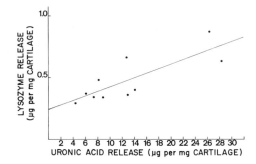

Fig. 4. Correlation of lysozyme extraction and uronic acid release from 10 cartilage specimens.

studies that cartilage lysozyme is not of serum origin. Greenwald *et al.* (14) compared the concentrations of lysozyme and a lysosomal enzyme β-glucuronidase, in joint cartilage and in synovium. No β-glucuronidase activity was found in cartilage. Synovial tissue, on the other hand, had significant levels of β-glucuronidase but no lysozyme. It was concluded that cartilage lysozyme was nonlysosomal. Sokoloff (15) showed that the physical properties of hyaline cartilage are altered by incubation with electrolyte solutions similar to those used in our study.

It is intriguing to speculate that the physical properties of cartilage and perhaps the alterations of these properties in pathologic states result from a change in such cationic constituents of cartilage as lysozyme. In addition, at least a portion of the elevated serum and synovial lysozyme found in rheumatoid arthritis (11) may result from cartilage breakdown.

References

1. Hollander, J. L., "Arthritis and Allied Conditions," 7th ed. Lea & Febiger, Philadelphia, Pennsylvania, 1966.
2. Schubert, M., and Hamerman, D., "A Primer on Connective Tissue Biochemistry." Lea & Febiger, Philadelphia, Pennsylvania, 1968.
3. Fleming, A., *Proc. Roy. Soc., Lond.* [*Biol.*] **93**, 306 (1922).
4. Meyer, K., and Hahnel, E., *J. Biol. Chem.* **163**, 723 (1946).
5. Kuettner, K. E., Guenther, H. L., Ray, R. D., and Schumacher, G. F. B., *Calcif. Tissue Res.* **1**, 298 (1968).
6. Kuettner, K. E., Soble, L. W., Eisenstein, R., and Yaeger, J. A., *Calcif. Tissue Res.* **2**, 93 (1968).
7. Kuettner, K. E., Soble, L. W., Guenther, H. L., Croxen, R. L., and Eisenstein, R., *Calcif. Tissue Res.* **5**, 56 (1970).
8. Eisenstein, R., and Kuettner, K. E., *Calcif. Tissue Res., Suppl.* p. 137 (1970).
9. Josephson, A. S., Greenwald, R. A., and Tsang, A., *Clin. Res.* **19**, 729 (1971).
10. Jollès, P., *Angew. Chem.* [*Engl.*] **3**, 28 (1964).

11. Pruzanski, W., Saito, S., and Ogryzlo, M. A., *Arthritis Rheum.* **13,** 389 (1970).
12. Thomas, L., *J. Exp. Med.* **104,** 245 (1956).
13. Weissman, G., Potter, J. L., McCluskey, R. T., and Schubert, M., *Proc. Soc. Exp. Biol. Med.* **102,** 584 (1959).
14. Greenwald, R. A., Josephson, A. S., Diamond, H. S., and Tsang, A., *J. Clin. Invest* **51,** 2264 (1972).
15. Sokoloff, L., *Science* **141,** 1055 (1963).

Lysozyme Production and Abnormalities in Rheumatic Diseases*

W. PRUZANSKI, M. A. OGRYZLO, and A. KATZ

Lysozyme (muramidase) has been detected in synovial fluid and identified as such by the lysoplate (1) and by the radial immunodiffusion technique, using specific anti-human lysozyme antisera (2–4).

Lysozyme activity was estimated in 235 sera and 258 synovial fluids from patients with various articular diseases. The mean values and standard deviations are shown in part in Tables I and II. In patients with rheumatoid arthritis the level of lysozyme in the serum varied from 4.5 to 48.2 μg/ml, being greater than 15 μg/ml in 34% of the cases. None of these patients had elevated blood urea nitrogen and none exhibited monocytosis. In contrast, serum lysozyme was much less frequently elevated in other articular diseases. Thus among 38 patients with nonrheumatoid forms of arthritis, serum lysozyme was greater than 15 μg/ml in only 11%. However, occasional high values were also encountered in other acute forms of arthritis not included in this report, notably in acute nonspecific synovitis and acute gout.

In rheumatoid synovial fluids, lysozyme activity ranged from 2.2 to 118 μg/ml, being higher than 15 μg/ml in 74% of the patients. Increased

*Supported by Grants-in-aid from the Medical Research Council of Canada and The Canadian Arthritis and Rheumatism Society.

TABLE I

Serum Lysozyme in Articular Diseases (Normal 9.7 ± 1.8 μg/ml)

Diagnosis	No. of estimations	Mean ± S.D. (μg/ml)
Rheumatoid arthritis	158	14.0 ± 5.5
Osteoarthritis	14	12.2 ± 3.3
Meniscectomy	6	8.9 ± 2.3
Post-traumatic	2	8.4 − 8.9
Ankylosing spondylitis	5	9.8 ± 1.3
Psoriatic arthritis	6	11.1 ± 1.8
Reiter's syndrome	5	8.2 ± 2.5

lysozyme activity was also noted in synovial fluids from patients with other forms of inflammatory arthritis, such as acute gout, ankylosing spondylitis, psoriasis, acute synovitis, and septic arthritis. Low lysozyme activity was invariably observed in synovial fluids from patients with osteoarthritis, tears of the semilunar cartilage, and traumatic nonsanguineous joint effusions.

Synovial fluid lysozyme was estimated in two simultaneously aspirated but unequally affected joints in 15 patients with rheumatoid arthritis and in one with psoriatic arthritis. The level of lysozyme was significantly higher in 11 of the more severely affected, paired joints. No such difference was noted in 13 other patients in whom two simultaneously aspirated joints were equally affected.

No correlation could be demonstrated between the activity of lysozyme in the serum and the number of circulating polymorphonuclears and monocytes, either in the patients with arthritis or in healthy individuals. Likewise

TABLE II

Synovial Fluid Lysozyme in Articular Diseases

Diagnosis	No. of estimations	Mean ± S.D. (μg/ml)
Rheumatoid arthritis	152	23.6 ± 13.5
Osteoarthritis	20	8.8 ± 3.5
Meniscectomy	9	6.0 ± 1.6
Post-traumatic	4	0.8 − 11.0
Ankylosing spondylitis	7	16.4 ± 9.7
Psoriatic arthritis	8	14.3 ± 4.5
Reiter's syndrome	6	9.3 ± 2.4

TABLE III

Correlation between RBAF and Lysozyme in Rheumatoid Synovial Fluid

Synovial fluid	No. of patients	Lysozyme (mean) (μg/ml)
RBAF (+)	35	21.2
RBAF (−)	35	16.5
p value	—	0.05

no correlation could be demonstrated between lysozyme concentrations and the number of cells found in the synovial fluids. There was also no correlation between the serum and synovial fluid lysozyme level.

Although no correlation was observed between the activity of lysozyme and the concentration of rheumatoid factor in the synovial fluid, a correlation could be shown (Table III) between lysozyme activity and rheumatoid biologically active factor (RBAF) activity ($p = 0.05$) (5, 6). There was no correlation between lysozyme and the concentration of IgG, IgM, or IgD, but it did correlate with the level of IgA ($p < 0.05$).

At the present time one can only speculate on the source of lysozyme in synovial fluid. Four possible sites of origin might be considered: lysozyme may accumulate in the synovial space as a result of (a) transudation or

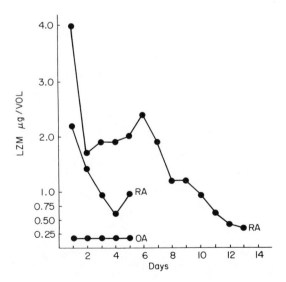

Fig. 1. Lysozyme activity in μg/vol of exchanged culture medium. RA: explants of rheumatoid synovium. OA: explant of synovium from osteoarthritic joint.

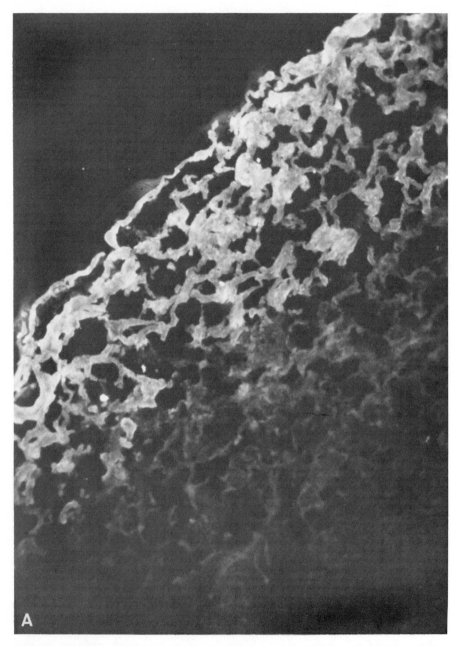

Fig. 2. (A) Rheumatoid synovium showing positive staining of lining cells with fluorescein-conjugated rabbit anti-human lysozyme antiserum. ($\times$ 100, UV light, 3 min exposure.) (B) Higher magnification ($\times$ 160) of (A) showing positive staining of the cytoplasm but not of the nuclei of lining cells.

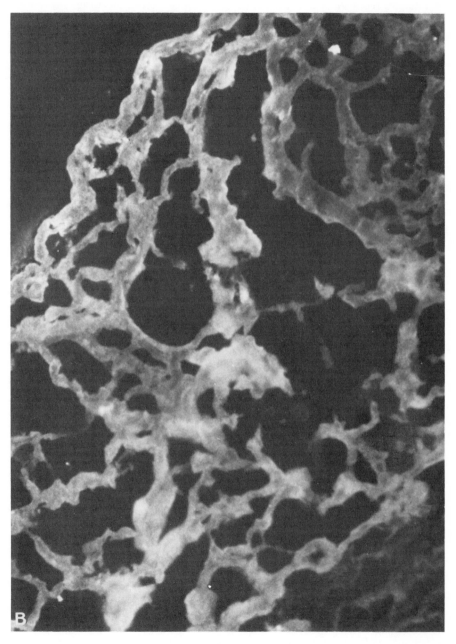

Fig. 2B

exudation from vessels in the synovial lining or actual bleeding, (b) release from polymorphonuclear and mononuclear cells in the synovial fluid and tissues, (c) liberation from damaged cartilage (7) and (d) production and discharge from the synovial lining cells.

In order to elucidate some of these possibilities, synovial tissue obtained from the joints of 6 patients with rheumatoid arthritis, 4 patients with osteoarthritis and 2 with injuries to the semilunar cartilage, were cultured *in vitro* for periods of 3–13 days. The volume of the culture medium was measured and changed daily. Lysozyme activity was calculated in $\mu g/ml$ and also per total volume of the exchanged medium. The explants were prepared in duplicate, and on each successive day one of the explants was examined histologically. The results showed that lysozyme was liberated into the medium from rheumatoid synovium in amounts 5–20 times greater than from synovium of patients with osteoarthritis or tears of the semilunar cartilage (Fig. 1).

Likewise, small explants of synovium grown in culture for different periods of time were incorporated into the lysoplate medium and the zone of lysis calculated. These studies showed that on the "0" day of culture, rheumatoid explants were 4–300 times more active, and on the fifth day of culture, 3–20 times more active in the lysis of *Micrococcus lysodeikticus,* than explants from osteoarthritic joints. Although the rheumatoid synovium was invariably infiltrated to varying degrees by polymorphonuclear and mononuclear cells, the concentration of lysozyme in the medium and the extent of lysis on the lysoplate did not appear to correlate with the extent of infiltration of the synovium. Histologically the explants were found to be viable with minimal degeneration of the synovial lining cells.

In an attempt to demonstrate the presence of lysozyme in various cells, synovium from 4 patients during meniscectomy, 3 patients with osteoarthritis, and 2 patients with rheumatoid arthritis were stained with fluorescein labeled anti-human lysozyme antiserum. By this technique lysozyme was shown to be located in the synovial lining (Fig. 2). Furthermore, the intensity of staining was greater in rheumatoid synovium and appeared to be present in all layers of the proliferating lining cells. Lysozyme was also demonstrated in the cytoplasm of polymorphonuclears and monocytic cells scattered within the synovial tissues and in the synovial fluid.

In order to investigate the possible damaging effect of lysozyme on synovial tissue, 2 synovial explants from osteoarthritic joints and 2 from rheumatoid joints were cultured for 24–48 hr in medium containing 1000 $\mu g/ml$ of purified human lysozyme. Synovial tissue from the same joints grown in media without lysozyme served as controls. After 24–48 hr the synovial explants were studied by electronmicroscopy. Preliminary results suggested that rheumatoid synovium exhibited more cellular degeneration of the lining cells when grown in the presence of lysozyme. These observations are presently being extended.

Summary

Lysozyme activity was found to be high in the majority of rheumatoid synovial fluids and in some rheumatoid sera. Moreover, the activity of synovial fluid lysozyme appeared to correlate with the severity of the articular damage. Lysozyme was also elevated in some synovial fluids in acute gout, septic arthritis, and in other arthritides. Lysozyme was invariably low in synovial fluid from degenerative joint diseases. The level of lysozyme did not correlate with the number of polymorphonuclear or monocytic cells.

Using an immunofluorescent staining technique, lysozyme was found to be located in the synovial lining cells, and was much more evident in rheumatoid arthritis than in osteoarthritis, and in normal synovium. Lysozyme was also detected in the neutrophiles and mononuclear cells within the synovium and in the synovial fluid. In tissue culture, lysozyme was liberated into the medium from explants of synovium for a period of at least 2 weeks. Much more lysozyme was liberated from rheumatoid than from degenerative joint disease synovial explants. These studies indicate that the synovial lining cells may be one of the sources of lysozyme found in synovial fluid. The possible damaging effect of lysozyme on synovial cells is presently being studied.

Acknowledgments

The excellent technical help of Mrs. S. Saito, Miss C. De la Torre, and Miss J. A. Jay is greatly appreciated.

References

1. Osserman, E. F., and Lawlor, D. P., *J. Exp. Med.* **124,** 921 (1966).
2. Pruzanski, W., Saito, S., and Ogryzlo, M. A., *Arthritis Rheum.* **13,** 389 (1970).
3. Pruzanski, W., Russell, M. L., and Ogryzlo, M. A., *Proc. Can. Conf. Res. Rheum. Dis., 1970* pp. 16–19 (1972).
4. Pruzanski, W., and Ogryzlo, M. A., *Semin. Rheumatol.* **1,** 361 (1972).
5. Baumal, R., and Broder, I., *Clin. Exp. Immunol.* **3,** 555 (1968).
6. Russell, M. L., Gordon, D. A., and Broder, I., unpublished data.
7. Greenwald, R. A., Josephson, A. S., Diamond, H. S., and Tsong, A., *J. Clin. Invest.* **51,** 2264 (1972).

Lysozyme in Human Genital Secretions

GEBHARD F. B. SCHUMACHER

General Considerations

Male and female sex organs, including those involved in the production of external secretions, such as the prostate gland, the seminal vesicles, and the Cowper's gland in the male, and the uterine cervix, the endometrium, and the fallopian tubes in the female, represent target tissues for sex hormones. Secretions of the male accessory glands, called the seminal plasma of the ejaculate, provide a vehicle for the transport of spermatozoa. These, once deposited in the posterior vaginal fornix, come in contact with the gel-like secretions of the cervical canal and enter the structural network of the cervical mucus by virtue of their intrinsic motility. However, cyclic changes of the semisolid gel component as well as of the soluble components of the cervical mucus permit sperm penetration into the cervical canal only for a short period of time at midcycle prior to ovulation and shortly thereafter. Midcycle mucus is very favorable for sperm survival. The morphologic configuration of the cervical crypts and clefts contributes to the storage and preservation of spermatozoa and their sustained and prolonged release to the uterine cavity and the oviducts.

Information on the presence of lysozyme in human genital secretions is scattered and incomplete. This may be due to the fact that the collection of the material presents some problems and that only small amounts are available. In addition, the biophysical properties of cervical mucus cause problems in handling and experimentation. Microradial diffusion methods in gel are most suitable for the quantitative assessment of soluble proteins and enzymes in genital secretions, especially cervical mucus (21, 24, 29, 30, 33). Radial diffusion in *Micrococcus lysodeikticus* containing agar gel as described by Osserman and Lawlor (17) and Schumacher and Wied (33) was used for the quantitative assessment of lysozyme in genital secretions.

Lysozyme in Human Seminal Plasma

Antibacterial substances are present in human semen and prostatic fluid. Taylor and Morgan (36) reported in 1952 that one of the two factors with antibacterial activity was lysozyme or a lysozyme-like substance. The other factor resisted heating at 100°C for 30 min and was suspected to be related to spermine (9). Lysozyme in human seminal plasma was demonstrated by Schill and Schumacher (22) and Hirschhäuser and Kionke (10) using the microradial diffusion technique in agar gel (33). These studies indicate a rather low level of lysozyme in human specimens as compared to rhesus monkey seminal plasma (21). Heat stability of the enzyme at pH 4.5 and 100°C for 1–2 min could be demonstrated. Traces of lysozyme can be found in acrosomal extracts (39) from washed human spermatozoa but not from washed rabbit or bull spermatozoa (33a).

The lysozyme content of human seminal plasma shows considerable individual variations, although the lysozyme level does not seem to be related to the count, motility, or morphology of spermatozoa in the ejaculate. Table I shows the results of lysozyme determinations in 87 specimens of "good," "fair," and "poor" quality. Specimens containing leukocytes or bacteria display a greater variation but not a typical pattern that could be related to the presence of bacteria and/or leukocytes.

The source of the lysozyme in genital secretions has not yet been established. Tissues of the seminal vesicles that produce, for instance, the coagulating proteins, trypsin inhibitors, prostaglandins, and fructose do not seem to contain lysozyme according to Hirschhäuser and Kionke (10). Table II shows some analytical data on a human "split-ejaculate" (5). The first portion of an ejaculate usually contains a large number of spermatozoa. The secretions derive from the epididymis and the ampulla of the vas deferens where the sperm cells are stored, from the Cowper's gland, and mainly from the prostatic gland. The last portion stems from the seminal vesicles. The results in Table II indicate clearly that most of the lysozyme derives from the Cowper's gland or from the prostate gland that also produces, for instance, acid phosphatase, plasminogen activator, and citric acid.

TABLE I

Lysozyme in Human Seminal Plasma[a] of Different Morphological Quality

Good	Fair	Poor	Azoospermia	Bacteria present	Bacteria and leukocyte; present
Count: >60 mill/ml	Count: 20–60 mill/ml	Count: <20 mill/ml	—	Good	Good
Motil: >60%	Motil: 40–60%	Motil: <40%	—	Fair	Fair
Morph: >80% oval forms	Morph: 60–80% oval forms	Morph: <60% oval forms	—	Poor	Poor
				Azoospermic	Azoospermic
M = 34.0	M = 51.6	M = 63.5	M = 42.0	M = 53.6	M = 212.6
S.D. = 34.5	S.D. = 70.3	S.D. = 82.6	S.D. = 33.9	S.D. = 72.3	S.D. = 331.4
Range = 4–140	Range = 3–280	Range = 3–300	Range = 10–100	Range = 11–280	Range = 11–800
N = 23	N = 18	N = 22	N = 6	N = 13	N = 5

[a] Microgram per ml egg-white lysozyme equivalent. There is no statistically significant difference between the groups on the 95% confidence level.

TABLE II

Lysozyme and Other Constituents in Human Split Ejaculates

Ejaculate portion	I	II	III
Sperm count (million/ml)	900	75	75
Motility	65%	25%	10%
Coagulation	Negative	+	+
Liquefaction	−	+	+
Lysozyme (μg/ml, egg-white lysozyme equivalent)	46	12	8
Fructose (qualitative)	Negative	±	+++
pH	7.4	8.2	8.4

Male accessory glands are influenced by sex hormones (15). Androgen deficiency leads to a decrease in concentration of certain substances such as fructose (23). The level of these substances can be raised by administration of testosterone. The effect of a progestational agent, Megestrol-acetate (Mead-Johnson Laboratories, Evansville, Indiana), on the composition of semen has been studied recently (32). The steroid was administered orally to 6 male volunteers for 4 weeks increasing the dosage from 0.5 mg per day during the first week to 1.0, 2.0, and 4.0 mg during the subsequent weeks. Ejaculates were obtained before, during, and after the treatment. Figure 1 shows the effect of this steroid on the lysozyme content of the ejaculates. The low dosage of 0.5 and 1.0 mg daily does not seem to have significant effects. After administration of 2.0 mg daily for a week, however, the lysozyme values *decrease* significantly. Further increase of the dosage is followed by an increase of lysozyme in 4 of the 6 cases. Withdrawal of the hormone results in a decrease again in 3 instances and an increase in 2 cases. Any attempt to explain these observations would be pure speculation at this point. However, it should be noted that administration of progestagens as oral contraceptives is followed by a significant *increase* of lysozyme and other soluble proteins in female genital secretions (see next section).

Lysozyme in Human Uterine Secretions

Lysozyme was found in vaginal secretions by Vecchietti in 1948 (37) and in cervical secretions by Pommerenke and Taylor in 1953 (19). Rozanski *et al.* (20) described in 1962 a lysozyme-like substance in human cervical mucus. It was the only antibacterial factor that could be demonstrated. However, the authors used midcycle mucus which is low in soluble proteins and enzymes (26–28).

During the course of microzone-electrophoretic studies on cervical secretions, a fraction migrating toward the cathode at pH 8.6 was observed
in the majority of specimens under investigation by the authors (24, 33).
This cathodic fraction is usually almost invisible after staining with Pon-
ceau-S or Lightgreen; however, it stains very well with Nigrosin (Colab
Laboratories, Inc., Chicago Heights, Illinois). The electrophoretic mobility of this fraction on cellulose acetate membranes was identical with the
cathodic fraction of lacrimal fluid, which is known to be lysozyme. Lytic
activity of the cathodic fraction in cervical secretions against *M. lysodeikticus* could be established by a simple sandwich technique. The cellulose
acetate membranes were cut after electrophoretic separation, one strip was
stained with Nigrosin (0.0025% in 2% acetic acid after precipitating the
proteins on the membrane by 3% trichloroacetic acid); the other strip was
placed on the surface of an agar gel (1% in 1/15 *M* phosphate buffer, pH
7.0) containing a suspension (0.03%) of dried *M. lysodeikticus* (Worthington
Biochemical Corp., Freehold, N.J.) as a specific substrate for lysozyme.
Lytic activity could be demonstrated after a few hours in an area which was
identical with the cathodic fraction of cervical mucus or lacrimal fluid.
Lytic activity on the cathodic side of normal serum electropherograms

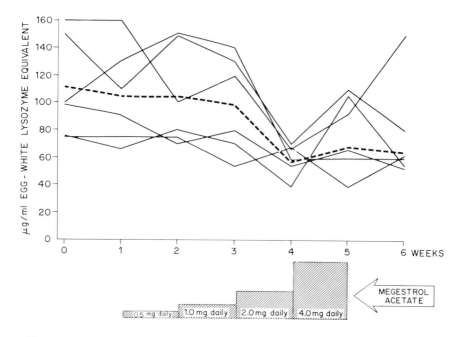

Fig. 1. Lysozyme in human seminal plasma during treatment with increasing doses of the progestational agent Megestrol-acetate. Dotted line indicates mean values.

could also be demonstrated; however, the concentration of lysozyme is very low in serum, and no staining reaction with Nigrosin could be observed.

A characteristic property of lysozyme is its heat stability at low pH. This treatment of native cervical mucus causes precipitation of lysozyme together with the mucus and other proteins (20). Sephadex G-50 gel filtration allows separation of soluble proteins since the mucoid material is trapped in the upper part of the column. Figure 2 shows the results of the microradial diffusion test in *M. lysodeikticus* containing agar gel (24, 33) on the cervical mucus fraction. The cervical mucus lysozyme can be seen to be heat stable at 100°C for more than 2 min under acid conditions (pH 4.2).

Microradial diffusion assay of lysozyme in agar gel (1% in 1/15 *M* phosphate buffer, pH 7.0) containing a suspension of dried *M. lysodeikticus* (0.03%) has proved to be a very useful micromethod for the quantitative assessment of this enzyme in uterine secretions (21, 24, 30, 33). Cervical mucus cannot be measured accurately with volumetric devices because of its peculiar rheologic properties (25). Under standardized conditions, the well in the agar gel can be utilized as a volumetric device (21). The LKB immunodiffusion equipment (LKB Instruments, Inc., Rockville, Maryland) has been used for these studies. A very similar technique was utilized by Osserman and Lawlor (17) for the determination of lysozyme in urine and

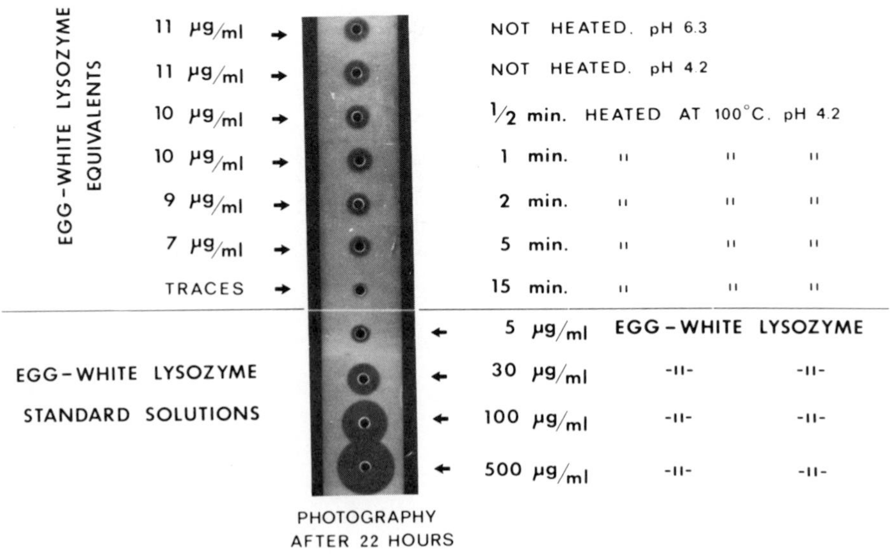

PHOTOGRAPHY
AFTER 22 HOURS

Fig. 2. Heat stability test at pH 4.2 of lysozyme in a fraction of human cervical mucus prepared by Sephadex G-50 gel filtration. The lysozyme activity is heat stable at 100°C for 1–2 min under these conditions.

other body fluids. The diameters of the transparent lysis zones around the wells in turbid agar measured in the NIL Universal Projector (National Instruments Laboratories, Inc., Rockville, Maryland) or other magnifying systems are proportional to the log of concentration of lysozyme. Crystallized egg-white (Worthington Biochemical Corporation, Freehold, New Jersey) or purified human lysozyme can be used for standard reference solutions.

Cervical mucus specimens from 20 normal healthy women during ovulatory cycles have been investigated. Increasing mucorrhea and Spinnbarkeit, positive fern test, low cellularity, and the presence of mobile sperms in cervical secretions are generally considered as results of increasing estrogen activity preceding ovulation. The increase of basal body temperature (BBT) is a characteristic symptom of progesterone activity which is usually the consequence of ovulation. Simultaneous effects are the decrease of Spinnbarkeit, the disappearance of the typical crystallization picture (fern test), the increase of the cell content of the mucus, the decrease of mucorrhea, and the decreased motility of sperm which no longer penetrate the cervical mucus (3).

Figure 3 shows the results of observations during 20 presumably ovulatory cycles (ovulation was induced in four cases by treatment with gonadotrophins). Only the courses of BBT, Spinnbarkeit, and lysozyme are recorded. Since there are remarkable individual differences in the time of ovulation, the curves have been arranged according to a conventional measure, namely, by determining the last day of low basal body temperature before the temperature rise as day 0 (the closest possible estimation of the time of ovulation). The lysozyme values are plotted in a semilogarithmic system because of the large-scale differences before and after ovulation and for the purpose of a better demonstration of changes in the low value range. The rather high post-menstrual values dropped very low during the last few days before the rise of the BBT. Lysozyme levels increased significantly together with the temperature rise. One to four days after ovulation, the lysozyme values increased to levels 10–50 times higher than the preovulatory values. The pre-ovulatory decrease is apparently concomitant with the increase of Spinnbarkeit, whereas the post-ovulatory decrease of Spinnbarkeit is apparently concomitant with the increase in lysozyme secretion. A comparison of the lysozyme concentration with all other parameters characterizing the estrogen and luteal phases indicates that the lowest lysozyme levels coincide with pre-ovulatory estrogen activity. Apparent post-ovulatory progesterone activity is characterized by an abrupt increase of the lysozyme levels in cervical secretions.

To ascertain whether sex steroids exert a regulating effect on the lysozyme level in cervical mucus, several cycles were studied in women under the sequential regimen of hormonal contraceptives. Figure 4 shows lysozyme,

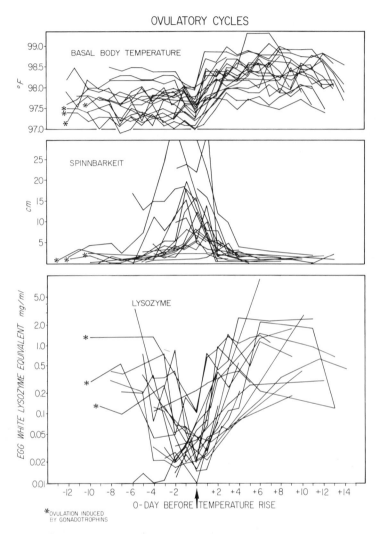

Fig. 3. Lysozyme in human cervical mucus during the menstrual cycle: Synopsis of 20 presumably ovulatory cycles. The curves are synchronized according to the last day of low basal body temperature (BBT). The changes in BBT, Spinnbarkeit of the cervical mucus, and lysozyme are recorded. Ovulation probably occurred in these cases on or shortly after day 0.

Spinnbarkeit, and BBT curves during 7 cycles [5 C-Quens (Eli Lilly and Co., Indianapolis, Indiana), 2 Oracon (Mead-Johnson Laboratories, Evansville, Indiana)]. The lysozyme level is low during estrogen administration and increases significantly under the influence of progestogens. The level of lysozyme during estrogen intake is not as low as the pre-ovulatory values during the ovulatory cycles.

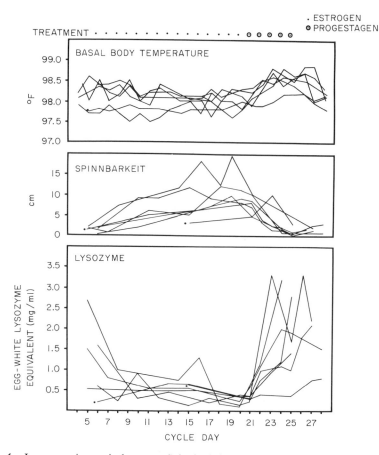

Fig. 4. Lysozyme in cervical mucus, Spinnbarkeit, and BBT during 7 cycles under the sequential regimen of oral contraception ["C-Quens" (E. Lilly), and "Oracon" (Mead-Johnson)]. The days of estrogen intake are indicated by dots, the days of administration of combined estrogen–progestogen pills are marked by dots in a circle on top of the graph. From G. F. B. Schumacher, *Advan. Biosci.* 95–119 (1969); reprinted with permission.

Since estrogens and progestogens appear to influence the concentration of lysozyme in cervical mucus, attempts were made to determine the site of secretion and to assess the distribution of diffusible lysozyme on the inner surface of the internal female genital tract. Frozen sections of the area of the squamo-columnar junction of uteri taken from patients in their reproductive age (hysterectomy) were laid on a slide covered with *M. lysodeikticus,* incubated for 30–60 min in a moist chamber and stained with Alcian Blue and Basic Fuchsin, according to Speece (34). Figure 5(A) shows a low power micrograph of the squamo-columnar junction. A narrow zone of bacteriolysis is visible along the lining of the mucus-producing tissue ele-

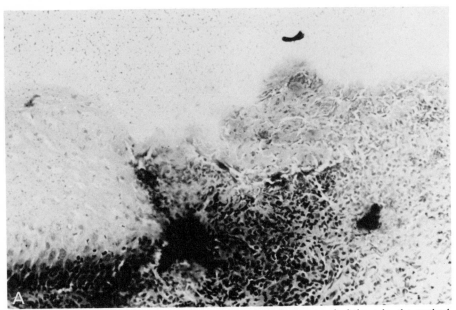

Fig. 5. Histochemical demonstration of lysozyme in human cervical tissue by the method of Speece (34). (A) Squamo-columnar junction (low power); (B–E) sections of different areas (high power).

Fig. 5. (B) Squamous epithelium.

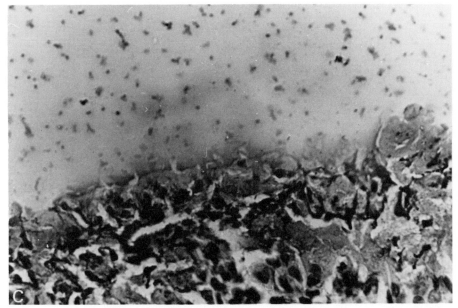

Fig. 5. (C) Columnar epithelium of the endocervix.

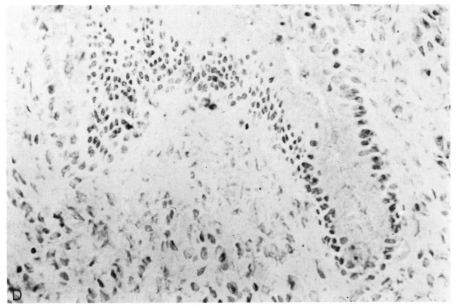

Fig. 5. (D) A glandular crypt of the cervical canal.

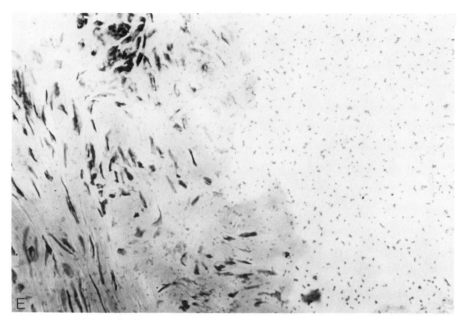

Fig. 5.   (E) Connective tissue from the peripheral areas of the cervix.

ments. This diffusion zone includes the adjacent part of the squamous epithelium. The other areas of the squamous epithelium show intact micrococci, whereas no micrococci are detectable over the columnar epithelial cells, their secretions, or underlying connective tissue. High power magnifications of these areas are shown in Fig. 5(B–E). Figure 5(B) represents the squamous epithelium with intact micrococci; (C) the columnar epithelium of the endocervix with lysed micrococci; (D) a glandular crypt of the cervical canal with lysed micrococci; and (E) connective tissue in the periphery of the cervix with some lysis. These results indicate that the lysozyme derives predominantly from the mucin-producing endocervical tissue components.

It is not known to what extent leucocytes (2) or macrophages (17) contribute to the lysozyme content of cervical secretions in humans. Polymorphonuclear leucocytes have been suspected to cause an increase in lysozyme in the infertile region of uterine lumen in rats and rabbits containing foreign bodies (18). However, it is still not known whether the polymorph invasion or changes in the composition of the uterine secretions induced by an intrauterine device (IUD) are the cause of the toxic effects on blastocysts (1).

The contribution of the endometrium to human uterine secretions, in the absence of foreign bodies, seems to be minor, as shown by the following

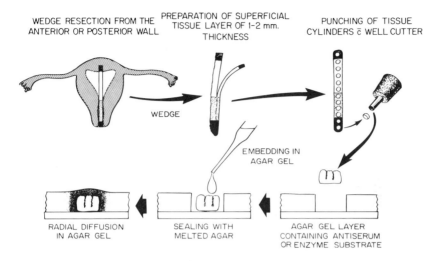

WEDGE RESECTION FROM THE ANTERIOR OR POSTERIOR WALL

PREPARATION OF SUPERFICIAL TISSUE LAYER OF 1-2 mm. THICKNESS

PUNCHING OF TISSUE CYLINDERS c̄ WELL CUTTER

WEDGE

EMBEDDING IN AGAR GEL

RADIAL DIFFUSION IN AGAR GEL

SEALING WITH MELTED AGAR

AGAR GEL LAYER CONTAINING ANTISERUM OR ENZYME SUBSTRATE

Fig. 6. Preparation of tissue cylinders from the endometrium and the cervix for radial gel diffusion. The tissue samples have been obtained in a similar way from the fallopian tubes, after removing the peritoneal covering and cutting the tube longitudinally. However, no attempts were made to separate mucosa and muscle layer. From G. F. B. Schumacher and M. J. Pearl, *Protides Biol. Fluids, Proc. Colloq.* **16**, 525–534 (1969); reprinted with permission.

studies of the distribution pattern of diffusible lysozyme in the tissues of the inner surface of the uterus (29, 30). Figure 6 demonstrates the procedures. Small tissue cylinders (of 5–10 mg wet weight) were taken from the posterior or anterior uterine wall immediately after hysterectomy and implanted in agar gel containing *M. lysodeikticus*. The lysozyme diffuses from the tissue into the agar and the quantity per mg wet weight of tissue can be determined from a standard curve. The patterns of two uteri, one taken on the fifth day of the cycle and the other taken on the 26th day of the cycle after a long-term treatment with Enovid (G. D. Searle and Co., Chicago, Illinois) are demonstrated in Fig. 7(A, B). These graphs show a remarkably high concentration of lysozyme in the cervical canal, particularly the lower part. The endometrium releases little lysozyme and uterine muscle even less. Figure 8 demonstrates the distribution of lysozyme in a fallopian tube (patient J. D. in Fig. 7). The values are in the same range as those of the endometrial tissues. Lysozyme activity seems higher in the ampulla of the tube, but this might be due to differences in the proportionate thickness of the mucoid membranes and muscle layers.

From these studies it may be concluded that lysozyme in cervical secretion derives predominantly from the secretory tissues of the uterine cervix.

Lysozyme most probably plays a major role in defense mechanisms against microbial invaders (7, 8, 11, 12, 35, 38). It is not known, at present,

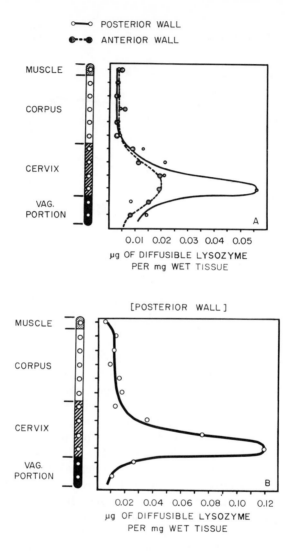

○——○ POSTERIOR WALL
●--● ANTERIOR WALL

μg OF DIFFUSIBLE LYSOZYME
PER mg WET TISSUE

[POSTERIOR WALL]

μg OF DIFFUSIBLE LYSOZYME
PER mg WET TISSUE

Fig. 7. (A) Patient J. P., 32 years, adhesions after 3 cesarean sections. Hysterectomy, 26th day of Enovid cycle. (B) Patient J. D., 37 years, uterus prolapse. Vaginal hysterectomy, 5th day of cycle. Distribution of diffusible lysozyme in the tissue of the inner surface of the uterus. Values for diffusible lysozyme were obtained as egg-white lysozyme equivalents (standard curve) per mg wet weight of tissue. The distribution patterns of two uteri removed from individuals in different hormonal situations are essentially the same. The values are generally lower in the tissues of patient J. P. The low values of the anterior wall of the cervix might be influenced by the scars after three cesarean sections. From G. F. B. Schumacher and M. J. Pearl, *J. Reprod. Med.* **3,** 171 (1969); reprinted with permission.

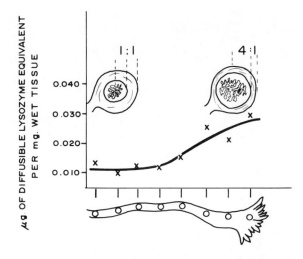

Fig. 8. Distribution pattern of diffusible lysozyme in a fallopian tube (patient J. D. in Fig. 7). The increased values per mg wet weight of tissue in the area of the ampulla might be due to the thicker mucosal layer in this part of the oviduct. From G. F. B. Schumacher and M. J. Pearl, *J. Reprod. Med.* **3,** 171 (1969).

whether the enzyme has functions in the mammalian organism although this would be conceivable, based on its character as a basic protein similar to protamines and histones (13, 14).

The observations presented in this and previous communications indicate that the level of lysozyme in human genital secretions is influenced by sex hormones. The mechanisms of steroid actions on lysozyme production or release are still obscure. Data on the biosynthesis of lysozyme in the target organs for sex steroids are lacking. It is also not known whether sex hormones effect the lysozyme level in other tissues, for instance, calcifying cartilage. This would be of interest with respect to the long-term effects of oral contraceptives.

Effects of Lysozyme on Spermatozoa

High concentrations of lysozyme may have a certain effect on spermatozoa that penetrate into the cervical mucus. The concentration of lysozyme in seminal plasma is relatively low and is only slightly higher than the concentration in pre-ovulatory midcycle cervical mucus. The lysozyme content of cervical secretions is many times higher during the luteal phase and in the early proliferative phase of the cycle. Lysozyme seems to react with surface structures of spermatozoa as indicated by our preliminary studies

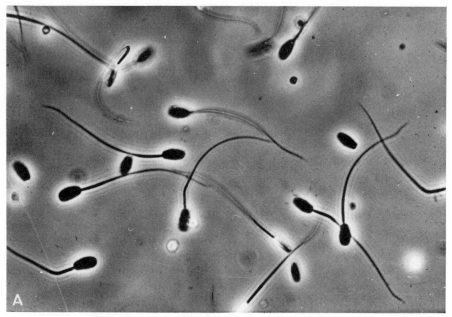

Fig. 9. Agglutination of washed rabbit spermatozoa. (A) Control in 2.5% glycine.

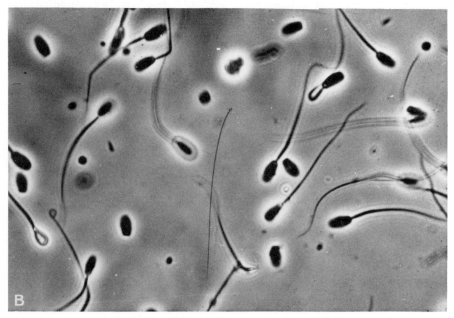

Fig. 9. (B) Control in 1% human albumin.

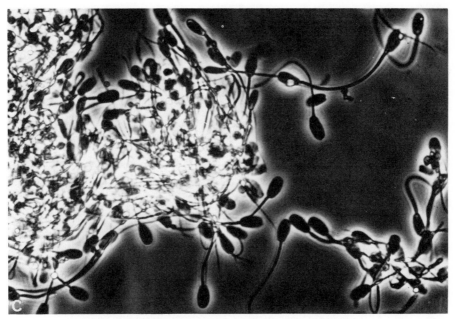

Fig. 9. (C) 1% Egg-white lysozyme (3 times cryst., Worthington).

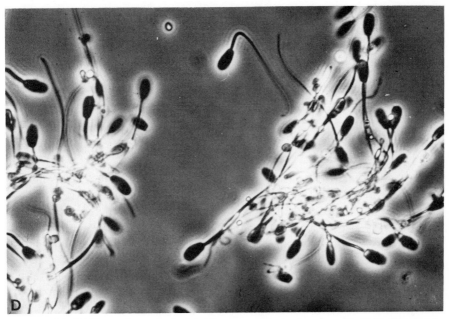

Fig. 9. (D) 1% Human lysozyme isolated from human urine (courtesy Dr. E. F. Osserman).

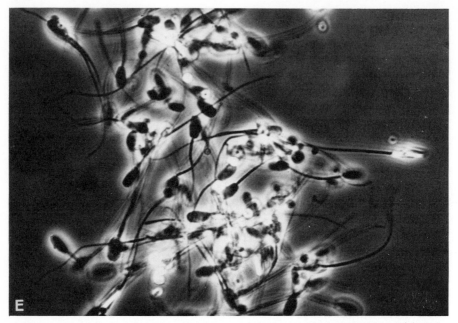

Fig. 9. (E) 1% Potato–trypsin inhibitor (courtesy Dr. V. G. Schwick, Marburg, Germany).

of rabbit and human spermatozoa treated with high concentration of egg-white and human lysozyme. Addition of 1% lysozyme in isotonic saline or glycine to a suspension of washed saline rabbit or human spermatozoa results in agglutination, as shown in Fig. 9(A–E). One percent human albumin does not produce this effect. Addition of lysozyme to spermatozoa suspended in seminal plasma has only a slight or no agglutinating effect. However, this is not a specific lysozyme effect. Other basic proteins or polylysine and polyornithine have similar or even stronger agglutinating effects. The strongest agglutination, even in the presence of seminal plasma, was found with a basic trypsin inhibitor from potatoes. Soybean trypsin inhibitor, a protein with a lower isoelectric point, does not possess agglutinating properties. It is known that polycations interact with cell surface structures and that terminal sialic acid residues are involved in this reaction (4). It is also known that a number of phytagglutinins, such as concanavalin A or influenza and Sendai viruses are capable of agglutinating spermatozoa on the basis of interaction with specific residues on the surface of the spermatozoa (6, 16). Attachment of Sendai virus particles to rabbit spermatozoa results in *in vitro* "capacitation" of the sperm cells, i.e., in obtaining the capacity to fertilize the ovum (6), a process that occurs under *in vivo* conditions in the uterus.

It remains to be determined whether the lysozyme in uterine secretions is of significance for sperm migration, capacitation, fertilization, and implantation of the blastocyst.

Summary

Lysozyme is present in human seminal plasma. The mean value of specimens considered to be of good quality according to morphological criteria was 34 μg/ml (egg-white lysozyme equivalent) with considerable individual variation. Mean values of ejaculates of lower quality, including pathological specimens, were higher, but the differences were statistically not significant. Preliminary observations indicate that the major portion of the lysozyme in seminal plasma probably derives from the prostate gland. Administration of a progestagen, Megestrol-acetate, results in a significant decrease of lysozyme in seminal plasma.

The level of lysozyme in the secretions of the human uterine cervix is strongly influenced by sex hormones. Lysozyme decreases during the estrogenic pre-ovulatory phase and shows a pronounced minimum at midcycle. Thereafter, the concentration increases rapidly during the luteal phase. Analogous changes can be observed after sequential administration of estrogens and progestagens. Lysozyme derives apparently from the mucus-producing glandular elements of the endocervix. Tissue samples from the lower part of the endocervix release 10 times more lysozyme than those from the endometrium of the corpus uteri. It is not known whether or not lysozyme is of biological significance for reproductive processes, although lysozyme appears to interact with the surface of spermatozoa similar to other basic proteins, as shown by agglutination.

Note Added in Proof

Observations by C. Hirschhäuser and R. Eliasson [*Life Sciences* **11,** 149 (1972)] indicating that lysozyme may be involved in the lysis of the seminal clot could *not* be confirmed in this laboratory using purified human lysozyme (Dr. E. Osserman) or crystallized egg-white lysozyme [P. F. Tauber, D. Propping, L. J. D. Zaneveld, and G. F. B. Schumacher, *Biol. Reprod.* **9,** 62 (1973) and unpublished results].

Acknowledgments

I wish to thank Dr. A. J. Wallingford, Jr., Albany Medical College, for preparing the microscopic slides, and Dr. A. P. Amarose, The University of Chicago, for the microphotography. The skillful technical assistance of Miss M. Feifel, Mrs. Jo Borg, Mrs. Margaret Crawford, Miss Elizabeth Hedlund, and Mr. B. Dragoje is gratefully acknowledged. These studies have been supported by USPHS Grants HD 2682 and HD 03696 and Ford Foundation Grant 690-0108 for research and training in reproductive biology.

References

1. Breed, W. G., Peplow, P. V., Eckstein, P., and Barker, S. A., *J. Endocrinol.* **52,** 575 (1972).
2. Briggs, R. S., Perillie, P. E., and Finch, S. C., *J. Histochem. Cytochem.* **14,** 167 (1966).
3. Carlborg, L., Johansson, E. D. B., and Gemzell, C., *Acta Endocrinol. (Kbh.)* **62,** 721 (1969).
4. Danon, D., Howe, C., and Lee, L. T., *Biochim. Biophys. Acta* **101,** 201 (1965).
5. Eliasson, R., and Lindholmer, C. A. R., *Fertil. Steril.* **23,** 252 (1972).
6. Ericsson, R. J., Buthala, D. A., and Norland, J. F., *Science* **173,** 54 (1971).
7. Glynn, A. A., and Milne, C. M., *Immunology* **12,** 639 (1967).
8. Grossgebauer, K., and Langmaack, H., *Klin. Wochenschr.* **46,** 1121 (1968).
9. Gurevitch, J., Rozansky, R., Weber, D., Brzezinsky, A., and Eckerling, B., *Am. J. Clin. Pathol.* **4,** 360 (1951).
10. Hirschhäuser, C., and Kionke, M., *Life Sci.* **10,** 333 (1971).
11. Jollès, P., *Angew. Chem. [Engl.]* **3,** 28 (1964).
12. Jollès, P., Charlemagne, D., Petit, J.-F., Marie, A.-C., and Jollès, J., *Bull. Soc. Chim. Biol. (Paris)* **47,** 2241 (1965).
13. Kuettner, K. E., Guenther, H. L., Ray, R. D., and Schumacher, G. F. B., *Calcif. Tissue Res.* **1,** 298 (1968).
14. Kuettner, K. E., Soble, L. W., Eisenstein, R., and Yaeger, J. A., *Calcif. Tissue Res.* **2,** 93 (1968).
15. Mann, T., *Ciba Found. Colloq. Endocrinol.* **16,** 233 (1967).
16. Nicolson, G. L., and Yanagimachi, R., *Science* **177,** 276 (1972).
17. Osserman, E. F., and Lawlor, D. P., *J. Exp. Med.* **124,** 921 (1966).
18. Parr, E. L., Schaedler, R. W., and Hirsch, J. G., *J. Exp. Med.* **126,** 523 (1967).
19. Pommerenke, W. T., and Taylor, P. W., Jr., *Ann. Ostet. Ginecol.* **75,** 891 (1953).
20. Rozanski, R., Persky, S., and Bercovici, B., *Proc. Soc. Exp. Biol. Med.* **110,** 876 (1962).
21. Schill, W.-B., and Schumacher, G. F. B., *Anal. Biochem.* **46,** 502 (1972).
22. Schill, W.-B., and Schumacher, G. F. B., *in* "The Biology of the Cervix" (R. J. Blandau and K. S. Moghissi, eds.), pp. 173–200. Univ. of Chicago Press, Chicago, Illinois, 1973.
23. Schirren, C., "Praktische Andrologie." Hartmann, Berlin, 1971.
24. Schumacher, G. F. B., *J. Reprod. Med.* **1,** 61 (1968).
25. Schumacher, G. F. B., *Adv. Biosci.* **4,** 95–119 (1969).
26. Schumacher, G. F. B., *Fertil. Steril.* **21,** 697 (1970).
27. Schumacher, G. F. B., *in* "Cervical Mucus in Human Reproduction" pp. 93–113. Scriptor, Copenhagen, 1973.
28. Schumacher, G. F. B., *in* "The Biology of the Cervix" (R. J. Blandau and K. S. Moghissi, eds.), pp. 201–231. Univ. of Chicago Press, Chicago, Illinois, 1973.
29. Schumacher, G. F. B., and Pearl, M. J., *Protides Biol. Fluids, Proc. Colloq.* **16,** 525–534 (1969).

30. Schumacher, G. F. B., and Pearl, M. J., *J. Reprod. Med.* **3,** 171 (1969).
31. Schumacher, G. F. B., Propping, D., Tauber, P., and Zaneveld, L. J. D., in preparation.
32. Schumacher, G. F. B., Schill, W.-B., Holschuh, L., and Newton, R., in preparation.
33. Schumacher, G. F. B., and Wied, G. L., *Proc. World Congr. Fertil. Steril., 5th, 1966* Int. Congr. Ser. No. 133, pp. 713–722. Excerpta Medica Foundation, Amsterdam (1967).
33a. Schumacher, G. F. B., and Zaneveld, L. J. D., unpublished data.
34. Speece, A. J., *J. Histochem. Cytochem.* **12,** 384 (1964).
35. Strominger, J. L., and Ghuysen, J. M., *Science* **156,** 213 (1967).
36. Taylor, P. W., and Morgan, H. R., *Surg. Gynecol. Obstet.* **94,** 662 (1952).
37. Vecchietti, G., *Quad. Clin. Ostet. Ginecol.* **3,** 73 (1948).
38. Wardlaw, A. C., *in* "Bacterial Endotoxins" (M. Landy and W. Braun, eds.), Rutgers Univ. Press, New Brunswick, New Jersey, 1964.
39. Zaneveld, L. J. D., Dragoje, B. M., and Schumacher, G. F. B., *Science* **177,** 702 (1972).

Lysozyme in Human Colostrum and Breast Milk*

NEWTON E. HYSLOP, JR., KATHRYN C. KERN, and
W. ALLAN WALKER

Introduction

Our own work with lysozyme originated from an interest in the mechanism of differentiation of the adult human breast into a secretory organ and the physiological regulation of its secretion. While both differentiation and secretion are controlled by hormones and although many of the consequences of their actions are well described, the identity of all the hormones involved and their modes of interaction in these two processes are as yet imperfectly understood. The reader interested in current hypotheses may refer to several recent reviews on the topic of hormonal controls over differentiation and secretion of the breast (1–3). A few general remarks on the consequences of their actions will serve to introduce the topic before we describe the results of our studies on lysozyme in human breast milk and colostrum.

*From the Departments of Medicine and Pediatrics, Harvard Medical School, and the Medical Service and Children's Service, Massachusetts General Hospital, Boston, Massachusetts. Newton Hyslop is an Investigator of the Howard Hughes Medical Institute and W. Allan Walker a Fellow of the Medical Foundation of Boston, Massachusetts.

The pattern of hormonal secretions affecting the breast is quite unlike the normally cyclical hormonal influences acting upon the nonpregnant female genital tract which results in cyclical alterations in local lysozyme concentrations as described by Schumacher (Chapter 39, this volume). The adult human breast responds to the sustained, "orchestrated hormonal symphony" accompanying pregnancy by differentiating from a fatty, non-secreting organ into a highly epithelialized structure capable of voluminous secretion in response to the appropriate biological signal.

With the completion of cellular proliferation during the last trimester of pregnancy, the prepartum breast begins to secrete a viscous, proteinaceous fluid called variously "biestings," "beestings," or "green milk" in medieval English. This thick milk is now more commonly referred to by its Latin name of colostrum.

The human female secretes 5–20 cm³ of colostrum per day before parturition. In the first few hours postpartum, colostrum continues to be the major product. However, the major hormonal changes accompanying parturition cause a pronounced increase in milk volume as well as changes in composition. During the first 72 hr postpartum the abrupt hormonal alterations affect the gross appearance and content of the breast's secretion, and the product is called "transitional milk." As responses to suckling establish a fixed pattern, the daily volume increases from 50- to 200-fold, and the stable product is referred to as "mature milk."

To recapitulate, there are three different products secreted by the breast; the prepartum secretion, small in volume, high in protein, called colostrum; the immediate postpartum secretion, a more watery product secreted in response to the major hormonal changes accompanying the loss of the placenta and other alterations associated with delivery, called transitional milk; and the product of established lactation, referred to as mature milk. Each of these secretions has a characteristic content of lysozyme and of other proteins, as we shall describe in this paper.

Our original goal was to examine the mechanism by which immuno-globulins gain access to the breast secretions. In order to interpret the changes in concentration of these proteins in the secretions, it was necessary to have standards of comparison, such as proteins which would be synthesized chiefly in the breast and proteins appearing in the milk but presumably derived from the serum. As examples of proteins of presumed serum origin we chose albumin and α_1-antitrypsin, both synthesized only in the liver. As examples of locally synthesized proteins, we chose α-lact-albumin and lysozyme. The latter, a minor constituent of milk, we assumed was made in the breast on the basis of earlier studies of Adinolfi and Glynn (4) who found lysozyme concentration in breast milk severalfold times higher than in serum. We were also intrigued by their suggestion that lysozyme interacted with complement and locally synthesized IgA anti-bodies to prepare a local defense against bacteria. Our choice of lysozyme and α-lactalbumin as markers of proteins both synthesized and secreted by the breast was not influenced at the time by knowledge of the remarkable structural homology between bovine α-lactalbumin and egg-white lysozyme. We were then unaware of those findings and also of the less striking ho-mology between human lysozyme and human α-lactalbumin of approxi-mately 28%, as reported by Hill, Steinman, and Brew (Chapter 4, this volume).

The following results are in the nature of a progress report on our work to date.

Method of Study

From our earlier experience in studying immunoglobulin concentrations in saliva (5), it was evident that sequential studies of individual donors would give more significant information than single samples from large numbers of women representing different stages of lactation. Consequently, the data presented are drawn from analyses of multiple samples of colos-trum and milk collected *seriatim* over a span of time ranging from 51 days prepartum to 245 days postpartum. The samples were donated by 5 normal women who had uncomplicated pregnancies and deliveries and who breast-fed their normal babies.

Both colostrum and milk were manually expressed by the donors and frozen until analysis. Total protein was determined on defatted colostrum and breast milk samples by the Lowry method. The Laurell–Clayman electroimmunodiffusion technique, as modified by us (6), was used to mea-sure albumin, α_1-antitrypsin, α-lactalbumin, IgA, and lysozyme. As little as 1 μg of protein per ml could be measured by this method. Antisera were prepared in rabbits to purified human lysozyme (a gift of Dr. Elliott Osser-man), secretory IgA (6) and α-lactalbumin (7). Commercial rabbit anti-sera (Behring Diagnostics) were used to measure albumin and α_1-anti-trypsin.

Results

The complexity of the changes in the protein secretions of the human breast under hormonal influence are indicated in Fig. 1 which is an agarose gel pattern obtained when *seriatim* samples of colostrum and milk are electrophoresed. Samples from 2 donors over a time span of 20 weeks are shown with their serums for comparison.

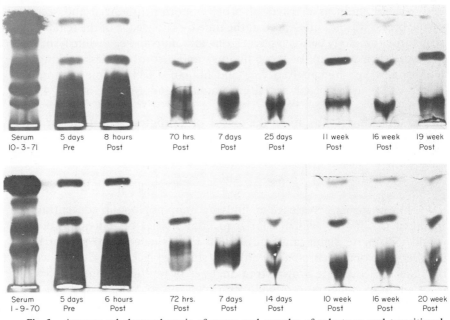

| Serum | 5 days | 8 hours | 70 hrs. | 7 days | 25 days | 11 week | 16 week | 19 week |
| 10-3-71 | Pre | Post | Post | Post | Post | Post | Post | Post |

| Serum | 5 days | 6 hours | 72 hrs. | 7 days | 14 days | 10 week | 16 week | 20 week |
| 1-9-70 | Pre | Post | Post | Post | Post | Post | Post | Post |

Fig. 1. Agarose gel electrophoresis of serum and samples of colostrum and transitional, early mature and late mature milk obtained *seriatim* from two donors. Electrophoresis was carried out in 1% agarose at pH (8.6) and ionic strength (0.075) in barbital buffer with calcium lactate. Proteins were stained with Amidoschwarz.

    The main bands visible in the prepartum and postpartum colostrums are the diffuse IgA band appearing near the origin, the intermediate α-lactalbumin band, and the anodal albumin band. Lysozyme is in such low concentration in both colostrum and milk that it cannot be seen even when sufficient cathodal space is provided on the gel for its banding position.

    During the first three days after delivery, the IgA band is replaced in transitional milk by a prominent casein band. At this time, the albumin band decreases, but with prolonged lactation it again becomes prominent. The early mature milk is, therefore, characterized by 2 significant bands representing locally produced proteins, casein, and α-lactalbumin. The late mature milk, seen with lactations lasting 2 months or longer, is characterized by the return of an easily visible albumin band.

    The multiple bands seen in the casein areas reflect variations in degree of phosphorylation of β-casein. The distinction between the alpha, beta, kappa, and other constituents of human casein can only be appreciated on starch-urea gel electrophoretic patterns.

The qualitative and quantitative changes seen on the stained electrophoretic patterns are more readily appreciated when the concentrations of specific components are examined as a function of time. One fact that cannot be appreciated from the electrophoretic patterns is that the major hormonal alterations accompanying delivery lead to marked changes in the water content of the breast's secretion. The change in total protein concentration, illustrated in Fig. 2, reflects the extent of dilution associated with the shift from production of colostrum to secretion of milk. As a result, measurements of the concentration of individual proteins in milk take on more significance when the results are expressed as a percentage of the total protein in the sample. For this reason, both absolute and relative concentrations of lysozyme as a function of time are shown in Figs. 3 and 4. The dilution of lysozyme content when colostrum is converted to milk is clearly

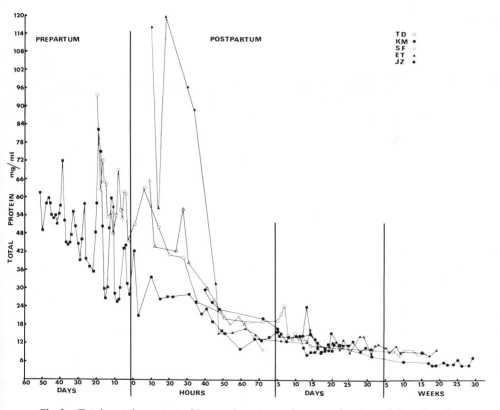

Fig. 2. Total protein content of human breast secretions as a function of time. Samples obtained *seriatim* from five donors. Note changes in time scale. Period encompasses prepartum and postpartum secretions.

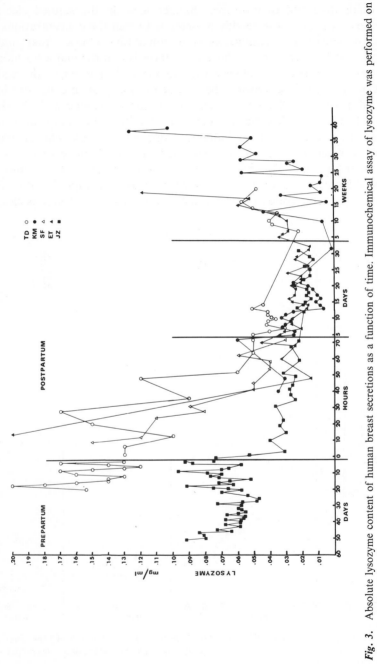

Fig. 3. Absolute lysozyme content of human breast secretions as a function of time. Immunochemical assay of lysozyme was performed on samples shown in Fig. 2.

evident in Fig. 3. The vertical lines divide the time scale into prepartum, immediate postpartum (first 72 hr), early mature milk (first 35 days), and late mature milk periods. Late mature milk was characterized by a rise in absolute lysozyme concentrations as well as by a rise in absolute albumin concentrations.

Figure 4 shows that the lysozyme proportion of total protein in colostrum, transitional milk, and early mature milk was remarkably constant for any given individual. However, late mature milk showed a substantial increase in the average lysozyme contribution to total protein although there were wide swings in content from sample to sample during the late period.

We made similar studies of the changes with time of both the absolute and relative contents of IgA, albumin, α_1-antitrypsin, and α-lactalbumin (8), and found that the absolute concentrations of all proteins fall with the changeover from colostrum to secretion of transitional milk and continue to fall with the coming of early mature milk. However, when concentrations were expressed as percentage of the total protein, it became apparent that all proteins did not behave identically. For example, for donor J. Z. the proportion of the total protein contributed by IgA fell from a massive 36% of colostrum protein to a relatively insignificant 5.2% of milk protein and never rose again. IgA was unique, however, among the proteins studied. The others followed the pattern of lysozyme and contributed proportionally more to the total protein content of late mature milk than to either transitional or early mature milk. These patterns were true irrespective of whether the proteins were originally of serum or breast origin. Albumin, although representing only 2% of total protein in colostrum, increased to as much as 4% of the protein in late milk, nearly equivalent to the IgA contribution. Similarly, α-lactalbumin, despite wide swings in content in both early and late milk, showed average values in late mature milk half again above its 10% contribution to the protein content of colostrum.

We next investigated whether any of the proteins we were examining were handled in similar ways by the breast. For example, were proteins like albumin and α_1-antitrypsin, both presumably transported from the serum, treated in the same way by the breast? Conversely, did proteins like lysozyme and α-lactalbumin, both presumably synthesized and exported by the breast, behave similarly? To answer these questions, we examined the relationships between the concentrations of specific proteins in individual samples. Plots of absolute concentrations, such as those shown in Figs. 5–7, were made for different pairs of proteins and the variance of a linear regression curve calculated by the least squares method. The range of concentrations exhibited by the *seriatim* samples dictated that high concentrations be plotted on one scale and low concentrations on another to minimize crowding of points. A natural division of the ranges of concentra-

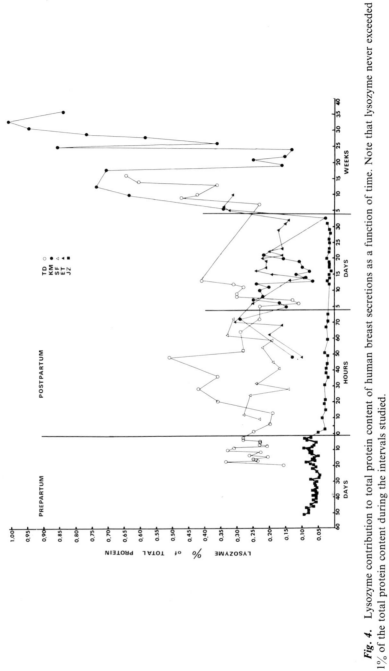

Fig. 4. Lysozyme contribution to total protein content of human breast secretions as a function of time. Note that lysozyme never exceeded 1% of the total protein content during the intervals studied.

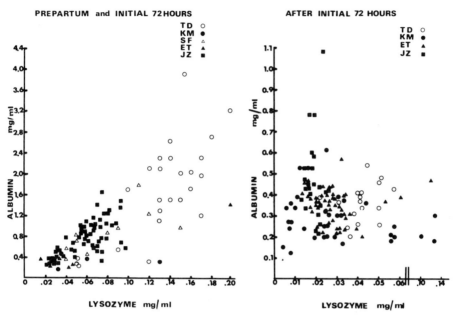

Fig. 5. Variance analysis of relation between changes in concentrations of albumin and lysozyme in human breast secretions from 5 donors. Lysozyme data shown in Fig. 2 are plotted against results of immunochemical assays of albumin in the same samples. Data are divided into 2 time segments and scale in right-hand side expanded to prevent crowding of values from period of mature milk production.

tions into two groupings occurred at 72 hr postpartum, coincident with the onset of production of early mature milk.

Direct comparison of concentrations from the prepartum period and first 72 hr was markedly affected by the dilution of all proteins during the changeover from colostrum to transitional milk. Therefore, variance analysis tended to be misleading when prepartum and first 72 hr data were analyzed because the concentrations of all proteins were markedly affected by dilution. Thus, there was excellent correlation between coincidental changes in absolute concentrations of lysozyme and albumin during the changeover from colostrum to milk, as shown on the left-hand side of Fig. 5. Only when the dilution factor was removed by the relatively steady state of protein concentration in mature milk was it possible to detect a substantial unrelatedness between the concentrations of albumin and lysozyme, as shown on the right-hand of Fig. 5.

Independent handling of lysozyme and proteins of serum origin was also suggested by the relationships between α-antitrypsin and lysozyme. Figure 6 shows the relationships between the absolute concentrations of lysozyme and albumin and lysozyme and α_1-antitrypsin measured in 41 samples from one donor and spanning late pregnancy to late postpartum. The left-hand

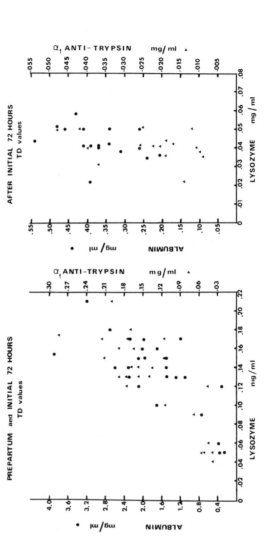

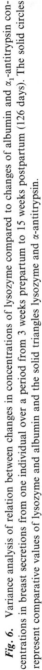

Fig. 6. Variance analysis of relation between changes in concentrations of lysozyme compared to changes of albumin and α_1-antitrypsin concentrations in breast secretions from one individual over a period from 3 weeks prepartum to 15 weeks postpartum (126 days). The solid circles represent comparative values of lysozyme and albumin and the solid triangles lysozyme and α-antitrypsin.

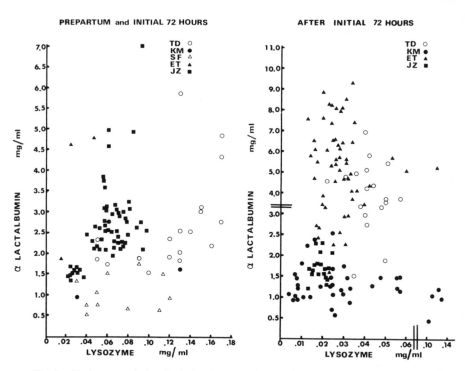

Fig. 7. Variance analysis of relation between changes in concentrations of α-lactalbumin and lysozyme in human breast secretions from five donors. Immunochemical assays of α-lactalbumin content are plotted against lysozyme concentrations in the same samples. Note change in scales.

graph again reflects the positive correlating effect of dilution, but it is evident that during production of early and late milk that the factors regulating breast secretion affected lysozyme, albumin and α_1-antitrypsin independently.

Alpha-lactalbumin was the only protein of those examined whose distinctive regulation during the prepartum and initial 72 hr periods was not obscured by the dilution factor. As shown in Fig. 7, there was a striking lack of proportionality between the concentrations of α-lactalbumin and lysozyme both before and after delivery. Correction of concentrations for percentage of total protein only magnified the lack of correlation.

The differences in the concentrations of lysozyme and α-lactalbumin are noteworthy. Alpha-lactalbumin, a major constituent of both colostrum and mature milk, comprised roughly 10% of the total protein of colostrum and approximately 15% of mature milk. Comparable figures for lysozyme were, respectively, 0.02% and 0.15%, a factor of nearly 1000-fold.

The increase in the volume of secretion accompanying the transition from colostrum to milk production is commonly of the order of a 200-fold

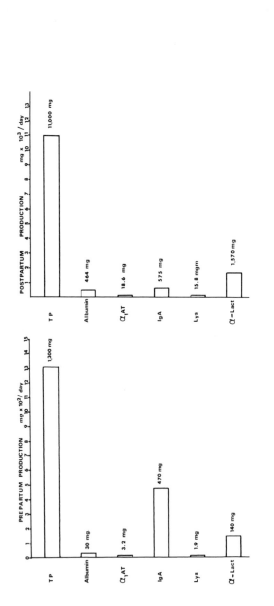

Fig. 8. Export of proteins into breast secretions: comparison of 24 hr production by one donor (both breasts) of colostral proteins with later production of mature milk proteins. Note difference in scale between prepartum and postpartum weight values (mg $\times$ 10^2/day vs. mg $\times$ 10^3/day). The amounts represented by the bars are given in numbers for each protein measured.

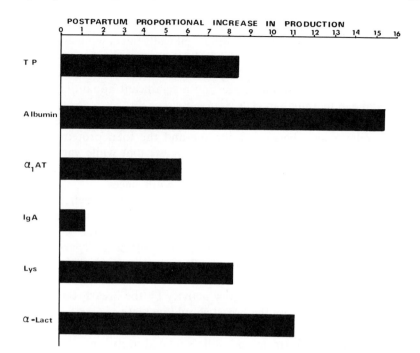

POSTPARTUM PROPORTIONAL INCREASE IN PRODUCTION

Fig. 9. Export of proteins into breast secretions: proportional changes in daily production accompanying transition from secretion of colostrum to secretion of mature milk. Donor J. Z. The data presented in Fig. 8 have been used to calculate the proportional increases in specific proteins.

increment in 24 hr volumes. Thus, when total production of proteins is studied one gains additional insight into the actual changes in amounts of proteins exported by the breast into the secretions. Total production studies for several proteins in donor J. Z. are shown in Figs. 8 and 9. These data were obtained from analyses of the entire output of colostrum (20 cm³) on day 51 prepartum and of the entire output of mature milk (928 cm³) on day 25 postpartum. The proportional differences between the prepartum and postpartum production values are shown in Fig. 9. While the absolute concentrations of all proteins fell, the 50-fold increase in total volume meant major increases in the elaboration of all proteins measured, with the exception of IgA. The increase in total lysozyme production was approximately 8-fold.

Discussion

The effects of pregnancy and lactation on the production of lysozyme by the human breast are clearly documented in the present studies, but the exact biological significance of these findings is unknown.

Several investigators have alluded to the possibility that the true substrate of lysozyme may not be bacterial cell walls but some mammalian cell glycoprotein or polysaccharide. The finding of lysozyme in cartilage (Kuettner, Eisenstein, and Sorgente, Chapter 36, this volume), far removed from bacteria, is but one example suggesting a significant nonantibacterial role for lysozyme in mammalian cells.

In line with these speculations, it may be significant that there is a rather constant relationship between lysozyme and the total protein content of colostrum and milk. This observation suggests that while various factors associated with synthesis and transport alter the composition of milk proteins, lysozyme remains independent of these influences.

By analogy with clinical studies of lysozyme in monocytic and monomyelocytic leukemia, it is possible to consider the appearance of lysozyme in milk as a spillover product from the epithelial cells of the breast. By this hypothesis, the relative constancy of the proportion of lysozyme to the total protein would be explained as reflecting the total activity of the secreting cell mass. Since the increase in total milk secretion following parturition requires enhanced secretory activity by the breast, the amount of lysozyme secreted daily by the breast should increase.

On the basis of present evidence it seems improbable that lysozyme's sole function in human colostrum and milk is related to its limited antibacterial activity, but other possible functions remain to be defined.

Acknowledgments

We are indebted to the LaLeche League of Boston for the opportunity to do this study and especially to the five dedicated volunteers who donated serial samples: T. D., S. F., K. M., E. T., and J. Z.

We acknowledge with pleasure the material support and encouragement given us by Kurt J. Isselbacher, M.D., and Morton N. Swartz, M.D. of the Massachusetts General Hospital.

This work was supported in part by a grant from the National Institutes of Health, AI-08070.

References

1. Topper, Y. J., *Recent Progr. Horm. Res.* **26**, 287 (1970).
2. Brew, K., *Essays Biochem.* **6**, 93 (1970).
3. Linzell, J. L., and Peaker, M., *Physiol. Rev.* **51**, 546 (1971).
4. Adinolfi, M., Glynn, A. A., Lindsay, M., and Milne, C. M., *Immunology* **10**, 517 (1965).
5. Hyslop, N. E., Jr., Lopez, M., and Fellarca, A., *Clin. Res.* **18**, 426 (1970).
6. Lopez, M., Tsu, T., and Hyslop, N. E., Jr., *Immunochemistry* **6**, 513 (1969).
7. Walker, W. A., Kern, K., and Hyslop, N. E., Jr., in preparation.
8. Hyslop, N. E., Jr., Kern, K., and Walker, W. A., unpublished data.

Agglutination of Rat Liver Mitochondria by Lysozyme

MATTEO ADINOLFI, JOHN N. LOEB, and
ELLIOTT F. OSSERMAN

As described in this and the following chapter (1), our laboratory has recently been investigating the effects of lysozyme (LZM) on mammalian cells and their constituent membranes and organelles. We were initially led to consider the possibility that LZM might specifically interact with mitochondria by previous observations on the kidneys of rats bearing a transplantable chloroleukemia which elaborates LZM (2). As shown in the electron micrographs in Fig. 1, the kidneys of these animals exhibit prominent droplet formation in the proximal tubular epithelial cells. These droplets have the appearance of complex phago-lysosomes containing osmiophilic, electron-dense aggregates which apparently represent the residue of degenerated lipid membranes. In addition many of these droplets contain structures with the appearance of degenerated mitochondria as suggested by residual cristae. It was further noted in these studies that mitochondria which were in immediate apposition to these droplets showed thinning of their outer membranes and in some instances apparent discontinuities. The present studies on isolated mitochondria were carried out in order to investigate the possibility that these morphologic changes might be due to a specific effect of LZM on mitochondrial membranes.

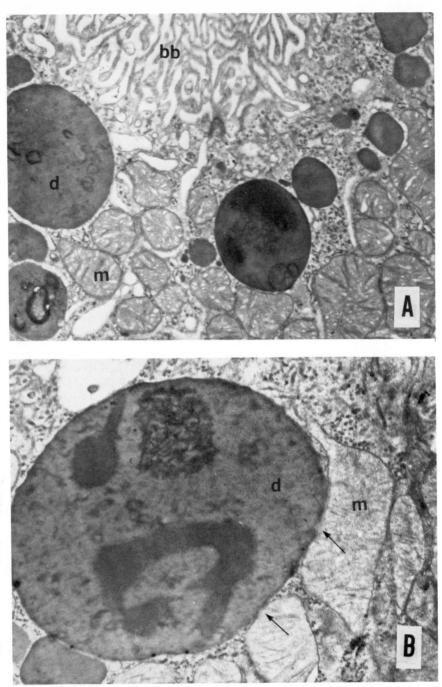

Fig. 1. Electron micrographs of the proximal tubular epithelial cells of a Wistar/Furth rat bearing the Shay chloroleukemia. Extensive droplet formation is evident. bb, brush border; d, droplets; m, mitochondria; arrows indicate points of discontinuity of mitochondrial membrane adjacent to a droplet. (A) × 5000; (B) × 10,000.

Materials and Methods

Mitochondria were isolated from the livers of adult Wistar/Furth rats by the method of Weinbach (3) with the exception that livers were directly homogenized in a loosely fitting hand-drive glass homogenizer in a medium (STE) containing 0.25 M sucrose, 1 mM Tris-HCl (pH 7.4), and 1 mM EDTA. Mitochondria which had been washed five times were resuspended in STE to form a stock suspension containing 2 to 4 mg of mitochondrial protein per ml; in the experiments which follow, the final concentration of mitochondrial protein in the various incubation mixtures was between 0.4 and 0.8 mg per ml. All steps prior to the actual incubations (see below) were carried out at 0.4°C.

Human LZM was isolated from the urine of patients with monocytic or monomyelocytic leukemia by the previously described bentonite method (4). Rat LZM was isolated by the same method from the urine of Wistar/ Furth rats bearing the Shay chloroleukemia. Hen egg-white LZM (LYSF 645, 2 × crystallized) and bovine pancreatic ribonuclease were obtained from Worthington Biochemical Corp., Freehold, N. J. Concanavalin A (2 × crystalized) and bovine α-lactalbumin were obtained from Nutritional Biochemicals Corp., Cleveland, Ohio.

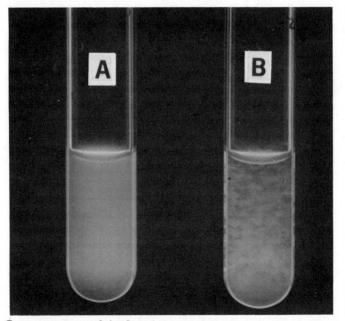

Fig. 2. Gross appearance of the flocculation–agglutination of rat liver mitochondria by lysozyme (LZM). (A) Control suspension in sucrose-Tris-EDTA (STE) buffer. (B) Same, 30 min after addition of human LZM, 100 μg/ml.

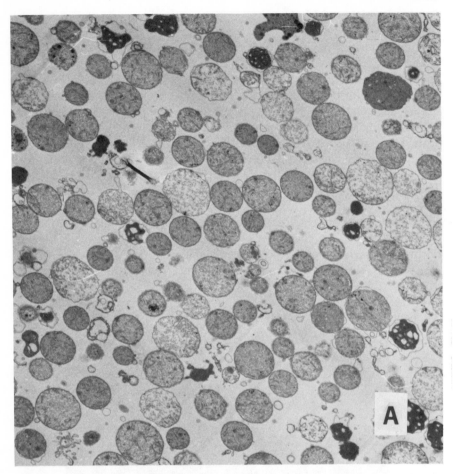

Fig. 3. Electron micrographs of control and LZM-treated rat liver mitochondria: (A) control; (B) 30 min after addition of human LZM, 100 μg/ml. Gluteraldehyde (3%) and osmium tetroxide (1%) fixation; Epon embedded (original magnification, × 1900).

Results

The addition of hen egg-white, rat, or human LZM to rat liver mitochondrial suspensions in STE medium caused flocculation which was visible within 10–30 min at room temperature. Both the rate and extent of mitochondrial agglutination were found to be dependent on the concentration of LZM in the reaction mixture, and agglutination could be observed with human LZM at concentrations as low as 20 μg/ml. Figure 2 shows the gross appearance of the reaction, and Fig. 3 shows electron micrographs

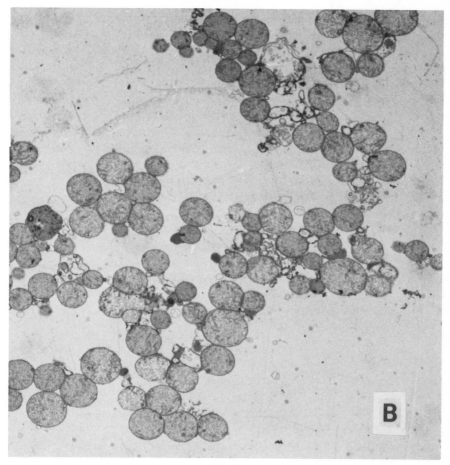

Fig. 3B

of the control and LZM-treated mitochondria. A striking agglutination or clumping of the mitochondria is evident after exposure to LZM, which can be conveniently monitored spectrophotometrically since it results in a progressive decrease in light scattering (Fig. 4).

Neither ribonuclease nor protamine, in concentrations of up to 300 μg/ml, agglutinated mitochondria, thus indicating that the reaction was not simply due to the presence of a low molecular-weight cationic protein. Similarly, there was no detectable mitochondrial agglutination with bovine α-lactalbumin (up to 400 μg/ml), a protein with considerable sequence homology and similar substrate binding affinities to those of LZM (5).

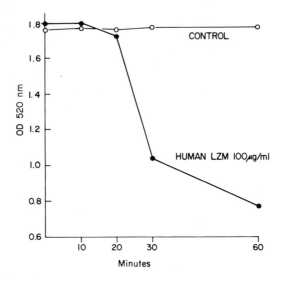

Fig. 4. Change in the optical density (520 nm) of a suspension of rat liver mitochondria associated with agglutination induced by the addition of human LZM, 100 μg/ml.

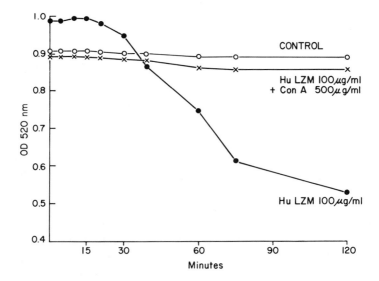

Fig. 5. Inhibition by concanavalin A (500 μg/ml) of the agglutination of rat liver mitochondria by human LZM (100 μg/ml).

In studies on the inhibition of the agglutination of mitochondria by LZM, it was found that the reaction was not inhibited by 1 mM EDTA (more than enough to chelate any traces of Ca^{2+} or Mg^{2+} in the LZM preparations) but was blocked by 2 mM EDTA and could actually be reversed by 4 mM EDTA. It was also observed that agglutination by LZM could be prevented by concanavalin A (Fig. 5).

In an effort to elucidate the mechanism of mitochondrial agglutination by LZM, studies were done to determine the effect of *N*-acetylglucosamine (NAG) on the reaction. Since NAG has been shown to be an inhibitor, albeit a relatively weak one, of the enzymatic cleavage of bacterial cell wall polysaccharides by LZM, an effect of NAG on the mitochondrial reaction might suggest a similar mechanism of action. It was found that the presence of 300 μg/ml of NAG delayed and partially blocked mitochondrial agglutination at 100 μg/ml of LZM; higher concentrations of NAG caused mitochondrial swelling and thus interfered with detection of the agglutination reaction.

Some preliminary studies have also been done to ascertain whether any acetylhexosamines are released from mitochondria by treatment with LZM. Although these studies have so far been negative, further studies are necessary before this possibility can be excluded.

Discussion

The present studies have demonstrated that LZM is capable of agglutinating isolated rat liver mitochondria. As yet the precise mechanism of this effect is unknown but the lack of a comparable effect by ribonuclease and protamine argue against its being attributable to a nonspecific action of low molecular weight cationic proteins. The fact that agglutination is blocked by concanavalin A and partially inhibited by *N*-acetylglucosamine suggests that the reaction may be related to constituent carbohydrate moieties of the mitochondrial membranes but this critical point needs further elucidation. Additional studies are also needed to determine the relevance of these *in vitro* observations to the *in vivo* structural changes of the mitochondria of the proximal renal tubular epithelium in patients and experimental animals with LZM-producing neoplasms and LZM-uria.

Acknowledgments

The authors gratefully acknowledge the technical assistance of Miss Lucy L. Young and Mrs. Harriett Ansari.

Studies supported by grant CA-02332 of the National Cancer Institute, and grant HD-05506 of the National Institute of Child Health and Human Development.

References

1. Osserman, E. F., Klockars, M., Halper, J., and Fischel, R. E., *in* "Lysozyme," (E. F. Osserman, R. E. Canfield, and S. Beychock, eds.), p. 471. Academic Press, New York (1973).
2. Osserman, E. F., and Azar, H. A., *Fed. Proc.* **28,** 619 (1969).
3. Weinbach, E. C., *Anal. Biochem.* **2,** 335 (1961).
4. Osserman, E. F., and Lawlor, D. P., *J. Exp. Med.* **124,** 921 (1966).
5. Hill, R. E., Steinman, H. M., and Brew, K., *in* "Lysozyme," (E. F. Osserman, R. E. Canfield, and S. Beychock, eds.), p. 55. Academic Press, New York (1973).

Studies of the Effects of Lysozyme on Mammalian Cells

ELLIOTT F. OSSERMAN, MATTI KLOCKARS, JAMES HALPER, and ROBERT E. FISCHEL

For the past 2 years, we have been investigating the possibility that certain polysaccharides, glycoproteins, or glycolipids of *mammalian* cells may be susceptible to enzymatic cleavage or substrate specific binding by lysozyme (LZM). Since these substances are major constituents of cell membranes, we have further postulated that LZM may serve an important function in the regulation of certain membrane-dependent cell functions and may participate in the surveillance of membrane abnormalities, particularly those associated with neoplastic transformation. In the preceding paper (1), we have presented evidence that LZM agglutinates isolated mitochondria and that this reaction can be blocked by concanavalin A. Presumably, this agglutination results from an action of LZM on a mitochondrial membrane constituent but as yet it has not been determined whether this effect is caused by specific enzymatic activity or by a nonenzymatic action, possibly related to LZM's low molecular weight and cationic properties.

In the present studies we have been concerned with examining the effects of LZM on the behavior of normal and transformed mammalian cells in culture, attempting thereby to gain some insights into the possible effects of endogenous LZM *in vivo*. For these studies of murine and human cells we have used homologous mammalian LZM's, i.e., human LZM isolated from the urine of patients with monocytic leukemia (2) and rat LZM isolated from the urine of rats bearing the transplantable chloroleukemia which elaborates LZM (3). The presently available evidence indicates that these LZM's are identical to their normal counterparts (2).

Materials and Methods

Tissue Culture Cell Lines

Normal mouse embryo fibroblasts (3T3) and Simian Virus transformed cells (SV101-3T3) were obtained from Dr. Claudio Basillico of the Department of Cell Biology of New York University School of Medicine. A rat liver cell line (RLB) which had apparently transformed spontaneously when grown in nutritionally deficient medium was obtained from Dr. Carmia Borek of the Department of Radiology of the College of Physicians and Surgeons (4). The 3T3 and SV-3T3 cells were grown in Dulbecco's Modified Eagle Medium supplemented with 10% calf serum and penicillin (100 U/ml) and streptomycin (50 U/ml). The rat liver (RLB) cells were grown in RPMI 1640 medium with 5% fetal calf serum and the same antibiotics.

All cell lines were maintained in plastic tissue culture flasks, 25 cm² (30 ml) (Falcon Plastics, Los Angeles). For the cell counting studies of 3T3 and SV-3T3 cells, plastic culture Petri dishes, 5 cm diameter, were used. The initial cell inocula were 5×10^4 or 1×10^5 cells per plate. All cultures were incubated at 37°C in a 7% CO_2:air atmosphere.

Cytologic studies and studies of [³H]thymidine uptake were carried out on cells cultured on 11×22 mm hanging glass coverslips according to the method described by Ryt maa (5). The coverslips were initially placed along the down side of sterile plastic culture tubes (16×125 mm) in the horizontal position. The volume (0.5 ml) between the lower surface of the coverslip and the test tube wall was then filled with a suspension of cells (10^3 to 10^4/ml) in tissue culture medium containing [³H]thymidine (1 μCi per ml). After filling each tube, it was quickly rotated 180° along its horizontal axis bringing the coverslips to the top side and thus permitting the cells to settle on its surface. In this position the tubes were closed with loose fitting plastic caps and placed in the incubator. In a separate experiment, it was determined that the majority of 3T3 and SV-3T3 cells were adherent to the coverslip surface within 30 min.

In all experiments, sufficient numbers of cultures were set up initially to permit harvesting 3 or 4 cultures after 1, 2, 3, and 4 days of incubation. At these intervals, the coverslips were removed from the culture tubes, air dried, fixed in 95% ethanol and washed with distilled water. They were then placed in scintillation vials containing 10 ml of Aquasol (New England Nuclear Co., Boston, Mass.) and counted in a Packard Tri-Carb liquid scintillation spectrometer. In previous studies (5) it was established that the channel ratios and energy spectra of [^{3}H]thymidine labeled bone marrow cells attached to coverslips was not significantly different from that of cells suspended directly in the scintillation fluid or from [^{3}H]thymidine in solution.

After counting, the coverslips were washed in 95% ethanol to remove the scintillation fluid and then stained with Wright Giemsa stain for cytologic examination.

Scanning and Transmission Electron Microscopic Studies of the Effects of Lysozyme on Human Liver Cells in Culture

Scanning and transmission electron microscopic studies of the effects of human and rat LZM on human liver cell cultures were carried out in collaboration with Prof. Marcel Bessis and Dr. Yves Le Charpentier at the Institute of Cellular Pathology, Paris (Bicetre). The cell line utilized for these studies was originally derived from a liver biopsy of a 4-year-old girl with hemolytic anemia and microspherocytosis. The tissue culture was started in January 1972 and showed restricted growth for 2 months (4 passages) at which time it apparently transformed spontaneously and proliferated more rapidly thereafter. When the present studies were done (August 1972), the cells had been passaged a total of 50 times. The cells were grown on glass coverslips in Leighton tubes using Eagle's medium supplemented with 20% fetal calf serum, glutamine, vitamins, heparin, and antibiotics. After dehydration, the coverslips were air dried ("critical point" method) and the cells coated with a conducting layer of vacuum evaporated gold. The coverslips were attached directly to the specimen holder of the microscope. The microscope (Cambridge-Stereoscan) was operated at 30 kV. Pictures were taken at 500–10,000 × magnification and at about a 40° angle to the monolayer surface. Parallel coverslip cultures were prepared and processed for transmission electron microscopy (Philips EM 200).

Lysozymes

Human LZM was isolated as previously described (2) from the urine of patients with monocytic or monomyelocytic leukemia by adsorption to bentonite followed by washing the bentonite five times with 5% aqueous pyridine and elution with 10% pyridine: sulfuric acid (pH 5.2). The lysozyme-containing eluate was dialyzed against distilled water until free of pyridine and then lyophilized. Rat LZM was isolated in the same manner from the urines of Wistar/Furth rats bearing the Shay chloroleukemia (3). This tumor was kindly provided in 1967 by Dr. William C. Moloney of Harvard Medical School, in its 108th transplant generation, and has been carried by serial intraperitoneal transplantations at 3–4 week intervals thereafter. Urine collections were made from animals housed in metabolic cages from 2 to 3 weeks after tumor transplantation when the excretion of LZM was maximal. The purity of these LZM preparations was ascertained by cellulose acetate and acrylamide gel electrophoresis and immunologic analyses, as previously described (2).

Results

Effects of Lysozyme on the Proliferation of 3T3 and SV-3T3 Cells in Culture

Figure 1 shows the effects of rat LZM (100 μg/ml) on the growth of normal (3T3) and transformed (SV-3T3) mouse embryo fibroblasts. In these experiments, LZM was added to the cultures at the time of initial plating. The initial cell inoculum was 5×10^4 cells in a 5 cm diameter culture plate. Under these conditions of a relatively low cell density and the presence of LZM throughout the culture period, the LZM appeared to effect a transient inhibition of the proliferation or killing of a certain number of both the 3T3 and the SV-3T3 cells, mainly evident during the first four days in culture. At the 5th day, particularly with the SV-3T3 cells, there was no significant difference in the total cell numbers in the control and LZM-treated cultures. As shown in Fig. 2, however, there was a significant *qualitative* difference between the control and LZM-treated SV-3T3 cultures on the 5th day. Thus, the control SV-3T3 culture showed a dense growth in the middle of the plate and sparse growth at the periphery, whereas the LZM-treated SV-3T3 cells were more uniformly distributed over the surface of the culture plate. Both the control and LZM-treated 3T3 cells were relatively uniformly distributed in their growth patterns.

Figure 3 shows the effects of adding rat LZM (100 μg/ml) to 3T3 and SV-3T3 cultures 48 hr after the initial plating of 1×10^5 cells, i.e., twice the inoculum as in the study shown in Fig. 1. Under these conditions it is evident that there was much less of an effect of LZM than in the prior experiments in which there was a lower cell density and the enzyme was present from the time of initial plating.

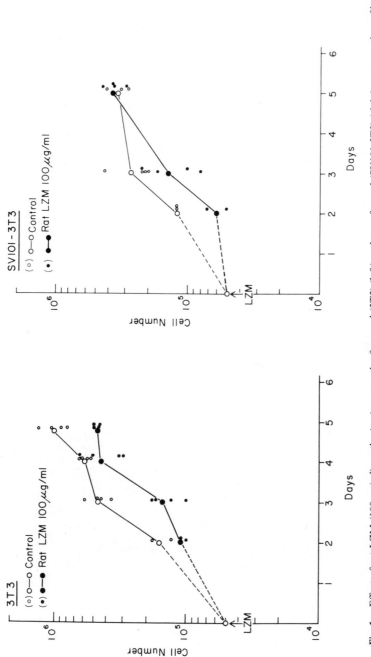

Fig. 1. Effect of rat LZM (100 μg/ml) on the *in vitro* growth of normal (3T3) (left) and transformed (SV101-3T3) (right) mouse embryo fibroblasts. The initial cell inoculum was 5 × 10⁴ cells in a 5 cm diameter plastic culture plate, and LZM was present in the medium throughout the entire period of culture. (Small open circle) and (small closed circle), individual plate counts; (large circles) mean of each group.

These results indicate that rat LZM at a concentration of 100 μg/ml transiently inhibits the proliferation or kills a certain number of both 3T3 and SV-3T3 mouse embryo fibroblasts if present from the time of initial plating. Presumably this reflects a particular sensitivity of these cells to LZM at the time of attachment and initiation of monolayer cell growth. More striking than the effects of LZM on the proliferation of these cells *in vitro,* however, was its *qualitative* effects on the pattern of growth of the SV-3T3 cells, vis, a more uniform distribution of the LZM-treated cells on the surface of the culture plate as compared to the control culture. When LZM was added to 2-day-old cultures with higher cell densities, it had no detectable effects on either 3T3 or SV-3T3 cells.

Effects of Lysozyme on the Uptake of [³H] Thymidine by 3T3 and SV-3T3 Cells in Culture

Figure 4 shows the effects of increasing concentrations of rat LZM (200 to 800 μg/ml) present from the time of initial plating on the uptake of [³H]-thymidine by normal (3T3) and transformed (SV-3T3) mouse embryo

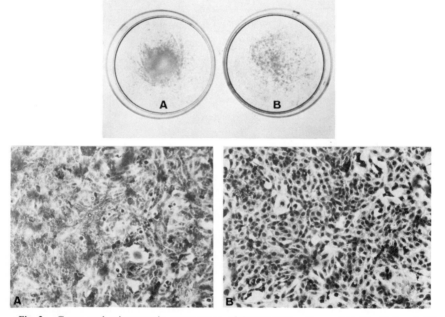

Fig. 2. Gross and microscopic appearance of SV101-3T3 cultures. (A) Control and (B) grown in the presence of rat LZM 100 μg/ml, on day 5 of the experiment shown in Fig. 1. The LZM-treated culture shows a more uniform distribution on the plate and a more regular cytology with less pleomorphism.

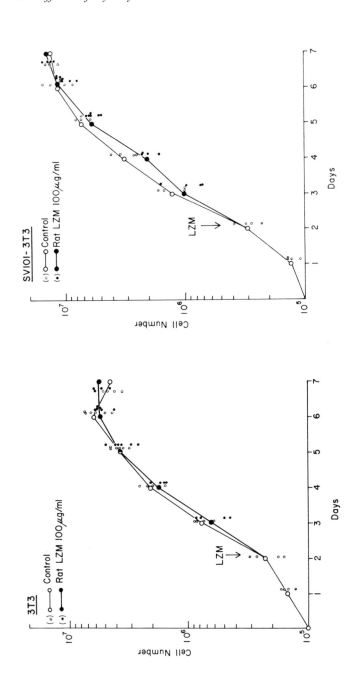

Fig. 3. Effect of rat LZM (100 μg/ml) on the *in vitro* growth of normal (3T3) (left) and transformed (SV101-3T3) (right) mouse embryo fibroblasts. The initial cell inoculum was 1 × 10⁵ cells in a 5 cm diameter plastic culture plate, and the LZM was added to the medium 48 hr after initiation of the culture.

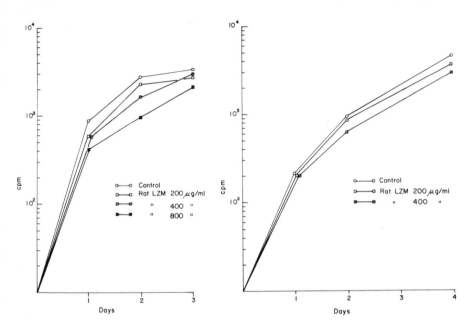

Fig. 4.   Effects of increasing concentrations of rat LZM (200–800 μg/ml) on the uptake of
³H-thymidine by normal (3T3) (left) and transformed (SV-3T3) (right) mouse embryo fibroblasts. Cells were grown as monolayers on hanging glass coverslips. The initial cell concentration was 10⁴/ml for the 3T3 cells and 4 × 10³/ml for the SV-3T3 cells. Each point represents
the mean of 3–4 cultures.

fibroblasts cultured on hanging glass coverslips. In both of these cell lines
it is evident that LZM produces a small but significant decrease in [³H]-
thymidine uptake, and this effect is apparently dose dependent. It is also
noteworthy, particularly with the 3T3 cultures, that the LZM inhibition of
[³H]thymidine uptake was most evident in the first 2 days *in vitro*, similar
to the pattern observed in the cell counting studies.

Effects of Lysozyme on the Uptake of [³H] Thymidine and Morphology of Transformed Rat Liver Cells in Culture

Figure 5 shows the effects of rat LZM (400 μg/ml) on the uptake of ³H-
thymidine by transformed rat liver (RLB) cells cultured on hanging glass
coverslips. The pattern and magnitude of the LZM effect is comparable
to that seen with the 3T3 and SV-3T3 cell lines, the maximal inhibition
being evident during the first 2 days in culture.

Figure 6 shows the cytology of the control and rat LZM-treated RLB
cells on the 2nd and 4th days *in vitro*. Whereas the control cultures exhibit
a relatively uniform epithelial hexagonal pattern, the LZM-treated cells

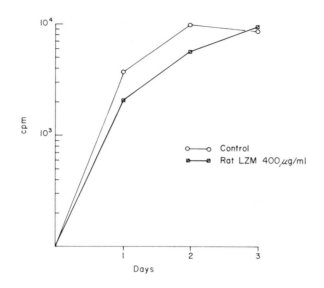

Fig. 5. Effect of rat LZM (400 μg/ml) on the uptake of [³H]thymidine by transformed rat liver cells (RLB) in hanging glass coverslip cultures. The initial cell concentration was 10⁴/ml. Each point represents the mean of 3–4 cultures.

show very marked nuclear and cytoplasmic irregularities. Thus, the nuclei of the LZM-treated cells vary considerably in size, shape, and staining intensity, and many appear distinctly pyknotic. The cytoplasmic changes in the LZM-treated cells are particularly striking and consist of a marked spreading, development of surface invaginations, and the extension of long, filamentous processes. It is also noteworthy that the control cells show numerous clear droplets, presumably lipid, which are relatively uniform in size and mainly perinuclear in distribution, whereas the LZM-treated cells show fewer droplets which are more irregular in size and scattered throughout the cytoplasm.

It should be emphasized that, although some of the cytologic changes associated with LZM treatment of these cells might suggest cell damage and degeneration due to a nonspecific toxicity of LZM, the [³H]thymidine incorporation data indicate that the net synthesis of DNA in these cultures was not markedly reduced by LZM.

Scanning and Transmission Electron Microscopic Studies of the Effects of Lysozyme on Human Liver Cells in Vitro

Figure 7 shows the effects of human LZM (200 μg/ml) on cultures of spontaneously transformed human liver cells after 20 hr *in vitro*, as seen by scanning electron microscopy (SEM). The LZM was added to the medium at the time of initial plating. After 20 hr *in vitro,* both the control and LZM-treated cells were firmly attached to the glass surface and the cell numbers were approximately the same. However, the appearance of the control and LZM-treated cells was strikingly different. The control cells were relatively more compact and high-contoured, and their cytoplasmic borders were sharply demarcated and had only a few short, filamentous projections. In contrast, the cells grown in the presence of LZM were markedly flattened and spread out on the glass surface, and their cytoplasmic borders showed numerous long, filamentous projections. Marked differences were also seen in the cell surfaces. The surfaces of the control cells appeared relatively smooth although a fine granularity was demonstrated by higher magnification [Fig. 7(C)], whereas the surfaces of the LZM-treated cells showed patches of verrucous nodularity plus scattered, smooth-surfaced nodules.

Cultures of these same human liver cells with rat LZM (200 μg/ml) showed similar effects although less marked than those produced by the homologous human LZM.

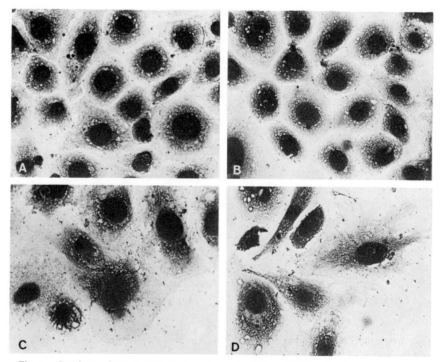

Fig. 6. Cytology of spontaneously transformed rat liver (RLB) cell cultures grown under control conditions and in the presence of rat LZM (400 μg/ml) for 2 and 4 days, respectively. (A) Control, 2 days; (B) control, 4 days; (C) rat LZM, 2 days; (D) rat LZM, 4 days.

Figure 8 shows transmission electron micrographs (TEM's) of these same human liver cell cultures transferred directly from the glass surfaces and embedded for sectioning. Consistent with the SEM observations, the control cells displayed relatively smooth cytoplasmic margins whereas the cytoplasmic margins of the LZM-treated cells showed numerous villous projections. In addition to these differences, the TEM's demonstrated that the control cells contained numerous large, fat droplets principally clustered around the nuclei and probably responsible for the fine surface granularity seen by SEM. These droplets were virtually absent from the LZM-treated cells. Furthermore, in the control cells, the mitochondria and ribosomes were irregular and poorly defined, whereas in the LZM-treated cells the mitochondrial membranes and cristae were more regular and sharply defined and the ribosomes more numerous and clustered into polysomes. Thus, with respect to certain features of internal structure the LZM-treated cells appeared more *normal* than the untreated control cells.

Discussion

The results of the present studies indicate that LZM produces very significant effects on a variety of mammalian cells *in vitro* and that these effects are not simply the expression of nonspecific toxicity. In the studies with normal and SV-transformed mouse embryo fibroblasts, LZM was shown to produce a relatively small and transient inhibition of both proliferation and thymidine uptake. These effects were particularly evident when the LZM was incorporated into the medium from the time of initiation of the cultures, and there was significantly less effect when LZM was added to the culture medium after 48 hr *in vitro*, and the initial cell density was greater.

The qualitative effects of LZM on the pattern of growth of mouse embryo fibroblasts, particularly those transformed by the 101 variant of SV40, were even more striking than its effects on cell proliferation and thymidine uptake. Thus, the SV-3T3 cells grown in the presence of LZM showed a more uniform and regular growth pattern and morphology than the control cultures suggesting a reacquisition of contact inhibition. It is also noteworthy that the LZM-treated SV-3T3 cells were more uniformly fibroblastic in their morphology and less pleomorphic than the control SV-3T3 cells.

The present studies further demonstrated that LZM affected the *in vitro* behavior and morphology of cultures of spontaneously transformed rat liver cells. In a manner similar to its effects on the mouse embryo fibroblasts, LZM produced a small but significant inhibition of thymidine by the transformed rat liver cells. However, the cytologic effects of LZM on the

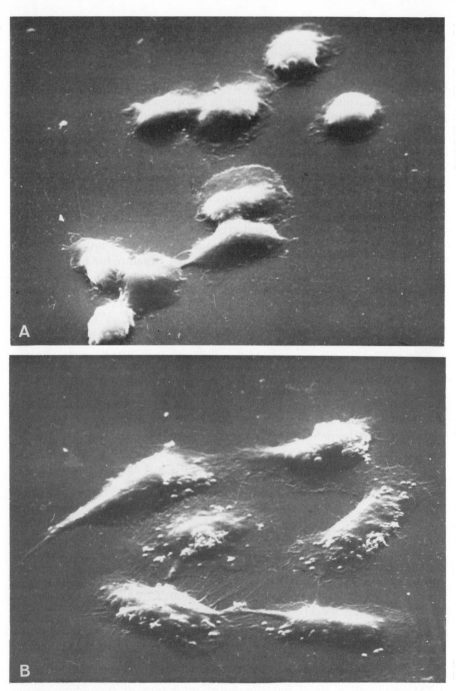

Fig. 7. Scanning electron micrographs of spontaneously transformed human liver cell cultures after 20 hr *in vitro*. (A) and (C), control; (B) and (D), human LZM (200 μg/ml). Original magnifications: (A) and (B), $\times$ 500; (C) and (D), $\times$ 2600.

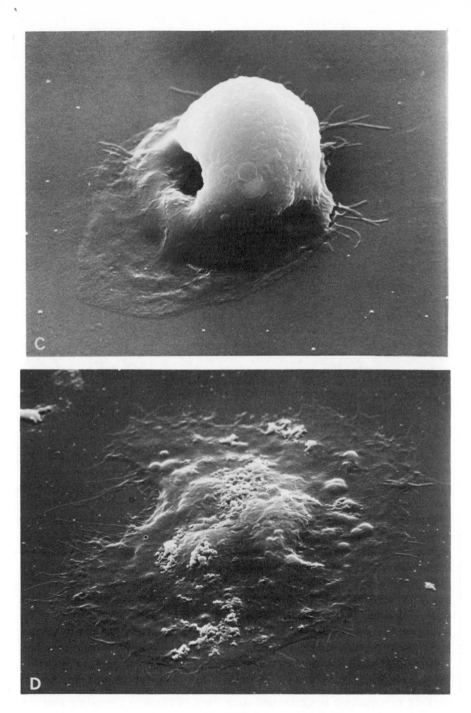

Fig. 7C and D

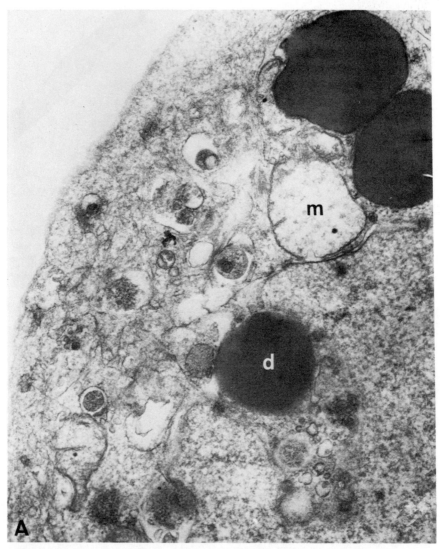

Fig. 8. Transmission electron micrographs of spontaneously transformed human liver cell cultures after 20 hr *in vitro*. (A) Control; × 61,500. m, mitochondrion; d, lipid droplet.

liver cells were quite different from those observed with the fibroblast cultures. Thus, in the presence of LZM, the liver cell cultures displayed very striking cytoplasmic irregularities consisting of marked spreading and the development of extensive invaginations and long, filamentous projections at the cell margins. It was also observed that the numerous perinuclear lipid

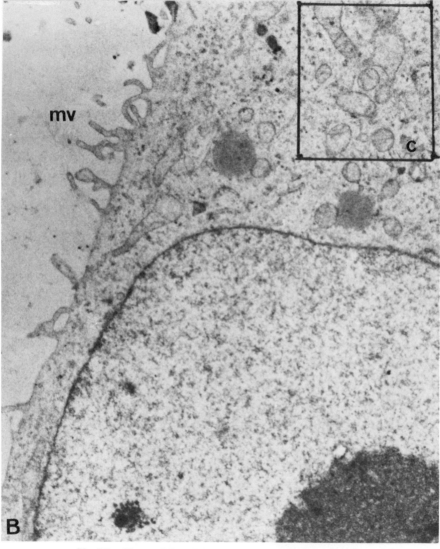

Fig. 8B.   Human LZM (200 µg/ml); × 15,000. mv, microvilli.

droplets in the control cultures were reduced in number and more irregularly distributed in the LZM-treated cells. Although certain of these effects might be indicative of nonspecific toxicity, the relatively insignificant inhibition of uptake of thymidine argues against this interpretation.

In our studies to date, probably the most convincing evidence of an effect of LZM on mammalian cells *in vitro* was that obtained by scanning

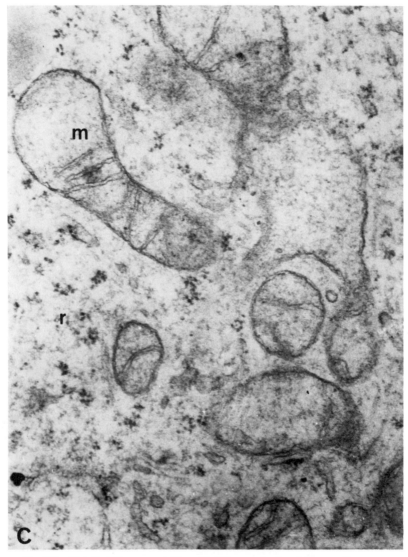

Fig. 8C.   Detail of B, upper right; × 61,500. m, mitochondrion; r, ribosomes.

and transmission electron microscopy of spontaneously transformed human liver cells. This particular cell line was employed because of its availability in the laboratory of Prof. Marcel Bessis at the Institute of Cellular Pathology in Paris where these studies were carried out. By scanning microscopy, it was found that the homologous human LZM at a concentration of 200 μg/ml produced major alterations in the contours and surfaces of these cells. These effects were already evident at 2 hr but were particularly striking after 20 hr *in vitro*. Whereas the control cells exhibited relatively compact and high contours, the LZM-treated cells were flattened and extensively spread out on the glass surfaces. They also displayed marked surface irregularities and long, filamentous cytoplasmic projections. These effects were apparently comparable to those observed by conventional microscopy with the rat liver cells grown in the presence of rat LZM. It is also noteworthy that the effects of the homologous human LZM on the cultured human liver cells were relatively more marked than that of heterologous rat LZM on these same cells, thus indicating an element of species specificity in the effects of LZM. It would also argue against the possibility that these effects are the nonspecific consequences of low molecular weight cationic proteins since the molecular weights and net electrical charges of the rat and human LZM are essentially the same.

The transmission electron microscopic studies of the control and LZM-treated human liver cells provided further evidence that the effects of LZM are not simply those of nonspecific toxicity. Thus, when compared to the control cells, the LZM-treated cells showed more normal-appearing mitochondria, more numerous and distinct polyribosomes, and a striking decrease in the number of lipid droplets. It also appeared that the outer ("plasma") membranes of the LZM-treated cells were thinner and more sharply defined than those of the control cells, consistent with a reduction in the amount of surface polysaccharide ("glycocalyx") by LZM.

In the presence of LZM, the changes in cell configurations from globular to flattened contours and the development of long, filamentous cytoplasmic projections would imply a significant increase in total surface area, and this, presumably, is accomplished either by the extension of existing membrane or the synthesis of new membrane, or both. The former process, i.e., the extension of existing membrane, could possibly be achieved by reorganization of a lattice structure due to an effect of LZM on membrane polysaccharides or a nonspecific polycation effect on the surface charge interactions of these cells with the supporting surfaces (6–10). It should be emphasized that even if these changes are due to polycation effects, they may still be of considerable biological significance in those situations where LZM is released from macrophages onto the surface of target cells. Alternatively or additionally, if LZM is causing these cells to produce more membrane, it is possible that the numerous mitochondria and polyribosomes and the reduced number of lipid droplets in the LZM-treated cells may be related to this synthesis.

The presence of macrophages and histiocytes in tumors of man (11–13) and experimental animals has been extensively documented (14–16), and there is considerable evidence, albeit mainly circumstantial, that these cells serve a major role in host defenses against neoplasia (14, 15, 17, 18). It is also probable that the antitumor effects of BCG vaccine and other adjuvants are related to their ability to stimulate or activate macrophages (19–25). At present, however, the precise mechanisms whereby macrophages carry out their putative antitumor functions and the specific biochemical mediators of these cellular activities are not known. The fact that LZM is elaborated by and released from macrophages is well documented (26, 27). An increased production and/or release of LZM by macrophages following BCG vaccination has also been demonstrated (27–29). This information in conjunction with the present evidence that LZM alters the structure and *in vitro* behavior of certain transformed mammalian cells leads us to postulate that LZM may in fact be one if not the major mediator of antitumor functions of macrophages. We also postulate that the cellular effects of LZM may be due to a combination of its cationic properties, facilitating interaction with the negatively charged cell surfaces (6, 7) and an enzymatic effect on a membrane polysaccharide(s). With respect to the first property, the polyanion heparin which strongly inhibits LZM activity against bacterial substrates (30, 31) has also been found to inhibit the immune cell-mediated killing of mastocytoma cells (32).

It should be emphasized that the *in vitro* effects of LZM shown in the present studies were obtained with homologous mammalian LZM's at concentrations which are probably within the range achieved *in vivo* in the sites where macrophages interact with target cells. These studies have also shown relatively little *in vitro* toxicity of LZM, and some preliminary animal studies (33) indicate that relatively large amounts of homologous LZM (up to 5 mg/gm body weight, administered intravenously or intraperitoneally, daily for several weeks) are very well tolerated by mice and rats. Thus, if the biological usefulness of LZM is established, its pharmacological use would seem to be feasible.

Summary

Homologous mammalian (rat and human) lysozymes (LZM's) have been found to alter the *in vitro* growth patterns and morphology of Simian virus-transformed mouse embryo fibroblasts (SV-3T3 cells) and spontaneously transformed rat and human liver cells. The more uniform growth pattern and cytology of the LZM-treated SV-3T3 cells suggests the reacquisition of contact inhibition. Scanning and transmission electron microscopy of the LZM-treated transformed human liver cells demonstrated marked alterations of cell surface contours presumed to be due to the effects of LZM on surface membranes. The mitochondria and polyribosomes of LZM-treated cells were more normal in appearance than in the control cells. This finding along with the relatively minor effects of LZM on the proliferation and uptake of [³H]thymidine indicates that LZM has very little if any nonspecific toxicity for these cells *in vitro*.

It is postulated that LZM may have specific effects on the membrane polysaccharides of normal and transformed mammalian cells and may serve an important role in the surveillance of membranes. It is further postulated that LZM may be a key mediator of the antitumor functions of macrophages.

Acknowledgments

We gratefully acknowledge the invaluable cooperation of Prof. Marcel Bessis and Dr. Yves Le Charpentier with the scanning electron microscopy studies and the excellent technical assistance of Miss Judith Weisfuse and Mrs. Carmen Colon.

These studies were supported by grant CA-02332 of the National Cancer Institute.

References

1. Adinolfi, M., Loeb, J. N., and Osserman, E. F., *in* "Lysozyme," (E. F. Osserman, R. E. Canfield, and S. Beychok, eds.), p. 463, Academic Press, New York (1973).
2. Osserman, E. F., and Lawlor, D. P., *J. Exp. Med.* **124,** 921 (1966).
3. Rosenthal, D. S., and Moloney, W. C., *Proc. Soc. Exp. Biol. Med.* **126,** 682 (1967).
4. Borek, C., *Proc. Natl. Acad. Sci. U.S.A.* **69,** 956 (1972).
5. Rytömaa, T., *In Vitro* **4,** 47 (1969).
6. Katchalsky, A., *Biophys. J.* **4,** 9 (1964).
7. Ambrose, E. J., *in* "The Biology of Cancer" (E. J. Ambrose and F. J. C. Roe, eds.), pp. 65–77. Van Nostrand-Reinhold, Princeton, New Jersey, 1966.
8. Moroson, H., *Cancer Res.* **31,** 373 (1971).
9. Weiss, L., and Zeigel, R., *J. Cell. Physiol.* **77,** 179 (1971).
10. Weiss, L., Zeigel, R., Jung, O. S., and Bross, I. D. J., *Exp. Cell Res.* **70,** 57 (1972).
11. Black, M. M., Kerpe, S., and Speer, F. D., *Am. J. Pathol.* **29,** 505 (1953).
12. Hamlin, I. M. E., *Br. J. Cancer* **22,** 383 (1968).
13. Silverberg, S. G., Chitale, A. R., Hind, A. D., Frazier, A. B., and Levitt, S. H., *Cancer* **26,** 1177 (1970).
14. Halpern, B. N., ed., "Rôle du Système Reticulo-Endothelial dans l'Immunité Anti-Bactérienne et Anti-Tumorale," No. 115. CNRS, Paris, 1953.
15. Thomas, L., *in* "Cellular and Humoral Aspects of the Hypersensitive States" (H. S. Lawrence, ed.), p. 529. Harper (Hoeber), New York, 1959.
16. Aoki, T., Teller, M. N., and Robitaille, M. -L., *J. Natl. Cancer Inst.* **34,** 255 (1965).
17. Old, L. J., Benacerraf, B., Clarke, D. A., Carswell, E. A., and Stockart, E., *Cancer Res.* **21,** 1281 (1961).
18. Old, L. J., Clarke, D. A., Benacerraf, B., and Goldsmith, M., *Ann. N. Y. Acad. Sci.* **88,** 264 (1960).
19. Old, L. J., Clarke, D. A., and Benacerraf, B., *Nature (Lond.)* **184,** 291 (1959).
20. Weiss, D. W., Bonhag, R. S., and Leslie, P., *J. Exp. Med.* **124,** 1039 (1966).

21. Mathé, G., Schwarzenberg, L., Amiel, J. L., Schneider, M., Cattan, A., and Schlumberger, J. R., *Cancer Res.* **27,** 2542 (1967).
22. Mathé, G., Pouillart, P., and Lapeyraque, F., *Br. J. Cancer* **23,** 814 (1969).
23. Mathé, G., Amiel, J. L., Schwarzenberg, L., Schneider, M., Cattan, A., Schlumberger, J. R., Hayat, M., and de Vassal, F., *Lancet* **1,** 697 (1969).
24. Bluming, A. Z., Vogel, C. L., Ziegler, J. L., Mody, N., and Kamya, G., *Ann. Intern. Med.* **76,** 405 (1972).
25. Zbar, B., and Tanaka, T., *Science* **172,** 271 (1971).
26. Myrvik, Q. N., Leake, E. S., and Fariso, B., *J. Immunol.* **86,** 133 (1961).
27. Cohn, Z. A., and Wiener, E., *J. Exp. Med.* **118,** 991 (1963).
28. Myrvik, Q. N., Leake, E. S., and Oshima, S., *J. Immunol.* **89,** 745 (1962).
29. Carson, M. E., and Dannenberg, A. M., Jr., *J. Immunol.* **94,** 99 (1965).
30. Nihoul, E., Massart, L., and Van Heel, G., *Arch. Int. Pharmacodyn. Ther.* **88,** 123 (1951).
31. Kerby, G. P., and Eadie, G. S., *Proc. Soc. Exp. Biol. Med.* **83,** 111 (1953).
32. Martz, E., and Benacerraf, B., *Fed. Proc.* **31,** 786 (1972).
33. Osserman, E. F., unpublished observations.

SECTION
4

LYSOZYME BIBLIOGRAPHY AND INDEX

Lysozyme Bibliography 1922–1972

BETTY R. MOORE AND ELLIOTT F. OSSERMAN

With the hope of assisting and stimulating ongoing and future lysozyme (LZM) research, we have compiled a Lysozyme Bibliography surveying the literature of the half-century since Fleming's original article. Comprised of approximately 2600 entries, the list is obviously extensive. It is also comprehensive, with 395 subject categories ranging from amoebae to X-ray crystallography. Nonetheless, it must be recognized that some contributions will inevitably have been overlooked or omitted. We regret and apologize for these omissions. Hopefully, they will not be too numerous or critical.

In assembling this bibliography, particular use has been made of the following indexing and abstracting resources: *Chemical Abstracts* (1922 to date); *Index Medicus* (1927–1954); *Current List of Medical Literature* (1956–1959); *Cumulative Index Medicus* and *Index Medicus* (1960 to date); *Biological Abstracts* and *Bioresearch Index* (1927 to date); and *Current Contents, Life Sciences* (1972–May 1973). We have also used previous review articles, particularly:

Thompson, R. Lysozyme and its relation to the antibacterial properties of various tissues and secretions (general review). *Arch. Pathol.* **30**, 1096–1134, 1940.

"Lysozyme, A Review." Armour and Co., Chicago, 1952.

International Symposium on Fleming's Lysozyme, Milan, 1959, Proceedings. Museo della Scienzae della Technica, Milan, 1959.

Colloque sur les lysozymes, organisé par P. Jollès. In *Exposés Annueles de Biochimie Médicale.* Paris, 1966.

Grossegebauer, K., and Langmaack, H. Ubersichten Lysozyme. *Klin. Wochenschr.* **46**, 1121–1127, 1968.

Phillips, D. C. On the stereochemical basis of enzyme action: lessons from lysozyme. *Harvey Lect.* **66,** 135–160, 1971–1972.
Imoto, T. *et al.* Vertebrate lysozyme. *In* "The Enzymes," 3rd edition (P. D. Boyer, ed.). Academic Press, New York, 1972, pp. 665–868.

Additional review articles are listed in the Lysozyme Bibliography.

Each reference in the bibliography is identified by a letter-number designation which appears to the left of the entry. The letter represents the first letter of the senior author's surname; the number represents the number of that reference within its letter group, e.g.,

M119. Meyer, K. The relationship of lysozyme to avidin. *Science* **99:**391–392, 1944.

In the index to the Lysozyme Bibliography under Interactions, avidin, M119 is found along with A24, G102, and L16, which identify other articles on this subject.

The majority of citations were indexed directly from the original articles, recognizing that article titles are frequently inadequate and occasionally misleading. Certain subject headings are very specific, e.g., Distribution, ear wax, whereas others are relatively broad, e.g., Bacteriophage LZM. With respect to the latter, it was not considered within the scope of this bibliography to include a detailed breakdown of phage LZM studies (i.e., gene mapping, biosynthesis, etc.), but a reasonable number of references were included to serve as a starting base for more detailed studies of this subject.

The experience of compiling this bibliography and accompanying index has been both illuminating and humbling—illuminating to appreciate the enormous volume and range of information and ideas which have emerged in 50 years of LZM research, and humbling to realize the limitations of one's own capacity to encompass even a small fraction of the total subject matter.

A

A1. Abeya, O. Lisozima; algunas observaciones sobre su acción en el tubo digestivo del lactante. *Sem. Med.,* **109:**785–792, 1956.
A2. Abraham, E. P. Some properties of egg-white lysozyme. *Biochem. Jour.,* **33:**622–630, 1939.
A3. Abraham, E. P., and Robinson, R. Crystallization of lysozyme. *Nature (London),* **140:**24, 1937.
A4. Acher, R., Jutisz, M., and Fromageot, C. Sur la structure du lysozyme. Etude de la liberation des groupes fonctionnels et des acides aminés au cours d'une hydrolyse acide ménagée. *Biochim. Biophysi. Acta,* **5:**493–498, 1950.
A4a. Acher, R., Chauet, J., Crocker, C., Laurila, U.-R., Thaureaux, J., and Fromageot, C. Isolement et caractérisation des peptides courts obtenus par hydrolyse acide. Etude de la structure du lysozyme et de la vasopressine. *Bull. Soc. Chim. Biol.,* **36:**167–179, 1954.

A4b. Acher, R., Jutisz, M., and Fromageot, C. Sur la structure du lysozyme. Etude de peptides basiques résultant d'une hydrolyse acide ménagée. *Biochim. Biophys. Acta,* **8:**442–449, 1952.

A4c. Acher, R., Thaureaux, J., Crocker, C., Jutisz, M., and Fromageot, C. Sur la structure du lysozyme. Quelques peptides de l'histidine et de la tyrosine resultant d'une hydrolyse ménagée. *Biochim. Biophys. Acta,* **9:**339–340, 1952.

A5. Achter, E. K., and Swan, I. D. On the conformation of lysozyme and alphalactalbumin in solution. *Biochemistry,* **10:**2976–2978, 1971.

A6. Acker, R. F., and Hartsell, S. E. Fleming's lysozyme. *Sci. Amer.,* **202:**132–142, 1960.

A7. Acton, H. C., and Myrvik, Q. N. Antigenicity of egg-white lysozyme. *J. Bacteriol.,* **84:**868–869, 1962.

A8. Adámek, L., Mison, P., Mohelská, H., Trnka, L. Ultrastructural organization of spheroplasts induced in *Mycobacterium sp. smegmatis* by lysozyme or glycine. *Arch. Mikrobiol.,* **69:**227–236, 1969.

A9. Adams, D. J., Evans, M. T. A., Mitchell, J. R., Phillips, M. C., and Rees, P. M. Adsorption of lysozyme and some acetyl derivatives at the air-water interface. *J. Polym. Sci.,* **34:**167, 1971.

A10. Adams, E. T., Jr. Sedimentation equilibrium in reacting systems. V. The application of sigma-ci-Mi2 to the analysis. *Biochemistry,* **6:**1864–1871, 1967.

A10a. Adams, E. T., and Filmer, D. L. Sedimentation equilibrium in reacting systems. IV. Verification of the theory. *Biochemistry,* **9:**2971–2990, 1966.

A11. Adams, G. E., Aldrich, J. E., Bisby, R. H., Cundall, R. B., Redpath, J. L., and Willson, R. L. Selective free radical reactions with proteins and enzymes: Reactions of inorganic radical anions with amino acids. *Radiat. Res.,* **49:**278–289, 1972.

A12. Adams, G. E., Willson, R. L., Aldrich, J. E., and Cundall, R. B. On the mechanism of the radiation-induced inactivation of lysozyme in dilute aqueous solution. *Int. J. Radiat. Biol.,* **16:**333–342, 1969.

A13. Adhikari, P. C., Raychaudhuri, C., and Chatterjee, S. N. The lysis of cholera and El Tor vibrios. *J. Gen. Microbiol.,* **59:**91–95, 1969.

A14. Adinolfi, M. Ontogeny of components of complement and lysozyme. *Ontogeny of Acquired Immunity, Ciba Found. Symp., 1971* pp. 65–85, 1972.

A15. Adinolfi, M., Glynn, A. A., Lindsay, M., and Milne, C. Serological properties of γA Antibodies to *Escherichia coli* Present in Human Colostrum. *Immunology,* **10:**517–526, 1966.

A16. Adinolfi, M., Martin, W., and Glynn, A. A. Ontogenesis of lysozyme in man and other mammals. *Protides Biol. Fluids, Proc. Colloq.,* **18:**91–93, 1971.

A16a. Adkins, B. J., and Yang, J. T. Difference spectropolarimetry as a probe for small conformational changes. *Biochemistry,* **7:**266–271, 1968.

A17. Afanas'eva, T. I., and Sheviakova, O. I. Lysozyme as an index of pathogenic properties of staphylococci. *Antibiotiki,* **15:**1036–1038, 1970. (Russ.)

A18. Agati, G., Stoppa, I. M., and Toscano, G. Ricerche sperimentali sull'effetto leucopenico conseguente all'associazione antimitotici-radiazioni ionizzanti. Influenza di farmaci ad azione protettiva e stimolante sulla leucopoiesi. *Minerva Radiol., Fisioter. Radiobiol.,* **10:**629–637, 1965.

A19. Akhmedov, Iu. D., Likhtenstein, G. I., Ivanov, L. V., and Kokhanov, Iu. V. Transglobular effects in lysozyme as investigated by the spin label method. *Dokl. Akad. Nauk SSSR,* **205:**372–374, 1972.

A20. Akinrimisi, E. O., and Ts'o, P. O. P. Interactions of purine with proteins and amino acids. *Biochemistry,* **3:**619–626, 1964.

A21. Albano, S. B., and Carollo, F. L'inibizione *in vivo* delle gonadotropine ipofisarie e corionica effettuata dal lisozima. *Boll. Soc. Ital. Biol. Sper.,* **36:**1142–1144, 1960.

A22. Albano, V. Action of lysozyme on the diptheria bacillus. *Boll. Ist. Sieroterap. Milan.,* **28**:244–251, 1949.

A23. Alderton, G., and Fevold, H. L. Direct crystallization of lysozyme from egg white and some crystalline salts of lysozyme. *J. Biol. Chem.,* **164**:1–5, 1946,

A24. Alderton, G., Lewis, J. C., and Fevold, H. L. The relationship of lysozyme, biotin and avidin. *Science,* **101**:151–152, 1945.

A25. Alderton, G., Ward, W. H., and Fevold, H. L. Isolation of lysozyme from egg white. *J. Biol. Chem.,* **157**:43–58, 1945.

A26. Aldrich, J. E., and Cundall, R. B. The radiation-induced inactivation of lysozyme. *Int. J. Radiat. Biol.,* **16**:343–358, 1969.

A27. Aldrich, J. E., Cundall, R. B., Adams, G. E., and Willson, R. L. Identification of essential residues in lysozyme: A pulse radiolysis method. *Nature (London),* **221**:1049–1050, 1969.

A28. Aldrich, K. M., and Sword, C. P. Methicillin-induced lysozyme-sensitive forms of staphylcocci. *J. Bacteriol.,* **87**:690–695, 1964.

A29. Alexander, J. W. Effect of thermal injury upon the early resistance to infection. *J. Surg. Res.,* **8**:128–137, 1968.

A30. Alexander, J. W. Serum and leukocyte lysosomal enzymes. Derangements following severe thermal injury. *Arch. Surg. (Chicago),* **95**:482–491, 1967.

A31. Alfonsini, L. Lytic activity of serum and saliva of tuberculous patients. *Boll. Soc. Ital. Biol. Sper.,* **24**:864–866, 1948.

A32. Allansmith, M. R., Drell, D., Anderson, R. P., and Newman, L. Comparison of electrophoretic mobility of tear lysozyme in 50 subjects. *Amer. J. Ophthalmolo.,* **71**:525–529, 1971.

A33. Allen, A. K., and Neuberger, A. The inhibition of goose lysozyme by oligosaccharides from the cell wall of micrococcus lysodeikticus. *Biochim. Biophys. Acta.,* **235**:539, 1971.

A34. Allerhand, A., Childers, R. F., and Oldfield, E. Natural-abundance carbon-13 nuclear magnetic resonance studies in 20-mm sample tubes. Observation of numerous single-carbon resonances of hen egg-white lysozyme. *Biochemistry,* **12**:1335–1341, 1973.

A35. Allerhand, J., Karelitz, S., Isenberg, H. D., Penbharkkul, S., and Ramos, A. The lacrimal proteins in Down's syndrome (mongolism). *J. Pediat.,* **62**:235–238, 1963.

A36. Allerhand, J., Karelitz, S., Penbharkkul, S., Ramos, A., and Isenberg, H. D. Electrophoresis and immunoelectrophoresis of neonatal tears. *J. Pediat.,* **62**:85–92, 1963.

A37. Allison, V. D. The antigenic properties of lysozyme-dissolved vaccines. *Brit. J. Expe. Patholo.,* **6**:99–108, 1925.

A38. Allison, V. D. The effect of the administration of vaccines on the lysozyme content of tissues and secretions. *Brit. J. Exp. Pathol.,* **5**:165–170, 1924.

A39. Altekar, W., and Gurnani, S. The nature of inactivation and interaction of lysozyme with guanidine hydrochloride *Indian J. Biochem. Biophys.,* **9**:293–296, 1972.

A40. Amano, T. The diverse nature of neutralizing antibodies of egg white lysozyme. *Jap. J. Bacteriol.,* **16**:636–637, 1961. (Jap.)

A41. Amano, T. , and Fujio, H. Antigenic determinants of lysozyme. *Tampakushitsu, Kakusan, Koso,* **13**:129–136, 1968. (Jap.)

A42. Amante, S. Azione del lisozima in alcune condizioni sperimentali di shock e di reazione d'allarme. *Minerva Chir.,* **13**:784–790, 1958.

A43. Amato, A., Costa, A. L., and Costa, A. Azione in vivo di policationi organici su un ceppo patogeno di Candida albicans. *Boll. Soc. Ital. Biol. Sper.,* **44**:729–731, 1968.

A44. Amato, A., Misefari, A., and Teti, D. Sull'azione eosinopenizzante del lisozima nel ratto. *Boll. Soc. Ital. Biol. Sper.*, **43**:629–632, 1967.

A45. Amato, A., Spadaro, M., and Costa, A. L. Comportamento della "sindrome enzimo-plasmatica" di topi infetti con virus EFH 120 e sottoposti a trattamento lisozimico. *Epatologia*, **12**:794–799, 1966.

A46. Amato, A. and Teti, D., Azione del lisozima su citostrutture di *M. Lysodeicticus*: Modificazioni dell'attività flogogena. *Boll. Soc. Ital. Biol. Sper.*, **43**:902–905, 1967.

A47. Ambrosio, L., and Mazza, V. Comportamento del lisozima nel siero di sangue e negli organi nelle intossicazioni sperimentali (piombo, benzolo, manganese). *Riv. Ist. Sieroter. Ital.*, **29**:252–262, 1954.

A48. Ambrosio, L., and Mazza, V. Lysozyme activity in diabetes superimposed upon fatigue. *Folia Med. (Naples)*, **33**:438–441, 1950.

A49. Ambrosio, L., and Mazza, V. Lysozyme of normal fatigued subjects. *Folia Med. (Naples)*, **33**:435–437, 1950.

A50. Ambrosio, L., and Mazza, V. Il comportamento del lisozima nei soggetti normali affaticati. *Folia Med. (Naples)*, **36**:372–381, 1953.

A51. Ananthanarayanan, V. S., and Bigelow, C. C. Unusual difference spectra of proteins containing tryptophan. II. Studies with proteins. *Biochemistry*, **8**:3723–3728, 1969.

A52. Ananthanarayanan, V. S., and Bigelow, C. C. Unusual difference spectra of proteins containing tryptophan. I. Studies with model compounds. *Biochemistry*, **8**:3717–3723, 1969.

A53. Andersen, O. Lysozymes in xerophthalmia. *Hospitalstidende*, **75**:1029–1037, 1932.

A54. Andersen, O. Untersuchungen über das Lysozym aus dem Micrococcus lysodeicticus. *Z. Immunitäetsforsch. Exp. Ther.*, **70**:90–103, 1931.

A55. Andersen, O. Uber die Verhältnisse des Lysozyms bei Xerophthalmie. *Acta Paediat., (Stockholm)*, **14**:81–91, 1932.

A56. Anderson, D. L., and Johnson, R. C. Electron microscopy of immune disruption of leptospires: Action of complement and lysozyme. *J. Bacteriol.*, **95**:2293–2309, 1968.

A57. Anderson, E. A., and Alberty, R. A. Homogeneity and the electrophoretic behavior of some proteins. II. Reversible spreading of steady-state boundary criteria. *J. Phys. Colloid Chem.*, **52**:1345–1364, 1948.

A59. Anderson, R. K., and Farmer, C. J. Studies on enzymatic digestion of gastric mucin. *Proc. Soc. Exp. Biol. Med.*, **32**:21–23, 1934.

A60. Ando, T., Fujioka, H., and Kawanishi, Y. The C-terminal sequence of lysozyme. *Biochim. Biophys. Acta*, **31**:553–555, 1959.

A61. Ando, E., and Gramignani, V. Sull'attività lisozimica e lipasica del siero di sangue di soggetti affetti da tbc. ossea ed osteoarticolare prima, durante e dopo trattamento con acido para-amino-salicilico. *Minerva Ortop.*, **4**:96–99, 1953.

A62. Andolsek, L. Il Deflamon nella terapia della trichomoniasi. *Minerva. Ginecol.*, **21**:1270–1278, 1969.

A63. Andreĭchin, M. A. Some indices of nonspecific reactivity of the organism in infectious hepatitis. *Vrach. Delo.*, **7**:141–143, 1971. (Russ.)

A64. Angelo, P. Ricerche sull'azione del lisozima nei confronti del virus del diftero-vaiolo dei polli. *Acta Med. Vet.*, **11**:135–140, 1965.

A65. Anikina, T. P., and Golosova, T. V. Effect of lysozyme on the protective function of microphages and macrophages. *Antibiotiki*, **12**:813–817, 1967. (Russ.)

A66. Anikina, T. P., Golosova, T. V., and Kokhanovskaia, T. M. The effect of lysozyme on the phagocytic capicity of leukocytes in experiments in vitro *Antibiotiki*, **11**:446–450, 1966. (Russ.)

A67. Araki, Y., Nakatani, T., Nakayama, K., and Ito, E. Occurrence of *N*-nonsubstituted glucosamine residues in peptidoglycan of lysozyme-resistant cell walls from *Bacillus cereus. J. Biol. Chem.,* **237**:6312–6322, 1972.

A68. Arcuri, F., Robert, L., and Piersantelli, N. Il comportamento dell'attività lisozimica in corso di epatite virale. *Epatologia,* **12**:410–413, 1966.

A69. Ardoino, L. A., Matracia, S., and Allegra, L. Incremento della urinazione ad opera del lisozima. *Boll. Soc. Ital. Biol. Sper.,* **36**:1374–1375, 1960.

A70. Arimura, H. Effect of lysozyme from human placenta on ectromelia virus. *Acta Virol. (Prague),* **17**:130–137, 1973.

A71. Armour and Company. Lysozyme: A review. Chicago, 1952.

A72. Arneberg, P. Quantitative determination of albumin, lysozyme, and amylase. A comparison of seven analytical methods. *Scand. J. Dent. Res.,* **78**:435–439, 1970.

A73. Arnheim, N., Jr., and Wilson, A. C. Quantitative immunological comparison of bird lysozymes. *J. Biol. Chem.,* **242**:3951–3956, 1967.

A74. Arnheim, N., Inouye, M., and Laudin, A. Chemical studies on the enzymatic specificity of goose egg white lysozyme. *J. Biol. Chem.,* **248**:233–236, 1973.

A75. Arnheim, N., Prager, E. M., and Wilson, A. C. Immunological prediction of sequence differences among proteins. Chemical comparison of chicken, quail, and pheasant lysozymes. *J. Biol. Chem.,* **244**:2085–2094, 1969.

A76. Arnheim, N., and Steller, R. Multiple genes for lysozyme in birds. *Arch. Biochem. Biophys.,* **141**:656–661, 1970.

A77. Arnheim, N., Sobel, J., and Canfield, R. Immunochemical resemblance between human leukemia and hen egg-white lysozyme and their reduced carboxymethyl derivatives. *J. Mol. Biol.,* **61**:237–250, 1971.

A78. Arnon, R. A selective fractionation of anti-lysozyme antibodies of different determinant specificities. *Eur. J. Biochem.,* **5**:583–589, 1968.

A79. Arnon, R., and Sela, M. Antibodies to a unique region in lysozyme provoked by a synthetic antigen conjugate. *Proc. Nat. Acad. Sci. U.S.,* **62**:163–170, 1969.

A80. Arnon, R., and Maron, E. Lack of immunological cross-reaction between bovine alpha-lactalbumin and hen's egg-white lysozyme. *J. Mol. Biol.,* **51**:703–707, 1970.

A81. Arnon, R., Maron, E., Sela, M., and Anfinsen, C. B. Antibodies reactive with native lysozyme elicited by a completely synthetic antigen. *Proc. Nat. Acad. Sci. U.S.,* **68**:1450–1455, 1971.

A82. Arnon, R., and Maron, E. An immunological approach to the structural relationship between hen egg-white lysozyme and bovine alpha-lactalbumin. *J. Mol. Biol.,* **61**:225–235, 1971.

A82a. Arsenis, C., Eisenstein, R., Soble, L. W., and Kuettner, K. E. Enzyme activities in chick embryonic cartilage. Their subcellular distribution in isolated chondrocytes. *J. Cell Biol.,* **49**:459–467, 1971.

A83. Arshavskii, I.A., Morachevskaya, E. V., and Shtamler, S. M. Lysozyme of the saliva during ontogenesis. *Byull. Eksp. Biol. Med.,* **21**:69–71, 1946.

A84. Arvidson, S., Holme, T., and Wadström, T. Formation of bacteriolytic enzymes in batch and continuous culture of *Staphylococcus aureus. J. Bacteriol.,* **104**:227–233, 1970.

A85. Asamer, H. Die Bedeutung der Muramidase-(Lysozym-)-Bestimmung für die Leukämiediagnostik. *Deut. Med. Wochenschr.,* **96**:829–830, 1971.

A86. Asamer, H. Schmalzl, F., and Braunsteiner, H. Der immunzytologische Lysozymnachweis in menschlichen Blutzellen. *Acta Haematol.,* **41**:49–54, 1969.

A87. Asamer, H., Schmalzl, F., and Braunsteiner, H. Immunocytological demonstration of lysozyme (muramidase) in human leukaemic cells. *Brit. J. Haematol.*, **20:**571–574, 1971.

A88. Asamer, H., Schmalzl, F. and Braunsteiner, H. Die diagnostische und prognostische Bedeutung der Muramidase-(Lysozym-) Bestimmung in Leukocytenlysaten, Serum und Harn von Leukämiepatienten. *Klin. Wochenschr.*, **49:**587–593, 1971.

A89. Atassi, M. Z., and Habeeb, A. F. S. A. Enzymic and immunochemical properties of lysozyme. I. Derivatives modified at tyrosine. Influence of nature of modification on activity. *Biochemistry*, **8:**1385–1393, 1969.

A90. Atassi, M. Z., Habeeb, A. F. S. A., and Ando, K. Enzymic and immunochemical properties of lysozyme. VII. Location of all the antigenic reactive regions. A new approach to study immunochemistry of tight proteins. *Biochim. Biophys. Acta*, **303:**203–209, 1973.

A91. Atassi, M. Z., Suliman, A. M., and Habeeb, A. F. S. A. Enzymic and immunochemical properties of lysozyme. IV. Conformation, enzymic activity and immunochemistry of derivatives modified at arginine residues. *Immunochemistry*, **9:**907–920, 1972.

A92. Atassi, M. Z., Habeeb, A. F. S. A., and Rydstedt, L. Lack of immunochemical cross-reaction between lysozyme and alpha-lactalbumin and comparison of their conformations. *Biochim. Biophys. Acta*, **200:**184–187, 1970.

A93. Atassi, M. Z., Perlstein, M. T., and Habeeb, A. F. S. A. Conformational studies on modified proteins and peptides. IV. Conformation of lysozyme derivatives modified at tyrosine or at tryptophan residues. *J. Biol. Chem.*, **246:**3291–3296, 1971.

A94. Atha, D. H., and Ackers, G. K. Calorimetric determination of denaturation enthalpy for lysozyme in guanidine hydrochloride. *J. Biol. Chem.*, **246:**5845–5848, 1971.

A95. Atkins, A. M., Barr, H. A., and Schofield, G. C. Localization of immunoglobulins and lysozyme in sheep lacrimal gland. *J. Anat.*, **110:**494, 1971.

A96. Audiffren, P., Vandekerkove, M., Faucon, R., and Nicol, J. Démasquage du substrat du lysozyme chez Neisseria meningitidis. *C. R. Soc. Biol.* **161:**394–396, 1967.

A97. Audran, R. Bactericidal and bacteriolytic immune reactions. Respective roles of complement and lysozyme. Importance in mechanisms of defense against infection. *Rev. Fr. Transfus.*, **15:**81–137, 1972.

A98. Aune, K. C., Salahuddin, A., Zarlengo, M. H., and Tanford, C. Evidence for residual structure in acid- and heat-denatured proteins. *J. Biol. Chem.*, **242:**4486–4489, 1967.

A99. Aune, K. C., and Tanford, C. Thermodynamics of the denaturation of lysozyme by guanidine hydrochloride. I. Dependence on pH at 25 degrees. *Biochemistry*, **8:**4579–4585, 1969.

A100. Aune, K. C., and Tanford, C. Thermodynamics of the denaturation of lysozyme by guanidine hydrochloride. II. Dependence on denaturant concentration at 25 degrees. *Biochemistry*, **8:**4586–4590, 1969.

A101. Aver'ianova, L. L. Effect of tetracycline on the content of lysozyme in the blood serum of rabbits and in human saliva. *Antibiotiki*, **15:**1034–1036, 1970. (Russ.)

A102. Aver'ianova, L. L., and Fidel'man, E. S. Effect of multiple intramuscular penicillin injections on the content of lysozyme in the rabbit blood. *Antibiotiki*, **9:**438–441, 1964. (Russ.)

A103. Aver'ianova, L. L., and Fidel'man, E. S. The effects of a single penicillin injection on the blood level of lysozyme in rabbits. *Antibiotiki*, **9:**38–41, 1964. (Russ.)

A104. Aver'ianova, L. L., Fidel'man, E. S., and Vol'man, I. B. Effect of bicillin therapy on the content of lysozyme in the saliva of rheumatic children. *Antibiotiki*, **10:**445–447, 1965. (Russ.)

A105. Avio, C. M., Polese, E., Rossi-Torelli, M., Imperato, S. Analisi cromatografica di lisozima radioattivo. *Nuovi Ann. Ig. Microbiol.*, **15:**353–356, 1964.

A106. Awataguchi, S., Kawabata, T., Watanabe, K., *et al.* Sinobronchitis in children and its treatment with anti-inflammatory enzyme. *Nippon Rinsho,* **28:**363–369, 1970. (Jap.)

A107. Axelsson, B., and Piscator, M. Renal damage after prolonged exposure to cadmium. *Arch. Environ. Health,* **12:**360, 1966.

A108. Azari, P. Action of sulfite on lysozyme. *Arch. Biochem. Biophys.,* **115:**230–232, 1966.

B

B1. Baba, S., Oashi, M., and Onoda, T. Evaluation of E-221-005 in inflammatory nasal diseases by double blind method. *Otolaryngology* (*Tokyo*), **44:**153–157, 1972. (Jap.)

B2. Baba, T. Studies on the immunospecific and non-specific substance of Bacillus anthracis, with special reference to antigenic substances used in Ascoli's reaction. 5. Changes in the serological properties of immunospecific and non-specific substance by enzymatic digestion. *Jap. J. Bacteriol.,* **19:**121–124, 1964. (Jap.)

B3. Babudieri, B., and Bietti, G. B. Electron microscopic observations on bacteriolysis produced by lysozyme of tears. *Arch. Ophthalmol.,* **39:**449–454, 1945.

B4. Baggiolini, M., Hirsch, J. G., and de Duve, C. Further biochemical and morphological studies of granule fractions from rabbit heterophil leukocytes. *J. Cell Biol.,* **45:**586–597, 1970.

B5. Baggiolini, M., Hirsch, J. G., and de Duve, C. Resolution of granules from rabbit heterophil leukocytes into distinct populations by zonal sedimentation. *J. Cell Biol.,* **40:**529–541, 1969.

B5a. Bail, O. Fleming's lysozyme. *Wien. Klin. Wochenschr.,* **36:**107–108, 1923. (Ger.)

B5b. Bail, O. Versuche über die Vielheit von Bakteriophagen. *Z. Immunitaetforsch. Exp. Ther.,* **38:**57–164, 1923–4.

B5c. Bainbridge, F. A. The action of certain bacteria on proteins. *J. Hyg.,* **11:**341–355, 1911.

B6. Baker, C. M. Molecular genetics of avian proteins. IX. Interspecific and intraspecific variation of egg white proteins of the genus *Gallus. Genetics,* **58:**211–226, 1968.

B7. Baker, C. M., and Manwell, C. Molecular genetics of avian proteins. VIII. Egg white proteins of the migratory quail, Coturnix coturnix—new concepts of "hybrid vigour." *Comp. Biochem. Physiol.,* **23:**21–42, 1967.

B8. Bakri, M., and Wolfe, F. H. Destabilization of casein micelles by lysozyme. *Can. J. Biochem.,* **49:**882–884, 1971.

B9. Balazs, T., and Roepke, R. R. Lysozymuria induced in rats by nephrotoxic agents. *Proc. Soc. Exp. Biol. Med.,* **123:**380–385, 1966.

B10. Balekjian, A. Y., Hoerman, K. C., and Berzinskas, V. J. Lysozyme of the human parotid gland secretion: Its purification and physicochemical properties. *Biochem. Biophys. Res. Commun.,* **35:**887–894, 1969.

B11. Balestreri, R., Pompei, A., and Jacopino, G. E. La corticosteroidogenesi in soggetti normali trattati con lisozima. *Arch. "Maragliano" Patol. Clin.* **21:**551–557, 1965.

B12. Ballardie, F. W., and Capon, B. 3,4-Dinitrophenyl Tetra-N-acetyl-Beta-chitotetraoside, a good chromophoric substrate for hens' egg-white lysozyme. *J. Chem. Soc. Chem. Commun.,* **14:**828–820, 1972.

B13. Banerjee, S. K., and Rupley, J. A. Binding of N-acetylglucosamine tetrasaccharide to lysozyme. *Arch. Biochem. Biophys.,* **155:**19–23, 1973.

B14. Banerjee, S. K., and Rupley, J. A. Temperature and pH dependence of the binding of oligosaccharides to lysozyme. *J. Biol. Chem.,* **248:**2117–2124, 1973.

B15. Barbel, I. E. The lysozyme content of tears after operations on the eyeball. *Sovet. Vestnik Oftalmol.,* **6:**446–452, 1935. (Russ.)

B16. Barbero, S., and Lorenzi, L. La nostra esperienza in tema di terapia dell'epatite virale dell'infanzia. *Epatologia,* **12:**476–480, 1966.

B17. Barbieri, D., and Genesi, M. Lisozima nel plasma e nei leucociti leucemici. *Haematologica,* **38:**891–899, 1954.

B18. Barbu, E., and Rampini, C. Le lysozyme des phages. *Exposes Annu. Biochim. Med.,* **27:**85–100, 1966.

B19. Barel, A. O., Dolmans, M., and Léonis, J. Spectroscopic studies on the reduction and reformation of the disulfide bonds of papaya lysozyme. *Eur. J. Biochem.,* **19:**488–495, 1971.

B20. Barel, A. O., and Léonis, J. Transitions thermiques dans le lysozyme. *Arch. Int. Physiol. Biochim.,* **71:**289–291, 1963.

B21. Barel, A. O., Prieels, J. P., Maes, E., Looze, U., and Léonis, J. Comparative physicochemical studies of human alpha-lactalbumin and human lysozyme. *Biochem. Biophys. Acta,* **257:**288–296, 1972.

B22. Barel, A. O., Turneer, M., and Dolmans, M. Spectral studies of the interaction of bovine alpha-Lactalbumin and egg-white lysozyme with 2-p-Toluidinylnaphthalene-6-sulfonate. *Eur. J. Biochem.,* **30:**26–32, 1972.

B23. Baretta, G. Modificazioni della cromogenesi e della fibrinolisi stafilococcica in vitro con il lysozima. *Fracastoro,* **59:**211–213, 1966.

B24. Baretta, G. Modificazioni dell'attività dell'alfa emolisina stafilococcica in vitro con il lisozima. *Riv. Ist. Sieroter. Ital.,* **39:**366–373, 1964.

B25. Baretta, G. Modificazioni dell'attività stafilo-coagulasica in vitro con il lisozima *Riv. Ist. Sieroter. Ital.,* **38:**290–295, 1963.

B26. Baricalla, A., and Giaccardi, P. Influenza della vitamina A sull'attività terapeutica del lisozima nella tricomoniasi vaginale. *Minerva Ginecol.,* **13:**458–460, 1961.

B27. Baricalla, A., and Giaccardi, P. Possibilità terapeutiche del lisozima nelle vaginite da trichomonas. *Minerva Ginecol.,* **12:**1062–1071, 1960.

B28. Barman, T. E. Reactivities of the tryptophan residues of alpha-lactalbumin and lysozyme to 2-hydroxy-5-nitrobenzyl bromide. *J. Mol. Biol.,* **52:**391–394, 1970.

B29. Barnes, J. M. The enzymes of lymphocytes and polymorphonuclear leucocytes. *Brit. J. Exp. Pathol.,* **21:**264–274, 1940.

B30. Barnes, K. P., Warren, J. R., and Gordon, J. A. Effect of urea on the circular dichroism of lysozyme. *J. Biol. Chem.,* **247:**1708–1712, 1972.

B31. Barondes, R. de R. Beneficial effects of egg-white in intestinal infections; lysozyme content of egg-white. *Med. Rec.,* **151:**138–139, 1940.

B32. Barondes, R., and Kane, T. J. Sickle cell anemia; sickle cell-accelerating factor and its relationship to lysozyme. *Mil. Surg.,* **103:**271–275, 1948.

B33. Baronenko, V. A., and Iskrzhitskaia, A. I. Effect of erythemic ultraviolet irradiation on the osmotic resistance of erythrocytes and lysozyme content in the blood serum. *Vop. Kurortol. Fizioter. Lech. Fiz. Kul't.,* **33:**448–450, 1968. (Russ.)

B34. Barratt, T. M., and Crawford, R. Lysozyme excretion as a measure of renal tubular dysfunction in children. *Clin. Sci.,* **39:**457–465, 1970.

B35. Barratt, T. M., and Crawford, R. Muramidase (lysozyme) excretion in children. *Arch. Dis. Childhood,* **44:**780–781, 1969.

B36. Barrett, J. T. Agglutination and lysis of erythrocytes by lysozyme. *Proc. Soc. Exp. Biol. Med.,* **97:**794–795, 1958.

B36a. Barstow, L. E., Hruby, V. J., Robinson, A. B., Rupley, J. A., Sharp, J. J., and Shimoda, T. Preparation of a protein with lysozyme activity—characterization. *Fed. Proc., Fed. Amer. Soc. Exp. Biol.,* **30:**1292, Abstr. No. 1292, 1971.

B37. Bartels, H. A., and Buchbinder, M. The presence of lysozyme in root canals. *Oral Surg., Oral Med. Oral Pathol.,* **10:**993–1001, 1957.

B37a. Barth, G., Records, R., Bunnenberg, E., Djerassi, C., and Voelter, W. Magnetic circular dichroism studies. XII. The determination of tryptophan in proteins. *J. Amer. Chem. Soc.*, **93**:2545–2547, 1971.

B38. Barzizza, C. M., and Ricardi, E. A. Mecanismo de acción de la lisozima. *Bol. Inst. Maternidad*, **14**:78–80, 1945.

B39. Barzizza, C. M., and Ricardi, E. A. Mecanismo de acción de la lisozima. *Obstet. Ginecol. Latino-Amer.*, **3**:556–558, 1945.

B40. Basilova, G. I., and Pilipenko, V. G. A possibility of the use of a reaction for C-reactive protein and the indices of lysozyme level in the blood serum for the assessment of reactogenicity and safety of live combined vaccines. *Zh. Mikrobiol., Epidemiol. Immunobiol.*, **9**:152, 1972.

B41. Bastman-Heiskanen, L. Lysozyme determination; clinically useful approximative method. *Nord. Med.*, **46**:1138–1139, 1951.

B42. Baumann-Grace, J. B., and Tomcsik, J. Lysozymempfindlichkeit der Spezies *B. megaterium. Experientia*, **13**:148, 1957.

B43. Baumgartner-Morf, R. Wirkung verschiedener Alkohole auf das Cytoplasma einer Bacillus-Spezies. *Pathol. Microbiol.*, **28**:313–331, 1965.

B44. Becker, M. E., and Hartsell, S. E. The synergistic action of lysozyme and trypsin in bacteriolysis. *Arch. Biochem.*, **55**:257–269, 1955.

B45. Beddell, C. R., and Blake, C. C. F. An x-ray study of the product of an unusual reaction in lysozyme. *Biochem. Soc. Symp.*, **31**:157–161, 1970.

B46. Beddell, C. R., Moult, J., and Phillips, D. C. Crystallographic studies of the active site of lysozyme. *Mol. Properties Drug Receptors, Ciba Found. Symp.*, 85–112, Churchill, London 1970.

B47. Bedoni, C., Felletti, V., and Zibordi, F. Il tasso serico del lisozima in alcuni processi infiammatori cronici di pertinenza otorinolaringoiatrica. *Arch. Ital. Otol., Rinol. Laringol.*, **72**:184–192, 1961.

B49. Bello, J. Thermal perturbation difference spectra of proteins containing tryptophyl residues. *Biochemistry*, **9**:3562–3568, 1970.

B50. Bellomo, G., and Muscolino, F. Lysozyme in the premature infant. *Pediatria (Naples)*, **69**:485–496, 1961.

B51. Benacchio, L. Azione combinata di lisozima e di un antidiarroico nel trattamento delle sindromi colitiche. *Minerva Gastroenterol.*, **12**:28–30, 1966.

B52. Benevento, L. Il lisozima nella rino-tracheo-bronchite influenzale del lattante. *Gazz. Med. Ital.*, **122**:401–402, 1963.

B53. Benisek, W. F., and Richards, F. M. Attachment of metal-chelating functional groups to hen egg white lysozyme. An approach to introducing heavy atoms into protein crystals. *J. Biol. Chem.*, **243**:4267–4271, 1968.

B54. Berengo, A., Manicardi, E., and Bussinello, E. Effetto della somministrazione di testosterone sull'attività lisozimica del siero nel coniglio. *Riv. Inst. Sieroter. Ital.*, **26**:319–325, 1951.

B55. Bergamini, L. Azione del lisozima sul *B. anthracis*. *Boll. Inst. Sieroter. Milan.*, **32**:256–260, 1953.

B56. Bergamini, L., and Ferrari, W. Inhibiting action of polysulfonate anticoagulants on the lytic action of lysozyme. *Boll. Ist. Sieroter. Milan.*, **27**:89–92, 1948.

B57. Berger, L. R., and Weiser, R. S. The beta-glucosaminidase activity of egg-white lysozyme. *Biochim. Biophys. Acta*, **26**:517–521, 1957.

B57a. Berger, L. R. and Weiser, R. S. The Beta-glucosaminidase activity of egg-white lysozyme. *Biochim. Biophys. Acta*, **26**:517–521, 1957.

B58. Bergonzi, F., and D'Amico, E. Prime osservazioni sul trattamento della timomegalia nel lattante con lisozima. *Minerva Pediat.,* **16**:1029–1030, 1964.

B59. Berlin, I., and Neujahr, H. Y. Studies of controlled lysis of washed cell suspensions of *Lactobacillus fermenti* and preparation of membrane-like fragments by a combined trypsin-lysozyme treatment. *Acta Chem. Scand.,* **22**:2972–2980, 1968.

B60. Berlin, R. D., and Wood, W. B., Jr. Studies on the pathogenesis of fever. XII. Electrolytic factors influencing the release of endogenous pyrogen from polymorphonuclear leukocytes. *J. Exp. Med.,* **119**:697–714, 1964.

B61. Berliner, L. J. A 6 A crystallographic study of a spin-labeled inhibitor complex with lysozyme. *J. Mol. Biol.,* **61**:189–194, 1971.

B62. Bernardi, L., and Mainoldi, F. L'impiego del lisozima nelle leucopenie da radiazioni ionizzanti. *Minerva Radiol.,* **12**:325–336, 1967.

B63. Bernath, F. R., and Vieth, W. R. Lysozyme activity in the presence of nonionic detergent micelles. *Biotechnol. Bioeng.,* **14**:737–752, 1972.

B64. Bernier, I., and Jolles, P. Action du N-bromosuccinimide sur le lysozym de blanc d'oeuf de poule. *C. R. Acad. Sci.,* **253**:745–747, 1961.

B65. Berthou, J., Laurent, A., and Jollès, P. Preliminary X-ray investigation of duck egg-white lysozyme II. *J. Mol. Biol.,* **71**:815–817, 1972.

B66. Bewley, T. A., and Li, C. H. The reduction of protein disulfide bonds in the absence of denaturants. *Int. J. Protein Res.,* **1**:117–124, 1969.

B67. Bewley, T. A., and Li, C. H. Reactivity of the tryptophan residues in lysozyme. *Nature (London),* **206**:624, 1965.

B68. Beychok, S. Side-chain optical activity in cystine-containing proteins: Circular dichroism studies. *Proc. Nat. Acad. Sci. U.S.,* **53**:999–1006, 1965.

B68a. Beychok, S., and Warner, R. C. Denaturation and electrophoretic behavior of lysozyme. *J. Amer. Chem. Soc.* **81**:1892–1896, 1959.

B69. Biantovskaya, E. T. Application of lysozyme in ophthalmology. *Vestn. Oftalmol.,* **14**:44–53, 1939. (Russ.)

B70. Biasotti, A., Messi, G., Ferrari, E. M., and Tiesi, P. A. Lysozyme therapy in grave ulcerative colitis. *Dia Med.,* **31**:2545–2548, 1959.

B71. Bibb, W. R., and Straughn, W. R. Formation of protoplasts from *Streptococcus faecalis* by lysozyme. *J. Bacteriol.,* **84**:1094–1098, 1962.

B72. Bierman, E. O., Ducey, E. F., Lobstein, O. E., and Ferrari, R. Lysozyme therapy in malignancy. *Filter,* **35**:109, 1963.

B73. Bijsterveld, O. P. van. Diagnostic tests in the Sicca syndrome. *Arch. Ophthalmol.,* **82**:10–14, 1969.

B74. Billitteri, A., and Bernardini, A. La produzione di lisozima da parte dello *Staph. aureus* in rapporto alla composizione del terreno di cultura. *Boll. Soc. Ital. Biol. Sper.,* **39**:1839–1843, 1963.

B76. Binder, R. A., and Gilbert, H. S. Muramidase in polycythemia vera. *Blood,* **36**:228–232, 1970.

B77. Bird, O. D., and Hoevet, B. The vitamin B_{12} - binding power of proteins. *J. Biol. Chem.,* **190**:181–189, 1951.

B78. Birdsell, D. C., and Cota-Robles, E. H. Production and ultrastructure of lysozyme and ethylenediaminetetraacetate-lysozyme spheroplasts of *Escherichia coli. J. Bacteriol.,* **93**:427–437, 1967.

B79. Bjurulf, C. Calorimetric titration of lysozyme. *Eur. J. Biochem.,* **30**:33–38, 1972.

B80. Bjurulf, C., Laynez, J., and Wadsö, I. Thermochemistry of lysozyme-inhibitor binding. *Eur. J. Biochem.,* **14**:47–52, 1970.

B81. Bjurulf, C., and Wadsö, I. Thermochemistry of lysozyme-inhibitor binding. *Eur. J. Biochem.,* 31:95–102, 1972.

B82. Black, L. W., and Hogness, D. S. The lysozyme of bacteriophage lambda. I. Purification and molecular weight. *J. Biol. Chem.,* 244:1968–1975, 1969.

B83. Black, L. W., and Hogness, D. S. The lysozyme of bacteriophage lambda. II. Amino acid and end group analysis. *J. Biol. Chem.,* 244:1976–1981, 1969.

B84. Black, L. W., and Hogness, D. S. The lysozyme of bacteriophage lambda. III. Ordering the cyanogen bromide peptides. *J. Biol. Chem.,* 244:1982–1987, 1969.

B85. Bladen, H., Hageage, G., Harr, R., and Pollock, F. Lysis of certain organisms by the synergistic action of complement and lysozyme. *J. Dent. Res.,* 52:371–376, 1973.

B86. Blake, C. C. F. A crystallographic study of the oxidation of lysozyme by iodine. *Proc. Roy. Soc., Ser. B* 167:435–438, 1967.

B86a. Blake, C. C. F. The preparation of isomorphous derivatives. *Advan. Protein Chem.,* 23:59–120, 1968.

B87. Blake, C. C. F., Fenn, R. H., North, A. C. T., Phillips, D. C., and Poljak, R. J. Structure of lysozyme. A Fourier map of the electron density at 6 angstrom resolution obtained by x-ray diffraction. *Nature (London),* 196:1173–1176, 1962.

B88. Blake, C. C. F., Johnson, L. N., Mair, G. A., North, A. C. T., Phillips, D. C., and Sarma, V. R. Crystallographic studies of the activity of hen egg-white lysozyme. *Proc. Roy. Soc., Ser. Biol.,* 167:378–388, 1967.

B89. Blake, C. C. F., Koenig, D. F., Mair, G. A., and North, A. C. T., Phillips, D. C., and Sarma, V. R. Structure of hen egg-white lysozyme. A three-dimensional Fourier synthesis at 2 angstrom resolution. *Nature (London),* 206:757–761, 1965.

B90. Blake, C. C. F., Mair, G. A., North, A. C. T., Phillips, D. C., and Sarma, V. R. On the conformation of the hen egg-white lysozyme. *Proc. Roy. Soc., Ser. B* 167:365–377, 1967.

B91. Blake, C. C. F., and Swan, I. D. A. X-ray analysis of structure of human lysozyme at 6 angstrom resolution. *Nature (London) New Biol.,* 232:12–15, 1971.

B92. Blasi, F., DeMasi, V. R., and Soria, M. The electron spin resonance spectra of chemically modified lysozyme irradiated in solid state. *Biophysik,* 3:335–338, 1967.

B93. Blatt, M. L., and Kessler, H. Human milk: Its lysozyme content and bacterial count. *Amer. J. Dis. Child.,* 53:768–784, 1937.

B94. Bleiweis, A. S., and Zimmerman, L. N. Formation of two types of osmotically fragile bodies from Streptococcus faecalis var. liquefaciens. *Can. J. Microbiol.,* 7:363–373, 1961.

B95. Blinova, M. I., Zhukovskaia, N. A., and Ermol'eva, Z. V. Experience in the prevention of influenza with lysozyme combined with ecmoline. *Antibiotiki,* 11:834–836, 1966. (Russ.)

B95a. Bloomfield, A. L. The fate of bacteria introduced into the upper air passages. *Bull. Johns Hopkins Hosp.,* 30:317–322, 1919.

B96. Blumberger, W., and Glatzel, H. Beeinflussung der Lysozymaktivität des Speichels durch landesübliche Würzmittel. *Deut. Zahnaerztl. Z.,* 23:666–674, 1968.

B97. Boasson, E. H. On the bacteriolysis by lysozyme. *J. Immunol.,* 34:281–293, 1938.

B98. Boasson, E. H. Lysozyme action. *Acta Brevia Neer. Physiol., Pharmacol., Microbiol.,* 6:59–63, 1936.

B99. Bobo, R. A., and Foster, J. W. The effect of enzyme treatments on *Brucella abortus* cell walls. *J. Gen. Microbiol.,* 34:1–8, 1964.

B100. Bodanszky, A., Bodanszky, M., Jorpes, E. J., Mutt, V., and Ondetti, M. A. Molecular architecture of peptide hormones optical rotatory dispersion of cholecystokinin-pancreozymin, bradykinin and 6-glycine bradykinin. *Experientia,* 26:948–950, 1970.

B101. Bogdanova, V. I., Panchenko, L. A., Bubliĭ, V. P., Belochkina, N. A. Blood serum lysozyme changes in patients with infectious hepatitis. *Vrach. Delo,* **7**:144–147, 1971. (Russ.)

B102. Bonaduce, A., Alosi, C., and Caccioni, I. Ricerche sul liquido idatideo bovino. IV. Determinazione della leucin-aminopeptidase, del lisozime e del glucosio. *Boll. Soc. Ital. Biol. Sper.,* **39**:31–32, 1963.

B103. Bonaduce, A., and Papparella, V. Emoagglutinazione crociata lisozima-virus Newcastle. *Boll. Soc. Ital. Biol. Sper.,* **40**:875–876, 1964.

B104. Bonavida, B. Structural basis for immune recognition of lysozymes. IV. Immunologically active peptide obtained by the action of cyanogen bromide on human lysozyme. *Immunochemistry,* **8**:829–839, 1971.

B105. Bonavida, B., Miller, A., and Sercarz, E. E. Structural basis for immune recognition of lysozymes. I. Effect of cyanogen bromide on hen egg-white lysozyme. *Biochemistry,* **8**:968–979, 1969.

B106. Bonavida, B., Sapse, A. T., and Sercarz, E. E. Rabbit tear proteins. I. Detection and quantitation of lysozyme in nonstimulated tears. *Invest. Ophtahlmol.,* **7**:435–440, 1968.

B107. Bonavida, B., and Sapse, A. T. Human tear lysozyme. II. Quantitative determination with standard Schirmer strips. *Amer. J. Ophthalmol.,* **66**:70–76, 1968.

B108. Bonavida, B., Sapse, A. T., and Sercarz, E. E. Human tear lysozyme. I. Purification, physicochemical, and immunochemical characterization. *J. Lab. Clin. Med.,* **70**:951–962, 1967.

B109. Bonavida, B., and Sercarz, E. Structural basis for immune recognition of lysozymes. II. Reactive but non-immunogenic epitopes. *Eur. J. Immunol.,* **1**:166–170, 1971.

B110. Bondar', Z. A., Finogenova, L. G., and Zolotnitskaia, R. P. Immunologic reactions in chronic liver disease. *Ter. Arkh.,* **43**:26–30, 1971. (Russ.)

B111. Bonfiglio, V., and Simonetti, N. Lisozima e filamentizzazione di B. megatherium. *Nuovi Ann. Ig. Microbiol.,* **9**:518–523, 1958.

B112. Bonfiglio, V., and Simonetti, N. Azione del lisozima au B. megatherium resistente alla streptomicina. *Riv. Ist. Sieroter. Ital.,* **33**:414–416, 1958.

B113. Bonifaci, E., Rattini, F. M., Baggio, P., and Gallo, E. Lisozima serico ed infezione da virus (polio 2 vivo) nei ratti timectomizzati alla nascita. *Minerva Pediat.,* **18**:490–491, 1966.

B114. Bonilla, C. A. Adsorption of lysozyme to polyacrylamide gel (Bio-Gel P-2). *J. Chromatogr.,* **47**:499–501, 1970.

B115. Bonnet, M. J., Calas, E., Fouquet, H., Florens, A., and Davin, A. Intérèt thérapeutique majeur du lysozyme en dermatologie. *Bull. Soc. Fr. Dermatol. Syphiligr.,* **79**:406, 1972.

B116. Borders, C. L., Jr. I. Purification and partial characterization of testicular hyaluronidase. II. Mechanistic studies of human lysozyme. Ph. D. Thesis, California Institute of Technology, Pasadena, 1968.

B117. Bordet, J., and Bordet, M. The bacteriolytic power of colostrum and of milk. *C. R. Acad. Sci.* **179**:1109–1113, 1924.

B118. Bordet, M. Essais d'extraction du lysozyme. *C. R. Soc. Biol.,* **99**:1254–1256, 1928.

B119. Bordet, M. Contribution a l'étude du lysozyme. *C. R. Soc. Biol.,* **99**:1252–1254, 1928.

B120. Borecký, L., and Lackovic, V. The reticuloendothelial system and virus infection. I. The reaction of mouse peritoneal mononucleate cells to infection with the mouse and egg line of influenza virus—persistence and dissemination of the virus in the mouse organism. *Acta Virol. (Prague), Engl. Ed.,* **8**:208–216, 1964.

B121. Bottino, S. Effetti del lisozima sulla mortalità di tipo intestinale da radiazioni ionzzanti. *Atti Accad. Fisiocr. Siena, Sez. Med.-Fis.*, **17**:1421–1428, 1968.

B122. Bradbury, J. H. The hydrazinolysis of insulin, lysozyme, wool proteins and wool. *Biochem. J.*, **68**:482–486, 1958.

B123. Bradbury, J. H., and King, N. L. R. Denaturation of proteins: Single or multiple step process? *Nature (London)*, **223**:1154–1156, 1969.

B123a. Bradbury, J. H., and Wilairat, P. Proton magnetic resonance spectroscopy of histidine residues in proteins. *Biochem. Biophys. Res. Commun.*, **29**:84–89, 1967.

B124. Bradford, W. L., and Roberts, J. B. The lysozyme content of blood. *J. Pediat.*, **8**:24–30, 1936.

B125. Bradshaw, R. A., and Deranleau, D. A. Use of *N*-methylnicotinamide chloride as a conformational probe in proteins. Identification of the binding sites in chicken egg-white lysozyme and a comparison with bovine alpha-lactalbumin. *Biochemistry*, **9**:3310–3315, 1970.

B126. Bradshaw, R. A., Kanarek, L., and Hill, R. L. The preparation, properties, and reactivation of the mixed disulfide derivative of egg white lysozyme and L-cystine. *J. Biol. Chem.*, **242**:3789–3798, 1967.

B127. Bragg, L. A discussion on the structure and function of lysozyme . . . London, The Royal Institution, 3 Feb. 1966. (Introduction and concluding remarks.) *Proc. Roy. Soc., Ser. B*, **167**:349 and 448, 1967.

B128. Brancadoro, G., and Docimo, R. Su alcuni poteri di difesa organica durante ibernazione artificiale. I. Il comportamento del tasso di lisozima ematico. (Ricerche sperimentali e cliniche.) *G. Ital. Chir.*, **9**:880–886, 1953.

B129. Brancato, U. Azione dell lisozima sui processi di guarigione delle ferite. *Riv. Patol. Clin.*, **14**:763–737, 1959.

B130. Brandtzaeg, P. The 'Lyso-Plate' method for quantification of muramidase activity. *Scand. J. Dent. Res.*, **80**:166–167, 1972.

B131. Brandtzaeg, P., and Mann, W. V. J. A comparative study of the lysozyme activity of human gingival pocket fluid, serum, and saliva. *Acta Odontol. Scand.*, **22**:441–455, 1964.

B132. Brant, D. A., and Schimmel, P. R. Analysis of the skeletal configuration of crystalline hen egg-white lysozyme. *Proc. Nat. Acad. Sci. U.S.*, **58**:428–435, 1967.

B133. Braun, O. H. Die Bedeutung enteralen Lysozyms für den Säugling. *Deut. Med. Wochenschr.*, **94**:1458–1462, 1969.

B134. Braun, O. H. Faekale Lysozymausscheidung bei Säuglingen. *Ann. Paediat.*, **205**:266–280, 1965.

B135. Braun, O. H. Fecal lysozyme excretion in infants and its effects on the intestinal flora. *Monatsschr. Kinderheilk.*, **109**:162–164, 1961. (Ger.)

B136. Braun, O. H. Der Einflub der Ernährung auf die fäkale lysozymausscheidung bei darmgesunden säuglingen. *Z. Kinderheilk.*, **83**:690–710, 1960.

B137. Braun, O. H. Lysozymausscheidung mit den Faeces bei Säuglingsenteritis. *Z. Kinderheilk.*, **81**:742–751, 1958.

B138. Braun, V., and Wolff, H. The murein-lipoprotein linkage in the cell wall of *Escherichia coli*. *Eur. J. Biochem.*, **14**:387–391, 1970.

B139. Brawerman, G., Revel, M., Salser, W., and Gros, F. Initiation factor requirements for the *in vitro* synthesis of T4 lysozyme. *Nature (London)*, **223**:957–958, 1969.

B140. Brew, K., Vanaman, T. C., and Hill, R. L. Comparison of the amino acid sequence of bovine alpha-lactalbumin and hens egg white lysozyme. *J. Biol. Chem.*, **242**:3747–3749, 1967.

B140a. Brewer, C. F., and Riehm, J. P. Evidence for possible nonspecific reactions between N-ethylmaleimide and proteins. *Anal. Biochem.*, **18**:248–255, 1967.

B141. Brey, W. S., Jr., Evans, T. E., and Hitzrot, L. H. Nuclear magnetic relaxation times of water sorbed by proteins. Lysozyme and serum albumin. *J. Colloid Interface Sci.*, **26**:306–316, 1968.

B142. Brey, W. S., Jr., Heeb, M. A., and Ward, T. M. Dielectric measurements of water sorbed on ovalbumin and lysozyme. *J. Colloid Interface Sci.*, **30**:13–20, 1969.

B143. Briggs, R. S., Perillie, P. E., and Finch, S. C. Lysozyme in bone marrow and peripheral blood cells. *J. Histochem. Cytochem.*, **14**:167–170, 1966.

B144. Brody, H. A., and Noble, R. E. Studies in recurrent oral aphthae. II. Lysozyme levels from the parotid and submaxillary gland. *J. Oral Med.*, **24**:17–19, 1969.

B144a. Brown, J. R. Disulphide bridges of lysozyme. *Biochem. J.*, **92**:13P, 1964.

B145. Brown, R. A., and West, G. B. The antihistaminic activity of lysozyme. *Int. Arch. Allergy Appl. Immunol.*, **26**:204–214, 1965.

B146. Brown, W. C., Sandine, W. E., and Elliker, P. R. Lysis of lactic acid bacteria by lysozyme and ethylenediaminetetraacetic acid. *J. Bacteriol.*, **83**:697–698, 1962.

B147. Browne, W. J., North, A. C. Phillips, D. C., Brew, K., Vanaman, T. C., and Hill, R. L. A possible three-dimensional structure of bovine alpha-lactalbumin based on that of hen's egg-white lysozyme. *J. Mol. Biol.*, **42**:65–86, 1969.

B148. Bruenn, J. Characterization of a recessive-lethal amber suppressor strain of *Salmonella typhimurium* by *in vitro* synthesis of T4 lysozyme. *Biochim. Biophys. Acta*, **269**:162–169, 1972.

B149. Brumfitt, W. The mechanism of development of resistance to lysozyme by some gram-positive bacteria and its results. *Brit. J. Exp. Pathol.*, **40**:441–451, 1959.

B150. Brumfitt, W., and Glynn, A. A. Intracellular killing of *Micrococcus lysodeikticus* by macrophages and polymorphonuclear leucocytes: A comparative study. *Brit. J. Exp. Pathol.*, **42**:408–423, 1961.

B151. Brumfitt, W., Wardlaw, A. C., and Park, J. T. Development of lysozyme-resistance in *Micrococcus lysodiekticus* and its association with an increased O-acetyl content of the cell wall. *Nature (London)*, **181**:1783–1784, 1958.

B152. Brundage, W. G. Bactericidal activity of antibody, complement, and lysozyme on *Micrococcus lysodeikticus* and their relationship to its chemical structure. Ph.D. thesis, Louisiana State University and Agriculture and Mechanics College, Baton Rouge, 1967.

B153. Brunelli, G., and Poggi, A. E. Il lisozima usato per via intraarticolare nelle artrosinoviti croniche. Osservazioni cliniche. *Minerva Med.*, **54**:1016–1021, 1963.

B154. Brunner, H., and Sussner, H. Raman scattering of native and thermally denatured lysozyme. *Biochim. Biophys. Acta*, **271**:16–22, 1972.

B155. Brusca, A. E., and Ardonino, A. L. Precipitazione di colloidi elettronegativi e di particelle virali per interazione con colloidi di carica opposta. *Riv. Ist. Sieroter. Ital.*, **40**:1–8, 1965.

B156. Brusca, A. E., and Ardonino, A. L. Interazione fra colloidi basici e sostanze Rivaltapositive di liquidi allantoidei di uova embrionate di pollo, inoculate con mixovirus. *Riv. Ist. Sieroter. Ital.*, **39**:477–485, 1964.

B157. Brusca, A. E., Comitini, E., Dionisi, O., *et al.* Protezione da lisozima delle lesioni flogistico-necrotiche del fegato di topini trattati con virus epatico EFH-120. *Epatologia*, **12**:150–158, 1966.

B158. Brusca, A. E., Guercio, V., and Ardoino, A. L. Precipitazione e concentrazione del virus della pseudo-peste aviaria a mezzo di una proteina elettropositiva. *Riv. Ist. Sieroter. Ital.*, **40**:9–17, 1965.

B159. Brusca, A. E., Mastroeni, P., and Rocca, A. Azione di colloidi basici su sospensioni di virus epatico EEF-120. *Epatologia,* 12:143–149, 1966.

B160. Brusca, A. E., Pasqualino, A., and Allegra, L. Effetti biologici del lisozima indipendenti dalla sua azione enzimatica. *Riv. Anat. Patol. Oncol.,* 17:509–516, 1960.

B161. Brusca, A. E., and Patrono, D. Interazione tra proteine basiche ed alcune strutture citologiche dei globuli rossi. II) Agglutinazione degli stromiematici e dei nuclei eritrocitici da parte di salmina, lisozima ed histone. *Boll. Soc. Ital. Biol. Sper.,* 38:736–739, 1962.

B162. Brusca, A. E., and Patrono, D. Interazione tra proteine basiche ed alcune strutture citologiche dei globuli rossi. 1) Riduzione della colorabilita' nucleare e citoplasmatica di eritrociti di pollo per trattamento con salmina, lisozima ed istone. *Boll. Soc. Ital. Biol. Sper.,* 38:734–736, 1962.

B163. Brusca, A. E., and Patrono, D. Interactive capacities of erythrocytic structures with basic proteins. I. Interactions between chicken erythrocyte nuclei and salmine, lysozyme and thymohistone. *Sicil. Sanit.,* 14:600–607, 1961. (Ital.)

B164. Brusca, A. E., and Patrono, D. Complessi elettrostatici insolubili tra lisozima e colloidi acidi batterici. *G. Batteriol. Virol. Immunol. Ann. Osped. Maria Vitteria Torino,* 53:211–220, 1960.

B165. Brusca, A. E., Patrono, D., and Ardoino, A. L. Interactive capacity of some erythrocytic structures with basic proteins. II. Interaction of bovine erythrocyte stromas and protein extracts of stromas with salmin, lysozyme and thymohistone. *Sperimentale,* 113:24–41, 1963. (Ital.)

B166. Brustad, T. Influence of H_2S on the radiosensitivity of enzymes and microorganisms. *Radiat. Res.,* 22:421–430, 1964.

B167. Brustad, T. The effects of radical scavengers on the radiosensitivity of lysozyme in dilute acqueous solutions of varying pH. *Radiat. Res.,* 27:456–473, 1966.

B168. Brustad, T. Study of the radiosensitivity of dry preparations of lysozyme, trypsin, and deoxyribonuclease, exposed to accelerated nuclei of hydrogen, helium, carbon, oxygen and neon. *Radiat. Res., Suppl.,* 2:65–74, 1960.

B169. Bruzzesi, M. R., Chiancone, E., and Antonini, E. Association-dissociation properties of lysozyme. *Biochemistry,* 4:1796–1800, 1965.

B170. Bucci, M. G. Zambrini's ptyalo-reaction and lysozyme in the tears and saliva of normal subjects and ocular patients. *Boll. Ocul.,* 40:268–288, 1961. (It.)

B171. Budetta, M. Sull'uso del lisozima associato agli antibiotici nella terapia delle sepsie delle infezioni virali del lattante ospedalizzato. *Clin. Pediat.,* 41:558–570, 1959.

B172. Budetta, M., and Benevento, L. Viral stomatitis and epidemic parotiditis treated with lysozyme. *Sem. Med. (Buenos Aires),* 119:1034–1038, 1961. (Span.)

B173. Buissière, J., Morélis, P., Chambon, P., and Fontanges, R. Aspect morphologique de la lyse des staphylocoques par *Bacillus alvei.* Role du lysozyme. *Ann. Inst. Pasteur, Paris,* 113:357–362, 1967.

B174. Bukharin, O. V., and Gerasimov, A. B. Use of lysozyme for the differentiation of Streptococcus viridans. *Antibiotiki,* 15:1039–1040, 1970. (Russ.)

B175. Bukharin, O. V., and Gerasimov, A. V. Effect of lysozyme on the production of bacteriocins in Streptococcus viridans. *Antibiotiki,* 14:608–610, 1969. (Russ.)

B176. Bukharin, O. V., and Iakovleva, Z. M. Non-specific prophylactic action of lysozyme in infections. *Antibiotiki,* 10:151–156, 1965. (Russ.)

B177. Bukharin, O. V., Solonina, I. P., Bel'skaia, V. I., *et al.* Prevention of influenza with lysozyme and folicobalamine. *Antibiotiki,* 11:836–838, 1966. (Russ.)

B178. Bukharin, O. V., Usviatsov, B., Frolov, B. A., Artischeva, L. I., and Smoliagen, A. I. Lysozyme activity of bacteria. *Antibiotiki,* 17:66–70, 1972. (Russ.)

B179. Bull, H. B. Determination of molecular weight of proteins in spread monolayers. *J. Biol. Chem.*, **185**:27–38, 1950.

B179a. Bull, H. B., and Breese, K. Ionization of proteins: analogue computer analysis. *Arch. Biochem. Biophys.*, **117**:106–109, 1966.

B180. Bultasová, H., Smahel, O., Svobodová, J., and Pinsker, P. Effect of lysozyme on the secretion and metabolism of cortisol. *Vnitr. Lek.*, **13**:833–837, 1967. (Czech.)

B181. Buontempo, U., Careri, G., and Faselia, P. Hydration water of globular proteins: The infrared band near 3300 cm -1. *Biopolymers*, **11**:519–521, 1972.

B182. Burchardt, K., Fenrych, W., and Wysocki, H. Muramidase activity in granulocytopenia. *Pol. Med. J. (Warsaw)*, **10**:876–881, 1971.

B183. Burchardt, K., Fenrych, W., and Wysocki, H. Muramidase activity in granulocytopenia. *Pol. Arch. Med. Wewn.*, **45**:617–623, 1970. (Pol.)

B184. Burchardt, K., and Wysocki, K. Renal elimination of mucopeptide N-acetyl-muramylhydrolase (muramidase, lysozyme). *Pol. Arch. Med. Wewn.*, **48**:305–309, 1972. (Pol.)

B185. Burghartz, N. Ergebnisse von Lysozymbestimmungen im Speichel bei Magengesunden und Magenkranken. *Klin. Wochenschr.*, **30**:284, 1952.

B186. Burghartz, N. Der Lysozymgehalt des Blutes und seine Beziehungen zym Ulcus pepticum. *Ver. Deut. Ges. Inn. Med.*, **57**:278–280, 1951.

B187. Burghartz, N., and Boosfeld, E. Untersuchungen zur Verbesserung der Lysozymbestimmungsmethode. *Klin. Wochenschr.*, **32**:181–182, 1954.

B188. Burghartz, N., and Boosfeld, E. Bericht über Lysozymuntersuchungen im Urin. *Klin. Wochenschr.*, **32**:182, 1954.

B189. Burghartz, N., and Quenzer, K. Lysozym und Allergie. Gleichzeitig ein Beitrag zum Problem der Ulcussentstehung. *Klin. Wochenschr.*, **32**:977–978, 1954.

B190. Bürgi, H., Regli, J., and Medici, T. Lysozym im Sputum bei chronischer Bronchitis. *Schweiz. Med. Wochenschr.*, **96**:1414–1415, 1966.

B191. Buss, D. H. Lysozyme activity in baboon milk. *Comp. Biochem. Physiol.*, **31**:783–787, 1969.

B192. Bussinello, E., Bajocchi, E., and Barbolini, G. Influenza della idrazide dell'acido isonicotinico (INI) sull' attivitá lisozimica del siero nell 'uomo tubercolotico. *Arch. Tisiol. Mal. App. Resp.*, **9**:967–971, 1954.

B193. Bustin, M., and Givol, D. Specific isolation of peptides containing modified tyrosine residues on insoluble antibody column. *Biochim. Biophys. Acta*, **263**:459–467, 1972.

B194. Butchard, C. G., Dwek, R. A., Ferguson, S. J., Kent, P. W., and Xavier, A. V. Mapping of the binding site of N-fluoroacetyl-D-glucosamine and analogues in hen egg lysozyme by ^{1}H and ^{19}F-NMR techniques with GD(III) and Mn(II) as paramagnetic probes. *FEBS Lett.*, **25**:91–93, 1972.

B195. Butchard, C. G., Dwek, R. A., Kent, P. W., Williams, R. J. P., and Xavier, A. V. A structural study by ^{19}F-nuclear-magnetic resonance of the binding of sugars to lysozyme. *Eur. J. Biochem.*, **27**:548–553, 1972.

B196. Butler, B., and Echols, H. Regulation of bacteriophage lambda development by gene N: properties of a mutation that bypasses N control of late protein synthesis. *Virology*, **40**:212–222, 1970.

B197. Butler, L. G., and Rupley, J. A. The binding of saccharide to crystalline and soluble lysozyme measured directly and through solubility studies. *J. Biol. Chem.*, **242**:1077–1078, 1967.

B198. Butovetskiĭ, L. D., Bukharin, O. V., and Bannikov, V. K. Immunological reactivity of patients with gonorrhea. *Vestn. Dermatol. Venerol.*, **45**:56–59, 1971. (Russ.)

B199. Buyanovskaya, I. S. Purified, dry preparations of lysozyme. *Microbe Variability Conf.*, 471–474, 1936.

C

C1. Cabezas, J., and Rubina, M. Lysozyme. *Ana. Quim. Farm.*, 9–17, 1942.
C2. Cabezas, J., and Vaccaro, H. Nonspecific factors in resistance. A biochemical and immunological study of lysozyme. *Proc. Int. Congr. Microbiol., 5th, 1950*, p. 196, 1950.
C3. Cabezas, J., Vaccaro, H., and Gonzalez, A. El "lisozima" en las mucosas nasal y paranasales. Su importancia como factor inespecifico de defensa. *Rev. Otorrinolaringol.*, **4**:37–55, 1944.
C4. Cahn-Bronner, C. E. The presence and action of lysozyme in nasal mucosa. *Ann. Otol., Rhinol., & Laryngol.*, **51**:250–252, 1942.
C5. Cali, G. Influenza del lisozima per aerosol sul titolo antistreptolisinico-O e sulla velocitá di eritrosedimentazione in soggetti sani. *Clin. Otorino-Laring.*, **16**:111–117, 1964.
C6. Califano, L. Action of lysozyme in liberating nucleic acid from bacteria. *Boll. Ist. Sieroter., Milan.*, **28**:366–371, 1949.
C7. Callerio, C. Studi sul "lysozym." *Boll. Soc. Med.-Chir. Pavia*, **46**:699–722, 1932.
C8. Callerio, C. Appearance of granules in the cytoplasm of tumour-cell cultures in contact with lysozyme. *Nature (London)*, **184**:202–203, 1959.
C9. Callerio-Babudieri, D. Lysozyme granules and lysosome structures in cell culture. *Nature (London)*, **212**:1274–1275, 1966.
C10. Callerio-Babudieri, D., and Callero, C. Ulteriori osservazioni siu granuli citoplasmatici lisosomo-simili ottenuti con lisozima fluorescente in cellule KB. *Nuovi Ann. Ig. Microbiol.*, **19**:265–268, 1968.
C11. Callerio-Babudieri, D., and Callerio, C. Electron microscope observations of tumour cell cultures in the presence of lysozyme. *Nature (London)*, **200**:693–694, 1963.
C12. Camba, R. Contributo all'interpretazione del diagramma elettroforetico dello sperma umano; dimostrazione del lisozima, della jaluronidasi e della fosfatasi acida. *Minerva Med.*, **77**:43–45, 1957.
C13. Campus, G. Lysozyme in the treatment of reactive stomatitis. *Boll. Mal. Orecchio, Gola Naso*, **79**:190–196, 1961.
C14. Candeli, A., and Mastrandrea, V. Azione del lisozima e del fluoruro di lisozima sugli sporigeni aerobi. *Boll. Ist. Sieroter. Milan.*, **43**:89–95, 1964.
C15. Candiani, G. B., and Bortolozzi, G. Aggiornamenti e prospettive nella terapia della trichomoniasi vaginale. *Minerva Ginecol.*, **21**:1259–1269, 1969.
C16. Canfield, R. E., and Liu, A. K. The disulfide bonds of egg white lysozyme (muramidase). *J. Biol. Chem.*, **240**:1997–2002, 1965.
C17. Canfield, R. E., and Anfinsen, C. B. Chromatography of pepsin and chymotrypsin digests of egg white lysozyme on phosphocellulose. *J. Biol. Chem.*, **238**:2684–2690, 1963.
C18. Canfield, R. E., and Anfinsen, C. B. Nonuniform labeling of egg white lysosyme. *Biochemistry*, **2**:1073–1078, 1963.
C19. Canfield, R. E. Peptides derived from tryptic digestion of egg white lysozyme. *J. Biol. Chem.*, **238**:2691–2697, 1963.
C20. Canfield, R. E. The amino acid sequence of egg white lysozyme. *J. Biol. Chem.*, **238**:2698–2707, 1963.
C21. Canfield, R. E., Kammerman, S., Sobel, J. H., and Morgan, F. J. Primary structure of lysozymes from man and goose. *Nature (London), (New Biol.)*, **232**:16–17, 1971.
C22. Canfield, R. E., and McMurry, S. Purification and characterization of a lysozyme from goose egg white. *Biochem. Biophys. Res. Commun.*, **26**:38–42, 1967.

C23. Cannefax, G. R., and Hanson, A. W. Ineffectiveness of lysozyme therapy in experimental syphilis in the rabbit. *Brit. J. Vener. Disease.*, **39**:192–194, 1963.

C24. Capon, B. Mechanism in carbohydrate chemistry. *Chem. Rev.* **69**:407–498, 1969.

C25. Cappuccino, J. G., Reilly, H. C., and Winston, S. Elevation of lysozyme in extracts of kidneys and spleens from tumor-bearing animals. *Cancer Res.*, **22**:850–856, 1962.

C26. Cappuccino, J. G., Winston, S., and Perri, G. C. Muramidase activity of kidney and spleen in Swiss mice challenged with B.C.G., zymosan and bacterial endotoxins. *Proc. Soc. Exp. Biol. Med.*, **116**:869–872, 1964.

C27. Caputi, F. Dissociazione microbica da lisozima (saggi sulla Mycotorula albicans). *G. Batteriol. Immunol.*, **44**:190–194, 1952.

C28. Caputo, A. Il lisozima nel canale alimentare. Ricerche sperimentali nel cane. *Arch. Sci. Biol. (Bologna)*, **36**:495–508, 1952.

C28a. Caputo, A. Caratteristiche del complesso pepsina-lisozima. *Giorn. Biochimica*, **4**: 310–331, 1955.

C29. Cardoso de Almeida, C. Lysozima. Fermento de defesa conjunctival existente na lagrima. Seu conceito actual. *Sao Paulo Med.*, **1**:207–219, 1941.

C29a. Carlström, D. The polysaccharide chain of chitin. *Biochim. Biophys. Acta*, **59**:361–364, 1962.

C30. Carmel, R., and Coltman, C. A., Jr. Serum vitamin B_{12}-binding capacity and muramidase changes with cyclic neutropenia induced by cytosine arabinoside. *Blood*, **37**:31–39, 1971.

C31. Carotenuto, A., Iannelli, O., and Ciardiello, A. Il lisozima nella terapia delle vasculopatie periferiche. *Rass. Int. Clin. Ter.*, **44**:517–520, 1964.

C32. Carotenuto, A., Iannelli, O., and Catanzano, C. Il lisozima nel trattamento postoperatorio delle malattie chirurgiche dell'addome. *Acta Med. Ital. Med. Trop. Subtrop. Gastroenterol.*, **18**:301–303, 1963.

C32a. Carr, C. W. Studies on the binding of small ions in protein solutions with the use of membrane electrodes. III. The binding of chloride ions in solutions of various proteins. *Arch. Biochem. Biophys.*, **46**:417–423, 1953.

C32b. Carr, C. W. Studies on the binding of small ions in protein solutions with the use of membrane electrodes. IV. The binding of calcium ions in solutions of various proteins. *Arch. Biochem.*, **46**:424–431, 1953.

C33. Carrera, G., and Galdiero, F. Lisozima e cloropromazina nella lisi bacterica. *G. Batteriol., Virol., Immunol. Ann. Osp. Maria Vittoria Torino*, **55**:330–335, 1962.

C34. Carson, K. J., and Eagon, R. G. Lysozyme sensitivity of the cell wall of Pseudomonas aeruginosa: Further evidence for the role of the non-peptidoglycan components in cell wall rigidity. *Can. J. Microbiol.*, **12**:105–108, 1966.

C35. Carson, M. E., and Dannenberg, A. M., Jr. Hydrolytic enzymes of rabbit mononuclear exudate cells. II. Lysozyme: Properties and quantitative assay in tuberculous and control inbred rabbits. *J. Immunol.*, **94**:99–104, 1965.

C36. Carat, R., and Cadeddu, N. Utile accorgimento per il prelievo di secrezioni per il dosaggio del lisozima. *Boll. Ocul.*, **32**:87–90, 1953.

C37. Caruso, P. Ulterio indagini sui rapporti tra properdina e lisozima. *Boll. Soc. Ital. Biol. Sper.*, **37**:526–529, 1961.

C38. Caruso, P., and Chimenz, B. Elettroshock e lisozima. *Boll. Soc. Ital. Biol. Sper.*, **35**:1443–1446, 1959.

C39. Casaglia, G., and Uncini-Manganelli, C. L'attività antiflogistica del lisozima in campo ginecologico. Ricerche sperimentali e cliniche. *Riv. Ital. Ginecol.*, **49**:428–439, 1965.

C40. Caselli, P. Experimental possibilities for the inactiviation of endogenous pyrogen by lysozyme. *C. R. Soc. Biol.*, **154**:84–86, 1960.

C41. Caselli, P. Analisi spettrofotometrica e potere litico di diverse preparazioni di lisozima purificato. *Arch. Sci. Biol. (Bologna)*, **33**:414–430, 1949.

C42. Caselli, P. Il comportamento del nucleo batterico nella lisi da lisozima. *Boll. Soc. Ital. Biol. Sper.*, **23**:1176–1179, 1947.

C43. Caselli, P. Apprezzamento dell'attività lisozimica. *Boll. Soc. Ital. Biol. Sper.*, **22**:57–59, 1946.

C44. Caselli, P., and Callerio, C. Inhibition of exogenous and endogenous pyrogens by lysozyme. *Nature (London)*, **188**:238, 1960.

C45. Caselli, P., and Cantelmo, P. Caratteri dell'attività inibente dell'urina umana normale lisa da lisozima. *Boll. Soc. Ital. Biol. Sper.*, **23**:564–565, 1947.

C46. Caselli, P., and Cavallo, G. Azione dei normali constituenti urinari sulla batteriolisi da lisozima. *Boll. Soc. Ital. Biol. Sper.*, **23**:566–567, 1947.

C47. Caselli, P., and Melucci, N. Attività lisozimica del siero di sangue di mammiferi, uccelli, rettili, anfibi e pesci. *Boll. Soc. Ital. Biol. Sper.*, **23**:568–569, 1947.

C48. Caselli, P., and Schumacher, H. Untersuchungen über den kolloidalen Zustand des Lysozyms im Nasensekret, Speichel und Magensaft. *Z. Gesamte Exp. Med.*, **126**:417–424, 1955.

C49. Caselli, P., and Schumacher, H. Mikroelektrophoretische Studien mit dem Serum-Lysozym. *Z. Gesamte Exp. Med.*, **124**:65–71, 1954.

C50. Caselli, P., and Tolone, S. L'attività lisozimica del liquor umano. *Boll. Soc. Ital. Biol. Sper.*, **22**:539–540, 1946.

C51. Caselli, P., and Tolone, S. L'attività antilisozimica del liquor umano. *Boll. Soc. Ital. Biol. Sper.*, **22**:537–539, 1946.

C52. Cassier, M., and Ryter, A. Sur un mutant de *Clostridium perfingens* donnant des spores sans tuniques à germination lysozyme-dépendante. *Ann. Inst. Pasteur, Paris*, **121**:717–732, 1971.

C53. Cassier, M., and Sebald, M. Germination lysozyme-dépendante des spores de *Clostridium perfringens* ATCC 3624 après traitement thermique. *Ann. Inst. Pasteur (Paris)*, **117**:312–324, 1969.

C54. Castaneda-Agullo, M. Nota preliminar sobre el aislamiento y purificación parcial de una lisozima de la leche de burra. *Gac. Med. Mex.*, **76**:165–169, 1946.

C55. Castermans, A. Complexation of transplantation antigens by lysozyme. *Transplant. Bull.*, **27**:98–100, 1961.

C56. Castrén, J. A., and Lavikainen, P. Animal experiments on the epithelization of corneal erosions when using lysozyme or steroids. *Acta Ophthalmol.*, **42**:33–37, 1964.

C57. Castriciano, N. Variazione della quota linfocitaria percentuale in silicotici dopo trattamento endovenoso con acido ascorbico e lysozyme. *Folia Med. (Naples)*, **51**:864–881, 1968.

C58. Catovsky, D., and Galton, D. A. G. Lysozyme activity and nitroblue-tetrazolium reduction in leukaemic cells. *J. Clin. Pathol.*, **26**:60–69, 1973.

C59. Catovsky, D., and Galton, D. A. Cytochemical demonstration of lysozyme and nitroblue tetrazolium reduction in leukaemia monocytes. *Brit. J. Haematol.*, **22**:634–635, 1972.

C60. Catovsky, D., Galton, D. A. G., and Griffin, C. The significance of lysozyme estimations in acute myeloid and chronic monocytic leukaemia. *Brit. J. Haematol.*, **21**:565–580, 1971.

C61. Catovsky, D., Galton, D. A. G., Griffin, C., Hoffbrand, A. V., and Szur, L. Serum lysozyme and vitamin B_{12} binding capacity in myeloproliferative disorders. *Brit. J. Haematol.*, **21**:661–672, 1971.

C62. Catovsky, D., Galton, D. A. G., and Robinson, J. Myeloperoxidase-deficient neutrophils in acute myeloid leukaemia. *Scand. J. Haematol.*, **9**:142–148, 1972.

C63. Cattan, D., and Bourgoin, D. Interactions between DNA and hen's egg white lysozyme. Effects of composition and structural changes of DNA, ionic strength and temperature. *Biochim. Biophys. Acta*, **161**:56–67, 1968.

C64. Cattan, D., Bourgoin, D., and Joly, M. Associations ADN-lysozyme. *Bull. Soc. Chim. Biol.*, **50**:923–924, 1968.

C65. Cattan, D., Bourgoin, D., and Joly, M. Association du lysozyme avec l'acide dé soxyribonucléique. I. Diagrammes de précipitation. *Bull. Soc. Chim. Biol.*, **44**:971–983, 1962.

C66. Cattan, D., and Dutheillet-Lamonthezie, N. Interaction préférentielle du lysozyme avec l'acide désoxyribonucléique de cellules d'ascite d'Ehrlich nouvellement synthétisé. *C. R. Acad. Sci., Ser. D*, **264**:1647–1650, 1967.

C67. Cattaneo, P. Potere lisozimico del liquido amniotico e potere antilisozimico del meconio. Ricerche sperimentali. *Clin. Ostet. Ginecol.*, **72**:220–222, 1949.

C68. Cattaneo, P., and Maggiora-Vergano, T. The lysozyme content of colostrum, early milk, meconium, and feces of the newborn during breast feeding or mixed artificial alimentation. *Rend. Ist. Super. Sanita (Ital. Ed.)*, **11**:994–1028, 1948.

C69. Cavaleri, M. Azione del lisozima sulla caseificazione del latte vaccino. *Minerva Nipiol.*, **7**:107–109, 1957.

C70. Cavaleri, M. Azione del lisozima sulla digestione gastrica del latte vaccino. *Minerva Nipiol.*, **7**:102–112, 1957.

C71. Cavalli, D., and Ottolenghi, G. Dell'azione antalgica del lisozima nelle forme neoplastiche dell'apparato genitale femminile in corso di trattamento radiante (radium, roentgen, gamma ray-therapy). *Riv. Ostet. Ginecol. Prat.*, **46**:438–461, 1964.

C72. Cavallo, G. Ricerche sul meccanismo della liberazione del mucopolisaccaride bacterico per azione del lisozima. *Boll. Ist. Sieroter. Milan.*, **30**:67–75, 1951.

C73. Cavallo, G., and Imperato, S. Caratteristiche di specie e di individuo nella produzione di sostanze naturali di difesa. *G. Mal. Infet. Parassit.*, **19**:77–85, 1967.

C74. Cavallo, G. Pontieri, G., and Imperato, S. Interaction between "liquoid" (sodium polyanethol-sulphonate) and lysozyme in the immune haemolytic reaction. *Experientia*, **19**:36–37, 1963.

C75. Cavka, V., and Prica, M. Über Lysozymwirkung in normalen und pathologischen Augensekreten. *Albrecht von Graefes Arch. Ophthalmol.*, **121**:740–755, 1929.

C75a. Cecil, R., and Wake, R. G. The reactions of inter- and intra-chain disulphide bonds in proteins with sulphite. *Biochem. J.*, **82**:401–406, 1962.

C76. Celestino, D., Crifo, S., and Cavanna, F. Lysozyme in the tissue of O.R.L. interest in panarteritic rats. Effects of parallel lysozyme treatment. *Valsalva*, **37**:253–257, 1961.

C77. Cernea, P., and Devien, C. V. Intérêt thérapeutique d'une nouvelle antibiothérapie associée. *Rev. Fr. Odonto-Stomatol.*, **14**:1541–1547, 1967.

C78. Cervini, C., Ciocci, A., and Lucherini, M. L'attività lisozimica serica nel Lupus eritematoso-sistemico. *Boll. Soc. Ital. Biol. Sper.*, **35**:766–768, 1959.

C79. Chakhava, O. V., and Goryunova, A. G. Formation of lysozyme in vitro by histiocyte-macrophage cultures. *Antibiotiki (Moscow)*, **10**:507–511, 1965. (Russ.)

C80. Chakhava, O. V., and Goryunova, A. G. In vitro lysozyme formation by culture of histiocytemacrophages. *Fed. Proc., Fed. Amer. Soc. Exp. Biol.*, **57**:Transl. Suppl., 707–709, 1966.

C81. Chaloupka, J., Křečkova, P., and Říhová, L. Changes in the character of the cell wall in growth of *Bacillus megaterium* cultures. *Folio Microbiol.*, **7**:269–274, 1962.

C82. Chaloupka, J., and Vereš, K. The nature and synthetic capacity of fragile cells of *Bacillus megatherium* partly deprived of cell walls. *Experientia,* **17**:562–573, 1961.

C83. Chaloupka, J., and Vereš, K. Formation of osmotically fragile rods by the action of lysozyme on *Bacillus megaterium* KM. *Folia Microbiol. (Prague),* **6**:379–385, 1961.

C84. Champagne, M., and Smith, D. B. Degradation by hydrazine of bovine plasma albumin and lysozyme. *Can. J. Biochem.,* **38**:715–726, 1960.

C85. Chandan, R. C., Parry, R. M., Jr., and Shahani, K. M. Purification and some properties of bovine milk lysozyme. *Biochim. Biophys. Acta,* **110**:389–398, 1965.

C86. Chandan, R. C., Shahani, K. M., and Holly, R. G. Lysozyme content of human milk. *Nature (London),* **204**:76–77, 1964.

C87. Chang, K. Y., and Carr, C. W. Studies on the structure and function of lysozyme. II. The similar activity of urea-treated and cyanate-treated lysozyme. *Biochim. Biophys. Acta,* **285**:377–382, 1973.

C88. Chang, K. Y., and Carr, C. W. Studies on the structure and function of lysozyme. I. The effect of pH and cation concentration of lysozyme activity. *Biochim. Biophys. Acta,* **229**:496–503, 1971.

C89. Chaput, A., and Gaboly, G. Enzymothérapie locale. Expérimentation clinque d'une nouvelle spécialité de thérapeutique dentaire. *Inform. Dent.,* **49**:210–214, 1967.

C90. Charachon, R., and Gagnon, J. Résultats cliniques de l'usage local d'une association lysozyme-papaïne en O.R.L. *Sem. Ther.,* **39**:562–565, 1963.

C91. Charlemagne, D., and Jollès, P. The action of various lysozymes on chitopentaose. *FEBS Lett.,* **23**:275–278, 1971.

C92. Charlemagne, D., Jollès, P. Inhibition par des polymères de la N-Acétylglucosamine de l'activite´ lysante à pH 6.2 de lysozymes d'origines differentes vis-à-vis de *Micrococcus lysodeikticus. C. R. Acad. Sci. Ser. D,* **270**:2721–2723, 1970.

C93. Charlemagne, D., and Jollès, P. La spécificité de divers lysozymes vis-à-vis de substrats de faible poids moléculaire provenant de la chitine. *Bull. Soc. Chim. Biol.,* **49**:1103–1113, 1967.

C94. Charlemagne, D., and Jollès, P. Les lysozymes des leucocytes et du plasma d'origine humaine. Hommes normaux et malades atteints de leucémie myéloïde chronique. *Nouv. Rev. Fr. Hematol.,* **6**:355–366, 1966.

C94a. Charlwood, P. A. Partial specific volumes of proteins in relation to composition and environment. *J. Amer. Chem. Soc.,* **79**:776–781, 1957.

C95. Charoki, C. Utilisation d'une médication enzymatique destinée au traitement local des affections de la bouche, du pharynx et du larynx. *Gaz. Med. Fr.,* **73**:142–144, 1966.

C96. Charoki, C. Utilisation d'une medication enzymatique destinee au traitement local des affections de la bouche, du pharynx et du larynx. *Rev. Laryngol., Paris,* **86**:1134–1135, 1965.

C97. Chatterjee, B. R., and Williams, R. P. Formation of spheroplasts from *Bacillus anthracis. J. Bacteriol.,* **89**:1128–1133, 1965.

C98. Chatterjee, B. R., and Williams, R. P. Preparation of spheroplasts from *Vibrio comma. J. Bacteriol.,* **85**:838–841, 1963.

C99. Chau, K. H., and Yang, J. T. Comparison of circular dichrometers: Normal and difference circular dichroism measurements. *Anal. Biochem.,* **46**:616–623, 1972.

C100. Chautin, S. M., and Zlatkina, E. I. Lysate therapy in ocular diseases. *Sov. Vestn. Optal'mol.,* **6**:713, 1935. (Russ.)

C101. Chen, H. C., and Craig, L. C. A dialysis study on the conformation of lysozyme and its binding properties with N-Acetyl-D-Glucosamine. *Bioorg. Chem.,* **1**:51–65, 1971.

C102. Chen, R. F. Limited reation of lysozyme with a flourescent labeling agent. *Biochem. Biophys. Res. Commun.,* **40**:1117–1124, 1970.

C103. Cherkasov, I. A., and Kravchenko, N. A. Structure-oriented addition of lysozyme to a polyacrylamide support. *Bull. Acad. Sci. USSR, Div. Chem. Sci., No. 10:*2374, 1972. (Russ.)

C104. Cherkasov, I. A., and Kravchenko, N. A. Determination of the heat of formation of the lysozyme-chitin enzyme-substrate complex by the chromatographic method. *Biokhimiya,* **35:**182–186, 1970. (Russ.)

C105. Cherkasov, I. A., and Kravchenko, N. A. Chromatographic approach to the study of the binding of lysozyme to substrate. *Biochim. Biophys. Acta,* **206:**289–294, 1970.

C106. Cherkasov, I. A., and Kravchenko, N. A. Improved method for the isolation of lysozymes by enzyme-substrate chromatography. *Biokhimiya,* **34:**1089–1091, 1969. (Russ.)

C107. Cherkasov, I. A., and Kravchenko, N. A. On some features of the sorbtion of lysozyme on chitin. *Biokhimiya,* **33:**761–765, 1968. (Russ.)

C108. Cherkasov, I. A., Kravchenko, N. A., and Kaverzneva, E. D. Chromatographic behavior of lysozyme in chitin containing column. *Dokl. Akad. Nauk SSSR,* **170:**213–215, 1966. (Russ.)

C109. Chesbro, W. R. Lysozyme and the production of osmotic fragility in enterococci. *Can. J. Microbiol.,* **7:**952–955, 1961.

C110. Chiancone, E., Bruzzesi, M. R., and Antonini, E. Studies on dextran and dextran derivatives. X. The interaction of dextran sulfate with lysozyme. *Biochemistry,* **5:**2823–2828, 1966.

C111. Chien, J. C. W., and Brandts, J. F. Natural abundance fourier transform of 13^c nuclear magnetic resonance spectra of lysozyme. *Nature (London), New Biol.,* **230:**209–210, 1971.

C112. Chipman, D. M. A kinetic analysis of the reaction of lysozyme with oligosaccharides from bacteria cell walls. *Biochemistry,* **10:**1714–1722, 1971.

C113. Chipman, D. M., Grisaro, V., and Sharon, N. The binding of oligosaccharides containing N-acetylglucosamine and N-acetylmuramic acid to lysozyme. The specificity of binding subsites. *J. Biol. Chem.,* **242:**4388–4394, 1967.

C114. Chipman, D. M., Pollock, J. J., and Sharon, N. Lysozyme-catalyzed hydrolysis and transglycosylation reactions of bacterial cell wall oligosaccharides. *J. Biol. Chem.,* **243:**487–496, 1968.

C115. Chipman, D. M., and Schimmel, P. R. Dynamics of lysozyme-saccharide interactions. *J. Biol. Chem.,* **243:**3771–3774, 1968.

C116. Chipman, D. M., and Sharon, N. Mechanism of lysozyme action. *Science,* **165:**454–465, 1969.

C117. Chistiakova, L. A., and Kravchenko, N. A. On the mechanism of salt activation of lysozyme. *Biokhimiya,* **37:**1126–1132, 1972. (Russ.)

C118. Choné, B., and Müller, D. Lysozymwirkung auf den Ehrlich-Aszites-Tumor der Maus. *Aerztl. Forsch.,* **22:**405–411, 1968.

C119. Chou, J. Y. UGA nonsense suppression assayed by T4 DNA-dependent *in vitro* synthesis of lysozyme. *Biochem. Biophys. Res. Commun.,* **41:**981–986, 1970.

C120. Chumachenko, N. V., and Sokolova, Z. M. Study of the effect of antibiotics on the activity of serum lysozyme and of aseptic exudate of experimental animals. *Antibiotiki (Moscow),* **12:**933–937, 1967. (Russ.)

C121. Chun, P. K. C., Wong, L., and Paik, Y. K. Siemsen, A. W., and Hokama, Y. Urinary C-reactive protein and lysozyme in renal homotransplantation. *Hawaii Med. J.,* **31:**262–265, 1972.

C122. Chun, P. W., Kim, S. J., Williams, J. D., Cope, W. T., Tang, L. H., and Adams, E. T., Jr. Graphical approach to non-ideal cases of self-associating protein systems. *Biopolymers,* **11:**197–214, 1972.

C123. Chung, Y. C., Upadhyay, J. M., and Garcia F. J. Purification and properties of a bacteriolytic enzyme from a soil amoeba *Hartmannella glebae*. *Arch. Biochem. Biophys.*, **135**:244–252, 1969.

C124. Churchich, J. E. Luminescence properties of muramidase and reoxidized muramidase. *Biochim. Biophys. Acta*, **92**:194–197, 1964.

C125. Churchich, J. E. The polarization of fluorescence of reoxidized muramidase (lysozyme). *Biochim. Biophys. Acta*, **65**:349–350, 1962.

C126. Churchich, J. E. Tryptophan residues in native and reoxidized muramidase: Luminescence properties. *Biochim. Biophys. Acta*, **120**:406–412, 1966.

C126a. Churchlich, J. E. Energy transfer in protein pyridoxamine-5-phosphate conjugates. *Biochemistry*, **4**:1405–1410, 1965.   –

C127. Churchich, J. E., and Irwin, R. Pyridoxyl-5-phosphate-lysozyme. Physical studies. *Biochim. Biophys. Acta*, **214**:157–167, 1970.

C128. Ciocci, A., and Violanti, A. Salivary lysozyme and Zambrini's ptyalo-reaction. *Progr. Med.*, **15**:Suppl., 771–778, 1959. (Ital.)

C129. Coice, C., and Zerneri, L. Attivita coagulante del lisozima nell'uomo. *Arch. Ital. Otol. Rinol. Laringol.*, **69**:271–278, 1958.

C130. Cline, M. J., Melmon, K. L., Davis, W. C., and Williams, H. E. Mechanism of endotoxin interaction with human leucocytes. *Brit. J. Haematol.*, **15**:539–547, 1968.

C131. Clough, P. W. Lysozyme in ulcerative colitis. *Ann. Intern. Med.*, **37**:813, 1952.

C132. Coassolo, M. Sulla presenza del lisozima nel cerume del condotto uditivo umano. *Minerva Otorinolaringol.*, **2**:505–507, 1952.

C133. Cocioba, L. Some data on lysozyme. *Rev. Stiint. Med.*, **27**:907–912, 1938.

C134. Cocuzza, S., and Ferrari, G. Comportamento dell'ornitin-carbamil-transferasi serica in soggetti affetti da epatite virale dopo trattamento con lisozima. *Epatologia*, **12**:342–344, 1966.

C135. Cocuzza, S., and Mignone, F. Influenza del lisozima sull'iperplasia timica. *Minerva Pediat.*, **18**:473–478, 1966.

C136. Cohen, J. S., and Jardetzky, O. Nuclear magnetic resonance studies of the structure and binding sites of enzymes. II. Spectral assignments and inhibitor binding in hen egg-white lysozyme. *Proc. Nat. Acad. Sci., U.S.*, **60**:92–99, 1968.

C137. Cohen, J. S. Proton magnetic resonance studies of human lysozyme. *Nature (London)*, **223**:43–46, 1969.

C137a. Cohen, L. A. Chemical modification as a probe of structure and function. *In* "The Enzymes" (P. D. Boyer, ed.), 3rd ed. Vol. 1, pp. 147–211. Academic Press, New York, 1970.

C138. Cohn, Z. A., and Hirsch, J. G. This isolation and properties of the specific cytoplasmic granules of rabbit polymorphonuclear leucocytes. *J. Exp. Med.*, **112**:983–1004, 1960.

C139. Cohn, Z. A., and Wiener, E. The particulate hydrolases of macrophages. I. Comparative enzymology, isolation, and properties. *J. Exp. Med.*, **118**:991–1008, 1963.

C140. Cohn, Z. A., and Wiener, E. The particulate hydrolases of macrophages. II. Biochemical and morphological response to particle ingestion. *J. Exp. Med.*, **118**:1009–1020, 1963.

C141. Cole, J. B., Bryan, M. L., and Bryan, W. P. Thermodynamics of solution of lysozyme crystals. *Arch. Biochem. Biophys.*, **130**:86–91, 1969.

C142. Coleman, S. E., van de Rijn, I., and Bleiweis, A. S. Lysis of grouped and ungrouped streptococci by lysozyme. *Infect. Immunity*, **2**:563–569, 1970.

C143. Collins, J. Relation between penicillin and N-acetylmuramic acid in the binding site of lysozyme. *Proc. Roy. Soc., Ser. B*, **167**:441–442, 1967.

C144. Colobert, L. Action du lysozyme sur les bactéries à gram négatif. *Exposes Annu. Biochim. Med.,* **27**:65–84, 1966.

C145. Colobert, L. On the nature of the membrane limiting protoplasts obtained by the action of lysozyme of Bacillus megaterium. *Ann. Inst. Pasteur, Paris,* **103**:37–52, 1961.

C146. Colobert, L., and Creach, O. Structure de la paroi ectoplasmique d'*Eberthella typhi*. Nature et localisation du substrat du lysozyme. *Ann. Inst. Pasteur, Paris,* **99**:672–694, 1960.

C147. Colobert, L. Lysozyme and autarcesis. *Pathol. Biol.,* **8**:1023–1038, 1960.

C148. Colobert, L. Etude de la lyse de salmonelles pathogènes provoquée par le lysozyme, après délipidation partielle de la paroi externe. *Ann. Inst. Pasteur, Paris,* **95**:156–167, 1958.

C149. Colobert, L. Cinétique de la phase spécifique de la lyse de Coccus P. (sp. Sarcina flava) par le lysozyme. Influence des ions. *Bull. Soc. Chim. Biol.,* **39**:1155–1162, 1957.

C150. Colobert, L. A propos de l'obtention prétendue de «protoplastes» par l'action contrôlée du lysozyme sur *Micrococcus lysodeikticus* et *Sarcina lutea*. *C. R. Soc. Biol.* **151**:114–116, 1957.

C151. Colobert, L. Obtention de protoplastes à partir de *Eberthella typhi* par action contrôlé e du lysozyme. *C. R. Soc. Biol.,* **151**:1904–1906, 1957.

C152. Colobert, L. Gonflement de la parot externe des Salmonelles par l'effet du lysozyme. *C. R. des Soc. Biol.,* **151**:1553–1555, 1957.

C153. Colobert, L. Destruction par le lysozyme, après délipidation de la paroi externe de Salmonelles pathogènes. *C. R. Acad. Sci.,* **245**:1674–1676, 1957.

C154. Colobert, L. Inactivation par le lysozyme de l'antigène somatique de Salmonelles pathogènes chauffées. *C. R. Acad. Sci.,* **244**:2863–2865, 1957.

C155. Colobert, L., and Creach, O. Obtention du substrat du lysozyme contenu dans la paroi ectoplasmique d' *Eberthella typhi*. *Biochim. Biophys. Acta,* **40**:167–169, 1960.

C156. Colobert, L., and Dirheimer, G. Le dosage de l'activité enzymatique du lysozyme. *Bull. Soc. Chim. Biol.,* **44**:141–147, 1962.

C157. Colobert, L., and Dirheimer, G. Action du lysozyme sur un substrat glycopeptidique isolé de *Micrococcus lysodeikticus*. *Biochim. Biophys. Acta,* **54**:455–468, 1961.

C158. Colobert, L., and Dirheimer, G. Obtention d'un substrat purifié du lysozyme à partir de *Micrococcus Lysodeikticus*. *C. R. Acad. Sci.,* **250**:423–442, 1960.

C159. Colobert, L., and Lenoir, J. Etude du mécanisme de la lyse de Coccus P (sp. Sarcina flava) par le lysozyme. *Ann. Inst. Pasteur, Paris,* **91**:Suppl., 74–88, 1956.

C160. Colobert, L., and Louisot, P. Formes filtrables d' *Eberthella typhi* obtenues par l'action du lysozyme. *Ann. Inst. Pasteur, Paris,* **97**:108–111, 1959.

C161. Colobert, L., and Servant, P. L'effet du lysozyme sur l'antigène somatique d' *Eberthella typhi*. *Ann. Inst. Pasteur, Paris,* **97**:549–559, 1959.

C162. Colucci, C. F., and Iacono, G. Ricerche sull'azione del lisozima su alcuni ceppi di virus influenzale e sui loro ricettori. *Acta Med. Ital. Mal. Infet. Parassit.,* **5**:431–436, 1950.

C162a. Colvin, J. R. The binding of anions by lysozyme, calf thymus histone sulphate, and protamine sulphate. *Can. J. Chem.,* **30**:320–331, 1952.

C162b. Colvin, J. R. The size and shape of lysozyme. *Can. J. Chem.,* **30**:831–834, 1952.

C162c. Colvin, J. R. Binding of anions by denatured proteins. *Can. J. Chem.,* **30**:973–984, 1952.

C162d. Colvin, J. R. The adsorption of methyl orange by lysozyme. *Can. J. Biochem. Physiol.,* **32**:109–119, 1954.

C162e. Colvin, J. R. A note on the stoichiometry of adsorption of anions by lysozyme. *Can. J. Biochem. Physiol.,* **33**:651–653, 1955.

C163. Colwell, C. A., Hess, A. R., and Tavaststjerna, M. Mononuclear cells from animals of divergent susceptibility to tuberculosis. I. Enzyme studies. *Amer. Rev. Resp. Dis.*, **88**:37–46, 1963.

C164. Combined Staff Clinics. Ulcerative colitis. *Amer. J. Med.*, **6**:481–494, 1949.

C165. Commission on Enzyme Nomenclature, International Union of Biochemistry. *Compr. Biochem.*, **14**:138, 1965.

C166. Commission on Enzyme Nomenclature, International Union of Biochemistry. *Compr. Biochem.*, **13**:106, 1964.

C167. Confalonieri, C., Suppa, G., and Grifoni, V. Un nuovo metodo per la determinazione della cronoresistenza leucocitaria in vitro. *Haematol. Lat.*, **4**:289–301, 1961.

C168. Conti, F., and Ricca, M. L'associazione cloramfenicolo-lisozima nel trattamento delle affezioni catarrali respiratorie acute dell'infanzia. *Omnia Med. Ther.*, **43**:885–891, 1965.

C169. Conti, U., Cortinovis, R., and Pedronetto, S. L'azione del lisozima sulla fibrinolisi attivata; studio tromboelastografico. *Farmaco, Ed. Sci.*, **12**:847–852, 1957.

C170. Corbetta, L. L'impiego del lisozima nel trattamento endobronchiale delle supurazioni broncopolmonari. *Arch. Ital. Otol. Rinol. Laringol.*, **70**:132–142, 1959.

C171. Corpe, W. A., and Winters, H. Hydrolytic enzymes of some periphytic marine bacteria. *Can. J. Microbiol.*, **18**:1483–1490, 1972.

C172. Corper, H. J. Lysozyme and tuberculosis. *Amer. Rev. Tuberc.*, **25**:59–68, 1932.

C173. Cortese, G. Sulle variazioni del contenuto in lisozima degli essudati purulenti. *G. Ital. Chir.*, **5**:378–387, 1949.

C174. Costa, A. L., Amato, A., and Vermiglio, G. Azione *in vitro* di policationi organici su un ceppo patogeno di *Candida albicans*. *Boll. Soc. Ital. Biol. Sper.*, **44**:731–735, 1968.

C175. Costa, A. L., Costa, A., and Spadaro, M. Sulla protezione di omoinnesti cutanei in animali sottoposti a pan-irradiazione e trattamento lisozimico. *Atti Soc. Peloritana Sci. Fis. Mat. Natur.*, **9**:371, 1963.

C176. Costa, A. L., Spadaro, M., and Mundo, A. Sul rapporti tra lisozima e tasso di colesterolo surrenale nei ratti pan-irradiati. *Boll. Soc. Ital. Biol. Sper.*, **42**:1280–1283, 1966.

C177. Costa, A. L., Spadaro, M., Orecchio, F., *et al.* Sul comportamento di alcuni fattori dell'immunità naturale in corso di epatite virale. *Epatologia*, **12**:800–805, 1966.

C178. Cottafavi, M. Lisozima e fattore di chiarificazione. *Minerva Ginecol.*, **11**:504–506, 1959.

C179. Cottafavi, M. Sull'influenza esercitata dal lisozima sulla coagulabilità ematica. *Minerva Ginecol.*, **11**:500–504, 1959.

C180. Cottafavi, M. Sull'attività lisozimica del liquor folliculi dell 'ovaio umano. *Minerva Ginecol.*, **8**:392–396, 1956.

C181. Cottini, G. B., and Randazzo, S. D. Wirkungsmöglichkeiten des immunisierenden Factors von Fleming (lysozym) bei Hautkrankheiten Beitrag zur ortlichen Behandlung der torpiden Unterschenkelgeschwüre. *Hautarzt*, **11**:414–418, 1960.

C182. Covelli, I., and Wolff, J. Iodination of the normal and buried tyrosyl residues of lysozyme. I. Chromatographic analysis. *Biochemistry*, **5**:860–866, 1966.

C183. Covey, W., Perillie, P., and Finch, S. C. The origin of tear lysozyme. *Proc. Soc. Exp. Biol. Med.*, **137**:1362–1363, 1971.

C184. Cowburn, D. A., Brew, K., and Gratzer, W. B. An analysis of the circular dichroism of the lysozyme-alpha-Lactalbumin group of proteins. *Biochemistry*, **11**:1228–1234, 1972.

C185. Cox, S. T., Jr., and Eagon, R. G. Action of ethylenediaminetetraacetic acid, tris (hydroxymethyl)-aminomethane, and lysozyme on cell walls of Pseudomonas aeruginosa. *Can. J. Biochem.*, **14**:913–922, 1968.

C186. Crescenzi, V., and Delben, F. Application of differential enthalpic analysis to the study of biopolymer solutions. *Int. J. Protein Res.,* **3:**57–62, 1971.

C187. Crescenzi, V., Quadrifoglio, F., and Cesàro, A. Interactions of hexamethylene tetramine with biopolymers and model compounds in aqueous solution. *Int. J. Protein Res.,* **3:**49–51, 1971.

C188. Crespi, H. L., Mandeville, S. E., and Katz, J. The action of lysozyme on several blue-green algae. *Biochem. Biophys. Res. Commun.,* **9:**569–573, 1962.

C189. Crifo, S., and Celestino, D. Lysozyme in tissues of otorhinolaryngoiatric interest in some laboratory animals. *Valsalva,* **37:**323–327, 1961.

C190. Crifo, S., Celestino, D., and Cavanna, F. Study of the distribution of lysozyme-I-131 in some tissues of the rat. Preliminary radioautographic research. *Boll. Mal. Orecchio, Gola. Nas,* **79:**578–587, 1961.

C191. Crifo, S., Celestino, D., and Pagni, B. Lysozyme of the upper respiratory tract in normal conditions and in induced pathology. Demonstration of a peculiar behavior of the laryngeal tissues. *Ann. Laringol., Otol., Rinol., Faringol.,* **60:**705–709, 1961. (Ital.)

C192. Cristofani, M. Effetti terapeutici del lisozima di Fleming in una epidemia di morbillo. *Minerva Med.,* **58:**2282–2285, 1967.

C193. Crohn, B. B., and Yarnis, H. Present trends in the treatment of ulcerative colitis. *N. Y. State J. Med.,* **51:**2129–2135, 1951.

C194. Crolle, G., and Bianco, S. Orientamenti attuali sulla terapia delle affezioni emofiliche con particolare riguardo all'azione del lisozima. *Progr. Med.,* **13:**296–303, 1957.

C195. Crombie, L. B., and Muschel, L. H. Quantitative studies on spheroplast formation by the complement system and lysozyme on gram negative bacteria. *Proc. Soc. Exp. Biol. Med.,* **124:**1029–1033, 1967.

C196. Crook, N. E., Stephen, J., and Smith, H. Recovery of bound homologous protein antigens from disulphide-linked immunosorbents at neutral pH. *Immunochemistry,* **9:**945–960, 1972.

C197. Crosato, M., and Ferrari, G. Comportamento elettroforetico delle lipoproteine seriche in soggatti affetti da epatite virale prima e dopo trattamento con lisozima. *Epatologia,* **12:**353–358, 1966.

C198. Crowder, J. G., Martin, R. R., and White, A. Release of histamine and lysosomal enzymes by human leukocytes during phagocytosis of staphylococci. *J. Lab. Clin. Med.,* **74:**436–444, 1969.

C199. Crowder, J. G., and White, A. C. Selective changes in white cell lysosomal enzymes in man. *Amer. J. Med. Sci.,* **255:**327–335, 1968.

C200. Cucinotta, U. Sull'uso di alcuni farmaci agenti sulla flora gastroenterica nell'occlusione intestinale acuta sperimentale. *Riv. Gastroenterol.,* **11:**1–11, 1959.

C201. Cunningham, F. E., and Cotterill, O. J. The influence of yolk on egg white lysozyme. *Poultry Sci.,* **50:**1013–1016, 1971.

C202. Currò, S., Deodato, G., Falcidia, E., *et al.* Ricerche sperimentali sull'azione antineoplastica del lisozima di Fleming. 2. Azione del lisozima sul sarcoma 180. *Chir. Patol. Sper.,* **12:**1022–1032, 1964.

C203. Currò, S., Deodato, G., Falcidia, E., Rasà, G., and Marchese, V. Ricerche sperimentali sull'attività antineoplastica del lisozima di Fleming. Nota 1. Azione del lisozima di Fleming. *Chir. Patol. Sper.,* **12:**1002–1021, 1964.

C204. Cusma, N., and Mastrandrea, V. Ricerche sull'attività antibatterica del lisozima. *Boll. Soc. Ital. Biol. Sper.,* **39:**572–576, 1963.

C205. Cutinelli, C. Relation between leucocytes and lysozyme. *Boll. Soc. Ital. Biol. Sper.,* **24:**449–452, 1948.

C206. Cutinelli, C. Sul contenuto in lisozima dei sieri immuni. *Riv. Ist. Sieroter. Ital., Sez. I*, 22:231–250, 1947.

C207. Cutinelli, C., and La Manna, N. Comportamento del nucleo dello pneumococco nella lisi da lisozima. *Riv. Inst. Sieroter. Ital.*, 25:1–8, 1950.

C208. Cutinelli, C., and La Manna, N. Tasso lisozimico del siero e leucociti. *Riv. Inst. Sieroter. Ital.*, 24:157–167, 1949.

C209. Cutinelli, C., and La Manna, N. Modificazioni del metabolismo glucidieo dello pneumococco dopo la lisi da lisozima. *Boll. Soc. Ital. Biol. Sper.*, 23:1157–1159, 1947.

C210. Cutinelli, C., and La Manna, N. Modificazioni morfologiche e chimiche dello pneumococco per azione del lisozima. *Boll. Soc. Ital. Biol. Sper.*, 23:1149–1153, 1947.

D

D1. Daghfous, T., Nabil, B., Fouche-Saillanfest, P. *et al.* Association d'antibiotiques et d'enzymes dans le traitement du trachome. *Rev. Int. Trachome*, 47:43–48, 1971.

D2. Dahlquist, F. W., Borders, C. L., Jr., Jacobson, G., and Raftery, M. A. The stereospecificity of human, hen, and papaya lysozymes. *Biochemistry*, 8:694–700, 1969.

D3. Dahlquist, F. W., Jao, L., and Raftery, M. On the binding of chitin oligosaccharides to lysozyme. *Proc. Nat. Acad. Sci. U.S.*, 56:26–30, 1966.

D3a. Dahlquist, F. W., and Raftery, M. Specificity of cleavage of chitotriose by lysozyme. *Nature*, 213:625–626, 1967.

D4. Dahlquist, F. W., and Raftery, M. A. A nuclear magnetic resonance study of association equilibria and enzyme-bound environments of N-acetyl-D-glucosamine anomers and lysozyme. *Biochemistry*, 7:3269–3276, 1968.

D5. Dahlquist, F. W., and Raftery, M. A. A nuclear magnetic resonance study of enzyme-inhibitor association. The use of pH and temperature effects to probe the binding environments. *Biochemistry*, 7:3277–3280, 1968.

D6. Dahlquist, F. W., and Raftery, M. A. Some properties of contiguous binding subsites on lysozyme as determined by proton magnetic resonance spectroscopy. *Biochemistry*, 8:713–725, 1969.

D7. Dahlquist, F. W., Rand-Meir, T., and Raftery, M. A. Application of secondary alpha-deuterium kinetic isotope effects to studies of enzyme catalysis. Glycoside hydrolysis by lysozyme and beta-glucosidase. *Biochemistry*, 8:4214–4221, 1969.

D8. Dahlquist, F. W., Rand-Meir, T., and Raftery, M. A. Demonstration of carbonium ion intermediate during lysozyme catalysis. *Proc. Nat. Acad. Sci. U.S.*, 61:1194–1198, 1968.

D9. Dainotto, F., Tognazzi, D., and Violanti, A. Blood coagulation and lysozyme action. *Policlin. Sez. Med.*, 68:380–389, 1961.

D10. Dainotto, F., Tognazzi, D., and Violanti, A. Lysozyme and blood coagulation in rheumatic disease in the 4th stage. *Policlin. Sez. Med.*, 68:271–283, 1961.

D11. Dalaly, B. K., Eitenmiller, R. R., and Vakil, J. R. Simultaneous isolation of human milk ribonuclease and lsysozyme. *Anal. Biochem.*, 37:208–211, 1970.

D12. Daly, S. Lysozyme of nasal mucus. Method of preparation and preliminary report on its effect on growth and virulence of common pathogens of paranasal sinuses. *Arch. Otolaryngol.*, 27:189–196, 1938.

D13. Dam, R. *In vitro* studies on the lysozyme-ovomucin complex. *Poultry Sci.*, 50:1824–1831, 1971.

D14. Dambly, C., Courturier, M., and Thomas, R. Control of development in termperate bacteriophages. II. Control of lysozyme synthesis. *J. Mol. Biol.*, 32:67–81, 1968.

D15. Dambly, C., Courturier, M., and Thomas, R. Le contrôle de la synthese du lysozyme chez les bactériophages tempérés. *Arch. Int. Physiol. Biochim.,* **75:**155–156, 1967.

D16. Daniel, R. S., and Brooks, M. A. Intracellular bacteroids: Electron microscopy of *Periplaneta americana* injected with lysozyme. *Exp. Parasitol.,* **31:**232–246, 1972.

D17. Daniels, J. C., Fukushima, M., Fish, J. C., Lindley, J. D., Remmers, A. R., Jr., Sarles, H. E., and Ritzmann, S. E. Studies on lysozyme (Muramidase). II. Serum and urine muramidase patterns in chronic uremia. *Tex. Rep. Biol. Med.,* **30:**9–22, 1972.

D18. Daniels, J. C., Fukushima, M., Larson, D. L., Abston, S., and Ritzmann, S. E. Studies on muramidase (lysozyme). I. Serum and urine muramidase activity in burned children. *Tex. Rep. Biol. Med.,* **29:**13–39, 1971.

D19. Das Gupta, B. R., and Boroff, D. A. Increased selectivity of photooxidation for amino acid residues. *Biochim. Biophys. Acta,* **97:**157–159, 1965.

D19a. Dasgupta, B. R., Rothstein, E., and Boroff, D. A. Method for quantitative determination of free and peptide-linked tryptophan after reaction with 2-hydroxy-5-nitrobenzyl bromide. *Anal. Biochem.,* **11:**555, 1965.

D20. Davidson, A. J. Lysozymuric response to contrast material in the dog. *Invest. Radiol.,* **3:**65–66, 1968.

D20a. Davidson, B. E., and Hird, F. J. R. The reactivity of the disulphide bonds of purified proteins in relationship to primary structure. *Biochem. J.,* **104:**473–485, 1967.

D21. Davidson, W. D., and Prockop, D. J. Lysozymuria: An enzymatic measure of renal tubular damage. *J. Clin. Invest.,* **43:**1274, 1964.

D22. David-West, T. S., Desalu, A. B. O., and Smith, J. A. Lysozymes in yellow fever infection. *Can. J. Microbiol.,* **18:**1829–1835, 1972.

D23. Davies, J. V., Ebert, M., and Shalek, R. J. The radiolysis of dilute solutions of lysozyme. II. Pulse radiolysis studies with cysteine and oxygen. *Int. J. Radiat. Biol.,* **14:**19–27, 1968.

D24. Davies, R. C., and Neuberger, A. Modification of lysine and arginine residues of lysozyme and the effect on enzymatic activity. *Biochim. Biophys. Acta,* **178:**306–317, 1969.

D25. Davies, R. C., Neuberger, A., and Wilson, B. M. The dependence of lysozyme activity on pH and ionic strength. *Biochim. Biophys. Acta,* **178:**294–305, 1969.

D26. Davis, C. S. Diagnostic value of muramidase. *Postgrad. Med.,* **49:**51–54, 1971.

D27. Davis, S. D., Gemsa, D., Iannetta, A., and Wedgwood, R. J. Potentiation of serum bactericidal activity by lysozyme. *J. Immunol.,* **101:**277–281, 1968.

D27a. Dayhoff, M. O. "Atlas of Protein Sequence and Structure," Vols. I–V and Supplements. Washington, National Biomed. Res. Fdn., Georgetown Univ. Med. Ctr. 1967–73.

D28. Dechaume, M., Laudenback, P., and Recoing, J. Expérimentation clinique de comprimés associant lysozyme, papäine et bacitracine. *Rev. Stomatol., Paris,* **68:**133–138, 1967.

D29. De Cecco, L. Natural immunity and uterine malignant neoplasms after roentgen therapy. *Quad. Clin. Ostet. Ginecol.,* **16:**953–962, 1961. (Ital.)

D30. De Cicco, N., Rolando, P., and Leonardi, L. Sull'incremento dell'immunità aspecifica in rapporto all'impiego del lisozima, e sua tolleranza nel lattante. *Pediatria (Naples),* **72:**493–512, 1964.

D31. Deckx, R. J., Vantrappen, G. R., and Parein, M. M. Localization of lysozyme activity in a Paneth cell granule fraction. *Biochim. Biophys. Acta,* **139:**204–207, 1967.

D32. De Duve, C. Les lysosomes. *Bull. Acad. Roy. Med. Belg.,* **23:**608–618, 1958.

D33. Delben, F., and Crescenzi, V. Thermal denaturation of lysozyme. A differential scanning calorimetry investigation. *Biochim. Biophys. Acta,* **194:**615–618, 1969.

D34. DelCampo, A., Castellani, G., and Franzoni, A. I fattori naturali di resistenza immunitaria nell'està sneile. *G. Gerontol.,* **14:**1123–1133, 1966.

D35. Del Campo, A., and Zardini, V. Contributo allo studio della fisiopatologia della vaccinazione antitifo-paratifica preventiva. II. Effetti della somministrazione del vaccino T.A.B. sui fattori naturali di resistenza immunitaria. *Riv. Ital. Ig.*, **25**:614–627, 1965.

D36. De Luca, R., Amato, A., and Parrinello, G. Effetti delle inoculazioni di immunoglobuline sui fattori umorali dell'immunità naturale nel bambino. *Pediatria (Naples)*, **78**:225–245, 1970.

D37. De Luca, R., Caruso, P., and Guzzetta, F. Lisozima e properdina. *Boll. Soc. Ital. Biol. Sper.*, **35**:601–603, 1959.

D38. de Montis, G. F. Ricerche sull'eventual azione antigerminativa sui miceti del lisozima puro o contenuto nel liquido articolare. *G. Clin. Med. (Bologna)*, **33**:756–762, 1952.

D39. Deonier, R. C., and Williams, J. W. Self-association of muramidase (lysozyme) in solution at 25 degrees, pH 7.0, and 1 = 0.20. *Biochemistry*, **9**:4260–4267, 1970.

D40. Deranleau, D. A., Bradshaw, R. A., and Schswzer, R. The use of n-methylnicotinamide chloride as a conformational probe for chicken egg-white lysozyme. *Proc. Nat. Acad. Sci. U.S.*, **63**:885–889, 1969.

D41. Derechin, M. Analysis of associating systems using the multinomial theory. *Biochemistry*, **10**:4981–4986, 1971.

D42. Derechin, M. Alternative equations for mass and molar equilibrium constants derived from the multinomial theory. *Biochemistry*, **11**:4153–4156, 1971.

D43. Desai, A. M., and Korgaonkar, K. S. Studies on the effects of cobalt-60 gamma rays on protamine sulfate, lysozyme and insulin by using monolayer technique. *Radiat. Res.*, **21**:61–74, 1964.

D44. De Serio, N., and Grimaldi, G. Comportamento del quadro sieroproteico elettroforetico nei conigli roentgenirradiati. 3. Protezione del danno da radiazione mediante prolungata lisozimoterapia endovenosa antedecente all'inizio del trattamento jonizzante. *Quad. Radiol.*, **33**:23–33, 1968.

D45. De Serio, N., and Grimaldi, G. Comportamento del quadro sieroproteico elettroforetico nei conigli roentgenirradiati. IV. Protezione del danno da radiazione mediante associazione della lisozimoterapia endovenosa al trattamento jonizzante. *Quad. Radiol.*, **33**:125–137, 1968.

D46. De Wijs, H., and Jollès, P. Cell walls of three strains of mycobacteria (*Nycobacterium phlei, Mycobacterium fortuitum, Mycobacterium kansasii*). Preparation, analysis and digestion by lysozymes of different origins. *Biochim. Biophys. Acta*, **83**:326–332, 1964.

D47. Dianoux, A. C., and Jollès, P. Nouvelles données concernant le lysozyme de blanc d'oeuf d'oie: Teneur en tryptophane, stabilitè et spécificité. *Helv. Chim. Acta*, **52**:611–616, 1969.

D48. Dianoux, A. C., and Jollès, P. Différences dans le comportement des lysozymes de blanc d'oeuf de poulte et d'ole vis-a-vis de Micrococcus lysodeikticus. *Bull. Soc. Chim. Biol.*, **51**:1559–1564; 1969.

D49. Dianoux, A. C., and Jollès, P. Etude d'un lysoyme pauvre en cystine et en tryptophane: Le lysozyme de blanc d'oeuf d'oie. *Biochim. Biophys. Acta*, **133**:472–479, 1967.

D50. Dickerson, R. E., Reddy, J. M., Pinkerton, M., and Steinrauf, L. K. A 6 angstrom model of triclinic lysozyme. *Nature (London)*, **196**:1178, 1962.

D50a. Dickman, S. R., Proctor, C. M. Factors affecting the activity of egg white lysozyme. *Arch. Biochem. Biophys.*, **40**:364–372, 1952.

D51. Dietrich, F. M. Inactivation of egg-white lysozyme by ultrasonic waves and protective effect of amino-acids. *Nature (London)*, **195**:146–148, 1962.

D52. Dietrich, F. M., and Bloch, H. Delayed hypersensitivity in tuberculous mice. I. The effect of tuberculin on the release of lysozyme in vivo and in vitro. *Int. Arch. Allergy Appl. Immunol.,* **20:**102–121, 1962.

D53. Di Nardo, A. U. Reazione immunologica crociata tra lisozima di planco d'uovo di pollo e lisozima di cavallo. *Ig. Mod.,* **63:**218–222, 1970.

D54. Dinelli, C. A., and Montani, E. Thrombodensitographic study of the procoagulant action of lysozyme in hepatic patients. *Sem. Med.,* **119:**1045–1046, 1961.

D55. Diomede-Fresa, V., and Fumarola, D. On chemical and biological protection against radiation injury. III. Effects of the administration of lysozyme in acute and chronic x-ray poisoning. *Ann. Med. Nav.,* **66:**777–784, 1961.

D56. Di Pietro, S., and Pellegris, G. Proprietà antalgiche del lisozima in soggetti portatori di neoplasie maligne. *Minerva Med.,* **54:**1882–1889, 1963.

D57. Dlabač, V. The sensitivity of smooth and rough mutants of *Salmonella typhimurium* to bactericidal and bacteriolytic action of serum, lysozyme and to phagocytosis. *Folia Microbiol. (Prague),* **13:**439–449, 1968.

D58. Dobraya, T. E. Lysozyme. *Vestn. Mikrobiol., Epidemiol., Parazitol.,* **16:**492–502, 1939.

D59. Dolby, J. M. The antibacterial effect of *Bordetella pertussis* antisera. *Immunology,* **8:**484–498, 1965.

D60. Dolmans, M., and Léonis, J. Dénaturation thermique du lysozyme: Effet des alcools et d'un inhibiteur compétitif. *Arch. Int. Physiol. Biochim.,* **75:**161–162, 1967.

D61. Donaldson, D. M., and Tew, J. G. Differentiation of muramidase and beta-lysin. *Proc. Soc. Exp. Biol. Med.,* **122:**46–49, 1966.

D62. Donin, I. K. Lysozyme in therapy of ulcerative keratitis. *Sov. Vestn. Oftalmol.,* **9:**638–642, 1936.

D63. Donovan, J. W., Laskowski, M., Jr., and Scheraga, H. A. Influenza of ionization of carboxyl groups on the ultraviolet absorption spectrum of lysozyme. *Biochim. Biophys. Acta,* **29:**455–456, 1958.

D63a. Donovan, J. W., Laskowski, M. Jr., and Scheraga, H. A. Abnormal ionizable groups in lysozyme. *J. Amer. Chem. Soc.,* **82:**2154–2163, 1960.

D63b. Donovan, J. W., Laskowski, M. Jr., and Scheraga, H. A. The effects of charged groups on the chromophores of lysozyme and of amino acids. *J. Amer. Chem. Soc.,* **83:** 2686–2694, 1961.

D64. Dorello, U. Sul comportamento dell'attività lisozimica nel siero di alcuni animali. *Boll. Soc. Ital. Biol. Sper.,* **23:**1107–1109, 1947.

D65. Dorfman, V. A. Physicochemical nature of bacteriolysis. *Nature (London),* **153:**169–170, 1944.

D66. Dorfman, V. A. Bacteriolysis. II. Kinetics of the lysozyme lysis. *Byull. Eksp. Biol. Med.,* **14:**112–114, 1943. (Russ.)

D67. Dorfman, V. A., and Karakash, E. I. Bacteriolysis. I. Study of the oxidation-reduction and electrokinetic potentials in the process of bacteriolysis. *Byull. Eksp. Biol. Med.,* **13:**48–52, 1942. (Russ.)

D68. Dorfman, V. A., Moldavskaya, E. A., Kastorskaya, T. L., and Zasypkina, P. S. The physiochemical nature of antibacterial action. *Biokhimiya,* **10:**407–422, 1945. (Russ.)

D69. Dorfman, V. A., and Moldavskaya, E. A. On the physiochemical nature of bacteriolysis. *Amer. Rev. Sov. Med.,* **2:**271, 1945.

D70. Dorfman, V. A., Moldavskaya, E. A., Kastorskaya, T. L., and Zasypkina, P. S. The physiocochemical nature of antibiotic action. *Amer. Rev. Sov. Med.,* **3:**500–510, 1946.

D71. Dorofeĭchuk, V. G. Determination of lysozyme activity by nephelometric method. *Lab. Delo,* **1**:28–30, 1968. (Russ.)

D72. Dose, K., and Caputo, A. Die Anwendung der Hochspannungspherographie bei der quantitativen Totalanalyse von Proteinhydrolysaten Untersuchungen am Lysozym. *Biochem. Z.,* **328**:376–385, 1956.

D73. Dose, K., Risi, S., and Rauchfuss, H. Veränderung von Aminosäuren und Aktivität bei Röntgenbestrahlung von krystallisiertem Lysozym. *Biophysik,* **3**:202–206, 1966.

D74. Douglas, H. W., and Parker, F. Electrophoretic studies on bacteria. 3. The growth cycle of *Bacillus megaterium,* the behaviour of cells and the changes produced by lysozyme. *Biochem. J.,* **68**:99–105, 1958.

D75. Dresden, M., and Hoagland, M. B. Polyribosomes from Escherichia coli; enzymatic method for isolation. *Science,* **149**:647–649, 1965.

D76. Driedger, A. A., and Grayston, M. J. Rapid lysis of cell walls of *Micrococcus radiodurans* with lysozyme: Effects of butanol pretreatment on DNA. *Can. J. Microbiol.,* **16**:889–893, 1970.

D77. Drożański, W. Bacteriolytic spectrum of the enzyme produced by Acanthamoeba castellanii. *Acta Microbiol. Pol., Ser. A,* **1**:169–174, 1969.

D78. Dubach, U. C. Diagnostik von Nierenkrankheiten mit Hilfe von Urienzymen. *Praxis,* **60**:527–528, 1971.

D79. Dubin, S. B., Feher, G., and Benedek, G. B. Study of the chemical denaturation of lysozyme by optical mixing spectroscopy. *Biochemistry,* **12**:714–720, 1973.

D80. Dubois-Prévost, R. The action of fluoride on salivary inhibitors. *Caries Res.,* **5**:11–12, 1971.

D81. Dubois-Prévost, R. Le lysozme, applications alimentaires. *Aliment. Vie,* **58**:44–49, 1970.

D82. Dubois-Prévost, R. Le lysozyme. *Inform. Dent.,* **49**:2845–2848, 1967.

D83. Dubois-Prevost, R. Le lysozyme et son application en diététique infantile. *Sem. Ther.,* **43**:612–615, 1967.

D84. Duckworth, D. H. Role of lysozyme in the biological activity of bacteriophage ghosts. *J. Virol.,* **3**:92–94, 1969.

D85. Duncan, C. L., Labbé, R. G., and Reich, R. R. Germination of heat- and alkali-altered spores of *Clostridium perfringens* type A by lysozyme and an initiation protein. *J. Bacteriol.,* **109**:550–559, 1972.

D86. Dunn, B. M., and Bruice, T. C. Physical organic models for the mechanism of lysozyme action. *Advan. Enzymol.,* **37**:1–60, 1973.

D87. Dunnill, P. Sequence similarities between hen egg-white and T4 phage lysozymes. *Nature (London),* **215**:621–622, 1967.

D88. Dunnill, P. The use of helical net-diagrams to represent protein structures. *Biophys. J.* **8**:865, 1968.

D89. Dwek, A., Kent, P. W., and Xavier, A. V. N-Fluoroacetyl-D-glucosamine as a molecular probe of lysozyme structure using (19^F) fluorine-nuclear-magnetic resonance techniques. *Eur. J. Biochem.,* **23**:343–348, 1971.

E

E1. Eagon, R. G., and Carson, K. J. Lysis of cell walls and intact cells of Pseudomonas aeruginosa by ethylenediamine tetraacetic acid and by lysozyme. *Can. J. Microbiol.,* **11**:193–201, 1965.

E2. Ebisuzaki, K. A test of a hypothesis concerning ultraviolet irradiation damage and phage functions. *Virology,* **26**:390–393, 1965.

E3.   Economou, J. S. Observations on a possible association between the acid mucopoly-saccharides and lysozyme. *J. Surg. Oncol.*, **3**:89–95, 1971.

E3a. Edelhoch, H. Spectroscopic determination of tryptophan and tyrosine in proteins. *Biochemistry*, **6**:1948–1954, 1967.

E4.   Edelhoch, H., and Steiner, R. F. Structural transitions of lysozyme in urea solution. *Biochim. Biophys. Acta*, **60**:365–372, 1962.

E5.   *Editorial.* Lysozyme and ulcerative colitis. *J. Amer. Med. Ass.*, **148**:653, 1952.

E6.   *Editorial.* Studies on muramidase in leukemia. *Rev. Clin. Espan.*, **111**:634–635, 1968.

E7.   *Editorial.* Lysozyme, ribonuclease and low molecular weight urinary proteins in renal diseases. *Rev. Clin. Espan.*, **111**:551–552, 1968.

E7a. Edman, P., and Josefson, L. Reversible enzyme inactivation due to *N,O*-Peptidyl shift. *Nature*, **178**:1189–1190, 1957.

E7b. Ehrenpreis, S. and Warner, R. C. The interaction of conalbumin and lysozyme. *Arch. Biochem. Biophys.*, **61**:38–50, 1956.

E8.   Eisenstein, R., Arsenis, C., and Kuettner, K. E. Electron microscopic studies of cartilage matrix using lysozyme as a vital stain. *J. Cell Biol.*, **47**:626–631, 1970.

E9.   Eisenstein, R., and Kuettner, K. C. An ultrastructural organization of cartilage matrix as revealed through the use of lysozyme. *Calcif. Tissue Res.*, **4**:Suppl., 137–138, 1970.

E10. Eisenstein, R., Soble, L. W., and Kuettner, K. E. Lysozyme in epiphyseal cartilage. III. Effects of protamine, toluidine blue and histamine on mouse embryonic cartilage in organ culture. *Amer. J. Pathol.*, **60**:43–56, 1970.

E11. Eisenstein, R., Sorgente, N., Arsenis, C., and Kuettner, K. E. Vitamin D effects on tissue and serum lysozyme. *Arch. Pathol.*, **94**:479–485, 1972.

E12. Eisler, D. M., and von Metz, E. Anti-*Pasteurella pestis* factor in the organs of normal mice and guinea pigs. I. Biologic characteristics. *J. Immunol.*, **91**:287–294, 1963.

E13. el-Gammal, M. Y., and Mostafa, M. S. Estimation of tears lysozyme in some eye diseases. *Bull. Opthalmol. Soc. Egypt*, **64**:285–297, 1971.

E14. Elias, A., Zalman, M. W., Sacaleanu, M., Strejaru, F. Rapport entre la capacité d'élaborer le lysozyme et certains caractères des staphylocoauques. *Arch. Roum. Pathol. Exp. Microbiol.*, **29**:395–400, 1970.

E15. Elkonin, L. G. On the problem of the interrelation of factors of natural and acquired immunity. I. Change in lysozyme and complement activity of blood serum under the influence of vaccines. *Zh. Mikrobiol. Epidemiol. Immunobiol.*, **40**:105, 1963. (Russ.)

E15a. Eley, D. D., and Hedge, D. G. Properties of biological membranes. The structure of films of proteins adsorbed on lipids. *Discuss. Faraday Soc.*, **21**:221–228, 1956.

E15b. Eley, D. D., and Hedge, D. G. The energetics of lipid-protein interactions. *J. Colloid. Science*, **12**:419–429, 1957.

E16. Emrich, J. Lysis of T4-infected bacteria in the absence of lysozyme. *Virology*, **35**:158–165, 1968.

E17. Emrich, J., and Streisinger, G. The role of phage lysozyme in the life cycle of phage T4. *Virology*, **36**:387–391, 1968.

E18. Endo, E. Kidney lysozyme in hypoxic radiation chimeras. *Exp. Hematol.* **15**:158, 1968.

E19. Epstein, C. J., and Goldberger, R. F. A study of factors influencing the reactivation of reduced egg white lysozyme. *J. Biol. Chem.*, **238**:1380–1383, 1963.

E20. Epstein, L. A., and Chain, E. Some observations on preparation and properties of substrate of lysozyme. *Brit. J. Exp. Pathol.*, **21**:339–355, 1940.

E21. Erickson, O. F. Drug influences on lacrimal lysozyme production. *Stanford Med. Bull.*, **18**:34–39, 1960.

E22. Erickson, O. F., Feeney, L., and McEwen, W. K. Filter-paper electrophoresis of tears.

2. Animal tears and presence of "slow-moving lysozyme." *AMA Arch. Ophthalmol.,* **55**:800–806, 1956.

E23. Ermol'eva, Z. V. Lysozymes. *Usp. Sovrem. Biol.,* **9**:68–80, 1938. (Russ.)

E24. Ermol'eva, Z. V., Blinova, M. I., Kagramanova, K. A. *et al.* Use of lysozyme and ekmolin for the prophylaxis of influenza. *Antibiotiki (Moscow),* **13**:368–371, 1968. (Russ.)

E25. Ermol'eva, Z. V., and Buyanovskaya, I. S. Le lysozyme, ses propriétés et ses applications. *Acta Med. URSS,* **1**:248–257, 1938.

E26. Ermol'eva, Z. V., and Buyanovskaya, I. S. Lysozyme and phagolysozyme in summer diarrheas of children. *Sov. Med.,* **4**:14–15, 1940.

E27. Ermol'eva, Z. V., and Buyanovskaya, I. S. Lysozyme, its properties and applications. *Microbe Variability Conf.,* pp. 459–470, 1936.

E28. Ermol'eva, Z. V., and Buyanovskaya, I. S. Ueber die Natur des Lysozyms. *Zentralbl. Bakteriol. Parasitenk. Infektionskr., Abt. 1,* **122**:267–270, 1931.

E29. Ermol'eva, Z. V., Furer, N. M., Ravich, I. V., Navashin, S. M., Braude, A. I., Fomina, I. P., Zhukovskaya, N. A., Balezina, T. I., Ved'mina, E. A., Golosova, T. V., Nemirovskaya, B. M., and Terent'eva, T. G. Experimental study and clinical application of lysozyme. *Fed. Proc., Fed. Amer. Soc. Exp. Biol.,* **23**:Transl. Suppl. 75–78, 1964.

E30. Ermol'eva, Z. V., Golosova, T. V., Ved'mina, E. A., Shenderovich, V. A., and Zhukovskaia, N. A. The use of lysozyme for the antisepsis of pathogenic Staphylo-cocci carriers. *Antibiotiki (Moscow),* **7**:359–361, 1962. (Russ.)

E31. Ermol'eva, Z. V., Ravich, I. V., Navashin, S. M., Braude, A. I., Fomina, I. P., Terentyeva, T. G., Vaisberg, G. E., Furer, N. M., Boiko, V. I., and Pokidova, N. V. Antitumor activity of certain substances of natural origin. *Acta Unio Int. Contra Cancrum,* **20**:295–296, 1964.

E32. Espey, L. L., and Lipner, H. Enzyme-induced rupture of rabbit Graafian follicles. *Amer. J. Physiol.,* **208**:208–213, 1965.

E33. Eudy, W. W., and Burrous, S. E. Effect of antibacterial therapy on renal lysozyme levels in rats developing bacterial pyelonephritis. *Antimicrob. Ag. Chemother.,* **1**:85–89, 1972.

E34. Eudy, W. W., and Burrous, S. E. Effects of aminoglycosides on renal lysozyme in rats. *Biochem. Med.,* **6**:419–425, 1972.

E35. Eudy, W. W., and Burrous, S. E. Renal lysozyme levels in animals developing *Proteus mirabilis*-induced pyelonephritis. *Appl. Microbiol.,* **21**:300–305, 1971.

E36. Eudy, W. W., Burrous, S. E., and Sigler, F. W. Renal lysozyme levels in animals developing "sterile pyelonephritis." *Infec. Immunity,* **4**:269–273, 1971.

F

F1. Fabricant, N. D. Relation of the pH of nasal secretions in situ to the activity of lysozyme. Report of a case of experimentally induced allergic rhinitis. *Arch. Otolaryn-gol.,* **41**:53–55, 1945.

F2. Faure, A., and Jollès, P. Etude de la réactivité immunologique entre alpha-lactalbu-mines humaine et bovine et plusieurs lysozymes d'origine humaine et de blancs d'oeufs d'oiseaux. *C. R. Acad. Sci. Ser. D,* **271**:1916–1918, 1970.

F3. Faure, M, and Maréchal, J. Complexes lysozyme-cardiolipide-lecithine. *Bull. Soc. Chim. Biol.,* **50**:1537–1546, 1968.

F4. Favre, H., Wauters, J. P., and Fabre, J. La lysozymurie dans le diagnostic de diverses néphropathies. *J. Urol. Nephrol.,* **76**:1017–1021, 1970.

F5. Fawcett, R. I., Limbird, T. J., Oliver, S. L., and Borders, C. L., Jr. Chemical modification of human lysozyme. Acetylation and nitration of tyrosine. *Can. J. Biochem.*, **49**:816–821, 1971.

F6. Fazleeva, L. K. Change in immunologic reactivity caused by corticosteroid therapy of dermatoses. *Zh. Mikrobiol., Epidemiol. Immunobiol.*, **49**:67–71, 1972. (Russ.)

F7. Fedorova, G. I. Production of protoplasts and spheroplasts in gram-positive organisms. *Antibiotiki (Moscow)*, **14**:880–993, 1969. (Russ.)

F8. Feeney, R. E., Ducay, E. D., and MacDonnell, L. R. Inactivation of lysozyme by trace elements. *Fed. Proc. Fed. Amer. Soc. Exp. Brol.*, **11**: No. 1, Part 1:209–20, 1952.

F9. Feiner, R. R., Meyer, K., and Steinberg, A. Bacterial lysis by lysozyme. *J. Bacteriol.*, **52**:375–384, 1946.

F10. Feingold, D. S., Goldman, J. N., and Kuritz, H. M. Locus of the action of serum and the role of lysozyme in the serum bactericidal reaction. *J. Bacteriol.*, **96**:2118–2126, 1968.

F11. Fellenberg, R. von, and Levine, L. Proximity of the enzyme active center and an antigenic determinant of lysozyme. *Immunochemistry*, **4**:363–365, 1967.

F12. Felsenfeld, H., and Handschumacher, R. E. The interation between lysozyme and penicillin. *Mol. Pharmacol.*, **3**:153–160, 1967.

F13. Ferina, F., and Giambanco, V. Adsorbimento del lisozima da parte dei vari tessuti. *G. Batteriol., Virol. Immunol. Ann. Osp. Maria Vittoria Torino* **53**:327–331, 1960.

F14. Ferina, F., Matracia, S., and Caradonna, D. Prevention of the death of rabbits and of the reduction of the complement titer of guinea pig serum caused by silica, by the action of histone, lysozyme and trypsin. *Med. Lav.*, **52**:494–497, 1961.

F15. Ferina, F., and Meldolesi, M. F. Interazione tra agglutinogeno somatico tifica e lisozima con riduzione della agglutinabilità da agglutinine antitifiche. *Riv. Ist. Sieroter. Ital.*, **35**:655–660, 1960.

F16. Ferina, F., and Meldolesi, M. F. Incremento della sensibilità dei batteri del gruppo tifo-paratifi agli antisieri agglutinanti, per pretrattameto dei germi con lisozima. *Boll. Ist. Sieroter. Milan.*, **39**:556–560, 1960.

F17. Ferlazzo, A., Bellomo, G., and Lombardo, G. Influenza del lisozima sul potere proteolitico delli feci. *Boll. Soc. Ital. Biol. Sper.*, **36**:222–226, 1960.

F18. Ferlazzo, A., and Lombardo, G. Das Enzym von Fleming (lysozym) und seine Bedeutung für die Säuglingsernahrung. *Ann. Paediat.*, **200**:305–318, 1963.

F19. Ferlazzo, A., Lombardo, G., and Scalisi, S. Lysozyme and cow's milk. *Minerva Nipiol.*, **10**:121, 1960. (Ital.)

F20. Ferlazzo, A., Minicuci, P., and Ricca, M. Lysozyme and cutaneous reactivity to tuberculin. *Arch. Tisiol. Mal. App. Resp.*, **17**:67–80, 1962. (Ital.)

F21. Ferrari, G., and Cocuzza, S. Comportamento dell'isocitrico deidrogenasi serica in soggetti affetti da epatite virale dopo trattamento con lisozima. *Epatologia*, **12**:349–352, 1966.

F22. Ferrari, R., Callerio, C., and Podio, G. Antiviral activity of lysozyme. *Nature (London)*, **183**:548, 1959.

F23. Ferrari, R., and Matracia, S. Interaction between lysozyme and a phospholipid of neoplastic tissue (Oncolipin.) *Nature (London)*, **192**:1187, 1961.

F24. Ferrari, R., and Podio, G. A proposito del meccanismo d'azione del lisozima di Fleming nella terapie delle virosi. *Minerva Dermatol.*, **34**:207–208, 1959.

F25. Ferraris, G. La terapia dell'infezione puerperale. *Minerva Ginecol.*, **18**:1018–1028, 1966.

F26. Ferrauto, A., and Calafato, M. Determinazione della attività lisozimica delle piastrine con colture essiccate e standardizzate di *Micrococcus lysodeicticus. Boll. Soc. Ital. Biol. Sper.*, **36**:1326–1329, 1960.

F27. Ferreira, C. B. Experimentação clinica com una associação de antibióticos: Tetraci-clina-cloranfenicol-lisozima. *Rev. Brasil. Med.,* **22**:526–535, 1965.

F28. Ferri, G. G., Pozza, F., and Calzavara, F. Influenza del lisozima di Fleming sulla leucopenia da radioterapia. Contributo casistico. *Quad. Radiol.,* **30**:759–766, 1965.

F29. Ferrini, U. Changes in amino acid content and inactivation of lysozyme following ultraviolet irradiation. *Arch. Biochem. Biophys.,* **107**:126–131, 1964.

F30. Ferrini, U., and Zito, R. Modification of histidine by ultraviolet irradiation of lysozyme. *J. Biol. Chem.,* **238**:pc3824–pc3825, 1963.

F31. Ferrini, U. Relazione tra struttura e funzione nel lisozima in seguito ad irradiazione ultravioletta. *Boll. Soc. Ital. Biol. Sper.,* **39**:51–54, 1963.

F32. Fevold, H. L. Egg proteins. *Advan. Protein Chem.,* **6**:187–252, 1951.

F33. Fevold, H. L., and Alderton, G. Lysozyme (Crystalline: From egg white). *Biochem. Prep.,* **1**:67–71, 1949.

F34. Fiaccavento, W. Modificazioni del potere litico del siero di sangue negli ustionati (ricerche sperimentali). *G. Batteriol. Immunol.,* **43**:302–312, 1951.

F35. Fiaccavento, W., and Natale, D. Potenziamento in virto del potere antibatterico dei sulfamidici con lisozima. *Arch. Sci. Med.,* **91**:378–380, 1951.

F36. Fidel'man, E. S., and Aver'ianova, L. L. On changes in the content of lysozyme in the blood of animals under the influence injections of Streptococcus and homologous tissue antigen. *Byull. Eksp. Biol. Med.,* **58**:39–41, 1964. (Russ.)

F37. Finch, S. C., Gnabasik, F. J., and Rogoway, W. Relations entre le lysozyme of serum et le "turnover" leucocytaire. *Med. Hyg.* **22**:972, 1964.

F38. Finch, S. C., Gnabasik, F. J., and Rogoway, W. Lysozyme and Leukopoiesis. *Proc. Int. Congr. Fleming's Lysozyme, 3rd, 1964.*

F39. Finch, S. C., Lamphere, J. P., and Jablon, S. The relationship of serum lysozyme to leukocytes and other constitutional factors. *Yale J. Biol. Med.,* **36**:350–360, 1964.

F40. Finch, S. C., Lamphere, J. P., and Jablon, S. Serum lysozyme determinations in the ABCC-JNIH adult health study sample, Hiroshima April–June, 1961. *ABCC Tech. Rep., 1959–1972, Hiroshima & Nagasaki, Japan,* No. 5–63, 1972.

F41. Findlay, G. M. A contribution to the etiology of experimental keratomalacia. *Brit. J. Exp. Pathol.,* **6**:16–21, 1925.

F42. Findlay, J. B. C., and Brew, K. The complete amino-acid sequence of human-lactalbumin. *Eur. J. Biochem.,* **27**:65, 1972.

F43. Findley, J. E., and Akagi, J. M. Lysis of *Desulfovibrio vulgaris* by ethylenediaminete-traacetic acid lysozyme. *J. Bacteriol.,* **96**:1427–1428, 1968.

F44. Fink, M. E., and Finch, S. C. Serum muramidase and granulocyte turnover. *Proc. Soc. Exp. Biol. Med.,* **127**:365–367, 1968.

F45. Finlayson, A. J. The performic acid oxidation of egg-white lysozyme. *Can. J. Biochem.,* **47**:31–37, 1969.

F46. Fiore, M. Comportamento del lisozima nelle neoplasie. II. Il lisozima nel tessuto neoplastico. *Tumori,* **38**:57–66, 1952.

F47. Fiorentini, S., and Capozzi, A. Determinazione della attività litica della saliva dopo somministrazione di lisozima per os. *Arcisp. S. Anna di Ferrara,* **18**:939–944, 1965.

F48. Firking, F. C. Serum muramidase in haematological disorders: Diagnostic value in neoplastic states. *Aust. N. Z. J. Med.,* **2**:28–33, 1972.

F49. Fischer, H., and Tronnier, H. Klinische Erfahrungen mit virostatischen Medikamen-ten. *Med. Klin. (Munich),* **61**:916–919, 1966.

F50. Fisher, D. B., and Pardee, A. B. The Luria-Latarjet effect studied by T4-Lysozyme production. *Virology,* **34**:91–96, 1968.

F51. Fitz-James, P. Fate of the mesosomes of *Bacillus megaterium* during protoplasting. *J. Bacteriol.,* **87**:1483–1491, 1965.

F52. Flanaga, P. and Lionetti, F. Lysozyme distribution in blood. *Blood,* **10**:497–501, 1955.

F53. Fleck, J. Etude du glycopeptide d'une forme L stable de Proteus. II. Action du lysozyme. *Ann. Inst. Pasteur, Paris,* **115**:57–61, 1968.

F54. Fleming, A. On a remarkable bacteriolytic element found in tissues and secretions. *Proc. Roy. Soc., Ser. B,* **93**:306–317, 1922.

F55. Fleming, A., and Allison, V. D. On the specificity of the protein of human tears. *Brit. J. Exp. Pathol.,* **6**:87–90, 1925.

F56. Fleming, A., and Allison, V. D. Further observations on a bacteriolytic element found in tissues and secretions. *Proc. Roy. Soc. London, Series B,* **44**:142–151, 1923.

F57. Fleming, A., and Allison, V. D. Observations on a bacteriolytic substance ("lysozyme") found in secretions and tissues. *Brit. J. Exp. Pathol.,* **3**:252–260, 1922.

F58. Fleming, A., and Allison, V. D. On the development of strains of bacteria resistant to lysozyme action and the relation of lysozyme action to intracellular digestion. *Brit. J. Exp. Pathol.,* **8**:214–218, 1927.

F59. Fleming, A. Arris and Gale Lecture on Lysozyme. A bacteriolytic ferment found normally in tissues and secretions. *Lancet,* **1**:217–220, 1929.

F60. Fleming, A. Lysozyme. (President's Address.) *Proc. Roy. Soc. Med.,* **26**:Part 1, 71–84, 1932–1933.

F61. Florey, H. The relative amounts of lysozyme present in the tissues of some mammals. *Brit. J. Exp. Pathol.,* **11**:251–261, 1930.

F62. Fluke, D. J. Temperature dependence of ionizing radiation effect on dry lysozyme and ribonuclease. *Radiat. Res.,* **28**:677–693, 1966.

F63. Foa, C., Foa, R., Muratore, R., and Olmer, J. Morphological and cytochemical aspects of intrasplenic red blood cell destruction in human hemolytic anemias. *J. Microsc. (Paris),* **14**:47A–48A, 1972.

F64. Fogelson, S. J., and Lobstein, O. E. A statistical comparison of the blood lysozyme activity of normal adults and of patients with localized and generalized carcinomatosis. *Amer. J. Dig. Dis.,* **21**:324–326, 1954.

F65. Fogelson, S. J., and Lobstein, O. E. Role of detergent complexes in experimental gastric and duodenal ulcers. *Proc. Soc. Exp. Biol. Med.,* **75**:334–337, 1950.

F66. Fogelson, S. J., Ross, A., and Lobstein, O. E. A method for the quantitative determination of lysozyme activity in blood. *Amer. J. Dig. Dis.,* **21**:327–331, 1954.

F67. Fontana, A., Veronese, F. M., and Scoffone, E. Sulfenyl halides as modifying reagents for polypeptides and proteins. III. Azobenzene-2-sulfenyl bromide, a selective reagent for cysteinyl residues. *Biochemistry,* **7**:3901–3905, 1968.

F68. Forsgren, A. Protein A from Staphylococcus aureus. VIII. Production of protein A by bacterial and L-forms of *S. aureus. Acta Pathol. Microbiol. Scand.,* **75**:481–490, 1969.

F69. Foss, J. G. Hydrophobic bonding and conformational transitions in lysozyme, ribonuclease and chymotrypsin. *Biochim. Biophys. Acta,* **47**:569–579, 1961.

F69a. Foss, J. G. Configurational transition in lysozyme. *Biochim. Biophys. Acta,* **43**:300–310, 1960.

F70. Foti, E., Rasá, G., and Mancuso, B. Sull'azione leucocitogena del lisozima di fleming in soggetti trattati con chemioantiblastici. *Minerva Chir.,* **21**:693–695, 1966.

F71. Fradkin, M. Y., Beketovskiy, N. P., and Levina, L. S. Lysozyme in postoperative ocular infections, *Vestn. Oftalmol. ,* **11**:496–500, 1937.

F72. Fradkin, M., Y., Levina, L., *et al.* Lysozyme in ocular infections. *Sov. Vestn. Ottal'mol.,* **6**:Part 3, 383, 1935. (Abstract in *Amer. J. Opthamol.,* **18**:375, 1935.)

F73. Fraenkel-Conrat, H. The essential groups of lysozyme, with particular reference to its reaction with iodine. *Arch. Biochem.,* **27**:109–124, 1950.

F73a. Fraenkel-Conrat, H. The reaction of proteins with ^{14}C-labeled N-carboxyleucine anhydride. *Biochim. Biophys. Acta,* **10**:180–182, 1953.

F74. Fraenkel-Conrat, H., Cooper, M., and Olcott, H. S. Reaction of CH$_2$O with proteins. *J. Amer. Chem. Soc.,* **67**:950–954, 1945.

F75. Fraenkel-Conrat, H., Mohammad, A., Ducay, E. D., and Mecham, D. K. Molecular weight of lysozyme after reduction and alkylation of the disulfide bonds. *J. Amer. Chem. Soc.,* **73**:625–627, 1951.

F76. Fraenkel-Conrat, H., and Olcott, H. S. Esterification of proteins by alcohols of low molecular weight. *J. Biol. Chem.,* **161**:259–268, 1945.

F77. Francesconi, F., and Resti, F. Clinical and experimental observations on the cicatricial action of lysozyme. *Gazz. Med. Ital.,* **118**:527–530, 1959.

F78. Frank, M. Uber die Beziehungen des Lysozyms zur A-Avitaminose. *Monatsschr. Kinderheilk.,* **60**:345–349, 1934.

F79. Fribourg-Blanc, A. Sensibilisation, par le lysozyme, du test d' immobilisation des tré ponémes. *Ann. Inst. Pasteur Paris,* **102**:460–468, 1962.

F80. Friedberg, F. Effect of irradiation on some lyophilized proteins. *Radiat. Res.,* **38**:34–42, 1969.

F81. Friedberg, I., and Avigad, G. High lysozyme concentration and lysis of *Micrococcus lysodeikticus. Biochim. Biophys. Acta,* **127**:532–535, 1966.

F82. Friedberger, E., and Hoder, F. Lysozyme and flocculation phenomenon of egg white. *Z. Immunitaetsforsch. Exp. Ther.,* **74**:429–447, 1932.

F82a. Frieden, E. H. "Enzymoid" properties of lysozyme methyl ester. *J. Amer. Chem. Soc.,* **78**:961–965, 1956.

F83. Friedland, B. R., Anderson, D. R., and Forster, R. K. Non-lysozyme antibacterial factor in human tears. *Amer. J. Ophthalmol.,* **74**:52–59, 1972.

F84. Friend, B. A., Eitenmiller, R. R., and Shahani, K. M. Reduction and reactivation of human and bovine milk lysozymes. *Arch. Biochem Biophys.,* **149**:435–440, 1972.

F85. Fromageot, C. The quantitative analysis of amino acids in proteins: Insulin and lysozyme. *Cold. Spring Harbor Symp. Quant. Biol.,* **14**:49–54, 1949.

F86. Fromageot, C. Lysozyme. *Congr. Chim. Biol., C. R. 7th, 1940,* Vol. 1:63–84, 1948; Suppl. to Bull. Soc. Chim. Biol. **30,** 1948.

F87. Fromageot, C., and Privat de Garilhe, M. La composition du lysozyme en acides aminés. II. Acides aminés totaux. *Biochim. Biophys. Acta,* **4**:509–517, 1950.

F88. Fromageot, C., and Privat de Garilhe, M. La composition du lysozyme en acides aminés. 1. Acides aromatiques, acides dicarboxyliques et bases hexoniques. *Biochim. Biophys. Acta,* **3**:82–91, 1949.

F89. Fromageot, C., and Schnek, G. Le spectre ultraviolet du lysozyme; avec des considé rations sur le spectre ultra-violet de divers acides aminés et de quelques- uns de leurs peptides. *Biochim. Biophys. Acta,* **6**:113–122, 1959.

F90. Fujio H., Imanishi, M. Nishioka, K., and Amano, T. Proof of independency of two antigenic sites in egg white lysozyme. *Biken J.,* **11**:219–223, 1968.

F91. Fujio, H., Imanishi, M., Nishioka, K., and Amano, T. Antigenic structures of hen egg white lysozyme. II. Significance of the N- and C-terminal region as an antigenic site. *Biken J.,* **11**:207–218, 1968.

F91a. Fujio, H., Kishiguchi, S., Shinka, S., Saiki, Y., and Amano, T. Immunochemical studies on lysozyme. I. Comparative studies of lysozyme and lysozyme methyl ester. *Biken J.* **2**:56–76, 1959.

F92. Fujio, H., Sakato, N., and Amano, T. The immunological properties of region specific antibodies directed to hen egg-white lysozyme. *Biken J.,* **14**:395–404, 1971.

F93. Funatsu, M. Chemical modification and enzymatic activities of lysozyme. *Tampak-ushitsu, Kakusan, Koso,* **13**:105–112, 1968. (Jap.)

G

G1. Gajdos, A. Biochimie des lysozymes. *Presse Med.*, **79**:351–354, 1971.

G1a. Gagen, W. L. The significance of the "partial specific volume" obtained from sedimentation data. *Biochemistry*, **5**:2553–2557, 1966.

G2. Galiazzo, G., Jori, G., and Scoffone, E. Selective and quantitative photochemical conversion of the tryptophyl residues to kynurenine in lysozyme. *Biochem. Biophys. Res. Commun.*, **31**:158–163, 1968.

G3. Galiazzo, G., Tamburro, A. M., and Jori, G. Photodynamic action of porphyrins on amino acids and proteins. Spectral studies on mono and di-sulphoxide derivatives of lysozyme. *Eur. J. Biochem.*, **12**:362–368, 1970.

G4. Galla, F., Chieregato, G. C., and Rossetti, C. L'attivita' lisozimico-serica nell'eritematode chronico. *Boll. Soc. Ital. Biol. Sper.*, **36**:936–938, 1960.

G5. Gallarate, L., Sandri, O., and Palmieri, R. Il potere lisozimico del succo gastrico in relazione alla curva dell'acidità sotto stimolo instaminico. *Boll. Ist. Sieroter. Milan.*, **31**:392–400, 1952.

G6. Gallarate, L., Sandri, O., and Palmieri, R. Il potere lisozimico nel gastropaziente. *Boll. Inst. Sieroter. Milan.*, **31**:274–283, 1952.

G7. Galliera, A., and Lazzari, G. Diagnosi di leucosi monocitica su piastra con micrococco lysodeikticus. *Arch. Sci. Med.*, **128**:91–96, 1971.

G8. Gallo, A. A., Swift, T. J., and Sable, H. Z. Magnetic resonance study of the Mn 2 + lysozyme complex. *Biochem. Biophys. Res. Commun.*, **43**:1232–1238, 1971.

G9. Galloni, O., and Castaldi, G. Contributo allo studio clinico terapeutico e esperimentale dei rapporti fra lisozima e timo. *Minerva Pediat.*, **18**:485–489, 1966.

G10. Gallop, R. G. C., Tozer, B. T., Stephen, J., and Smith, H. Separation of antigens by immunological specificity. Use of cellulose-linked antibodies as immunosorbents. *Biochem. J.*, **101**:711–716, 1966.

G11. Galyean, R. D., Cotterill, O. J., and Cunningham, F. E. Yolk inhibition of lysozyme activity in egg white. *Poultry Sci.*, **51**:1346–1353, 1972.

G12. Gambetti, G. Azione del Lisozima de Fleming comparativamente all'attività antiblastica della ciclofosfamide nei confronti del carcinoma di Ehrligh sperimentato nel cavo orale. *Mondo Odontostomatol.*, **9**:660–671, 1967.

G13. Garagnani, A., and Facchini, G. Influenza del lisozima sull'azione anticoagulante della eparina. *Arch. Patol. Clin. Med.*, **34**:41–46, 1957.

G14. Gardner, F. A., and Nikoopour, H. Growth characteristics of *Pseudomonas fluorescens* on conalbumin and lysozyme substrates. *Poultry Sci.*, **48**:43–48, 1969.

G14a. Garfinkel, D., and Edsall, J. T. Raman spectra of amino acids and related compounds. X. The Raman spectra of certain peptides and of lysozyme. *J. Amer. Chem. Soc.*, **80**: 3818–3823, 1958.

G15. Gasparetto, A., Corradi, C., and Matarazzi, R. Azione leucocitogena della muramidasi. Osservazioni cliniche. *Minerva Radiol., Fisioter. Radiobiol.*, **10**:176–181. 1965.

G16. Gebicki, J. M., and James, A. M. Effect of some lytic agents on Aerobacter aerogenes. *Nature (London)*, **182**:725–726, 1958.

G17. Gemsa, D., Davis, S. D., and Wedgwood, R. J. Lysozyme and serum bactericidal action. *Nature (London)*, **210**:950–951, 1966.

G18. Genazzani, E., and Miele, E. Radiazioni ionizzantie lisozima. - II) Influenza di vari farmaci sulla inattivazione biologica del lisozima da parte dei R.X. *Boll. Soc. Ital. Biol. Sper.*, **35**:1798–1801, 1959.

G19. Genazzani, E., and Miele, E. Radiazioni ionizzantie lisozima. 1. Inattivazione biologica del lisozima da raggi X. *Boll. Soc. Ital. Biol. Sper.*, **35**:1796–1798, 1959.

G20. Georgescu, M. Lizozimul factor al rezistentei naturale a organismului. *Rev. Med.-Chir. Soc. Med. Natur. Iasi,* **74:**13–20, 1970.

G21. Gerlsma, S. Y., and Stuur, E. R. The effect of polyhydric and monohydric alcohols on the heat-induced reversible denaturation of lysozyme and ribonuclease. *Int. J. Peptide Protein Res.,* **4:**377–383, 1972.

G22. Germana, G., and Galletti, C. Il comportamento dell'attività lisozimica della saliva in rapporto all'anestesia di superficie. *Oto-Rino-Laringol. Ital.,* **31:**104–109, 1962.

G23. Gerwing, J., and Thompson, K. Studies on the antigenic properties of egg-white lysozyme. I. Isolation and characterization of a tryptic peptide from reduced and alkylated lysozyme exhibiting haptenic activity. *Biochemistry,* **7:**3888–3892, 1968.

G24. Geschwind, I. I., and Li, C. H. The guanidination of some biologically active proteins. *Biochim. Biophys. Acta,* **25:**171–178, 1957.

G25. Gesteland, R. F., and Salser, W. Bacteriophage T4 lysozyme mRNA. *Genetics,* **61:**Suppl., 429–437, 1969.

G26. Gesteland, R. F., Salser, W., and Bolle, A. In vitro synthesis of T4 lysozyme by suppression of amber mutations. *Proc. Nat. Acad. Sci. U.S.,* **58:**2036–2042, 1967.

G26a. Gelewitz, E. W., Riedman, W. L., and Klotz, I. M. Some quantitative aspects of the reaction of diazonium compounds with serum albumin. *Arch. Biochem. Biophys.,* **53:** 411–424, 1954.

G27. Geyer, G. Lysozyme in Paneth cell secretions. *Acta Histochem.,* **45:**126–132, 1973.

G28. Ghoos, Y., and Vantrappen, G. The cytochemical localization of lysozyme in Paneth cell granules. *Histochem. J.,* **3:**175–178, 1971.

G29. Ghoos, Y., and Vantrappen, G. The cytochemical localization of lysozyme activity in leucocytes. *Histochem. J.,* **2:**11–16, 1970.

G30. Ghosh, B. K., and Murray, R. G. Fine structure of *Listeria monocytogenes* in relation to protoplast formation. *J. Bacteriol.,* **93:**411–426, 1967.

G31. Ghuysen, J. M. Use of bacteriolytic enzymes in determination of wall structure and their role in cell metabolism. *Bacteriol. Rev.,* **32:**Suppl., 425–464, 1968.

G32. Ghuysen, J. M. Précisions sur la structure des complexes disaccharide-peptide libérés des parois de Micrococcus lysodeikticus sous l'action des beta (1–4) N-acetylhexosa-minidases. *Biochim. Biophys. Acta,* **47:**561–568, 1961.

G33. Ghuysen, J. M. Acetylhexosamine compounds enzymically released from *Micrococcus lysodeikticus* cell walls. II. Enzymic sensitivity of purified acetylhexosamine and acetylhexosamine-peptide complexes. *Biochim. Biophys. Acta,* **40:**473–480, 1960.

G34. Ghuysen, J. M., and Salton, M. R. J. Acetylhexosamine compounds enzymically released from *Micrococcus lysodeikticus* cell walls. I. Isolation and composition of acetylhexosamine and acetylhexosamine-peptide complexes. *Biochim. Biophys. Acta,* **40:**462–472, 1960.

G35. Ghuysen, J. M., Petit, J. F., Muñoz, E., Legh-Bouille, M., and Dierickx, L. Action des enzymes bacteriolytiques. Autres que le lysozymes sur le peptidoglycane des parois cellulaires bactériennes. *Exposes Annu. Biochim. Med.,* **27:**101–110, 1966.

G36. Giacometti, C. Comportamento della attività lisozimica del liquido di bolla provocata zona cutanea sottoposta all' azione di irradiazioni Roentgen in dose eritema; ricerche sperimentali. *Ann. Ital. Dermatol. Sifil.,* **11:**91–98, 1956.

G37. Giacometti, C., and Anselmi, L. Ricerche sulla attività lisozimica in soggetti piodermitici comparativamente nel siero di sangue e nel liquido di bolla provocata in sede di alterazioni piococciche e in zone sane. *Rass. Dermatol. Sifilogr.,* **10:**104–115, 1957.

G38. Giannoni, E., and Bernicchi, L. Ricerche sul lisozima nel muco nasale della zona olfattoria e della zona respiratoria. *Minerva Otorinolaringol.,* **5:**302–305, 1955.

G39. Gibbons, R. J., de Stoppelaar, J. D., and Harden, L. Lysozyme insensitivity of bacteria indigenous to the oral cavity of man. *J. Dent. Res.,* **45:**877–881, 1966.

G40. Gildemeister, E. Untersuchungen über das Lysozym. *Zentralbl. Bakteriol., Parasitenkunde, Infektionskrankheiten, (Abt. 1: Orig.,* **136:**408–412, 1936.

G40a. Gill, T. J. III, McLaughlin, E. M., and Omenn, G. S. Studies of polypeptide structure by fluorescence techniques. III. Interaction between dye and macromolecule in fluorescent conjugates. *Biopolymers,* **5:**297–311, 1967.

G41. Girard, J. P. Propriétés antianaphylactiques du 9-(N-méthylpiperidyliden-4') thioxanthen (BP 400) et du lysozyme, comparées à celles de la prednisone. *Med. Exp.,* **9:**400–408, 1963.

G42. Girfanova, Kh. N. Changes of bactericidal properties of natural protein and lysozyme of chicken egg in course of incubation. *Dokl. Akad. Nauk SSSR,* **68:**1105–1108, 1949. (Russ.)

G43. Givol, D., Fuchs, S., and Sela, M. Isolation of antibodies to antigens of low molecular weight. *Biochim. Biophys. Acta,* **63:**222–224, 1962.

G44. Glad, B. W., Spikes, J. D., and Kumagai, L. F. Tyrosine and thyronine analogs and inhibitors of the dye-sensitized photoinactivation of lysozyme. *Proc. Soc. Exp. Biol. Med.,* **131:**1278–1280, 1969.

G45. Gladstone, G. P., and Johnston, H. H. The effect of cultural conditions on susceptibility of *Bacillus anthracis* to lysozyme. *Brit. J. Exp. Pathol.,* **36:**363–372, 1955.

G46. Glass, G. B. J. Gastric mucin and its constituents. *Gastroenterology,* **23:**636–658, 1953.

G47. Glass, G. B. J., Pugh, B. L., Grace, W. J., and Wolf, S. Observations on the treatment of human gastric and colonic mucus with lysozyme. *J. Clin. Invest.,* **29:**12–19, 1950.

G48. Glass, G. B. J., and Pugh, B. L. Effect of lysozyme on gastric and colonic mucus of man in vitro. *Amer. J. Med.,* **7:**422, 1949.

G49. Glassman, H. N., and Molnar, D. M. Precipitation and inhibition of lysozyme by surface-active agents. *Arch. Biochem. Biophys.,* **32:**170–180, 1951.

G50. Glazer, A. N., Barel, A. O., Howard, J. B., and Brown, D. M. Isolation and characterization of fig lysozyme. *J. Biol. Chem.,* **244:**3583–3589, 1969.

G50a. Glazer, A. N., and Simmons, N. S. Structure-function relationships in lysozyme by circular dichroism. *J. Amer. Chem. Soc.* **88:**2335–2336, 1966.

G51. Glick, A. D., Ranhand, J. M., and Cole, R. M. Degradation of Group A streptococcal cell walls by egg-white lysozyme and human lysosomal enzymes. *Infec. Immunity,* **6:**403–413, 1972.

G52. Glickson, J. D., McDonald, C. C., and Phillips, W. D. Assignment of tryptophan indole NH proton resonances of lysozyme. *Biochem. Biophys. Res. Commun.,* **35:**493–498, 1969.

G53. Glickson, J. D., Phillips, W. D., and Rupley, J. A. Proton magnetic resonance study of the indole NH resonances of lysozyme. Assignment, deuterium exchange kinetics, and inhibitor binding. *J. Amer. Chem. Soc.,* **93:**4031–4038, 1971.

G54. Glynn, A. A. The complement lysozyme sequence in immune bacteriolysis. *Immunology,* **16:**463–471, 1969.

G55. Glynn, A. A. Role et répartition du lysozyme intracellulaire et extracellulaire. *Exposes Annu. Biochim. Med.,* **27:**111–128, 1966.

G55a. Glynn, A. A. Lysozyme: antigen, enzyme and antibacterial agent. *Sci. Basis Med.,* 31–52, 1968.

G56. Glynn, A. A., and Brumfitt, W. Lysozyme intracellulaire et resistance aux infections bactériennes. *C. R. Soc. Biol.,* **156:**995–1000, 1962.

G57. Glynn, A. A., Brumfitt, W., and Salton, M. R. J. The specific activity and specific inhibition of intracellular lysozyme. *Brit. J. Exp. Pathol.*, **47**:331–336, 1966.

G58. Glynn, A. A., Martin, W., and Adinolfi, M. Levels of lysozyme in human foetuses and newborns. *Nature (London)*, **225**:77–78, 1970.

G59. Glynn, A. A., and Milne, C. M. A kinetic study of the bacteriolytic and bactericidal action of human serum. *Immunology*, **12**:639–653, 1967.

G60. Glynn, A. A., and Milne, C. M. Lysozyme and immune bacteriolysis. *Nature (London)*, **207**:1309–1310, 1965.

G61. Glynn, A. A., and Parkman, R. Studies with an antibody to rat lysozyme. *Immunology*, **7**:724–729, 1964.

G62. Gohar, M. A. Lysozym. *Zentralbl. Bakteriol. Parasitenk. Infektionskrankh., (Abt. 1: Orig.*, **119**:240–244, 1930.

G63. Gold, L. M., and Schweiger, M. Control of beta-glucosyltransferase and lysozyme synthesis during T4 deoxyribonuculeic acid-dependent ribonucleic acid and protein synthesis *in vitro. J. Biol. Chem.*, **245**:2255–2258, 1970.

G64. Goldbach, W., Haenel, H., and Grütte, F. K. Uber die pH-Abhäglgkeit des faekalen Lysozymgehaltes beim Erwachsenen. *Ernaehrungsforschung*, **9**:295–299, 1964.

G65. Goldbach, W., and Haenel, H. Uber Lysozym- und Proteasenbildung bei Staphylokokken. *Zentralbl. Bakteriol. Parasitenk. Infektionskrankh. Hyg. Abt. 1:* **195**:201–205, 1964.

G66. Goldbach, W., Haenel, H., and Schmidt, F. Uber den lysozymgehalt von malignen Rattentumoren im Vergleich zu normalen Rattengewben. *Acta Biol. Med. Ger.*, **10**:579–583, 1963.

G67. Goldberg, M. E., and Rossi, G. V. The turbidometric determination of lysozyme in animal gastric juice. *Amer. J. Pharm.*, **130**: 78–81, 1958.

G68. Goldberg, W. M., Chakrabarti, S., and Filipich, R. Urinary lactic dehydrogenase, alkaline phosphatase and lysozyme studies in renal disease. *Can. Med. Ass. J.*, **94**:1264–1268, 1966.

G69. Goldberger, R. F., and Epstein, C. J. Characterization of the active product obtained by oxidation of reduced lysozyme. *J. Biol. Chem.*, **238**:2988–2991, 1963.

G70. Goldberger, R. F., Epstein, C. J., and Anfinsen, C. B. Purification and properties of a microsomal enzyme system catalyzing the reactivation of reduced ribonuclease and lysozyme. *J. Biol. Chem.*, **239**:1406–1410, 1964.

G71. Goldschmidt, M. C., Kuhn, C. R., Perry, K., and Johnson, D. E. EDTA and lysozyme lavage in treatment of pseudomonas and coliform bladder infections. *J. Urol.*, **107**:969–972, 1972.

G72. Goldschmidt, M. C., and Wyss, O. The role of tris in EDTA toxicity and lysozyme lysis. *J. Gen. Microbiol.*, **47**:421–431, 1967.

G73. Goldsworthy, N. E., and Florey, H. Some properties of mucus, with special reference to its antibacterial functions. *Brit. J. Exp. Pathol.*, **11**:192–208, 1930.

G74. Golosova, T. V., Skurkovich, G. V., Shenderovich, V. A., and Anikina, T. P. The lysozyme titer in patients with various otorhinolaryngological diseases. *Antibiotiki (Moscow)*, **10**:447–450, 1965. (Russ.)

G75. Gonet, A., Scherer, J., and Girard, J. P. Antalgic properties of lysozyme in patients with neoplasms with bone metastases. Preliminary notes. *Ther. Umsch.*, **19**:158–159, 1962.

G76. Gonzales, R. M. Importancia del lysozyma en pediatria. *Rev. Chil. Pediat.*, **15**:320–326, 1944.

G77. Gordon, J. A. Denaturation of globular proteins. Interaction of guanidinium salts with three proteins. *Biochemistry*, **11**:1862–1870, 1972.

G78. Gordon, J. A. Optical rotatory dispersion studies of globular proteins, including alphachymotrypsin. *J. Biol. Chem.,* **243:**4615–4625, 1968.

G79. Gorin, G., Fulford, R., and Deonier, R. C. Reaction of lysozyme with dithiothreitol and other mercaptans. *Experientia,* **24:**26–27, 1968.

G80. Gorin, G., Papapavlou, L., and Tai, L. W. X-ray inactivation of lysozyme in dilute solution. *Int. J. Radiat. Biol.,* **15:**33–41, 1969.

G81. Gorin, B., Wang, S. F., and Papapavlou, L. Assay of lysozyme by its lytic action on M. lysodeikticus cells. *Anal. Biochem.,* **39:**113–127, 1971.

G81a. Gorini, L., and Felix, F. Influence du manganèse sur la stabilité du lysozyme. I. Influence du manganèse sur la vitresse d'inactivation irréversible du lysozyme par la chaleur. *Biochim. Biophys. Acta,* **10:**128–135, 1953.

G81b. Gorini, L., Felix, F., and Fromageot, C. Influence du manganèse sur la stabilité du lysozyme. II. Rôle protecteur du manganèse lors de l'hydrolyse du lysozyme par la trypsine. *Biochim. Biophys. Acta,* **12:**283–288, 1953.

G82. Goriunova, A. G. Dynamics of lysozyme titer changes in guinea pigs under the action of ionizing radiations. *Med. Radiol.,* **8:**77–79, 1963. (Russ.)

G83. Goriunova, A. G., and Chakhava, O. V. Determination of formation of lysozyme in different cell cultures of human and animal origin. *Antibiotiki (Moscow),* **15:**28–30, 1970. (Russ.)

G84. Gorokhoff, M., and Charoki, K. Apropos of enzyme medication with physiological action adapted to the mucous membranes of otorhinolaryngological region. *Rev. Laryngol. Paris,* **85:**88–92, 1963.

G85. Goryunova, A. G., and Chakhava, O. V. Determination of muramidase production in various cell cultures of humans and animals. *Antibiotiki (Moscow),* **15:**28–30, 1970.

G86. Gorzynski, E. A., Neter, E., and Cohen, E. Effect of lysozyme on the release of erythrocyte-modifying antigen from staphylococci and *Micrococcus lysodeikticus. J. Bacteriol.,* **80:**207–211, 1960.

G87. Gould, G. W., Georgala, D. L., and Hitchins, A. D. Fluorochrome-labelled lysozyme: Reagent for the detection of lysozyme substrate in cells. *Nature (London),* **200:**385–386, 1963.

G88. Gould, G. W., and Hitchins, A. D. Sensitization of bacterial spores to lysozyme and to hydrogen peroxide with agents which rupture disulphide bonds. *J. Gen. Microbiol.,* **33:**413–423, 1963.

G89. Grace, W. J., Seton, P. H., Wolf, S., and Wolff, H. G. Studies of human colon; variations in concentration of lysozyme with life situation and emotional state. *Amer. J. Med. Sci.,* **217:**241–251, 1949.

G90. Grand, J. P., Scherer, J., and Fischer, G. Therapeutic properties of lysozyme in pain connected with generalized neoplasms and various virus diseases. *Ther. Umsch.,* **19:**503–506, 1962. (Fre.)

G91. Grando, F., and Baratella, B. O lisozima de Fleming como profilático da morbilidade genérica e da perturbações da nutrição do lactente. *Hospital (Rio de Janeiro),* **68:**252–255, 1965.

G92. Gratia, A., and Gorecsky, L. The ultracentrifugation of immune bodies and of normal bactericidal substances. *Z. Immunitaetsforsch.,* **93:**18–26, 1938.

G93. Gray, S. J., and Reifenstein, R. W. Antilysozyme and steroid therapy in ulcerative colitis. *Ann. N. Y. Acad. Sci.,* **58:**474–486, 1954.

G94. Gray, S. J., Reifenstein, R. W., Benson, J. A., Jr., and Young, J. C. G. Treatment of ulcerative colitis and regional enteritis with ACTH; significance of fecal lysozyme. *AMA Arch. Intern. Med.,* **87:**646–662, 1951.

G95. Gray, S. J., Reifenstein, R. W., Connolly, E. P., Spiro, H. M., and Young, J. C. G. Studies on lysozyme in ulcerative colitis. *Gastroenterology,* **16**:687–696, 1950.

G96. Gray, S. J., Reifenstein, R. W., Young, J. C. G., Spiro, H. M., and Connolly, E. P. The source of gastric lysozyme. *J. Clin. Invest.,* **29**:1595–1600, 1950.

G97. Graziani, G., Imbasciati, E., Goi, A., and Brancaccio, D. Sul significato della lisozimuria quale indice di alterazioni tubulari in alcune nefropatie. *Arch. Ital. Urol. Nefrol.,* **42**:389–400, 1969.

G98. Grebner, E. E., and Neufeld, E. F. Stimulation of a protein glycosylation reaction by lysozyme. *Biochim. Biophys. Acta,* **192**:337–349, 1969.

G99. Green, F. C., and Schroeder, W. A. A terminal amino acid residue of lysozyme as determined with 2,4-Dinitrofluorobenzene. *J. Amer. Chem. Soc.,* **73**:1385, 1951.

G100. Green, G. E. Properties of a salivary bacteriolysin and comparison with serum beta lysin. *J. Dent. Res.,* **45**:882–889, 1966.

G101. Green, M. N., Kulczycki, L. L., Nemer, M., Blume, S., and Shwachman, H. Lysozyme levels in cystic fibrosis of the pancreas. *AMA J. Dis. Child.,* **97**:303–307, 1959.

G102. Green, N. M. Evidence for a genetic relationship between avidins and lysozymes. *Nature (London),* **217**:254–256, 1968.

G103. Greenberger, J. S., Rosenthal, D. S., and Moloney, W. C. Studies on hypermuramidasemia in the normal and chloroleukemic rat: The role of the kidney. *J. Lab. Clin. Med.,* **81**:116–121, 1973.

G104. Greenblatt, I. J., Jacobi, M., and Cohen, T. D. Effect of a lysozyme inactivating anion exchange polymer in treatment of peptic ulcer: Experimental and clinical study. *Amer. J. Dig. Dis.,* **18**:362–368, 1951.

G105. Greenfield, J., and Bigland, C. H. Detection of lysozyme in developing avian embryos. *Poultry Sci.,* **50**:1748–1753, 1971.

G105a. Greenfield, N., and Fasman, G. D. Computed circular dichroism spectra for the evaluation of protein conformation. *Biochemistry,* **8**:4108–4116, 1969.

G106. Greenwald, R. A., Josephson, A. S., Diamond, H. S., and Tsang, A. Human cartilage lysozyme. *J. Clin. Invest.,* **51**:2264–2270, 1972.

G107. Greenwald, R. A., and Sajdera, S. W. Extracellular localization of lysozyme in rachitie rat cartilage. *Proc. Soc. Exp. Biol. Med.,* **142**:924–927, 1973.

G108. Grekova, N. A. Factors of natural bodily immunity to brucellosis. I. The effect of the vaccinal and infectious processes in brucellosis on the lysozyme concentration in the body of the animal. *Zh. Mikrobiol., Epidemiol. Immunobiol.,* **44**:42–44, 1967. (Russ.)

G109. Grignolo, A. Ricerche su possibili interferenze tra azione antibatterica della penicillina e lisi da lysozyme lacrimale. *Boll. Ocul.,* **26**:699–709, 1947.

G110. Grigorian, G. S. Treatment of subclinical bovine mastitis with lysozymic milk. *Veterinariya (Moscow),* **44**:94–95, 1967. (Russ.)

G111. Gros, C., and Labouesse, B. Study of the dansylation reaction of amino acids, peptides and proteins. *Eur. J. Biochem.,* **7**:463–470, 1969,

G112. Grossgebauer, K., and Langmaack, H. Wechselwirkungen zwischen Lysozym und Influenzaviren. *Z. Naturforschu. B,* **23**:952–961, 1968.

G113. Grossgebauer, K. Ein Hämagglutinationstest zum Nachweis von Lysozym. *Blut,* **14**:93–96, 1966.

G114. Grossgebauer, K., and Langmaack, H. Lysozyme. Ergebnisse und Probleme. *Klin. Wochenschr.,* **46**:1121–1127, 1968.

G115. Grossgebauer, K., Pohle, H. D., and Langmaack, H. Vermehrtes Auftreten von Lysozym im Liquor und Urin bei Fällen von Meningitis. *Klin. Wochenschr.,* **46**:1127–1131, 1968.

G116. Grossgebauer, K., Raettig, H., Langmaack, H., and Küchler, R. Zur unspezifischen Schutzwirkung von Lysozym und Aristologhiasäure bei bakteriellen und viralen Infektionen. *Zentralbl. Bakteriol., Parasitenk., Infektionskr., Hyg.,* **213**:401–415, 1970.

G117. Grossgebauer, K., Schmidt, B., and Langmaack, H. Ein In-vitro-Test zum Nachweis von hämagglutinierenden Eigenschaften einiger Antibiotika sowie von Antibiotikum-Lysozym-Wechselwirkungen. *Blut,* **19**:63–72, 1969.

G118. Grossgebauer, K., Schmidt, B., and Langmaack, H. Lysozyme production as an aid for identification of potentially pathogenic strains of staphylococci. *Appl. Microbiol.,* **16**:1745–1747, 1968.

G119. Grossowicz, N., and Ariel, M. Mechanism of protection of cells by spermine against lysozyme-induced lysis. *J. Bacteriol.,* **85**:293–300, 1963.

G120. Grossweiner, L. I., and Usui, Y. Flash photolysis and inactivation of aqueous lysozyme. *Photochem. Photobiol.,* **13**:195–214, 1971.

G121. Gruen, L. C., and Nicholls, P. W. Improved recovery of tryptophan following acid hydrolysis of proteins. *Anal. Biochem.,* **47**:348–355, 1972.

G121-A. Grula, E. A., and Hartsell, S. E. Lysozyme in the bacteriolysis of gram-negative bacteria. II. Factors influencing clearing during the Nakamura treatment. *Can. J. Microbiol.,* **3**:23–34, 1957.

G122. Grula, E. A., and Hartsell, S. E. Lysozyme in the bacteriolysis of gram-negative bacteria. I. Morphological changes during use of Nakamura's technique. *Can. J. Microbiol.,* **3**:13–21, 1957.

G123. Guerrisi, G. Le modifiche vascolari nella genesi del lisozima a seconda delle varie forme di flogosi tonsillari, tonsilliti ipertrofiche. *Acta Med. Ital. Mal. Infet. Parassit.,* **10**:92–95, 1955.

G124. Guerrisi, G. Le modifiche vascolari nella genesi del lisozima a seconda delle varie forme di flogosi tonsillari; tonsilliti focale. *Acta Med. Ital. Mal. Infet. Parassit.,* **10**:63–69, 1955.

G125. Guffanti, A. Prime ricerche cliniche sull'ampicillinato di lisozima in oftalmologia. *Minerva Oftalmol.,* **10**:139–147, 1968.

G126. Guffanti, A. Sull'interesse biologico del lisozima in oculistica. *Arch. Sci. Med.,* **124**:553–565, 1967.

G127. Guire, P. Differences in tryptophan exposure between chicken egg-white lysozyme and bovine alpha-lactalbumin. *Biochim. Biophys. Acta,* **221**:383–386, 1970.

G128. Gulik-Krzywicki, T., Shechter, E., Iwatsubo, M., Ranck, J. L., and Luzzati, V. Correlations between structure and spectroscopic properties in membrane model systems. Tryptophan and I-anilino-8-naphthalene sulfonate fluorescence in protein-lipid-water phases. *Biochim. Biophys. Acta,* **219**:1–10, 1970.

G129. Gurnani, S., Arifuddin, M., and Bhargava, P. M. Studies on egg white lysozyme: Effect of trichloroacetic acid. *Indian J. Biochem.,* **5**:37–43, 1968.

H

H1. Haas, D. J. Preliminary studies on the denaturation of cross-linked lysozyme crystals. *Biophys. J.,* **8**:549–555, 1968.

H2. Haas, D. J. Preliminary x-ray data for two new forms of hen egg-white lysozyme. *Acta Crystallogr.,* **23**:666, 1967.

H3. Habbal, Z., and Durr, I. F. Osmotic fragility of lysozyme- and thiol-treated *Lactobacillus plantarum. J. Bacteriol.,* **100**:1409–1410, 1969.

H4. Habeeb, A. F. S. A. Quantitation of conformational changes on chemical modification of proteins: Use of succinylated proteins as a model. *Arch. Biochem. Biophys.,* **121:**652–664, 1967.

H5. Habeeb, A. F. S. A. The reaction of sulphuric acid with lysozyme and horse globin. *Can. J. Biochem. Physiol.,* **39:**31–43, 1961.

H6. Habeeb, A. F. S. A., and Atassi, M. Z. Enzymic and immunochemical properties of lysozyme. V. Derivatives modified at lysine residues by guanidination, acetylation, succinylation or maleylation. *Immunochemistry,* **8:**1047–1059, 1971.

H7. Habeeb, A. F. S. A., and Atassi, M. Z. Enzymic and immunochemical properties of lysozyme. IV. Demonstration of conformational differences between alpha-lactalbumin and lysozyme. *Biochim. Biophys. Acta,* **236:**131–141, 1971.

H8. Habeeb, A. F. S. A., and Atassi, M. Z. Enzymic and immunochemical properties of lysozyme. Evaluation of several amino group reversible blocking reagents. *Biochemistry,* **9:**4939–4944, 1970.

H9. Habeeb, A. F. S. A., and Atassi, M. Z. Enzymic and immunochemical properties of lysozyme. II. Conformation, immunochemistry and enzymic activity of a derivative modified at tryptophan. *Immunochemistry,* **6:**555–566, 1969.

H10. Hachimori, Y., Horinishi, H., Kurihara, K., and Kazuo, S. States of amino acid residues in proteins. V. Different reactivities with H_2O_2 of tryptophan residues in lysozyme, proteinases and zymogens. *Biochim. Biophys. Acta,* **93:**346–360, 1964.

H11. Haenel, H. On the significance of antimicrobial substances in food. *Ernaehrungsforschung,* **7:**680–692, 1963. (Ger.)

H12. Haenel, H., Goldbach, W., and Grutte, F. K. Nutrition and fecal lysozyme activity in infants. *Ernaehrungsforschung,* **8:**282–289, 1963. (Ger.)

H13. Haenel, H., Goldbach, W., Hoffmann, G., *et al.* Zum Nachweis von Hefen, Staphylokokken, aeroben Sporenbildern und Pseudomonas sowie von Lysozym im Stuhl gesunder Erwachsener. *Zentralb. Bakteriol., Parasitenk, Infektionskr. Hyg., Abt. 1: Orig.,* **197:**244–255, 1965.

H13a. Haggis, G. H. Proton-deuteron exchange in protein and nucleoprotein molecules surrounded by heavy water. *Biochim. Biophys. Acta,* **23:**494–503, 1957.

H14. Haldar, D., and Chatterjee, A. N. Action of lysozyme on Vibrio cholerae. *Ann. Biochem. Exp. Med.,* **22:**255–260, 1962.

H15. Haley, E. E., Corcoran, B. J., Dorer, F. E., and Buchanan, D. L. Beta-aspartyl peptides in enzymatic hydrolysates of protein. *Biochemistry,* **5:**3229–3235, 1966.

H16. Hall, L. D., and Grant, C. W. M. An n.m.r. study of the effects of fluorine substituents on the association between lysozyme and derivatives of 2-amino-2-deoxy-D-glucose. *Carbohyd. Res.,* **24:**218–220, 1972.

H17. Hallauer, C. Klinische und experimentelle Untersuchungen über den Lysozymgehalt im Bindehaustsack und in der Tränenflüssigkeit. *Arch. Augenheilk.,* **103:**199–215, 1930.

H18. Hallauer, C. Ueber das Lysozym. *Zentralbl. Bakteriol., Parasitenk. Infektionskrankh., Abt. 1: Orig.,* **114:**519–529, 1929.

H19. Halper, J. P., Latovitzki, N., Bernstein, H., and Beychok, S. Optical activity of human lysozyme. *Proc. Nat. Acad. Sci. U.S.,* **68:**517–522, 1971.

H19a. Halwer, M., Nutting, G. C., and Brice, B. A. Molecular weight of lactoglobulin, ovalbumin, lysozyme and serum albumin by light scattering. *J. Amer. Chem. Soc.* **73:** 2786–2793, 1951.

H20. Hamaguchi, K. Binding of protomer enzyme with substrates and inhibitors—conformational changes by binding of inhibitors. *Tampakushitsu Kakusan, Koso,* **15**:1271–1280, 1970. (Jap.)

H20a. Hamaguchi, K. Studies on protein denaturation by surface chemical method. I. The relationship between monolayer properties and urea denaturation of lysozyme. *J. Biochem. (Tokyo),* **42**:449–459, 1955.

H20b. Hamaguchi, K. Studies on protein denaturation by surface chemical method. II. On the mechanism of surface denaturation of lysozyme. *J. Biochem. (Tokyo),* **42**:705–714, 1955.

H20c. Hamaguchi, K. Studies on protein denaturation by surface chemical method. III. The interaction of urea with lysozyme monolayer. *J. Biochem. (Tokyo)* **43**:83–92, 1956.

H20d. Hamaguchi, K. Studies on protein denaturation by surface chemical method. IV. On the structure of lysozyme monolayer. *J. Biochem. (Tokyo),* **43**:355–367, 1956.

H20e. Hamaguchi, K. Studies on the denaturation of lysozyme. I. Kinetics of heat denaturation. *J. Biochem. (Tokyo),* **44**:695–706, 1957.

H20f. Hamaguchi, K. Studies on the denaturation of lysozyme. II. Urea denaturation. *J. Biochem. (Tokyo),* **45**:79–88, 1958.

H20g. Hamaguchi, K., and Funatsu, M. On the action of egg white lysozyme on glycol chitin. *J. Biochem. (Tokyo),* **46**:1659–1660, 1959.

H20h. Hamaguchi, K., Rokkaku, K., Funatsu, M. and Hayashi, K. Studies on the structure and enzymatic function of lysozyme. I. Enzymatic action of lysozyme on glycol chitin. *J. Biochem. (Tokyo),* **48**:351–357, 1960.

H20i. Hamaguchi, K., and Rokkaku, K. Studies on the structure and enzymatic function of lysozyme. II. Effect of urea on lysozyme. *J. Biochem. (Tokyo),* **48**:358–362, 1960.

H21. Hamaguchi, K. Conformation and enzymatic activities of lysozyme. *Tampakushitsu Kakusan, Koso,* **13**:2–8, 1968. (Jap.)

H22. Hamaguchi, K. Secondary and tertiary structures of lysozyme. *Seikagaku,* (Jap.) **38**:387–401, 1966.

H23. Hamaguchi, K. Structure of muramidase (lysozyme). VIII. Effect of dimethylsulfoxide on the stability of muramidase. *J. Biochem. (Tokyo),* **56**:441–449, 1964.

H24. Hamaguchi, K. Structure of muramidase (lysozyme). V. Effect of N,N-dimethylformamide and the role of disulfide bonds in the stability of muramidase. *J. Biochem. (Tokyo),* **55**:333–339, 1964.

H25. Hamaguchi, K., Hayashi, K., Imoto, T., and Funatsu, M. Structure of muramidase (lysozyme). IV. Effects of ethylene glycol, polyethylene glycol and sucrose on the physico-chemical and enzymatic properties of muramidase. *J. Biochem. (Tokyo),* **55**:24–29, 1964.

H26. Hamaguchi, K., and Imahori, K. Structure of muramidase (lysozyme). VI. Effect of amide derivatives and the structure of internal fold of the muramidase molecule. *J. Biochem. (Tokyo),* **55**:388–393, 1964.

H27. Hamaguchi, K., and Kurono, A. Structure of muramidase (lysozyme). III. Effect of 2-chloroethanol, ethanol and dioxane on the stability of muramidase. *J. Biochem. (Tokyo),* **54**:497–505, 1963.

H28. Hamaguchi, K., and Kurono, A. Structure of muramidase (lysozyme). I. The effect of guanidine hydrochloride on muramidase. *J. Biochem. (Tokyo),* **54**:111–112, 1963.

H29. Hamaguchi, K., Kurono, A., and Goto, S. Structure of muramidase (lysozyme). II. Effect of lithium chloride and bromide on the stability of muramidase. *J. Biochem. (Tokyo),* **54**:259–266, 1963.

H30. Hamaguchi, K., and Sakai, H. Structure of lysozyme. IX. The effect of temperature on the conformation of lysozyme. *J. Biochem. (Tokyo)*, **57**:721–732, 1965.

H31. Hambleton, P. The sensitivity of gram-positive bacteria, recovered from aerosols, to lsyozyme and other hydrolytic enzymes. *J. Gen. Microbiol.*, **61**:197–204, 1970.

H32. Hamdy, M. K. Effect of trauma and infection on lysozyme in poultry tissue. *Proc. Soc. Exp. Biol. Med.*, **131**:409–415, 1969.

H33. Hampe, O. G. Conformation of lysozyme in aqueous solution. Effect of ionic strength and protein concentrations. *Eur. J. Biochem.*, **31**:32–37, 1972.

H34. Hanke, N., Prager, E. M., and Wilson, A. C. Quantitative immunological and electrophoretic comparison of primate lysozymes. *J. Biol. Chem.*, **248**:2824–2828, 1972.

H35. Hannoüz, M. Antigénicité et pureté d'une préparation de lysozyme. *Pathol. Microbiol.*, **28**:139–146, 1965.

H36. Hanschke, M., and Heilmann, H. H. Über die Lysozymaktivität in Knochen und Knorpel. *Beitr. Orthop. Traumatol. (Berlin)*, **17**:618–620, 1970.

H37. Hansen, N. E., and Karle, H. Blood and bone-marrow lysozyme in neutropenia: An attempt toward pathogenetic classification. *Brit. J. Haematol.*, **21**:261–270, 1971.

H38. Hansen, N. E., and Karle, H. Clinical significance of muramidase (lysozyme) in plasma and urine. *Nord. Med.*, **85**:325–329, 1971. (Dan.)

H39. Hansen, N. E., Karle, H., and Andersen, V. Muramidase activity of bone marrow plasma. Studies in haematologically normal individuals and in granulocytopenic patients. *Acta Med. Scand.*, **185**:387–392, 1969.

H40. Hansen, N. E., Karle, H., Andersen, V., and Olgaard, K. Lysozyme turnover in man. *J. Clin. Invest.*, **51**:1146–1155, 1972.

H41. Hansen, N. E., Karle, H., and Andersen, V. Lysozyme turnover in the rat. *J. Clin. Invest.*, **40**:1473–1477, 1971.

H42. Hansen, N. E., Karle, H., and Axelsen, N. H. Post-gamma protein in urinary protein. *Acta Med. Scand.*, **190**:175–178, 1971.

H43. Hansen, N. E., and Weeke, E. Renal transplantation and muramidase activity. Urinary muramidase as indicator of tubular damage in patients with renal transplants. *Acta Med. Scand.*, **188**:317–321, 1970.

H44. Hara, S. The substrate specificity of egg white lysozyme. *Seikagaku*, **44**:146, 1972. (Jap.)

H45. Hara, S., and Matsushima, Y. Studies on the substrate specificity of egg white lysozyme. IV. A comparative study of the substrate specificities of lysozymes from different sources. *J. Biochem. (Tokyo)*, **72**:993–1000, 1972.

H46. Hara, S., and Matsushima, Y. Studies on the substrate specificity of egg white lysozyme. I. The N-acyl substituents in the substrate mucopolysaccharides. *J. Biochem. (Tokyo)*, **62**:118–125, 1967.

H47. Hara, S., Nakagawa, Y., and Matsushima, Y. Studies on the substrate specificty of egg white lysozyme. III. The mode of the enzymatic action on partially O-hydroxyethylated chitin. (glycol chitin). *J. Biochem. (Tokyo)*, **68**:53–62, 1970.

H48. Hara, Y. On the lysozyme activity in the chronic tonsillitic patient. *Kumamoto Igakkai Zasshi*, **44**:986–990, 1970. (Jap.)

H49. Harrison, J. F., and Barnes, A. D. The urinary excretion of lysozyme in dogs. *Clin. Sci.*, **38**:533–547, 1970.

H50. Harrison, J. F., Barnes, A. D., and Blainey, J. D. Lysozymuria and renal transplantation. *Transplantation*, **13**:372–377, 1972.

H51. Harrison, J. F., Barnes, A. D., and Blainey, J. D. Proteinuria, lysozymuria, and renal transplantation. *Brit. J. Surg.*, **58**:299–300, 1971.

H52. Harrison, J. F., Lunt, G. S., Scott, P., and Blainey, J. D. Urinary lysozyme, ribonuclease, and low-molecular-weight protein in renal disease. *Lancet,* **1**:371–375, 1968.

H53. Harrison, J. F., and Swingler, M. The effect of serum macromolecules on the lysis of Micrococcus lysodeikticus cells by lysozyme. *Clin. Chim. Acta,* **31**:149–154, 1971.

H54. Hartdegen, F. J., and Rupley, J. A. The oxidation by iodine of tryptophan 108 in lysozyme. *J. Amer. Chem. Soc.,* **89**:1743–1745, 1967.

H55. Hardegen, F. J., and Rupley, J. A. Inactivation of lysozymes by iodine oxidation of a single tryptophan. *Biochim. Biophys. Acta,* **92**:625–627, 1964.

H56. Harte, R. A., and Rupley, J. A. Three-dimensional pictures of molecular models— lysozyme. *J. Biol. Chem.,* **243**:1663–1669, 1968.

H57. Hartsell, S. E. Lysozyme activity of rehydrated, spray-dried whole-egg powder. *Food Res.,* **13**:136–142, 1948.

H58. Hartsell, S. E. The newer knowledge of lysozyme and bacteria. *Proc. Indiana Acad. Sci.,* **57**:44–53, 1948.

H59. Hartsell, S. E., and Caldwell, J. J. Lysozyme and the differentiation of group D streptococci. *Sem. Med.,* **120**:195–204, 1962. (Span.)

H60. Harvey, S. C., and Hoekstra, P. Dielectric relaxation spectra of water adsorbed on lysozyme. *J. Phys. Chem.,* **76**:2987–2994, 1972.

H61. Hasebe, K. Biochemical studies on synovial fluid. II. Effect of intra-articular administration of aurothiomalate on mucopolysaccharase activities in rheumatoid synovial fluid. *Fukushima J. Med. Sci.,* **15**:45–53, 1968.

H62. Hasebe, K. Biochemical studies on synovial fluid. I. Mucopolysaccharase activities in synovial fluid of rheumatoid arthritis. *Fukushima J. Med. Sci.,* **15**:35–44, 1968.

H63. Hash, J. H. Measurement of bacteriolytic enzymes. *J. Bacteriole,* **93**:1201–1202, 1967.

H64. Hash, J. H., and Rothlauf, M. V. The N,O-diacetylmuramidase of *Chalaropsis* species. *J. Biol. Chem.,* **242**:5586–5590, 1967.

H64a. Hashizume, H., Shiraki, M., and Imahori, K. Study of circular dichroism of proteins and polypeptides in relation to their backbone and side conformations. *J. Biochem. (Tokyo),* **62**:543–551, 1967.

H65. Hatano, H. Experimental studies on the effects of chloramphenicol-enzyme combinations in the field of ophthalmology. *Nippon Ganka Gakkai Zasshi,* **71**:1225–1238, 1967. (Jap.)

H66. Hattori, T. Malignant tumor and lysozyme. *Saishin Igaku,* **26**:930–936, 1971. (Jap.)

H67. Haupt, H., and Heimburger, N. Humanserumproteine mit hoher Affinität zu carboxymethyl celluose, I. Isolierung von lysozym, Clq und zwei bisher unbekannten aplha-Globulinen. *Hoppe-Seyler's Z. Physiol. Chem.,* **353**:1125–1132, 1972.

H68. Hawiger, J. Frequency of staphylococcal lsyozyme production tested by plate method. *J. Clin. Pathol.,* **21**:390–393, 1968.

H69. Hawiger, J. Purification and properties of lysozyme produced by *Staphylococcus aureus. J. Bacteriol.,* **95**:376–384, 1968.

H70. Hawiger, J. Staphylococcal lysozyme. 4. The lysozyme activity against various substrates. *Exp. Med. Biol.,* **20**:118–124, 1968.

H71. Hawiger, J. Med: Staphylococcal lysozyme. 3. Factors that influence the course of reaction of staphylococcal lysozyme. *Exp. Med. Microbiol.,* **20**:107–117, 1968.

H72. Hawiger, J. Staphylococcal lysozyme. II. Purification of lysozyme produced by Staphylococcus aureus 524. *Exp. Med. Microbiol.,* **20**:1–8, 1968.

H73. Hawiger, J. Staphylococcal lysozyme. I. Occurrence of lysozyme in comparison with the other staphylococcal factors. *Exp. Med. Microbiol.,* **19**:337–344, 1967.

H74. Hawthorne, J. R. Note on the products formed in lysis of *Micrococcus lysodeikticus* by egg white lysozyme. *Biochim. Biophys. Acta,* **6**:94–96, 1959.

H75. Hawthorne, J. R. The action of egg white lysozyme on ovomucoid and ovomucin. *Biochim. Biophys. Acta,* **6**:28–35, 1950.

H76. Hayashi, H., Amano, K.-I., Araki, Y., and Ito, E. Action of lysozyme on oligosaccharides from peptidoglycan n-unacetylated at glucosamine residues. *Biochem. Biophys. Res. Commun.,* **50**:641–648, 1973.

H77. Hayashi, H., Araki, Y., and Ito, E. Occurrence of glucosamine residues with free amino groups in cell wall peptidoglycan from bacilli as a factor responsible for resistance to lysozyme. *J. Bacteriol.,* **113**:592–598, 1973.

H78. Hayashi, K. Reaction mechanism of lysozyme catalyzed hydrolysis. *Tanpakushitsu Kakusan, Koso,* **13**:121–128, 1968. (Jap.)

H79. Hayashi, K., Fujimoto, N., Kugimiya, M., and Funatsu, M. The enzyme-substrate complex of lysozyme with chitin derivatives. *J. Biochem. (Tokyo),* **65**:401–405, 1969.

H80. Hayashi, K., Hamaguchi, K., and Funatsu, M. Heat activation of muramidase. *J. Biochem. (Tokyo),* **53**:374–380, 1963.

H81. Hayashi, K., Imoto, T., and Funatsu, M. Crystallization of the lysozyme-product complex. *J. Biochem. (Tokyo),* **63**:550–552, 1968.

H82. Hayashi, K., Imoto, T., and Funatsu, M. The enzyme-substrate complex in a muramidase (lysozyme) catalyzed reaction. II. Evidence for the conformational changes in the enzyme. *J. Biochem. (Tokyo),* **55**:516–521, 1964.

H83. Hayashi, K., Imoto, T., Funatsu, G., and Funatsu, M. The position of the active tryptophan residue in lysozyme. *J. Biochem. (Tokyo),* **58**:227–235, 1965.

H84. Hayashi, K., Kugimiya, M., and Funatsu, M. Heat stability of the lysozyme-substrate complex. *J. Biochem. (Tokyo),* **64**:93–97, 1968.

H85. Hayashi, K., Imoto, T., and Funatsu, M. The enzyme-substrate complex in a muramidase catalyzed reaction. Difference spectrum of complex. *J. Biochem. (Tokyo),* **54**:381–387, 1963.

H85a. Hayashi, K., Imoto, T., and Funatsu, M. The enzyme-substrate complex in a muramidase (lysozyme) catalyzed reaction. *J. Biochem. (Tokyo),* **55**:516–521, 1964.

H86. Hayashi, K., Kugimiya, M., Imoto, T., Funatsu, M., and Bigelow, C. C. The inhibitory interaction of cationic detergents with the active center of lysozyme. I. Site of interaction. *Biochemistry,* **7**:1461–1466, 1968.

H86a. Hayashi, K., Kugimiya, M., Imoto, T., Funatsu, M. and Bigelow, C. C. The inhibitory interaction of cationic detergents with the active center of lysozyme. II. The pH dependence of the interaction. *Biochemistry,* **7**:1467–1472, 1968.

H87. Hayashi, K., Shimoda, T., Funatsu, M., and Imoto, T. Iodination of lysozyme. II. Reactivity and position of tyrosine residues. *J. Biochem. (Tokyo),* **64**:365–370, 1968.

H88. Hayashi, K., Shimoda, T., Yamada, K., Kumai, A., and Funatsu, M. Iodination of lysozyme. I. Differential iodination of tyrosine residues. *J. Biochem. (Tokyo),* **64**:239–245, 1968.

H89. Hayes, J. P. Estimation of serum and urinary muramidase with the eel aggregometer. *J. Clin. Pathol.,* **25**:454–455, 1972.

H90. Hayslett, J. P., Perillie, P. E., and Finch, S. C. Urinary muramidase and renal disease. Correlation with renal histology and implication for the mechanism of enzymuria. *N. Eng. J. Med.,* **279**:506–512, 1968.

H91. Heise, E. R., and Myrvik, Q. N. Secretion of lysozyme by rabbit alveolar macrophages *in vitro. Res. J. Reticuloendothel. Soc.,* **4**:510–523, 1967.

H92. Heise, E. R., Myrvik, Q. N., and Leake, E. S. Effect of bacillus Calmette-Guérin on the levels of acid phosphatase, lysozyme and cathepsin in rabbit alveolar macrophages. *J. Immunol.*, **95**:125–130, 1965.

H93. Hendrich, F., Máchovà, L., Horecká, J., and Pospíšil, L. Uber die Aktivität des Lysożyms im Serum von Kranken mit Hepatitis epidemica. *Wien. Med. Wochenschr.*, **119**:231–233, 1969.

H94. Hendrich, F., Pospíšil, L. Lysozyme. *Acta Fac. Med. Univ. Brun.*, **00**:000, 1971.

H95. Hendrich, F., Pospíšil, L., Vitulová, V. Kuthan, F., Máchová, L. Lysozyme in joint effusions. *Vnitr. Lek.*, **14**:433–436, 1968. (Czech.)

H96. Hendrich, F., Houbal, V., Horecká, J., Máchová, L., Pospíšil, L. Serum lysozyme. I. Lysozyme activity in the serum of patients with acute hepatitis. *Vnitr. Lek.*, **13**:838–841, 1967. (Czech.)

H97. Heneine, I. F., and Kimmel, J. R. The sulfhydryl groups of papaya lysozyme and their relation to biological activity. *J. Biol. Chem.*, **247**:6589–6596, 1972.

H98. Hénon, M., and Delaunay, A. Influence exercée par la phagocytose sur le lysozyme leucocytaire chez des animaux normaux ou intoxiqués par une endotoxine bactérienne. *Ann. Inst. Pasteur, Paris*, **108**:723–735, 1965.

H99. Hénon, M., and Delaunay, A. Modifications apportées au lysozyme leucocytaire par la phagocytose. *C. R. Acad. Sci.*, **260**:1807–1809, 1965.

H100. Hénon, M., Delaunay, A., and Bazin, S. Modifications enzymatiques produites dans les leucocytes et le sérum de cobayes intoxiqués par une endotoxine bactérienne. *Ann. Inst. Pasteur, Paris*, **107**:604–617, 1964.

H101. Henriksen, T. Free radicals induced in enzymes by electrons and heavy ions. *Radiat. Res.*, **32**:Suppl. 7, 87–101, 1967.

H102. Henriksen, T. Effect of the irradiation temperature on the production of free radicals in solid biological compounds exposed to various ionizing radiations. *Radiat. Res.*, **27**:694–709, 1966.

H103. Henrikson, R. L. Selective S-methylation of cysteine in proteins and peptides. *Biochem. Biophys. Res. Commun.*, **41**:967–972, 1970.

H104. Heppel, L. A. Selective release of enzymes from bacteria. *Science*, **156**:1451–1455, 1967.

H105. Hermann, J., Jollès, J., and Jollès, P. Multiple forms of duck-egg white lysozyme. Primary structure of two duck lysozymes. *Eur. J. Biochem.*, **24**:12–17, 1971.

H106. Hermann, J., and Jollès, J. The primary structure of duck egg-white lysozyme. II. *Biochim. Biophys. Acta*, **200**:178–179, 1970.

H107. Hermans, J., Jr., Puett, D., and Acampora, G. On the conformation of denatured proteins. *Biochemistry*, **8**:22–30, 1969.

H108. Herrlich, P., Schweiger, M., and Sauerbier, W. Host- and phage-RNA polymerase mediated synthesis of T_7 lysozyme in vivo. *Mol. Gen. Genet.*, **112**:152–160, 1971.

H109. Herrlich, P., and Schweiger, M. T3 and T7 bacteriophage deoxyribonucleic acid-directed enzyme synthesis in vitro. *J. Virol.*, **6**:750–753, 1970.

H110. Herz, F., Kaplan, E., and Stevenson, J. H., Jr. Acetylcholinesterase inactivation of enzyme-treated erythrocytes. *Nature (London)*, **200**:901–902, 1963.

H110a. Herzig, D. J., Rees, A. W., and Day, R. A. Bifunctional reagents and protein structure determination. The reaction of phenolic disulfonyl chlorides with lysozyme. *Biopolymers*, **2**:349–360, 1964.

H111. Hess, G. P. Observations on the structure and function of two hydrolytic enzymes, chymotrypsin and lysozyme. *Brookhaven Sym. Biol.*, **21**:155–171, 1968.

H112. Hewitt, L. F. Effects of lysozyme on the oxidation-reduction potentials of *M. lysodeikticus* cultures. *Biochem. J.*, **25**:1452–1457, 1931.

H112a. Hexner, P. E., Radford, L. E., and Beams, J. W. Achievement of sedimentation equilibrium. *Proc. Nat. Acad. Sci. U.S.*, **47**:1848–1852, 1961.

H113. Heymann, H., Ginsberg, T., Gulick, Z. R., and Mayer, R. L. Inhibition of pepsin and lysozyme by acidic polymers. *Proc. Soc. Exp. Biol. Med.*, **100**:279–282, 1959.

H114. Heymann, H., Manniello, J. M., and Barkulis, S. S. Structure of streptococcal cell walls. 3. Characterization of an alanine-containing glucosaminylmuramic acid derivative liberated by lysozyme from streptococcal glycopeptide. *J. Biol. Chem.*, **239**:2981–2985, 1964.

H115. Heymans, M. Contribution à l'étude du lysozyme. *Arch. Int. Med. Exp.*, **3**:223–235, 1927.

H116. Hiatt, R. B., Engle, C., Flood, C., and Karush, A. The role of the granulocyte as a source of lysozyme in ulcerative colitis. *J. Clin. Invest.*, **31**:721–726, 1952.

H117. Hilding, A. C. Lysozyme: One of the natural defense factors of the eye. *Minn. Med.*, **18**:360–365, 1935.

H118. Hilding, A. Changes in lysozyme content of nasal mucus during colds. *Arch. Otolaryngol.*, **20**:38–46, 1934.

H119. Hilding, A. Studies on common cold and nasal physiology. *Trans. Amer. Laryngol. Ass.*, **56**:253–271, 1934.

H120. Hill, J., and Leach, S. J. Hydrogen exchange at carbon-hydrogen sites during acid or alkaline treatment of proteins. *Biochemistry*, **3**:1814–1818, 1964.

H121. Hillar, M., and Rzeczycki, W. Interaction of macrocations and macroanions with mitochondria. *Acta Biochim. Pol.*, **12**:133–141, 1965.

H122. Hine, S. The use of Neuzym (lysozyme) in diseases of the ear, nose and throat. *Jibiinkoka Rinsho*, **37**:887–890, 1965. (Jap.)

H122a. Hiremath, C. B., and Day, R. A. Introduction of covalent cross-linkages into lysozyme by reaction with alpha,alpha′-dibromo-p-xylenesulfonic acid. *J. Amer. Chem. Soc.*, **86**:5027–5028, 1964.

H122b. Hirsch, J. G. Antimicrobial factors in tissues and phagocytic cells. *Bacteriol. Rev.*, **24**:133–140, 1960.

H123. Hirschhäuser, C., and Eliasson, R. Origin and possible function of muramidase (lysozyme) in human seminal plasma. *Life Sci., Part II*, **11**:149–154, 1972.

H124. Hirschhäuser, C., and Kionke, M. Demonstration of muramidase (lysozyme) in human seminal plasma. *Life Sci., Part II*, **10**:333–335, 1971.

H124a. Hnojewyj, W. S., and Reyerson, L. H. The sorption of H_2O and D_2O by lyophilized lysozyme. *J. Phys. Chem.*, **63**:1653–1654, 1959.

H124b. Hnojewyj, W. S., and Reyerson, L. H. Further studies on the sorption of H_2O and D_2O vapors by lysozyme and the deuterium-hydrogen exchange effect. *J. Phys. Chem.*, **65**:1694–1698, 1961.

H124c. Hoare, D. G., and Koshland, D. E., Jr. A method for the quantitative modification and estimation of carboxylic acid groups in proteins. *J. Biol. Chem.*, **242**:2447–2453, 1967.

H125. Hobart, C. Lysozyme in treatment of ophthalmic conditions. *Amer. J. Ophthalmol.*, **33**:1409–1416, 1950.

H126. Hoder, F. Steht die bakterienauflösende Wirkung des Lysozyme in einer Beziehung zur Wirkung der Immunkörper? *Immunitäetsforsch. Exp. Ther.*, **70**:457–458, 1931.

H127. Hodgson, C. F., McVey, E. B., and Spikes, J. D. The effect of oxygen concentration on the quantum yields of the dye-sensitized photoinactivation of trypsin, alpha-chymotrypsin and lysozyme. *Experientia*, **25**:1021–1022, 1969.

H128. Hoerman, K. C., Englander, H. R., and Shklair, I. L. Lysozyme: Its characteristics in human parotid and submaxillo-lingual saliva. *Proc. Soc. Exp. Biol. Med.,* **92**:875–878, 1956.

H129. Holden, J. T., and Van Balgooy, J. N. A. Effect of nutritional and physiological factors on the reaction between *Lactobacillus plantarum* and muramidase. *Biochem. Biophys. Res. Commun.,* **19**:401–406, 1965.

H130. Holladay, L. A., and Sophianopoulos, A. J. Association sites of lysozyme in solution. II. Concentration dependence of the near-ultraviolet circular dichroism. *J. Biol. Chem.,* **247**:1976–1979, 1972.

H131. Holler, E., Rupley, J. A., and Hess, G. P. Kinetics of lysozyme-substrate interactions. *Biochem. Biophys. Res. Commun.,* **37**:423–429, 1969.

H132. Holler, E., Rupley, J. A., and Hess, G. P. Kinetics of lysozyme-substrate interactions. *Biochem. Biophys. Res. Commun.,* **40**:166–170, 1970.

H133. Holt, L. A., and Milligan, B. The labeling of proteins by irradiation in tritiated water. *Biochim. Biophys. Acta,* **264**:432–439, 1972.

H134. Holt, L. A., Milligan, B., and Rivett, D. E. Tritiation of tryptophyl residues in proteins. *Biochemistry,* **10**:3559–3564, 1971.

H135. Holt, R. J. Lysozyme production by staphylococci and micrococci. *J. Med. Microbiol.,* **4**:375–379, 1971.

H136. Holyoke, E. D., and Johnson, F. H. The influence of hydrostatic pressure and pH on the rate of lysis of *Micrococcus lysodeikticus* by lysozyme. *Arch. Biochem. Biophys.,* **31**:41–48, 1951.

H137. Holzknecht, E., and Poli, G. Il lisozima nel bambino. *Minerva Pediat.,* **3**:380–386, 1951.

H138. Holzknecht, E., and Poli, G. Il dosaggio del lisozima su piastre insemenzate con Micrococcus lysodeicticus. *Boll. Ist. Sieroter. Milan.,* **29**:194–201, 1950.

H139. Holzman, R. S., Gardner, D. E., and Coffin, D. L. In vivo inactivation of lysozyme by ozone. *J. Bacteriol.,* **96**:1562–1566, 1968.

H140. Hook, W. A., Carey, W. F., and Muschel, L. H. Alterations in serum lysozyme and properdin titers of mice following x-irradiation or treatment with zymosan or endotoxin. *J. Immunol.,* **84**:569–575, 1960.

H141. Hook, W. A., Toussaint, A. J., and Muschel, L. H. Antibody, lysozyme and complement levels in animals chemically protected against x-irradiation. *J. Immunol.,* **91**:100–106, 1963.

H142. Hopkins, T. R., and Spikes, J. D. Conformational changes of lysozyme during photodynamic inactivation. *Photochem. Photobiol.,* **12**:175–184, 1970.

H143. Hopkins, T. R., and Spikes, J. D. Inactivation of crystalline enzymes by photodynamic treatment and gamma radiation. *Radiat. Res.,* **37**:253–260, 1969.

H144. Horinishi, H., Hachimori, Y., Kurihara, K., and Shibata, K. States of amino acid residues in proteins. 3. Histidine residues in insulin, lysozyme, albumin and proteinases as determined with a new reagent of diazo-I-H-tetrazole. *Biochim. Biophys. Acta,* **86**:477–489, 1964.

H144a. Horinishi, H., Nakaya, K., Tani, A., and Shibata, K. States of amino acid residues in proteins. XV. Ethyl-morpholinylpropyl-carbodiimide for modification of carboxyl groups in proteins. *J. Biochem. (Tokyo),* **63**:41–50, 1968.

H145. Howard, J. B., and Glazer, A. N. Papaya lysozyme. Terminal sequences and enzymatic properties. *J. Biol. Chem.,* **244**:1399–1409, 1969.

H146. Howard, J. B., and Glazer, A. N. Studies of the physiochemical and enzymatic properties of papaya lysozyme. *J. Biol. Chem.,* **242**:5715–5723, 1967.

H147. Howes, E. L., Howes, E. L., Jr., and Armitage, C. Enzymatic digestion of mucoproteins. *Proc. Soc. Exp. Biol. Med.*, **113**:216–221, 1963.
H148. Howlett, G. J., Jeffrey, P. D., and Nichol, L. W. The effects of pressure and thermodynamic nonideality on the sedimentation equilibrium of chemically reacting systems: Results with lysozyme at pH 6.7 and 8.0. *J. Phys. Chem.*, **76**:777–783, 1972.
H149. Howlett, G. J., and Nichol, L. W. Computer simulation of sedimentation equilibrium distributions for systems involving heterogeneous associations. The interaction of N-acetylglucosamine with lysozyme. *J. Biol. Chem.*, **247**:5681–5685, 1972.
H150. Howlett, G. J., and Nichol, L. W. A sedimentation equilibrium study of the interaction between ovalbumin and lysozyme. *J. Biol. Chem.*, **248**:619–621, 1973.
H151. Hunkeler, F. Le Mécanisme bactériolytique du lysozyme. *Pharm. Acta Helv.*, **45**:717–727, 1970.
H152. Hurst, A., and Kruse, H. The effect of thioglycolic acid or propanol pretreatment on lysing of *Streptococcus lactis* with lysozyme and salt. *Can. J. Microbiol.*, **18**:945–947, 1972.
H153. Hurst, D. J., and Coffin, D. L. Ozone effect on lysosomal hydrolases of alveolar macrophages in vitro. *Arch. Intern. Med.*, **127**:1059–1063, 1971.
H154. Hurst, D. J., Gardner, D. E., and Coffin, D. L. Effect of ozone on acid hydrolases of the pulmonary alveolar macrophage. *Res., J. Reticuloendothel. Soc.*, **8**:288–301, 1970.
H155. Hussels, H., Langmaack, H., Grossgebauer, K., and Förster, D. Über die Lysozym-und Toxinbildung von Tierstaphylokokken. *Zentralbl. Bakteriol., Parasitenk., Infektionskr. Hyg., Abt. 1: Orig.*, **214**:203–206, 1970.
H156. Hvidt, A. The deuterium exchange of lysozyme as followed by an infrared spectrophotometric method. *C. R. Trav. Lab. Carlsberg*, **33**:475–495, 1963.
H157. Hvidt, A., and Kanarek, L. The deuterium exchange of lysozyme as followed by the gradient method. *C. R. Trav. Lab. Carlsberg*, **33**:463–474, 1963.
H157a. Hvidt, A. A discussion of the pH dependence of the hydrogen-deuterium exchange of proteins. *C. Rend. Travaux Lab. Carlsberg*, **34**:299–317, 1963–1964.
H157b. Hvidt, A., and Nielsen, S. O. Hydrogen exchange in proteins. *Advan. Protein Chem.*, **21**:287–386, 1966.

I

I1. Iavorkovskiĭ, L. L., and Grant, Kh. Ia. Blood serum lysozymes in leukemia. *Probl. Gematol. Pereliv. Krovi*, **14**:39–42, 1969. (Russ.)
I2. Igelmann, J. M., Rotte, T. C., Schecter, E. Blaney, D. Exposure of enzymes to laser radiation. *Ann. N.Y. Acad. Sci.*, **122**:790–801, 1965.
I3. Iizuka, T. Recent advances in ocular therapeutics with special reference to enzymatic preparations. *Jap. J. Clin. Ophthalmol.*, **18**:1217–1225, 1964. (Jap.)
I4. Ikari, N. S., and Donaldson, D. M. Serum beta-lysin and muramidase levels in germfree and conventional rats. *Proc. Soc. Exp. Biol. Med.*, **133**:49–52, 1970.
I5. Ikeda, K., and Hamaguchi, K. Interactions of Mn^{2+}, Co^{2+}, and Ni^{2+} ions with hen egg-white lysozyme and with its N-acetyl-chitooligosaccharide complexes. *J. Biochem.* (*Tokyo*), **73**:307–322, 1973.
I6. Ikeda, K., and Hamaguchi, K. A tryptophyl circular dichroic band at 305 mu of hen egg-white lysozyme. *J. Biochem. (Tokyo)*, **71**:265–273, 1972.

I7. Ikeda, K., and Hamaguchi, K. Interaction of alcohols with lysozyme. I. Studies on circular dichroism. *J. Biochem. (Tokyo),* **68**:785–794, 1970.

I8. Ikeda, K., and Hamaguchi, K. The binding of N-acetylglucosamine to lysozyme. Studies on circular dichroism. *J. Biochem. (Tokyo),* **66**:513–520, 1969.

I9. Ikeda, K., Hamaguchi, K., Imanishi, M., and Amano, T. Effect of pH on the ultraviolet optical rotatory dispersion and circular dichroism of lysozyme. *J. Biochem. (Tokyo),* **62**:315–320, 1967.

I10. Ikeda, K., Hamaguchi, K., Miwa, S., and Nishina, T. Circular dichroism of human lysozyme. *J. Biochem. (Tokyo),* **71**:371–378, 1972.

I11. Il'iashenko, B. N. Infection of lysozyme spheroplasts with phi-X174 DNA phage. *Vop. Virusol.,* **8**:667–676, 1963. (Russ.)

I12. Il'iashenko, B. N., and Ditiatkin, S. Ia. Study of E. coli lysozyme spheroplast competence with respect to phage infectious DNA. *Mikrobiologiya,* **35**:122–127, 1966. (Russ.)

I13. Imada, M., and Inouye, M., Eda, M., and Tsugita, A. Frameshift mutation in the lysozyme gene of bacteriophage T4: Demonstration of the insertion of four bases and the preferential occurrence of base addition in acridine mutagenesis. *J. Mol. Biol.,* **54**:199–217, 1970.

I14. Imai, K., Takagi, T., and Isemura, T. Recovery of the intact structure of muramidase (lysozyme) after reduction of all disulfide linkages in 8 M urea. *J. Biochem. (Tokyo),* **53**:1–6, 1963.

I15. Imam, S. A., and Dutta, G. P. Effect of emetine hydrochloride on cysts of amoebae in presence of trypsin and lysozyme. *Indian J. Biochem. Biophys.,* **6**:233–235, 1969.

I16. Imanaga, Y., and Park, J. T. Studies on the cell walls of *Micrococcus lysodeikticus.* Fractionation of the nondialyzable components from a lysozyme digest of cell walls. *Biochemistry,* **11**:4006–4012, 1972.

I17. Imanishi, M., Miyagawa, N., and Fujio, Hajime, Amano, T. Highly inhibitory antibody fraction against enzymic activity of egg white lysozyme on a small substrate. *Biken J.,* **12**:85–96, 1969.

I18. Imanishi, M., Miyagawa, N., and Amano, T. A rapid microassay of lysozyme activity with reduced chitin oligosaccharides as substrate. *Biken J.,* **12**:25–30, 1969.

I19. Imanishi, M., Shinka, S., Miyagawa, N., Amano, T., and Tsugita, A. Amino acid composition of duck and turkey egg-white lysozymes. *Biken J.,* **9**:107–114, 1966.

I20. Imbriano, A. E., and Mazzucco, P. The synergic activity of lysozyme and methylene blue of various antibiotics. *Sem. Med. (Buenos Aires),* **119**:1129–1132, 1961. (Span.)

I21. Imbriano, A. E., and Mazzucco, P. The biological activity of lysozyme. Experimental study. *Sem. Med. (Buenos Aires)* **118**:446–448, 1961. (Span.)

I22. Imoto, T., Doi, Y., Hayashi, K., and Funatsu, M. Characterization of enzyme-substrate complex of lysozyme. II. Effects of pH and salts. *J. Biochem. (Tokyo),* **65**:667–671, 1969.

I23. Imoto, T., Forster, L. S., Rupley, J. A., and Tanaka, F. Fluorescence of lysozyme: Emissions from tryptophan residues 62 and 108 and energy migration. *Proc. Nat. Acad. Sci. U.S.,* **69**:1151–1155, 1972.

I24. Imoto, T., Hayashi, K., and Funatsu, M. Characterization of enzyme-substrate complex of lysozyme. I. Two types of complex. *J. Biochem. (Tokyo),* **64**:387–392, 1968.

I25. Imoto, T., Johnson, L. N., North, A. C. T., Phillips, D. C., and Rupley, J. A. Vertebrate lysozymes. *In* "The Enzymes" (P. D. Boyer, ed.), 3rd. ed., Vol. 7, pp. 665–868. Academic Press, New York, 1972.

126. Imoto, T., and Yagishita, K. Chitin coated cellulose as an adsorbent of lysozyme-like enzymes. Preparation and properties. *Agr. Biol. Chem.*, **37**:465–470, 1973.

127. Imperato, S., Riccardino, N., and Imperato, A. Interazione tra virus e sostanze basiche. I. Effetto del metillisozima sul virus influenzale: Attività infettante en antigenica del complessi. *Ig. Mod.*, **59**:429–437, 1966.

128. Inada, T., Sawanishi, K., Yoshida, O., Okada, K., and Miyake, Y. Clinical use of lysozyme in the field of urology. *Hinyokika Kiyo*, **12**:713–722, 1966. (Jap.)

129. Inada, Y. States of tyrosine residues in the molecules of insulin, lysozyme and catalase. *J. Biochem. (Tokyo)*, **49**:217–225, 1961.

130. Inagaki, J., Taniguchi, T., Niitani, H., and Kimura, K. Lysosome and lysozyme of tumor cells. *Igaku to Seibutsugaku*, **77**:181–186, 1968. (Jap.)

131. Inai, S., Hirao, F., Kishimoto, S., and Takahashi, H. Studies on serum lysozyme. II. Lysozyme activity in normal human serum and in serum of patients with various diseases. *Med. J. Osaka Univ.*, **9**:33, 1958.

132. Inoue, K., Yomenasu, K., Takamizawa, A., and Amano, T. Studies on the immune bacteriolysis. XIV. Requirement of all nine components of complement for immune bacteriolysis. *Biken J.*, **11**:203–206, 1968.

133. Inouye, M., Okada, Y., and Tsugita, A. The amino acid sequence of T4 phage lysozyme. I. Tryptic digestion. *J. Biol. Chem.*, **245**:3439–3454, 1970.

134. Inouye, M., Imada, M., and Tsugita, A. The amino acid sequence of T4 phage lysozyme. IV. Dilute acid hydrolysis and the order of tryptic peptides. *J. Biol. Chem.*, **245**:3479–3484, 1970.

135. Inouye, M., Imada, M., Akaboshi, E., and Tsugita, A. The amino acid sequence of T4 phage lysozyme. II. Chymotryptic digestion. *J. Biol. Chem.*, **245**:3455–3466, 1970.

136. Inouye, M., Akaboshi, E., Kuroda, M., and Tsugita, A. Replacement of all tryptophan residues in T4 bacteriophage lysozyme by tyrosine residues. *J. Mol. Biol.*, **50**:71–81, 1970.

137. Inouye, M., and Tsugita, A. Amino acid sequence of T2 phage lysozyme. *J. Mol. Biol.*, **37**:213–223, 1968.

138. Inouye, M., Akaboshi, E., Tsufita, A., Streisinger, G., and Okada, Y. A frame-shift mutation resulting in the deletion of two base pairs in the lysozyme gene of bacteriophage T4. *J. Mol. Biol.*, **30**:39–47, 1967.

139. Inouye, M., and Tsugita, A. The amino acid sequence of T4 bacteriophage lysozyme. *J. Mol. Biol.*, **22**:193–196, 1966.

140. Inouye, M., Lorena, M. J., Tsugita, A. The amino acid sequence of T4 phage lysozyme. III. Peptic digestion. *J. Biol. Chem.*, **245**:3467–3478, 1970.

141. Inouye, M., Yahata, H., Ocada, Y., and Tsugita, A. Change of tryptophan residue at the 158th position of T4 bacteriophage lysozyme into a tyrosine residue by suppression and spontaneous reversion of an amber mutant. *J. Mol. Biol.*, **33**:957–961, 1968.

142. Irwin, R., and Churchich, J. E. Rotational relaxation time of pyridoxyl 5-phosphate lysozyme. *J. Biol. Chem.*, **246**:5329–5334, 1971.

142a. Isemura, T., Takagi, T., Maeda, Y., and Imai, K. Recovery of enzymatic activity of reduced Taka-amylase A and reduced lysozyme by air-oxidation. *Biochem. Biophys. Res. Commun.*, **5**:373–377, 1961.

143. Iser-Vincentelli, J. B., and Léonis, J. Etude des fonctions carboxyles anormales du lysozyme. *Arch. Int. Physiol. Biochim.*, **75**:170–171, 1967.

144. Iser-Vincentelli, J., and Léonis, J. Effet des alcools sur la dénaturation thermique du lysozyme. *Arch. Int. Physiol. Biochim.*, **72**:327–328, 1964.

145. Ishiwa, H., and Yokokura, T. A lytic enzyme(s) acting on *Lactobacilli*. *Jap. J. Microbiol.*, **15**:539–541, 1971.

146. Ishizeki, C., and Iwahara, S. Effect of endotoxin on serum lysozyme level in rabbit. *Eisei Shikenjo Kokoku,* **86**:64–68, 1968. (Jap.)

147. Issinger, O. G., and Hausmann, R. Synthesis of bacteriophage-coded gene products during infection of Escherichia coli with amber mutants of T3 and T7 defective in gene 1. *J. Virol.,* **11**:465–472, 1973.

148. Ito, M. Polarographic studies on proteins and enzymes. IV. Protein double waves of protein mixtures with special reference to protein interaction. *Mie Med. J.,* **16**:173–192, 1967.

149. Ito, M. Polarographic studies on proteins and enzymes. III. Effect of protein denaturation and reduction on the protein double wave of lysozyme. *Mie Med. J.,* **16**:105–117, 1966.

150. Ito, M. Polarographic studies on proteins and enzymes. I. Polarographic behavior of various proteins. *Mie Med. J.,* **13**:149–178, 1964.

151. Ito, M. Polarographic studies of lysozyme. II. The catalytic protein wave of lysozyme in ammonia buffer containing hexamminecobalt (III) chloride. *Nippon Nogei Kagaku Kaishi,* **33**:238–243, 1959.

152. Ito, M. Polarographic studies of lysozyme. I. The catalytic protein wave of lysozyme in amonia buffer containing cobaltous chloride. *Nippon Nogei Kagaku Kaishi,* **33**:233–238, 1959.

153. Ito, T. Photoinactivation of solid dye-enzyme system: Requirement of water molecules. *Curr. Mod. Biol.,* **3**:233–236, 1970.

154. Ivanova, L. N., and Perekhval'skaia, T. V. The effect of retrograde introduction of a preparation of hyaluronidase and lysozyme on the reabsorption of osmotically free water in the dog kidney. *Dokl. Akad. Nauk SSSR,* **181**:1013–1016, 1968. (Russ.)

155. Iwahara, S., Ishizeki, C., and Koshinuma, K. Lysozyme activity of the serum and organs of the rabbit. *Igaku to Seibutsugaku,* **77**:5–7, 1968. (Jap.)

156. Iwamoto, Y., Nakamura, R., Watanabe, T., and Tsunemitsu, A. Purification and amino acid analysis of human parotid saliva lysozyme. *J. Dent. Res.,* **49**:1104–1110, 1970.

157. Iwamoto, Y., Watanabe, T., Tsunemitsu, A., Fukui, K., and Moriyama, T. Lysis of streptococci by lysozyme from human parotid saliva and sodium lauryl sulfate. *J. Dent. Res.,* **50**:1688, 1971.

158. Iwasawa, K. A clinical study of mucopolysaccharide dissolving enzyme, lysozyme (neuzym) in otorhinolaryngology. Report 1. Effect of neuzym in chronic sinusitis of children—especially on the basis of changes of one-layer X-ray pictures. *Oto-Rhino-Laryngol.,* **11**:50, 1968.

159. Iwata, K., Eda, T., and Tomiyama, T. Studies on the extracellular enzymes of staphylococci. 4. Identification of pathogenic staphylococci by means of the lysozyme activity test. *Nippon Saikingaku Zasshi,* **23**:700–707, 1968. (Jap.)

161. Iwata, K., and Eda, T. Studies on the extracellular enzymes of staphylococci. 3. Conditions for determination of lysozyme activity for the identification of pathogenic staphylococci. *Nippon Saikingaku Zasshi,* **23**:465–472, 1968. (Jap.)

162. Iwazawa, T. Clinical application of a lysozyme, Leftose, in chronic sinusitis. *Jibi Inkoka Rinsho,* **39**:767–776, 1967. (Jap.)

163. Izaka, K. I., Shirakawa, H., Yamada, M., and Suyama, T. Method for isolation of human placental lysozyme. *Anal. Biochem.,* **42**:299–309, 1971.

J

J1. Jacob, A., Jorre, J., Vokral, N., Palou, A. M., and Thelier, J. Dosage du lysozyme par une méthode de diffusion. *Ann. Pharm. Fr.,* **22**:301–306, 1964.

J2. Jakubowski, A., and Pospíšil, L. Lysozymaktivität des Serums im Verlaufe von experimenteller Syphilis des Kaninchens. 2. Quantitative Veränderungen des Lysozyms. *Z. Immunitaetsforsch. Allerg. Klin. Immunol.,* **131**:449–452, 1966.

J3. James, L. K., Jr., and Hilborn, D. A. Lysozyme adsorption and activity in urea solutions. *Biochim. Biophys. Acta,* **151**:279–281, 1968.

J4. James, W. M. The lysozyme content of tears. *Amer. J. Ophthalmol.,* [3], **18**:660, 1935.

J5. James, W. M. The lysozyme content of tears. *Amer. J. Ophthalmol.* [3], **18**:1109–1113, 1935.

J6. Janicke, B. W., and Patnode, R. A. Increase in circulating lysozyme-like enzyme following sensitization of guinea pigs with *Mycobacterium tuberculosis. Amer. Rev. Resp. Dis.,* **83**:872–877, 1961.

J7. Jáurequi-Adell, J. Purification of lysozymes by retention on Bio Gel CM 30. *Biochimie,* **53**:1167–1173, 1971.

J8. Jáuregui-Adell, J., Cladel, G., Ferraz-Pina, C., and Rech, J. Isolation and partial characterization of mare milk lysozyme. *Arch. Biochem. Biophys.,* **151**:353–355, 1972.

J9. Jauregui-Adell, J., Jollès, J., and Jollès, P. The disulfide bridges of hen's egg-white lysozyme. *Biochim. Biophys. Acta,* **107**:97–111, 1965.

J10. Jauregui-Adell, J., and Jollès, P. Contribution à l'étude du centre actif du lysozyme de blanc d'oeuf de poule: Action de l'acide iodacétique á pH 5.5. *Bull. Soc. Chim. Biol.,* **46**:141–147, 1964.

J11. Jaworkowsky, L., Bersin, W., Grant, H., Rudens, J. Das Serumlysozym in der Differentialdiagnose akuter Leukosen. *Folia Haematol. (Leipzig),* **97**:265–278, 1972.

J12. Jaworkowski, L., Bersin, W., Grant, H., and Turgel, E. Das Serumlysozym bei chronischer Myeloleukose. *Folia Haematol. (Leipzig),* **93**:184–195, 1970.

J13. Jay, J. M. Production of lysozyme by Staphylococci and its correlation with three other extracellular substances. *J. Bacteriol.,* **91**:1804–1810, 1966.

J14. Jayaraman, R. Transcription of bacteriophage T4 DNA by Escherichia coli RNA polymerase in vitro: Identification of some immediate-early and delayed-early genes. *J. Mol. Biol.,* **70**:253–263, 1972.

J15. Jeanloz, R. W. Substrat du lysozyme structure chimique. *Exposes Annu. Biochim. Med.,* **27**:45–63, 1966.

J16. Jeffries, L. Sensitivity to novobiocin and lysozyme in the classification of micrococcaceae. *J. Appl. Bacteriol.,* **31**:436–442, 1968.

J17. Jelliffe, D. B. Active anti-infective properties of human milk. *Lancet,* **2**:167–168, 1971.

J18. Jensen, H. B., and Kleppe, K. Effect of ionic strength, pH, amines and divalent cations on the lytic activity of T4 lysozyme. *Eur. J. Biochem.,* **28**:116–122, 1972.

J19. Jensen, H. B., and Kleppe, K. Studies on T4 lysozyme. Affinity for Chitin and the use of chitin in the purification of the enzyme. *Eur. J. Biochem.,* **26**:305–312, 1972.

J20. Jensen, R. S., Tew, J. G., and Donaldson, D. M. Extracellular beta-lysin and muramidase in body fluids and inflammatory exudates. *Proc. Soc. Exp. Biol. Med.,* **124**:545–549, 1967.

J21. Jeronimidis, G., and Damiani, A. Interpretation of some conformational data in myoglobin and lyszoyme. *Nature (London), New Biol.,* **229**:150–151, 1971.

J22. Jirgensons, B. Optical rotatory dispersion and conformation of various globular proteins. *J. Biol. Chem.,* **238:**2716–2722, 1963.

J22a. Jirgensons, B. Optical rotation and viscosity of native and denatured proteins. I. Influence of pH and the stability of various proteins toward several denaturing reagents. *Arch. Biochem. Biophys.,* **39:**261–270, 1952.

J22b. Jirgensons, B., and Straumanis, L. Optical rotation and viscosity of native and denatured proteins. VIII. Rotatory dispersion studies. *Arch. Biochem. Biophys.,* **68:** 319–329, 1957.

J22c. Jirgensons, B. Optical rotation and viscosity of native and denatured proteins. XIII. Further studies on enzyme proteins. *Arch. Biochem. Biophys.,* **92:**216–220, 1961.

J23. Johansson, B. G., and Malmquist, J. Quantitative immunochemical determination of lysozyme (muramidase) in serum and urine. *Scand. J. Clin. Lab. Invest.,* **27:**255–261, 1971.

J24. Johansson, B. G., and Ravnskov, U. The serum level and urinary excretion of alpha$_2$-Microglobulin, Beta$_2$-microglobin and lysozyme in renal disease. *Scand. J. Urol. Nephrol.,* **6:**249–256, 1972.

J25. Johnson, K. G., and Campbell, J. N. Effect of growth conditions on peptidoglycan structure and susceptibility to lytic enzymes in cell walls of *Micrococcus sodonensis. Biochemistry,* **11:**277–286, 1972.

J26. Johnson, L. N. An interaction between lysozyme and penicillin. *Proc. Roy. Soc, Ser. B.,* **167:**439–440, 1967.

J27. Johnson, L. N. The structure and function of lysozyme. *Sci. Progr. (London),* **54:**367–385, 1966.

J28. Johnson, L. N. The three dimensional structure of hen egg white lysozyme as determined by x-ray diffraction. *Exposes Annu. Biochim. Med.,* **27:**31–33, 1966.

J29. Johnson, L. N., and Phillips, D. C. Structure of some crystalline lysozyme-inhibitor complexes determined by X-ray analysis at 6 angstrom resolution. *Nature (London),* **206:**761–763, 1965.

J30. Johnson, L. N., Phillips, D. C., and Rupley, J. A. The activity of lysozyme: An interim review of crystallographic and chemical evidence. *Brookhaven Symp. Biol.,* **21:**120–138, 1968.

J31. Jollès, G., and Jollès, P. Isolement et caractérisation d'un facteur antihistaminique présent dans certains lots de lysozyme. *Bull. Soc. Chim. Fr.* [5], Fasc. **11:**3300–3301, 1965.

J32. Jollès, J., Bernier, I., Jauregui, J., and Jollès, P. Amino acid composition of 3 samples of lysozyme from chicken egg: New determination of the number of cystine residues. *C. R. Acad. Sci.,* **250:**413–414, 1960.

J33. Jollès, J., Dianoux, A.-C., Hermann, J., Niemann, B., and Jollès, P. Relationship between the cystine and tryptophan contents of 5 different lysozymes and their heat stability and specific activity. *Biochim. Biophys. Acta,* **128:**568–570, 1966.

J34. Jollès, J., Hermann, J., Niemann, B., and Jollès, P. Differences between the chemical structures of duck and hen egg-white lysozymes. *Eur. J. Biochem.,* **1:**344–346, 1967.

J35. Jollès, J., Jauregui-Adell, J., Bernier, I., and Jollès, P. La structure chimique du lysozyme de blanc d'oeuf de poule: Etude détaillée. *Biochim. Biophys. Acta,* **78:**668–689, 1963.

J36. Jollès, J., Jauregui-Adell, J., and Jollès, P. Preliminary contribution to the study of the disulfide bonds in hen's egg-white lysozyme. *Biochim. Biophys. Acta,* **71:**488–490, 1963.

J37. Jollès, J., and Jollès, P. Comparision between human and bird lysozymes: Note concerning the previously observed deletion. *FEBS Lett.,* **22:**31–33, 1972.

J38. Jollès, J., and Jollès, P. Human milk lysozyme: Unpublished data concerning the establishment of the complete primary structure; comparison with lysozymes of various origins. *Helv. Chim. Acta,* **54**:2668–2675, 1971.

J39. Jollès, J., and Jollès, P. La structure chimique primaire du lysozyme du lait de femme: Establissement d'une formule dévelopée provisoire. *Helv. Chim. Acta,* **52**:2671–2675, 1969.

J40. Jollès, J., and Jollès, P. Sur la structure chimique du lysozyme de lait de femme. *Bull. Soc. Chim. Biol.,* **50**:2543–2551, 1968.

J41. Jollès, J., and Jollès, P. Human tear and human milk lysozymes. *Biochemistry,* **6**:411–417, 1967.

J42. Jollès, J., and Jollès, P. Isolation and purification of the biologically active deslysylvalyl phenylalanine-lysozyme. *Biochem. Biophys. Res. Commun.,* **22**:22–25, 1966.

J43. Jollès, J., and Jollès, P. Structure chimique du lysozyme de blanc d'oeuf de poule; la formule dévelopée. *C. R. Acad. Sci.,* **253**:2773–2775, 1961.

J44. Jollès, J., and Jollès, P. Structure chimique de la moitie N-terminale du lysozyme de blanc d'oeuf de poule. *Biochim. Biophys. Acta,* **45**:407–408, 1960.

J45. Jollès, J., Jollès, P., and Jauregui-Adell, J. Etablissement d'une formule provisoire du lysozyme de blanc d'oeuf de poule. *Bull. Soc. Chim. Biol.,* **42**:1319–1329, 1960.

J46. Jollès, J., Spotorno, G., and Jollès, P. Lysozymes characterized in duck egg-white: Isolation of a histidine-less lysozyme. *Nature (London),* **208**:1204–1205, 1965.

J47. Jollès, J., Van Leemputten, E., Mouton, A., and Jollès, P. Amino acid sequence of guinea-hen egg-white lysozyme. *Biochim. Biophys. Acta,* **257**:497–510, 1972.

J48. Jollès, P. Die evolution von primarstrukturen. *Chimia,* **25**:1, 1971.

J49. Jollès, P. Lysozymes: A chapter of molecular biology. *Angew. Chem. Int. Ed. Engl.,* **8**:227–239, 1969.

J50. Jollès, P. Lysozymes. *Biochem. J.,* **110**:25P, 1968.

J51. Jollès, P. Rapports entre la structure et l'activité de quelques lysozymes. *Bull. Soc. Chim. Biol.,* **49**:1001–1012, 1967.

J52. Jollès, P. Relationship between chemical structure and biological activity of hen egg-white lysozyme and lysozymes of different species. *Proc. Roy. Soc., Ser. B,* **167**:350–364, 1967.

J53. Jollès, P. Chimie et biochimie des lysozymes. I. Le lysozyme de blanc d'oeuf de poule: Structure chimique, ponts disulfures, relations entre structure et activité. II. Etudes comparée de quelques lysozymes d'origines diverses. *Exposes Annu. Biochim. Med.,* **27**:1–30, 1966.

J54. Jollès, P. Recent developments in the study of lysozymes. *Angew. Chem., Int. Ed. Engl.,* **3**:28–36, 1964.

J55. Jollès, P. La structure chimique du lysozyme de blanc d'oeuf de poule. *Med. Hyg.,* **22**:965, 1964.

J56. Jollès, P. Action comparée de trois lysozymes d'origine différente sur des substrats glycopeptidiques isolés de *Micrococcus lysodeikticus. Biochim. Biophys. Acta,* **69**:505–510, 1963.

J57. Jollès, P. Lysozyme. *In* "Enzymes" (P. D. Boyer, H. Lardy, and K. Myrback, eds.), Vol. 4, p. 431. Academic Press, New York, 1960.

J58. Jollès, P., and Berthou, J. High temperature crystallization of lysozyme: An example of phase transition. *FEBS Lett.,* **23**:21–23, 1972.

J59. Jollès, P., Bonnafé, M., Mouton, A., and Schwarzenberg, L. Dosage automatique du lysozyme dans l'urine de divers leucémiques: Caratérisation et purification d'un lysozyme chez les seuls malades atteints de leucémie myéloblastique aiguë. *Rev. Fr. Etud. Clin. Biol.,* **12**:996–998, 1967.

J60. Jollès, P., Charlemagne, D., Petit, J.-F., Marie, A. C., and Jollès, J. Biochimie comparée des lysozymes. *Bull. Soc. Chim. Biol.,* **47:**2241–2259, 1965.

J61. Jollès, P., Jauregui-Adell, J., and Jollès, J. Le lysozyme de blanc d'oeuf de poule: Disposition des ponts disulfures. *C. R. Acad. Sci.,* **258:**3926–3928, 1964.

J62. Jollès, P., and Jollès, J., Synthèse par la méthode de Merrifield d'un octapeptide faisant partie du lysozyme de blanc d'oeuf de poule (Enchaînement Cys[64]—Gly[71]). *Helv. Chim. Acta,* **51:**980–986, 1968.

J63. Jollès, P., and Jollès, J. Lysozyme from human milk. *Nature (London),* **192:**1187–1188, 1961.

J64. Jollès, P., and Jollès, J. Hydrolyse trypsique du lysozyme de blanc d'oeuf de poule ré duit: Isolement de peptides contenant du tryptophane et de la cystine. *C. R. Acad. Sci.,* **246:**1109–1111, 1958.

J65. Jollès, P., Jollès, J., and Jauregui, J. Structure du lysozyme d'oeuf de poule. III. Etude des peptides de l'hydrolysat chymotrypsique du lysozyme dénaturé par la chaleur. *Biochim. Biophys. Acta,* **31:**96–99, 1959.

J66. Jollès, P., Jollès-Thaureaux, J., and Fromageot, C. Structure du lysozyme d'oeuf de poule. II. Etude de quelques peptides de l'hydrolysat pepsique. *Biochim. Biophys. Acta,* **27:**439–442, 1958.

J67. Jollès, P., Jollès-Thaureaux, J., and Fromageot, C. Recherche de lysozymes chez quelques invertébrés; lysozyme de *Nephthys hombergi. C. R. Soc. Biol.,* **151:**1368–1369, 1957.

J68. Jollès, P., and Ledieu, M. Comparaison entre les structures chimiques du lysozyme d'oeuf de poule et du lysozyme de la rate de chien. *Biochem. Biophys. Acta,* **36:**284–285, 1959.

J69. Jollès, P., and Ledieu, M. Le lysozyme de rate de chien. II. Composition en acides aminès et résidus N- et C-terminaux. *Biochim. Biophys. Acta,* **31:**100–103, 1959.

J70. Jollès, P., Petit, J. F., Charlemagne, D., Salmon, S., and Jollès, J. Etude comparée de quelques lysozymes humains. *Proc. Int. Symp. Lysozyme, 3rd, Milan 1964.*

J71. Jollès, P., Saint Blancard, J., Charlemagne, D., Dianoux, A.-C., Jollès, J., and Le Baron, J. L. Comparative behaviour of six different lysozymes in the presence of an inhibitor. *Biochim. Biophys. Acta,* **151:**532–537, 1968.

J72. Jollès, P., Sternberg, M., and Mathé, G. The relationship between serum lysozyme levels and the blood leukocytes. *Isr. J. Med. Sci.,* **1:**445–447, 1965.

J73. Jollès, P., Sternberg, M., and Mathé, G. Etude de la teneur en lysozyme du serum chez des patients atteints de leucemies et hematosarcomes. *Proc. Symp. Int. Lysozyme, 3rd, Milan 1964.*

J74. Jollès, P., and Theareaux, J. Hydrolyse trypsize du lysozyme d'oeuf de poule; séparation par chromatographie sur colonne et analyse des peptides de l'hydrolysat. *C. R. Acad. Sci.,* **243:**1685–1688, 1956.

J75. Jollès, P., Thaureaux, J., and Fromageot, C. L'enchaînement du C-terminal du lysozyme d'oeuf de poule. *Arch. Biochem. Biophys.,* **69:**290–294, 1957.

J76. Jollès, P., and Zuili, S. Purification et étude comparée de nouveaux lysozymes; extraits du poumon de poule et de *Nephthys Hombergi. Biochim. Biophys. Acta,* **39:**212–217, 1960.

J77. Jollès, P., Zowall, H., Jauregui-Adell, J., and Jollès, J. Nouvelle méthode chromatographique de préparation des lysozymes. *J. Chromatogr.,* **8:**363–368, 1962.

J78. Jollès-Thaureaux, J., Jollès, P., and Fromageot, C. Structure du lysozyme d'oeuf de poule. Etude des peptides de l'hydrolysat trypsique. *Biochim. Biophys. Acta,* **27:**298–310, 1958.

J79. Jonas, A., and Weber, G. Strong binding of hydrophobic anions by bovine serum albumin peptides covalently linked to lysozyme. *Biochemistry,* **10:**4492–4496, 1971.

J80. Jones, F. T. Optical and crystallographic properties of lysozyme chloride. *J. Amer. Chem. Soc.,* **68**:854–857, 1946.

J81. Jones, R. H., Nevin, T. A., Guest, W. J., and Logan, L. C. Lytic effect of trypsin, lysozyme, and complement on *Treponema pallidum. Brit. J. Vener. Dis.,* **44**:193–200, 1968.

J82. Jori, G., Galiazzo, G., Marzotto, A., and Scoffone, E. Selective and reversible photo-oxidation of the methionyl residues in lysozyme. *J. Biol. Chem.,* **243**:4272–4278, 1968.

J83. Jori, G., Galiazzo, G., and Scoffone, E. Photodynamic action of porphyrins on amino acids and proteins. III. Further studies on the hematoporphyrin-sensitized photooxidation of lysozyme. *Experientia,* **27**:379–380, 1971.

J84. Jori, G., Galiazzo, G., and Scoffone, E. Photodynamic action of porphyrins on amino acids and proteins. I. Selective photooxidation of methionine in aqueous solution. *Biochemistry,* **8**:2868–2875, 1969.

J85. Jori, G., Galiazzi, G., and Scoffone, E. Dye-sensitized selective photo-oxidation of cysteine. *Int. J. Protein Res.,* **1**:289–298, 1969.

J86. Jori, G., Gennari, G., Galiazzo, G., and Scoffone, E. Paramagnetic metal ions as protectors of selected regions of protein molecules from photodynamic action. *Biochim. Biophys. Acta,* **236**:749–766, 1971.

J87. Josefsson, L. Number of formyl groups in the formic acid inactivated protein enzymes. *Acta Chem. Scand.,* **19**:2421–2422, 1965.

J88. Josefsson, L. Changes in the physico-chemical properties of proteins during the formic acid-induced transformation. *Biochim. Biophys. Acta,* **59**:128–136, 1962.

J89. Josefsson, L., and Edman, P. Reversible inactivation of lysozyme due to N,O-peptidyl shift. *Biochim. Biophys. Acta,* **25**:614–623, 1957.

J90. Josephson, A. S., and Wald, A. Enhancement of lysozyme activity by anodal tear protein. *Proc. Soc. Exp. Biol. Med.,* **131**:677–679, 1969.

J91. Josslin, R. Physiological studies on the t gene defect in T4-infected *Escherichia coli. Virology,* **44**:101–107, 1971.

J92. Joynson, M. A., North, A. C. T., Sarma, V. R., Dickerson, R. E., and Steinrauf, L. K. Low-resolution studies on the relationship between the triclinic and tetragonal forms of lysozyme. *J. Mol. Biol.,* **50**:137–142, 1970.

K

K1. Kábátová, A., and Zeman, V. Azione del Deflamon sul Trichomonas vaginalis. *Minerva Ginelcol.,* **22**:837–841, 1970.

K2. Kábátová, A., Zeman, V., Jakubowski, A., and Pospišil, L. L. Der lysozymgehalt im Scheidenmilieu. *Gynaecologia,* **164**:336–342, 1967.

K3. Kagitomi, T. Histo-pathological studies of mucopolysaccharidase treatment on chronic sinusitis. *Otolaryngology (Tokyo),* **38**:175–185, 1966. (Jap.)

K4. Kagramanova, K. A., Isolation and composition of lysozyme substrate from bacterial cells. *Antibiotiki (Moscow),* **12**:407–410, 1967. (Russ.)

K5. Kagramanova, K. A., and Ermol'eva, Z. V. Comparative characteristics of the methods for determination of the activity of lysozyme. *Antibiotiki (Moscow),* **11**:917–919, 1966. (Russ.)

K5a. Kaiser, E. Inhibition and activation of lysozyme. *Nature,* **171**:607–608, 1953.

K6. Kamaya, T. Lytic action of lysozyme on *Candida albicans. Mycopathol. Mycol. Appl.,* **42**:197–207, 1970.

K7. Kamaya, T. Flocculation phenomenon of *Candida albicans* by lysozyme. *Mycopathol. Mycol. Appl.,* **37**:320–330, 1969.

K8. Kanarek, L., Bradshaw, R. A., and Hill, R. L. Regeneration of active enzyme from the mixed disulfide of egg white lysozyme and cystine. *J. Biol. Chem.,* **240**:pc2755–pc2757, 1965.

K9. Kanarek, L., Hvidt, A., Leonis, J., and Ottesen, M. Deuterium-hydrogen exchange between lysozyme and water. III. Various recent results. *Arch. Int. Physiol. Biochim.,* **70**:408–409, 1962.

K10. Kanarek, L., Hvidt, A., Léonis, J., and Ottesen, M. Echange deutérium-hydrogéne entre des lysozymes altérés et l'eau. *Arch. Int. Physiol. Biochim.,* **69**:596–597, 1961.

K11. Kanarek, L., Hvidt, A., Leonis, J., and Ottesen, N. Effet du pH et de la température sur l' échange deutérium-hydrogène entre le lysozyme et l'eau. *Arch. Int. Physiol. Biochim.,* **69**:387–388, 1961.

K12. Kaneda, M., Kato, I., Tominaga, N., Titani, K., and Narita, K. The amino acid sequence of quail lysozyme. *J. Biochem. (Tokyo),* **66**:747–749, 1969.

K13. Kaplan, J. H. Anion diffusion across artificial lipid membranes. The effects of lysozyme on anion diffusion from phospholipid liposomes. *Biochim. Biophys. Acta,* **290**:339–347, 1972.

K14. Kapp, W., Klunker, W., and Fellmann, N. Klinisch-statische Prüfung der analgetisch-antiphlogistischen Eigenschaften von Lysozym bei primär chronischer Polyarthritis. *Praxis,* **57**:1465–1466, 1968.

K15. Kartasheva, A. L., and Dubovik, B. V. Immune bacteriolysis in plague and the role of cellular-humoral factors in its development. *Zh. Mikrobiol., Epidemiol., Immunobiol.,* **49**:116–124, 1972. (Russ.)

K15a. Kashiba, S., Niizu, K., Tanaka, S., Nozu, H., and Amano, T. Lysozyme, an index of pathogenic staphylococci. *Biken's J.* **2**:50–55, 1959.

K16. Kasymova, Kh. A., and Chernikova, Z. S. State of natural immunity in humans following multiple vaccinations with live brucellosis vaccine BA-19. II. Effect of vaccine strains on lysozyme contents in animals and in immunized humans. *Zh. Mikrobiol., Epidemiol. Immunobiol.,* **48**:91–95, 1971. (Russ.)

K17. Kato, K., Kotani, S., Imanishi, M., and Amano, T. Soluble lysozyme-substrate isolated from *Micrococcus lysodeikticus* cell walls by *Flavobacterium* L$_{11}$ enzyme digestion. *Biken J.,* **6**:223–227, 1963.

K18. Kato, K., and Murachi, T. Chemical modification of tyrosyl residues of hen egg-white lysozyme by diisopropylphosphorofluoridate. *J. Biochem. (Tokyo),* **69**:725–737, 1971.

K19. Kato, K., Suginaka, H., Kishida, H., Takamasa, K., Yanagida, I., Matsuda, T., Hirata, T., and Murayama, Y. Study of the structure of *Micrococcus lysodeikticus* cell walls with Flavobacterium L-11 enzyme. *Biken J.,* **11**:59–65, 1968.

K20. Kato, T., Yakushiji, M., Tsuda, H., *et al.* Lysozyme, a mucopolysaccharase, in the field of obstetrics and gynecology. 2. Serum lysozyme activity in healthy and pregnant women. *Sanfujinka No Jissai,* **20**:1273–1279, 1971. (Jap.)

K21. Kato, T., Yakushiji, M., Tsuda, H. *et al.* Studies on lysozyme (N-acetylmuramide glycanohydrolase) determination in obstetrics and gynecology. I. On serum lysozyme activity. *Sanfujinka No Jissai,* **17**:827–832, 1968. (Jap.)

K22. Katz, S., and Miller, J. E. Structure-volume relationships for proteins. Comparative dilatometric study of acid-base reactions of lysozyme and ovalbumin in water and denaturing media. *Biochemistry,* **10**:3569–3574, 1971.

K23. Katz, W. Vergleichende Untersuchungen an teilchenge-bundenem und freiem T2-Lysozym und Kristallisierung beider Enzyme. *Z. Naturforsch, B,* **19**:129–133, 1964.

K24. Katz, W., Katz, W., Berger, D., and Martin, H. H. Action of an N-acetylmuramidase from a Limax amoeba on the murein of Proteus mirabilis. *Biochim. Biophys. Acta,* **244**:47–57, 1971.

K25. Katz, W., and Weidel, W. Isolation and characterization of the lysozyme bound to Ts phage. *Z. Naturforsch. B,* **16**:363–368, 1961. (Ger.)

K26. Kaverzneva, E. D., Maksimov, V. I., and Osipov, V. I. Structural disorders of lysozyme and ribonuclease following gamma-irradiation in a dry state. *Biofizika,* **16**:581–588, 1971. (Russ.)

K27. Keeler, R., The effect of bilateral nephrectomy on the production and distribution of muramidase (lysozyme) in the rat. *Can. J. Physiol. Pharmacol.,* **48**:131–138, 1970.

K28. Keeler, R. Evidence for a renal-dependent factor in the control of muramidase (lysozyme) formation. *Can. J. Physiol. Pharmacol.,* **47**:831–832, 1969.

K29. Keller, F., and Maurer, H. Enzymbehandlung entzündlicher Erkrankungen des Mund- und Rachenraumes. *Ther. Gegenw.,* **106**:663–672, 1967.

K30. Kemall, D. Il comportamento del lisozima ematico nei distimici. *Acta Neurol.,* **4**:527–533, 1949.

K31. Kent, J. F., and De Weerdt, J. B. Enhancement by lysozyme of the sensitivity of *Treponema pallidum* immobilization tests. *Brit. J. Vener. Dis.,* **39**:37–40, 1963.

K32. Kent, P. W., and Dwek, R. A. N-fluoroacetyl-alpha-D-glucosamine as a molecular probe of lysozyme structure by using [¹⁹F] fluorine Nuclear-magnetic-resonance techniques. *Biochem. J.,* **121**:11P-12P, 1971.

K33. Kerby, G. P. Methods for the study of surviving leukocytes. D. Lysozyme technic for measuring cell damage. *Methods Med. Res.,* **7**:140–141, 1958.

K34. Kerby, G. P. A method for detection of leukocyte injury based on release of a lysozyme-like enzyme. *Proc. Soc. Exp. Biol. Med.,* **81**:129–131, 1952.

K35. Kerby, G. P., and Chaudhuri, S. N. Plasma levels and the release of a lysozyme-like enzyme from tuburculin-exposed leukocytes of tuberculous and nontuberculous human beings. *J. Lab. Clin. Med.,* **41**:632–636, 1953.

K36. Kerby, G. P., and Eadie, G. S. Inhibition of lysozyme by heparin. *Proc. Soc. Exp. Biol. Med.,* **83**:111–113, 1953.

K37. Kerby, G. P., and Taylor, S. M. Enzymatic activity in human synovial fluid from rheumatoid and non-rheumatoid patients. *Proc. Soc. Exp. Biol. Med.,* **126**:865–868, 1967.

K38. Kern, R. A., Kingkade, M. J., Kern, S. F., and Behrens, O. K. Characterization of the action of lysozyme on *Staphylococcus aureus* and on *Micrococcus lysodeikticus. J. Bacteriol.,* **61**:171–178, 1951.

K39. Kertész, D., and Caselli, P. Sur le spectre d'absorption du lysozyme oxydé. *Bull. Soc. Chim. Biol.,* **32**:583–586, 1950.

K40. Kertész, D., and Teti, M. Reversible inactivation of lysozyme by the system, polyphenol oxidase plus an o-Diphenol (so-called Tyrosinase). *C. R. Soc. Biol.,* **144**:923–924, 1950.

K41. Khechinashvili, N. N., Privalov, P. I., and Tiktopula, E. I. Calorimetric investigation of lysozyme thermal denaturation. *FEBS Lett.,* **30**:57–60, 1973.

K42. Khorlin, A. Ya., Shashkova, E. A., and Zurabyan, S. E. Interaction of lysozyme with isomeric p-nitrophenyl 0-2-acetamido-2-deoxy-Beta-d-glucopyranosyl)-0-(2-acetamido-2-deoxy-Beta-D-glycopyranosyl)-2-acetamido-2-dexoy-Beta-D-glucopyranosides. *Carbohyd. Res.,* **21**:269–273, 1972.

K43. Khristiuk, V. M. Problem of allergic reactivity and nonspecific immunity in children with neurodermatitis and eczema. *Vestn. Dermatol. Venerol.,* **45**:21–26, 1971. (Russ.)

K44. Khurgin, Iu. I., Rosliakov, V. Ia., Kliachko-Gurvich, A. L., and Brueva, T. P. Adsorption of water vapors by alpha-chymotrypsin and lysozyme. *Biokhimiya,* **37**:485–492, 1972. (Russ.)

K45. Kigasawa, T. Antigene Eigenschaften der Lysozyme. *Z. Immunitäetsforsch. Exp. Ther.,* **57**:146–152, 1928.

K46. Kigasawa, T. Lysozymwirkungen des Eiereiweisses. *Z. Immunitäetsforsch. Exp. Ther.*, **54**:155–180, 1927–1928.

K46a. King, G. S. D. An x-ray investigation of lysozyme iodide and nitrate. *Acta Cryst.*, **12**:216–219, 1959.

K47. King, J. R., and Gooder, H. Reversion to the streptococcal state of enterococcal protoplasts, spheroplasts, and L-forms. *J. Bacteriol.*, **103**:692–696, 1970.

K48. King, J. R., and Gooder, H. Induction of enterococcal L-forms by the action of lysozyme. *J. Bacteriol.*, **103**:686–691, 1970.

K49. King, T. P. Selective chemical modification of arginyl residues. *Biochemistry*, **5**:3454–3459, 1966.

K49a. King, T. P., and Craig, L. C. Countercurrent distribution studies with ribonuclease and lysozyme. *J. Amer. Chem. Soc.*, **80**:3366–3370, 1958.

K50. Klemparskaya, N. N. Flocculation of bacteria by lysozyme. *Z. Microbiol., Epidemiol. Immunitetsforsch.*, **2–3**:124–128, 1939.

K51. Kleňha, J., and Kras, V. Lysozyme in mouse and human skin. *J. Invest. Dermatol.*, **49**:396–399, 1967.

K52. Klimova, K. N., and Ivanova, N. M. Changes in certain plasma factors of natural defense system in dogs under the influence of bacterial lipopolysaccharide. *Byull. Eksp. Biol. Med.*, **62**:80–82, 1966. (Russ.)

K53. Klishin, V. V. The activity of lysozyme in the blood serum of patients suffering from El-Tor Cholera, and of vibrio carriers. *Zh. Mikrobiol., Epidemiol. Immunobiol.*, **49**:52–54, 1972. (Russ.)

K54. Klotz, I. M., Griswold, P., and Gruen, D. M. Infrared spectra of some proteins and related substances. *J. Amer. Chem. Soc.*, **71**:1615–1620, 1949.

K55. Klotz, I. M., Stellwagen, E. C., and Stryker, V. H. Ionic equilibria in protein conjugates: Comparison of proteins. *Biochem. Biophys. Acta*, **86**:122–129, 1964.

K56. Klotz, I. M., Urquhart, J. M., and Weber, W. W. Penicillin—protein complexes. *Arch. Biochem.*, **26**:420–435, 1950.

K57. Klotz, I. M., and Walker, F. M. Complexes of lysozyme. *Arch. Biochem.*, **18**:319–325, 1948.

K58. Knox, J. R. Protein molecular weight by x-ray diffraction. *J. Chem. Educ.*, **49**:476–479, 1972.

K59. Kobayashi, T., Kiyono, K., Sakamoto, Y. *et al.* Herpes zoster in radiotherapy for malignant tumors—report of 5 cases and significance of lysozyme preparations. *Jap. J. Cancer Clin.*, **13**:714–718, 1967. (Jap.)

K59a. Koch, A. L., Lamont, W. A., and Katz, J. J. The effect of anhydrous strong acids on ribonuclease and lysozyme. *Arch. Biochem. Biophys.*, **63**:106–117, 1956.

K60. Kocka, F., Magoc, T., and Searcy, R. L. Action of sulfated polyanions used in blood cultures lysozyme, complement and antibiotics. *Ann. Clin. Lab. Sci.*, **2**:470–473, 1972.

K61. Koeva-Slavkova, N., Klain, S. B., and Leparskii, E. A. Nonspecific immunity of mature and premature children. *Vop. Okhr. Mater. Detstva*, **14**:13–19, 1969. (Russ.)

K62. Kohn, A. Lysis of frozen and thawed cells of *Escherichia coli* by lysozyme, and their conversion into spheroplasts. *J. Bacteriol.*, **79**:697–706, 1960.

K63. Kohoutek, J. Herpes simplex of the cornea. (Critical review.) *Cesk. Oftalmol.*, **20**:208–214, 1964. (Czech.)

K64. Kol'man, A. E. Lysozymes. *Antibiotiki (Moscow)*, **12**:740–750, 1967. (Russ.)

K65. Kolomnikov, I. S., Svoboda, P., and Vol'pin, M. E. Splittig of C—C-bond by complexes of transition metals. *Izv. Akad. Nauk SSSR, Ser. Khim.*, **12**:2818, 1972.

K66. Kondo, L. Hanna, L., and Keshishyan, H. Reduction in Chlamydial infectivity by lysozyme. *Proc. Soc. Exp. Biol. Med.*, **142**:131–136, 1973.

K67. Kondo, Y. Study of lysozyme activity in feces of infants. 3. Serum lysozyme activity in normal infants and variations of stool lysozyme activity in infants with diarrhea. *Acta Paediat. Jap.*, **71**:1452–1460, 1967. (Jap.)

K68. Kondo, Y. Study of lysozyme activity in feces of infants. *Acta Paediat. Jap.*, **71**:1433–1440, 1967. (Jap.)

K68a. Konigsberg, W. H., and Becker, R. R. The preparation of C^{14}-Polypeptidyl-proteins. *J. Amer. Chem. Soc.*, **81**:1428–1431, 1959.

K69. Kopeć, W., and Metzger, M. Muramidase from tissue fluid of syphilitic rabbit testes. II. Comparative study of the enzyme and hen egg white muramidase. *Arch. Immunol. Ther. Exp.*, **16**:533–545, 1968.

K70. Kopeć, W. Muramidase from tissue fluid of syphilitic rabbit testes. I. Purification and properties of the enzyme. *Arch. Immunol. Ther. Exp.*, **15**:161–175, 1967.

K71. Kopeloff, N., Harris, M. M., and McGinn, B. Lysozyme in saliva. *Amer. J. Med. Sci.*, **184**:632–367, 1932.

K72. Korzhenkova, M. P., Mikhaïlov, Z. M., and Vologodskaia, N. A. Age peculiarities of nonspecific immunologic reactivity in acute respiratory diseases in children. *Pediatriya (Moscow)*, **49**:3–7, 1970. (Russ.)

K72a. Koshland, D. E., and Neet, K. E. The catalytic and regulatory properties of enzymes. *Ann. Rev. Biochem.*, **37**:359–410, 1968.

K73. Kotani, S., Harada, S., Kitaura, T., Hashimuto, Y., Matsubara, T., and Chimori, M. Lysis of isolated BCG cell walls with enzymes. *Biken J.*, **5**:117–119, 1962.

K74. Kotani, S., Hashimoto, S., Matsubara, T., Kato, K., Harada, K., Kogami, J., Kitaura, T., and Tanaka, A. Lysis of isolated BCG cell walls with enzymes. 2. Demonstration of 'bound wax D' as a component of BCG cell walls. *Biken J.*, **6**:181–196, 1963.

K75. Kotani, S., Kato, K., Matsubara, T., Mori, Y., Chimori, M., Kishida, H. Changes in susceptibility of *Staphylococcus aureus* and *Corynebacterium diphtheriae* cell walls to egg white lysozyme, the L3- and L-11-enzymes caused by trichloroacetic acid treatment. *Biken J.*, **6**:317–320, 1964.

K76. Kotelchuck, D., and Scheraga, H. A. The influence of short-range interactions on protein conformation. II. A model for predicting the alpha-helical regions of proteins. *Proc. Nat. Acad. Sci. U.S.*, **62**:14–21, 1969.

K77. Kotelchuck, D., and Scheraga, H. A. The influence of short-range interactions on protein conformation. I. Side chain-backbone interactions within a single peptide unit. *Proc. Nat. Acad. Sci. U.S.*, **61**:1163–1170, 1968.

K78. Kovanyi, G., and Holzner, H. Lysozymaktivität im flüssigen Inhalt von Ovarialzysten. *Wien. Klin. Wochenschr.*, **82**:375–378, 1970.

K79. Kovanyi, G., and Karobath, H. Lysozymaktivität in Serum und Harn nach Herzmuskelinfarkt. *Verh. Deut. Ges. Kreislaufforsch.*, **35**:398–401, 1969.

K80. Kovanyi, G., and Letnansky, K. Urine and blood serum muramidase (lysozyme) in patients with urogenital tumors. *Eur. J. Cancer*, **7**:25–31, 1971.

K81. Kowal-Gierczak, B., Falowska-Adamczyk, W., and Hańczyc, H. Muramidase activity in blood serum of patients with allergic bronchial asthma. *Pol. Med. J.*, **10**:1136–1141, 1971.

K82. Kowal-Gierczak, B., Falowska-Adamczyk, W., and Hańczyc, H. Muramidase activity in blood serum of patients with allergic bronchial asthma. *Pol. Tyg. Lek.*, **25**:2018–2020, 1970.

K83. Kowal-Gierczak, B., Falowska-Adamczyk, W., and Hańczyc, H. Activity of salivary muramidase in allergic bronchial asthma. *Pol. Med. J.*, **8**:584–588, 1969.

K84. Kowal-Gierczak, B., Falowska-Adamczyk, W., and Hanczyc, H. Sialine muramidase activity in allergic bronchial asthma. *Pol. Tyg. Lek.*, **23**:1718, 1968.

K85. Kowal-Gierczak, B., Falowska-Adamczyk, W., Wrzyszcz, M. *et al.* Effect of vitamin A on the level of seromucoid and fibrinogen and the activity of muramidase in the serum of patients with bronchial asthma. *Pol. Tyg. Lek.,* **27**:169–171, 1972. (Pol.)

K86. Kowal-Gierczak, B., Hańczycowa, H., and Lubczyńska-Kowalska, W. Muramidase activity in saliva, gastric juice and serum in patients having undergone gastrectomy for gastric or duodenal ulcer. *Pol. Tyg. Lek.,* **27**:1071–1073, 1972. (Pol.)

K87. Kowalski, C. J., and Schimmel, P. R. Interaction of lysozyme with alpha-N-acetyl-D-glucosamine. *J. Biol. Chem.,* **244**:3643–3646, 1969.

K88. Kraus, F. W., and Mestecky, J. Immunohistochemical localization of amylase, lysozyme and immunoglobulins in the human parotid gland. *Arch. Oral Biol.,* **16**:781–789, 1971.

K89. Kraus, F. W., Mestecky, J., Hammack, W. J., and Ward, J. L. Immunofluorescent localization of lysozyme in human leukocytes. *Ala. J. Med. Sci.,* **6**:417–421, 1969.

K89a. Krause, S., and O'Konski, C. T. Electric properties of macromolecules. VIII. Kerr constants and rotational diffusion of some proteins in water and in glycerol-water solutions. *Biopolymers,* **1**:503–515, 1963.

K90. Kravchenko, N. A. Lysozyme as a transferase. *Proc. Roy. Soc., Ser. B,* **167**:429–430, 1967.

K91. Kravchenko, N. A., Chentsova, T. V., and Kaverzneva, E. D. Studies on esterification lysozyme and the influence of reaction conditions on the composition of the ester mixture. *Biokhimiya,* **33**:355–359, 1968. (Russ.)

K92. Kravchenko, N. A., Kagramanova, K. A., and Kuznetsov, Iu. D. Effect of salts and other factors on lysozyme activity. *Biokhimiya,* **32**:618–623, 1967. (Russ.)

K93. Kravchenko, N. A., Kleopina, C. V., and Kaverzneva, E. D. Studies on an active center of lysozyme using the carboxymethylation method. *Biokhimiya,* **30**:713–720, 1965. (Russ.)

K94. Kravchenko, N. A., Kleopina, G. V., and Kaverzneva, E. D. Isolation and desalting of the products formed in treatment of lysozyme by iodoacetic acid. *Biokhimiya,* **30**:534–542, 1965. (Russ.)

K95. Kravchenko, N. A., Kleopina, G. V., and Kaverzneva, E. D. A study of carboxymethylation of lysozyme by iodoacetic acid. *Biokhimiya,* **30**:195–202, 1965. (Russ.)

K96. Kravchenko, N. A., Kléopina, G. V., and Kaverzneva, E. D. Investigation of the active sites of lysozyme. Carboxymethylation of the imidazole group of histidine and of the epsilon-aminogroup of lysine. *Biochim. Biophys. Acta,* **92**:412–414, 1964.

K97. Kravchenko, N. A., and Lapuk, V. Kh. The pathway of the reaction of selective photooxidation of tryptophan in lysozyme. *Biokhimiya,* **35**:64–71, 1970. (Russ.)

K98. Kravchenko, N. A., and Lapuk, V. Kh. Isolation and characteristics of the products of selective photooxidation of tryptophan in lysozyme. *Biokhimiya,* **34**:832–838, 1969. (Russ.)

K99. Kravchenko, N. A., Zrelov, V. P., and Klabunovskii, E. I. On modified fermentative and optic activities of lysozyme following its electron and proton irradiation. *Dokl. Akad. Nauk SSSR,* **155**:1449–1451, 1964. (Russ.)

K100. Kregzde, J., Lambert, L. L., and Davidson, W. D. Lysozymuria in renal calculosis following spinal cord injury. *Urol. Int.,* **24**:310–317, 1969.

K101. Kretchmar, A. L., and Comas, F. V. Correlation between kidney muramidase activity and plasma iron removal in rats with Walker-256 carcinosarcoma. *Proc. Soc. Exp. Biol. Med.,* **116**:275–277, 1964.

K102. Krigbaum, W. R., and Kügler, F. R. Molecular conformation of egg-white lysozyme and bovine alpha-lactalbumin in solution. *Biochemistry,* **9**:1216–1223, 1970.

K103. Krisch, K. Zur papierchromatographischen Bausteinanalyse des Lysozym-Substrates. *Hoppe Seyler's Z. Physiol. Chem.,* **311**:131–135, 1958.

K104. Krishnamoorthy, R. V. Increased lysozyme activity in muscular atrophy of denervated frogs. *Enzymologia,* **43**:353–358, 1972.
K105. Kriss, A. E. The lysozyme in actinomycetes. *Microbiology (USSR),* **9**:32–38, 1940. (Russ.)
K106. Kronman, B. S., Carson, D. A., and Osserman, E. F. *In vitro* cultivation of a rat monocytic leukemia with lysozyme production. *Fed. Proc., Fed. Amer. Soc. Exp. Biol.* **29**:502, 1970 (abstr.).
K107. Kronman, M. J. Similarity in backbone conformation of egg white lysozyme and bovine gamma lactalbumin. *Biochem. Biophys. Res. Commun.,* **33**:535–541, 1968.
K108. Kronman, M. J., Robbins, F. M., and Andreotti, R. E. Reaction of N-bromosuccinimide with lysozyme. *Biochim. Biophys. Acta,* **147**:462–472, 1967.
K109. Krull, L. H., and Friedman, M. Reduction of protein disulfide bonds by sodium hydride in dimethyl sulfoxide. *Biochem. Biophys. Res. Commun.,* **29**:373–377, 1967.
K110. Kubát, Z., Rokos, J., Procházka, P., and Liebl, V. Experimental interaction opacity of the cornea. *Acta Ophthalmol.,* **43**:219–223, 1965.
K111. Kubát, Z., Rokos, J., Procházka, P., and Liebl, V. Interaction of corneal polysaccharides with basic macromolecules. *Casopis Led. Cesk.,* **103**:909–913, 1964. (Czech.)
K112. Kubelka, V. Influence du lysozyme sur différents virus et la possible utilization de cet enzyme dans le laboratoire virologique. *Epatologia,* **12**:588–591, 1966.
K113. Kuettner, K. E., Eisenstein, R., Soble, L. W., and Arsenis, C. Lysozyme in epiphyseal cartilage. IV. Embryonic chick cartilage lysozyme—its localization and partial characterization. *J. Cell Biol.,* **49**:450–458, 1971.
K114. Kuettner, K. E., Soble, L. W., Gunether, H. L., and Eisenstein, R. Lysozyme in epiphyseal cartilage. I. The nature of the morphologic response of cartilage in cluture to exogenous lysozym. *Calcif. Tissue Res.,* **5**:56–63, 1970.
K115. Kuettner, K. E., Soble, L. W., Ray, R. D., Croxen, R. L., Passovoy, M., and Eisenstein, R. Lysozyme in epiphyseal cartilage. II. The effect of egg white lysozyme on mouse embryonic femurs in organ cultures. *J. Cell Biol.,* **44**:329–339, 1970.
K116. Kuettner, K. E., Soble, L. W., Eisenstein, R., and Yaeger, J. A. The influence of lysozyme on the appearance of epiphyseal cartilage in organ culture. *Calcif. Tissue Res.,* **2**:93–105, 1968.
K117. Kuettner, K. E., Sorgente, N., Arsenis, C., and Eisenstein, R. Cartilage lysozyme, an extracellular basic (cationic) protein. *Isr. J. Med. Sci.,* **7**:407–409, 1971.
K118. Kuettner, K. E., Wezeman, F. H., Simmons, D. J., Lisk, P. Y., Croxen, R. L., Soble, L. W., and Eisenstein, R. Lysozyme in preosseous cartilage. V. The response of embryonic chick cartilage to antilysozyme antibodies in organ culture. *Lab. Invest.,* **27**:324–330, 1972.
K119. Kuettner, K. E., Guenther, H. L., Ray, R. D., Schumacher, G. F. B. Lysozyme in preosseous cartilage. *Calcif. Tissue Res.,* **1**:298–305, 1968.
K120. Kuhlmann, F. The nervous ulcer, its treatment and prevention. *Hippokrates,* **34**:135–138, 1963. (Ger.)
K121. Kühnemund, O., Köhler, W., and Prokop, O. Untersuchungen über angebliches lysozymähnliches Verhalten des Protektins Anti-A_{hp} aus der Eiweibdrüse der Weinbergschnecke *(Helix pomatia). Z. Immunitäetsforsch., Exp. Klin. Immunol.,* **144**:344–351, 1972.
K122. Kunze, M. Der Nachweis der Staphylokokken-Muramidase (Lysozym). *Zentralbl. Bakteriol., Parasitenk. Infektionskr. Hyg., Erste Abt. 1: Orig., Reihe A.,* **219**:50–55, 1972.
K123. Kunze, M. Muramidase-Bildung durch *Listeria monocytogenes. Zentralbl. Bakteriol., Parasitent., Infektionskr. Hyg., Abt. 1: Orig., Reihe A,* **218**:249–250, 1971.

K124. Kuramitsu, S., Ikeda, K., Hamaguchi, K., Miwa, S., and Nishina, T. Interactions of N-Acetyl-chitooligosaccharides with human and hen lysozymes. *J. Biochem., (Tokyo),* **72:**1109–1115, 1972.

K125. Kurata, M., Otodani, S., and Kawada, J. Immunochemical studies on hen egg white lysozyme. II. Some properties of acetyl-lysozymes. *Yakwgakwzasshi,* **92:**64–68, 1972.

K126. Kurata, M., Otodani, S., and Kawada, J. Immunochemical studies on hen egg white lysozyme. I. Acetylation of lysozyme. *J. Pharm. Soc. Jap.,* **91:**494–498, 1971. (Jap.)

K127. Kurihara, K., Horinishi, H., and Shibata, K. Reactions of cyanuric halides with proteins. I. Bound tyrosine residues of insulin and lysozyme as identified with cyanuric fluoride. *Biochim. Biophys. Acta,* **74:**678–687, 1963.

K128. Kurono, A., and Hamaguchi, K. Structure of muramidase (lysozyme). VII. Effect of alcohols and the related compounds on the stability of muramidase. *J. Biochem. (Tokyo),* **56:**432–440, 1964.

K129. Kürti, V. Lysozyme and its therapeutic use in pediatrics. *Cesk. Pediat.,* **20:**640–643, 1965 (Czech.)

K130. Kusnetzova, N. P., Glikina, M. V., Andreeva, N. A., Illarionova, N. G., Volkova, L. A., and Samsonov, G. V. Investigation of interaction of lysozyme with polymethacrylic acid. *Mol. Biol.,* **7:**42, 1973.

K131. Kuznetsova, L. L. Indices of immunologic reactivity in cerebral circulatory disorders. *Zh. Mikrobiol., Epidemiol., Immunobiol.,* **72:**10–13, 1972. (Russ.)

K132. Kyle, R. A., and Linman, J. W. Index of granulocyte turnover. *N. Engl. J. Med.,* **280:**109, 1969.

K133. Kyle, R. A., Pierre, R. V., and Bayrd, E. D. Multiple myeloma and acute myelomonocytic leukemia. *N. Eng. J. Med.,* **283:**1121–1125, 1970.

L

L1. Lalli, P. Il comportamento della carie sperimentale nel ratto trattato con fluoruro di lisozima. *Mondo Odontostomatol.,* **10:**37–39, 1968.

L2. Landman, O. E., and Halle, S. Enzymically and physically induced inheritance changes in *Bacillus subtilis. J. Mol. Biol.,* **7:**721–738, 1963.

L3. Landon, A. Les enzymes vers une utilisation spécifique en thérapeutique dentaire. *Inform. Dent.,* **48:**5153–5156, 1966.

L4. Landucci, L., and Biagini, R. Il contenuto e le variazioni fisiologiche del lisozima nel siero di sangue del bambino. *Riv. Clin. Pediat.,* **52:**293–299, 1953.

L5. Landureau, J. C., and Jollès, P. Lytic enzyme produced *in vitro* by insect cells: Lysozyme or Chitinase? *Nature (London),* **225:**968, 1970.

L6. Lapanje, S., and Wadsö, I. A calorimetric study of the denaturation of lysozyme by guanidine hydrochloride and hydrochloric acid. *Eur. J. Biochem.,* **22:**345–349, 1971.

L7. Lapuk, V. Kh., Chistyakova, L. A., and Kravchenko, N. A. On the tryptophan determination in lysozyme and its photoxidation products. *Anal. Biochem.,* **24:**80–89, 1968.

L8. Lark, C., Bradley, D., and Lark, K. G. Further studies on the incorporation of *D*-methionine into the bacterial cell wall. Its incorporation into the R-layer and the structural consequences. *Biochim. Biophys. Acta,* **78:**278–288, 1963.

L9. LaRue, J. N., and Speck, J. C., Jr. Turkey egg white lysozyme. Preparation of the crystalline enzyme and investigation of the amino acid sequence. *J. Biol. Chem.,* **245:**1985–1991, 1970.

L10. Lashchenko, P. The effect of egg albumin in killing spores and in retarding development. *Z. Hyg. Infektionskr.,* **64:**419–427, 1909.

L11. Laterza, G. Rilievi clinici ed ematochimici nel trattamento della epatite virale con lisozima. *Epatologia,* **12**:488–496, 1966.

L12. Lato, M., and Antonelli, G., L'influenza del lisozima negli stati emofilici. *Clin. Pediat.,* **44**:36–44, 1962.

L13. Lato, M., and Silvana, C. Lysozyme in the first stages of (human) growth. *Riv. Biol.,* **41**:477–488, 1949.

L14. Latovitzki, N., Halper, J. P., and Beychok, S. Spectrophotometric titration of tyrosine residues in human lysozyme. *J. Biol. Chem.,* **246**:1457–1460, 1971.

L15. Laufman, H., Ross, A., and Lobstein, O. E. Quantitation of lysozyme activity in experimental intestinal strangulation obstruction. *Quart. Bull. Northwest. Univ. Med. Sch.,* **27**:99–110, 1953.

L16. Laurence, W. L. On the possible identity of "avidin" and eggwhite lysozyme. *Science,* **99**:392–393, 1944.

L17. Laurentaci, G., and Scupola, G. Sulla temporanea sopravvivenza degli omoinnesti di cute favorita dal lisozima. Ricerche sperimentali. *Acta Chir. Ital.,* **19**:1225–1257, 1963.

L18. Lawrence, C. A., and Klingel, H. Effects of azo-sulfonamides upon lysozyme. *Proc. Soc. Exp. Biol. Med.,* **52**:129–130, 1943.

L19. Layne, J. S., and Johnson, E. J. Resistant properties of Azotobacter cysts induced in response to mineral deficiencies. *J. Bacteriol.,* **88**:956–959, 1964.

L20. Lazowski, J. Studies on localization of the activity of certain lysozyme enzymes in human sperm. *Ginekol. Pol.,* **39**:1299–1303, 1968.

L21. Lazzerini, S. Evoluzione della ferite cutanee trattate con lisozima. *Minerva Chir.,* **14**:469–474, 1959.

L21a. Leach, S. J., Meschers, A., and Swanepoel, O. A. The electrolytic reduction of proteins. *Biochemistry,* **4**:23–27, 1965.

L21b. Leach, S. J., Némethy, G., and Scheraga, H. A. Intramolecular steric effects and hydrogen bonding in regular conformations of polyamino acids. *Biopolymers,* **4**: 887–904, 1966.

L22. Leake, E. S., Gonzalez-Ojeda, D., and Myrvik, Q. N. Enzymatic differences between normal alveolar macrophages and oil-induced peritoneal macrophages obtained from rabbits. *Exp. Cell Res.,* **33**:553–561, 1964.

L23. Leake, E. S., and Myrvik, Q. N. Changes in morphology and in lysozyme content of free alveolar cells after the intravenous injection of killed BCG in oil. *Res. J. Reticuloendothel. Soc.,* **5**:33–53, 1968.

L24. Leake, E. S., and Myrvik, Q. N. Differential release of lysozyme and acid phosphatase from sub-cellular granules of normal rabbit alveolar macrophages. *Brit. J. Exp. Pathol.,* **45**:384–392, 1964.

L25. Lee, B., and Richards, F. M. The interpretation of protein structures: Estimation of static accessibility. *J. Mol. Biol.,* **55**:379–400, 1971.

L26. Leemhuis, M. P., Spijkerman, M. G., and Swanen-Sterag, D. H. Muramidase and acute rejection of kidney grafts. *Lancet,* **2**:1309, 1972.

L27. Leemhuis, M. P., Spijkerman, M. G., and Swanen-Sterag, D. H. Muramidase and acute rejection of kidney grafts. *Lancet,* **2**:387–388, 1972.

L28. Leffell, M. S., and Spitznagel, J. K. Association of lactoferrin with lysozyme in granules of human polymorphonuclear leukocytes. *Infec. Immunity,* **6**:761–765, 1972.

L29. Lehrer, S. S. Solute perturbation of protein fluorescence. The quenching of the tryptophyl fluorescence of model compounds and of lysozyme by iodide ion. *Biochemistry,* **10**:3254–3263, 1971.

L30. Lehrer, S. S. Deuterium isotope effects on the fluorescence of tryptophan in peptides and in lysozyme. *J. Amer. Chem. Soc.,* **92**:3459–3462, 1970.

L31. Lehrer, S. S. The selective quenching of tryptophan fluorescence in proteins by iodide ion: Lysozyme in the presence and absence of substrate. *Biochem. Biophys. Res. Commun.*, **26**:767–772, 1967.

L32. Lehrer, S. S., and Fasman, G. D. Fluorescence of lysozyme and lysozyme substrate complexes. Separation of tryptophan contributions by fluorescence difference methods. *J. Biol. Chem.*, **242**:4644–4651, 1967.

L33. Lehrer, S. S., and Fasman, G. D. The fluorescence of lysozyme and lysozyme substrate complexes. *Biochem. Biophys. Res. Commun.*, **23**:133–138, 1966.

L34. Lemperle, G., Müller, E., Michaelis, W., and Wieczorek, Z. Die Bestimmung des Lysozyms in Serum und Urin nach Nierentransplantation. *Langenbecks Arch. Chir.*, **325**:719–725, 1969.

L34a. Lenk, H. P., Wenzel, M., and Schütte, E. Darstellung von Oligosacchariden des Glucosamins und Acetylglucosamins und deren Spaltung durch Lysozyme. *Naturwissenschaften*, **47**:516–517, 1960.

L35. Lenti, G. Determination of lysozyme. Minimal lytic unit. *Boll. Soc. Ital. Biol. Sper.*, **22**:847–848, 1946. (Ital.)

L36. Leoncini, G., and D'Angelo, P. Trattamento terapeutico e profilattico della epatite virale con lisozima. *Epatologia*, **12**:208–214, 1966.

L37. Léonis, J. Preliminary investigation of the behavior of lysozyme in urea solutions. *Arch. Biochem. Biophys.*, **65**:182–193, 1956.

L38. Le Pecq, J. B., Bourgoin, D., and Joly, M. Association du lysozyme avec l'acide désoxyribonucléique. II. Modalité de l'association a pH constant. *Bull. Soc. Chim. Biol.*, **44**:985–996, 1962.

L39. Lesobre, R., and Aubin-Hesse, H. L'association propionyl-érythromycine, tétracycline, lysozyme dans le traitement des affections bronchiques et pulmonaires. *Presse Med.*, **74**:2935–2936, 1966.

L40. Letnansky, K., and Hübl, P. Die lysozymaktivität des menschlichen Magensaftes. *Wien. Klin, Wochenschr.*, **74**:46–48, 1962.

L41. Levi, J. A., Speden, J. B., Vincent, P. C., and Gunz, F. W. Studies on muramidase in hematologic disorders. I. Serum muramidase and serum lactic dehydrogenase in leukemia. *Cancer*, **31**:939–947, 1973.

L42. Levi, J. A., Speden, J. B., Vincent, P. C., and Gunz, F. W. Studies of muramidase in haematological disorders: Serum and marrow muramidase in leukaemia. *Pathology*, **5**:59–68, 1973.

L42a. Levitt, M., and Lifson, S. Refinement of protein conformations using a macromolecular energy minimization procedure. *J. Mol. Biol.*, **46**:269–279, 1969.

L43. Lévy, J. M., Stoll, C., and Francfort, J. J. L'hypertrophie thymique du nourrisson. Effet du lysozyme. *Ann. Pediat.*, **18**:138–145, 1971.

L44. Lewis, J. C., Snell, N. S., Hirschmann, D. J., and Fraenkel-Conrat, H. Amino acid composition of egg proteins. *J. Biol. Chem.*, **186**:23–35, 1950.

L45. Lewis, P. N., and Scheraga, H. A. Prediction of structural homology between bovine alpha-lactalbumin and hen egg white lysozyme. *Arch. Biochem. Biophys.*, **144**:584–588, 1971.

L46. Libman, B. G., Kagramanova, K. A., and Ermol'eva, Z. V. Use of lysozyme in medicine. *Sov. Med.*, **34**:34–40, 1971. (Russ.)

L47. Libman, B. G., Kagramanova, K. A., and Ialysheva, N. I. Lysozyme of the mucosa of the respiratory pathways, their microbial flora and early post-intubation complications. *Eksp. Khir. Anesteziol.*, **14**:72–75, 1969. (Russ.)

L48. Lieberman, J. Enzymatic disolution of pulmonary secretions. An *in vitro* study of sputum from patients with cystic fibrosis of pancreas. *Amer. J. Dis. Child.*, **104**:342–348, 1962.

L48a. Lienhard, G. E., Secemski, I. I., Koehler, K. A., and Lindquist, R. N. Enzymatic catalysis and the transition state theory of reaction rates: transition state analogs. *Cold Spring Harbor Symp. Quant. Biol., 36*:45–51, 1971.

L49. Light, R. A. The nature of toxic material in closed loop intestinal obstruction; identification of lysozyme. *Surgery, 30*:195–205, 1951.

L50. Lin, T.-Y., and Koshland, D. E., Jr. Carboxyl group modification and the activity of lysozyme. *J. Biol. Chem., 244*:505–508, 1969.

L51. Linder, P. W., and Bryan, W. P. Chitotriose, a "tritium exchange probe" of the active cleft of lysozyme. *J. Amer. Chem. Soc., 93*:3061–3062, 1971.

L52. Linz, R. Some properties of purified lysozyme. *C. R. Soc. Biol., 126*:1281–1282, 1937.

L53. Linz, R. Purification of the lysozyme of egg white. *C. R. Soc. Biol., 126*:1279–1280, 1937.

L54. Liotet, S., and Reveilleau, J. Etude des larmes humaines par electrophorese et par immuno-electrophorese. *Ann. Ocul., 198*:12, 1965.

L55. Liotet, S., and Rouchy, J. P. Etude bactériologique de 2,000 conjonctivites. *Arch. Ophtalmol.* [N.S.], *31*:887–894, 1971.

L56. Lippman, M. E., and Finch, S. C. A quantitative study of muramidase distribution in normal and nitrogen mustard-treated rats. *Yale J. Biol. Med., 45*:463–470, 1972.

L57. Lis, J., and Schleif, R. Lambda lysozyme synthesis in the absence of N protein. *Virology, 45*:532–533, 1971.

L58. Liso, V. Lysozyme activity of monocytes during acute hemopathies. Preliminary study. *Haematologica, 57*:535–540, 1972. (Ital.)

L59. Liso, V., Troccoli, G., and Schena, F. P. La valutazione del lisozima nelle leucemie acute mielogene. *Haematologica, 57*:241–245, 1972.

L60. Litwack, G. Effect of thyroid hormone on the bacteriolytic action of muramidase (lysozyme) *in vitro. Biochim. Biophys. Acta, 67*:501–504, 1963.

L61. Litwack, G. Thyroid gland potency and kidney lysozyme state. *J. Biol. Chem., 231*:175–181, 1958.

L62. Litwack, G. Chromatographic studies of rat kidney lysozyme. *Proc. Soc. Exp. Biol. Med., 98*:408–412, 1958.

L63. Litwack, G. Development of *Micrococcus lysodeikticus* resistant to lysozyme. *Nature (London), 181*:1348–1350, 1958.

L64. Litwack, G. Effect of induced states on tissue lysozyme activity. *Proc. Soc. Exp. Biol. Med., 94*:764–767, 1957.

L65. Litwack, G. Photometric determination of lysozyme activity. *Proc. Soc. Exp. Biol. Med., 89*:401–403, 1955.

L66. Litwack, G., Chakrabarti, S. G., and Wojciechowski, V. Human kidney lysozyme. *Arch. Biochem. Biophys., 83*:566–567, 1959.

L67. Litwack, G., and Prasad, A. L. N. Development of lysozyme resistance in *Micrococcus lysodeiktricus* and *Sarcina lutea. Nature (London), 196*:543–545, 1962.

L68. Litwack, G., and Sears, M. L. Precipitation of lysozyme with thyroxine or thyroglobulin. *J. Biol. Chem., 240*:674–678, 1965.

L69. Litwack, G., and Wojciechowski, V. Modification of lysozyme thermostability by cytochrome c preparations. *Proc. Soc. Exp. Biol. Med., 103*:431–433, 1960.

L70. Lobstein, O. E. The determination of lysozyme (muramidase) by rate reduction. *Clin. Chem., 18*:720, 1972.

L71. Lobstein, O. E. Atoxic lysozyme. Determination of the activity of lysozyme. Massive doses of lysozyme by intravenous route. Histological examination of the organs and demonstration of the atoxic nature of lysozyme. *Sem. Med., 119*:1024–1034, 1961.

L72. Lobstein, O. E., and Bierman, E. O. The story of lysozyme. *J. Amer. Coll. Med. Technol.*, **1**:7, 1962.

L73. Lobstein, O. E., and Fogelson, S. J. Lysozyme activity in gastric juice of normal adults: Preliminary report. *Amer. J. Dig. Dis.*, **18**:282–283, 1951.

L74. Lobstein, O. E., and Fogelson, S. J. The effect of chloride concentration in the determination of lysozyme activity. *Amer. J. Dig. Dis.*, **18**:298–299, 1951.

L75. Lobstein, O. E., and Fogelson, S. J. The effect of a detergent complex on the deactivation and rate of reappearance of lysozyme activity in gastric juice fractional samples. *Amer. J. Dig. Dis.*, **18**:214–215, 1951.

L76. Lobstein, O. E., and Fogelson, S. J. Bacteriolytic (turbidimetric) determination of lysozyme in physiological fluids. *Quart. Bull. Northwest. Univ. Med. Sch.*, **25**:89–92, 1951.

L77. Lobstein, O. E., Jopes, H. S., and Fogelson, S. J. Effect of time and temperature on the susceptibility of lyophilized *Microccocus lysodeikticus* to lysozyme. *Quart. Bull. Northwest. Univ. Med. Sch.*, **26**:19–21, 1952.

L78. Locquet, J. P., Saint-Blancard, J., and Jollès, P. Apparent affinity constants of lysozymes from different origins for *Micrococcus lysodeikticus* cells. *Biochim. Biophys. Acta*, **167**:150–153, 1968.

L79. Lodinová, R., Wagner, V., and Jouja, V. Effect of artificial colonization of the intestines with non-pathogenic strain of *E.coli* 083 and of lysozyme administration on the formation of immunoglobulins and coproantibodies in infants. *Cesk. Pediat.*, **27**:495–598, 1972. (Czech.)

L80. Loe, K. G., Henriksen, S. D., and Berdal, P. Lysozyme content of tear fluid from ozaena patients. *Acta Pathol. Microbiol. Scand.*, **40**:521–525, 1957.

L81. Loew, G. H., and Thomas, D. D. Molecular orbital calculations of the catalytic effect of lysozyme. 1. Glu 35 as general acid catalyst. *J. Theor. Biol.*, **36**:89–104, 1972.

L82. Lombardo, G., and Londrillo, A. Ricerche sui tranquillanti. III. Comportamento del potere lisozimico e della proteina C reattiva in animali trattati con meprobamato. *Boll. Soc. Ital. Biol. Sper.*, **35**:318–319, 1959.

L83. Lominski, I., and Gray, S. Inhibition of lysozyme by 'Suramin.' *Nature (London)*, **192**:683, 1961.

L84. Longsworth, L. G., Cannan, R. K., and MacInnes, D. A. An electrophoretic study of the proteins of egg white. *J. Amer. Chem. Soc.*, **62**:2580–2590, 1940.

L84a. Longworth, J. W. Tyrosine phosphorescence of proteins. *Biochem. J.*, **81**:23P–24P, 1961.

L84b. Longworth, J. W. Conformations and interactions of excited states. II. Polystyrene, polypeptides, and proteins. *Biopolymers*, **4**:1131–1148, 1966.

L85. Looze, Y., Barel, A. O., and Léonis, J. Propriétés physico-chimiques du composant II du lysozyme de cane. *Arch. Int. Physiol. Biochim.*, **78**:1004–1006, 1970.

L86. Looze, Y., Vincentelli, J. B., Dolmans, M., and Léonis, J. Comparison conformationnelle des lysozymes d'oie et de poule. *Arch. Int. Physiol. Biochim.*, **77**:386–388, 1969.

L87. Looze, Y., Vincentelli, J. B., Lavinha, F., and Léonis, J. Etude de l'effet dénaturant d'alcools polyhydroxyliques sur le lysozyme à l'état dissous. *Arch. Int. Physiol. Biochim.*, **79**:837–838, 1971.

L88. Lord, R. C., and Mendelsohn, R. Laser Raman spectra of aqueous lysozyme denatured by lithium bromide. *J. Amer. Chem. Soc.*, **94**:2133–2135, 1972.

L89. Lord, R. C., and Yu, N. T. Laser-excited Raman spectroscopy of biomolecules. I. Native lysozyme and its constituent amino acids. *J. Mol. Biol.*, **50**:509–524, 1970.

L90. Lorena, M., Inouye, M., and Tsugita, A. Studies on the lysozyme from the bacteriophage T4 eJD7eJD4, carrying two frame shift mutations. *Mol. Gen. Genet.*, **102**:69–78, 1968.

L91. Lostia, A. Disattivazione operata da alcuni dermatofiti sul lisozima aggiunto al substrato colturale. *Rass. Med. Sarda,* **67**:217–221, 1964.

L92. Lostia, A. On the effects of lysozyme on the development in vitro of certain dermatomycetes. *Rass. Med. Sarda,* **62**:335–342, 1960. (Ital.)

L93. Lostia, A. Observations on the use of lysozyme in some skin diseases of known or presumed viral origin. *Rass. Med. Sarda,* **62**:329–334, 1960.

L94. Lowe, G. Possible hydrolytic mechanisms. *Proc. Roy. Soc., Ser. B,* **167**:431–434, 1967.

L95. Lowe, G., Sheppard, G., Sinnott, M. L., and Williams, A. Lysozyme-catalysed hydrolysis of some beta-Aryl di-N-acetylchitobiosides. *Biochem. J.,* **104**:893–899, 1967.

L96. Loza-Tulimowska, M. Rola lizozymu w skórze. *Przeg. Dermatol.,* **57**:107–109, 1970.(Pol.)

L97. Ludovici, L. J. "Fleming. Discoverer of Penicillin" pp. 94–121. Andrew Dakers, Ltd., London, 1952.

L98. Luk'ianenko, V. I. Comparative characteristics of humoral factors of natural immunity in dolphin (*Phocaena phocaena*). *Fed. Proc., Fed. Amer. Soc. Exp. Biol.,* **25**:Transl. Suppl., 337–338, 1966.

L99. Luk'ianenko, V. I. Interspecies differences in the incidence of detection and rate of serum lysozyme in fish. *Izv. Akad. Nauk SSSR, Ser. Biol.,* **3**:409–413, 1965. (Russ.)

L100. Lundblad, G., and Hultin, E. Human serum lysozyme (muramidase). I. Viscosimetric determination with glycol chitin and purification by selective adsorption. *Scand. J. Clin. Lab. Invest.,* **18**:201–208, 1966.

L101. Lundblad, G., Vesterberg, O., Zimmerman, R., and Lind, J. Studies on lysozyme from human leucemic urine by isoelectric focusing. *Acta Chem. Scand.,* **26**:1711–1713, 1972.

L102. Lungarotti, F., Vassallo, C., and Zampogna, S. Lysozyme in association with vitamins C and K in the treatment of some hemorrhagic conditions in surgery. *Ann. Laringol., Otol., Rinol., Faringol.,* **60**:710–720, 1961. (Ital.)

L103. Lunghetti, R., Blanco, G., Di Paolo, N. *et al.* Considerazioni clinico terapeutiche sull'associazione herpes zoster e leucemia linfatica cronica. *Atti Accad. Fisiocrit. Siena, Sez. Med.-Fis.,* **17**:941–953, 1968.

L103a. Luzzati, A., Champagne, M. Masse moléculaire et dimensions de la molécule de lysozyme en solution. *C. Rend. Acad. Sci.,* **244**:2930–2932, 1957.

L104. Lupo, M. Primi tentativi di protezione leucocitaria con lisozima nelle tele-panirradiazioni. *Minerva Radiol., Fisioter. Radiobiol.,* **10**:164–165, 1965.

L105. Luvoni, R. Due casi mortali da prima iniezione intramuscolare di lisozima. *Minerva Medicoleg.,* **83**:130–134, 1963.

L106. Luzzati, V., Witz, J., and Nicolaieff, A. Détermination de la masse et des dimensions des protéines en solution par la diffusion centrale des rayons x mesurée à l'échelle absolue: Exemple du lysozyme. *J. Mol. Biol.,* **3**:362–378, 1961.

L107. Lycke, E., Lund, E., and Strannegård, O. Enhancement by lysozyme and hyaluronidase of the penetration by *Toxoplasma gondii* into cultured host cells. *Brit. J. Exp. Pathol.,* **46**:189–199, 1965.

M

Mc1. McBride-Warren, P. A., and Mueller, D. D. Hydrogen-deuterium exchange of lysozyme. I. Rate constants and pH dependence. *Biochemistry,* **11**:1785–1792, 1972.

Mc2. McCabe, W. C., and Fisher, H. F. Measurement of the excluded volume of protein molecules by differential spectroscopy in the near infra-red. *Nature* (London), 207:1274–1276, 1965.

Mc3. McCoubrey, A., and Smith, M. H. Retention of enzymatic activity by egg white lysozyme bearing alkylamidino substituents. *Biochem. Pharmacol.*, 15:1623–1625, 1966.

Mc4. McDade, J. E., and Tripp, M. R. Lysozyme in oyster mantle mucus. *J. Invertebr. Pathol.*, 9:581–582, 1967.

Mc5. McDonald, C. C., Phillips, W. D., and Glickson, J. D. Nuclear magnetic resonance study of the mechanism of reversible denaturation of lysozyme. *J. Amer. Chem. Soc.*, 93:235–246, 1971.

Mc6. McDonald, C. C., and Phillips, W. D. Perturbation of the PMR spectrum of lysozyme by Co^{+2}. *Biochem. Biophys. Res. Commun.*, 35:43–51, 1969.

Mc7. McDonald, C. C., and Phillips, W. D. Manifestations of the tertiary structures of proteins in high-frequency nuclear magnetic resonance. *J. Amer. Chem. Soc.*, 89:6332–6231, 1967.

Mc8. McDonald, H. C., Odstrchel, G., and Maurer, P. H. Hyperimmune response to a protein and a polypeptide antigen coated on Pneumococcus R36A. *J. Immunol.*, 109:881–883, 1972.

Mc9. McElroy, L. J., and Casida, L. E., Jr. An evaluation of rhodamine-labeled lysozyme as a fluorescent stain for *in situ* soil bacteria. *Can. J. Microbiol.*, 18:933–936, 1972.

Mc10. McEwen, W. K., and Kimura, S. J. Filter-paper electrophoresis of tears. Lysozyme and its correlation with keratoconjunctivitis sicca. *Amer. J. Ophthalmol.*, 39:Part 2, 200–202, 1956.

Mc11. McGhee, J. R., and Freeman, B. A. Effect of lysosomal enzymes on *Brucella*. *Res. J. Reticuloendothel. Soc.*, 8:208–219, 1970.

Mc11a. McKelvy, J. F., Eshdat, Y., and Sharon, N. Action of specific and irreversible inhibitors of hen egg white lysozyme on lysozymes from four other sources. *Fed. Proc., Fed. Amer. Soc. Exp. Biol.*, 29:532, Abstr. No. 1598, 1970.

Mc12. McMahon, L. C., Zanger, B., and Desforges, J. G. Serum muramidase in newborns. *Biol. Neonate*, 17:24–29, 1971.

Mc12a. MacRitchie, F., and Alexander, A. E. Kinetics of adsorption of proteins at interfaces. Part I. The role of bulk diffusion in adsorption. *J. Colloid. Sci.*, 18:453–457, 1963.

M1. Maack, T. Changes in the activity of acid hydrolases during renal reabsorption of lysozyme. *J. Cell Biol.*, 35:268–273, 1967.

M2. Maack, T., Brentani, R., and Rabinovitch, M. Increase in 'alkaline' ribonuclease of rat kidney cortex after lysozyme and egg-white administration. *Nature (London)*, 186:158, 1960.

M3. Maack, T., and Kinter, W. B. Transport of protein by flounder kidney tubules during long-term incubation. *Amer. J. Physiol.*, 216:1034–1043, 1969.

M4. Maack, T., Mackensie, D. D. S., and Kinter, W. B. Intracellular pathways of renal reabsorption of lysozyme. *Amer. J. Physiol.*, 221:1609–1616, 1971.

M5. Maass, D., and Weidel, W. Final proof for the identity of enzymic specificities of egg-white lysozyme and phage T2 enzyme. *Biochim. Biophys. Acta*, 78:369–370, 1963.

M6. Macbeth, W. A. A. G., McKenzie, R. G., and Taverner, P. L. The effect of cholecystectomy upon renal excretion rates of ribonuclease, lysozyme and total protein. *N. Z. Med. J.*, 75:359–361, 1972.

M7. Maccari, M., and Gatti, M. T. Influenza del lisozima sull'assorbimento intestinale della tetraciclina. *Boll. Soc. Ital. Biol. Sper.*, 44:1071–1073, 1968.

M8. Macchi, L., and Acerboni, F. La lisozimuria nella gravidanza normale e nella sindrome edemo nefrosica ipertensiva. *Ann. Ostet. Ginecol.*, 88:814–820, 1966.

M9. Machmer, P. Uber einen charge-Transfer-Komplex des Lysozyms. *Naturwissenschaften*, **53**:405–406, 1968.

M10. Madecka-Borkowska, I., and Deptuch, A. "Crossing" paper electrophoresis for the detection of complexes of enzymes with their inhibitors. *J. Chromatog.*, **13**:251–254, 1964.

M11. Madon, E., and Cocuzza, S. Comportamento dlettroforectico delle glicoproteine seriche in soggetti affetti da epatite virale prima e dopo trattamento con lisozima. *Epatologia*, **12**:359–364, 1966.

M12. Maeda, H., Glaser, C. B., and Meienhofer, J. Facile assignment of disulfide bonds in proteins through detection of cystine-containing peptides in peptide maps. *Biochem. Biophys. Res. Commun.*, **39**:1211–1218, 1970.

M13. Maes, E. D., Dolmans, M., Vincentelli, J. B., and Léonis, J. Bovine alpha-lactalbumin: Its structure in solution as compared to hen lysozyme. *Arch. Int. Physiol. Biochim.*, **77**:388–390, 1969.

M14. Magaldi, R. Sull'activitá lisozimica della saliva nella patogenesi della parotite postoperatoria. *Atti Accad. Fisiocrit. Siena*, **3**:95–106, 1956.

M15. Maggi, M., and Fossati, G. Considerazioni sull'impiego dei cortisonici e del lisozima nella terapia della epatite virale nell'infanzia. *Minerva Pediat.*, **18**:1958–1960, 1966.

M16. Maggio, E., and Vingiani, A. Il lisozima ed i suoi effetti clinici e batteriologici nel trattamento locale dell'otite media prurulenta cronica. *Acta Med. Ital. Mal. Infet. Parassit.*, **10**:312–321, 1955.

M17. Magliulo, E., Petrocini, S., Fossati, G. C., and Sprovieri, G. Effetto di una associazione lisozima-antibiotici sulla attività fagocitaria dei granulociti neutrofili umani. *Arch. Sci. Med.*, **125**:329–334, 1968.

M18. Maglietta, C., Toschi, G. P., and Parenti, G. F. Behavior of the serum lysozyme level in pulmonary tuberculosis. *Riv. Patol. Clin.*, **35**:135–150, 1962.

M19. Magnuson, J. A., and Magnuson, N. S. A new NMR technique to study disulfide reduction: Comparison of lysozyme and alpha-lactalbumin. *Biochem. Biophys. Res. Commun.*, **45**:1513–1517, 1971.

M20. Maini, G., and Sozzi, F. Il contenuto in lisozima del siero di sangue nei bambini delle varie età. *Lattante*, **20**:9–19, 1949.

M21. Mainoldi, F. La muramidasi nelle leucopenie da citostatici. Nota preventiva. *Minerva Radiol. Fisioter. Radiobiol.*, **10**:169–172, 1956.

M22. Makhlina, M. S. Lysozyme in therapy of summer diarrhea in children. *Sov. Med.*, **4**:9–10, 1940.

M23. Maksimov, I. Effect of egg lysozyme on disaccharides. *Dokl. Akad. Nauk SSSR*, **172**:210–213, 1967. (Russ.)

M24. Maksimov, V. I. Investigation of the mechanism of aggregation of gamma-irradiated lysozyme employing a turbidimetric method. *Biokhimiya*, **32**:835–842, 1967. (Russ.)

M25. Maksimov, V. I., Kaverzneva, E. D., and Kravchenko, N. A. On the mode of lysozyme action on oligosaccharide fragments of chitin. *Biokhimiya*, **30**:1007–1014, 1965. (Russ.)

M26. Maksimov, V. I., Kaverzneva, E. D., and Osipov, V. I. Mechanism of radiation damage of lysozyme irradiated in dry state. *Biofizika*, **17**:978–985, 1972. (Russ.)

M27. Maksimov, V. I., and Mosin, V. A. Chromatography of native and gamma-irradiated lysozymes on dense gels. *J. Chromatog.*, **47**:361–368, 1970.

M28. Maksimov, V. I., and Osipov, V. I. Study of the ability of gamma-irradiated lysozyme to form aggregates (solubility test and gel-filtration method.) *Biokhimiya*, **32**:98–105, 1067. (Russ.)

M29. Maksimov, V. I., Osipov, V. I., and Kaverzneva, E. D. Suppression of the tendency toward aggregation of gamma-irradiated lysozyme after treatment by mercaptoethylamine. *Radiobiologiya,* **9**:355–358, 1969. (Russ.)

M30. Maksimov, V. I., Osipov, V. I., and Kaverzneva, E. D. A study on the hydrodynamic size of gamma-irradiated lysozyme by gel filtration under different conditions. *Biokhimiya,* **33**:451–458, 1968. (Russ.)

M31. Maksimov, V. I., Osipov, V. I., and Kaverzneva, E. D. The use of low molecular weight fragments of chitin for determining the activity of gamma-irradiated lysozyme. *Biokhimya,* **32**:1169–1174, 1967. (Russ.)

M32. Maksimov, V. I., Osipov, V. I., and Kaverzneva, E. D. Investigation of gamma-irradiated ribonuclease by gel-filtration. *Biokhimiya,* **32**:403–408, 1967. (RUss.)

M33. Maksimov, V. I., and Tepelina, O. M. Chitodextran as substrate and inducer in the reaction of lysozyme with oligosaccharides. *Biokhimiya,* **31**:918–923, 1966. (Russ.)

M34. Makulu, D. R., and Wagner, M. Lysozyme activity in the serum, saliva and tears of germfree and conventional rats and mice. *Proc. Indiana Acad. Sci.,* **76**:183, 1966.

M35. Malamy, M. H., and Horecker, B. L. Release of alkaline phosphatase from cells of Escherichia coli upon lysozyme spheroplast formation. *Biochemistry,* **3**:1889–1893, 1964.

M36. Malik, S. R., Gupta, A. K., and Berry, R. Electrophoretic analysis of lacrimal fluid proteins and demonstration of lysozyme in normal human tears. *J. All-India Ophthalmol. Soc.,* **18**:106–110, 1970.

M37. Malke, H. Uber das Vorkommen von Lysozym in Insekten. *Z. Allg. Mikrobiol.,* **5**:42–47, 1965.

M38. Malke, H. Wirkung von Lysozym auf die Symbionten der Blattiden. *Z. Allg. Mikrobiol.,* **4**:88–91, 1964.

M39. Malke, H. Production of aposymbiotic cockroaches by means of lysozyme. *Nature (London),* **204**:1223–1224, 1964.

M40. Malke, H., and Schwartz, W. Untersuchungen über die Symbiose von Tieren mit Pilzen und Bakterien. XI. Die Rolle des Wirtslysozyms in der Blattidensymbiose. *Arch. Mikrobiol.,* **53**:17–32, 1966.

M41. Malmquist, J. Serum and urinary lysozyme in leukaemia and polycythaemia vera. *Scand. J. Haematol.,* **9**:258–266, 1972.

M42. Maltman, J. R., and Webb, S. J. The action of hydrolytic enzymes and vapor rehydration on semidried cells of *Klebsiella pneumoniae. Can. J. Microbiol.,* **17**:1443–1450, 1971.

M43. Mandel, M. Proton magnetic resonance spectra of some proteins. I. Ribonuclease, oxidized ribonuclease, lysozyme, and cytochrome c. *J. Biol. Chem.,* **240**:1586–1592, 1965.

M43a. Mandel, M. The effect of temperature on the proton magnetic resonance spectra of ribonuclease, oxidized ribonuclease, and lysozyme. *Proc. Nat. Acad. Sci. U.S.,* **52**: 736–741, 1964.

M44. Mandeles, S., and Ducay, E. D. Site of egg white protein formation. *J. Biol. Chem.,* **237**:3196–3199, 1962.

M45. Manfredi, G., and Salvioli, G. P., Jr. Contributo alla conoscenza della neuromielite ottica o malattia di Devic-Gault nell 'eta' infantile. Raro decorso favorevole da trattamento antidisreattivo locale. *Clin. Pediat. (Bologna),* **44**:191–198, 1962.

M46. Manwell, C. Molecular palaeogenetics: Amino acid sequence homology in ribonuclease and lysozyme. *Comp. Biochem. Physiol.,* **23**:383–406, 1967.

M47. Marchand, J. Utilisation en O.R.L. des propiétés anti-hémorragiques du lysozyme. *Ann. Oto-Laryngol. Chir. Cervico-Faciale,* **78**:860–862, 1961.

M48. Marciani, D. J., and Tolbert, B. M. Effects of gamma-irradiation on the enzymatic properties of lysozyme. *Biochim. Biophys. Acta,* **302**:376–381, 1973.

M49. Marciani, D. J., and Tolbert, B. M. Analytical studies of fractions from irradiated lysozyme. *Biochim. Biophys. Acta,* **271**:262–273, 1972.

M50. Marcolongo, R., and Contu, L. Behavior of lysozyme activity in some blood diseases. *Haematol. Lat.,* **5**:141–173, 1962.

M51. Margulis, L. A. Lysozyme and phagolysozyme in prophylaxis and treatment of acute intestinal diseases of children. *Sov. Med.,* **4**:15–17, 1940.

M52. Marinova-Pemrova, R., and Pereverzev, N. A. The transmission of multiple drug resistance between penicillin and lysozyme spheroplasts of E. coli K-12 substrains. *Zh. Mikrobiol., Epidemiol. Immunbiol.,* **46**:80–84, 1969. (Russ.)

M53. Mark, K. K. A lysozyme assay method for low activity. *Anal. Biochem.,* **37**:447–450, 1970.

M54. Mark, K. K. The relationship between the synthesis of DNA and the synthesis of phage lysozyme in *Escherichia coli* infected by bacteriophage T4. *Virology,* **42**:20–27, 1970.

M55. Mark, K.-K., and Chen, I. Correlation between the reduction of phage lysozyme level and the time of lysis in *Escherichia coli* infected by bacteriophage T4. *Biochem. Biophys. Res. Commun.,* **46**:1102–1105, 1972.

M56. Maron, E., Arnon, R., and Bonavida, B. Sequential appearance of antibodies directed against different antigenic determinants of hen egg-white lysozyme. *Eur. J. Immunol.,* **1**:181–185, 1971.

M57. Maron, E., Arnon, R., Sela, M., Perin, J.-P., and Jollès, P. Immunological comparison of bird and human lysozymes and of their "loop" regions. *Biochim. Biophys. Acta,* **214**:222–224, 1970.

M58. Maron, E., and Bonavida, B. A sensitive immunoassay for human lysozyme in biological fluids. *Biochim. Biophys. Acta,* **229**:273–275, 1971.

M59. Maron, E. Eshdat, Y., and Sharon, N. Immunological studies of affinity labelled hen egg-white lysozyme and of the active site region of related lysozymes. *Biochim. Biophys. Acta,* **278**:243–249, 1972.

M60. Maron, E., Shiozawa, C., Arnon, R., and Sela, M. Chemical and immunological characterization of a unique antigenic region in lysozyme. *Biochemistry,* **10**:763–771, 1971.

M61. Maron, E., Webb, C., Teitelbaum, D., and Arnon, R. Cell-mediated vs humoral response in the cross-reaction between hen egg-white lysozyme and bovine alpha-lactalbumin. *Eur. J. Immunol.,* **2**:294–297, 1972.

M62. Marras, G., and Salis, B. Inhibiting action of lysozyme on the activity of some local anesthetics. *Arch. Ital. Sci. Farmacol.,* **9**:413–418, 1959.

M63. Marras, G., and Salis, B. Il lisozima inibisce l'azione anestetica locale della lidocaina. *Boll. Soc. Ital. Biol. Sper.,* **35**:794–796, 1959.

M64. Marri, R., and Linari, G. On a photometric method of measuring the biological activity of lysozyme. *Farmaco, Ed. Prat.,* **17**:663–666, 1962.

M65. Marshall, M. E., and Deutsch, H. F. Clearances of some proteins by the dog kidney. *Amer. J. Physiol.,* **163**:461–467, 1950.

M66. Martin, R. R., Crowder, J. G., White, A., and Jackson, G. Human reactions to staphylococcal antigens. A possible role of leucocyte lysosomal enzyme. *J. Immunol.,* **99**:269–275, 1967.

M67. Marzotto, A. Circular dichroism of acetoacetylated proteins. *Experientia,* **26**:1084–1085, 1970.

M68. Marzotto, A., and Galzigna, L. On the enzymatic hydrolysis of carboxymethlychitin by lysozyme. *Hoppe-Seyler's Z Physiol. Chem.*, **350**:427–430, 1969.

M69. Marzotto, A., Pajetta, P., Galzinga, L., and Scoffone, E. Reversible acetoacetylation of amino groups in proteins. *Biochim. Biophys. Acta*, **154**:450–456, 1968.

M70. Massaro, D. Synthesis of proteins by alveolar cells. *Nature (London)*, **215**:646–647, 1967.

M71. Mastrandrea, V., and Candell, A. Su alcuni caratteri delle varianti "S" e "R" di un ceppo di Bacillus anthracis. *Bull. Soc. Ital. Biol. Sper.*, **44**:170–173, 1968.

M72. Matracia, S., and Brusca, A. L'attivita lisozimica nella fatica e nella setticemia stafilococcica sperimentale. *G. Batteriol. Immunol.*, **49**:402–410, 1956.

M73. Matracia, S., Gargano, F., and Salerno, A. Valutazione a mezzo del blu di toluidina della riduzione delle sostanze Rivalta: Positive, operata da trattamento di essudati sierosi con ACTH, lisozima e salmina. *Boll. Soc. Ital. Biol. Sper.*, **36**:820–822, 1960.

M74. Matracia, S., Salerno, A., and Caradonna, D. Indagini sulla interazione di fosfatidi e glicolipidi del sarcoma ascite di Yoshida con il lisozimo. *Boll. Soc. Ital. Biol. Sper.*, **37**:548–550, 1961.

M75. Matsunaga, F., and Kubo, A. Ulcerative colitis and lysozyme activity. *Naika (Tokyo)*, **11**:229–240, 1963. (Jap.)

M76. Matsushima, Y., Miyazaki, T., and Kasai, S. Chemical evidence for the Phillips model of the action of hen egg white lysozyme. *Nature (London)*, **219**:265, 1968.

M76a. Matsushima, A., Hachimori, Y., Inada, Y. and Shibata, K. States of amino acid residues in proteins. XII. Amino groups with different reactivities toward Naphthoquinone. *J. Biochem. (Tokyo)*, **63**:328–336, 1967.

M77. Matsushita, K., Tani, Y., and Miyaura, K. Clinical evaluation of lysozyme ophthalmic solution (MT-L). *Folia Ophthalmol. Jap.*, **20**:34–40, 1969. (Jap.)

M78. Matthews, K. S., and Cole, R. D. Interaction of lysozyme with f2 bacteriophage. *J. Mol. Biol.*, **68**:173–176, 1972.

M79. Matthyssens, G., and Kanarek, L. Antigenic and structural studies on peptic fragments of hen egg-white lysozyme. I. *Arch. Int. Physiol. Biochim.*, **79**:1032–1033, 1971.

M80. Matthyssens, G., Simons, G., and Kanarek, L. Antigenic and structural studies on peptic fragments of hen egg-white lysozyme. II. *Arch. Int. Physiol. Biochim.*, **79**:1033–1034, 1971.

M81. Matthyssens, G. E., Simons, G., and Kanarek, L. Study of the thermal-denaturation mechanism of hen egg-white lysozyme through proteolytic degradation. *Eur. J. Biochem.*, **26**:449–454, 1972.

M82. Mauel, J., and Defendi, V. Infection and transformation of mouse peritoneal macrophages by Simian Virus 40. *J. Exp. Med.* **134**:335–350, 1971.

M83. Maurois, A. "The Life of Sir Alexander Fleming; Discoverer of Penicillin" (Translated by G. Hopkins), Dutton, New York, 1959. pp. 109–122.

M84. Mazzola, G. Trattamento degli itteri gravi neonatali con corticosteroidi e lisozima. *Minerva Pediat.*, **17**:1710–1712, 1965.

M85. Mazzucco, P. S. Lysozyme in sinusitis, sinus diseases and sinutosis. *Sem. Med.*, **119**:1050–1052, 1961.

M86. Mazzucco, P. Physiopathological treatment of inflammations of the nasal and paranasel mucosa. *Omnia Ther.*, **10**:135–144, 1959.

M87. Meadows, D. H., Markley, J. L., Cohen, J. S., and Jardetzky, O. Nuclear magnetic resonance studies of the structure and binding sites of enzymes. I. Histidine residues. *Proc. Nat. Acad. Sci. U.S.*, **58**:1307–1313, 1967.

M88. Meadows, E. C., and Levison, J. The effect of lysozyme on canine gastrointestinal mucosa; further study. *Gastroenterology,* 21:569–573, 1952.

M89. Means, G. E., and Feeney, R. E. Reductive alkylation of amino groups in proteins. *Biochemistry,* 7:2192–2201, 1968.

M90. Meier, H., and Hoag, W. G. Activity of lysozyme in inbred mice. *J. Bacteriol.,* 83:689–690, 1962.

M90a. Mejbaum-Katzenellenbogen, W. Insoluble protein-tannin compounds. *Acta Biochim. Polon.,* 6:341–349, 1959. (Engl.)

M90b. Mejbaum-Katzenellenbogen, W. Studies on regeneration of protein from insoluble protein-tannin compounds. I. Removal of tannin from the protein-tannin compounds by caffeine. *Acta Biochim. Polon.,* 6:350–363, 1959. (Engl.)

M91. Melsom, A., and Weiser, R. S. The therapeutic action of lysozyme on pneumococcal infection of mice. *J. Infec. Dis.,* 102:203–213, 1958.

M92. Merigan, T. C., and Dreyer, W. J. Studies on the antigenic combining sites in bacteriophage lysozyme. *Ann. N. Y. Acad. Sci.,* 103:765–772, 1963.

M93. Merlini, M., Gusmano, R., and Galletti, I. Nota preliminare su di un nuovo indirizzo terapeutico della mallattia reumatica (m.r.) *Minerva Pediat.,* 16:1402–1403, 1964.

M94. Merlini, M., Vallarino, G., and Gusmano, R. L'impiego del lisozima nella terapia dell'epatite virale. *Epatologia,* 12:215–221, 1966.

M95. Mesrobeanu, I., Mesrobeanu, L., Mitrica, N., and Papazian, E. Gli enzimi leucocitari ed il problema della sopravvivenza degli stafilococchi nei leucociti dopo fagocitosi. *Minerva Med.,* 48:1375–1379, 1957.

M96. Mesrobeanu, L., and Noeppel, O. Revue des travaux sur le lysozyme. *C. R. Soc. Biol.,* 131:1265–1267, 1939.

M97. Mesrobeanu, L., and Noeppel, O. Review of work on lysozyme. *Arch. Roum. Pathol. Exp. Microbiol.,* 11:247–272, 1938.

M98. Messner, B., and Mohrig, W. Zum gemeinsamen Vorkommen van Hämagglutininen und Lysozymem bei Pflazen und Tieren. Bemerkungen zu den Prokopschen Protektinen. *Acta Biol. Med. (Gdansk),* 4:891–903, 1970.

M99. Městecký, J. Detection of lysozyme production in individual cells by the plaque technique. *Folia Microbiol. (Prague),* 12:175–176, 1967.

M100. Městecký, J., Jílek, M., and Marěcková, M. Variances in the levels of serum lysozyme in x-irradiated mice. *Folia Microbiol. (Prague),* 11:179–183, 1966.

M101. Městecký, J., Kraus, F. W., Williams, R. C., and Mims, M. C. Rivanol-ethanol fractionation of parotid fluid and colostrum. *Experientia,* 25:892–893, 1969.

M102. Městecký, J., and Medlín, J. Role of lysozyme and character of antibodies in experimental viral infection. *Acta Virol. (Prague), Eng. Ed.,* 11:206–215, 1967.

M103. Městecký, J., Tlaskalová, H., Mandel, L., and Jilek, J. Serum lysozyme in germ-free piglets. *Folia Microbiol. (Prague),* 12:406, 1967.

M104. Metcalf, R. H., and Deibel, R. H. Effect of lysozyme on enterococcal viability in low ionic environments. *J. Bacteriol.,* 113:278–286, 1973.

M105. Metcalf, R. H., and Deibel, R. H. Growth of *Streptococcus faecium* in the presence of lysozyme. *Infec. Immunity,* 6:178–183, 1972.

M106. Metcalf, R. H., and Deibel, R. H. Differential lytic response of enterococci associated with addition order of lysozyme and anions. *J. Bacteriol.,* 99:674–680, 1969.

M107. Metzger, M. The lysozyme level in the blood serum and tissue fluids of rat testes in syphilis. *Postephy Hig. Med. Dosw.,* 17:719–724, 1963. (Pol.)

M108. Metzger, M. A study of the role of serum and tissue lysozyme upon the treponeme immobilization reaction. *Amer. J. Hyg.,* 76:267–275, 1962.

M109. Metzger, M. A one-day diagnostic TPI test with lysozyme. *Arch. Immunol. Ter. Dosw.,* **9:**733–744, 1961.

M110. Metzger, M., Hardy, P. H., Jr., and Nell, E. Influence of lysozyme upon the treponeme immobilization reaction. *Amer. J. Hyg.,* **73:**236–244, 1961.

M111. Metzger, M., and Kopeć, W. Staining bacterial cells with fluorescent lysozyme. *Arch. Immunol. Ther. Exp.,* **12:**473–482, 1964.

M112. Metzger, M., and Podwińska, J. Studies of the mechanism of development of agglutinability of pathogenic *Treponema pallidum*. I. Accelerating effect of some proteolytic and mucopolysaccharide-splitting enzymes. *Arch. Immunol. Ther. Exp.,* **13:**516–524, 1965.

M113. Metzger, M., and Ruczkowska, J. Influence of lysozyme upon the reactivity of Treponema pallidum in the fluorescent antibody reaction. *Arch. Immunol. Ther. Exp.,* **12:**702–708, 1964.

M114. Metzger, M., and Ruczkowska, J. Application of the oil technique in the one-day TPI test with lysozyme. *Arch. Immunol. Ther. Exp.,* **10:**870–876, 1962.

M115. Metzger, M., and Szulga, T. Lysozyme levels in the sera of guinea pigs and rabbits in the course of experimental tuberculosis. *Arch. Immunol. Ther. Exp.,* **11:**467–475, 1963.

M116. Metzger, M., and Szulga, T. The lysozyme activity of the blood serum of guinea pigs and rabbits in experimental tuberculosis. *Postepy Hig. Med. Dosw.,* **17:**725–729, 1963. (Pol.)

M117. Meyer, K. Lysozyme. *Bull. N.Y. Acad. Med.,* **30:**995–996, 1954.

M118. Meyer, K. Mycolytic enzymes. "Currents in Biochemical Research," pp. 277–290. Wiley (Interscience), New York, 1946.

M119. Meyer, K. The relationship of lysozyme to avidin. *Science,* **99:**391–392, 1944.

M120. Meyer, K. *et al.* Lysozyme activity in chronic ulcerative colitis. *Rev. Gastroenterol.,* **16:**476–479, 1949.

M121. Meyer, K., Gellhorn, A., Prudden, J. F., Lehman, W. L., and Steinberg, A. Lysozyme activity in ulcerative alimentary disease II. Lysozyme activity in chronic ulcerative colitis. *Amer. J. Med.,* **5:**496–502, 1948.

M122. Meyer, K., Gellhorn, A., Prudden, J. F., Lehman, W. L., and Steinberg, A. Lysozyme in chronic ulcerative colitis. *Proc. Soc. Exp. Biol. Med.,* **65:**221–222, 1947.

M123. Meyer, K., and Hahnel, E. The estimation of lysozyme by a viscosimetric method. *J. Biol. Chem.,* **163:**723–732, 1946.

M124. Meyer, K., Hahnel, E., and Steinberg, A. Lysozyme of plant origin. *J. Biol. Chem.,* **163:**733–740, 1945.

M125. Meyer, K., Palmer, J. W., Thompson, R., and Khorazo, D. On mechanism of lysozyme action. *J. Biol. Chem.,* **113:**479–486, 1936.

M126. Meyer, K., Prudden, J. F., Lehman, W. L., and Steinberg, A. Lysozyme activity in ulcerative alimentary disease; lysozyme in peptic ulcer. *Amer. J. Med.,* **5:**482–495, 1948.

M127. Meyer, K., Prudden, J. F., Lehman, W. L., and Steinberg, A. Lysozyme content of the stomach and its possible relationship to peptic ulcer. *Proc. Soc. Exp. Biol. Med.,* **65:**220–221, 1947.

M128. Meyer, K., Thompson, R., Palmer, J. W., and Khorazo, D. The purification and properties of lysozyme. *J. Biol. Chem.,* **113:**303–309, 1936.

M129. Meyer, K., Thompson, R., Palmer, J. W., and Khorazo, D. Nature of lysozyme action. *Science,* **79:**61, 1934.

M130. Meyer, O. T., and Dannenberg, A. M., Jr. Radiation, infection, and macrophage function. II. Effect of whole body radiation on the number of pulmonary alveolar macrophages and their levels of hydrolytic enzymes. *Res. J. Reticuloendothel. Soc.,* **7**:79–90, 1970.

M131. Michaels, G. B., and Eagon, R. G. The effect of ethylenediaminetetraacetate and lysozyme on isolated lipopolysaccharide from *Pseudomonas aeruginosa. Proc. Soc. Exp. Biol. Med.,* **122**:866–868, 1966.

M132. Miglior, M., and Orzalesi, N. F. Indagini in vitro, sulla capacità difissare il lisozima da parte dei tessuti oculari. *Arch. Vecchi Anat. Patol. Med. Clin.,* **34**:737–746, 1961.

M133. Miglior, M., and Pirodda, A. Sul contento di lisozima dell'uovo di pollo incubato in rapporto alla cultura del virus erpetico. *Riv. Ist. Sieroter. Ital.,* **29**:115–121, 1954.

M134. Miglior, M., and Pirodda, A. Attività lisozimica del tessuto corneale dell'uomo e di alcuni mammiferi domestici. *G. Ital. Oftalmol.,* **6**:494–497, 1953.

M135. Mikhaïlova, Z. M., and Mikheeva, G. A. Indices of nonspecific immunologic reactivity in healthy children during various age periods. *Pediatriya (Moscow),* **46**:47–52, 1967. (Russ.)

M136. Mikhaïlova, Z. M., Sokolova, A. F., and Derechinskaia, Sh. L. Dynamics of complement, its components and serum lysozyme activity in respiratory viral and enteroviral infections in children. *Vop. Okhr. Mater. Detstva (Moscow),* **12**:9–14, 1967. (Russ.)

M137. Mikulewicz, W., Kubik, K., and Bednarski, W. Studies on proteins of nasal mucous membrane secretion in fibrinolytic aspect. *Acta Oto-Laryngol.,* **70**:379–382, 1970.

M138. Miller, A., Bonavida, B., Stratton, J. A., and Sercarz, E. Cross-reactivity of some rabbit anti-human lysozyme sera with gallinaceous lysozymes. *Biochim. Biophys. Acta,* **243**:520–524, 1971.

M139. Miller, T. E. Killing and lysis of gram-negative bacteria through the synergistic effect of hydrogen peroxide, ascorbic acid, and lysozyme. *J. Bacteriol.,* **98**:949–955, 1969.

M140. Miller, T. E., Cameron, C. M., and North, J. D. K. Distribution of lysozyme in the rat kidney and the role of this enzyme in experimental pyelonephritis. *Proc. Soc. Exp. Biol. Med.,* **128**:749–752, 1968.

M141. Miller, W. N., and Casida, L. E., Jr. Microorganisms in soil as observed by staining with rhodamine-labeled lysozyme. *Can. J. Microbiol.,* **16**:305–307, 1970.

M142. Millett, F., and Raftery, M. A. An NMR method for characterizing conformation changes in proteins. *Biochem. Biophys. Res. Commun.,* **47**:625–632, 1972.

M143. Millett, F., and Raftery, M. A. A ^{19}F nuclear magnetic resonance study of the binding of trifluoroacetylglucosamine oligomers to lysozyme. *Biochemistry,* **11**:1639–1643, 1972.

M144. Mil'man, M. Sh. Lysozyme in nasal secretions and its significance in atrophic rhinitis. *Vestn. Otorinolaringol,* **22**:50–54, 1960. (Russ.)

M145. Minagawa, K. Significance of lysozyme in infant nutrition. Synthesis of lysozyme by *Lactobacillus bifidus. Acta Paediat. Jap.,* **74**:761–767, 1970. (Jap.)

M146. Minagawa, K. Study of lysozyme in infants. 2. Study of various factors influencing feces lysozyme activity in health infants. *Acta Paediat. Jap.,* **72**:1324–1333, 1968. (Jap.)

M147. Minagawa, K. Study of lysozyme in infants. (1) Effect of oral administration of lysozyme on serum lysozyme activity. *Acta Paediat. Jap.,* **71**:1484–1490, 1967. (Jap.)

M148. Minton, L. R. Paralimbal ring keratitis and absence of lysozyme in lupus erythematosus. *Amer. J. Ophthalmol.,* **60**:532–535, 1965.

M149. Mirelman, D. S., and Sharon, N. Isolation and study of the chemical structure of low molecular weight glycopeptides from *Micrococcus lysodeikticus* cell walls. *J. Biol. Chem.,* **242**:3414–3427, 1967.

M150. Mirisola, F., and Dolci, G. Ricerche sperimentali sull'azione ipercoagulante del lisozima. *Ann. Stomatol.,* **14**:445–462, 1965.

M151. Mitchell, W. M., and Hash, J. H. The N,O-diacetylmuramidase of *Chalaropsis species*. II. Physical properties. *J. Biol. Chem.,* **244**:17–21, 1969.

M152. Miyama, A., Plescia, O. J., and Braun, W. Comparison of bactericidal and hemolytic serum systems. II. Analysis of inhibitors in normal serum fractions. *Proc. Soc. Exp. Biol. Med.,* **117**:386–389, 1964.

M153. Miyazaki, T., and Matsushima, Y. Studies on the substrate specificity of egg white lysozyme. II. Mode of the enzymatic action on partially-O-carboxymethylated chitin. *Bull. Chem. Soc. Jap.,* **41**:2754–2757, 1968.

M154. Mizuguchi, Y., and Tokunaga, T. Spheroplasts of Mycobacteria. 1. Spheroplast formation with glycine and lysozyme. *Med. Biol. (Tokyo),* **76**:301–305, 1968. (Jap.)

M155. Mizuno, J., Matsuoka, Y., Miyata, H. *et al.* Measurement of lysozyme activity values in the normal nasal secretion. *Otolaryngology (Tokyo),* **39**:539–543, 1967. (Jap.)

M156. Mizunoe, K., and Dannenberg, A. M., Jr. Hydrolases of rabbit macrophages. III. Effect of BCG vaccination, tissue culture, and ingested tubercle bacilli. *Proc. Soc. Exp. Biol. Med.,* **120**:284–290, 1965.

M157. Modéer, T., and Söder, P. O. A diffusion method for determination of lysozyme activity. *Scand. J. Dent. Res. (Copenhagen),* **79**:533–535, 1971.

M158. Moeller, H. C., Klotz, A. P., and Kirsner, J. B. Lack of effect of crystalline lysozyme on the isolated intestinal pouch of the dog. *Gastroenterology,* **20**:604–608, 1952.

M159. Moeller, H., C., Marshall, H. C., and Kirsner, J. B. Lysozyme production in response to injury of gastrointestinal tract in dogs. *Proc. Soc. Exp. Biol. Med.,* **76**:159–161, 1951.

M160. Mogilenko, A. F. Lysozyme level in calves in acute broncopneumonia and in normal conditions. *Veterinariya (Moscow),* **48**:88–89, 1972. (Russ.)

M161. Mohrig, W., and Messner, B. Significance of lysozyme for the antibacterial immunity of insects. *Zh. Obshch. Biol.,* **30**:62–71, 1969. (Russ.)

M162. Mohrig, W., and Messner, B. Lysozym als antibakterielles Agens im Bienenhonig und Bienengift. und Bienengift. *Acta Biol. Med. Ger.,* **21**:85–95, 1968.

M163. Moncalvo, F. Action of lysozyme, dextran and an aspecific lipoprotein complex on blood levels of properdin. *Riv. Emoter. Immunomatol. (Pavia),* **6**:189–200, 1959.

M164. Monier, R., and Fromageot, C. Quelques peptides résultant de l'hydrolyse partielle du lysozyme. *Biochim. Biophys. Acta,* **5**:224–227, 1950.

M165. Monier, R., and Jutisz, M. Etude de l'hydrolyse acide du lysozyme. Possibilité d'eviter la destruction du tryptophane au cours de cette hydrolyse. *Bull. Soc. Chim. Biol.,* **32**:228–233, 1950.

M166. Monodane, T., Hara, S., and Matsushima, Y. Studies on the substrate specificity of egg white lysozyme. V. subsites C and D in the active site. *J. Biochem. (Tokyo),* **72**:1175–1183, 1972.

M167. Montague, M. D. The enzyme degradation of cell walls of *Streptococcus faecalis*. *Biochim. Biophys. Acta,* **86**:588–595, 1964.

M168. Montanari, G. Determination del tasso lisozimico in soggetti pemfigomatosi prima e dopo l'ablazione di foci oro-dentali presunti attivi. *Mondo Odontostomatol. (Bologna),* **9**:679–687, 1967.

M169. Montero Rodriguez, A. Tratamiento de las corizas de los lactantes, con una asociación de antihistamínicos, un protector capilary lisozima. *Rev. Espan. Pediat.,* **21**:291–303, 1965.

M170. Monteverde, A., Uglietti, A., and Grazioli, C. Valutazione delle disponibilità della riserva granulocitaria e della resistenza meccanica leucocitaria in soggetti normali dopo somministrazione di lisozima. *Minerva Radiol., Fisioter. Radiobiol.,* **10:**173–176, 1965.

M171. Monti, P. C. Lysozyme therapy and autohemotherapy in otorhinolayrngology. *Sem. Med. (Buenos Aires),* **119:**1038–1042, 1961. (Span.)

M172. Moore, G. L., and Day, R. A. Protein conformation in solution: Cross-linking of lysozyme. *Science,* **159:**210–211, 1968.

M173. Moreno, A., and Pagliano Sassi, L. Il lisozima nella terapia dell'epatite virale. *Epatologia,* **12:**505–507, 1966.

M174. Moreno, A., and Pagliano Sassi, L. Livelli lisozimici e transminasici nel decorso dell'epatite virale. *Epatologia,* **12:**288–291, 1966.

M175. Morgan, W. T., and Riehm, J. P. The isolation and some enzymatic properties of des-arginylleucine-lysozyme. *Biochem. Biophys. Res. Commun.,* **30:**50–56, 1968.

M176. Morioka, T., Nishimura, M., and Matsumura, T. Anti-inflammatory effect of lysozyme paste on experimental gingivitis induced by various chemical mediators in guinea pig. *J. Periodontol.,* **41:**341–348, 1970.

M177. Moss, J. N., and Martin, G. J. The inhibition of lysozyme activity. *Amer. J. Dig. Dis.,* **15:**412–414, 1948.

M178. Moult, J., Eshdat, Y., and Sharon, N. The identification by X-ray crystallography of the site of attachment of an affinity label to hen egg-white lysozyme. *J. Mol. Biol.,* **75:**1–4, 1973.

M179. Mourzinn, A. N., and Souchkowa, G. Lysozyme of tears in tracoma. *Rev. Int. Trachome,* **12:**1–15, 1935.

M180. Mouton, A., and Jollès, J. On the identity of human lysozymes isolated from normal and abnormal tissues or secretions. *FEBS Lett.,* **4:**337, 1969.

M181. Mozes, E., Maron, E., Arnon, R., and Sela, M. Strain-dependent differences in the specificity of antibody responses toward lysozyme. *J. Immunol.,* **106:**862–864, 1971.

M182. Muggia, F. M., Heinemann, H. O., Farhangi, M., and Osserman, E. F. Lysozymuria and renal tubular dysfunction in monocytic and myelomonocytic leukemia. *Amer. J. Med.,* **47:**351–366, 1969.

M183. Murachi, T., Miyake, T., and Kato, K. Spectral studies on hen egg-white lysozyme modified with diisopropylphosphorofluoridate. *J. Biochem. (Tokyo),* **69:**209–217, 1971.

M184. Murachi, T., Miyake, T., and Yamasaki, N. Alkylphosphorylation of hen egg-white lysozyme by diisopropylphosphorofluoridate. *J. Biochem. (Tokyo),* **68:**239–244, 1970.

M185. Murachi, T. A general reaction of diisopropylphosphorofluoridate with proteins without direct effect on enzymic activities. *Biochim. Biophys. Acta,* **71:**239–241, 1963.

M186. Muschel, L. H., and Jackson, J. E. Activity of the antibody-complement system and lysozyme against rough gram negative organisms. *Proc. Soc. Exp. Biol. Med.,* **113:**881–884, 1963.

M187. Myrvik, Q. N. Serum and tissue lysozyme levels associated with granulomatous reactions. *Tuberculology,* **18:**91–94, 1960.

M188. Myrvik, Q. N., Leake, E. S., and Fariss, B. Lysozyme content of alveolar and peritoneal macrophages from the rabbit. *J. Immunol.,* **86:**133–136, 1961.

M189. Myrvik, Q. N., Leake, E. S., and Oshima, S. A study of macrophages and epitheloid-like cells from granulomatous (BCG-induced) lungs of rabbits. *J. Immunol.,* **89:**745–751, 1962.

M190. Myrvik, Q., and Weiser, R. S. A tuberculostatic serum substance possessing lysozyme-like properties. *Amer. Rev. Tuberc.,* **64:**669, 1951.

M191. Myrvik, Q. N., Weiser, R. S., and Agar, H. D. Lethal and cytologic effects of lysozyme on tubercle bacilli. *Amer. Rev. Tuberc.,* **67**:217–231, 1953.
M192. Myrvik, Q. N., Weiser, R. S., and Kelly, M. C. Studies on the mode of action of lysozyme on mycobacteria. *Amer. Rev. Tuberc.,* **68**:564–574, 1953.

N

N1. Nace, G. W., Suyama, T., and Iwata, T. The relationship between a lysozyme-like enzyme and frog adenocarcinoma. *Ann. N. Y. Acad. Sci.,* **126**:204–221, 1965.
N2. Nagai, H., Usui, T., Hayakawa, T., and Usui, N. Congenital nonspecific defense factors in vivo. III. Behavior of congenital non-specific defense factors in experimental streptococcal infection. *Ann. Paediat. Jap.,* **7**:321–329, 1961. (Jap.)
N3. Nagai, H., Usui, T., Okubo, Y., Tanaka, M. The levels of non-specific defense factors in maternal and umbilical cord blood. *Ann. Paediat. Jap.,* **9**:130–134, 1963. (Jap.)
N4. Nagasawa, K., Nishizaki, S., and Yokota, I. On the National Institute of Hygienic Sciences Standard "lysozyme standard." *Bull. Nat. Inst. Hyg. Sci., (Tokyo),* **88**:84–85, 1970. (Jap.)
N5. Nagase, H. A study on the clinical significance of serum and urinary muramidase activity in leukemics. *Nagoya J. Med. Sci.,* **34**:13–26, 1971.
N6. Nagy, J., and Straub, F. B. Electrolytic reduction of disulfide bonds and biological activity of some proteins. *Acta Biochim. Biophys. Acad. Sci. Hung. (Budapest),* **4**:15–25, 1969.
N7. Naha, P. M. Intracistronic mapping of the structural gene for lysozyme in coliphage lambda. *Virology,* **29**:676–678, 1966.
N8. Naithani, V. K., and Dhar, M. M. Synthetic substitute lysozymes. *Biochem. Biophys. Res. Commun.,* **29**:368–372, 1967.
N9. Najarian, J. S., Noble, R. E., Braby, P., and Brainerd, H. D. Lysozyme determination as a measure of rejection of kidney homotransplants. *Surg. Forum,* **16**:258–260, 1965.
N10. Nakae, Y., Ikeda, K., Azuma, T., and Hamaguchi, K. Circular dichroism of hen egg-white lysozyme modified with N-acetylimidazole. *J. Biochem. (Tokyo),* **72**:1155–1162, 1972.
N11. Nakamura, K. Plasma lysozyme in the childhood. *Acta Paediat. Jap.,* **74**:751–760, 1970. (Jap.)
N12. Nakamura, M. Studies on the biological properties of Mycobacterium leprae murium. 4. Effect of the administration of various kinds of enzymes on the virulence of Mycobacterium leprae murium. *Jap. J. Bacteriol.,* **17**:387–390, 1962. (Jap.)
N13. Nakamura, O. Actions of lysozyme. *Wien. Klin. Wochenschr.,* **36**:322–323, 1923. (Ger.)
N14. Nakamura, O. Ueber Lysozymwirkung. *Z. Immunitäetsforsch. Exp. Ther., 1,* **38**:425–447, 1923.
N15. Nakanishi, M., Tsuboi, M., and Ikegami, A. Fluctuation of the lysozyme structure. II. Effects of temperature and binding of inhibitors. *J. Mol. Biol.,* **75**:673–682, 1973.
N16. Nakanishi, M., Tsuboi, M., and Ikegami, A. Fluctuation of the lysozyme structure. *J. Mol. Biol.,* **70**:351–361, 1972.
N17. Nakano, M., Okitsu, F., Okumura, Y. *et al.* On changes in the enzyme activities of the host peritoneal cells in the immune process in experimental typhoid. *Jap. J. Bacteriol.,* **19**:110–116, 1964. (Jap.)
N17a. Nakaya, K., Horinishi, H., and Shibata, K. States of amino acid residues in proteins. XIII. Monochloroquinone as a new reagent for discrimination of amino groups. *J. Biochem. (Tokyo),* **63**:337–344, 1967.

N18. Nakazawa, S., Itagaki, M., Yokota, T., Otani, Y., Miwa, M., Onitake, J., Nakay-ama, T., and Fusaoka, N. Fundamental studies on the antibiotic action of lysozyme. *Jap. J. Antibiot., Ser. B,* **19**:34–47, 1966. (Jap.)

N19. Nakazawa, S., Yamamoto, A., Yokota, Y., Mitsutake, T., Miyoshi, H., and Yamashina, Y. Basic studies on the antibiotic activity of lysozyme. II. Combined effects with aminobenzyl penicillin. *Jap. J. Antibiot.,* **21**:10–14, 1968. (Jap.)

N20. Narabayaski, S. Transfer of lysozyme chloride into nasal, paranasal mucosa and surrounding tissues. *J. Otolaryngol. Jap.,* **73**:473–484, 1970. (Jap.)

N21. Narducci, U. Il potere lisozimico del liquido lacrimale del neonato; possibili rapporti fra lisozima lacrimale e profilassi oculare neonatale. *Minerva Ginecol.,* **6**:117–122, 1954.

N21a. Narita, K. Reaction of anhydrous formic acid with proteins. *J. Amer. Chem. Soc.,* **81**: 1751–1756, 1959.

N22. Natanson, M. S. Lysozyme in therapy of blepharitis. *Vestn. Oftalmol'.,* **14**:22–23, 1939.

N23. Natanson, D. M. Lysozyme in ophthalmology. *Vestnik Oftalmol.,* **11**:501–506, 1937.

N24. Nemes, J. L., and Wheatcroft, M. G. Action of salivary lysozyme on *Micrococcus lysodeikticus. Oral Surg., Oral Med. Oral Pathol.,* **5**:653–658, 1952.

N25. Némethy, G., Phillips, D. C., Leach, S. J., and Scheraga, H. A. A second right-handed helical structure with the parameters of the Pauling-Corey alpha-helix. *Nature (London),* **214**:363–365, 1967.

N26. Nermut, M. V., and Svoboda, A. Reversion of spheroplasts produced by lysozyme into rods in *Proteus vulgaris. Nature (London),* **193**:396–397, 1962.

N27. Neter, E. Action of sulfamido compounds upon *M. lysodeikticus* and lytic and bactericidal activities of lysozyme. *Proc. Soc. Exp. Biol. Med.,* **48**:106–109, 1941.

N28. Neu, H. C., Dreyfus, J., 3rd., and Canfield, R. E. Effect of human lysozyme on gram-positive and gram-negative bacteria. *Antimicrob. Ag. Chemother.,* **8**:442–444, 1968.

N28a. Neuberger, A., Davies, R. C. Modification of lysine and arginine residues of lysozyme and the effect on enzymatic activity. *Biochim. Biophys. Acta,* **178**:306–317, 1969.

N29. Neuberger, A., and Wilson, B. M. Inhibition of lysozyme by derivatives of D-glucosamine. I. *Biochim. Biophys. Acta,* **147**:473–486, 1967.

N30. Neuberger, A., and Wilson, B. M. Inhibition of lysozyme by N-acyl-D-glucosamine derivatives. *Nature (London),* **215**:524–525, 1967.

N31. Neuman, M. Le lysozyme. Acquisitions récentes. *Concours Med.,* **86**:6438–6444, 1964.

N32. Neumann, H., Goldberger, R. F., and Sela, M. Interaction of phosphorothioate with the disulfide bonds of ribonuclease and lysozyme. *J. Biol. Chem.,* **239**:1536–1540, 1964.

N33. Neurath, A. R., and Brunner, R. Fractionation of proteins with different isoelectric points by Rivanol. *Experientia,* **25**:668–671, 1969.

N34. Neville, W. M., and Eyring, H. Hydrostatic pressure and ionic strength effects on the kinetics of lysozyme. *Proc. Nat. Acad. Sci. U. S.,* **69**:2417–2419, 1972.

N35. Nevin, T. A., and Guest, W. J. Complement and lysozyme requirements for spirochetolysis in guinea pig serum. *J. Bacteriol.,* **94**:1388–1393, 1967.

N36. Nichol, L. W., Ogston, A. G., and Winzor, D. J. Evaluation of gel filtration data on systems interacting chemically and physically. *Arch. Biochem. Biophys.,* **121**:517–521, 1967.

N37. Nickel, W. F., Jr., Gordon, G. M., and Andrus, W. DeW. Studies on lysozyme as an etiologic agent in ulcerative colitis. *Gastroenterology,* **17**:406–408, 1951.

N38. Niemann, B., Hermann, J., and Jollès, J. Structures chimiques des moités N-terminales de deux lysozymes de blanc d'oeuf de cane. *Bull. Soc. Chim. Biol.*, **50**:923–924, 1968.

N39. Nigro, N., Bonenti, G., and Benso, L. Variazioni del tasso serico di fosfoesosoisomerasi e di leucino-amino-peptidasi nel corso di trattamento lisozimico delle epatiti virali dell'infanzia. *Epatologia*, **12**:318–325, 1966.

N40. Ninni, M., and Barbui, T. Recenti progressi in tema di terapia della epatite virale con particolare riguardo all'impiego del lisozima. *Epatologia*, **12**:444–456, 1966.

N41. Nishihara, K., Isoda, K., and Homma, M. Fecal bacterial flora in the use of dry milk treated with lysozyme. *Acta Paediat. Jap.*, **71**:95–102, 1967. (Jap.)

N42. Niutta, R., Paternò, M., Urcioli, A., and Elifani, G. L'associazione metaciclina-lisozima nella cura della brucellosi umana. *G. Mal. Infet. Parassit.*, **21**:131–133, 1969.

N43. Nizegorodcew, M., Kaminska-Mecner, M., and Deler, A. Comparison between the effect of alcohol extract of chicken eggs and the effect of lysozyme on acid-fast bacilli. *Gruzlica Choroby Pluc*, **40**:485–488, 1972. (Pol.)

N44. Nobile, F., and Andrei, A. Intra-arterial therapy with lysozyme in varicose ulcers of the lower extremity. *Atti Accad. Fisiocrit. Siena, Sezione Medico-Fisica*, **8**:648–650, 1960. (Ital.)

N45. Nobile, F., and Andrei, A. Anatomo-pathological study of guinea pigs treated with parenteral injections of lysozyme. *Atti Accad. Fisiocrit. Siena*, **8**:645–647, 1960. (Ital.)

N46. Noble, R. E., and Brainerd, H. D. Urine and serum lysozyme alterations in a case of acute renal failure. *J. Urol.*, **96**:852–853, 1966.

N47. Noble, R. E., and Fudenberg, H. D. Leucocyte lysozyme activity in myelocytic leukemia. *Blood*, **30**:465–473, 1967.

N48. Noble, R. E., and Koch-Weser, D. *In vitro* inhibition of leukocyte uptake of radioactive endotoxin by components of normal serum. *Proc. Soc. Exp. Biol. Med.*, **121**:541–545, 1966.

N49. Noble, R. E., Najarian, J. S., and Brainerd, H. D. Urine and serum lysozyme measurement in renal homotransplantation. *Proc. Soc. Exp. Biol. Med.*, **120**:737–740, 1965.

N50. Noguer, S., Borrell, J., and Noguer, D. S. Treatment of recurrent herpes with antipoliomyelitis vaccine and lysozyme. First trials. *Actas Dermo-Sifiliogr.*, **51**:25–30, 1960. (Span.)

N51. Noller, E. C., and Hartsell, S. E. Bacteriolysis of *Enterobacteriaceae*. II. Pre- and co-lytic treatments potentiating the action of lysozyme. *J. Bacteriol.*, **81**:492–499, 1961.

N52. Noller, E. C., and Hartsell, S. E. Bacteriolysis of *Enterobacteriaceae*. 1. Lysis by four lytic systems utilizing lysozyme. *J. Bacteriol.*, **81**:482–491, 1961.

N53. Nomura, M., Hosoda, J., and Nishimura, S. Enzyme formation in lysozyme lysate of *Bacillus subtilis*. *Biochim. Biophys. Acta*, **28**:161–167, 1958.

N54. Nord, C.-E., Modéer, T., Söder, P.-O., and Bergstrom, J. Enzyme activities in experimental gingivitis in man. *Scand. J. Dent. Res. (Copenhagen)*, **79**:510–514, 1971.

N55. Nord, C. E., Söder, P. O., and Lindqvist, L. Lysozyme activity of dental plaque material. *Sv. Tandlaek. Tidskri*, **62**:493–500, 1969.

N56. Nord, C. E., and Wadström, T. Chitinase activity and substrate specificity of three bacteriolytic endo-beta-N-acetylmuramidases and one endo-beta-N-acetylglucosaminidase. *Acta Chem. Scand.*, **26**:653–660, 1972.

N57. Norregaard, S. Lysozyme. *Nord. Med.*, **44**:1923–1925, 1950.

N57a. North, A. C. T., and Phillips, D. C. X-ray studies of crystalline proteins. *Prog. Biophys. Mol. Biol.* **19** (Pt. 1): 5–132, 1969.

N58. Nossel, H. L., Rubin, H., Drillings, M., and Hsieh, R. Inhibition of Hageman factor activation. *J. Clin. Invest.*, **47**:1172–1180, 1968.
N59. Novikova, V. A., Razzhivina, R. V., and Shakurova, T. K. Lysozyme content in the blood serum of children with acute pneumonia. *Vop. Okhrany Mater. Detstva (Moscow)*, **13**:84–85, 1968. (Russ.)
N60. Novokreschchenov, M. V. Apropos of the nervous regulation of the lysozyme activity of the blood serum. *Patol. Fiziol. Eksp. Ter. (Moscow)*, **8**:80–81, 1964. (Russ.)
N61. Nunnari, A., Belfiore, F., and Calafato, M. L'attività lisozimica dei leucociti normali e leucemici. *Boll. Soc. Ital. Biol. Sper.*, **39**:714–718, 1963.
N62. Nunziata, B., and Rolando, P. Rilievi sull'introduzione per via intradermica nella cavia di linfa vaccinica associata a lisozima e rispettivamente a prednisone. *Pediatria (Naples)*, **72**:980–988, 1964.

O

O1. Ocada, Y., Amagase, S., and Tsugita, A. Frameshift mutation in the lysozyme gene of bacteriophage T4: Demonstration of the insertion of five bases, and a summary of *in vivo* codons and lysozyme activities. *J. Mol. Biol.*, **54**:219–246,1970.
O1a. Offord, R. E. Protection of peptides of biological origin for use as intermediates in the chemical synthesis of proteins. *Nature* **221**:37–40, 1969.
O2. Ogasahara, K., and Hamaguchi, K. Structure of lysozyme. XII. Effect of pH on the stability of lysozyme. *J. Biochem. (Tokyo)*, **61**:199–210, 1967.
O3. Ogawa, H., and Miyazaki, H. Immunochemical studies on the human skin lysozyme. *J. Invest. Dermatol.*, **58**:59–62, 1972.
O4. Ogawa, H., Miyazaki, H., and Kimura, M. Isolation and characterization of human skin lysozyme. *J. Invest. Dermatol.*, **57**:111–116, 1971.
O5. Ogreba, V. I., Vasil'ev, N. V., and Nemirovskaia, L. Ia. On the method of determination of blood serum lysozyme activity. *Lab. Delo*, **2**:77–79, 1969. (Russ.)
O6. Ohashi, N. A study of urinary and serum lysozyme activity in the urological field. *Jap. J. Urol.*, **60**:1033–1052, 1969. (Jap.)
O7. Ohta, H., and Nagase, H. Lysozyme activities in various leukemias—with special reference to monocytic leukemia. *Saishin Igaku*, **26**:2386–2393, 1971. (Jap.)
O8. Ohta, H., and Nagase, H. Serum, urine, and leukocyte muramidase (lysozyme) activity in monocytic leukemia and other hematologic malignancies. *Acta Haematol. Japo.*, **34**:498–512, 1971.
O9. Ohta, H., and Osserman, E. F. Hydrolytic enzymes of human blood monocytes and neutrophils with special reference to lysozyme (muramidase). *Tohoku J. Exp. Med.*, **107**:229–240, 1972.
O10. Ohta, H., and Nagase, H. Serial estimation of serum, urine, and leukocyte muramidase (lysozyme) in monocytic leukemia. *Acta Haematol.*, **46**:257–266, 1971.
O10a. Ohta, Y., Gill, T. J. III, and Leung, C. S. Volume changes accompanying the antibody-antigen reaction. *Biochemistry*, **9**:2708–2713, 1970.
O11. Oka, T., and Schimke, R. T. Progesterone antagonism of estrogen-induced cytodifferentiation in chick oviduct. *Science*, **163**:83–85, 1969.
O12. Oka, M., and Seppää, O. Muramidase (lysozyme) in joint fluid and serum of rheumatic patients. *Acta Rheumatol. Scand.*, **16**:223–230, 1970.
O13. Okabe, N., and Takagi, T. Effect of labelling with dansyl group on the renaturation of lysozyme. *Biochim. Biophys. Acta*, **229**:484–495, 1971.
O14. Okada, Y., Streisinger, G., Emrich, J., Newton, J., Tsugita, A., and Inouye, M.

Frame shift mutations near the beginning of the lysozyme gene of bacteriophage T4. *Science,* **162**:807–808, 1968.

O15. Okada, Y., Streisinger, G., Emrich, J., Tsugita, A., and Inouye, M. The lysozyme of a triple frame-shift mutant strain of bacteriophage. *J. Mol. Biol.,* **40**:299–304, 1969.

O16. Okada, Y., Streisinger, G., Owen, J., Newton, J., Tsugita, A., and Inouye, M. Molecular basis of a mutational hot spot in the lysozyme gene of bacteriophage T4. *Nature (London),* **236**:338–341, 1972.

O17. Okada, Y., Terzaghi, E., Streisinger, G., Emrich, J., Inouye, M., and Tsugita, A. A frame-shift mutation involving the addition of two base pairs in the lysozyme gene of Phage T4. *Proc. Nat. Acad. Sci. U.S.,* **56**:1692–1698, 1966.

O18. Okubo, Y. Studies on the localization of serum lysozyme and properdin, using cold ethanol fractionation, method 10 of Cohn. *Ann. Paediat. Jap.,* **7**:195–210, 1961. (Jap.)

O19. Olivetti, L. Il tasso lisozimico nel siero e nel liquido bolla nelle dermatosi piogeniche. *G. Ital. Dermatol. Sifilol.,* **91**:386–398, 1951.

O20. Olsson, I., and Venge, P. Cationic proteins of human granulocytes. I. Isolation of the cationic proteins from the granules of leukaemic myeloid cells. *Scand. J. Haematol.,* **9**:204, 1972.

O21. Oram, J. D., and Reiter, B. Phage-associated lysins affecting group N and group D streptococci. *J. Gen. Microbiol.,* **40**:57–70, 1965.

O22. Orecchio, F., Amato, A., and Misefari, A. Modificazioni del midollo osseo di ratto per trattamento intensivo con cloruro di lisozima. *Boll. Soc. Ital. Biol. Sper.,* **42**:1261–1265, 1966.

O23. Orecchio, F., Amato, A., and Misefari, A. Sul rapporti tra lisozima e distribuzione dei leucociti nei tessuti e nel sangue periferico del ratto. *Boll. Soc. Ital. Biol. Sper.,* **42**:1037–1040, 1966.

O24. O'Reilly, J. M., and Karasz, F. E. Heat of denaturation of lysozyme. *Biopolymers,* **9**:1429–1435, 1970.

O25. Orfei, Z. Ricerche su di un eventuale azione inattivante del lisozima sul virus erpetico e vaccinico. *Rend. Ist. Super. Sanita,* **18**:426–429, 1955.

O26. Orfila, G., and Rampini, C. Quelques aspects des variations physio-pathologiques du lysozyme. *Exposes Annu. Biochim. Med.,* **27**:155–159, 1966.

O27. Orru, A. First trials of the use of lysozyme in the treatment of lichen ruber planus. *Rass. Med. Sarda,* **64**:589–593, 1962.

O28. Orzalesi, F. Primi tentativi di terapia di cheratiti erpetiche con lisozima. *G. Ital. Oftalmol.,* **6**:210–216, 1953.

O29. Orzalesi, F., and Miglior, M. Neutralisation "in vitro" du virus herpétique par le lisozyme. *Bull. Mem. Soc. Fr. Ophtalmol.,* **65**:433–437, 1952.

O30. Osawa, T., and Nakazawa, Y. Lysozyme substrates. Chemical synthesis of p-Nitrophenyl 0-(2-Acetamido-2-Deoxy-β-D-Glucopyranosyl)-(1 → 4)-0-(2-Acetamido-2-Deoxy-β-D-Glucopyranosyl)-(1 → 4)-2-Acetamido-2-Deoxy-β-D-Glucopyranoside and its reaction with lysozyme. *Biochim. Biophys. Acta,* **130**:56–63, 1966.

O30a. Osawa, T. Lysozyme substrates. Synthesis of p-Nitrophenyl 2-acetamido-4-0(2-acetamido-2-deoxy-Beta-D-glupyranosyl)-2-deoxy-Beta-D-glucopyranoside and its Beta-D-(1 to 6)isomer. *Carbohydr. Res.* **1**:435–443, 1965–6.

O31. Oshima, S., Myrvik, Q. N., and Leake, E. S. The demonstration of lysozyme as a dominant tuberculostatic factor in extracts of granulomatous lungs. *Brit. J. Exp. Pathol.,* **42**:138–144, 1961.

O32. Oshima, S., Myrvik, Q. N., and Leake, E. S. Studies on the antitubercular substance lysozyme in extract of extravasated lung cells of the sensitized rabbit. *Kyoto Daigaku Kekkaku Kenkyusho Kiyo,* **9**:154–163, 1961. (Jap.)

O33. Osipov, V. I., Maksimov, V. I., and Kaversneva, E. D. Properties of lysozyme and ribonuclease A, irradiated in the dry state, following treatment with beta-mercaptoethylamine. *Radiobiologiya,* **12**:26–32, 1972. (Russ.)

O34. Osserman, E. F. Monocytic and monomyeloctyic leukaemia with increased serum and urine lysozyme as a late complication in plasma cell myeloma. *Brit. Med. J.,* **2**:327, 1971.

O35. Osserman, E. F. Clinical and biochemical studies of plasmacytic and monocytic dyscrasias and their interrelationships. *Trans. & Stud. Coll. Physicians Philadelphia,* 4, **36**:134–146, 1969.

O36. Osserman, E. F. Association between plasmacytic and monocytic dyscrasias in man: Clinical and biochemical studies. *Gamma Globulins, Proc. Nobel Symp., 3rd, 1967,* p. 573, 1967.

O37. Osserman, E. F. Crystallization of human lysozyme. *Science,* **155**:1536–1537, 1967.

O38. Osserman, E. F. Lysozymuria in renal and non-renal disease. *In* "Proteins in Normal and Pathological Urine" (Y. Manuel, J. P. Revillard, and H. Betuel, eds.), pp. 260–270. Karger, Basel, 1970.

O39. Osserman, E. F., Cole, S. J., Swan, I. D. A., and Blake, C. C. F. Preliminary crystallographic data on human lysozyme. *J. Mol. Biol.,* **46**:211–212, 1969.

O40. Osserman, E. F., and Lawlor, D. P. Serum and urinary lysozyme (muramidase) in monocytic and monomyelocytic leukemia. *J. Exp. Med.,* **124**:921–952, 1966.

O41. Ota, Y., Hibino, Y., Asaba, K., Sugiura, K., and Samejima, T. On the conformational change of hen egg-white lysozyme by chemical scission. *Biochim. Biophys. Acta,* **236**:802–805, 1971.

O42. Otson, R., Reyes-Zamora, C., Tang, J. Y., and Tsai, C. S. Interaction of beta-aryl di-N-acetylchitobiosides with lysozyme. *Can. J. Biochem.,* **51**:1–6, 1973.

O43. Ottolenghi, G., and Cavalli, D. Terapia delle leucopenie iatrogene da associazione chemio-radioterapica mediante lisozima per via orale. *Minerva Radiol., Fisioter. Radiobiol.,* **10**:181–183, 1965.

O44. Ovchinnikov, N. M., and Podvinska, Ia. Ia. Effect of lysozyme and temperature on acceleration of the *Trepanema pallidum* immobilization test. *Vestn. Dermatol., Venerol.,* **42**:67–69, 1968. (Russ.)

O45. Oyake, H. Salivary lysozyme and its relation to angular stomatitis in arboflavinosis. III. Morphological changes of epithelial cells of lips after incubation with lysozyme. *Tohoku J. Exp. Med.,* **75**:234–237, 1961.

O46. Oyake, H. Salivary lysozyme: Its relation to angular stomatitis in ariboflavinosis. Part II. Lytic action of lysozyme upon *Micrococcus lysodeikticus* cultured with or without riboflavin. *Tohoku J. Exp. Med.,* **75**:197–200, 1961.

O47. Oyake, H. Salivary lysozyme: Its relation to angular stomatitis in ariboflavinosis. I. Salivary lysozyme activity in ariboflavinosis. *Tohoku J. Exp. Med.,* **75**:190–196, 1961.

O48. Ozaki, M., Higashi, Y., Amano, T., Miyama, A., and Kashiba, S. Nature of spheroplasting agent, leucozyme C, in guinea pig leucocytes. *Biken J.,* **8**:175–188, 1965.

P

P1. Padayatty, J. D. Size of polysome involved in the synthesis of 54 Phage lysozyme. *Indian J. Biochem. & Biophys.,* **9**:210–212, 1972.

P2. Padgett, G. A., and Hirsch, J. G. Lysozyme: Its absence in tears and leukocytes of cattle. *Aust. J. Exp. Biol. Med. Sci.,* **45**:569–570, 1967.

P3. Pagnes, P., and Galla, F. Test di immobilizzazione del treponema pallido (T.P.I. test) e lisozima. *Minerva Dermatol., 40*:286–291, 1965.

P4. Paikina, S. Sh. The stimulating effect of casein on lysozyme. *Z. Mikrobiol., Epidemiol., Immunitäetsforsch.,* No. 2:100–104, 1941.

P5. Palmer, K. J., Ballantyne, M., and Galvin, J. A. The molecular weight of lysozyme determined by the X-ray diffraction method. *J. Amer. Chem. Soc., 70*:906–908, 1948.

P6. Palmiter, R. D. Regulation of protein synthesis in chick oviduct. 1. Independent regulation of ovalbumin, conalbumin, ovomucoid, and lysozyme induction. *J. Biol. Chem., 247*:6450–6461, 1972.

P7. Palmiter, R. D., and Gutman, G. A. Appendix. Fluorescent antibody localization of ovalbumin, conalbumin, ovomucoid, and lysozyme in chick oviduct magnum. *J. Biol. Chem., 247*:6459–6461, 1972.

P8. Panconesi, E., Vallecchi, C., and Giannotti, B. Cutaneous reactivity to 48/80 in subjects treated with lysozyme. *Rass. Dermatol. Sifilogra., 13*:52–59, 1960.

P9. Panebianco, G., and Baravelli, P. Variazioni del contenuto in lisozima del muco nasale nei laringectomizzati. *Oto-Rino-Larinool. Ital., 27*:211–222, 1959.

P10. Panebianco, G., and Motta, G. Il lisozima della saliva parotidea e sottomascellare; modificazione indotte dalla pilocarpina e dalla sezione della corda del timpano. *Oto-Rino-Laringol. Ital., 24*:350–366, 1956.

P11. Paris, J. Recent data on the role of lysozyme in digestive pathology. *Echo Med. Nord., 21*:305–315, 1950.

P12. Parrot, J.-L., and Nicot, G. Antihistaminic action of lysozyme. *Nature (London), 197*:496, 1963.

P13. Parrot, J.-L., Nicot, G., Laborde, C., and Canu, P. Inhibition de diverses actions biologiques de l'histamine par le lysozyme. *J. Physiol. (Paris), 54*:739–748, 1962.

P14. Parry, R. M., Jr., Chandan, R. C., and Shahani, K. M. A rapid and sensitive assay of muramidase. *Proc. Soc. Exp. Biol. Med., 119*:384–386, 1965.

P15. Parry, R. M., Jr., Chandan, R. C., and Shahani, K. M. Isolation and characterization of human milk lysozyme. *Arch. Biochem. Biophys., 130*:59–65, 1969.

P16. Parsons, S. M., and Raftery, M. A. Ionization behavior of the cleft carboxyls in lysozyme-substrate complexes. *Biochemistry, 11*:1633–1638. 1972.

P17. Parsons, S. M., and Raftery, M. A. Ionization behavior of the catalytic carboxyls of lysozyme. Effects of temperature. *Biochemistry, 11*:1630–1633, 1972.

P18. Parsons, S. M., and Raftery, M. A. Ionization behavior of the catalytic carboxyls of lysozyme. Effects of ionic strength. *Biochemistry, 11*:1623–1629, 1972.

P19. Parsons, S. M., and Raftery, M. A. Ionization behavior of the catalytic carboxyls of lysozyme. *Biochem. Biophys. Res. Commun., 41*:45–49, 1970.

P20. Parsons, S. M., and Raftery, M. A. The identification of aspartic acid residue 52 as being critical to lysozyme activity. *Biochemistry, 8*:4199–4205, 1969.

P21. Parsons, S. M., Jao, L., Dahlquist, F. W., Borders, C. L., Groff, T., Racs, J., and Raftery, M. A. The nature of amino acid side chains which are critical for the activity of lysozyme. *Biochemistry, 8*:700–712, 1969.

P22. Pascual, R. S., Perillie, P. E., Sulavik, S., Donadio, J. A., Gee, J. B. L., and Finch, S. C. Lysozyme studies in sarcoidosis. *Ann. Intern. Medi., 76*:880, 1972.

P23. Pasynskii, A. G., and Kastorskaya, T. Electrophoresis of lysozyme by the Tiselius method. *C. R. Acad. Sci. Russ., 49*:504–507, 1945.

P24. Pasynskiĭ, A. G. Diffusion and molecular weight of lysozyme. *C. R. Acad. Sci. Russ., 48*:579–581, 1945.

P25. Patterson, D., Weinstein, M., Nixon, R., and Gillespie, D. Interaction of ribosomes and the cell envelope of *Escherichia coli* mediated by lysozyme. *J. Bacteriol.,* **101**:584–591, 1970.

P26. Pavone Macaluso, M., and Ardoino, L. A. Azione di un colloide elettropositivo (lisozima) sulla escrezione urinaria di acqua e di elettroliti, nel coniglio. *Boll. Soc. Ital. Biol. Sper.,* **37**:15–18, 1961.

P27. Pavone Macaluso, M., and Ciofalo, G. Aumento del numero degli eosinofili circulanti dopo somministrazione di proteine basiche (lisozima, protamina, tripsina e chimotripsina) nell' uomo. *Boll. Soc. Ital. Biol. Sper.,* **39**:150–152, 1963.

P28. Pavoni, P., Semprebene, L., Sedati, P., and Mancini, L. Comparative distribution of protein radioactivity in the body fluids and tissues of rabbits treated with lysozyme labeled with I-131, with human serum albumin and with elemental I-131. *Rass. Fisiopatol. Clin. Ter.,* **34**:67–80, 1962. (Ital.)

P29. Pecht, I., Maron, E., Arnon, R., and Sela, M. Specific excitation energy transfer from antibodies to dansyl-labeled antigen. Studies with the "loop" peptide of hen egg-white lysozyme. *Eur. J. Biochem.,* **19**:368–371, 1971.

P30. Pellegrini, R., and Vertova, P. Synergism between penicillins and lysozyme. Data which led to the synthesis of a new antibiotic salt: Lysozyme ampicillinate. Part I. *Arzneim. Forsch.,* **19**:110–112, 1969.

P31. Pellegrini, R., and Vertova, P. Synergism beyween penicillins and lysozyme. Data which led to the synthesis of a new antibiotic salt: Lysozyme ampicillinate. Part II. *Arzneim.-Forsch.,* **19**:149–153, 1969.

P32. Pellegrini, R., and Vertova, P. Synergism between penicillins and lysozyme. Data which led to the synthesis of a new antibiotic salt: Lysozyme ampicillinate. Part 3. *Arzneim.-Forsch.,* **19**:375–379, 1969.

P33. Pelletier, M., Delaunay, A., and Bazin, S. Sur l'agglutination bactérienne produite par un complexe polypeptidique extrait du thymus de veau avec ou sans tratiement pré alable par le lysozyme, la trypsine, le chlorure de calcium ou le polypeptide P II. *Pathol. Biol.,* **13**:46–53, 1965.

P34. Pelzer, H. The chemical structure of two mucopeptides released from *Escherichia coli* B cell walls by lysozyme. *Biochim. Biophys. Acta,* **63**:229–234, 1962.

P35. Pène, J. J. A sensitive assay for egg-white and bacteriophage-induced lysozyme. *Biochem. Biophys. Res. Commun.,* **28**:365–373, 1967.

P36. Pentimalli, L. Attività lisozimica del siero de sanque del bambino. *Arch. Ital. Pediat.,* **13**:93–109, 1949.

P37. Penrose, L. S. Lysozyme content of saliva in psychotics. *Lancet,* **2**:689–699, 1930.

P38. Penrose, M., and Quastel, J. H. Cell structure and cell activity. *Proc. Roy. Soc., Ser. B,* **107**:168–181, 1930.

P39. Penzikova, G. A., and Mikhaĭlova, G. R. Lysozyme-induced lysis of *Actinomyces fradiae* cells. *Mikrobiologiya,* **32**:465–470, 1963. (Russ.)

P40. Perillie, P. E., and Finch, S. C. Lysozyme in leukemia. *Med. Clin. N. Amer.,* **47**:395–407, 1973.

P41. Perillie, P. E., and Finch, S. C. Muramidase studies in Philadelphia-chromosome-positive and chromosome-negative chronic granulocytic leukemia. *N. Engl. J. Med.,* **283**:456–459, 1970.

P42. Perillie, P. E., Kaplan, S. S., and Finch, S. C. Significance of changes in serum muramidase activity in megaloblastic anemia. *N. Engl. J. Med.,* **277**:10–12, 1967.

P43. Perillie, P. E., Kaplan, S. S., Lefkowitz, E., Rogaway, W., and Finch, S. Studies of muramidase (lysozyme) in leukemia. *J. Amer. Med. Ass.,* **203**:317–322, 1968.

P44. Périn, J.-P., and Jollès, P. The lysozyme from *Nephthys hombergi* (Annelid). *Biochim. Biophys. Acta,* **263**:683–689, 1972.

P45. Perin, J.-P., and Jollès, P. Etude comparée des lysozymes de leucocytes de sujets sains et de malades atteints de leucémie myéloïde chronique. *Clin. Chim. Acta,* **42:**77–84, 1972.

P46. Periti, P. F., Quagliarotti, G., and Liquori, A. M. Recognition of alpha-helical segments in proteins of known primary structure. *J. Mol. Biol.,* **24:**313–322, 1967.

P47. Perkins, H. R. Inhibition of lysozyme by positively charged groups in the mucopeptide of the bacterial cell wall. *Proc. Roy. Soc., Ser. B,* **167:**443–445, 1967.

P48. Perkins, H. R. Substances reacting as hexosamine and as N-acetyl-hexosamine liberated from bacterial cell walls by lysozyme. *Biochem. J.,* **74:**186–192, 1960.

P49. Perkins, H. R. The structure of a disaccharide liberated by lysozyme from the cell walls of *Micrococcus lysodeikticus. Biochem. J.,* **74:**182–186, 1960.

P50. Perkins, H. R. The action of hot formamide on bacterial cell walls. *Biochem. J.,* **95:**876–882, 1965.

P51. Perkins, H. R., and Allison, A. C. Cell-wall constituents of rickettsiae and psittacosis-lymphogranuloma organisms. *J. Gen. Microbiol.,* **30:**469–480, 1963.

P52. Perri, G. C., Cappuccino, J. G., Faulk, M., Mellors, J., and Stock, C. C. Variations of the content of lysozyme in normal rats and rats bearing Jensen sarcoma following surgery. *Cancer Res.,* **23:**431–435, 1963.

P53. Perri, G. C., and Faulk, M. Lysozyme levels in the kidneys of cancer patients. *Proc. Soc. Exp. Biol. Med.,* **113:**245–247, 1963.

P54. Perri, G. C., Faulk, M., Mellors, J., and Stock, C. C. Crystallization of a basic protein (lysozyme) from kidneys of tumour-bearing rats (Jensen sarcoma). *Nature (London),* **193:**649–651, 1962.

P55. Perri, G. C., Faulk, M., and Money, W. L. Thyroid function and kidney muramidase. *Proc. Soc. Exp. Biol. Med.,* **115:**185–188, 1964.

P56. Perri, G. C., Faulk, M., Shapiro, E., and Money, W. L. Role of the kidney in accumulation of egg white muramidase in experimental animals. *Proc. Soc. Exp. Biol. Med.,* **115:**189–192, 1964.

P56a. Person, P. and Fine, A. S. Reversible inhibitions of cytochrome system components by macromolecular polyions. *Arch. Biochem. Biophys.,* **94:**392–404, 1961.

P57. Perugini, S. L'action du lysozyme sur la coagulation du sang. *Presse Med.,* **65:**719–720, 1957.

P58. Perugini, S., Gobbi, F., and Ghisleri, G. Lisozima e coagulazione del sangue. II. Ricerche clinico-terapeutiche sull' attività antiemorragica del lisozima. *Haematologica,* **42:**855–877, 1957.

P59. Perugini, S., Gobbi, F., and Ghisleri, G. Lisozima e coagulazione del sangue. I. Ricerche sperimentali sul meccanismo dell'azione coagulante del lisozima. *Haematologica,* **42:**831–854, 1957.

P60. Perusi, A., and Falcone, F. Obsservazioni sui valori del lisozima sinoviale, serico ed urinario in alcune artropatie. *Arch. Ortop.,* **65:**249–254, 1952.

P60a. Pessen, H., Kumosinski, T. F., and Timasheff, S. N. The use of small-angle X-ray scattering to determine protein conformation. *J. Agric. Food Chem.* **19:**698–702, 1971.

P61. Peters, J. H. Immunogenic and antigenic modification of trachoma Bedsoniae. *Amer. J. Ophthalmol.,* **63:**1506–1512, 1967.

P62. Peterson, R. G., and Hartsell, S. E. Lysozyme spectrum of the gram-negative bacteria. *J. Infec. Dis.,* **96:**75–81, 1955.

P63. Petit, J.-F., and Jollès, P. Purification and analysis of human saliva lysozyme. *Nature (London),* **200:**168–169, 1963.

P64. Petit, J.-F., Panigel, M., and Jollès, P. Purification et analyse d'un lysozyme extrait du placenta humaine. *Bull. Soc. Chim. Biol.,* **45:**211–217, 1963.

P65. Petrakis, N. L., Doherty, M., Lee, R. E., Smith, S. C., and Page, N. L. Demonstration and implications of lysozyme and immunoglobulins in human ear wax. *Nature (London)*, **229**:119–120, 1971.

P66. Pette, J. Le lysozyme. Ulilisation thérapeutique et prophylactique. *Inform. Den.*, **48**:1245–1248, 1966.

P67. Phillips, D. C. On the stereochemical basis of enzyme action: Lessons from lysozyme. *Harvey Lect.*, **66**:135–160, 1972.

P68. Phillips, D. C. The hen egg-white lysozyme molecule. *Proc. Nat. Acad. Sci. U.S.*, **57**:484–495, 1967.

P69. Phillips, D. C. The three-dimensional structure of an enzyme molecule. *Sci. Amer.*, **215**:78–90, 1966.

P70. Phillips, D. C. The structure and function of lysozyme. *Proc. Roy. Inst. Gt. Brit.*, **40**:530, 1965.

P71. Phillips, G. O., Power, D. M., Robinson, C., and Davies, J. V. Ion binding of penicillins to proteins. *Biochim. Biophys. Acta*, **215**:491–502, 1970.

P72. Piazza, B., and Lattuca, C. Influenza del lisozima e della protamina sulla funzione renale. Ricerche sperimentali. *Sicil. Sanit.*, **4**:60–67, 1966.

P73. Piccaluga, A. Azione delle diverse proteine di albume d'uovo sulle Mastzellen. *Arch. "Vecchi" Anat. Patol. Med. Clin.*, **24**:1103–1113, 1956.

P74. Piliero, S. J., and Colombo, C. Action of antiinflammatory drugs on the lysozyme activity and "turbidity" of serum from rats with adjuvant arthritis or endocrine deficiency. *J. Pharmacol. Exp. Ther.*, **165**:294–299, 1969.

P75. Pinner, M., and Voldrich, M. Lysozyme; occurrence of an apparently sterile phase; relation of lysis to dissociation. *J. Infec. Dis.*, **60**:6–14, 1937.

P76. Piomelli, S., Bruzzese, L., and Schettini, L. Osservazioni sulla attività del lisozima and processo dell'emocoagulazione: Ricerche in vitro ed in vivo in soggetti normali. *Progr. Med. (Naples)*, **12**:681–685, 1956.

P77. Pipino, G., and Marchiodi, C. Influenza del lisozima nelle leucopenie da antimitotici. Osservazioni cliniche preliminari. *Minerva Radiol., Fisioter. Radiobiol.*, **10**:166–169, 1965.

P78. Pipitone, V., Russo, R., and Dailly, L. Effect of heparin and of ultraviolet rays on plasma lysozyme and on extracorporeal fractions. *Boll. Ist. Sieroter.*, **38**:450–455, 1959. (Ital.)

P79. Pipitone, V., and Russo, R. Study of the influence of urea and phenol on the enzymatic activity of lysozyme. Research "in vivo" and "in vitro." *Boll. Ist. Sieroter.*, **38**:391–399, 1959. (Ital.)

P80. Pipitone, V., Russo, R., and Dailly, L. Primi dati sul comportamento dell'attivitá lisozimica del siero in soggetti iperazotemici. *Boll. Soc. Ital. Biol. Sper.*, **35**:291–294, 1959.

P81. Pipitone, V., Russo, R., and Triggiani, G. Behavior of serum protein fractions in subjects with increased lysozyme activity of the serum. *Boll. Ist. Sieroter.*, **38**:400–412, 1959. (Ital.)

P82. Piscator, M. Proteinuria in chronic cadmium poisoning. 3. Electrophoretic and immunoelectrophoretic studies in urinary proteins from cadmium workers, with special reference to the excretion of low molecular weight proteins. *Arch. Environ. Health*, **12**:335, 1966.

P83. Piszkiewicz, D., and Bruice, T. C. The identification of histidine-15 as part of an esteratic site of hen's egg white lysozyme. *Biochemistry*, **7**:3037–3047, 1968.

P83a. Piszkiewicz, D., and Bruice, T. C. Glycoside hydrolysis. II. Intramolecular carboxyl and acetamido group catalysis in Beta-glycoside hydrolysis. *J. Amer. Chem. Soc.*, **90:** 2156–2163, 1968.

P84. Pirie, A. The effect of lysozyme on the union between a phage and the susceptible *Bacillus megatherium. Brit. J. Exp. Pathol.*, **21:**125–132, 1940.

P85. Piszkiewicz, D., and Bruice, T. C. Interaction of cellodextrins with lysozyme: The necessity of the 2-acetamido group for binding and hydrolysis. *Arch. Biochem. Biophys.*, **129:**317–320, 1969.

P86. Pivnik, L. E. On the importance of Black's reaction and the iodine test, and determining lysozyme activity in the blood serum in some skin diseases. *Vestn. Dermatol. Venerol.*, **40:**31–34, 1966. (Russ.)

P87. Pletsityĭ, D. F. Immunogenesis and non-specific factors of natural resistance. 3. Effect of active immunization on the lysozyme content of the blood in animals. *Byull, Eksp. Biol. Med.*, **55:**69–73, 1963. (Russ.)

P88. Pletsityĭ, D. F., Gorshunova, L. P. and Fidel'man, E. S. Immunogenesis and nonspecific factors of natural resistance. II. Effect of anti-rabies vaccination on the lysozyme content of human saliva and blood. *Zh. Mikrobiol., Epidemiol. Immunobiol.*, **40:**38–42, 1963. (Russ.)

P89. Pletsityĭ, D. F., and Magaéva, S. V. Sur le rôle du système limbique du cerveau dans la régulation des réactions immunologiques spécifiques et non-spécifiques. *Rev. Immunol.*, **35:**43–48, 1971.

P90. Pletsityĭ, D. F., and Mirismailov, M. I. Immunogenesis and non-specific factors of natural resistance. V. Changes in the content of lysozyme in the serum and saliva of people under the influence of typhoid-paratyphoid-tetanus vaccine. *Zh. Mikrobiol., Epidemiol. Immunobiol.*, **45:**105–108, 1968. (Russ.)

P91. Pletsityĭ, D. F., Monaenkov, A. M., Ostrovskii, Iu. B., and Boinik, P. T. Immunogenesis and non-specific factors of natural resistance. I. Effect of active immunization on the content of lysozyme in animal saliva. *Zh. Mikrobiol., Epidemiol. Immunobiol.*, **33:**112–117, 1962. (Russ.)

P92. Pletsityĭ, D. F., and Shaganov, L. N. Immunogenesis and nonspecific natural resistance factors. IV. On changes in the lysozyme content of the blood serum of horses after hyperimmunization. *Zh. Mikrobiol., Epidemiol. Immunobiol.*, **42:**19–21, 1965. (Russ.)

P93. Pokiodova, N. V., Babaian, S. S., and Ermol'eva, Z. V. Isolation of lysozyme from human placenta. *Antibiotiki (Moscow)*, **16:**456–458, 1971. (Russ.)

P94. Poli, G., and Holzknecht, E. Il lisozima nell'alimentazione de lattante. *Minerva Pediat.*, **3:**386–393, 1951.

P95. Poljak, R. J. Heavy-atom attachment to crystalline lysozyme. *J. Mol. Biol.*, **6:**244–246, 1963.

P96. Pollock, J. J., Chipman, D. M., and Sharon, N. Glycosyl transfer to acceptor saccharides catalyzed by lysozyme. *Arch. Biochem. Biophys.*, **120:**235–238, 1967.

P97. Pollock, J. J., Chipman, D. M., and Sharon, N. The active site of lysozyme: Some properties of subsites *E* and *F. Biochem. Biophys. Res. Commun.*, **28:**779–784, 1967.

P98. Pollock, J. J., and Sharon, N. Studies on the acceptor specificity of the lysozyme-catalyzed transglycosylation reaction. *Biochemistry*, **9:**3913–3925, 1970.

P99. Pollock, J. J., and Sharon, N. Formation and cleavage of $1 \rightarrow 2$ glycosidic bonds by hen's egg white lysozyme. *Biochem. Biophys. Res. Commun.*, **34:**673–680, 1969.

P100. Ponder, E. Effect of basic proteins on the adhesiveness of red cells. *Nature (London)*, **209:**307–308, 1966.

P101. Ponnuswamy, P. K., Warme, P. K., and Scheraga, H. A. Role of medium-range interactions in proteins. *Proc. Nat. Acad. Sci. U.S.*, **70:**830–833, 1973.

P102. Ponomareva, O. I. Lysozyme in surgery. *Amer. Rev. Sov. Med.*, 3:432–442, 1946.

P103. Ponomareva, O. I. Experimental studies of lysozyme. (In vivo effects on Staphylococci.) *Amer. Rev. Sov. Med.*, 3:426–431, 1946.

P104. Ponomareva, O. I. Changes of bacteria *in vivo* following application of lysozyme in therapeutic experiments. *Vestn. Mikrobiol., Epidemiol., Parazitol.*, 19:538–544, 1940.

P105. Ponomareva, O. I. Application of lysozyme in surgical practice. *Vestn. Mikrobiol., Epidemiol., Parazitol.*, 18:293–306, 1940.

P106. Pontieri, G. M., Cotrufo, M., Ciccimarra, F., and Tolone, G. Attempts to isolate C'_3 activity from pig serum. *Experientia*, 21:75–76, 1965.

P107. Pontoni, L., and Paoletti, A. Aumento del lisozima serico per azione della adrenalina. *Acta Med. Ital. Mal. Infet. Parassit.*, 4:85–90, 1949.

P108. Pontorieri, N., and Lista, A. The action of lysozyme in staphylococcal sepsis in the infant. *Sem. Med.*, 119:1049–1050, 1961. (Span.)

P109. Poortmans, J. R. Effect of exercise on the renal clearance of amylase and lysozyme in humans. *Clin. Sci.*, 43:115–120, 1972.

P110. Pop, A., Gorcea, V., Elias, A. *et al.* The action of the fluoride ion and lysozyme on the microbial flora in the dental plaque and saliva. *Stomatologia*, 18:527–533, 1971. (Rum.)

P111. Portis, S. A. Idiopathic ulcerative colitis: Newer concepts concerning its cause and management. *J. Amer. Med. Ass.*, 139:208–214, 1949.

P112. Pospíšil, L. Current state of specific syphilis serology and its future perspectives. *Cesk. Dermatol.*, 44:102–107, 1969. (Czech.)

P113. Pospíšil, L., and Jakubowski, A. Lysozymaktivität des Serums im Verlaufe von experimenteller Syphilis des Kaninchens. 1. Veränderungsdynamik der Lysozymaktivität. *Z. Immunitaetsforsch., Allerg. Klin. Immunol.*, 131:444–448, 1966.

P114. Potkin, V. E. The influence of prolonged transversely directed radial accelerations on the secretion of intestinal juice and enzymes in dogs. *Byull. Eksp. Biol. Med.*, 61:43–47, 1966. (Russ.)

P115. Powning, R. F., and Irzykiewicz, H. Effect of lysozyme on chitin oligosaccharides. *Biochim. Biophys. Acta*, 124:218–220, 1966.

P115a. Powning, R. F., and Irzykiewicz, H. Detection of chitin oligosaccharides on paper chromatograms. *J. Chromatog.* 17:621–623, 1965.

P116. Prager, E. M., Arnheim, N., Mross, G. A., and Wilson, A. C. Amino acid sequence studies on bobwhite quail egg white lysozyme. *J. Biol. Chem.*, 247:2905–2916, 1972.

P117. Prager, E. M., and Wilson, A. C. Comparison of multiple duck lysozymes. *Biochem. Genet.*, 7:269–272, 1972.

P118. Prager, E. M., and Wilson, A. C. Dependence of immunological cross-reactivity upon sequence resemblance among lysozymes. II. Comparison of precipitin and microcomplement fixation results. *J. Biol. Chem.*, 246:7010–7017, 1971.

P119. Prager, E. M., and Wilson, A. C. Dependence of immunological cross-reactivity upon sequence resemblance among lysozymes. I. Microcomplement fixation studies. *J. Biol. Chem.*, 246:5978–5989, 1971.

P120. Prager, E. M., and Wilson, A. C. Multiple lysozymes of duck egg white. *J. Biol. Chem.*, 246:523–530, 1971.

P121. Praissman, M., and Rupley, J. A. Comparison of protein structure in the crystal and in solution. III. Tritium-hydrogen exchange of lysozyme and a lysozyme-saccharide complex. *Biochemistry*, 7:2446–2450, 1968.

P122. Prasad, A. L. N., and Litwack, G. Growth and biochemical characteristics of *Micrococcus lysodeikticus*, sensitive or resistant to lysozyme. *Biochemistry*, 4:496–501, 1965.

P123. Prasad, A. L. N., and Litwack, G. Measurement of the lytic activity of lysozymes (muramidases). *Anal. Biochem.*, **6**:328–334, 1963.

P124. Prasad, A. L. N., and Litwack, G. Relationship of lysozyme resistance to carotenogenesis in *Micrococcus lysodeikticus. Biochim. Biophys. Acta*, **46**:452–456, 1961.

P125. Preiss, J. W. Complexing of lysozyme with poly C and other homopolymers. *Biophys. J.*, **8**:1201–1210, 1968.

P126. Preiss, J. W., and Stevenson, D. A. Some parallelisms in the behavior of pancreatic ribonuclease and chicken lysozyme toward homopolyribonucleotides. *Biophys. J.*, **12**:80–91, 1972.

P127. Pretolani, E. Inibizione *in vitro* ed *in vivo* dell'elastasi pancreatica da parte del lisozima. *Boll. Soc. Ital. Biol. Sper.*, **37**:1223–1225, 1961.

P128. Previero, A., Coletti-Previero, M. A., and Axelrud-Cavadore, C. Prevention of cleavage next to tryptophan residues during the oxidative splitting by N-bromosuccinimide of tryosyl peptide bonds in proteins. *Arch. Biochem. Biophys.*, **122**:434–438, 1967.

P129. Previero, A., Coletti-Previero, M. A., and Jollès, P. Localization of non-essential tryptophan residues for the biological activity of lysozyme. *J. Mol. Biol.*, **24**:261–268, 1967.

P130. Prickett, P. S., Miller, N. J., and McDonald, F. G. Lysozyme studies of tissues from animals deficient in vitamin A. *J. Bacteriol.*, **33**:39, 1937.

P131. Prince, W. R., and Garren, H. W. An investigation of the resistance of white leghorn chicks to *Salmonella gallinarum. Poultry Sci.*, **45**:1149–1153, 1966.

P132. Pringle, B. H., DePaulis, D. C., and Pemrick, T. D. Estimation of lysozyme in certain biologic materials. *Amer. J. Clin. Pathol.*, **21**:1039–1044, 1951.

P133. Prixováa, J. Cytoplasmic bactericidal factors in polynuclear leukocytes of the rabbit. *Sb. Ved. Pr. Lek. Fak. Karlovy Univ. Hradei Kralove*, **11**:Suppl, 161–195, 1968. (Czech.)

P134. Prixová, J., and Kotýnek, O. Bactericidal substances in cytoplasm of leucocytes. An attempt to separate lysozyme by gel filtration. *Folia Microbiol. (Prague)*, **13**:468–471, 1968.

P135, Prockop, D. J., and Davidson, W. D. Study of urinary and serum lysozyme in patients with renal disease. *N. Engl. J. Med.*, **270**:269–274, 1964.

P136. Prosperi, P., Borselli, L., and Vitale, S. Modificazioni istologiche indotte dal trattamento prolungato con lisozima nelle ghiandole a secrezione interna di ratti. *Minerva Pediat*, **17**:1008–1014, 1965.

P137. Protass, J. J., and Korn, D. Inhibition of lysozyme synthesis by actinomycin D in bacteriophage T4-infected cells of *Escherichia coli. Proc. Nat. Acad. Sci. U.S.*, **55**:832–835, 1966.

P138. Protein chemistry. Denaturation and refolding. *Nature (London)*, **221**:317, 1969.

P139. Proux, C., and Toubiana, C. G. Le traitement des radiomucites de la sphère O.R.L. par une médication d'emploi facile: La lysopaïne O.R.L. comprimés. *Sem. Ther.*, **43**:182–185, 1967.

P140. Prozorov, A. A. Effect of egg lysozyme on B. subtilis cell permeability to transforming DNA. *Dokl. Akad. Nauk SSSR*, **160**:472–474, 1965. (Russ.)

P141. Prudden, J. F., and Lane, N. Studies on the mechanism of alimentary lysozyme production. Absence of parasympathetic and histamine control. *Gastroenterology*, **15**:104–109, 1950.

P142. Prudden, J. F., Lane, N., and Levison, J. Lysozyme titres in regional enteritis, miscellaneous tissues, micro-organisms, and excreta. *Proc. Soc. Exp. Biol. Med.*, **72**:220–222, 1949.

P143. Prudden, J. F., Lane, N., and Meyer, K. Effect of orally and intra-arterially administered lysozyme on canine gastrointestinal mucosa. *Amer. J. Med. Sci.,* **219**:291–300, 1950.

P144. Purdden, J. F., Lane, N., and Meyer, K. Lysozyme content of granulation tissue. *Proc. Soc. Exp. Biol. Med.,* **72**:38–39, 1949.

P145. Prudden, J. F., and Meadows, E. C. Studies on the mechanism of alimentary lysozyme production, II. *Gastroenterology,* **23**:403–418, 1953.

P146. Pruitt, K. M., Caldwell, R. C., Jamieson, A. D., and Taylor, R. E. Interaction of salivary proteins with tooth surface. *J. Dent. Res.,* **48**:818–823, 1969.

P147. Pruzanski, W. Serum and urinary proteins and renal function in chronic myelogenous leukemia. *Amer. J. Med. Sci.* **263**:163–171, 1972.

P148. Pruzanski, W., Leers, W.-D., and Wardlaw, A. C. Bacteriolytic and bactericidal activity in monocytic and myelomonocytic leukemia with hyperlysozymemia. *Cancer Res.,* **33**:867–873, 1973.

P149. Pruzanski, W., Leers, W.-D., and Wardlaw, A. C. Bacteriolytic and bactericidal activity of maternal and cord sera. Relationship to complement, lysozyme, transferrin, and immunoglobulin levels. *Can. J. Microbiol.,* **18**:1551–1555, 1971.

P150. Pruzanski, W., and Ogryzlo, M. A. Lysozyme in rheumatic diseases. *Semin. Arthritis and Rheum.,* **1**:361–381, 1972.

P151. Pruzanski, W., and Platis, M. E. Serum and urinary proteins, lysozyme (muramidase), and renal dysfunction in mono- and myelomonocytic leukemia. *J. Clin. Invest.,* **49**:1694–1708, 1970.

P152. Pruzanski, W., and Saito, S. G. Diagnostic value of lysozyme (muramidase) estimation in biological fluids. *Amer. J. Med. Sci.,* **258**:405–415, 1969.

P153. Pruzanski, W., Saito, S., and Ogryzlo, M. A. Significance of lysozyme (muramidase) in rheumatoid arthritis. I. Levels in serum and synovial fluid. *Arthritis Rheum.,* **13**:389–399, 1970.

P154. Pryme, I. F., and Berentsen, S. A. Lysozyme synthesis in *Escherichia coli* B cells infected with Phage T7. *Biochim. Biophys. Acta,* **204**:630–632, 1970.

P155. Pryme, I. F., Joner, P. E., and Jensen, H. B. Appearance of phage associated lysozyme in E. coli B cells immediately after infection with phage T2. *Biochem. Biophys. Res. Commun.,* **36**:676–681, 1969.

P156. Pulatova, M. K., Pasoian, V. G., Kaiushin, L. P. *et al.* E.P.R. spectra of a gamma-irradiated monocrystal of lysozyme chloride. *Dokl. Akad. Nauk SSSR,* **194**:711–714, 1970. (Russ.)

P157. Puskás, M., Antoni, F., Staub, M., Farkas, Gy. and Piffkó, P. Lysozyme activity of the human tonsil. *ORL, J. Oto-Rhino-Laryngol. Borderlands,* **34**:160–000, 1972.

P158. Pusztai, Z., Csizér, Z., and Joó, I. Effect of different bactericidal systems on B. pertussis and B. parapertussis organisms. *Ze. Immunitaetsforsch., Exp. Klin. Immunol.,* **141**:1–13, 1970.

P159. Pusztai, Z., Joo, I., and Eckhardt, E. Isolation of antigenic fractions from *Bordetella pertussis* by means of lysozyme. *Nature (London),* **196**:296–297, 1962.

Q

Q1. Quash, G. A. The effect of antibodies anti-polyamine on the growth of sarcoma T.R.V.L. 53886 in mice. *West Indian Med. J.,* **18**:1–4, 1969.

Q2. Quattrocchi, G., and Armellini, G. Influenza della gonadotropina corionica su alcuni indici immunitari. *Arch. Sci. Med.,* **115**:183–191, 1963.

Q3. Quelin-Zuili, S. Etude comparée de nouveaux lysozymes. Purification et action sur deux substrats purifiés. *Pathol. Biol.,* **11**:875–884, 1963.

Q4. Quevauviller, A., and Maziére, M. Le lysozyme augmente l'activité des anesthésiques locaux. *C. R. Soc. Biol.,* **160**:2276–2278, 1966.

Q5. Quie, P. G., and Hirsch, J. G. Antiserum to leucocyte lysosomes. Its cytotoxic, granulolytic, and hemolytic activities. *J. Exp. Med.,* **120**:149–160, 1964.

R

R1. Rabassini, A. Fleming's lysozyme in the treatment of multiple sclerosis. *Riv. Sper. Freniat. Med. Leg. Alienazioni Ment.,* **84**:1095–1100, 1960.

R2. Rabassini, A., and Bernardi, S. Terapia multifocale della sclerosi in placche, con particolare riferimento alla terapia con lisozima. *Minerva Med.,* **57**:1039–1041, 1966.

R3. Rabe, E. F., and Curnen, E. C. The occurrence of lysozyme in the cerebrospinal fluid and serum of infants and children. *J. Pediat.,* **38**:147–153, 1951.

R4. Radha, E., and Krishna-Moorthy, R. V. Differentiation of lysozyme activity in the fast, slow and cardiac muscles of chick. *Curr. Sci.,* **41**:183, 1972.

R5. Radici, M., and Galloni, O. La nostra esperienza in tema di terapia dell'epatite virale dell'infanzia. *Epatologia,* **12**:365–372, 1966.

R6. Raeste, A.–M. Lysozyme (muramidase) activity of leukocytes and exfoliated epithelial cells in the oral cavity. *Scand. J. Dent. Res.,* **80**:422–427, 1972.

R7. Raftery, M. A., and Dahlquist, F. W. The chemistry of lysozyme. *Fortschr. Chem. Org. Naturst.,* **27**:340–381, 1969.

R8. Raftery, M. A., Dahlquist, F. W., Chan, S. I., and Parsons, S. M. A proton magnetic resonance study of the association of lysozyme with monosaccharide inhibitors. *J. Biol. Chem.,* **243**:4175–4180, 1968.

R9. Raftery, M. A., Dahlquist, F. W., Parsons, S. M., and Wolcott, R. G. The use of nuclear magnetic resonance to describe relative modes of binding to lysozymes of homologous inhibitors and related substrates. *Proc. Nat. Acad. Sci. U.S.,* **62**:44–51, 1969.

R10. Raftery, M. A., Huestis, W. H., and Millett, F. Use of ^{19}F-nuclear magnetic resonance spectroscopy for detection of protein conformation changes: application to lysozyme, ribonuclease, and hemoglobin. *Cold Spring Harbor Symp. Quant. Biol.,* **36**:541–550, 1971.

R11. Raftery, M. A., and Rand-Meir, T. On distinguishing between possible mechanistic pathways during lysozyme-catalyzed cleavage of glycosidic bonds. *Biochemistry,* **7**:3281–3289, 1968.

R11a. Raftery, M. A., Rand-Meir, T., Dahlquist, F. W., Parsons, S. M., Borders, C. L. Jr., Wolcitt, R. G., Beranek, W. Jr., and Jao, L. Separation of glycosaminoglycan saccharide and glycoside mixtures by gel filtration. *Anal. Biochem.* **30**:427–435, 1969.

R12. Ragnunathan, R., and Gurnani, S. Differential association of nuclear lysozyme in rat tissues. *Indian J. Biochem. & Biophys.,* **9**:166–167, 1972.

R13. Raghunathan, R., and Gurnani, S. Comparative studies on nuclear lysozyme in rat tissues. *Arch. Biochem. Biophys.,* **147**:527–533, 1971.

R14. Ralston, D. J. Staphylococcal sensitization: Specific biological effects of phage K on the bacterial cell wall in lysis-from-without. *J. Bacteriol.,* **85**:1185–1193, 1963.

R15. Ralston, D. J., and Elberg, S. S. Intramonocytic destruction of Brucella: Potentiating effect of glycine on intracellular lysozyme activity. *J. Infec. Dis.,* **109**:71–80, 1961.

R15a. Ramachandran, L. K., and Witkop, B. Selective cleavage of C-tryptophyl peptide bonds in protein and peptides. *J. Amer. Chem. Soc.,* **81**:4028–4032, 1959.

R16. Ramunni, M. Su alcuni rapporti tra attività lisozimica e potere complementare e battericida del sangue. *Riv. Inst. Sieroter. Ital.*, **29**:329–336, 1954.

R17. Rand-Meir, T., Dahlquist, F. W., and Raftery, M. A. Use of synthetic substrates to study binding and catalysis by lysozyme. *Biochemistry*, **8**:4206–4214, 1969.

R18. Rao, G. R. K., and Burma, D. P. Purification and properties of phage P22-induced lysozyme. *J. Biol. Chem.*, **246**:6474–6479, 1971.

R19. Rao, D. N., and Gurnani, S. Enzymatic and physico-chemical properties of DNP derivatives of lysozyme obtained at pH 9.1. *Indian J. Biochem. & Biophys.*, **9**:49–51, 1972.

R20. Rao, D. N., and Gurnani, S. DNP derivatives of lysozyme obtained at pH 7.2—effect of modification on physico-chemical properties of the enzymes. *Indian J. Biochem. & Biophys.*, **9**:43–48, 1972.

R21. Rao, D. N., and Gurnani, S. Studies on dinitrophenylation of lysozyme at pH 7.2 — effect of modification on enzyme activity. *Indian J. Biochem. & Biophys.*, **9**:36–42, 1972.

R22. Rao, G. J. S., and Ramachandran, L. K. The role of tryptophan residues in the enzymic activity of lysozyme. *Biochim. Biophys. Acta*, **59**:507–508, 1962.

R22a. Rasper, J., and Kauzmann, W. Volume changes in protein reactions. I. Ionization reactions of proteins. *J. Amer. Chem. Soc.* **84**:1771–1777, 1962.

R23. Ravnskov, U. Muramidase and acute rejection of kidney graft. *Lancet*, **2**:716, 1972.

R24. Rawitch, A. B. The rotational diffusion of bovine alpha-Lactalbumin: A comparison with egg white lysozyme. *Arch. Biochem. Biophys.*, **151**:22–27, 1972.

R25. Rawitch, A. B., and Weber, G. The reversible association of lysozyme and thyroglobulin. Cooperative binding by near-neighbor interactions. *J. Biol. Chem.*, **247**:680–685, 1972.

R26. Raykher, Z. A., and Romanova, M. A. Lysozyme in therapy of dyspepsia. *Sov. Med.*, **4**:11–12, 1940.

R27. Razin, S., and Argaman, M. Comparative studies of Mycoplasma (PPLO), bacterial protoplasts and L-forms. *Harefuah*, **62**:259–263, 1962. (Heb.)

R28. Redner, P. Lisozima e microfungos ceratofilicos patogênicos. *Hospital (Rio de Janeiro)*, **71**:575–578, 1967.

R29. Reed, H., and Tuschingham, G. The antibacterial property of vitreous. *AMA Arch. Ophthalmol.*, **62**:780–781, 1959.

R30. Rees, A. R., and Offord, R. E. X-ray crystallography and the amino acid sequence of lysozyme. *Biochem. J.*, **130**:965–968, 1972.

R31. Regan, E. The lysozyme content of tears. *Amer. J. Ophthalomol.*, **33**:600–605, 1950.

R32. Reichert, R., Hochstrasser, K., and Eisenmann, M.: Simultane Bestimmung von Peroxydase, Lysozym und niedermolekularem Proteaseninhibitor im menschlicehn Nasensekret. *Arch. Klin. Exp. Ohren-Nasen- Kehlkopfheilk.*, **203**:109–114, 1972.

R33. Reifenstein, R. W., and Gray, S. J. The effect of adrenocorticotropic hormone upon the fecal lysozyme titer in ulcerative colitis. *Gastroenterology*, **19**:547–557, 1951.

R34. Reifenstein, R. W., Gray, S. J., Spiro, H. M., Young, C. G., and Connolly, E. P. Relationship of lysozyme to other components of gastric secretion in peptic ulcer. *Gastroenterology*, **16**:387–403, 1950.

R35. Reimer, S. M., and Litwack, G. Environmental effects of sodium chloride on *Micrococcus lysodeikticus* and its lysis by lysozyme. *Enzymologia*, **22**:373–383, 1961.

R36. Repaske, R. Lysis of gram-negative organisms and the role of versene. *Biochim. Biophys. Acta*, **30**:225–232, 1958.

R37. Repaske, R. Lysis of gram-negative bacteria by lysozyme. *Biochim. Biophys. Acta*, **22**:189–191, 1956.

R38. Resti, M. Attività lisozimica del siero di sangue e del liquido endperitoneale in corso di occlusione intestinale sperimentale del conglio. *Riv. Patol. Clin.*, **17**:1061–1068, 1962.

R39. Resti, M., and Andreis, G. Comportamento del tasso ematico di lisozima dopo emorragia acuta nel coniglio. *Riv. Patol. Clin.*, **17**:933–940, 1962.

R40. Rettger, L. F., and Sperry, J. A. The antiseptic and bactericidal properties of egg white. *J. Med. Res.*, **26**:55–64, 1912.

R41. Revis, G. J. Immunoelectrophoretic identification of lysozyme in saliva. *Proc. Soc. Exp. Biol. Med.*, **137**:90–96, 1971.

R42. Ribble, J. C. Increase of plasma lysozyme activity following injections of typhoid vaccine. *Proc. Soc. Exp. Biol. Med.*, **107**:597–600, 1961.

R43. Ribble, J. C., and Bennett, I. L., Jr. Lysozyme release as an indication of *in vivo* leukocyte injury by endotoxin. *Clin. Res.*, **7**:41, 1959.

R44. Ribble, J. C., and Hook, E. W. Inactivation of T2r(plus) bacteriophage by lysozyme. *Bull. Johns Hopkins Hosp.*, **107**:108–124, 1960.

R45. Richmond, M. H. A technique for the location of egg-white lysozyme on zone electrophoresis papers using plates containing *Micrococcus lysodeikticus*. *Biochim. Biophys. Acta*, **27**:209–210, 1958.

R46. Ridley, F. Lysozyme: Antibacterial body present in great concentration in tears, and its relation to infection of human eye. *Proc. Roy. Soc. Med. (Lond.)*, **21**:55–66, 1928.

R47. Rinaldi, L. Action of lysozyme on blood coagulation time in normal subjects. *Boll. Mal. Orecchio, Gola Naso*, **78**:253–271, 1960.

R48. Rindello, S. Ricerche sul "lysozym" in rapporto ad alcune questioni interessanti l'oftalmologia. *Boll. Ocul.*, **15**:1215–1231, 1936.

R49. Ristow, S. S., and Wetlaufer, D. B. Evidence for nucleation in the folding of reduced hen egg lysozyme. *Biochem. Biophys. Res. Commun.*, **50**:544–550, 1973.

R50. Riblet, R. J., and Herzenberg, L. A. Mouse lysozyme production by a monocytoma: Isolation and comparision with other lysozymes. *Science*, **168**:1595–1597, 1970.

R51. Riley, M., and Perham, R. N. The reversible reaction of protein amino groups with exo-cis-3,6-endoxo-delta4-tetrahydrophthalic anhydride. The reaction with lysozyme. *Biochem. J.*, **118**:733–739, 1970.

R51a. Ritland, H. N., Kaesberg, P., and Beeman, W. W. An X-ray investigation of the shapes and hydrations of several protein molecules in solution. *J. Chem. Phys.*, **18**:1237–1242, 1950.

R52. Ritzerfeld, W. von Enzymatische Beeinflussung der Antibiotikawirkung. *Arzneim.-Forsch.*, **19**:674–676, 1969.

R53. Ritzerfeld, W. von, and Kleining, R. Untersuchungen zur antibakteriellen Wirkung von Chloramphenicol und Nitrofurantoin unter Lysozymeinfluss. *Arch. Hyg. Bakteriol.*, **152**:477–481, 1968.

R54. Robel, E. J., and Crane, A. B. An accurate method for correcting unknown amino acid losses from protein hydrolyzates. *Anal. Biochem.*, **48**:233–246, 1972.

R55. Roberts, E. A. H. The preparation and properties of purified egg-white lysozyme. *Quart. J. Exp. Physiol.*, **27**:89–98, 1937.

R56. Roberts, E. A. H., Maegraith, B. G., and Florey, H. W. A comparison of lysozyme preparations from egg-white, cat and human saliva. *Quart. J. Exp. Physiol.*, **27**:381–391, 1938.

R57. Robinson, D. S., and Monsey, J. B. A reduced ovomucin-reduced lysozyme complex from egg white. *Biochem. J.*, **115**:64P, 1969.

R58. Robinson, D. S., and Monsey, J. B. A quantitative study of a reduced ovomucin-natural lysozyme interaction. *Biochem. J.*, **115**:64P–65P, 1969.

R59. Robyt, J. F., Ackerman, R. J., and Chittenden, C. G. Reaction of protein disulfide groups with Ellman's reagent: A case study of the number of sulfhydryl and disulfide groups in *Aspergillus oryzaeapha*-amylase,papain, and lysozyme. *Arch. Biochem. Biophys.*, **147**:262–269, 1971.

R59a. Rodney, G., Swanson, A. L., Wheeler, L. M., Smith, G. N., and Worrell, C. S. The effect of a series of flavonoids on hyaluronidase and some other related enzymes. *J. Biol. Chem.*, **183**:739–747, 1950.

R60. Rogers, J. C., and Kornfeld, S. Hepatic uptake of proteins coupled to fetuin glycopeptide. *Biochem. Biophys. Res. Commun.*, **45**:622–629, 1971.

R61. Rognoni, F. Sull'alto contenuto di lisozima nel timo. *Boll. Soc. Ital. Biol. Sper.*, **40**:64–66, 1964.

R62. Rognoni, F., and Bergonzi, F. Contributi sperimentali sul meccanismo d'azione del lisozima in rapporto alla sua azione sul timo. *Minerva Pediat.*, **18**:482–483, 1966.

R63. Rognoni, F., and Poggi Longostrevi, G. Modificazioni dell'attività lisozimica nell' intestino di animali irradiati. *Minerva Radiol., Fisioter. Radiobiol.*, **10**:303–305, 1965.

R64. Rolicka, M. The mechanism of action of penicillin. On spheroplastic rods of *Escherichia coli* obtained with the aid of penicillin and lysozyme. *Postepy Hig. Med. Dosw.*, **16**:735–743, 1962. (Pol.)

R65. Rolla, G. Bacteriostatic activity in the saliva. *Nor. Tandlaegeforen. Tid.*, **75**:101–110, 1965. (Nor.)

R66. Romanenko, A. A. On the question of the state of natural immunity in mental patients. *Zh. Nevropatol. Psikhiat. S. S. Korsakova*, **68**:1825–1828, 1968. (Russ.)

R67. Romanoff, A. L., and Romanoff, A. J. "The Avian Egg." Wiley, New York, 1949.

R68. Romeo, D., and De Bernard, B. Formation *in vitro* of a lipoprotein membrane masking lysozyme activity. *Nature (London)*, **212**:1491–1492, 1966.

R70. Rosenthal, D. S., Maglio, R., and Moloney, W. C. Muramidasuria and hyperkaluria in the chloroleukemic rat. *Proc. Soc. Exp. Biol. Med.*, **141**:499–500, 1972.

R71. Rosenthal, D. S., and Moloney, W. C. Muramidase activity in leukemic rats. *Proc. Soc. Exp. Biol. Med.*, **126**:682–685, 1967.

R72. Rosenthal, L., and Lieberman, H. The role of lysozyme in the development of intestinal flora of the newborn infant. *J. Infect. Dis.*, **48**:226–235, 1931.

R73. Ross, V. The transitional pH's and equivalent combining weights of lysozyme, cytochrome c, alpha-lactalbumin and mouse milk casein in their reactions with salmine. *Arch. Biochem. Biophys.*, **72**:1–7, 1957.

R74. Ross, V. Precipitation of insulin by lysozyme and effect of the complex on the blood sugar level. *Proc. Soc. Exp. Biol. Med.*, **72**:465–468, 1949.

R75. Rossi, G.-L., Holler, E., Kumar, S., Rupleg, John A., and Hess, G. P. Labelling of the catalytic site of lysozyme. *Biochem. Biophys. Res. Commun.*, **37**:757–766, 1969.

R76. Rossi, R. The action of lysozyme in artificial feeding and in dyspepsia. *Sem. Med.*, **119**:1019–1023, 1961. (Span.)

R77. Rossi, R., Frangini, V., and Moggi, C. Il lisozima nella terapia dell'epatite virale. *Epatologia*, **12**:763–770, 1966.

R78. Rouit, H. Enzymothérapie des affections pulpo-apicales. Le lysozyme. *Inform. Dent.*, **50**:3845–3848, 1968.

R79. Rowland, J. "The Penicillin Man." Roy Publ., New York, 1957.

R80. Roxby, R., and Tanford, C. Hydrogen ion titration curve of lysozyme in 6 M guanidine hydrochloride. *Biochemistry*, **10**:3348–3352, 1971.

R81. Royce, P. C. Role of renal uptake of plasma protein in compensatory renal hypertrophy. *Amer. J. Physiol.*, **212**:924–930, 1967.

R82. Rozgonyi, F., and Redai, I. The effect of methicillin in combination with trypsin and lysozyme on the growth of methicillin sensitive and resistant *Staphylococcus aureus* strains. *Acta Microbiol.*, **19**:133–143, 1972.

R83. Rozgonyi, F., and Rédai, I. The effect of lysozyme and meticillin on the growth of meticillin resistant and sensitive staphylococcus aureus strains. *Acta Microbiol.*, **17**:95–103, 1970.

R84. Rozhanskiy, V. I. Lysozyme in therapy of purulent wounds. *Nov Khir. Arkh.*, **44**:3–10, 1939.

R85. Rubenstein, K. E., Schneider, R.S., and Ullman, E. F. "Homogeneous" enzyme immunoassay. A new immunochemical technique. *Biochem. Biophys. Res. Commun.*, **47**:846–851, 1972.

R86. Ruczaj, Z., Sawnor-Korszyńska, D., Paśś, L., and Raczyńska-Bojanowska, K. On the release of enzymes upon lysozyme treatment of Streptomyces. *Acta Biochim. Pol.*, **16**:371–378, 1969.

R87. Rudders, R. A., and Bloch, K. J. Myeloma renal disease: Evaluation of the role of muramidase (lysozyme). *Amer. J. Med. Sci.*, **262**:79–85, 1971.

R88. Ruiz Sanchez, A., Ponce de Leon, E., Ramirez, A., and Ruiz Sanchez, F. La lisozima. *Medicina (Mexico City)*, **29**:221–228, 1949.

R89. Rupley, J. A. The binding and cleavage by lysozyme of *N*-acetyl-glucosamine oligosaccharides. *Proc. Roy. Soc., Ser. B*, **167**:416–428, 1967.

R90. Rupley, J. A. The hydrolysis of chitin by concentrated hydrochloric acid, and the preparation of low-molecular-weight substrates for lysozyme. *Biochim. Biophys. Acta*, **83**:245–255, 1964.

R91. Rupley, J. A., Butler, L., Gerring, M., Hartdegen, F. J., and Pecoraro, R. Studies on the enzymic activity of lysozyme. 3. The binding of saccharides. *Proc. Nat. Acad. Sci. U.S.*, **57**:1088–1095, 1967.

R92. Rupley, J. A., and Gates, V. Studies on the enzymic activity of lysozyme, II. The hydrolysis and transfer reactions of N-acetylglucosamine oligosaccharides. *Proc. Nat. Acad. Sci. U.S.*, **57**:496–510, 1967.

R93. Rupley, J. A., Gates, V., and Bilbrey, G. R., Lysozyme catalysis. Evidence for a carbonium ion intermediate and participation of glutamic acid 35. *J. Amer. Chem. Soc.*, **90**:5633–5635, 1968.

R94. Rybicka, I. Sensitivity of the PW8 strain of *Corynebacterium diphtheriae* to the effect of lysozyme obtained from chicken egg white. *Med. Dosw. Mikrobiol.*, **15**:113–124, 1963. (Pol.)

R95. Rydnik, R. I., and Agranovich, S. M. Lysozyme and phagolysozyme in therapy of diarrhea in infants and children. *Sov. Med.*, **6**:24–26, 1942.

R96. Rytel, M. W., and Lytle, R. I. Alterations in Humoral Factors in the Course of Acute Respiratory Infections," Rep. MF 022.03.07-4002 RU 66.14. Naval Med. Res. Unit No. 4, U.S. Naval Hospital, Great Lakes, Illinois, 1966.

R97. Ryter, A., Frehel, C., and Ferrandes, B. Comportement des mésosmes lors de l'attaque de *Bacillus subtilis* par le lysozyme en milieu hyper-ou hypotonique. *C. R. Acad. Sci. Ser. D*, **265**:1259–1262, 1967.

S

S1. Sabbag, Y. Ovservacões acèrca do uso de uma associacão de Cloranfenicol, Tetraciclina e Lisozima no Tratamento de Anexites. *Rev. Brasil. Med.*, **22**:544–545, 1965.

S2. Saint-Blancard, J., Allary, M., and Jollès, P. Influence du Bierbrich Scarlet sur la cinétique de lyse de *Microccus lysodeikticus* par plusiers lysozymes. *Biochimie,* **54:**1375–1376, 1972.

S3. Saint-Blancard, J., Capbern, A., Ducassé, D., and Jollès, P. Influence de la N-acétylglucosamine, du chitobiose et du chitotriose sur la fluorescence, en fonction du pH, de 6 lysozymes d'origines différentes. *C. R. Acad. Sci., D,* **269:**858–861, 1969.

S4. Saint-Blancard, J., Chuzel, P., Mathieu, Y., Perrot, J., and Jollès, P. Influence of pH and ionic strength on the lysis of *Micrococcus lysodeikticus* cells by six human and four avian lysozymes. *Biochim. Biophys. Acta,* **220:**300–306, 1970.

S5. Saint-Blancard, J., Hénique, Y., Ducassé, D., Dianoux, A.-C., and Jollès, P. Etude cinétique de l'inhibition de lysozymes d'origines diverses par la N-acétylglucosamine. *Bull. Soc. Chim. Biol.,* **50:**1783–1790, 1969.

S6. Saint-Blancard, J., and Jollès, P. Comportement cinétique de plusieurs lysozymes en fonction du pH et de la force ionique vis-à-vis de *Micrococcus lysodeikticus.* *Biochimie,* **54:**7–15, 1972.

S7. Saint-Blancard, J., Le Baron, J. L., Monteux, J. *et al.* Evolution de l'activité lysozymique du plasma dans le sang conservé. *Rev. Fr. Transfus.,* **12:**259–269, 1969.

S8. Saint-Blancard, J., Le Baron, J. L., Monteaux, J. *et al.* Activité lysozymique du plasma citrate. *Transfusion (Paris),* **9:**225–228, 1966.

S9. Saitta, G. Ricerche sul comportamento del potere lisozimico e della colesterolemia nel saturinismo professionale. *Folia Med. (Naples),* **38:**718–726, 1955.

S10. Sakakibara, R., and Hamaguchi, K. Structure of lysozyme. XIV. Acid-base titration of lysozyme. *J. Biochem. (Tokyo),* **64:**613–618, 1968.

S11. Sakato, N., Fujio, H., and Amano, T. Antigenic structures of hen egg-white lysozyme. III. The antigenic stie close to the catalytic site. *Biken J.,* **15:**135–152, 1972.

S12. Sakato, N., Fujio, H., and Amano, T. The electric charges of antibodies directed to unique regions of hen egg-white lysozyme. *Biken J.,* **14:**405–411, 1971.

S13. Sakurai, M. A study on protein fractions in tears. *Acta Soc. Ophthalmol. Jap.,* **73:**826–841, 1969. (Jap.)

S14. Salosi, C., Londrillo, A., and Mazzaglia, E. Comportamento di alcuni indici di immunità aspecifica in conigli sottoposti a trattamento con mucoproteina gastrica e vaccinati. 3. *Arch. Sci. Med.,* **125:**13–16, 1968.

S15. Salser, W., Gesteland, R. F., and Bolle, A. *In vitro* synthesis of bacteriophage lysozyme. *Nature (London),* **215:**588–591, 1967.

S16. Salser, W., Gesteland, R. F., and Ricard, B. Chacterization of lysozyme messenger and lysozyme synthesized in vitro. *Cold Spring Harbor Symp. Quant. Biol.,* **34:**771–780, 1969.

S17. Salton, M. R. J. Action of lysozyme on gram-positive bacteria and the structure of cell walls. *Exposes Annu. Biochim. Med.,* **27:**35–43, 1966.

S17a. Salton, M. R. "The Bacterial Cell Wall." Elsevier, Amsterdam, 1964.

S18. Salton, M. R. J. The lysis of micro-organisms by lysozyme and related enzymes. *J. Gen. Microbiol.,* **18:**481–490, 1958.

S19. Salton, M. R. J., and Pavlik, J. G. Studies of the bacterial cell wall. VI. Wall composition and sensitivity to lysozyme. *Biochim. Biophys. Acta,* **39:**398–407, 1960.

S20. Salton, M. R. J., and Ghuysen, J. M. Acetylhexosamine compounds enzymically released from *Micrococcus lysodeikticus* cell walls. III. The structure of DI- and tetra-saccharides released from cell walls by lysozyme and Streptomyces F_1 enzyme. *Biochim. Biophys. Acta,* **45:**355–363, 1960.

S21. Salton, M. R. J., and Ghuysen, J. M. The structure of di- and tetra- saccharides released from cell walls by lysozyme and Streptomyces F_1 enzyme and the beta(1 to 4) N-acetylhexosaminidase activity of these enzymes. *Biochim. Biophys. Acta,* **36:**552–554, 1959.

S22. Salton, M. R. J. The properties of lysozyme and its action on microorganisms. *Bacteriol. Rev.,* **21**:82–100, 1957.

S23. Salton, M. R. J. Studies of the bacterial cell wall. V. The action of lysozyme on cell walls of some lysozyme-sensitive bacteria. *Biochim. Biophys. Acta,* **22**:495–506, 1956.

S23a. Salton, M. R. J. Chemistry and function of amino sugars and derivatives. *Ann. Rev. Biochem.,* **34**:143–174, 1965.

S24. Salvi, G. L. Ricerche intorno all'attività del cortisone sul potere lisozimico delle lagrime. *Boll. Ocul.,* **32**:349–356, 1953.

S25. Sammons, H. G. Mucinases in ulcerative colitis. *Lancet,* **2**:239–240, 1951.

S26. Sandholzer, L. A. Bacteriophage and lysozyme in relation to oral infection. *J. Amer. Dent. Ass. Dent. Cosmos,* **25**:1399–1405, 1938.

S27. Sangiorgi, M. Il lisozima come antiprotozoario. *G. Batteriol. Immunol.,* **39**:429–432, 1948.

S28. Sankaran, K., and Gurnani, S. On the variation in the catalytic activity of lysozyme in fishes. *Indian J. Biochem. & Biophys.,* **9**:162–165, 1972.

S29. Santacroce, L. Radiation sickness in its relation with the lysozyme activity of the blood. *Riv. Ostet. Ginecol. Prat.,* **42**:730–738, 1960. (Ital.)

S30. Santacroce, L. On the behavior of C-reactive protein and lysozyme activity in pelvic inflammations. *Riv. Ostet. Ginecol. Prat.,* **42**:189–194, 1960. (Ital.)

S31. Sapse, A. T., Bonavida, B., Stone, W., Jr., and Sercarz, E. E. Human tear lysozyme. III. Preliminary study on lysozyme levels in subjects with smog eye irritation. *Amer. J. Ophthalmol.,* **66**:76–80, 1968.

S32. Sapuppo, A. Specific and aspecific action of lysozyme on Treponema pallidum. *Nuovi Ann. Ig. Microbiol.,* **12**:234–239, 1961. (Ital.)

S33. Sapuppo, A. On the action of lysozyme on Treponema pallidum. *Nuovi Ann. Ig. Microbiol.,* **12**:72–78, 1961. (Ital.)

S34. Sasaki, Y., Hosokawa, K., Inago, M. *et al.* Drug allergy from lysozyme preparations. *Otolaryngology (Tokyo),* **41**:207–212, 1969. (Jap.)

S35. Sato, H., Diena, B. B., and Greenberg, L. Spheroplast induction and lysis of BCG strains by glycine and lysozyme. *Can. J. Microbiol.,* **12**:255–261, 1966.

S36. Satomura, K., Yokota, M., and Shiraha, S. Clinical study of lysozyme in surgery. *Arch. Jap. Chir.,* **34**:1076–1091, 1965. (Jap.)

S37. Sauerbier, W., and Bräutigam, A. R. Control of gene function in bacteriophage T4. II. Synthesis of messenger ribonucleic acid and protein after interrupting deoxyribonucleic acid replication and glucosylation. *J. Virol.,* **5**:179–187, 1970.

S38. Sauter, E. A., and Petersen, C. F. Quality characteristics of eggs from hens indexed for lysozyme content of egg white. *Poultry Sci.,* **51**:957–960, 1972.

S39. Savini, R., and Mercurelli, D. Sopra l'azione opsonica del lisozima. *Boll. Soc. Ital. Biol. Sper.,* **23**:764–765, 1947.

S40. Saxena, V. P., and Wetlaufer, D. B. Formation of three-dimensional structure in proteins. I. Rapid nonenzymic reactivation of reduced lysozyme. *Biochemistry,* **9**:5015–5023, 1970.

S41. Scano, V., and Garau, A. Sugli aspetti nipiofarmacologici di alcuni enzimi. *Minerva Nipiol.,* **17**:363–367, 1967.

S42. Schill, W.-B., and Schumacher, G. F. B. Radial diffusion in gel for micro determination of enzymes. I. Muramidase, alpha-amylase, DNase I, RNase A, acid phosphatase, and alkaline phosphatase. *Anal. Biochem.,* **46**:502–533, 1972.

S43. Schmalzl, F., and Braunsteiner, H. Zur cytochemie der monocytenleukamie. *Klin. Wochenschr.,* **46**:1185, 1968.

S44. Schmalzl, F., Huber, H., Asamer, H., Abrederis, K., and Braunsteiner, H. Cytochemical and immunohistologic investigations on the source and the functional changes of mononuclear cells in skin window exudates. *Blood,* **34**:129–140, 1969.

S45. Schmidt, H., and Rosenkranz, P. On the participation of singlet oxygen in the acridine orange sensitized photoinactivation of lysozyme. *Z. Naturforsch. B,* **27**:1436–1438, 1972.

S46. Schnaitman, C. A. Effect of ethylenediaminetetraacetic acid, Triton X-100, and lysozyme on the morphology and chemical composition of isolated cell walls of *Escherichia coli. J. Bacteriol.,* **108**:553–563, 1971.

S47. Schnek, A. G., Ledoux, L. Léonis, J., and Charles, P. Préparation de lysozyme et de protéines du blanc d'oeuf marqué au ¹⁴C et a ³H. *Arch. Int. Physiol. Biochim.,* **69**:601–602, 1961.

S48. Scholnik, A. P., and Kass, L. A direct cytochemical method for the identification of lysozyme in various tissues. *J. Histochem. Cytochem.,* **21**:65–72, 1973.

S49. Schrodt, M. J., Eisenstein, R., Ray, R. D., and Kuettner, K. E. Lysozyme in embryonic cartilage: Ontogenic studies. *Surg. Forum,* **19**:461–462, 1968.

S49a. Schubert, M., and Franklin, E. C. Interaction in solution of lysozyme with chondroitin sulfate and its parent protein polysaccharide. *J. Amer. Chem. Soc.,* **83**:2920–2925, 1961.

S50. Schumacher, G. F. B. Protein analysis of secretions of the female genital tract. *Lying-in. J. Reprod. Med.,* **1**:61, 1968.

S51. Schumacher, G. F. B., and Pearl, M. J. Muramidase (lysozyme) in cervical secretions. *Protides Biol. Fluids, Proc. Colloq.,* **16**:525–534, 1969.

S52. Schumacher, G. F. B., and Wied, G. L. Semiquantitative microanalysis of proteins in cervical mucus. *Fert. Steril., Proc. World Congr., 5th, 1966, Excerpta Medica,* Congr. Ser. No. 133, pp. 713–722, 1966.

S53. Schumacher, H. Neue Erkenntnisse über die Biochemie des Lysozyms und ihre klinische Bedeutung für die Augenheilkunde. *Bibl. Ophthalmol.,* **52**:142–211, 1958.

S54. Schumacher, H. Die Bedeutung des Lysozyms für die Pathogenese von Magen- und Darmkrankheiten. *Aerztl. Forsch.,* **9**:534–539, 1955.

S55. Schumacher, H. Experimentelle Untersuchungen über die Adsorption von Lysozym an Mineralstaub. *Beitr. Silikoseeforsch.,* **23**:59–000, 1953.

S56. Schumacher, H. Über die Bestimmung der Lysozymaktivität in biologischem Material, mit besonderer Berücksichtigung der Tranen. *Z. Immunitaetsforsch. Exp. Ther.,* **110**:389–394, 1953.

S57. Schumacher, H., and Albano, A. Lysozymwirkung auf Gewebezellen in vitro. *Klin. Wochenschr.,* **31**:768, 1953.

S58. Schütte, E., and Krisch, K. Uber die Darstellung und Eigenschaften des sogenannten Lysozyme-Substrates. *Hoppe Seyler's Z. Physiol. Chem.,* **311**:121–130, 1958.

S59. Schweiger, M., and Gold, L. M. Bacteriophage T4 DNA-dependent in vitro synthesis of lysozyme. *Proc. Nat. Acad. Sci. U.S.,* **63**:1351–1358, 1969.

S60. Schweiger, M., and Gold, L. M. DNA-dependent in vitro synthesis of bacteriophage enzymes. *Cold Spring Harbor Symp. Quant. Biol.,* **34**:763–766, 1969.

S61. Schweiger, M., Herrlich, P., and Millette, R. L. Gene expression *in vitro* from deoxyribonucleic acid of bacteriophage T7. *J. Biol. Chem.,* **246**:6707–6712, 1971.

S62. Schweighofer, D., and Starlinger, P. Zur Protoplastierung von E. coli B mit Lysozym und Versen. *Arch. Mikrob.* (Berlin), **32**:219–223, 1959.

S63. Scibienski, R., Fong, S., and Benjamini, E. Cross-tolerance between serologically non-cross-reacting forms of egg white lysozyme. *J. Exp. Med.,* **136**:1308–1312, 1972.

S64. Scibienski, R., and Sercarz, E. Suppression and selection of B Cells during lysozyme tolerance in the mouse. *J. Immunol.,* **110**:540–545, 1973.

S65. Scoffone, E., Jori, G., and Galiazzo, G. Selective photo-oxidation of amino acids in proteins. *Biochem. Soc. Symp.*, **31**:163–170, 1970.

S66. Scolari, E. G. Il lisozima in dermatologia. *Minerva Med.*, **56**:457–459, 1965.

S67. Scolari, É. G. Il lisozima in dermatologia. *Minerva Dermatol.*, **39**:281–283, 1964.

S68. Scolari, E. G. New treatment with lysozyme against the pain of cancer patients. *Sem. Med.*, **119**:1012–1019, 1961. (Span.)

S69. Scolari, E. G. Nuovo terapia col lisozima contro il dolore dei cancerosi. *Minerva Med.*, **52**:1825–1830, 1961.

S70. Scolari, E. G. New therapy with lysozyme for pain in cancer patients. *Rass. Dermatol. Sifilogr.*, **14**:1–14, 1961. (Ital.)

S71. Scolari, E. G., Nannelli, M., and Vallecchi, C. The intravenous use of lysozyme in the pain caused by cancer and first observations concerning the pharmacological combinations. *Rass. Dermatol. Sifilogra.*, **14**:283–294, 1961. (Ital.)

S72. Secemski, I. I., Lehrer, S. S., and Lienhard, G. E. A transition state analog for lysozyme. *J. Biol. Chem.*, **247**:4740–4748, 1972.

S73. Secemski, I. I., and Lienhard, G. E. The role of strain in catalysis by lysozyme. *J. Amer. Chem. Soc.*, **93**:3549–3550, 1971.

S74. Segagni, E. La terapia delle epatiti virali acute alla luce di una pluriennale esperienza pediatrica. *Epatologia*, **12**:687–692, 1966.

S75. Segre, A. Sôbre o emprêgo de uma nova droga no tratamento da tricomoniase vaginal. *Hospital (Rio de Janeiro)*, **75**:1483–1492, 1969.

S76. Segre, A. Sull'impiego di un nuovo farmaco (Deflamon) nel trattamento della trichomoniasi vaginale. *Minerva Ginecol.*, **20**:52–59, 1968.

S77. Sekiguchi, M., and Cohen, S. S. The synthesis of messenger RNA without protein synthesis. II. Synthesis of phage-induced RNA and sequential enzyme production. *J. Mol. Biol.*, **8**:638–659, 1964.

S78. Sela, M. Antigens and antigenicity. *Naturwissenschaften*, **56**:206–211, 1969.

S79. Sela, M. Antigenicity: Some molecular aspects. *Science*, **166**:1365–1374, 1969.

S80. Sela, M., and Steiner, L. A. Inhibition of lysozyme by some copolymers of amino acids. *Biochemistry*, **2**:416–421, 1963.

S80a. Sela, M., White, F. H., and Anfinsen, C. B. The reduction cleavage of disulfide bonds and its application to problems of protein structure. *Biochim. Biophys. Acta*, **31**:417–426, 1959.

S81. Seleste, E. Keeping quality of cow's milk and mother's milk; investigations with reference to temperature, heat treatment, changes in flavour and pH, and lysozyme content. *Ann. Med. Exp. Biol. Fenn.*, **31**:Suppl. 8, 3–78, 1953.

S82. Seligman, B. R., Rosner, F., Parise, F., and Lee, S. L. Serum muramidase levels in acute leukemia. *Amer. J. Med. Sci.*, **264**:69–82, 1972.

S83. Seneca, H., and Peer, P. Effect of antibacterials, antibiotics, enzymes and steroids on phagocytosis. *J. Amer. Geriat. Soc.*, **14**:187–199, 1966.

S84. Senn, H. J., Chu, B., O'Malley, J., and Holland, J. F. Experimental and clinical studies on muramidase (lysozyme). I. Muramidase activity of normal human blood cells and inflammatory exudates. *Acta Haematol.*, **44**:65–77, 1970.

S85. Senn, H. J., and Rhomberg, W. U. Muramidaseaktivität in Serum und Urin bei akuten und chronischen Leukämien. *Schweiz. Med. Wochenschr.*, **100**:1993–1995, 1970.

S86. Sergeev, P. V., Kovalev, I. E., Ionov, I. D., Burkin, A. A., and Imambaev, S. E. Effect of exogenous lysozyme on skin homograft rejection and production of circulating antibodies in mice. *Farmakol. Toksikol. (Moscow)*, **34**:729–730, 1971. (Russ.)

S87. Sermann, R. Effetti della gonadotropina corionica sul testicolo di rana esculenta in relazione ad una eventuale azione antigonadotropa del lisozima di Fleming. *Minerva Ginecol.,* **13**:1266–1270, 1961.

S88. Serra, G. E., and Chisalè, E. Il lisozima per uso topico nella chirurgia plastico ginecologica. *Minerva Ginecol.,* **13**:1282–1284, 1961.

S89. Serra, G. E., and Chisalé, E. Il lisozima per via parenterale nelle cervicopatio infiammatorie. *Minerva Ginecol.,* **13**:1134–1136, 1961.

S90. Serra, G. E., and Vallerino, V. Il lisozima di fleming nel trattamento delle cervidopatie. *Minerva Ginecol.,* **11**:854–856, 1959.

S91. Serre, A., Chalmin, C., and Romieu, C. Etudes de certains facteurs non spécifiques de l'immunité au cours des processus néoplasiques. *Rev. Fr. Etud. Clin. Biol.,* **14**:1025–1028, 1969.

S92. Serrou, B., Dubois, J. B., Balmes, J. L., and Girard, C. Estimation of serum lysozyme in liver cirrhosis. *Nouv. Presse Med.,* **1**:2324, 1972.

S93. Sessa, G., and Weissmann, G. Incorporation of lysozyme into liposomes. A model for structure-linked latency. *J. Biol. Chem.,* **245**:3295–3301, 1970.

S94. Shabalina, S. V., Rubtsov, I. V., and Danilov, V. I. Dynamics of serum titers of lysozyme in infectious hepatitis and acute dysentery. *Zh. Mikrobiol., Epidemiol. Immunobiol.,* **48**:89–93, 1971. (Russ.)

S95. Shabordin, D. A. Photometric determination of lysozyme activity. *Arch. Biol. Nauk,* **60**:No.2, 83–86, 1940.

S96. Shah, S. B., and King, H. K. The action of lysozyme on bacterial electron transport systems. *J. Gen. Microbiol.,* **44**:1–13, 1966.

S97. Shalek, R. J., and Smith, C. E. The radiolysis of dilute solutions of lysozyme. I. Enzymatic studies relating to the oxygen effect. *Int. J. Radiat. Biol.,* **14**:9–17, 1968.

S98. Sharon, N. The chemical structure of lysozyme substrates and their cleavage by the enzyme. *Proc. Roy. Soc., Ser. B,* **167**:402–415, 1967.

S99. Sharon, N., Jollès, J., and Jollès, P. Contribution à l'étude du mécanisme d'action de lysozymes d'origines diverses. *Bull. Soc. Chim. Biol.,* **48**:731–732, 1966.

S100. Sharon, N., and Seifter, S. A transglycosylation reaction catalyzed by lysozyme. *J. Biol. Chem.,* **239**:PC2398–PC2399, 1964.

S100a. Sharp, J. J., Robinson, A. B., and Kamen, M. D. Preparation of a protein with lysozyme activity synthesis by Merrifield procedure. *Fed. Proc., Fed. Amer. Soc. Exp. Biol.,* **30**:1273, Abstr. No. 1287, 1971.

S101. Shashkova, E. A., Vikha, I. V., Vichutinskii, A. A., and Khorlin, A. Y. Interaction of lysozyme with low-molecular inhibitors and modified substrates containing beta-(1-4)- and beta-(1-6)-glucosaminidic bonds. *Biokhimiya,* **35**:124–131, 1970. (Russ.)

S101a. Shaw, E. Chemical modification by active-site-directed reagents. *In* "The Enzymes" (P. D. Boyer, ed.), 3rd. ed. Vol. 1, pp. 91–146. Academic Press, New York, 1970.

S102. Shechter, Y., Burstein, Y., and Patchornik, A. Sulfenylation of tryptophan-62 in hen egg-white lysozyme. *Biochemistry,* **11**:653–660, 1972.

S102a. Shechter, E., and Blout, E. R. An analysis of the optical rotatory dispersion of polypeptides and proteins, II. *Proc. Nat. Acad. Sci. U.S.,* **51**:794–800, 1964.

S103. Shechter, E., Gulik-Krzywicki, T., Azerad, R., and Gros, C. Correlations between structure and spectroscopic properties in membrane model systems. Fluorescence of dansylated protein and dansylated lipid in protein-lipid-water phases. *Biochim. Biophys. Acta,* **241**:431–442, 1971.

S104. Shehadeh, I. H., Carpenter, C. B., Monterio, C. H., and Merrill, J. P. Renal allograft rejection. An analysis of lysozymuria, serum complement, lymphocyturia, and heterophil antibodies. *Arch. Intern. Med.,* **125**:850–857, 1970.

S105. Shelton, K. R. Dodecyl sulfate-polyacrylamide gel electrophoresis of N'-dimethylami-nonaphthalene-5-sulfonyl-proteins: A covalently attached fluorescent label. *Biochem. Biophys. Res. Commun.*, **43**:367–371, 1971.

S106. Shibko, S., and Tappel, A. L. Rat-kidney lysozymes: Isolation and properties. *Biochem. J.*, **95**:731–741, 1965.

S107. Shih, J. W.-K., and Hash, J. H. The N,O-diacetylmuramidase of *Chalaropsis* species. III. Amino acid composition and partial structural formula. *J. Biol. Chem.*, **246**:994–1006, 1971.

S108. Shimaki, N., Ikeda, K., and Hamaguchi, K. Interaction of urea and guanidine hydrochloride with lysozyme. *J. Biochem. (Tokyo)*, **70**:497–508, 1971.

S109. Shimaki, N., Ikeda, K., and Hamaguchi, K. Interaction of alcohols with lysozyme. II. Studies on difference absorption spectra. *J. Biochem. (Tokyo)*, **68**:795–803, 1970.

S110. Shimizu, M. *et al.* Serum and urine lysozyme activity in various blood disorders and hypopotassemia in leukemia. *Jap. J. Clin. Hematol.*, **13**:549–559, 1972. (Jap.)

S111. Shinitzky, M., Grisaro, V., Chipman, D. M., and Sharon, N. Influence of inhibitory sugars on the fluorescence of lysozyme. *Arch. Biochem. Biophys.*, **115**:232–233, 1966.

S112. Shinitzky, M., Katchalski, E., Grisaro, V., and Sharon, N. Inhibition of lysozyme by imidazole and indole derivatives. *Arch. Biochem. Biophys.*, **116**:332–343, 1966.

S113. Shinka, S., Imanishi, M., Miyagawa, N., Amano, T., Inouye, M., and Tsugita, A. Chemical studies on antigenic determinants of hen egg white lysozyme. I. *Biken J.*, **10**:89–107, 1967.

S114. Shirko, G. N., Zakharova, M. S., Lapaeva, I. A., Pereverzev, N. A. The kinetics of the formation and morphology of B. pertussis spheroplasts. *Zh. Mikrobiol., Epidemiol. Immunobiol.*, **49**:85–88, 1972. (Russ.)

S115. Shockman, G. D., and Cheney, M. C. Autolytic enzyme system of *Streptococcus faecalis*. V. Nature of the autolysin-cell wall complex and its relationship to properties of the autolytic enzyme of *Streptococcus faecalis*. *J. Bacteriol.*, **98**:1199–1207, 1969.

S116. Shockman, G. D., and Martin, J. J. Autolytic enzyme system of *Streptococcus faecalis*. IV. Electron microscopic observations of autolysin and lysozyme action. *J. Bacteriol.*, **96**:1803–1810, 1968.

S117. Shockman, G. D., Pooley, H. M., and Thompson, J. S. Autolytic enzyme system of *Streptococcus faecalis*. III. Localization of the autolysin at the sites of cell wall synthesis. *J. Bacteriol.*, **94**:1525–1530, 1967.

S118. Shugar, D. The measurement of lysozyme activity and the ultra-violet inactivation of lysozyme. *Biochim. Biophys. Acta*, **8**:302–309, 1952.

S119. Shukla, O. P., and Krishna Murti, C. R. Preparation of lysozyme from egg white by precipitation with hexametaphosphate. *J. Sci. Ind. Res., Sec. C*, **20**:105–108, 1961.

S120. Shulman, J. A., and Petersdorf, R. G. Relationship of endogenous pyrogen to lysozyme. *Proc. Soc. Exp. Biol. Med.*, **114**:376–379, 1963.

S121. Siboulet, A. Les urétrites à virus; thérapeutiques actuelles. *Sem. Ther.*, **39**:422–423, 1963.

S122. Simkin, R. D., Cole, S. A., Ozawa, H., Magdoff-Fairchild, B., Eggena, P., Rudko, A., and Low, B. W. Precipitation and crystallization of insulin in the presence of lysozyme and salmine. *Biochim. Biophys. Acta*, **200**:385–394, 1970.

S123. Simmonds, H., Troup, G. M., Hsiu, J. J., Walford, R. L. Renal lysozyme levels in mice thymectomized at birth. *Proc. Soc. Exp. Biol. Med.*, **126**:879–882, 1967.

S124. Simon, L. D. Infection of *Escherichia coli* by T2 and T4 bacteriophages as seen in the electron microscope: T4 head morphogenesis. *Proc. Nat. Acad. Sci. U.S.*, **69**:907, 1972.

S125. Sindoni, G., and Restivo, O. I vaccini lisozimizzati nella terapia immunologica della brucellosi. *G. Mal. Infet. Parassit.*, **21**:116–119, 1969.

S126. Sjöquist, J., Movitz, J., Johannson, I.-B., and Hjelm, H. Localization of protein A in the bacteria. *Eur. J. Biochem.*, **30**:190–194, 1972.

S127. Skarin, A. T., Matsuo, Y., and Moloney, W. C. Muramidase in myeloproliferative disorders terminating in acute leukemia. *Cancer*, **29**:1336–1342, 1972.

S128. Skarin, A. T., and Moloney, W. C. Muramidase levels in leukemia. *N. Engl. J. Med.*, **280**:1360, 1969.

S128a. Skarnes, R. C., and Watson, D. W. Antimicrobial factors of normal tissues and fluids. *Bacteriol. Rev.*, **21**:273–294, 1957.

S129. Skerjanc, J., and Lapanje, S. A dilatometric study of the denaturation of lysozyme by guanidine hydrochloride and hydrochloric acid. *Eur. J. Biochem.*, **25**:49–53, 1972.

S130. Skripkin, Iu. K., Somov, B. A., and Butov, Iu. S. Current problems of studying allergic reactivity and nonspecific immunity in patients with eczema and neurodermatitis. *Vestn. Dermatol. Venerol.*, **45**:16–21, 1971. (Russ.)

S131. Smarda, J. Microscopic picture of lysozyme and glycine spheroplasts of *Escherichia coli* exposed to colicin. *Folia Microbiol. (Prague)*, **10**:179–181, 1965.

S132. Smékal, F. Lysis of lyophilized *Escherichia coli* cells with egg-white lysozyme without ethylenediaminetetra-acetic acid. *Folia Microbiol. (Prague)*, **18**:146–148, 1973.

S132a. Smith, D. B., Wood, G. C., and Charlwood, P. A. Application of the Archibald ultracentrifuge procedure to lysozyme and apurinic acid. Evaluation using a mechanical integrator. *Can. J. Chem.*, **34**:364–370, 1956.

S133. Smith, E. L., Kimmel, J. R., Brown, D. M., Thompson, E. O. P. Isolation and properties of a crystalline mercury derivative of a lysozyme from papaya latex. *J. Biol. Chem.*, **215**:67, 1955.

S133a. Smith, E. L. Evolution of enzymes. *In* "The Enzymes" (P. D. Boyer, ed.), 3rd ed. Vol. 1, pp. 267–339. Academic Press, New York, 1970.

S133b. Smith, G. N., and Stocker, C. Inhibition of crystalline lysozyme. *Arch. Biochem.*, **21**:383–394, 1949.

S134. Smith, I., Smith, M., and Roberts, L. Models for tertiary structures. Myoglobin and lysozyme. *J. Chem. Educ.*, **47**:302–305, 1970.

S135. Smolelis, A. N., and Hartsell, S. E. Factors affecting the lytic activity of lysozyme. *J. Bacteriol.*, **63**:665–674, 1952.

S136. Smolelis, A. N., and Hartsell, S. E. Occurrence of lysozyme in bird egg albumens. *Proc. Soc. Exp. Biol. Med.*, **76**:455–456, 1951.

S137. Smolelis, A. N., and Hartsell, S. E. The determination of lysozyme. *J. Bacteriol.*, **58**:731–736, 1949.

S138. Smolens, J., and Charney, J. The antigenicity of crystalline lysozyme. *J. Bacteriol.*, **54**:101–107, 1947.

S139. Snapper, I., and Seld, D. A Screening test for the presence of urinary lysozyme (muramidase). *Blood*, **31**:516–517, 1968.

S140. Söder, P.-Ö., Nord, C.-E., and Linder, L. Separation of enzymes from 2-day-old dental plaque material. *Scand. J. Dent. Res. (Copenhagen)*, **79**:515–517, 1971.

S141. Söder, P.-Ö., Nord, C.-E., Lundblad, G., and Kjellman, O. Investigation on lysozyme, protease and hyaluronidase activity in extracts from human leukocytes. *Acta Chem. Scand.*, **24**:129–136, 1970.

S142. Sohler, A., Romano, A. H., and Nickerson, W. J. Biochemistry of the actinomycetales. III. Cell wall composition and the action of lysozyme upon cells and cell walls of the actinomycetales. *J. Bacteriol.*, **75**:283–290, 1958.

S143. Sonnino, F. R., Gazzaniga, P. P., and Mastroeni, P. Immunoelectrophoretic characterization of *Micrococcus lysodeikticus* components. *Riv. Ist. Sieroter. Ital.,* **41**:277–282, 1966.

S144. Sophianopoulos, A. J. Association sites of lysozyme in solution. I. The active site. *J. Biol. Chem.,* **244**:3188–3193, 1969.

S144a. Sophianopoulos, A. J., and Sasse, E. A. Isoelectric point of proteins by differential conductimetry. *J. Biol. Chem.,* **240**:PC1864–PC1866, 1965.

S145. Sophianopoulos, A. J., and Van Holde, K. E. Physical studies of muramidase (lysozyme). II. pH-dependent dimerization. *J. Biol. Chem.,* **239**:2516–2524, 1964.

S146. Sophianopoulos, A. J., Rhodes, C. K., Holcomb, D. N., and Van Holde, K. E. Physical studies of lysozyme. I. Characterization. *J. Biol. Chem.,* **237**:1107–1112, 1962.

S147. Sophianopoulos, A. J., and Van Holde, K. E. Evidence for dimerization of lysozyme in alkaline solution. *J. Biol. Chem.,* **236**:PC82–PC83, 1961.

S148. Sophianopoulos, A. J., and Weiss, B. J. Thermodynamics of conformational changes of proteins. I. pH-dependent denaturation of muramidase. *Biochemistry,* **3**:1920–1928, 1964.

S149. Sorgente, N., Hascall, V. C., and Kuettner, K. E. Extractability of lysozyme from bovine nasal cartilage. *Biochim. Biophys. Acta,* **284**:441–450, 1972.

S150. Spadaro, M., Costa, A. L., and Mundo, A. Sui rapporti tra lisozima e tasso di colesterolo surrenale nel ratto. *Boll. Soc. Ital. Biol. Sper.,* **42**:1040–1044, 1966.

S151. Spadaro, M., Costa, A. L., Orecchio, F. *et al.* Sul comportamento di alcuni fattori dell'immunità naturale in corso di infezione sperimentale da virus EFH 120. *Epatologia,* **12**:548–554, 1966.

S152. Spanier, P., and Deribas, D. Nucleolytic nature of lysozyme. *Rev. Microbiol. Appl. Agr., Hyg., Ind.,* **3**:61–66, 1937.

S153. Speck, J. C., Jr., and Rynbrandt, D. J. A convenient method for isolating the disaccharide and tetrasaccharide in muramidase digests of *Micrococcus lysodeikticus* cell walls. *Anal. Biochem.,* **19**:426–433, 1967.

S154. Speece, A. J. Histochemical distribution of lysozyme activity in organs of normal mice and radiation chimeras. *J. Histochem. Cytochem.,* **12**:384–391, 1964.

S155. Spitznagel, J. K. Normal serum cytotoxicity for P_{32}-labeled smooth *Enterobacteriaceae*. III. Isolation of a gamma-g normal antibody and characterization of other serum factors causing P_{32} loss. *J. Bacteriol.,* **91**:401–408, 1966.

S156. Srivastava, V. K., and Bigelow, C. C. On the oxidation of lysozyme by N-Bromosuccinimide. *Biochim, Biophys. Acta,* **285**:373–376, 1972.

S157. Srivastava, S., Mathur, K. B., and Dhar, M. M. Synthetic substitute enzymes. Part III. Glutamic acid copolymers with lysozyme-like activity. *Experientia,* **26**:11–12, 1970.

S158. Stafani, M., Corradi, G., and D'Amico, D. Sulla presunta attività antiblastica del lisozima. *Acta Chir. Ital. (Padua),* **22**:Suppl. 1, 59–67, 1966.

S159. Staffieri, M. Sull'attività lisozimica ed antilisozimica del cerume e sulla probabile natura lisozimica del suo potere battericida. *Arch. Ital. Laringol.,* **58**:295–302, 1950.

S160. Staffierei, M., and Zorzi, M. Comportamento della attività lisozimica della secrezione nasale in presenza di terpeni, oli essenziali, istamina, antistaminici, sali d'argento, sulfamidici, antibiotici (ricerche sperimentali "in vitro"). *Arch. Ital. Otol., Rinol. Laringol.,* **62**:173–186, 1951.

S161. Staffieri, M., and Zorzi, M. Attività lisozimica della saliva e della secrezione nasale in rapporto alla anestesia di superficie; ricerche sperimentali in vivo e in vitro. *Arch. Ital. Otol. Rinol. Laringol.,* **61**:Suppl. 3, 138–142, 1950.

S162. Staffieri, M., Zorzi, M., and Sburlati, L. Il comportamento della batteriolisi da lysozyma nelle lacrime, saliva e secreto nasale del nato a termine e del prematuro. *Boll. Ist. Sieroter., Milan.*, **29**:139–147, 1950.

S163. Stanford, R. H., Jr., Marsh, R. E., and Corey, R. B. An x-ray investigation of lysozyme chloride crystals containing complexions of niobium and tantalum: Three-dimensional Fourier plot obtained from data extending to a minimum spacing of 5 angstroms. *Nature (London)*, **196**:1176–1178, 1962.

S164. Stasiw, R. O., Zaun, J. W., Patel, A. B., and Brown, H. D. Microcalorimetric study of binding by lysozyme of N-acetyl glucosamine oligomers. *Int. J. Peptide Protein Res.*, **5**:11–00, 1973.

S165. Stefanachi, L., and de Natale, L. Therapeutic results with the intramuscular and intraspinal administration of lysozyme obtained in subjects afflicted with Sydenham's chorea. *Rass. Neuropsichiat. Sci. Affini.*, **15**:206–208, 1961. (Ital.)

S166. Stefanovic, J., and Absolonová, O. Serum lysozyme and trypsin inhibitor levels and leukocyte count in rabbits following irradiation. *Bratislav. Lek. Listy*, **54**:618–625, 1970.(Slov.)

S166a. Steim, J. M. Differential thermal analysis of protein denaturation in solution. *Arch. Biochem. Biophys.*, **112**:599–604, 1965.

S166b. Steinberg, D., Vaughan, M., Anfinsen, C. B. and Gorry, J. Preparation of tritiated proteins by the Wilzbach method. *Science*, **126**:447–448, 1957.

S167. Steiner, R. F. Structural transitions of lysozyme. *Biochim. Biophys. Acta*, **79**:51–63, 1964.

S168. Steiner, R. F. Reversible association processes of globular proteins. II Electrostatic complexes of plasma albumin and lysozyme. *Arch. Biochem. Biophys.*, **47**:56–75, 1953.

S168a. Steiner, R. F. Reversible association processes of globular proteins. IV. Fluorescence methods in studying protein interactions. *Arch. Biochem.*, **46**:291–311, 1953.

S168b. Steiner, R. F. Structural transitions of lysozyme. *Biochim. Biophys. Acta*, **79**:51–63, 1964.

S168c. Steiner, R. F., and Edelhoch, H. Influence of pH and urea on the ultra-violet fluorescence of several globular proteins. *Nature*, **192**:873–874, 1961.

S169. Steinrauf, L. K. Preliminary x-ray data for some new crystalline forms of alpha-lactoglobulin and hen egg white lysozyme. *Acta Crystallogr.*, **12**:77, 1959.

S170. Steere, R. L., and Ackers, G. K. Restricted-diffusion chromatography through calibrated columns of granulated agar gel; a simple method for particle-size determination. *Nature (London)*, **196**:475–476, 1962.

S171. Stephen, J., Gallop, R. G., and Smith, H. Separation of antigens by immunological specificity. Use of disulphide-linked antibodies as immunosorbents. *Biochem. J.*, **101**:717–720, 1966.

S172. Sternberger, L. A., Osserman, E. F. and Seligman, A. M. Lysozyme and fibrinogen in normal and leukemic blood cells: A quantitative electron immunocytochemical study. *Johns Hopkins Med. J.*, **126**:188–209, 1970.

S173. Sternlicht, H., and Wilson, D. Magnetic resonance studies of macromolecules. I. Aromatic-methyl interactions and helical structure effects in lysozyme. *Biochemistry*, **6**:2881–2892, 1967.

S174. Stevens, C. O., and Bergstrom, G. R. The multiple nature of crystalline egg-white lysozyme. *Proc. Soc. Exp. Biol. Med.*, **124**:187–191, 1967.

S175. Stevens, C. O., Henderson, L. E., and Tolbert, B. M. Radiation chemistry of proteins. II. Enzymic activity and deuterium exchange properties of lysozyme and alpha-chymotrypsin. *Arch. Biochem. Biophys.*, **107**:367–373, 1964.

S176. Stevens, C. O., and Long, J. L. 4,4'-Difluoro-3,3'-dinitrophenylsulfone as a cross-linking reagent for lysozyme. *Proc. Soc. Exp. Biol. Med.*, **131**:1312–1316, 1969.

S177. Stevens, C. O., Long, J. L., and Upjohn, D. Radiation produced aggregation in crystalline preparations of ribonuclease, lysozyme, and trypsin. *Proc. Soc. Exp. Biol. Med.*, **132**:951–956, 1969.

S178. Stevens, C. O., and Sauberlich, H. E. Proteolytic digestion rates for altered enzymes. *Radiat. Res.*, **41**:362–374, 1970.

S179. Stevens, C. O., Sauberlich, H. E., and Bergstrom, G. R. Radiation-produced aggregation and inactivation in egg white lysozyme. *J. Biol. Chem.*, **242**:1821–1826, 1967.

S180. Stevens, C. O., Tolbert, B. M., and Bergstrom, G. R. Effects of oxygen on irradiated solid egg-white lysozyme. *Radia. Res.*, **42**:232–243, 1970.

S181. Stevens, C. O., Tolbert, B. M., and Reese, F. E. Radiation chemistry of proteins. I. Modified lysozyme substances. *Arch. Biochem. Biophys.*, **102**:423–429, 1963.

S182. Steytler, J. G. Die Belang van Lisosiem (muramidase) in hematologiese studies. *S. Afr. Med. J.*, **45**:1202–1205, 1971.

S183. Stollerman, G. H., Rytel, M., and Ortiz, J. Accessory plasma factors involved in the bactericidal test for type-specific antibody to group A Streptococci. II. Human plasma cofactor(s) enhancing opsonization of encapsulated organisms. *J. Exp. Med.*, **117**:1–17, 1963.

S184. Stone, J. L. The effect of lysozyme on the production of tetanus toxin. II. Mouse M.L.D. *Yale J. Biol. Med.*, **25**:239–244, 1953.

S185. Stone, J. L. The effect of lysozyme on the production of tetanus toxin; 1. Studies with flocculation. *J. Bacteriol.*, **64**:299–303, 1952.

S186. Stoppa, I. M. Influenza di farmaci ad azione protettiva sull'effetto leucopenico conseguente all'associazione chemio-radioterapica. *Minerva Radiol., Fisioter. Radiobiol.*, **10**:183–185, 1965.

S168a. Stracher, A., and Becker, R. R. Polyvalyl-proteins. *J. Amer. Chem. Soc.*, **81**:1432–1435, 1959.

S187. Strani, G. F. It trattametno dello zoster con lisozima ad alte dosi per os. *Minerva Dermatol.*, **43**:288–295, 1968.

S188. Stratton, K. Electron spin resonance studies on proton-irradiated ribonuclease and lysozyme. *Radiat. Res.*, **32**:Suppl. 7, 102–115, 1967.

S189. Straus, J. H., Gordon, A. S., and Wallach, D. F. H. The influence of tertiary structure upon the optical activity of three globular proteins: Myoglobin, hemoglobin and lysozyme. *Eur. J. Biochem.*, **11**:201–212, 1969.

S190. Straus, W. Changes in "droplet" fractions from rat kidney cells after intraperitoneal injection of egg white. *J. Biophys. Biochem. Cytol.*, **3**:933, 1957.

S191. Streisinger, G., Emrich, J., Okada, Y., Tsugita, A., and Inouye, M. Direction of translation of the lysozyme gene of bacteriophage T4 relative to the linkage map. *J. Mol. Biol.*, **31**:607–612, 1968.

S192. Streisinger, G., Mukai, F., Dreyer, W., Miller, B., and Horiuchi, S. Mutations affecting the lysozyme of phage T4. *Cold Spring Harbor Symp. Quant. Biol.*, **26**:25–30, 1961.

S193. Streisinger, G., Okada, Y., Emrich, J., Newton, J., Tsugita, A., Terzaghi, E., and Inouye, M. Frameshift mutations and the genetic code. *Cold Spring Harbor Symp. Quant. Biol.*, **31**:77–84, 1966.

S194. Stringa, L., and Bellone, F. On the behavior of lysozyme, as index of the aspecific immunological status of the organism, in gynecological phlogosis subjected to resolvent treatment. *Boll. Ist. Sieroter. Milan.*, **40**:65–70, 1961.

S195. Stringa, L., and Brusotti, A. Lysozyme in antiblastic chemotherapy in gynecology. *Quad. Clin. Ostet. Ginecol.,* **16**:751–756, 1961.

S196. Strominger, J. L., and Ghuysen, J.-M. Mechanisms of enzymatic bacteriolysis. Cell walls of bacteri are solubilized by action of either specific carbohydrases or specific peptidases. *Science,* **156**:213–221, 1967.

S197. Strosberg, A. D., and Kanarek, L. Immunochemical studies on hen's egg-white lysozyme. The role of the lysine, the histidine and the methionine residues. *Eur. J. Biochem.,* **14**:161–168, 1970.

S198. Strosberg, A. D., and Kanarek, L. Etude de l'antigenicite du lysozyme. IV. Rôle des résidus méthionine. *Arch. Int. Physiol. Biochim.,* **76**:950–951, 1968.

S199. Strosberg, A. D., and Kanarek, L. Etude de l'antigénicité du lysozyme. II. Rôle des ré sidus amino- et carboxy-terminaux. *Arch. Int. Physiol. Biochim.,* **76**:202–203, 1968.

S200. Strosberg, A. D., and Kanarek, L. Etude de l'antigénicité du lysozyme. I. Rôle du ré sidu histidine. *Arch. Int. Physiol. Biochim.,* **75**:898–900, 1967.

S201. Strosberg, A. D., Van Hoeck, B., and Kanarek, L. Immunochemical studies on hen's egg-white lysozyme Effect of selective nitration of the three tyrosine residues. *Eur. J. Biochem.,* **19**:36–41, 1971.

S202. Strosberg, A. D., Nihoul-Deconinck, C., and Kanarek, L. Weak immunological cross-reaction between bovine alpha-lactalbumin and hen's egg-white lysozyme. *Nature (London),* **227**:1241–1242, 1970.

S203. Strukelj, L. Muramidasi nei liquidi ascitice dei tumori ascite e nei liquidi delle tasche carcino-matose. *Friuli Med.,* **20**:969–974, 1965.

S204. Strukelj, L. Ricerche sulla tasca carcinomatosa e granulomatosa. I. Diffusione del lisozima ricavato da blanco d'uovo nella tasca carcinomatosa. *Friuli Med.,* **20**:915–922, 1965.

S205. Studebaker, J. F., Sykes, B. D., and Wien, R. A nuclear magnetic resonance study of lysozyme inhibition. Effects of dimerization and pH on saccharide binding. *J. Amer. Chem. Soc.,* **93**:4579–4585, 1971.

S206. Sud, I. J. Use of crude egg white lysozyme for preparing protoplasts of *B. megaterium. Indian J. Pathol. Bacteriol.,* **9**:313–315, 1966.

S207. Sullivan, N. P., and Manville, I. A. Relationship of the diet to the self-regulatory defense mechanism. II. Lysozyme in vitamin A and in uronic acid deficiencies. *Amer. J. Pub. Health Nat. Health,* **27**:1108–1115, 1937.

S208. Sulsenti, G. Lysozyme in various syndromes in O.R.L. *Boll. Mal. Orecchio, Gola Naso,* **79**:1–11, 1961.

S209. Sulsenti, G., Nucci, C., and Camprini, C. Modivicazioni del taso di lisozima in alcune dermopatie. *Arch. Ital. Otol., Rinol. Laringol.,* **73**:805–842, 1962.

S210. Summers, W. C., and Jakes, K. Phage T7 lysozyme mRNA transcription and translation *in vivo* and *in vitro. Biochem. Biophys. Res. Commun.,* **45**:315–320, 1971.

S211. Sundar, C. V., and Wu, H. C. Escherichia coli mutants permissive for T4 bacteriophage with deletion of e gene (phage lysozyme). *J. Bacteriol.,* **114**:656–665, 1973.

S212. Suponitskaia, V. M. Apropos of bactericidal factors in the blood serum I. Bactericidal factors in the serum of normal animals and healthy human subjects. *Zh. Mikrobiol., Epidemiol. Immunobiol.,* **41**:107–111, 1964. (Russ.)

S213. Suranyi, L. Lysozyme. *Magy. Orvosi Arch.,* **27**:577–585, 1926.

S214. Suranyi, L. Ueber das Lysozym. *Z. Immunitäetsforsch. Exp. Ther.,* **49**:166–179, 1926.

S215. Sussman, M., Asscher, A. W., and Jenkins, J. A. S. The intrarenal distribution of lysozyme (muramidase). *Invest. Urol.,* **6**:148–152, 1968.

S216. Suu, V.-T., Congdon, C. C., and Kretchmar, A. L. Lysozyme activity in radiation chimeras. *Proc. Soc. Exp. Biol. Med.,* **115**:825–829, 1964.

S217. Suu, V.-T., Congdon, C. C., and Kretchmar, A. L. Increase in lysozyme activity in kidneys of irradiation chimeras. *Proc. Soc. Exp. Biol. Med.,* **113**:481–485, 1963.

S218. Suyama T., Oguro, Y., Doi, T., and Izaka, K.-I. Protective effect of human placental lysozyme on experimental infections in mice induced by *Clostridium tetani, Staphylococcus aureus* and *Diplococcus pneumoniae. Jap. J. Antibiot., Ser. B,* **22**:287–298, 1969. (Jap.)

S219. Suzuki, H., Kitahara, T., Yamagishi, M. *et al.* Studies on the effects of lysozyme on the growth of Bifudus bacilli. (2). Effects on intestinal flora and putridity in rats. *Acta Paediat. Jap.,* **70**:861–866, 1966. (Jap.)

S220. Suzuki, H., Kitahara, T., Yamagishi, M. *et al.* Studies on the effect of lysozyme on the development of Bifidus in milk. *Acta Paediat. Jap.,* **70**:368–372, 1966. (Jap.)

S220a. Suzuki, S. Ueber die Wirkungsweise der Leukozyten auf saprophytische Keime. *Arch. Hyg.,* **74**:345–378, 1911.

S221. Suzuki, Y., and Rode, L. J. Effect of lysozyme on resting spores of *Bacillus megaterium. J. Bacteriol.,* **98**:238–245, 1969.

S222. Swan, I. D. A. The inhibition of hen egg-white lysozyme by imidazole and indole derivatives. *J. Mol. Biol.,* **65**:59–62, 1972.

S223. Sweeney, E. A., and Shaw, J. H. The effect of dietary lysozyme supplements on caries incidence in rats. *Arch. Oral Biol.,* **8**:775–776, 1963.

S224. Sykes, B. D. A transient nuclear magnetic resonance study of the kinetics of methyl *N*-acetyl-D-glucosaminide inhibition of lysozyme. *Biochemistry,* **8**:1110–1116, 1969.

S225. Sykes, B. D., and Dolphin, D. The role of distortion in the lysozyme-catalysed hydrolysis of glucosides. *Nature (London),* **233**:421–422, 1971.

S226. Sykes, B. D., and Parravano, C. A nuclear magnetic resonance study of the inhibition of lysozyme by *N*-acetyl-D-glucosamine and di-N-acetyl-D-glucosamine. *J. Biol. Chem.,* **244**:3900–3904, 1969.

S227. Sykes, B. D., Patt, S. L., and Dolphin, D. The role of distortion in the lysozyme mechanism. *Cold Spring Harbor Symp. Quant. Biol.,* **36**:29–33, 1971.

S228. Syrén, E., and Raeste, A.-M. Identification of blood monocytes by demonstration of lysozyme and peroxidase activity. *Acta Haematol.,* **45**:29–35, 1971.

S229. Szeszák, F., Szabó, G., Erdei, J., and Müller, F. Changes of resistance to lysozyme and ultrasonic disintegration of the mycelium of *Streptomyces griseus* under the influence of chelating agents and polyvalent cations. *Can. J. Microbiol.,* **14**:769–773, 1968.

S230. Szpilman, H. A. Diagnostic and prognostic value of lysozyme (muramidase) determinations in leukemias. *Pol. Arch. Med. Wewn.,* **45**:723–725, 1970. (Pol.)

S231. Szturm-Rubinsten, S. Pouvoir pathogène expérimental et sensibilité au lysozyme des cultures de Shigella. *Ann. Inst. Pasteur, Paris,* **94**:508–511, 1958.

S232. Szymczyk, T., Gratkowska, H., and Kozlowska, I. Lysozyme. I. Structure and properties. *Czas. Stomatol.,* **22**:213–219, 1969. (Pol.)

T

T1. Takahashi, S., and Schmid, K. Basic polypeptides of human plasma. *Biochim. Biophys. Acta,* **82**:627–629, 1964.

T2. Takahashi, T., Hamaguchi, K., Hayashi, K., Imoto, T., and Funatsu, M. Structure of lysozyme. X. On the structural role of tryptophan residues. *J. Biochem. (Tokyo),* **58**:385–387, 1965.

T2a. Tabachnick, M., and Giorgio, N. A. Jr. Interaction of thyroid hormone with histone from calf thymus nuclei. *Nature,* **212:**1610–1611, 1966.

T3. Takaoka, T., Katsuta, H., Ishiki, S., and Konishi, T. Effects of lysozyme on the proliferation of fibroblasts in tissue culture. *Jap. J. Exp. Med.,* **42:**221–232, 1972.

T4. Takasuka, N. The action of lysozyme upon inflammatory sinus mucous membrane. *Otolaryngólogy (Tokyo),* **38:**895–901, 1966. (Jap.)

T5. Takeda, H., Strasdine, G. A., Whitaker, D. R., and Roy, C. Lytic enzymes in the digestive juice of Helix pomatia; chitinases and muramidases. *Can. J. Biochem.,* **44:**509–518, 1966.

T5a. Takenaka, O., and Shibata, K. States of amino acid residues in proteins. XX. Fluorescence of Stilbene dyes adsorbed on hydrophobic regions of protein molecules. *J. Biochem. (Tokyo),* **66:**805–814, 1969.

T6. Takikawa, K., and Ohta, H. On the nature of neutrophilic granules. *Acta Haematol. Jap.,* **29:**571–577, 1966. (Jap.)

T7. Takumi, K., Kawata, T., and Hisatsune, K. Autolytic enzyme system of *Clostridium botulinum.* II. Mode of action of autolytic enzymes in *Clostridium botulinum* type A. *Jap. J. Microbiol.,* **15:**131–141, 1971.

T7a. Tallan, H. H., and Stein, W. H. Chromatographic studies on lysozyme. *J. Biol. Chem.* **200:**507–514, 1953.

T8. Tallei, F. Ricerche sull'attività lisozimica ed antilisozimica del vitreo e della retina del coniglio. *G. Ital. Oftalmol.,* **3:**464–467, 1950.

T9. Tamaki, S., and Matsuhashi, M. Increase in sensitivity to antibiotics and lysozyme on deletion of lipopolysaccharides in *Escherichia coli* strains. *J. Bacteriol.,* **114:**453–454, 1973.

T10. Tan, K. H., and Lovrien, R. Enzymology in aqueous-organic cosolvent binary mixtures. *J. Biol. Chem.,* **247:**3278–3285, 1972.

T11. Tanford, C., and Aune, K. C. Thermodynamics of the denaturation of lysozyme by guanidine hydrochloride. III. Dependence on temperature. *Biochemistry,* **9:**206–211, 1970.

T12. Tanford, C., Aune, K. C., and Ikai, A. Kinetics of unfolding and refolding of proteins. III. Results for lysozyme. *J. Mol. Biol.,* **73:**185–197, 1973.

T13. Tanford, C., Pain, R. H., and Otchin, N. S. Equilibrium and kinetics of the unfolding of lysozyme (muramidase) by guanidine hydrochloride. *J. Mol. Biol.,* **15:**489–504, 1966.

T13a. Tanford, C. and Wagner, M. L. Hydrogen ion equilibria of lysozyme. *J. Amer. Chem. Soc.,* **76:**3331–3336, 1954.

T13b. Tanford, C. The interpretation of hydrogen ion titration curves of proteins. *Advances Protein Chem.,* **17:**69–165, 1962.

T14. Tanford, C., and Roxby, R. Interpretation of protein titration curves. Application to lysozyme. *Biochemistry,* **11:**2192–2198, 1972.

T15. Tani, I., Uchida, K., Takeshita, T. *et al.* Evaluation of the effect of lysozyme on chronic sinusitis. (2) Histological examination of the maxillary sinal mucosa treated by combined use with antibiotics. *Otolaryngology (Tokyo),* **38:**759–763, 1966. (Jap.)

T16. Tarlovskaya, S. I. Lysozyme in therapy of corneal ulcers. *Vestn. Oftalmol.,* **14:**20–21, 1939.

T17. Tassoni, G. Ulteriori ricerche in vitro sull'azione fluidificante di enzimi e sostanze riducenti e della loro associazione. *Valsalva,* **46:**281–285, 1970.

T18. Tassoni, G. Studio dell'attività fluidificante in vitro del lisozima e della papaina. *Valsalva,* **45:**242–246, 1969.

T19. Taylor, P. W., and Morgan, H. R. Antibacterial substances in human semen and prostatic fluid. *Surg. Gynecol. Obstet.*, **94**:662–668, 1952.

T19a. Teale, F. W. J. The ultraviolet fluorescence of proteins in neutral solution. *Biochem. J.*, **76**:381–388, 1960.

T20. Tecilazich, F. La terapie delle epatiti virali dell'infanzia. *Epatologia*, **12**:514–516, 1966.

T21. Tecilazich, F. Terapia dell'ipertrofia timico del lattante. *Minerva Pediat.*, **18**:489–490, 1966.

T22. Tecilazich, F. The prophylactic action of lysozyme in measles infection. *Sem. Med. (Buenos Aires)*, **119**:1046–1057, 1961. (Span.)

T23. Teichberg, V. I., Kay, C. M., and Sharon, N. Separation of contributions of tryptophans and tyrosines to the ultraviolet circular dichroism spectrum of hen egg-white lysozyme. *Eur. J. Biochem.*, **16**:55–59, 1970.

T24. Teichberg, V. I., Plasse, T., Sorell, S., and Sharon, N. A spectrofluorimetric study of human lysozyme. *Biochim. Biophys. Acta*, **278**:250–257, 1972.

T24a. Teichberg, V. I., and Sharon, N. A spectrofluorometric study of tryptophan 108 in hen egg-white lysozyme. *FEBS Lettr.*, **7**:171–174, 1970.

T25. Teichberg, V. I., and Shinitzky, M. Fluorescence polarization studies of lysozyme and lysozyme-saccharide complexes. *J. Mol. Biol.*, **74**:519–531, 1973.

T26. Teicher, E., Maron, E., and Arnon, R. The role of specific amino acid residues in the antigenic reactivity of the loop peptide of lysozyme. *Immunochemistry*, **10**:265–271, 1973.

T27. Terry, J. M., Blainey, J. D., and Swingler, M. C. An automated method for lysozyme assay. *Clin. Chim. Acta*, **35**:317–320, 1971.

T28. Terzaghi, E., Okada, Y., Streisinger, G., Inouye, M., and Tsugita, A. Changes of a sequence of amino acids in phage T4 lysozyme by acridine-induced mutations. *Proc. Nat. Acad. Sci. U.S.*, **56**:500–507, 1966.

T29. Teti, M. I fattori immunitari in corso di epatite virale sperimentale. *Epatologia*, **12**:83–88, 1966.

T30. Teti, M. Variazioni dei fattori umorali dell'immunità naturale. *Clin. Eur. (Rome)*, **4**:333–337, 1965.

T31. Teti, M. Sui rapporti tra leucociti polimorfonucleati e concentrazione di policationi organici in vivo. *Minerva Radiol., Fisioter. Radiobiol.*, **10**:161–164, 1965.

T32. Teti, M. Isolamento e caratterizzazione di una ribonucleoproteina nel mezzo liquido della bacteriolisi lisozimica. *Riv. Ist. Sieroter. Ital.*, **27**:279–281, 1952.

T33. Teti, M., and Falcone, G. Ricerche manometriche sulle modificazioni della respirazione propria ed in presenza di substrato di alcuni stipiti di S. typhi sottoposti all'azione del lisozima. *Riv. Ist. Sieroter. Ital.*, **27**:462–466, 1952.

T34. Teti, M., and Fiore, M. Morphological modification induced by lysozyme in microorganisms of varying sensitivity. *Boll. Ist. Sieroter. Milan.*, **28**:160–169, 1949.

T35. Tets, V. L. Lysozyme and its use in the food industry. *Vop. Pitan.*, **7**:No. 3, 80–90, 1938.

T36. Teuber, M. Lysozyme-dependent production of spheroplast-like bodies from polymyxin B-treated *Salmonella typhimurium*. *Arch. Mikrobiol.*, **70**:139–146, 1970.

T37. Tew, J. G., Hess, W. M., and Donaldson, D. M. Lysozyme and beta-lysin release stimulated by antigen-antibody complexes and bacteria. *J. Immunol.*, **102**:743–750, 1969.

T38. Tew, J. G., Scott, R. L., and Donaldson, D. M. Plasma beta-lysin and lysozyme following endotoxin administration and the generalized Shwartzman reaction. *Proc. Soc. Exp. Biol. Med.*, **136**:473–478, 1971.

T39. Thacore, H., and Willett, H. P. The formation of spheroplasts of Mycobacterium tuberculosis in tissue culture cells. *Amer. Rev. Resp. Dis.*, **93**:786–796, 1965.

T40. Thacore, H. and Willett, H. P. Formation of spheroplasts of *Mycobacterium tuberculosis* by lysozyme. *Proc. Soc. Exp. Biol. Med.*, **114**:43–47, 1963.

T41. Thomas, E. W. Interaction between diacetylchitobiose methyl glycoside and lysozyme as studied by NMR spectroscopy. *Biochem. Biophys. Res. Commun.*, **29**:628–634, 1967.

T42. Thomas E. W. Interaction between lysozyme and acetamido sugars as detected by proton magnetic resonance spectroscopy. *Biochem. Biophys. Res. Commun.*, **24**:611–615, 1966.

T43. Thomas, E. W., McKelvy, J. F., and Sharon, N. Specific and irreversible inhibition of lysozyme by 2',3'-epoxypropyl beta-glycosides of N-acetyl-D-glucosamine oligomers. *Nature (London)*, **222**:485–486, 1969.

T44. Thompson, A. R. The C-terminal residue of lysozyme. *Nature (London)*, **169**:495–496, 1952.

T44a. Thompson, A. R. Amino acid sequence in lysozyme. I. Displacement chromatography of peptides from a partial hydrolysate on ion-exchange resins. *Biochem. J.*, **60**:507–515, 1955.

T45. Thompson, E. O. P. The C-terminal amino-acid sequence of lysozyme. *Biochim. Biophys. Acta*, **25**:210, 1957.

T45a. Thompson, E. O. P., O'Donnell, I. J. Quantitative reduction of disulphide bonds in proteins using high concentrations of mercaptoethanol. *Biochim. Biophys. Acta*, **53**:447–449, 1961.

T46. Thompson, K. E., and Levy, J. G. Effect of sequential degradation on the haptenic activity of a tryptic peptide isolated from reduced and alkylated egg-white lysozyme and the haptenic properties of its amino-terminal residues. *Biochemistry*, **9**:3463–3468, 1970.

T47. Thompson, R. Lysozyme and antibacterial properties of tears. *Arch. Ophthalmol.*, N.S. **25**:491–509, 1941.

T48. Thompson, R. Lysozyme and its relation to the antibacterial properties of various tissues and secretions. *Arch. Pathol.*, **30**:1096–1134, 1940.

T49. Thompson, R., and Gallardo, E., Jr. The susceptibility of the antigenic types of staphylococci to several protective enzymes. *J. Bacteriol.*, **36**:302, 1938.

T50. Thompson, R., and Gallardo, E. The concentration of lysozyme in the tears in acute and chronic conjunctivitis. With a note on the source of the lysozyme of tears. *Amer. J. Ophthalmol.*, 3, **19**:684–685, 1936.

T51. Thompson, R., and Khorazo, D. Susceptibility to lysozyme of staphylococci. *Proc. Soc. Exp. Biol. Med.*, **33**:299–302, 1935.

T52. Thomsen, J., Lund, E. H., Kristiansen, K., Brunfeldt, K., and Malmquist, J. A val-Val sequence found in a human monocytic leukemia lysozyme. *FEBS Lett.*, **22**:34–36, 1972.

T53. Thorsell, W., Björkman, N., and Wittander, G. Studies on the action of some enzymes on the cyst wall of isolated metacerkariae from the liver fluke *Fasciola hepatica* L. *Experientia*, **2**:587–589, 1965.

T54. Tichý, P., Rytír, V., and Kohoutová, M. Genetic transformation and transfection of *Bacillus subtilis* spheroplasts. *Folia Microbiol. (Prague)*, **13**:510–514, 1968.

T55. Tichý, P., and Kohoutová, M. Transformation of lysozyme spheroplasts of Bacillus subtilis. *Folia Microbiol. (Prague)*, **13**:317–323, 1968.

T56. Tietze, F. Enzymic release of amino acids from the carboxyl terminus of native and modified egg-white lysozyme. *Arch. Biochem. Biophys.*, **87**:73–80, 1960.

T57. Timasheff, S. N., and Inoue, H. Preferential binding of solvent components to proteins in mixed water-organic solvent systems. *Biochemistry,* **7**:2501–2513, 1968.

T58. Timasheff, S. N., and Rupley, J. A. Infrared titration of lysozyme carboxyls. *Arch. Biochem. Biophys.,* **150**:318–323, 1972.

T59. Tinozzi, C. C. L'associazione lisozima-corticosteroidi in dermatologia. *Minerva Dermatol.,* **40**:209–212, 1965.

T60. Tinozzi, G. C., and Bruni, L. Sulla eventuale utilità terapeutica dell'associazione tra lisozimae cortisonici in dermatologia. Contributo clinico preliminare. *Minerva Dermatol.,* **38**:34–37, 1963.

T61. Tischendorf, F. W., and Heckner, F. Atypisches plasmozytom mit G3K- MK-doppelparaproteinamie Bence-Jones-protein (Type K) und lysozymurie. *Haematol. Bluttransfus.,* **8**:162, 1969.

T62. Tischendorf, F. W., Ledderose, G., Müller, D., Orywall, D., and Wilmanns, W. Chronische Myelosen mit massiver Lysozymurie unter Milzbestrahlung. *Klin. Wochenschr.,* **50**:250–257, 1972.

T63. Tischendorf, F. W., Ledderose, G., Müller, D., and Wilmanns, W. Heavy lysozymuria after X-irradiation of the spleen in human chronic myelocytic leukaemia. *Nature (London),* **235**:274–275, 1972.

T64. Tischendorf, F. W., and Osserman, E. F. Immunochemical analyses of human lysozyme in monocytic dyscrasias. *Protides Biol. Fluids, Proc. Colloq.,* **19**:197–204, 1969.

T65. Tischendorf, F. W., and Tischendorf, M. M. Immunochemische und enzymatische Analyse des menschlichen Lysozyms aus dem Urin von Monocytenleukämien. *Verh. Deut. Ges. Inn. Med.,* **75**:515–518, 1969.

T65a. Tiselius, A., Hjertén, S., and Levin, O. Protein chromatography on calcium phosphate columns. *Arch. Biochem. Biophys.,* **65**:132–155, 1956.

T66. Tobin, M. C. Raman spectra of crystalline lysozyme, pepsin, and alpha chymotrypsin. *Science,* **161**:68–69, 1968.

T67. Toder, V. A. The concentration of lysozyme in the macrophages of the peritoneal exudate of guinea pigs sensitized to streptococci. *Zh. Mikrobiol., Epidemiol. Immunobiol.,* **47**:115–118, 1970. (Russ.)

T68. Toder, V. A. Increased level of lysozyme in macrophages of guinea pigs sensitized by streptococcal antigens in the administration of specific antigen. *Byull. Eksp. Biol. Med.,* **68**:75–78, 1969. (Russ.)

T69. Tojo, T., Hamaguchi, K., Imanishi, M., and Amano, T. Structure of lysozyme. XI. Spectrophotometric titration of tyrosyl groups of hen and duck egg-white lysozyme. *J. Biochem. (Tokyo),* **60**:538–542, 1966.

T70. Tolone, S. Lysozyme in nervous tissue. *Boll. Soc. Ital. Biol. Sper.,* **24**:692–693, 1948.

T71. Tolone, S. Blood lysozyme and nicotinic acid. *Acta Neurol.,* **2**:296–305, 1947.

T72. Tolone, S. Indagini sull'attività lisozimica del liquor umano patologico. *Boll. Soc. Ital. Biol. Sper.,* **23**:381–382, 1947.

T73. Tolone, S. Natura del potere antibatterico liquorale. *Boll. Soc. Ital. Biol. Sper.,* **23**:369–370, 1947.

T74. Tolone, S. Aspetti del potere antibatterico liquorale. *Boll. Soc. Ital. Biol. Sper.* **23**:367–368, 1947.

T75. Tolone, S. Alterazioni morfologiche del *Micrococcus lysodeikticus* per azione del liquor normale e patologico. *Boll. Soc. Ital. Biol. Sper.,* **23**:364–366, 1947.

T76. Tolone, S. Ricerche sull'attività antibatterica del liquido cefalorachidiano. -Aggiunte preliminari alla morfologia del *Micrococcus lysodeikticus.* Nuovo metodo di colorazione Autolisi. *Boll. Soc. Ital. Biol. Sper.,* **23**:362–364, 1947.

T77. Tomcsik, J., and Guex-Holzer, S. Änderung der Struktur der Bakterienzelle im Verlauf der Lysozym-Einwirkung. *Schweiz. Z. Allg. Pathol. Bakteriol.,* **15**:517–525, 1952.

T77a. Tomimatsu, Y., and Gaffield, W. Optical rotatory dispersion of egg proteins. I. Ovalbumin, conalbumin, ovomucoid, and lysozyme. *Biopolymers,* 3:509–517, 1965.

T78. Tortorici, G. Sul potere lisozimico delle colecisti (ricerche cliniche e sperimentali). *Boll. Ist. Sieroter. Milan.,* **31**:51–54, 1952.

T79. Tosolini, G. C., and Galliera, A. Il potere lisozimico nel sangue materno e fetale. *Friuli Med.,* **18**:927–935, 1963.

T80. Toussi, T. Nouvelle thérapeutique de la douleur chez les cancéreux, au moyen du lysozyme (Antalzyme). *Schweiz. Med. Wochenschr.,* **94**:1463–1465, 1964.

T81. Tramer, Z., and Shugar, D. Deuteron and gamma-irradiation of dried preparations of lysozyme and ribonuclease. *Acta Biochim. Pol.,* **9**:281–293, 1962.

T82. Trimarchi, F., and Teti, D. Azione del lisozima cloruro sui serbatoi leucocitari nel ratto. *Boll. Soc. Ital. Biol. Sper.,* **42**:1272–1276, 1966.

T83. Tropeano, L., Cordaro, S., and Fichera, G. Sul comportamento del tasso delle agglutinine del siero e del quadro sieroproteico dopo somministrazione di lisozima parenterale in soggetti affeti da febbre tifoidea e da brucellosi. *Acta Med. Ital. Mal. Infet. Parasit.,* **12**:85–93, 1957.

T84. Troup, G. M., Wagner, I., and Walford, R. L. Liver pigment, liver histidase, and renal lysozyme changes in relation to age in normal and irradiated Syrian hamsters. *Radiat. Res.,* **29**:489–498, 1966.

T85. Troup, G. M., and Walford, R. L. Transplantation disease, renal lysozyme, and aging. *Transplantation,* **5**:43–50, 1967.

T86. Tsai, C. S., and Matsumoto, K. Acetamido group in the binding and catalysis of synthetic substances by lysozyme. *Biochem. Biophys. Res. Commun.,* **39**:864–869, 1970.

T87. Tsai, C. S., Reyes-Zamora, C., and Otson, R. Effect of glycosidic isologs in lysozyme catalysis. *Biochim. Biophys. Acta,* **250**:172–181, 1971.

T87a. Tsai, C. S., Tang, J. Y., and Subbarao, S. C. Substituent effect on lysozyme-catalysed hydrolysis of some beta-Aryl Di-N-acetylchitobiosides. *Biochem. J.,* **114**:529–534, 1969.

T88. Tsugita, A. Phage lysozyme and other lytic enzymes. *In* "The Enzymes" (P. D. Boyer, ed.), 3rd ed., Vol. 5, Academic Press, New York, 1971.

T89. Tsugita, A., and Inouye, M. Complete primary structure of phage lysozyme from *Escherichia coli* T4. *J. Mol. Biol.,* **37**:201–212, 1968.

T90. Tsugita, A., Inouye, M., Terzaghi, E., and Streisinger, G. Purification of bacteriophage T4 lysozyme. *J. Biol. Chem.,* **243**:391–397, 1968.

T91. Tsugita, A., Inouye, M., Imagawa, T., Nakanishi, T., Okada, Y., Emrich, J., and Streisinger, G. Frameshift mutations resulting in the changes of the same amino acid residue (140) in T4 bacteriophage lysozyme and *in vivo* codons for Trp, Tyr, Met, Val and Ile. *J. Mol. Biol.,* **41**:349–364, 1969.

T92. Tsukiori, N., Sasaki, S., Takahashi, T. *et al.* Synergic activity of lysozyme in combination with antituberculous agents. I. Inhibitory activity against M. tuberculosis in vitro and synergic effect in vivo measured by S.S.A.A.T. method. *Kekkaku,* **40**:555–559, 1965. (Jap.)

T93. Tumanian, M. A., Izvekova, A. V., Levitan, M. Kh. *et al.* Blood lyszoyme contents in patients with nonspecific ulcerous colitis as an index of natural resistance of the organism to infection. *Sov. Med.,* **32**:88–90, 1969. (Russ.)

T94. Tyeryar, F. J., Jr., and Doetsch, R. N. Protoplasts of the giant bacterium, *Caryophanon latum* Peshkoff. *Nature (London),* **195**:1327–1328, 1962.

U

U1. Uchida, Y., and Hirai, T. A simplified method for measurement of tear lysozyme. *Acta Soc. Ophthalmol. Jap.*, **73**:1184–1188, 1969. (Jap.)

U2. Uhlenbruck, G. Uber eine Reaktion von Lysozym mit Gangliosiden. *Naturwissenschaften*, **54**:286, 1967.

U3. Uhlig, H., Lehmann, K., Salmon, S., Jollés, J., and Jollés, P. Partieller enzymatischer Abbau von Lysozym und Ribonuclease mit einer mikrobiellen Aminopeptidase. *Biochem. Z.*, **342**:553–556, 1965.

U3a. Urnes, P., and Dotty, P. Optical rotation and the conformation of polypeptides and proteins. *Advan. Proteins Chem.*, **16**:401–544, 1961.

U4. Utevska, S. L., and Tamarina, A. E. Role of lysozyme of tonsils in local immunity. *Eksp. Med. (Kharkov)*, pp. 91–99, 1936. (Russ.)

U5. Utrilla, A. Jimenez, J., and Bravo, J. Influence of lysozyme on the *Treponema pallidum* immobilization test performed by the Utrilla-Bravo technic. *Amer. J. Clin. Pathol.*, **44**:709–711, 1965.

U6. Utrilla, A., Jimenez, J., and Bravo, J. Influence of lysozyme on the Treponema pallidum immobilization test perfomred by utrilla-bravo technic. *Tech. Bull. Registry Med. Technol. (Baltimore)*, **35**:197–199, 1965.

V

V1. Vaccaro, H., Cabezas, J., and Berrios, H. Factores inespecificos de defensa en ginecologia y obstetrica. Importancia del lisozima. *Bol. Soc. Chil. Obstet. Ginecol.*, **10**:233–258, 1945.

V2. Vaccaro, H., Cabezas, J., and Copaja, D. El lisozima en el tracto digestivo del recién nacido, y lactante. *Rev. Chil. Pediat.*, **15**:597–623, 1944.

V3. Vadehra, D. V., Baker, R. C., and Naylor, H. B. Distribution of lysozyme activity in the exteriors of eggs from *Gallus gallus*. *Comp. Biochem. Physiol., B*, **43**:503–508, 1972.

V4. Vaglio, N. Vitamin A e lisozima. Sul potere della vit. A di esaltare l'attività lisozimica di alcuni liquidi organici; ricerche nel campo, ostetrico. *Arch. Ostet. Ginecol.*, **54**:183–194, 1949.

V5. Vaglio, N. Vitamin C and A and lysozyme. The ability of vitamin A to increase the lysozyme activity of blood serum. 2. Vitamin A and lysozyme. The ability of Vitamin A to increase the lysozyme activity of some organic fluids. *Clin. Obstet. Ginecol.*, **50**:9–10, 1948.

V6. Vallecchi, C. Possibility of the use of lysozyme in the treatment of radiation sickness. III. *Rass. Dermatol. Sifilogra.*, **12**:35–41, 1959.

V7. Vallecchi, C. Observations on the behavior of the lysozyme activity of blood in toto in bearers of malignant neoplams subjected to ionizing therapy. II. *Rass. Dermatol. Sifilogra.*, **12**:27–34, 1959.

V8. Vallet Armengol, F. J. Antistaphylococcal immunoenzymotherapy. *Rev. Clin. Espan.*, **81**:419–421, 1961. (Span.)

V9. Van Berkel, T. J. C., Koster, J. F., and Hulsmann, W. C. Distribution of L- and M-type pyruvate kinase between parenchymal and Kupffer cells of rat liver. *Biochim. Biophys. Acta*, **276**:425–429, 1972.

V10. Vandoni, G., and Bertè, F. Attività lisozimica del liquido amniotico e delgi organi di feto bovino in diversi periodi di gravidanza. *Boll. Soc. Ital. Biol. Sper.*, **37**:743–745, 1961.

V11. van Eikeren, P., and Chipman, D. M. Substrate distortion in catalysis by lysozyme.

Interaction of lysozyme with oligosaccharides containing N-acetylxylosamine. *J. Amer. Chem. Soc.,* **94:**4788–4790, 1972.

V12. Varshavskaya, R. R. Lysozyme in diseases of eye. *Vestn. Oftalmol.,* **16:**471, 1940.

V13. Vasario, U., Riccardino, N., and Perinetti, G. Ricerche sull'eliminazione biliare del lisozima. *Arch. Sci. Med.,* **120:**54–57, 1965.

V14. Vasil'chenko, V. G. Effect of removable dental prosthesis made of plastic AKR-7 on the lysozyme content of saliva. *Stomatologiya (Moscow),* **44:**82–84, 1965. (Russ.)

V15. Vatteroni, M. Studio su alcuni fattori di immunità naturale nelle neoplasie maligne, con particolari riferimenti alle localizzazioni genitali femminili. *Quad. Clin. Ostet. Ginecol.,* **19:**168–177, 1964.

V16. Velican, D., and Velican, C. Lysozyme and the pericapillary reticulin network. *Nature (London),* **215:**889–890, 1967.

V17. Velican, D., and Velican, C. Observations on the action of lysozyme on reticulin. *Rev. Roum. Med. Interne,* **4:**209, 1967.

V18. Velican, D., and Velican, C. Observations on the action of lysozyme on reticulin. *Stud. Cercet. Med. Interna,* **8:**245–247, 1967. (Rum.)

V19. Venco, L. Ricerche sul "lysozym" nelle lacrime. *Rass. Ital. Ottalmol.,* **2:**519–552, 1933.

V19a. Venkatachalam, C. M. Stereochemical criteria for polypeptides and proteins. V. Conformation of a system of three linked peptide units. *Biopolymers,* **6:**1425–1436, 1968.

V20. Venkatappa, M. P., and Steinrauf, L. K. Sedimentation equilibrium studies on heavy atom derivatives of lysozyme. *Indian J. Biochem.,* **5:**28–30, 1968.

V21. Venkatasubramanian, K., Vieth, W. R., and Wang, S. S. Studies of lysozyme immobilized on collagen. *J. Ferment. Technol.,* **50:**600, 1972.

V22. Verhoeven, J., and Schwyzer, R. Charge transfer as a molecular probe in systems of biological interest. VI. Interactions between lysozyme and the methyl 2-acetamido-6-O (N-methyl-isonicotinylium) 2-deoxy-Beta-D-glucopyranoside ion. *Helv. Chim. Acta,* **55:**2572–2581, 1972.

V23. Vernon, C. A. The mechanisms of hydrolysis of glycosides and their relevance to enzyme-catalysed reactions. *Proc. Roy. Soc., Ser. B,* **167:**389–401, 1967.

V24. Veronese, F. M., Boccú, E., and Fontana, A. Modification of tryptophan 108 in lysozyme by 2-nitro-4-carboxyphenylsulfenyl chloride. *FEBS Lett.,* **21:**277, 1972.

V25. Veronova, M. S. Lysozyme in the therapy of diarrhea of infants. *Pediatriya, (Moscow),* **3:**67, 1941.

V26. Vershigora, A. E., Begunova, T. I., Sidorenko, E. N. *et al.* Indices of specific and non-specific immunological reactivity in the desensitization process. *Zh. Ushn. Nos. Gorl. Bolez.,* **28:**85–89, 1968. (Russ.)

V27. Vichtinskii, A. A., Zaslavskiĭ, B. Yu., and Khorlin, A. Ya. Investigation of interactions of lysozyme with Serum albumins by the method of microcalorimetry. *Biokhimiya,* **36:**1294–0000, 1971. (Russ.)

V28. Vichtinskii, A. A. Zaslavskiĭ, B. Y., and Khorlin, A. V. Study of interactions of lysozyme with serum albumins by a microcalorimetric method. *Biokhimia,* **36:**1294–1296, 1971. (Russ.)

V29. Vietzke, W. M., Perillie, P. E., and Finch, S. C. Serum muramidase in patients with neutropenia. *Yale J. Biol. Med.,* **45:**457–462, 1972.

V30. Villanueva, J. R., Gascón, S., and Garcia Acha, I. Lytic activity on yeast cell walls as a useful character for the separation of *Streptomyces* and *Nocardia. Nature (London),* **198:**911–912, 1963.

V31. Vincentelli, J. B., and Léonis, J. Etude physico-chimique du lysozyme à l'état cristallin. II. Effet de l'urée sur la dénaturation thermique. *Arch. Int. Physiol. Biochim.,* **78:**1014–1015, 1970.

V32. Vincentelli, J. B., and Léonis, J. Etude physico-chemique du lysozyme a l'état cristalin. I. Effet des alcools sur la dénaturation thermique. *Arch. Int. Physiol. Biochim.,* **77:**561–562, 1969.

V33. Vincentelli, J. B., Looze, Y., and Léonis, J. Etude physico-chimique du lysozyme à l'état crystallin. 3. Effet des alcools polyhydroxyliques sur la denaturation thermique, *Arch. Int. Physiol. Biochim.,* **79:**855–856, 1971.

V34. Violle, H. Contribution à l'étude du lysozyme. *Presse Med.,* **61:**846–848, 1953.

V35. Viscidi, E., Consiglio, E., and Roche, J. Binding of thyroid hormones by lysozyme: Fluorescence quenching studies. *Biochim. Biophys. Acta,* **121:**424–426, 1966.

V36. Vitale, S., and Borselli, L. Sul comportamento della fosfatasi alcalina e dell'acido ribonucleinico in alcune ghiandole a secrezione interna di ratti in corso di trattamento protratto con lisozima. *Minerva Pediat.,* **17:**1015–1019, 1965.

V37. Vitetta, M., Costa, A. L., and Saitta, T. La condizione biologica dell'immaturità affettiva esplorata con le varizioni del lisozima serico e leucocitario. *Rass. Neuropsichiat. Sci. Affini,* **19:**633–638, 1966.

V38. Vitetta, M., and Saitta, T. La valutazione della reazione alla frustrazione nella immaturità affettiva in rapporto ad alcuni fattori biologici. *Rass. Neuropsichiat. Sci. Affini,* **19:**785–791, 1965.

V39. Vittadini, G. Ulteriore contributo alla terapia dell'iperplasia timica con lisozima. *Minerva Pediat.,* **18:**478–479, 1966.

V40. Vogliazzo, U., and Gheis, F. Attivita' lisozimica nel siero di sangue di 200 silicotici. *Med. Lav.,* **52:**262–270, 1961.

V41. Volpe, A. Evaluation of the modifications of the neutrophil leukocytosis in some acute peritonites treated with lysozyme. *Rass. Int. Clin. Ter.,* **42:**1129–1137, 1962.

V42. Volpe, A., and Santoro, S. Il lisozima nella prevenzione delle aderenze peritoneali post-operatorie. *G. Ital. Chir.,* **18:**231–241, 1962.

V43. Voronova, M. S. Lysozyme in therapy of diarrhea in infants. *Pediatriya, (Moscow),* **3:**67, 1941.

V44. Voss, J. G. Lysozyme lysis of gram-negative bacteria without production of spheroplasts. *J. Gen. Microbiol.,* **35:**313–317, 1964.

W

W1. Wadström, T. Bacteriolytic enzymes from *Staphylococcus aureus.* Properties of the endo-β-N/acetylglycosaminidase. *Biochem. J.,* **120:**745–752, 1970.

W2. Wadström, T., and Hisatsune, K. Bacteriolytic enzymes from *Staphylococcus aureus.* Specificity of action of endo-β-N/acetylglucosaminidase. *Bichem. J.,* **120:**735–744, 1970.

W3. Wagner, K., and Katz, W. Assoziationsverhalten von T2-Lysozym. *Z. Naturforsch. B.* **19:**230–234, 1964.

W4. Wahl, P., and Lami, H. Etude du déclin de la fluorescence du lysozyme-1-dimé thylaminonaphtalène-5-sulfonyl. *Biochim. Biophys. Acta,* **133:**233–242, 1967.

W5. Walker, J. B., and Hnilica, V. S. Developmental changes in arginine: X amidinotransferase activity in streptomycin-producing strains of Streptomyces. *Biochim. Biophys. Acta,* **89:**473–482, 1964.

W6. Wang, K. J., Grant, R., Janowitz, H. D., and Grossman, M. I. Action of lysozyme on gastrointestinal mucosa. *Arch. Pathol.,* **49:**298–308, 1950.

W7. Wardlaw, A. C. The complement-dependent bacteriolytic activity of normal human serum. I. The effect of pH and ionic strength and the role of lysozyme. *J. Exp. Med.,* **115:**1231–1249, 1962.

W8. Warner, N., Moore, M. A. S., and Metcalf, D. A transplantable myelomonocytic leukemia in BALB-c mice: Cytology, karyotype, and muramidase content. *J. Nat. Cancer Inst.,* **43:**963–982, 1969.

W9. Warren, G. H., and Gray, J. Effect of sublethal concentrations of penicillins on the lysis of bacteria by lysozyme and trypsin. *Proc. Soc. Exp. Biol. Med.,* **120:**504–511, 1965.

W10. Warren, G. H., and Gray, J. Production of a polysaccharide by *Staphylococcus aureus.* III. Action of penicillins and polysaccharides on enzymic lysis. *Proc. Soc. Exp. Biol. Med.,* **116:**317–323, 1964.

W11. Warren, G. H., Gray, J., and Bartell, P. The lysis of *Pseudomonas aeruginosa* by lysozyme. *J. Bacteriol.,* **70:**614–619, 1955.

W12. Warren, J. R., and Gordon, J. A. Denaturation of globular proteins. II. The interaction of urea with lysozyme. *J. Biol. Chem.,* **245:**4097–4104, 1970.

W13. Warth, A. D., and Strominger, J. L. Structure of the peptidoglycan of bacterial spores: Occurrence of the lactam of muramic acid. *Proc. Nat. Acad. Sci. U.S.,* **64:**528–535, 1969.

W14. Watanabe, K., and Takesue, S. Inhibitory effect of some gaseous hydrocarbons on the cell-lysis of *Micrococcus lysodeikticus* by egg-white lysozyme. *Agri. Biol. Chem.,* **36:**825–830, 1972.

W15. Watanabe, K., and Takesue, S. Effect of some hydrocarbon gases on egg-white lysozyme activities on different substrates. *Enzymologia,* **41:**99–111, 1971.

W16. Wauters, J. P., and Favre, H. L'intérèt de la mesure du lysozyme urinaire dans le diagnostic des méphropathies. *Schweiz. Med. Wochenschr.,* **100:**1903–1907, 1970.

W17. Webb, M. The action of lysozyme on heat-killed gram-positive microorganisms. *J. Gen. Microbiol.,* **2:**260–274, 1948.

W17a. Weber, G. Fluorescence-polarization spectrum and electronic-energy transfer in proteins. *Biochem. J.,* **75:**345–352, 1960.

W18. Weberschinke, J., and Kittnar, E. A shortened and simplified treponema test in diagnostic practice. *J. Hygi., Epidemiol., Microbiol., Immunol.,* **10:**483–487, 1966.

W19. Wedgwood, R. J., and Davis, S. D. Potentiation of serum bactericidal activity by lysozyme. *Fed. Proc., Fed. Amer. Soc. Exp. Biol.,* **27:**370, 1968.

W20. Weibull, C., Zacharias, B., and Beckman, H. Affinity of lysozyme to structural elements of the bacterial cell as studied with enzyme labelled with radioactive iodine. *Nature (London),* **184:**Suppl 22, 1744–1745, 1959.

W20a. Weidel, W., and Pelzer, H. Bagshaped macromolecules—a new outlook on bacterial cell walls. *Advan. Enzymol.,* **26:**193–232, 1964.

W21. Weil, E. Untersuchungen über die keimtötende Kraft der weifsen Blutkörperchen. *Arch. Hyg.,* **74:**289–344, 1911.

W21a. Weil, L., and Buchert, A. R. Photoöxidation of crystalline beta-lactoglobulin in the presence of methylene blue. *Arch. Biochem. Biophys.,* **34:**1–15, 1951.

W21b. Weil, L., Buchert, A. R., Mahler, J. Photoöxidation of crystalline lysozyme in the presence of methylene blue and its relation to enzymatic activity. *Arch. Biochem. Biophys.,* **40:**245–252, 1952.

W22. Weinbaum, G., Rich, R., and Fischman, D. A. Enzyme-induced formation of spheres from cells and envelopes of *Escherichia coli. J. Bacteriol.,* **93:**1693–1698, 1967.

W23. Weiser, R. S. Antibacterial action of lysozyme on tubercle bacilli. *Tuberculology,* **14:**217–219, 1954.

W24. Weiser, R. S., Youmans, G. P., Youmans, A. S., and Myrvik, Q. Lysozyme treatment of tuberculous mice. *J. Infec. Dis.,* **102**:53–59, 1958.

W25. Welsch, M. Lysis and transformation by lysozyme and actinomycin in "spheroplasts" of Streptomyces. *C. R. Soc. Biol.,* **154**:453–456, 1960.

W26. Welsch, M. Le «lysozyme» des staphylocoques. *C. R. Soc. Biol.,* **153**:2080–2083, 1959.

W27. Welsch, M. Formation de protoplastes d'Escherichia coli sous l'influence de la glycine et d'autres acides aminés. *Schweiz. Z. Allg. Pathol. Bakteriol.,* **21**:741–768, 1958.

W28. Welshimer, H. J., and Robinow, C. F. The lysis of Bacillus megatherium by lysozyme. *J. Bacteriol.,* **57**:489–499, 1949.

W29. Wenzel, M., Lenk, H.-P., and Schütte, E. Herstellung von Tri-[N-acetyl]-chitotriose-[²H] und deren Spaltung durch lysozym. *Hoppe-Seyler's Z. Physiol. Chem.,* **327**:13–20, 1961.

W29a. Wetlaufer, D. B., and Stahmann, M. A. Solubility and mechanism of dye-uptake in protein-dye salts. *J. Amer. Chem. Soc.,* **80**:1493–1500, 1958.

W30. Wetter, L. R., and Deutsch, H. F. Immunological studies on egg white proteins. IV. Immunochemical and physical studies of lysozyme. *J. Biol. Chem.,* **192**:237–242, 1951.

W31. Wharton, D. R. Lysozyme retention by cockroach *Periplaneta americana* L. *Science,* **163**:183–184, 1969.

W32. Wharton, D. R. A., and Lola, J. E. Lysozyme action on the cockroach, *Periplaneta americana,* and its intracellular symbionts. *J. Insect Physiol.,* **15**:1647–1658, 1969.

W32a. Whitaker, J. Determination of molecular weights of proteins by gel filtration on Sephadex. *Anal. Chem.,* **35**:1950–1953, 1963.

W33. White, F. H., Jr. Tritiation of lysozyme by the free-radical interceptor method and determination of the tritium distribution among individual residues of the chromatographically homogeneous protein. *Radiat. Res.,* **36**:470–482, 1968.

W34. White, P. B. Lysogenic strains of V. cholerae and the influence of lysozyme on cholera phage activity. *J. Pathol. Bacteriol.,* **44**:276–278, 1937.

W35. Wieczorek, Z., Czajka, M., and Kowalczyk, H. Assay of serum lysozyme by the diffusion method. *Arch. Immunol. Ther. Exp.,* **15**:829–832, 1967.

W36. Wieczorek, Z., Skurski, A., and Siemek, R. The effect of lysozyme on the phagocytosis of tubercle bacilli. *Arch. Immunol. Ther. Exp.,* **13**:197–203, 1965.

W37. Wien, R. W., Morrisett, J. D., and McConnell, H. M. Spin-label-induced nuclear relaxation. Distances between bound saccharides, histidine-15, and tryptophan-123 on lysozyme in solution. *Biochemistry,* **11**:3707–3716, 1972.

W38. Wiener, E., and Levanon, D. The *in vitro* interaction between bacterial lipopolysaccharide and differentiating monocytes. *Lab. Invest.,* **19**:584–590, 1968.

W39. Wiernik, P. H., and Serpick, A. A. Clinical significance of serum and urinary muramidase activity in leukemia and other hematologic malignancies. *Amer. J. Med.,* **46**:330–343, 1969.

W39a. Wilcox, F. H. Jr., and Daniel, L. J. Reduced lysis at high concentrations of lysozyme. *Arch. Biochem. Biophys.,* **52**:305–312, 1954.

W40. Wilhelm, J. M., and Haselkorn, R. The chain growth rate of T4 lysozyme in vitro. *Proc. Nat. Acad. Sci. U.S.,* **65**:388–394, 1970.

W41. Wilhelm, J. M., and Haselkorn, R. In vitro synthesis of T4 proteins: Lysozyme and the products of genes 22 and 57. *Cold Spring Harbor Symp. Quan. Biol.,* **34**:793–798, 1969.

W42. Wilkison, E. E., and Lehman, E. P. Relationship of lysozyme to mechanism and cause of death in intestinal obstruction. *Surg. Forum,* 94–100, 1953.

W43. Willett, H. P., and Thacore, H. Formation of spheroplasts of Mycobacterium tuberculosis by lysozyme in combination with certain enzymes of rabbit peritoneal monocytes. *Can. J. Microbiol.*, **13**:481–488, 1967.

W44. Willett, H. P., and Thacore, H. The induction by lysozyme of an L-type growth in Mycobacterium tuberculosis. *Can. J. Microbiol.*, **12**:11–16, 1966.

W45. Williams, E. J., Herskovits, T. T., and Laskowski, M., Jr. Location of chromophoric residues in proteins by solvent perturbation. 3. Tryptophyls in lysozyme and in alpha-chymotrypsinogen and its derivatives. *J. Biol. Chem.*, **240**:3574–3579, 1965.

W45a. Williams, E. J., and Laskowski, M. Jr. A method for distinguishing between complete and partial exposure of tryptophyls in proteins. alpha-Chymotrypsinogen and lysozyme. *J. Biol. Chem.*, **240**:3580–3584, 1965.

W46. Willoughby, W. F. Pulmonary arteritis induced by cationic proteins. *Amer. Rev. Resp. Dis.*, **105**:50–59, 1972.

W47. Wilson, A. T. Urinary lysozyme. Identification and measurement. *J. Pediat.*, **36**:39–44, 1950.

W48. Wilson, A. T., and Hadley, W. P. Urinary lysozyme. III. Lysozymuria in children with nephrotic syndrome. *J. Pediat.*, **36**:199–211, 1950.

W49. Wilson, A. T., and Hadley, W. P. Urinary lysozyme. II. Lysozymuria in healthy children and in children with miscellaneous diseases (sex difference). *J. Pediat.*, **36**:45–50, 1950.

W50. Wilson, G. A., and Bott, K. F. Effects of lysozyme on competence for *Bacillus subtilis* transfection. *Biochim. Biophys. Acta*, **199**:464–475, 1970.

W51. Wilson, L. A., and Spitznagel, J. K. Molecular and structural damage to *Escherichia coli* produced by antibody, complement, and lysozyme systems. *J. Bacteriol.*, **96**:1339–1348, 1968.

W52. Wiseman, D. A common antigen in the cell walls of three lysozyme-sensitive bacteria. *J. Pharm. Pharmacol.*, **15**:Suppl., 182T–184T, 1963.

W53. Wolff, J., and Covelli, I. Iodination of the normal and buried tyrosyl residues of lysozyme. II. Spectrophotometric analysis. *Biochemistry*, **5**:867–871, 1966.

W53a. Wolff, J., and Covelli, I. Factors in the iodination of histidine in proteins. *Eur. J. Biochem.*, **9**:371–377, 1969.

W54. Wolff, L. K. Researches on lysozyme. *Ned. Tijdschr. Geneesku.*, **00**:2303–2309, 1927.

W55. Wolff, L. K. Untersuchungen über das Lysozym. *Z. Immunitäetsforsch. Exp. Ther.*, **54**:188–198, 1927–1928.

W56. Wolff, L. K. Untersuchungen über das Lysozym. *Z. Immunitäetsforsch. Exp. Ther.*, **50**:88–100, 1927.

W57. Wolff, L. K. Researches on lysozyme. *Ned. Tijdschr. Geneesk.*, **70**:2340–2348, 1926.

W58. Wolin, M. J. Lysis of Vibrio succinogenes by ethylenediamine-tetraacetic acid or lysozyme. *J. Bacteriol.*, **91**:1781–1786, 1966.

W59. Wolinsky, I., and Cohn, D. V. Bone lysozyme: Partial purification, properties and depression of activity by parathyroid extract. *Nature (London)*, **210**:413–414, 1966.

W60. Wollman, E., and Wollman, E. Phases of the lysogenic function. Successive action of lysozyme and trypsin on lysogenic bacteria. *C. R. Soc. Biol.*, **131**:442–445, 1939.

W61. Wollman, E., and Wollman, E. Mise en liberté des bactériophages d'une souche spontanément lysogène par l'action du lysozyme. Application à la détermination du rapport numérique entre bactéries, et bactériophages. *C. R. Soc. Biol.*, **119**:47–50, 1935.

W62. Work, E. Factors affecting the susceptibility of bacterial cell walls to the action of lysozyme. *Proc. Roy. Soc., Ser. B*, **167**:446–447, 1967.

W63. Wright, D. G., and Malawista, S. E. The mobilization and extracellular release of granular enzymes from human leukocytes during phagocytosis. *J. Cell Biol.,* **53:**788–797, 1972.

W64. Wright, G. G. Influence of spermine and related substances on susceptibility of *Bacillus anthracis* to lysozyme. *Proc. Soc. Exp. Biol. Med.,* **108:**740–742, 1961.

Y

Y1. Yamada, G. Clinical studies on high-molecular constituents of gastric juice—clinical significance of changes of gastric juice lysozyme activity in various stomach diseases. *Sapporo Med. J.,* **35:**235–250, 1969. (Jap.)

Y2. Yamamoto, K. Effects of cobalt-60 gamma rays on lysozyme in acqueous solution. 3. *J. Radiat. Res.,* **12:**133–137, 1971.

Y3. Yamamoto, K. Effects of cobalt-60 gamma rays on lysozyme in aqueous solution. *J. Radiat. Res.,* **11:**85–91, 1970.

Y4. Yamamoto, K., and Matsushima, Y. Studies on amino-hexoses. XI. A synthetic substrate of egg white lysozyme: phenyl 6-0-(2-acetamido-2-deoxy-beta-D-glucopyranosyl)-2-acetamido-3-0-(D-1-carboxyethyl)-2-deoxy-beta-D-glucopyranoside. *Bull. Chem. Soc. Jap.,* **40:**194–196, 1967.

Y5. Yamashita, T., and Bull, H. B. Films of lysozyme adsorbed at air-water surfaces. *J. Colloid Interface Sci.,* **27:**19–24, 1968.

Y6. Yamashita, T., and Bull, H. B. Spread monolayers of lysozyme. *J. Colloid Interface Sci.,* **24:**310–316, 1967.

Y7. Yamazaki, Y. Enzymatic activities on cell walls in bacteriophage T4. *Biochim. Biophys. Acta,* **178:**542–550, 1969.

Y7a. Yang, J. T., and Foster, J. F. Intrinsic viscosity and optical rotation of proteins in acid media. *J. Amer. Chem. Soc.,* **77:**2374–2378, 1955.

Y8. Yashinsky, G. Y. Fluorescence decay of lysozyme and of iodine oxidized lysozyme. *FEBS Lett.,* **26:**123–126, 1972.

Y9. Yasunobu, K. T., and Wilcox, P. E. Differences in the susceptibility of lysozyme and alpha-lactalbumin to the action of tyrosinase. *J. Biol. Chem.,* **231:**309–313, 1958.

Y10. Yoshimoto, T., Tsuru, D. Studies on bacteriolytic enzymes. II. Purification and some properties of two types of staphylolytic enzymes from Streptomyces griseus. *J. Biochem. (Tokyo),* **72:**379–390, 1972.

Y11. Yoshimura, T. Imanishi, A., and Iseumura, T. Preparation and properties of poly-DL-alanyl-lysozyme. *J. Biochem. (Tokyo),* **63:**730–738, 1968.

Y12. Youman, J. D., 3d, Saarni, M. I., and Linman, J. W. Diagnostic value of muramidase (Lysozyme) in acute leukemia and preleukemia. *May. Clin. Proc.,* **45:**219–228, 1970.

Y13. Young, E. T., 2nd., and Van Houwe, G. Control of synthesis of glucosyl transferase and lysozyme messengers after T4 infection. *J. Mol. Biol.,* **51:**605–619, 1970.

Y14. Young, J. D., and Leung, C. Y. Immunochemical studies on lysozyme and carboxy-methylated lysozyme. *Biochemistry,* **9:**2755–2762, 1970.

Y15. Young, M. On the titration behavior of dimethylaminonaphthalene-protein conjugates. *Biochim. Biophys. Acta,* **71:**206–208, 1963.

Y16. Yutani, K., Yutani, A., Imanishi, A., and Isemura, T. The mechanism of refording [refolding] of the reduced random coil form of lysozyme. *J. Biochem. (Tokyo),* **64:** 449–455, 1968.

Z

Z1. Zabirov, I. Sh., and Schepetkina, L. V. Benzylpenicillin effect on lysozyme activity in blood serum, leucocytes and organs of albino rats. *Antibiotiki (Moscow)*, **17**:730–734, 1972.

Z2. Zäh, K. Das Lysozym des Speichels. *HNO Hlas-, Nasen-, Ohrenaerzte.*, **29**:57–65, 1938.

Z3. Zanca, A., and Benatti, M. Action of lysozyme on the reactivity and diffusion capacity of the skin. *Arch. Ital. Dermatol. Venerol. Sessuol.*, **29**:341–342, 1959.

Z4. Zangara, A., Lombardo, V., and Gazzano, A. Ricerche sull'azione del lisozima sulla mucosa gastrica. *Minerva Gastroenterol.*, **5**:239–245, 1959.

Z5. Zanni, Giberti, A. Treatment of cutaneous and cutaneomucosal viral diseases with Fleming's lysozyme. *Sem. Med. (Buenos Aires)*, **119**:1043–1044, 1961. (Span.)

Z6. Zanussi, C., Berengo, A., and Passerini, A. Influenza delle intossicazioni sperimentali da esotossine sul lisozima del siero. *Boll. Ist. Siertoer. Milan.*, **30**:55–59, 1951.

Z7. Zanussi, C., Lusiani, G. B., and Berengo, A. Effetto della somministrazione di colesterolo sul potere battericida verso l'E. typhi e sul lisozima del siero nel coniglio. *Riv. Ist. Sieroter. Ital., Sez. I*, **25**:175–183, 1950.

Z8. Zanussi, C., and Prati, G. Behavior of lysozyme of serum in various morbid conditions. *Boll. Ist. Siertoer. Milan.*, **28**:89–95, 1949.

Z9. Zaoli, G. Attività lisozimica della saliva, del secreto nasale, del siero di sangue in alcuni ammalati di tbc polmonare con quadri anatomo-clinici disparati. *Arch. Tisiol Mal. App. Resp.*, **11**:453–463, 1956.

Z10. Zaoli, G. Attività lisozimica della saliva in corso di terapia con aureomicina, cloroamfenicolo, terramicina. *Oto-Rino-Laringol. Ital.*, **22**:175–182, 1954.

Z11. Zaslavskii, B. Iu, Vichutinskiĭ, A. A., and Khorlin, A. Ia. Study of the interactions of hen's egg white lysozyme with Mn + 2 and Ca + 2 cations by the method of reaction microcalorimetry. *Biofizica*, **17**:536–538, 1972. (Russ.)

Z12. Zaslavskii, B. Iu., Vichutinskiĭ, A. A., and Khorlin, A. Ia. Study of the thermodynamics of the interaction of lysozyme with neutral sugars by the method of reaction microcalorimetry. *Biofizica*, **17**:385–389, 1972. (Russ.)

Z13. Zaslavsky, B. Yu., Vichutinsky, A. A., and Khorlin, A. Ia. Microcalorimetric studies on the interactions of hen's egg white lysozyme with cations Mn^2 and Ca^2 +. *Biofizika*, **17**:536, 1972.

Z14. Zaslavsky, B. Yu., Vichutinsky, A. A., and Khorlin, A. Ya. Microcalorimetric studies on thermodynamics of interactions of lysozyme with some neutral saccarides. *Biofizika*, **17**:385, 1972. (Russ.)

Z15. Zehavi, U. Modification by low temperature of the transfer function of lysozyme. *Biochim. Biophys. Acta*, **194**:526–531, 1969.

Z15a. Zehavi, U. and Jeanloz, R. W. Hydrolysis of Benzyl 2-acetamido-4-0-(2-acetamido-2-deoxy-Beta-D-Glucopyranosyl)-2-deoxy-Beta-D-Glucopyranoside by egg white lysozyme. *Carbohyd. Res.*, **6**:129–137, 1968.

Z16. Zehavi, U., and Lustig, A. On the reversibility of substrate-induced dissociation of lysozyme aggregates. Native and reacted lysozyme. *Biochim. Biophys. Acta*, **236**:127–130, 1971.

Z17. Zehavi, U., and Lustig, A. Substrate-induced dissociation of lysozyme dimers. *Biochim. Biophys. Acta*, **194**:532–539, 1969.

Z18. Zehavi, U., Pollock, J. J., Teichberg, V. I., and Sharon, N. Oligosaccharides containing glucose as substrates for hen's egg white lysozyme. *Nature (London)*, **219**:1152–1154, 1968.

Z19. Zeya, H. I., and Spitznagel, J. K. Antibacterial and enzymic basic proteins from leukocyte lysosomes: Separation and identification. *Science,* **142**:1085–1087, 1963.

Z20. Zeya, H. I., Spitznagel, J. K., and Schwab, J. H. Antibacterial action of PMN lysomal cationic proteins resolved by density gradient electrophoresis. *Proc. Soc. Ex. Biol. Med.,* **121**:250–253, 1966.

Z21. Zhitova, E. I., Kostina, V. V., and Kudriashova, K. I. Lysozyme and the functional activity of phagocytes in acute, protracted and chronic pneumonia. *Ter. Arkh.,* **40**:46–51, 1968. (Russ.)

Z22. Zhukovskaia, N. A., and Likina, T. N. On the question of nonspecific protective action of lysozyme on the organism. *Antibiotiki (Moscow),* **11**:920–924, 1966. (Russ.)

Z23. Zickler, F. Die Wirkung freier Phagenrezeptoren auf Proteus-Phagen in Gegenwart von Polykationen. *Z. Naturforsch. B,* **22**:418–421, 1967.

Z24. Zinder, N. D., and Lyons, L. B. Cell lysis: Another function of the coat protein of the bacteriophage f2. *Science,* **159**:84–86, 1968.

Z25. Zinder, N. D., and Arndt, W. F. Production of protoplasts of *Escherichia coli* by lysozyme treatment. *Proc. Nat. Acad. Sci. U.S.,* **42**:586–590, 1956.

Z26. Zini, F. Il lisozima risolve la Gram-positività della Candida albicans; parassitante una membrana embrionaria. *G. Mal. Infet. Parassit.,* **6**:21–24, 1954.

Z27. Zubairov, D. M., Popova, L. G., and Akhmetshina, M. Kh. Effect of lysozyme, ribonuclease and cytochrome C on activation of one of the factors of coagulation—the Hageman's Factor—by adrenaline. *Ukr. Biokhim. Zh.,* **45**:86–90, 1973. (Russ.)

Z28. Zucker, S., Hanes, D. J., Vogler, W. R., and Eanes, R. Z. Plasma muramidase: A study of methods and clinical applications. *J. Lab. Clin. Med.,* **75**:83–92, 1970.

Z29. Zuili, S., and Jollès, P. Comparative study of the action of 3 lysozymes of different origin on *Bacillus megatherium* and *Micrococcus lysodeikticus. C. R. Acad. Sci.,* **250**:3521–3523, 1960.

Z30. Zwart-Voorspuij, A. J., Gootjes, J., and Nauta, W. Th. An antilysozyme factor in radix liquiritiae. *Arzneim. Forsch.,* **10**:604–606, 1960.

Index to the Lysozyme Bibliography

B

Testes, *see* Distribution
Testosterone, *see* Interactions, androgens
Tetracyclines, *see* Interactions
Therapy, *see also* Disease states, Pharmacological effects, Virus interactions
  arthritis, B153, K14
  brucellosis, N42, S125, T83
  cancer, B72, C77, C202, D56, E31, G75, L46, L72, S33, S68, S69, S70, S71, S195, S204, T80
  dental infections, C89, L1, L3, M176, P66, R78, S223
  dermatological conditions, B115, C181, F6, L92, L93, O27, S66, T59, T60, Z5
  experimental infections, A45, B157, B159, B176, C23, C76, C118, C189, M91, M115, M176, N12, N17, P102, P103, P105, R84, S151, T29, W24
  experimental tumors, C120, C202, C203, G12, I30, L72, Q1, S158
  gastrointestinal diseases, A1, B31, B51, B70, B171, C32, C200, D81, E26, F27, G6, G91, K129, L46, L79, M22, M51, P105, P108, R26, R76, R95, S41, V25, V41, V42, V43
  gynecological diseases, C39, F27, K21, S1, S30, S88, S89, S90, S194
  hepatitis, A45, A68, B16, B157, B159, C134, C177, C197, D54, D81, F21, F27, L11, L36, M11, M15, M94, M173, N39, N40, R5, R77, S74, S151, T20, T29
  leukopenia, X-ray and drug-induced, A18, B62, C57, C71, C175, D44, D45, F28, F70, G15, L104, M21, M170, O43, P77, S186, T82, V6, V7
  neonatal jaundice, B190
  neurological disorders, M45, R1, R2
  ophthalmological diseases, B71, B170, C56, C100, D1, D62, F41, F71, F72, G125, H65, H125, I3, K63, K110, M77, N22, N23, O28, S24, T16, V12
  oral diseases, B172, C13, C77, C90, D28, G84, K29, M176, P139, S26
  otitis, B52, C90, G84, H122, M16, M171, S208
  pediatrics, general, D83, H11, V25
  puerperal infections, F25
  radiation injury, B121, D44, D45, D55, P139

  respiratory infections, A106, B1, B52, B95, B177, C95, C168, C170, C191, F27, H122, I58, L39, M86, M169, M171, N20, S208, T15, T18
  sinusitis, A106, B1, C90, C96, H56, I58, K3, M85, M86, M171, N20, S208, T4, T15
  trachoma, D1
  Trichomonas vaginitis, A62, B26, B27, F29, K1, P88, P151, S75, S76, S89, S90
  urinary tract infections, G71, I28, S121
  viral infections, B95, B159, B171, B172, B177, C96, C192, E24, E29, F49, G90, K59, K112, L93, M169, N50, S121, S187, T22, Z5
  wounds, B129, F77, L21, N44, S36, V8, V42
Thermal denaturation, *see* Denaturation, temperature
Thermal perturbation difference spectra, A16a, A51, A52, B49, F69, H27, H29, H85a, M43a, S109, S167, W45a
Thymus, thymectomy, *see also* Distribution B58, B113, C135, G9, K30, L43, R61, R62, S123, T21, V39
Thyroid function: thyroxine and thyroglobulin interactions, L60, L61, L64, L68, P55, T2a
Tissue culture *in vitro* synthesis, C79, C80
Titration curves, B68a, B179a, D63a, D63b, H124c, R22a, T13a, T13b
Tolerance, *see* Immunochemical studies
Toluidinylnaphthalene, *see* Interactions
Toxoplasma, see Protozoa
Trachoma, *see* Therapy
Transglycosylation, C116, D2, G98, K90, P96, P98, P99, R11, R91, R92, S72, S100, S101, Z15
Transition states, F69a, L48a
Transplantation antigens, *see* Interactions
Trauma, *see* Disease states
Treponema, C23, F79, J81, K31, K69, K70, M108, M109, M110, M112, M113, M114, N35, O44, P3, P112, S32, U5, U6, W18
Trichloracetic acid, *see* Interactions
Trichomonas vaginitis, *see* Disease states, vaginitis; Therapy
Tritiation, H133, H134, P121, S166b
Triton, *see* Surfactants
Trypsin, *see* Interactions

Subject Index